The Gulf of Mexico Sedimentary Basin

The Gulf of Mexico basin is one of the most prolific hydrocarbon-producing basins in the world, with an estimated endowment of 200 billion barrels of oil equivalent. This book provides a comprehensive overview of the basin, spanning the USA, Mexico, and Cuba. Topics covered include conventional and unconventional reservoirs, source rocks and associated tectonics, basin evolution from the Mesozoic to Cenozoic Era, and different regions of the basin from mature onshore fields to deepwater subsalt plays. Cores, well logs, and seismic lines are all discussed, providing local, regional, and basin-scale insights. The scientific implications of seminal events in the basin's history are also covered, including sedimentary effects of the Chicxulub impact. Containing over 200 color illustrations and 50 stratigraphic cross-sections and paleogeographic maps, this is an invaluable resource for petroleum industry professionals, as well as graduate students and researchers interested in basin analysis, sedimentology, stratigraphy, tectonics, and petroleum geology.

John W. Snedden is a senior research scientist at the Institute for Geophysics, the University of Texas at Austin. He directs the Gulf Basin Depositional Synthesis project, an industry research consortium investigating the depositional history of the Gulf of Mexico. He worked in the oil industry for more than 25 years.

William E. Galloway is Professor Emeritus in the Department of Geological Sciences and a research professor at the Institute for Geophysics, the University of Texas at Austin. He co-authored the reference book *Terrigenous Clastic Depositional Systems* (Springer-Verlag 1983) and founded the Gulf Basin Depositional Synthesis project.

The Gulf of Mexico Sedimentary Basin

Depositional Evolution and Petroleum Applications

John W. Snedden
The University of Texas at Austin
William E. Galloway
The University of Texas at Austin

CAMBRIDGE
UNIVERSITY PRESS

University Printing House, Cambridge CB2 8BS, United Kingdom

One Liberty Plaza, 20th Floor, New York, NY 10006, USA

477 Williamstown Road, Port Melbourne, VIC 3207, Australia

314–321, 3rd Floor, Plot 3, Splendor Forum, Jasola District Centre, New Delhi – 110025, India

79 Anson Road, #06–04/06, Singapore 079906

Cambridge University Press is part of the University of Cambridge.

It furthers the University's mission by disseminating knowledge in the pursuit of education, learning, and research at the highest international levels of excellence.

www.cambridge.org
Information on this title: www.cambridge.org/9781108419024
DOI: 10.1017/9781108292795

First published 2019

Printed in Singapore by Markono Print Media Pte Ltd.

A catalogue record for this publication is available from the British Library.

ISBN 978-1-108-41902-4 Hardback

Additional resources for this publication at www.cambridge.org/gomsb.

Dedicated to our wives, Peggy and Rosemary, for their patience, support, and love, both while we wrote this book and during the many years that brought us to this endeavor.

Contents

Part II Mesozoic Depositional Evolution

Preface

A "superbasin" is defined as having a prolific petroleum system (remaining recoverable reserves > 5 BBOE; past production > 5 BBOE) and a well-established surface infrastructure that allows exploration in a low-risk setting. With an estimated endowment of 200 BBOE and cumulative production of 60 BBOE, the Gulf of Mexico (including the USA and Mexico) easily qualifies. Yet few recent books have attempted to illuminate one of this superbasin's key success factors: the depositional systems that produce reservoir, source, and seal rock necessary to entrap hydrocarbons in conventional and unconventional plays.

Investigating the evolution of depositional systems in the Gulf of Mexico superbasin is also scientifically important, as it represents a 200-million-year-plus sedimentary archive, well-documented by oil and gas wells, Deep Sea Drilling Project (DSDP) and Integrated Ocean Drilling Program (IODP) sites, thousands of academic and industry seismic surveys, outcrop and core descriptions, academic and company publications, and an increasing number of advanced provenance studies (e.g., detrital zircon geochronology). Studies of modern Gulf of Mexico depositional systems continue to provide important insights on climate change (land loss due to sea-level rise, hurricane impact, river flooding) which are better informed by review of preceding Neogene depositional patterns. In fact, the immense volume of geological and geophysical data and archive of technical publications is, in its own way, a daunting challenge for both new and seasoned explorationists and scientists working in the Gulf.

The Gulf of Mexico basin has also served historically as a test bed to establish and refine ideas on depositional systems, beginning with Fisher and McGowen's (1967) study of the onshore Wilcox and extending to the mapping of deepwater Wilcox abyssal fans by Galloway and others following major Wilcox deepwater discoveries. One could argue that seismic stratigraphy may not have evolved as smoothly into sequence stratigraphy without the addition of the depositional systems tract concept developed by Brown and Fisher and documented by Gulf Coast examples published in AAPG Memoir 26 in 1977.

We also feel that there is a pressing need for a more comprehensive, basin-scale review of the Gulf of Mexico with the opening of Mexico to international exploration. Until recently, many investigations stopped at the USA–Mexico border, a political boundary that no natural depositional system or tectonic domain recognizes. In fact, the new 2D and 3D seismic surveys shot in Mexico have revealed much of the southern half of the basin and caused us to reconsider models based strictly upon the northern portion. This is coupled with new ideas on the deep crustal structure emerging from seismic refraction studies, tectonic uplift history from detrital zircon geochronology, and the new thoughts on the basin-wide effects of the Chicxulub impact event that ended the Mesozoic.

In total, we have conducted full-time research into the Gulf of Mexico for over 80 years, in both academia and industry. The Gulf Basin Depositional Synthesis (GBDS) project in the Institute for Geophysics at the University of Texas at Austin, supported continuously for over 22 years, provides the primary database of reflection seismic data and well data (e.g., logs, biostratigraphy), and associated rock data (e.g., cores, cuttings). It also enables the opportunity to collaborate with other researchers, students, and oil and gas company personnel with a mutual interest in deciphering this superb natural laboratory of ancient sedimentary processes. The most important products of this multi-decade effort are the paleogeographic reconstructions, restored to original plate tectonic and structural positions, which are summarized in the full-color maps for each depositional unit in this book.

While our book deals primarily with the depositional history from pre-salt to Pleistocene, much space is devoted to the structural trends that control and influence stratal accommodation, including salt tectonics. In fact, the overall Mesozoic and Cenozoic stratigraphic framework is a tectonostratigraphic scheme reflecting deep crustal processes (e.g., sea floor spreading), hinterland climate, uplift and sediment generation, extrabasinal transport systems, and processes within the depositional sink like salt evacuation, paleobathymetry, mass transport and slope failures, and more.

The introductory Part I begins with a foundational description of the unique tectonic setting that is necessary to understand how depositional trends emerge and evolve within the basin. This includes 10 basin-scale cross-sections across the

USA, Mexico, and Cuba, onshore to offshore. It is important to note that here and elsewhere in the book, a large number of figures are cross-sections based upon or showing reflection seismic data, the primary tool for exploration in the USA and Mexico. This is followed by detailed discussion of our tectonostratigraphic framework, including stratigraphic terminology for the Mesozoic and Cenozoic of the northern Gulf of Mexico (USA), southern Gulf of Mexico (Mexico), and Cuba. Robust explanation of depositional systems classifications we use for carbonate and siliciclastic domains, including key concepts like submarine fans, ramps, and aprons, and shelf edge recognition criteria are provided. Our database of seismic data and well information is illustrated at the end of this preface.

The main portion of the book (Parts II and III) follows a chronologic pathway from basin precursors to basin opening, nascent basin, and basin evolution to the end of Pleistocene time. In Chapter 2, which examines the poorly understood pre-salt section, we offer a new alternative model to the conventional view espoused by Amos Salvador in his seminal 1991 work on the basin, published in the GSA *Decade of North American Geology* (DNAG) volume. Our model of a post-orogenic successor basin-fill (versus rift system) for the central northern Gulf of Mexico is based on new results from detrital zircon geochronology, revised plate tectonic reconstructions, and reinterpretation of onshore seismic reflection data. Following this, in Chapter 3 is a discussion of the tectonostratigraphic phase of sea floor spreading and crustal cooling that continued to 140 Ma. Emerging new models on the Louann Salt origin and source marine water are described here. Chapter 3 continues with a tour through eolian systems of the Norphlet Sandstone, marine flooding and marine microbial development in the Smackover, rise of platform margin reefs in the Kimmeridgian, organic source facies enrichment in the Tithonian, and westward expansion of the Cotton Valley–Knowles reef system. The important Kimmeridgian shelf grain shoals and patch reefs that form important reservoirs in Mexico are also described. It should be noted that considerable text is devoted to description of eolian paleo-environments, given the relative unfamiliarity of such reservoirs to even experienced Gulf of Mexico interpreters. Discovery of the Norphlet deepwater play, with estimated recoverable resources of 1 BBOE, has generated considerable interest in such dryland systems. Periods of reduced bottom circulation result in at least two phases of source rock development, in the Oxfordian and Tithonian stages, that are linked to petroleum generation for both conventional and unconventional plays. The chapter finishes with development of formidable platform margin reefs at the end of the Jurassic, possibly positioned against the maximum fetch of this new marine basin. This largely prevented sands of updip fluvial–deltaic systems from passing beyond the shelf margin, a classic case of "reef blocking."

The Late Mesozoic Local Tectonic and Crustal Heating Phase, the subject of Chapter 4, follows the end of sea floor spreading and is marked by local tectonic uplifts, beginning with a major Early Cretaceous siliciclastic influx in the eastern Gulf of Mexico, likely from uplift of the Peninsular Arch, as indicated by detrital zircon geochronology of the Hosston Sandstone. The younger Tuscaloosa Sandstone marks the first major entry of siliciclastics into the central northern Gulf of Mexico deepwater basin in the Ceno-Turonian. The Eagle Ford Shale, a world-class unconventional play, forms in restricted shelf basins in south Texas. A reduction of siliciclastic input, combined with globally high sea level, results in pervasive deep marine sedimentation culminating in chalk deposition in the latest Cretaceous. The end of the Mesozoic Chicxulub impact event generated mass transport deposits, breccia, and hybrid flows related to seismic shaking and catastrophic slope failures, greatly modifying the land- and seascape of the basin and paving the way for long-lived source-to-sink transport systems routing sediment from the Laramide Orogenic Belt into the deep Gulf basin.

Part III completes the depositional history and structural evolution of the Gulf. The 62-million-year Cenozoic history is divided into three tectonostratigraphic phases. Chapter 5 discusses the Laramide Phase, which records the Paleocene–Middle Eocene interval, when the Gulf was directly impacted by compressional tectonism along the Laramide front and indirectly by transport of great volumes of terrigenous sediment from interior uplands through the newly evolving river networks. Depositional loading both depressed the sub-Gulf crust and initiated massive mobilization and redistribution of the Louann Salt. Chapter 6 reviews the Middle Cenozoic Gulf record of resurgent sediment supply from western uplifts and volcanic centers driven by regional crustal heating. Chapter 7 describes the final phase, presaged in the Early Miocene and clearly dominating the Middle Miocene–Pliocene, that was dominated by rejuvenation of Appalachian and mid-continental sources and consequent eastward migration of sediment supply and loading. Salt structures expanded in both diversity and complexity as earlier-formed salt canopies were loaded and overrun. The phase culminated with onset of montane glaciation and formation of the North American ice sheet. Finally, Chapter 8 is a synthesis of the long-term patterns of sediment supply, paleogeographic themes, and continental margin growth and evolution.

Part IV (Chapter 9) transitions from foundational science and paleogeographic considerations to discussion of the petroleum habitat associated with each depositional unit or unit aggregate (supersequence) described in Parts II and III. This includes known production trends in the USA, Mexico, and to a lesser degree in Cuba. Beginning with an overview of the current Gulf of Mexico resource size and spatial distribution, ensuing sections cover frontier (pre-salt), emerging (deepwater Tuscaloosa and Norphlet), existing (deepwater Wilcox), and mature (Plio-Pleistocene minibasin) conventional exploration plays. This chapter includes close examination of the unconventional plays that are well documented (Eagle Ford, Haynesville), reemerging (Navarro–Taylor, Austin Chalk) or currently technically challenged (Bexar–Pine Island,

Tuscaloosa Marine Shale). It finishes with a section on the seismic technology evolution that underpins current success in the subsalt of the US sector and will undoubtedly impact exploration in the underexplored Campeche salt province of Mexico.

Within chapters of the book are special sections devoted to topics that are germane to the Gulf of Mexico and related technologies that support scientific investigation and industry decision-making. These "boxes" cover a spectrum from special depositional nomenclature, detrital zircon geochronology, the history of scientific research into the Chicxulub impact, and the significance of the Alaminos Canyon BAHA wells for deepwater Wilcox exploration in the Gulf of Mexico.

We would be remiss if we did not acknowledge the very important contributions of others in our efforts to write this book. Foremost among these is Jon Virdell, who not only managed the GBDS project during the most important phases of the book's construction but also drafted many figures himself, coordinated with our publisher Cambridge University Press, oversaw student support, edited text, and obtained permissions for use of previously published figures, among many other tasks. We also thank Jeff Horowitz for patiently drafting many figures in the book. Patricia Ganey-Curry, who managed the creation, organization, and funding and budgeting of the GBDS project for the first 20 years of its history, is also thanked. Reviews and contributions to sections of the book by UT-Austin Institute of Geophysics researchers Christopher Lowery, Robert Cunningham, Ian Norton, William Fisher, and Craig Fulthorpe are noted. We are especially grateful to Michael Hudec of the UT Bureau of Economic Geology Applied Geodynamic Laboratory for reviews of our basin cross-sections and his insights on Gulf of Mexico salt tectonics that helped shape interpretations made by both of us. Finally, Dr. Timothy Whiteaker, a master GIS wizard, organized the continually expanding database and found ways to make and interpret the myriad array of maps that are the foundation of our synthesis.

Technical contributions and insights by others, including Angela McDonnell, Shirley Dutton, Jake Covault, Frank Peel, Martin Jackson (deceased), Robert Loucks (all of UT-Austin Bureau of Economic Geology), Mark Rowan, Tom Ewing, Art Waterman (PaleoData), Daniel Stockli (UT-Austin Department of Geological Sciences), Gary Kinsland (U. Louisiana), Michael Blum, Bruce Fredericks, Brad Prather (all of U. Kansas), Erik Scott (EOG), and Sean Gulick (UT-Austin Institute of Geophysics) were very helpful.

The theses of the following UT students also formed an important foundation for our interpretations captured in this book:

Snedden was the main supervisor, co-supervisor, or thesis committee member for the following students: Jie Xu, Jason Sanford, Caroline Bovay, Luciana de la Roche Tinker, Enrique Arce, Fernando Apango, Colin White, Keelan Umbarger, Harry Hull, Will Pinkston, of UT-Austin; other students worked with GBDS data and contributed to our understanding of the basin: Drew Eddy (formerly UT) and Kody Shellhouse (formerly U. Louisiana).

Galloway supervised numerous students who worked on various aspects of Gulf depositional history, including: Liangqing Xue, Holly Hoel, Richard Paige, William Dingus, Thomas Williams, Scott Hamlin, Janet Coleman, Ed Duncan, Scott Spradlin, Lawrence Meckel III, Qing Fang, Ricardo Combellas-Bigott, and Xinxia Wu.

Of course, financial support of the GBDS Industrial Associates Program members, past and present, includes the following companies:

Apache, Anadarko Petroleum Corporation, BHPB Pet (Deepwater) Inc., Bureau of Ocean Energy Management, BP America Production Company, Chesapeake Operating, L.L.C, Chevron North America Exploration & Production Company, Cobalt, ConocoPhillips, Devon, Ecopetrol, Eni, Equinor, Encana, ExxonMobil Exploration Company, Freeport McMoRan, GulfSlope, Hess Corporation, Hilcorp Energy Company, INPEX Corporation, Japan Oil, Gas and Metals National Corporation, LUKOIL, Maersk, Marathon, Mitsui E&P USA LLC, Murphy Exploration and Production Co., Nexen Petroleum U.S.A. Inc., Noble Energy Inc., Pemex, Petrobras, Pioneer, Repsol Services Company, Ridgewood Energy, Samson, Shell Exploration & Production Company, Stone Energy, Suncor, Talos Energy, TOTAL E&P Research & Technology USA, LLC, Venari, Woodside.

The seismic companies who provided seismic figures for this book and/or provided seismic data to the GBDS project include ION (and partners SEI, GPI), Spectrum, Fugro, Dynamic Global Associates, Multi-Client Geophysical, PGS, TGS, and WesternGeCO.

Financial support by UTIG (Mrinal Sen) is much appreciated.

Finally, we wish to acknowledge Cambridge University Press, including Emma Kiddle who commissioned our book, Zoë Pruce who efficiently coordinated production, and Gary Smith who copy-edited the manuscript.

Abbreviations

AAPG	American Association of Petroleum Geologists
AB	Alabama basin
AC	Austin Chalk
AE	Apalachicola Embayment
ANB	Anahuac Block
AO	Appalachian Orogen (Cretaceous limit)
AU	Arbuckle Uplift
AVO	Amplitude versus offset
BB	Burgos basin
BEG	Bureau of Economic Geology
BOEM	Bureau of Ocean Energy Management
BSEE	Bureau of Safety and Environmental Enforcement
BU	Burro Uplift
CAMP	Central Atlantic magmatic province
CCP	Clarke County Platform
CIE	Carbon isotope excursion
CNH	Commision de Nacional Hydrocarbons
CP	Coahuila Platform
CVB	Cotton Valley–Bossier
CVK	Cotton Valley–Knowles
DSDP	Deep Sea Drilling Project
DSSB	DeSoto salt basin
EFT	Eagle Ford–Tuscaloosa
EGoM	Eastern GoM Embayment
EM	Eagle Mills
EMARC	Energy and Minerals Applied Research Center
EP	Edwards Platform
ETB	East Texas basin
FWB	Fort Worth basin
GBDS	Gulf Basin Depositional Synthesis
GBR	Great Barrier Reef
GoM	Gulf of Mexico
GR	Glen Rose
GRG	Greater Rio Grande Embayment
HPHT	High-pressure/high-temperature
HVB	Haynesville–Buckner
IODP	Integrated Ocean Drilling Program
JD	Jackson Dome
KC	Keathley Canyon
LAD	last appearance datum
LCLA	low-continuity, low-amplitude
LEF	Lower Eagle Ford
LOC	limit of oceanic crust
LOM	level of organic metamorphism
LPB	La Popa basin
LS	Louann Salt
LU	Llano Uplift
LW	Lower Wilcox
MA	Muenster Arch
MAU	Marathon Uplift
MB	Mississippi Basin
MCU	Middle Cretaceous unconformity
MD	measured depth
ME	Mississippi Embayment
MSB	Mississippi salt basin
MTC	mass transport complex
MTD	mass transport deposit
MU	Monroe Uplift
MW	Middle Wilcox
NAZ	narrow azimuth
NLSB	North Louisiana salt basin
NOR	Norphlet
NT	Navarro–Taylor
OAE	oceanic anoxic events
OBS	on-bottom sensors
OCS	outer continental shelf
ODP	Ocean Drilling Program
OF	Oligocene Frio
OM	Ouachita Mountains
OU	Ocala Uplift
PB	Parras basin
PETM	Paleocene–Eocene Thermal Maximum
PH	Peyotes High
PTA	Pleistocene Trim A
PW	Paluxy–Washita
QFL	quartz–feldspar–lithic
RD	Rodessa
RGR	Rio Grande rift
SAP	Sarasota Platform
SEGE	Southeast Georgia Embayment
SFB	South Florida basin
SFL	shallow-focused resistivity log
SGR	South Georgia rift
SH	Sligo–Hosston

SJ San Juan volcanic field
SMK Smackover
SMO Sierra Madre Occidental volcanic field
SN Smackover–Norphlet
SP Southern Platform
SP spontaneous potential
SU Sabine Uplift
TB Tyler basin
TD total depth
TE Tampa Embayment
TLP tension leg platform
TMM Tampico–Misantla–Magiscatzin
TMS Tuscaloosa Marine Shale
TOC total organic carbon
TP Tuxpan Platform
TSCA Tamaulipas/San Carlos Arch
TSR thermochemical sulfate reduction
TVT true vertical thickness
UEF Upper Eagle Ford
USGS US Geological Survey
UTRR undiscovered technically recoverable resources
UW Upper Wilcox
WA Wiggins Arch
WAZ wide azimuth
WB Winnfield basin
XRF X-ray fluorescence

Part I

Introduction

Part I is the foundational introduction to the Gulf of Mexico basin. It provides a detailed description of the unique tectonic setting that is necessary to understand how depositional trends emerge and evolve within the basin. This includes analysis of 10 basin-scale cross-sections across the USA, Mexico, and Cuba, onshore to offshore. What follows is a robust discussion of the Gulf of Mexico tectonostratigraphic framework, including stratigraphic terminology for the Mesozoic and Cenozoic strata and explanation of depositional systems classifications for the ancient carbonate and siliciclastic domains.

Chapter 1 Introduction
Tectonic and Stratigraphic Framework

1.1 General Setting

In this book, we describe the greater Gulf of Mexico (GoM) basin as extending from the coastal plain in the southern USA to the coastal plain of southern Mexico, the Chiapas and Tabasco region, and east across the Yucatán Platform to Cuba, the Florida Straits, and the Florida onshore area (Figure 1.1). The Gulf basin has a central abyssal plain that generally lies at 13 km depth (Bryant *et al.* 1991). The eastern Gulf floor is dominated by the morphology of the Late Quaternary Mississippi Fan.

The continental slope of the northern Gulf margin displays a bathymetrically complex morphology that terminates abruptly in the Sigsbee Escarpment to the west and merges into the Mississippi Fan to the east (Steffens *et al.* 2003). The hallmark of the central Gulf continental slope is the presence of numerous closed to partially closed, equi-dimensional, **slope**

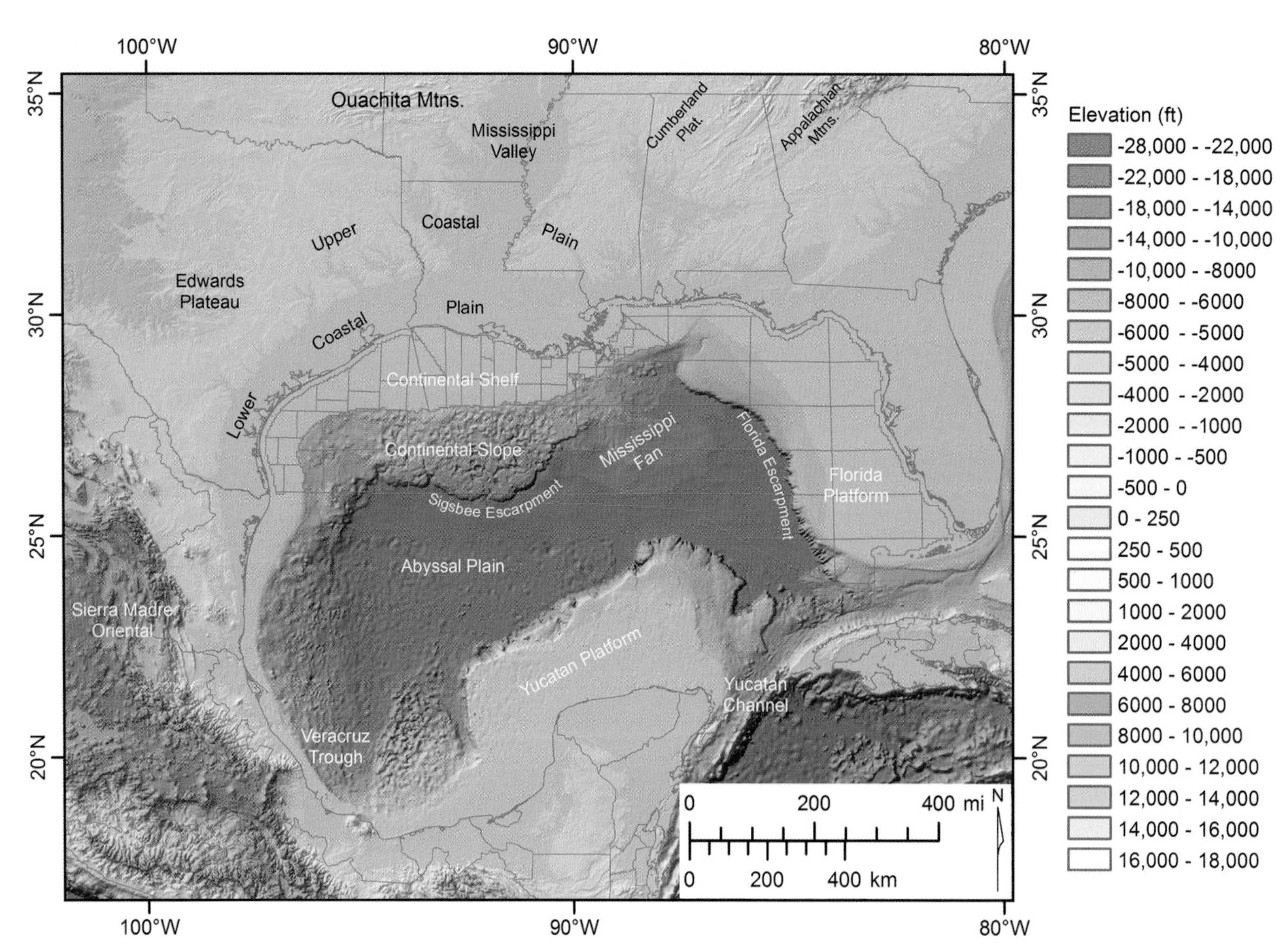

Figure 1.1 Location map for greater GoM basin, including important geographic and bathymetric features.

minibasins. In contrast, the Florida Platform forms a broad ramp and terrace that terminates at depth into the nearly vertical Florida Escarpment. The western Gulf margin displays intermediate width, and it too is quite bathymetrically complex. Here, numerous contour-parallel ridges and swales dominate the mid- to lower-slope morphology. The modern **shelf margin**, as reflected by a well-defined increase in basinward gradient, generally lies at a depth of 100–120 m. Landward, the northwestern, northern, and eastern GoM is bounded by broad, low-gradient shelves that range from 100 to 300 km in width (Figure 1.1). Today, and throughout its history, the Florida and Yucatán Platforms, which bound the basin on the east and south, persist as sites of carbonate deposition.

On shore, the northern and northwestern Gulf margins display a broad coastal plain (Figure 1.1). The lower coastal plain, a flat, low-relief surface, is underlain by Neogene and Quaternary strata. The upper coastal plain displays modest relief of less than about 100 m (328 ft) created by Quaternary incision into older Neogene, Paleogene, and Upper Cretaceous strata by numerous large and small rivers. The basin is bounded by a variety of Cenozoic, Mesozoic, and remnant Paleozoic uplands, including the Sierra Madre Oriental of Mexico, the Trans-Pecos mountains of west Texas, the Lower Cretaceous limestone-capped Edwards Plateau, Ouachita Mountains of southern Arkansas, and the Cumberland Plateau and southern Appalachian Mountains of northern Mississippi and Alabama. The northeast Gulf basin merges into the southern Atlantic coastal plain across northern Florida; however, the structural basin boundary is generally placed near the current west coast of the Florida peninsula.

Mexico's onshore topography strongly reflects the Sierra Madre Oriental in the north and the Chiapas deformational belts in the south of the country. The eastern onshore portion of Mexico is marked by short but steep gradient rivers that carry modern sediments toward a wave-dominated shoreline, a narrow shelf, and steep slope that terminates abruptly at the abyssal plain. Offshore, bathymetric maps show the sea floor complexity resulting from recent tectonic events: (1) the elongate, generally north–south oriented structures called the Mexican Ridges; and (2) the recent salt inflation and compression evidenced in the rugose hydrography of the Campeche and Yucatán salt provinces.

Across the Bay of Campeche lies the Yucatán carbonate platform, with equally steep margins that circumscribe the platform and its border with the adjacent Caribbean basin. The Yucatán channel separates Yucatán from Cuba, a tectonically complex mélange of various microplates that merged over 100 million years. Cuba lies across the Florida Straits from the South Florida basin, a short distance, but a world away in terms of its geological evolution.

1.2 Structural Framework

In order to understand the depositional evolution of the GoM, it is necessary to consider the structural framework that underpins and influences the sedimentary loading history of this immense natural repository. This extends to the deep crystalline crust and even mantle that can, in some cases, be detected by modern seismic reflection and refraction data. The accumulated sediment mass, including both siliciclastics and carbonates, also drove **gravity tectonics**, particularly where **evaporites** like salt respond in a ductile fashion at burial depths attainable by modern wells.

1.2.1 Deep Crustal Types

For many years, the form and lithology of the deep structure in the GoM was a matter of conjecture and inferences based upon rare penetrations of **basement** rock or sometimes-equivocal gravity and magnetic data. Recently, seismic refraction studies have greatly illuminated the form of the mantle and overlying crystalline and sedimentary crust (Van Avendonk *et al.* 2013, 2015; Christeson *et al.* 2014; Eddy *et al.* 2014). In addition, new plate tectonic models have altered previous suppositions on timing of basin opening and emplacement of **oceanic crust** (Norton *et al.* 2016). Alternative models, particularly for the pre-spreading rift phase, show convergence toward a consensus solution.

In general, these studies agree that the Gulf basin is largely surrounded by normal continental crust of the North American plate. Most of the structural basin is underlain by **transitional crust** that consists of **continental crust** that was stretched and attenuated by Middle to Late Jurassic rifting (Hudec *et al.* 2013a). Two types of transitional crust are differentiated (Figure 1.2). The basin margin is underlain by a broad zone of thick transitional crust, which displays modest thinning and typically lies at depths between 2 and 12 km subsea (Sawyer *et al.* 1991). The area of thick transitional crust

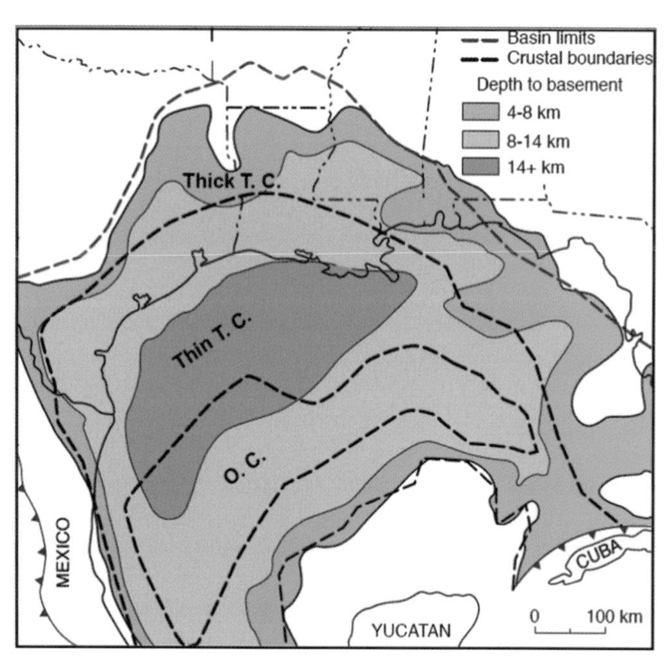

Figure 1.2 GoM crustal types. Modified from Galloway (2008).

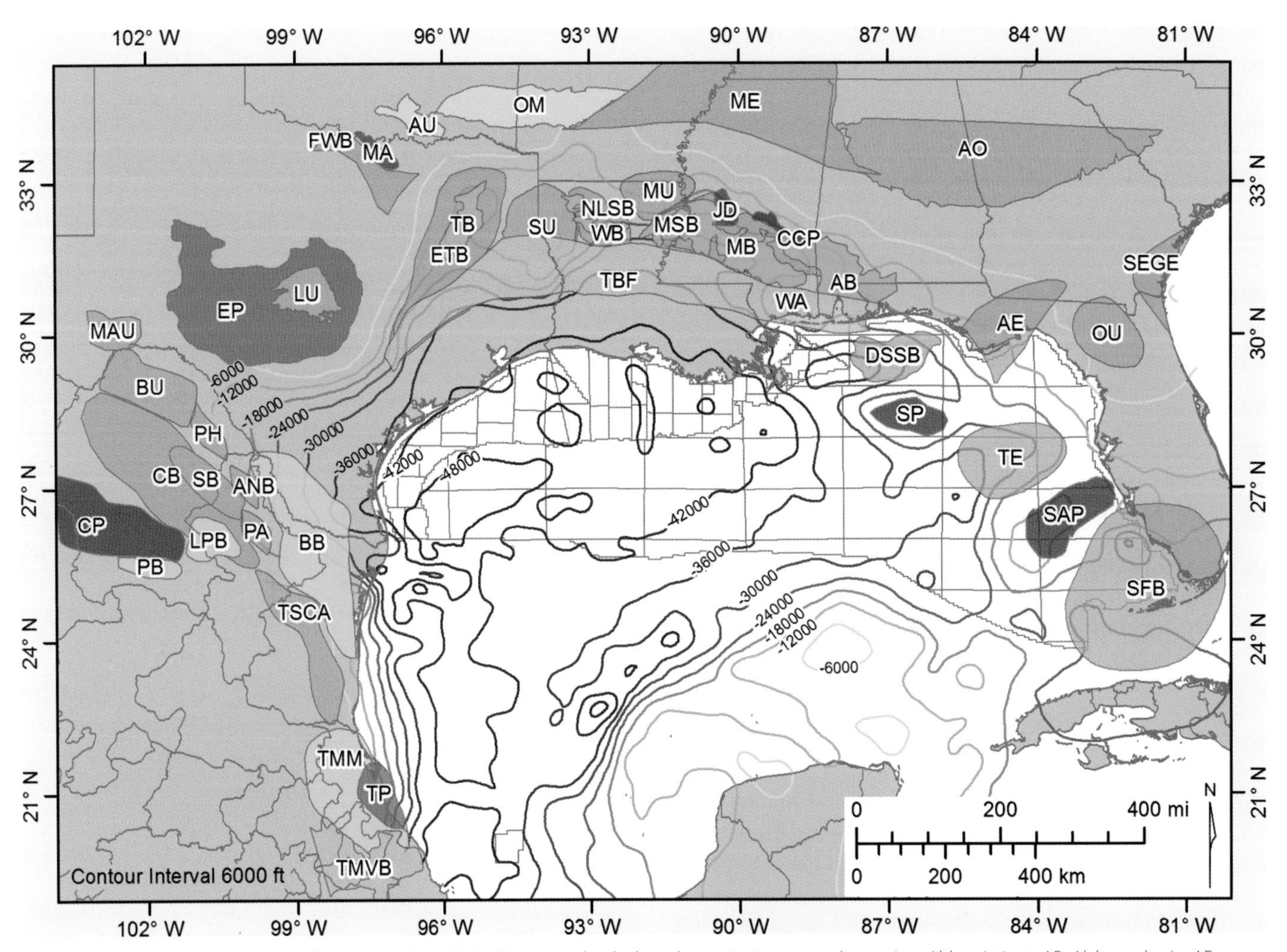

Figure 1.3 Key tectonostratigraphic features, northern GoM. Basement depths based on seismic structural mapping. Abbreviations: AB, Alabama basin; AE, Apalachicola Embayment; ANB, Anahuac Block, BB, Burgos basin; AO, Appalachian Orogen (Cretaceous limit); AU, Arbuckle Uplift; BU, Burro Uplift; CCP, Clarke County Platform; CP, Coahuila Platform; DSSB, DeSoto salt basin; EP, Edwards Platform; ETB, East Texas basin; FWB, Fort Worth basin; JD, Jackson Dome; LPB, La Popa basin; LU, Llano Uplift; MA, Muenster Arch; MAU, Marathon Uplift; MB, Mississippi Basin; ME, Mississippi Embayment; MSB, Mississippi salt basin; MU, Monroe Uplift; NLSB, North Louisiana salt basin; OM, Ouachita Mountains; OU, Ocala Uplift; PB, Parras basin; PH, Peyotes High; SAP, Sarasota Platform; SEGE, Southeast Georgia Embayment; SFB, South Florida basin; SP, Southern Platform; SU, Sabine Uplift; TB, Tyler basin; TE, Tampa Embayment; TMM, Tampico–Misantla–Magiscatzin; TP, Tuxpan Platform; TSCA, Tamaulipas/San Carlos Arch; WA, Wiggins Arch; WB, Winnfield basin. Terminology from various public sources, including Ewing and Lopez (1991).

consists of blocks of near-normal thickness continental crust separated by areas of stretched crust that has subsided more deeply. The result is a chain of named arches and intervening embayments and salt basins around the northern periphery of the Gulf basin (Figure 1.3).

Much of the present inner coastal plain, shelf, and continental slope is underlain by relatively homogeneous thin transitional crust, which is generally less than half of the 35 km thickness typical of continental crust and is buried to depths of 10–16 km below sea level. Reconstructions of deep seismic traverses (Peel *et al.* 1995; Radovich *et al.* 2007, 2011; Hudec *et al.* 2013b) indicate that basement may lie below 20 km in the central **depocenter** beneath the south Louisiana coastal plain and adjacent continental shelf. The deep, central Gulf floor is underlain by an arcuate belt of basaltic oceanic crust that was intruded during Late Jurassic through Early Cretaceous sea floor spreading (Hudec *et al.* 2013a; Norton *et al.* 2016). Surprisingly, the central Gulf crust generally lacks the magnetic signature typical of oceanic crust (Figure 1.4), which compounds interpretation difficulties, but recent gravity mapping (Sandwell *et al.* 2014) confirm earlier models of the location of the updip or landward limit of oceanic crust (LOC).

1.2.2 Seismic Refraction Studies of Deep Crust

The majority of data obtained for petroleum exploration is **seismic reflection data**, which allows both imaging through common depth point solutions and measurement of compressional seismic velocities to depths approaching 40,000 ft (12.2 km), depending on the energy source and cable. **Seismic refraction** data involves measurement of the compressional seismic velocities at much greater depths, approaching 40 km (25 miles). These velocities are a function of density in the deep earth and allow one to differentiate between mantle,

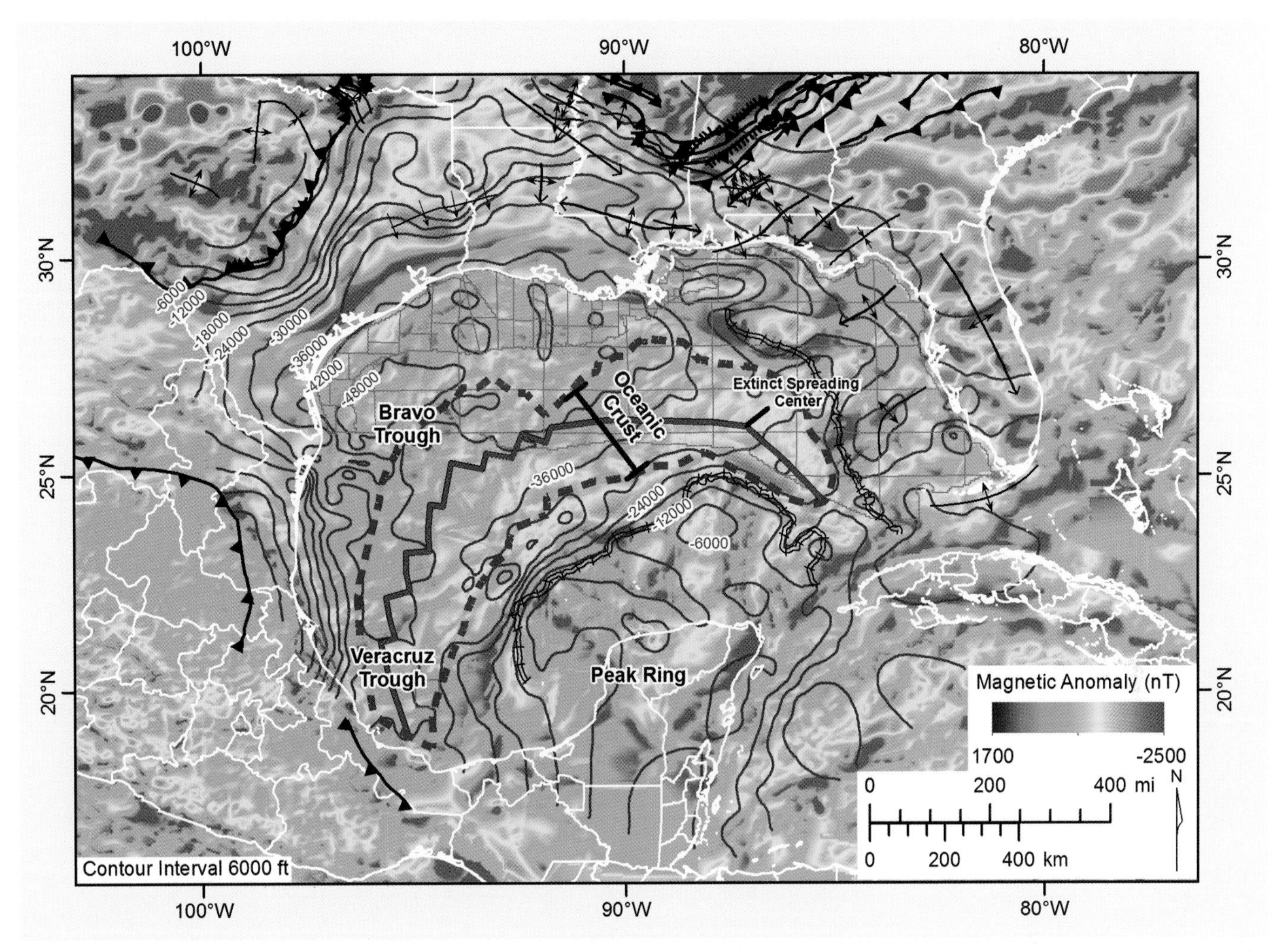

Figure 1.4 Mapped top of seismically defined basement with overlay of EMAG2 magnetic anomaly (Sandwell *et al.* 2014). Key tectonic features are discussed in the text. The limit of oceanic crust (red dashed line) is based on Hudec *et al.* (2013a, 2013b).

crystalline crust, and sedimentary crust, even where buried below thick intervals of salt and sedimentary rocks (Figure 1.5). In the northern GoM, a series of long (>500 km) seismic refraction lines were collected using bottom sensors (Figure 1.5). A line across the eastern GoM revealed the top of the mantle to shallow from about 34 km (21 miles) below the thick transitional crust below the Florida Platform to depths as shallow as 15 km (9 miles) in the area where oceanic crust is known to be present (Christeson *et al.* 2014; Figure 1.5). Above the mantle here lies a crystalline crust interval with unusually low velocities (in comparison to other areas), suggesting moderately attenuated continental crust. The sedimentary interval has compressional velocities in the range of 5.0 km/s (carbonate-dominated platform) to 3.0 km/s, where Miocene and younger strata are known to be present from well penetrations. The seismic refraction data also allow locating the boundaries of the LOC, here at a distance of 350–400 km from the start of the line just offshore of Florida. An intriguing observation is higher-than-expected seismic velocities at the LOC, suggestive of massive basalt emplacement associated with sea floor spreading (Christeson *et al.* 2014).

In the western GoM, seismic refraction data (Gumbo Line 1) revealed an unusual interval between high compressional velocity mantle and penetrated sedimentary crust (Van Avendonk *et al.* 2013). Below base of salt lies an unknown interval with considerable lateral crustal heterogeneity, thought to be rifted (attenuated) sedimentary crust with igneous intrusions. This interval ranges from 10–12 km at the top to as deep as 28 km depth above mantle rock. The lateral velocities variations that suggest igneous intrusions are documented in the shallow **pre-salt** interval of onshore areas, to be discussed in Section 2.2. The LOC is located inboard of the present-day Sigsbee Escarpment, though there is some uncertainty, given the thick salt canopy here (Van Avendonk *et al.* 2013). The presence of a pre-salt (Late Triassic[?] to Middle Jurassic[?]) interval in the deep northern GoM is consistent with observations from seismic reflection data in a pre-salt province

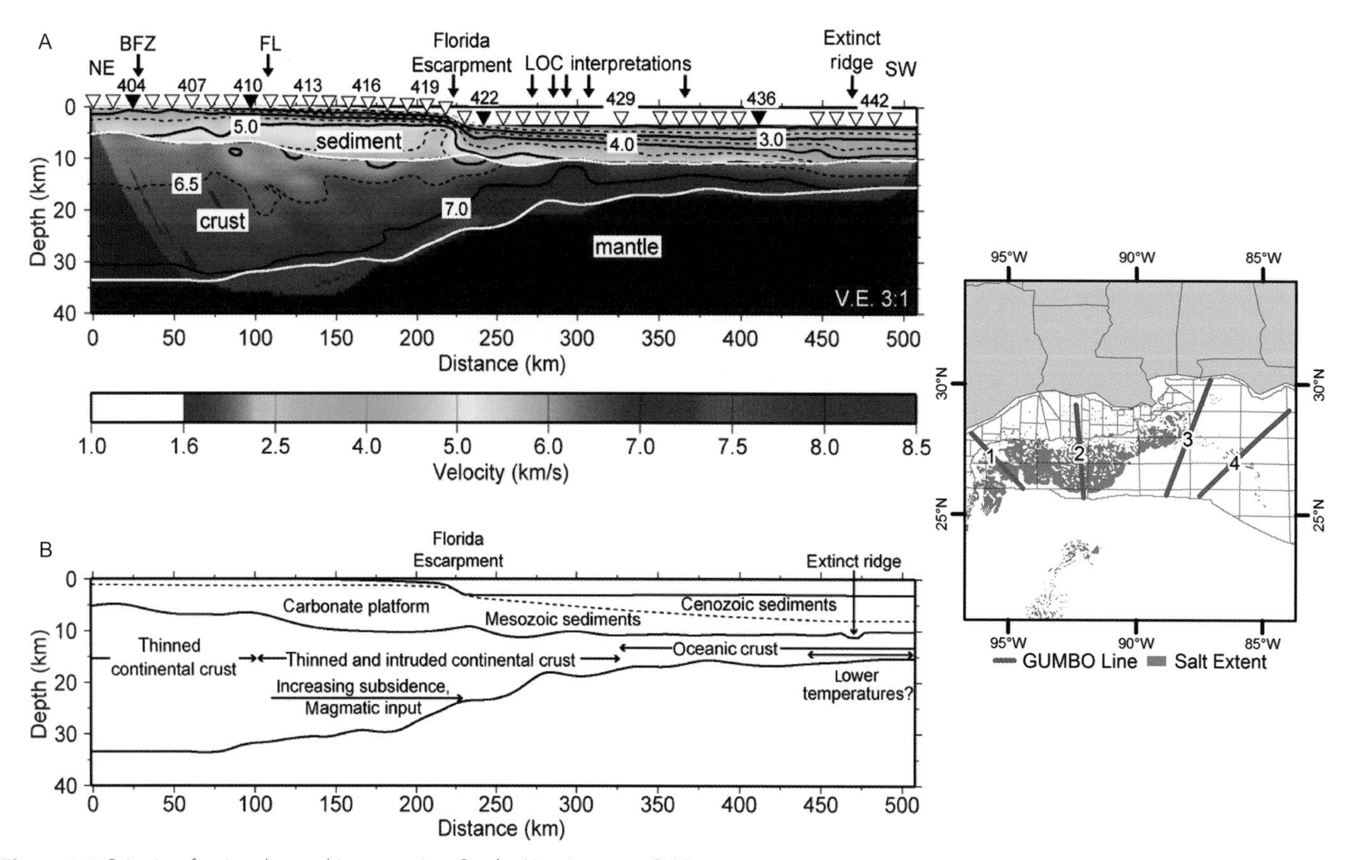

Figure 1.5 Seismic refraction data and interpretation, Gumbo Line 4, eastern GoM. Modified from Christeson *et al.* (2014).

offshore of Yucatán Province (Williams-Rojas *et al.* 2012; Miranda Peralta *et al.* 2014; Saunders *et al.* 2016).

1.2.3 Seismic Reflection Studies of Deep Crust

Seismic reflection surveys shot for oil and gas exploration provide some corroboration of seismic refraction interpretations, particularly for the eastern GoM where the salt canopy is absent. Here the general position of a Jurassic–Early Cretaceous **spreading center** in the eastern GoM has been suggested for many years, yet the precise location was not precisely known until Snedden *et al.* (2014) used several seismic criteria to define its location (Figure 1.6). Lin *et al.* (2019) subsequently refined its structure and evolution using newer vintage seismic reflection and gravity data. The extinct spreading center here displays morphological characteristics associated with slow-spreading mid-ocean ridges (rates of 1–4 cm/year; Perfit and Chadwick 1998): (1) large and wide axial valleys, 5–20 km wide; (2) deep axial valleys, often over 2 km deep; (3) normal faults that dip toward axial valleys; and (4) discontinuous, isolated basement highs, with elevations over 1 km above regional oceanic basement depth. Using seismic refraction data, Christeson *et al.* (2014) calculated a full spreading rate of 2.2 cm/year on a profile (Figure 1.5) in the same area. This estimate falls squarely in the slow spreading rate range globally and specifically for the comparable Mid-Atlantic Ridge system (McDonald 1982). Slow-spreading ridges express wide variety in tectonic and volcanic character, reflecting relatively unfocused magmatism (Sempere *et al.* 1993).

Structural-balanced restorations of the eastern Gulf further confirm the LOC location and timing of sea floor spreading (Curry *et al.* 2018). Upper Jurassic (Smackover and Norphlet) strata downlap onto oceanic crust, suggesting oceanic crust formation contemporaneous with deposition (Figure 1.6; see also Section 3.3.4). Latest Upper Jurassic (Haynesville-equivalent) and Cotton Valley intervals extend across all oceanic crust, constraining the end of sea floor spreading at about 155 Ma. These units are also contemporaneous with post-Smackover rafting in the eastern Gulf, suggesting a genetic relationship, as will be explored in Section 3.3.4.

1.2.4 Magnetic Data

Early attempts at mapping the extinct spreading center and LOC (Figure 1.4) were challenged by the generally indistinct character on magnetic data collected from the northern Gulf (e.g., Imbert and Phillippe 2005). This can be partly attributed to the low paleolatitude of the Gulf during the Jurassic, resulting in shallow magnetization vectors that subdued magnetic intensity at the surface but also the poor resolution of older surveys. Newer aeromagnetic data acquired for hydrocarbon exploration in Mexico have better constrained the location of oceanic crust, particularly when integrated with

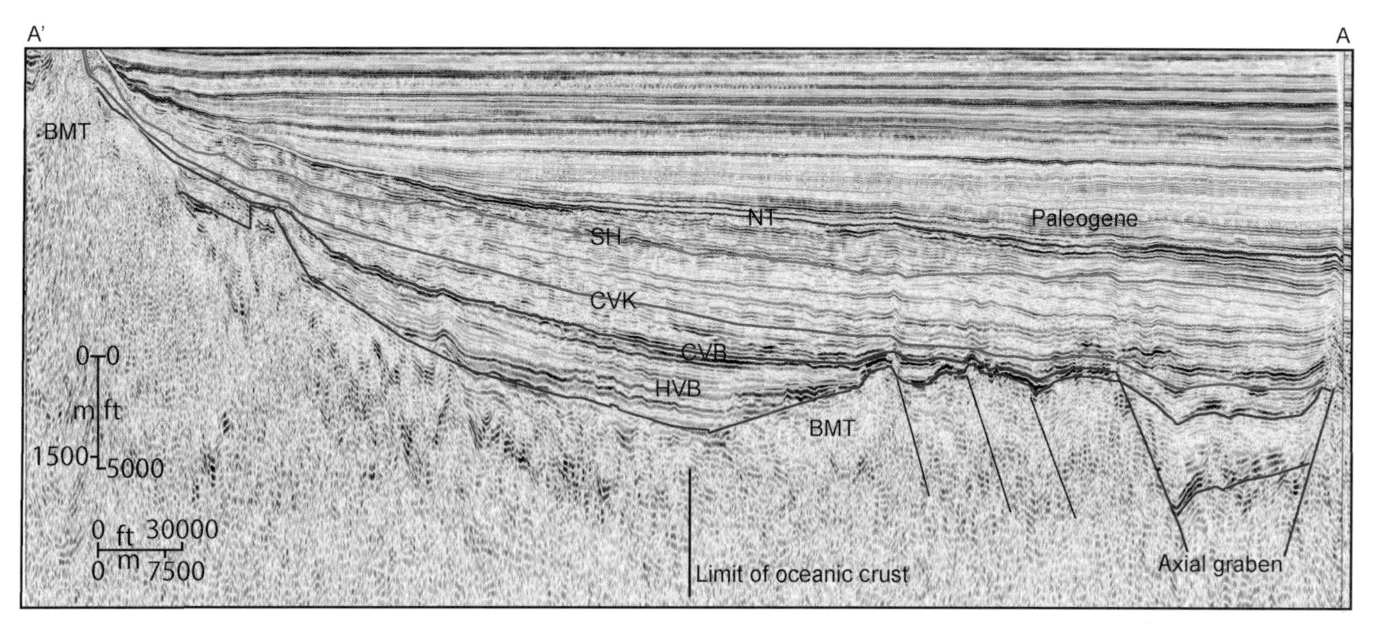

Figure 1.6 Seismic line interpretation in eastern GoM, extending from the Florida Platform across the inferred axial graben of the extinct spreading center showing lapout of HVB, CVB, and CVK supersequences onto oceanic crust. Other correlated horizons are SH, NT, and Paleogene (Wilcox) supersequences. Modified from Snedden *et al.* (2014). Seismic line courtesy of Spectrum. Abbreviations HVB, Haynesville–Buckner; CVB, Cotton Valley–Bossier; CVK, Cotton Valley–Knowles; SH, Sligo–Hosston; NT, Navarro–Taylor; BMT, basement.

comparable vintage northern Gulf data (Pindell *et al.* 2016). One prominent magnetic anomaly located in the central GoM has a distinctive pattern of orthogonally cross-cutting linear features superimposed upon an elongate margin parallel magnetic anomaly, thought to indicate the location of the youngest oceanic crust and thus the position of the extinct spreading center (Pindell *et al.* 2016). The calculated full spreading rates of 1–3.6 cm/year for the entire GoM are comparable to the slow spreading rates (2.2 cm/year) estimated for the eastern GoM (Christeson *et al.* 2014). Another trend, called the Campeche magnetic anomaly, is located downslope of the Yucatán **Platform margin** and constrains the Yucatán (Mayan) block position at the start of pre-salt deposition here, as discussed in Chapter 3.

1.2.5 Gravity Data

Sandwell gravity maps (Sandwell *et al.* 2014) also provide further documentation of the present-day crustal types and their position. Continental crust is generally indicated by gravity highs (e.g., Yucatán block) and oceanic crust by gravity lows, but local variations can occur as a function of igneous intrusions, salt, and depth variations along prominent escarpments.

1.3 Gravity Tectonics

Above the crystalline basement in the greater GoM basin, a thick sedimentary interval exists, deposited largely in the Mesozoic and Cenozoic. Beginning in the Jurassic, robust depositional systems delivered sediment into the basin, the siliciclastic systems fed by rivers draining a variety of source terranes in the northern Rockies, southern Rockies, Appalachians, Quachita Mountains (USA), and Sierra Madres and other areas of Mexico. Siliciclastic systems are particularly prominent in the Cenozoic, but Mesozoic systems of the Jurassic and Cretaceous were, at times, equally impressive in terms of accumulated thickness and caliber of sediment grade. Cenozoic deposition, which extended past the rigid Mesozoic carbonate margins, induced significant basinward translation due to gravitational loading. Shelf margin sediment loading and faulting created accommodation space and, where the Louann Salt was encountered, major salt evacuation. The resulting sedimentary accumulations were unusually thick (often $>$25,000 ft) but barely kept pace in the northern GoM with sediment influx from numerous continental-scale rivers. Loading onto salt also created complex salt mobilization and salt–sediment interaction that set up a wide diversity of trap types, heat flow variations, pathways for hydrocarbon migration, **depositional architectures**, and seal rock distributions.

As will be discussed in Section 9.4, improvements in imaging and illumination of the **subsalt** structure has vastly enhanced our understanding of the early basin history in the slope and **abyssal plain**. Regional to basinal scale seismic analysis has led to recognition of both extensional and contractional tectonics (and even raft tectonics) throughout the Mesozoic and Cenozoic. The extensive seismic and well control means that the structures here are well-imaged and thus studied (Worrall and Snelson 1989; Nelson 1991; Jackson *et al.* 1994; Diegel *et al.* 1995; Peel *et al.* 1995; Watkins *et al.* 1996a; Rowan *et al.* 2000, 2016; Radovich *et al.* 2007).

It is therefore worthwhile to describe some of the important structural styles that have been identified to date. It is also useful to view these tectonic features in the context of structural domains (Section 1.4) and 10 basin-scale cross-sections (Section 1.5).

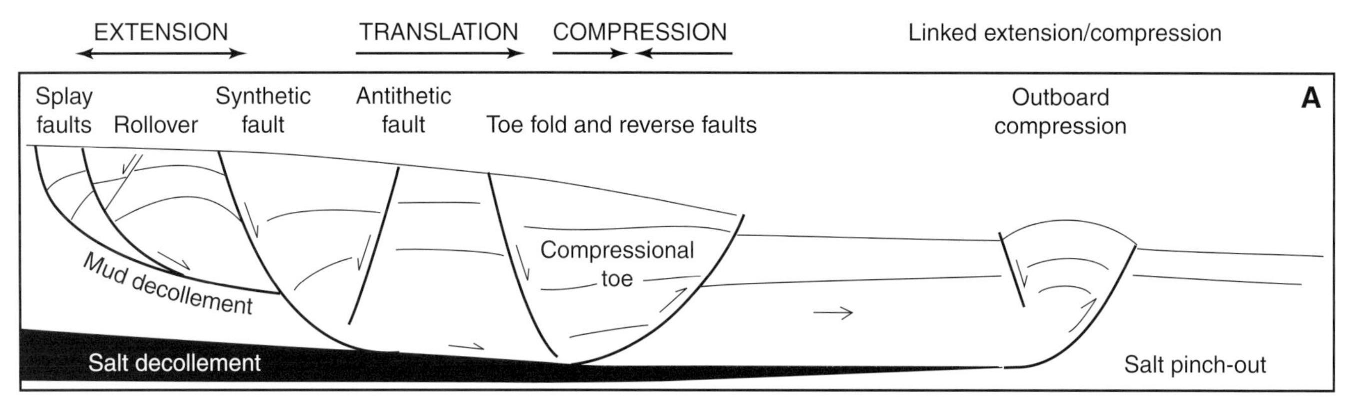

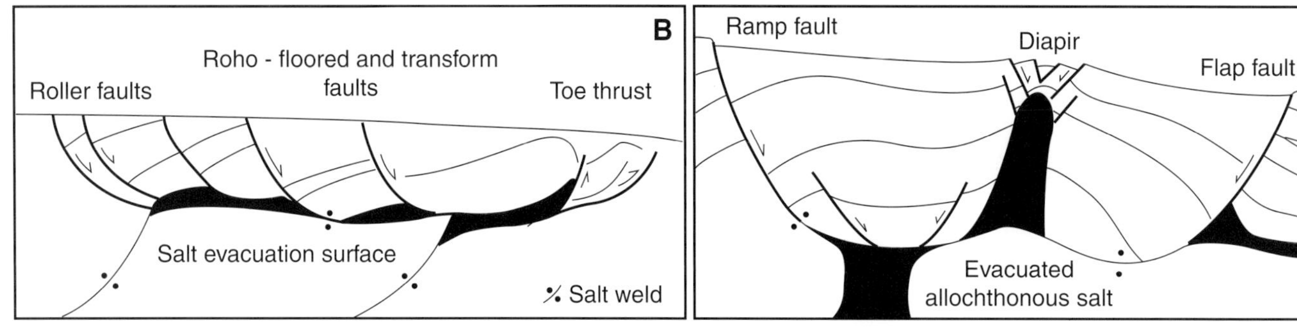

Figure 1.7 GoM gravity tectonics. (A) Linked extension and compression. (B) Roho salt detachment. (C) Salt withdrawal minibasin. From Galloway (2008).

Several pre-conditions set up the complex and diverse assemblages of GoM basin gravity tectonic structures. The combination of a thick, basin-floor Louann Salt substrate, rapid sediment loading, and offlap of a high-relief, continental margin sediment prism has resulted in mass transfer of salt and overpressured mud upward and basinward throughout Gulf history.

1.3.1 Growth Fault Families and Related Structures

Growth faults tend to nucleate and grow during active deposition at the **continental margin** (Winker 1982; Watkins *et al.* 1996b; Jackson and Hudec 2017). Here, extension results from basinward gravitational gliding or translation of the sediment wedge along a **detachment** zone, typically found within salt or overpressured deep marine mud (Rowan *et al.* 2004). Extension creates a family of features, including primary synthetic growth faults, splay faults, antithetic faults, and rollover anticlines (Figure 1.7A). In many parts of the GoM, updip extension is more or less balanced by a similar degree of contraction in downdip areas, as discussed in the following sections.

1.3.2 Basin-Floor Contractional Fold Belts

Basinward gravity spreading or gliding along a detachment zone, and resultant updip extension, requires compensatory compression at the toe of the displaced sediment body (Weimer and Buffler 1992; Hall *et al.* 1993; Fiduk *et al.* 1999; Trudgill *et al.* 1999). Contractional features include anticlinal toe folds and reverse faults (Figure 1.7A). They commonly form at the base of the slope, but can also extend onto the basin plain where a stepped discontinuity or termination of the decollement layer occurs. The deepwater fold belts (Atwater, Mississippi, etc.) are thought to represent adjustments to significant updip extension (Radovich *et al.* 2007). In other areas, extension may be balanced by squeezing salt bodies or salt weld development (Jackson and Hudec 2017; see Section 1.3.6).

1.3.3 Allochthonous Salt Bodies, Including Salt Canopies and Salt Sheets

Loading of the Louann Salt has resulted in regional extrusion of salt basinward and upward (Diegel *et al.* 1995; Fletcher *et al.* 1995; Peel *et al.* 1995). **Allochthonous salt** canopies typically develop beneath the continental slope, where salt rises as a series of coalescing diapirs or as injected tongues. Salt may also be extruded to the surface, forming salt sheets, or nappes, which move basinward, much like salt glaciers (Jackson and Hudec 2017).

1.3.4 Roho Fault Families

Lateral salt extension by gravity spreading creates a linked assemblage of extensional faults and compensating, downslope compressional toe faults, anticlines, and salt injections in the overlying sedimentary cover (Rowan 1995; Schuster 1995). In some cases, the top of allochthonous salt can acts as a decollement surface for faults (Figure 1.7B), as does autochthonous salt previously described. These are called **roho systems** and often occur in stratigraphically distinct fault groups or **fault families**.

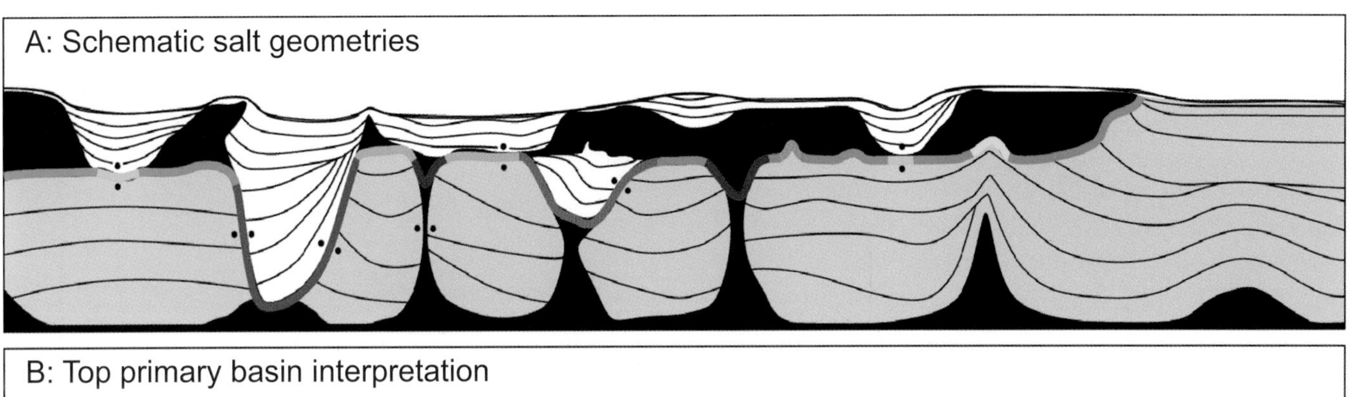

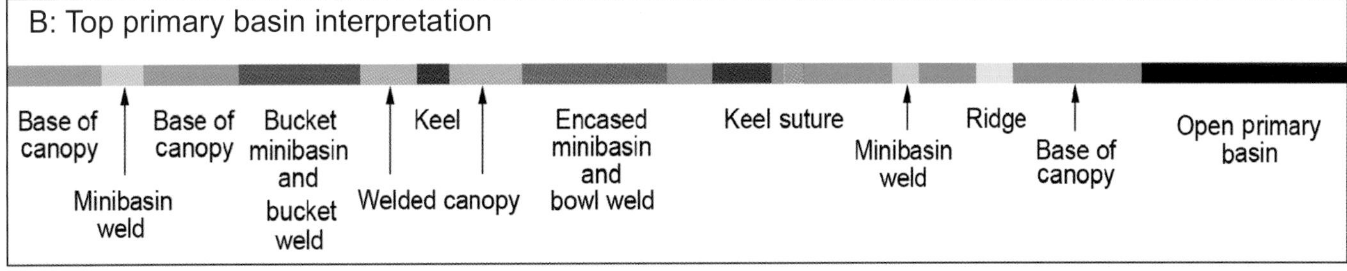

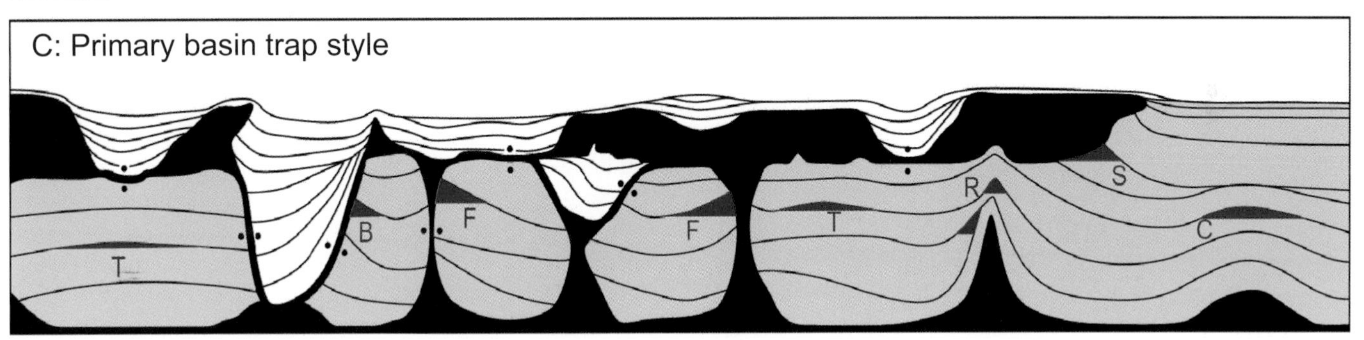

Figure 1.8 Schematic cross-sections of the bucket weld province. (A) Schematic salt geometries based on seismic interpretation. (B) Primary top basin interpretation. (C) Primary basin trap style. Letters indicate different trap styles in subsalt domain. Modified from Pilcher *et al.* (2014).

1.3.5 Salt Diapirs and Their Related Withdrawal Synclines and Minibasins

In the Gulf margin basins and embayments, salt **diapirs** rise directly from the autochthonous Louann "mother" salt (Seni and Jackson 1983; Fletcher *et al.* 1995; Rowan 1995; Rowan and Weimer 1998). Basinward, depositional loading of salt canopies and sheets beneath shelf and slope areas also causes renewed salt stock evacuation, creating high-relief salt diapirs and intervening depressions (Figure 1.7C). Progressive **salt evacuation** creates shifting, localized sites of extreme subsidence and sediment accumulation. Resulting features include withdrawal synclines created by local evacuation of salt from diapir flanks, bathymetric depressions, called minibasins, that form local depocenters, turtle structures, and local fault families, including down-to-basin ramp faults, counter-regional flap faults, and crestal faults above salt bodies.

1.3.6 Salt Welds

Salt welds are surfaces or zones that join strata originally separated by either autochthonous or allochthonous salt (Hudec and Jackson 2011). These are present where nearly complete expulsion of salt from stock feeders, dikes, salt tongues, or salt canopies has occurred (Jackson and Cramez 1989; Jackson *et al.* 1994; Figure 1.7B,C). Because the welds form some time after the deposition of adjacent strata, these juxtapose discordant stratigraphic intervals, sometimes with significant angularity of converging reflections (Hudec and Jackson 2011). Primary, secondary, and tertiary welds can be identified on the basis of the type of salt body that was welded (Jackson and Hudec 2017). Welds can also serve as detachment surfaces for younger listric faults.

Younger (secondary) sedimentary minibasins may be welded against older (primary) minibasins, resulting in drastically different ages, lithologies, and subsurface pressures (Pilcher *et al.* 2011; Figure 1.8). These are particularly prominent in a portion of the central GoM, the so-called "**bucket weld**" province. These bucket welds can act as lateral boundaries to hydrocarbon traps.

Salt welds can also act as regional decollement surfaces even when obvious linkage to downdip contraction is lacking. Regional decollements at welds are also known to be significant horizontal pressure barriers, with a significant increase in pressure in sub-weld intervals and attendant increase in risk, uncertainty, and well costs (see Section 9.20.1 on the Wilcox deep shelf play). Reverse faults can occur along welds (thrust welds), as observed in Campeche.

1.3.7 Rollovers and Expulsion Rollovers

Thickening and bending of strata toward a listric normal fault is commonly observed in the GoM Cenozoic and Mesozoic intervals. If expulsion of salt occurs to cause stratal thickening and rotation, with or without a fault, this structure is referred to as an **expulsion rollover** (Ge *et al.* 1997; Jackson and Hudec 2017). Large expulsion rollover structures have been identified in the Mississippi Canyon protraction block and represent some of the largest undrilled prospects in the basin (Harding *et al.* 2016). The orientation of these expulsion rollovers may indicate the general direction of sediment transport and loading (McDonnell *et al.* 2008), though these features are several orders of magnitude larger than depositional clinoforms and should not be used to indicate the location of paleo-shelf margins.

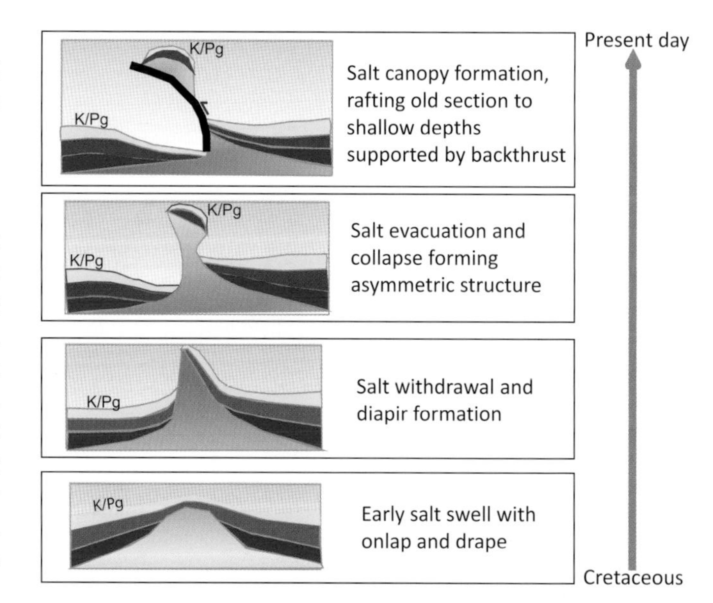

Figure 1.9 Development of a salt-related carapace structure. Modified from M. Rowan (pers. comm.).

1.3.8 Carapaces and Rafts

When moving salt carries roof material that is not firmly attached to surrounding strata, stratigraphic discontinuities can occur. Transported roof material can be tens of kilometers in lateral extent and sometimes as thick as the salt body (Jackson and Hudec 2017). The term carapace is used here in a restrictive sense to describe detached blocks above salt that have moved vertically relative to the surrounding strata, either actively by diapir rise or passively as younger sediments are deposited around the salt-supported blocks (Figure 1.9). Early drilling at or around the allochthonous salt canopy encountered blocks which tended to be older, thinner, and/or more stratigraphically condensed than the adjacent non-carapace interval (Hart *et al.* 2004). Carapaces are often structurally much higher than the regional level of coeval strata. For example, the Norton well (GB 754 #1) penetrated a carapace block where Top Cretaceous was encountered at 7180 ft (2189 m), much shallower than the regional depths of Cretaceous, closer to 30,000 ft (9.1 km; Cunningham *et al.* 2016). Initially, stratigraphic discontinuities within carapaces caused considerable confusion, including the misinterpreted Middle Cretaceous unconformity (MCU), which later analyses proved was actually the Cretaceous–Paleogene boundary (K–Pg; Dohmen 2002).

Carapaces do accumulate sediment above a diapir documenting that diapir's history, but also record information on older strata that is relevant to regional or basin reconstructions. Carapaces containing organically enriched intervals within both the Tithonian and Ceno-Turonian intervals provide critical evidence in characterization of these source rocks (Cunningham *et al.* 2016).

Rafts are more complicated salt tectonic features that are defined in two different ways. First, we recognize rafts as stratigraphic blocks formed as part of raft tectonic processes. Raft tectonics is a form of thin-skinned extension, with unusually large degrees of extension such that the footwall and hanging wall are often not in contact, unlike growth faults (Jackson and Hudec 2017). Raft gaps are filled in by synkinematic (syn-extensional) strata. Raft tectonics is well-documented in the Albian interval of Angola and the Oxfordian interval of the DeSoto Canyon protraction block (Pilcher *et al.* 2014).

A second use of the term raft applies to stratigraphic blocks that have been moved considerable distances downslope by allochthonous salt. For example, it is established from 3D seismic analysis that the salt canopy in the deepwater northern GoM has transported over 20 raft blocks across the Alaminos Canyon, Keathley Canyon, Walker Ridge, and Green Canyon protraction blocks, with distances ranging from less than 3 km to more than 80 km from their original positions (Fiduk *et al.* 2014). Over 3100 km^2 of rafted strata was identified, largely accumulating near the terminus of the salt canopy.

Primary or secondary minibasins (terminology of Pilcher *et al.* 2011) can become encased in salt as allochthonous salt flows over the minibasin subsiding onto a deeper salt level (Hudec and Jackson 2011). In some cases, salt evacuation continues, and the minibasin is instead surrounded by welds (Rowan and Inman 2011).

Thus, it is very important to consider the tectonic history of vertical and lateral salt transport when analyzing stratigraphic information from carapaces, rafts, and encased minibasins. Stratigraphic discontinuities are common, and in areas of poor seismic imaging are only revealed by drilling and biostratigraphic analysis. Some wells have penetrated salt-overturned intervals, where biostratigraphic datums are encountered in reverse order, resulting in major drilling "surprises" (Box 1.1).

Box 1.1 Stratigraphic Surprises Caused by Salt Tectonics

Seismic correlations in the deepwater GoM are often challenging due to the complexity of salt tectonics, limits on illumination below the thick and continuous allochthonous salt canopy, and imaging constraints around parautochthonous salt. In some areas, seismic imaging has failed to reveal the true stratal geometries, resulting in unanticipated structural interpretation problems encountered while drilling (Olson *et al.* 2015).

A prime example is a well drilled in the Green Canyon protraction block 639 (GC 639 #1), drilled in 2009 (Figure 1.10). After drilling through a normal Pleistocene to Pliocene stratal interval, and then a thick allochthonous salt body, the well began to encounter Cretaceous strata in reverse stratigraphic order. The quality of biostratigraphic tops was reasonably good, with most referred to as "definite" (DEF). Plotting the absolute ages of the various biohorizons indicates some variation but an overall downward younging of the interval (Figure 1.10A). This must have caused some concern, particularly if the trend was unanticipated. The well reached total depth (TD) near the top of the overturned Cretaceous. It is likely that the Mesozoic interval penetrated by GC 639 #001 was overturned by salt (Figure 1.10B).

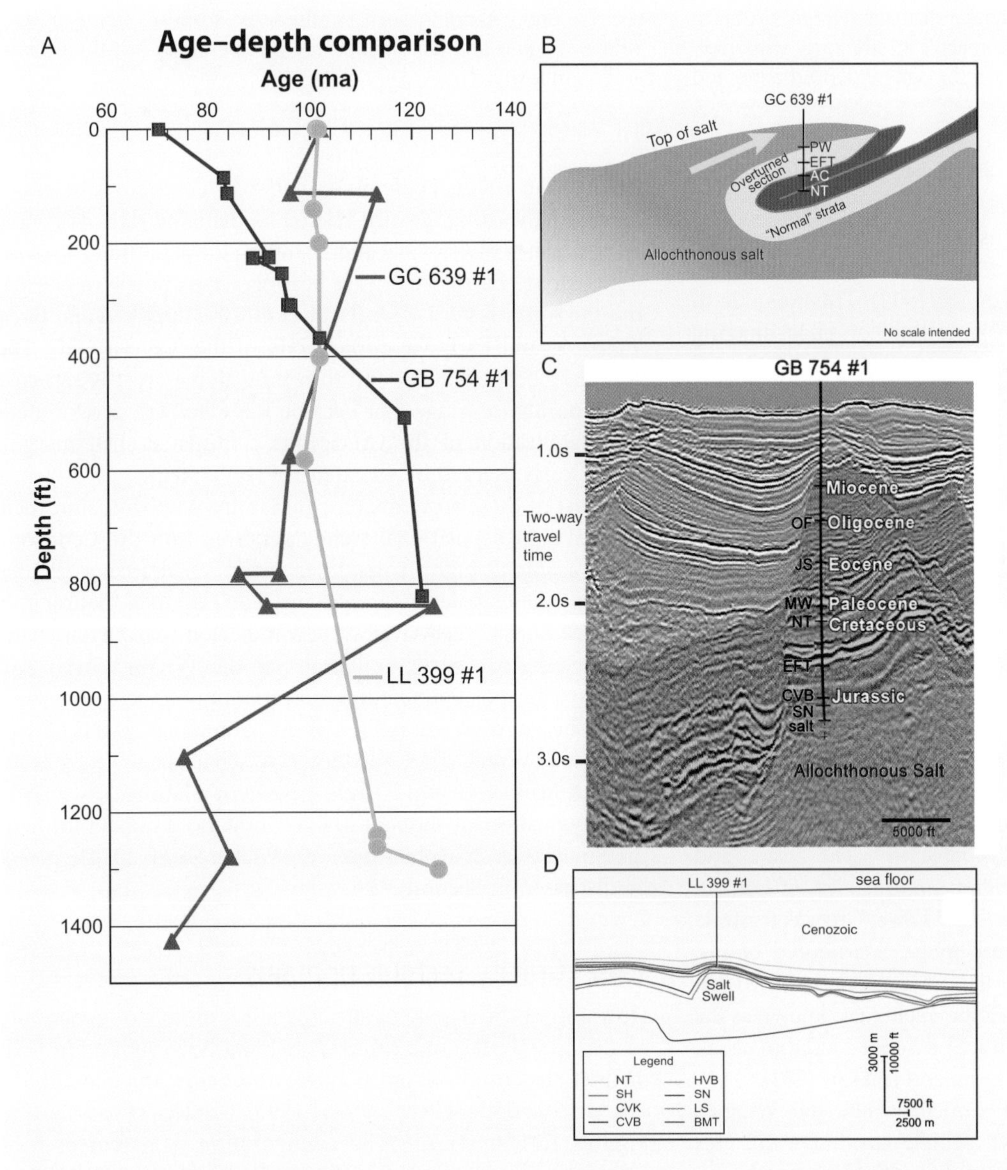

Figure 1.10 (A) Age–depth comparison of three deepwater GoM wells with depths (in feet) normalized to shallowest Mesozoic horizon penetrated in each well. Diversity of Mesozoic well penetrations in the deep GoM basin are illustrated by (B). (B) Well GC 639 #1 shown on a schematic structural configuration drawn from the original 2D seismic line. (C) Well GB 754 #1 (modified from Hart *et al.* 2004) is shown with stratigraphic horizons used in this book shown to the left of the well bore. (D) Well LL 399 #1 on a schematic structural configuration drawn from the original 2D seismic line. Sediment accumulation rates and structural geology related to the wells are discussed in the text.

Box 1.1 *(cont.)*

The GC 639 #1 well contrasts with the normal stratigraphic order encountered by two other wells penetrating the same stratigraphic interval elsewhere in the basin (Figure 1.10). GB 754 #1 (Norton **Prospect**) drilled through a stratigraphically condensed interval above a shallow salt structure or carapace feature (Figure 1.10C). LL 399 #1 (Cheyenne Prospect) tested a deep (parautochthonous) salt structure (Figure 1.10D).

GB 754 #1 has a relatively continuous, but low-sloping trend in comparison with most of the LL 399 #1 interval penetrated (the exception being the bottom 30.48 m [100 ft] just above the salt). The estimated sediment accumulation rates (<16 ft/my [4.8 m/my]) of GB 754 #1 are far lower than those of LL 399 #1 (>55 ft/my [16.7 m/my]), consistent with the former well having penetrated a condensed interval on a salt–carapace structure (Figure 1.10C). LL 399 #1 well penetrated a lower relief salt structure, with high sediment accumulation rates above the salt-influenced zone and lower rates just above the salt (Figure 1.10D). Seismic data confirm these structural differences. Taking overturning of the Mesozoic interval by salt into account, calculated sediment accumulation rates of >47 ft/my (14.3 m/my) for GC 639 #1 is more comparable with that of LL 399 #1 than with GB 754 #1, consistent with the idea that the interval penetrated in GC 639 #1 was originally within a subsiding, deepwater, primary basin until salt emplacement overturned the Mesozoic interval, rather than a modified carapace structure or other structural oddity.

Numerous other wells have encountered either reverse stratigraphic intervals, thin and condensed intervals on carapaces, or intervals that are spatially out of place due to salt rafting (Hart *et al.* 2004; Fiduk *et al.* 2014). Biostratigraphic analyses are a key tool for understanding such stratigraphic surprises, particularly where imaging and illumination are hampered by salt thickness and complexity.

1.4 Structural Domains

Original basin-scale cross-sections of the greater GoM largely used well logs to define structural provinces of the basin (e.g., Morton *et al.* 1988; Morton and Ayers 1992). This was due to the lack of long, regional 2D seismic lines or poor imaging around salt or various complex structures. Nonetheless, broad structural domains were defined and have, for the most part, been confirmed by new seismic interpretations. The exceptions are subsalt structural provinces, for obvious reasons, and areas with limited well control.

1.4.1 Basement Structural Province

The periphery of the greater GoM basin, underpinned by thick transitions to the continental crust, is segmented by a series of prominent basement structures (Figure 1.3; Ewing and Lopez 1991). Their influence on overlying stratigraphy has long been known as early exploration efforts targeted these, based on gravity, magnetics, or early single- or multi-fold seismic reflection. Established structures include a halo of embayments (epicratonic basins that open to the central Gulf) and closed basins and intervening arches and uplifts (Figure 1.3; Ewing 1991). The basins and embayments typically contain a significant thickness of Louann Salt and thicker sequences of Jurassic and Lower Cretaceous strata relative to the adjacent arches and uplifts. Salt-floored basins, including the East Texas basin, North Louisiana salt basin, Mississippi salt basin, and Apalachicola Embayment (also known as the DeSoto Canyon salt basin) contain well-described families of salt domes and related structures (e.g., Seni and Jackson 1983).

Basement highs, arches, and anticlines like the Wiggins Arch, Sabine Uplift, Llano Uplift, Middle Ground Arch, etc., are known to have been reactivated multiple times during multiple Mesozoic and Cenozoic tectonic events (Ewing 1991). Several of these marginal highs, including the San Marcos Arch, Sabine Arch, and Monroe Uplift display short pulses of uplift of as much as a few hundred meters, creating angular unconformities in Middle Cretaceous and Lower Eocene strata (Laubach and Jackson 1990). These pulses generally correlate to stages of Laramide thrusting, in turn related to changing rates of Pacific margin plate convergence and changing intracratonic compressional stress. Extensive crustal heating across northern Mexico and the southwestern USA (Gray *et al.* 2001) uplifted and tilted Mesozoic and Early Cenozoic strata of the western Gulf. The boundary between thick and thin transitional crust is reflected by a subsidence hinge that became the focus for development and stabilization of the Cretaceous continental shelf margin, most clearly marked by an extensive reef system.

Most of these structures formed in the Mesozoic, but their influence on depositional trends persisted into the Cenozoic. The basement highs sometimes acted as drainage divides between major paleo-river systems. For example, mapping of the Tuscaloosa paleo-river system indicated that fluvial channels avoided the structural highs of the Wiggins Arch and surrounding positive features, terminating at the coeval lowstand shelf margin where the Ceno-Turonian interval is greatly expanded (Woolf 2012; Snedden *et al.* 2016b).

The Middle Ground Arch (Southern Platform), which is entirely offshore, is thought to have influenced radial rafting of the Smackover–Norphlet interval (Pilcher *et al.* 2014), as will be discussed in Section 3.3.4.

1.4.2 Gravity Tectonic Domains

Downdip of the basement structures is a mosaic of genetically related gravity tectonic features that can be grouped into two distinct structural domains, above and below the allochthonous salt canopy (Peel *et al.* 1995; Hudec *et al.* 2013b; Figures 1.11 and 1.12). The configuration of basement rock below the canopy controlled the original Louann Salt distribution and its immediate post-depositional downdip migration onto contemporaneous or newly formed oceanic crust. The near-end of Cretaceous canopy architecture in turn influenced Cenozoic depositional patterns and faulting, detachment, and further basinward translation of salt and sediments.

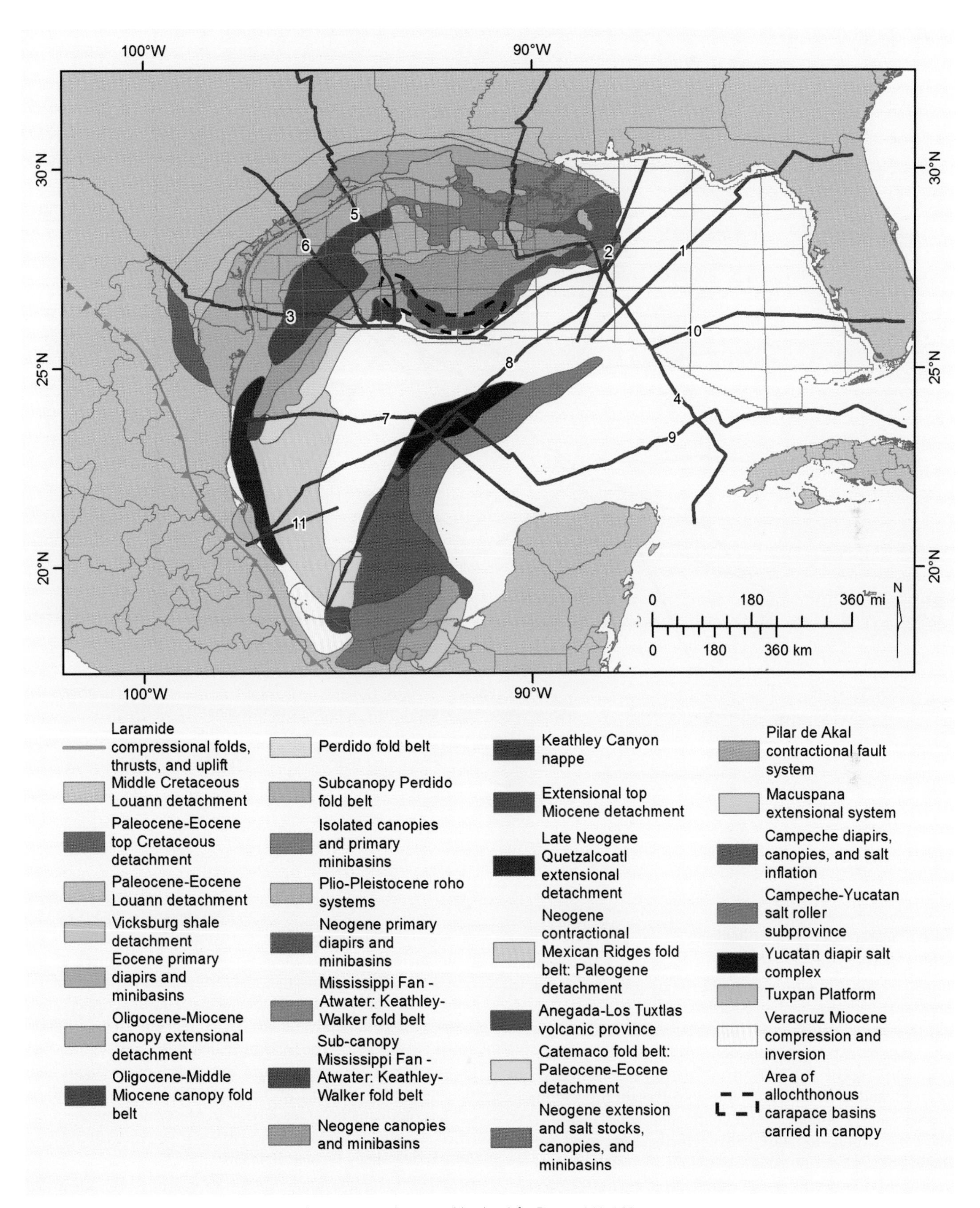

Figure 1.11 Tectonostratigraphic provinces with cross-section locations (blue lines) for Figures 1.13–1.22.

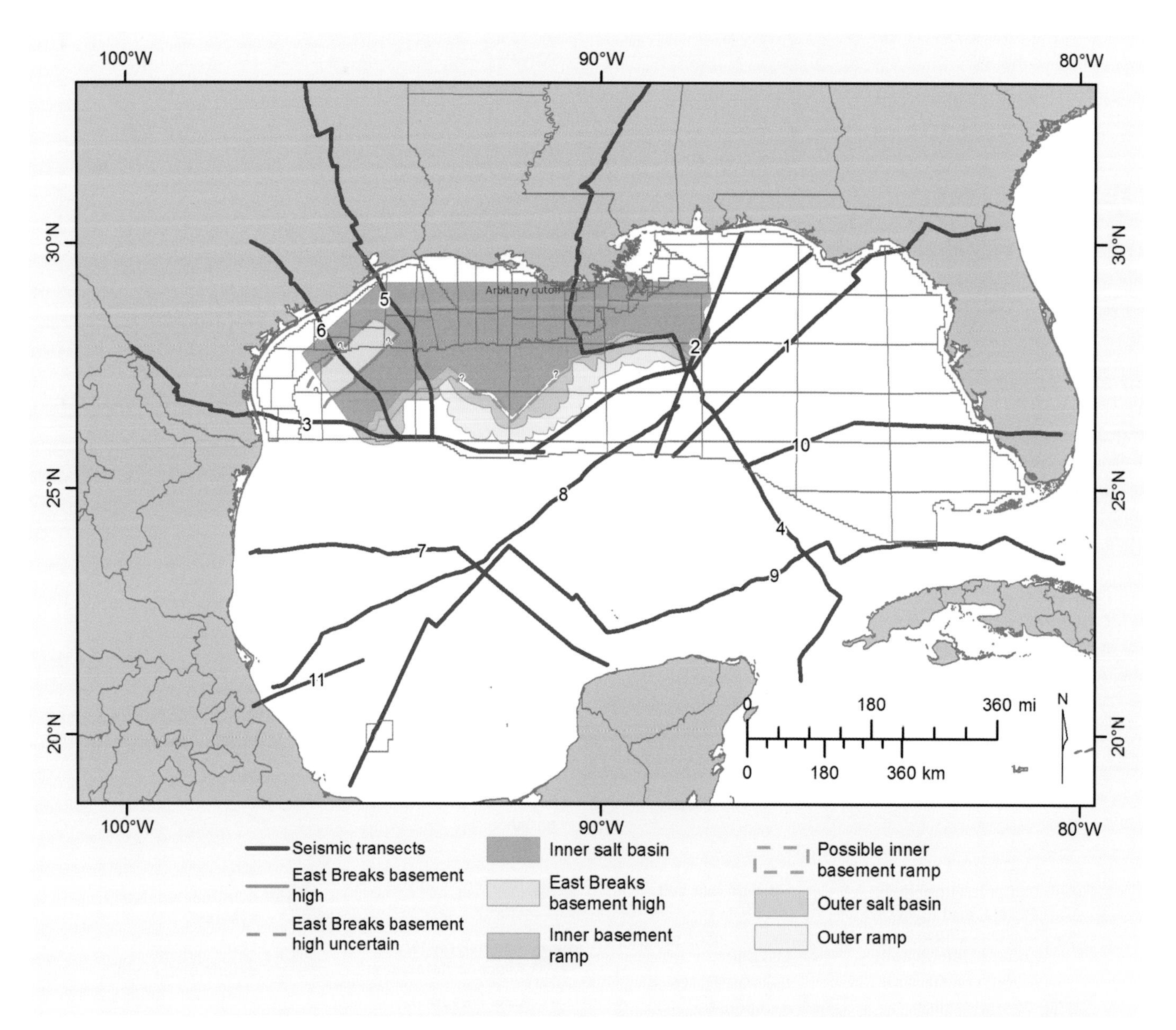

Figure 1.12 Subcanopy structural domain with cross-section locations (Figures 1.13–1.22). Modified from Hudec *et al.* (2013b)

1.4.2.1 Supracanopy Tectonic Domains

With a few exceptions, the northern GoM suprasalt structural domains had a finite time span of primary growth that can be associated with one or more successive episodes of Cenozoic siliciclastic sediment accumulation. Suprasalt domains generally become younger basinward, beginning with the Paleocene–Eocene detachment (at the top of the Cretaceous interval) and culminating in the Plio-Pleistocene minibasin and salt canopy domains of the continental slope (Figure 1.11). These structural domains will be further discussed in the context of the 10 regional cross-sections in Section 1.5.

The Middle Cretaceous Louann detachment, Oligocene–Lower Miocene, and Miocene compressional domains are exceptions to this general pattern. In addition, the full array of gravity tectonic structure domains of the northern Gulf basin includes the salt diapirs and related structures of the East Texas, North Louisiana, Mississippi, and DeSoto Canyon salt basins, which lie around the northern basin periphery, and a series of peripheral grabens, including the Luling–Mexia–Talco, State Line, and Pickins–Gilberton fault zones that delimit the landward extent of autochthonous Louann Salt. As mentioned earlier, growth of structures within these inboard domains occurred largely in Mesozoic time.

1.4.2.2 Subcanopy Tectonic Domains

Subsalt structural domains have only recently been identified due to the thick salt canopy that resisted illumination and imaging (Figure 1.12). Here, Louann Salt rests largely upon the "acoustic" basement, with limited coherent seismic data below this point. Seismic surveys, first in the Campeche basin of Mexico and then in the deepwater northern GoM, indicated a pronounced landward-dipping step in the acoustic basement, termed the inner ramp (Hudec *et al.* 2013b; Figure 1.12). With an estimated elevation of 1–4 km (depending on area and the local velocity model), this change in the base of salt is thought to represent the limit of the oceanic crust. This implies in turn that inboard transitional crust must be thinner or denser than the outboard oceanic crust (Hudec *et al.* 2013b). It also marks an important boundary for the original limit of the Louann Salt prior to extrusive sea floor spreading. Post-salt depositional creep onto oceanic crust varied as a function of ramp dip and depth, with greater salt advances in the area of the Walker Ridge salient, bounded to the west by the Brazos transfer fault (Figure 1.12). Original salt thicknesses of 3–4 km (1.7–2.5 miles) are thought to progressively decrease from the broad inner basin to the outer basin to the thinnest interval in the outer ramp perched on the oceanic crust (Hudec *et al.* 2013a). Salt canopy feeders are concentrated in the inner basin, where original source salt thicknesses were largest.

The concentration of contractional fold belts in the outer ramp and outer basin is not coincidental. Parts of the outer ramp were reactivated as thrusts during Miocene shortening of the Atwater fold belt (Hudec *et al.* 2013b). Large compressional anticlines of the Perdido fold belt, dated as Oligo-Miocene, are formed in the outer basin of the Alaminos Canyon area (Rowan *et al.* 2000). The Timbalier fold belt is found in the inboard subsalt or sub-weld region, but is thought genetically unrelated but similar in timing, reflecting uplift and seaward tilting of the onshore northern GoM during the Miocene (Jackson *et al.* 2011).

Another prominent subcanopy feature is the East Breaks basement high, where the acoustic basement is thought to be as shallow as 48 km (29 miles), based on new 3D wide azimuth (WAZ) seismic surveys (M. Hudec, pers. comm.). This structure likely effects local heat flow and depositional patterns of sediments as young as Oligo-Miocene age.

1.5 Basin-Scale Cross-Sections

Basin-scale cross-sections (Figures 1.13–1.22) illustrate the fundamental sedimentary and structural architecture of the greater GoM basin. All cross-sections are based upon newer vintage or recently reprocessed 2D depth-imaged seismic lines across the basin. Seismic horizons are correlated from well penetrations, which constrain the age, lithologic character, and paleo-environment of the Mesozoic and Cenozoic stratigraphic units.

These 10 representative cross-sections have been selected to illuminate several important observations: (1) change in deep crustal types (oceanic, transitional, continental); (2) important structural domains; (3) topographic/bathymetric features; (4) notable gravity tectonic features; (5) prominent basement structures; and (6) the large-scale, progressive shift in age of deposition and paleo-environment of the basin-fill. Shelf platform margins (Mesozoic) and shelf–slope interfaces (Cenozoic) are important paleophysiographic features that usually mark the transition from contemporaneous shelfal processes of waves, currents, and tides to the sedimentary gravity flows and mass transport/failures-dominated slope to abyssal plain. Note the boundary between thick and thin transitional crust, which became a subsidence hinge point, and often marks the position of the Mesozoic shelf platform margins, which in turn influenced Cenozoic expanded intervals due to increased accommodation.

Interpretation of subcrustal structure, such as the top of the mantle, was not attempted due to the use of 2D seismic reflection data that rarely permits unequivocal selection of the Moho boundary. Seismic refraction data, discussed in Section 1.2.2 provides guidance on cross-sections 1, 2, and 6 (see Van Avendonk *et al.* 2013, 2015; Christeson *et al.* 2014; Eddy *et al.* 2014), but recent vintage data was not available along the other sections.

It is important to note that in the last five years we have learned a lot more about the GoM through the effort to link onshore and offshore seismic data by seismic companies like ION. The reprocessing of older onshore data and merging with offshore data has allowed the first truly basinal cross-sections to be developed. For example, the presence of multiple, linked extensional–contractional structural belts of Early Paleogene and Oligo-Miocene age became evident (Radovich *et al.* 2007). From Cretaceous to Pleistocene, there is a repeated basinward migration with expansion as each interval fills the space in front of it created by extension and salt withdrawal. Radovich *et al.* (2011) estimated 100+ miles (161 km) of progradation and over 15,000 ft (4570 m) of aggradation. Without the dedicated efforts of seismic companies like ION, our understanding of this complex basin would not have been possible.

1.5.1 Cross-Section 1: Sigsbee Abyssal Plain to Peninsular Arch

Cross-section 1 is a transect from the northeastern Gulf, passing from the abyssal plain at the USA–Mexico international border to onshore northern Florida (Figure 1.13). Continental crust rises to depths as shallow as <1500 m (4920 ft) on the Peninsular Arch, as crystalline basement has been drilled in a number of onshore wells (Jordan *et al.* 1949). Penetrations include granites dated at 159 ± 3 Ma and basalts and diabases as young as 183 ± 5 Ma in the exotic Suwannee terrane of south Florida (Heatherington and Mueller 2003). While the Cenozoic interval is relatively thin in comparison to the central Gulf (reflecting limited fluvial input), the Mesozoic interval thickens substantially to the southwest, into the area of the Florida Middle Ground Arch and Tampa Embayment. The physicographic slope marks an abrupt termination of many Mesozoic units at the Florida Escarpment. Cretaceous strata have actually been dredged from the sea floor, suggesting that

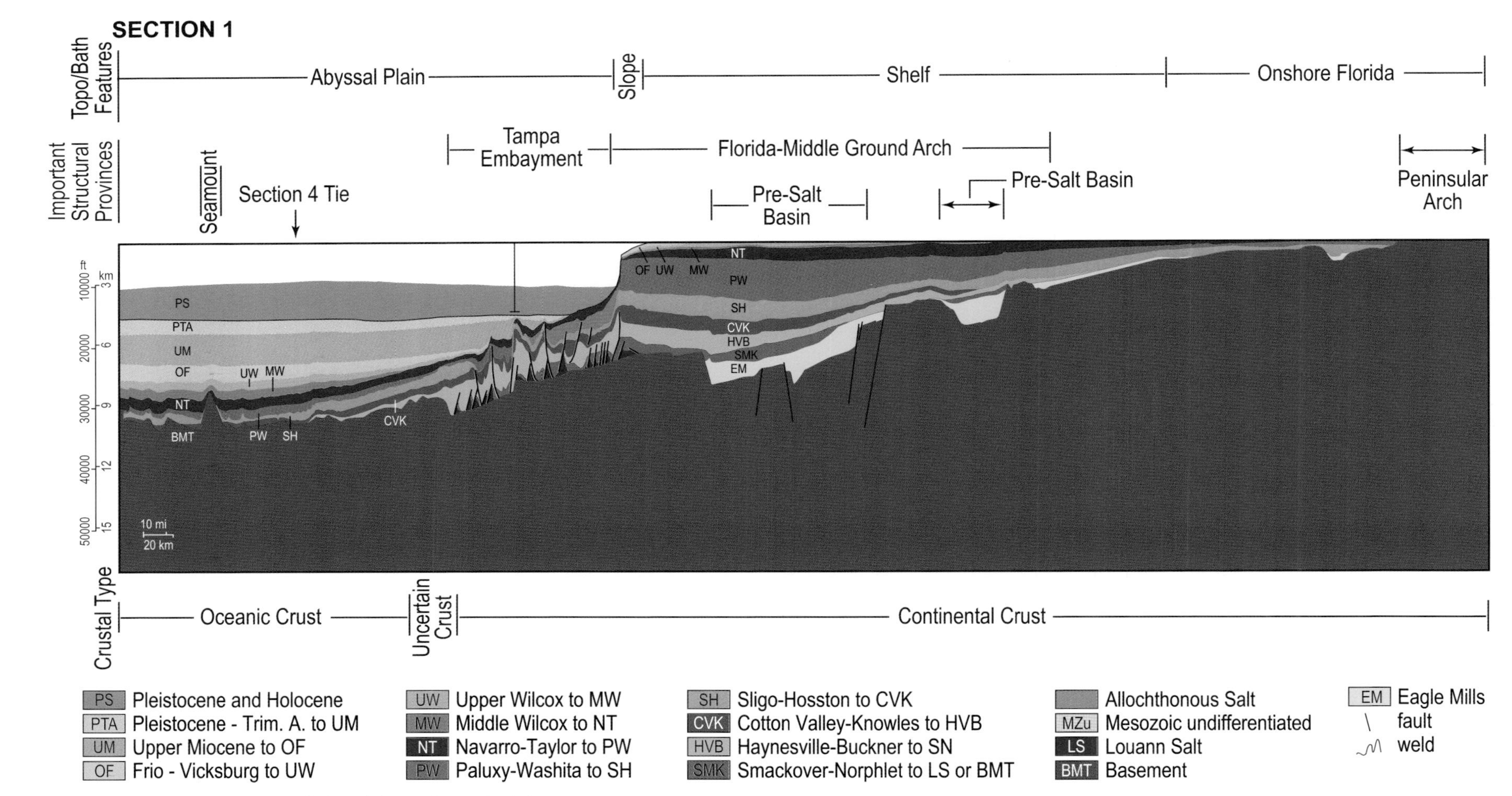

Figure 1.13 GoM cross-section 1: US abyssal plain to Peninsular Arch.

the escarpment is an erosional remnant inherited from the K–Pg impact event and subsequent slope failures and adjustments (Freeman-Lynde 1983).

The shelf portion of the cross-section includes an interpreted pre-salt interval, sedimentary rocks likely of Triassic–Middle Jurassic age, known in onshore areas as the Eagle Mills. The interpreted seismic structure is that of a horst/graben, possibly a continuation of the east coast rift system documented in the coastal plain of South Carolina, Georgia, and other states (Heffner 2013; Goggin and Rine 2014; Rine *et al.* 2014). The nearby well GV-707 penetrated a poorly dated siliciclastic interval between the Cretaceous (Aptian–Valanginian) Sligo–Hosston and Paleozoic carbonates.

Further seaward is the basinal portion of the Tampa Embayment where Louann Salt diapirs and Mesozoic rafts of Smackover and Norphlet are thought to be present, as observed in the DeSoto Canyon area (see discussion of cross-section 2, Figure 1.14). No salt canopy formed here, reflecting the general thinning of original salt toward the southeast.

A pronounced step up in basement, a change in elevation of several kilometers, is coincident with the termination of Louann Salt. This marks the seaward limit of transitional continental crust. A short segment of uncertain crust gives way seaward to oceanic crust, documented by magnetic and gravity data, as discussed in Section 1.2.4).

The downlap of Mesozoic stratigraphic units onto transitional and oceanic crust provides some indication of timing of oceanic crust emplacement (Snedden *et al.* 2013). Oxfordian Norphlet and Smackover rafts appear to have glided onto the oceanic or uncertain crust, suggesting that salt was present during the initial stages of sea floor spreading in order to provide a decollement. In other areas, locally thick minibasins (primary basins) are thought to contain Norphlet- or Smackover-equivalent strata (M. Hudec, pers. comm.).

The Haynesville (Kimmeridgian) and basal Cotton Valley–Bossier (Tithonian) strata continue across the oceanic crust before lapping out onto the oceanic crust near the extinct spreading center. Subsequent depositional units continue across the section, though distal thinning is observed on the seismic sections.

A pronounced structural feature, located at the basement step, fits the established characteristics of a seamount, a basement feature with an elevation greater than 1 km (3280 ft) above the regional basement level (Snedden *et al.* 2014). Such seamounts are relatively common across this area of oceanic crust (Stephens 2009, 2010). As mentioned in Section 1.2.4, the slow spreading rates associated with the GoM opening are thought to be associated with unfocused magmatism and in turn the poorly organized distribution of seamounts like this. Cenozoic strata from Cretaceous upward to Middle Wilcox drape the seamount, and compactional related features extend upward to the Oligocene. A number of these structural features have been leased in recent years, yet no drilling plans have yet been filed with the = Bureau of Ocean Energy Management (BOEM) or Bureau of Safety and Environmental Enforcement (BSEE).

As mentioned, Cenozoic deposition on the Middle Ground Arch and adjacent onshore Florida is thin, with limited accommodation on the Mesozoic platform here. By contrast, Neogene strata, particularly the Pleistocene interval of 5000 ft (1524 m), thicken substantially over the abyssal plain. This marks the Pleistocene Mississippi River input, but also the Miocene contributions by the paleo-Tennessee system. Paleogene deposition thins toward the platform margin, reflecting general western (Laramide) sources and linked drainage networks.

1.5.2 Cross-Section 2: Florida Shoreline to USA–Mexico International Border

Cross-section 2 is located further to the west, extending from the USA international border to just seaward of the Florida shoreline (Figure 1.14). Mesozoic strata, particularly Jurassic and Early Cretaceous intervals, thicken dramatically into the DeSoto salt basin, also known as the Appalachicola Embayment. The Florida Middle Arch is prominent and its extension into the deepwater is the site of significant industry exploration efforts and nearby Norphlet reservoir discoveries such as the Appomattox Prospect (see Section 9.6).

The Norphlet raft province is well illustrated here (Figure 1.14). As described in Section 3.3.4, raft tectonics is a form of thin-skinned extension, with unusually large degrees of extension such that the footwall and hanging wall are often not in contact (Jackson and Hudec 2017). The dismembering of stratigraphic units occurs as blocks glide downslope on a detachment surface, in this case the top of the Louann Salt. Intervening troughs between raft blocks are filled with younger units, providing age control on the timing of rafting. Rafting apart of the Norphlet–Smackover interval must have been contemporaneous with deposition of the Haynesville–Buckner and Cotton Valley–Bossier as these fill in the gaps between rafts. In some areas Cotton Valley–Knowles and even basal Sligo–Hosston also fill raft gaps. The timing of sea floor spreading and rafting is similar enough to consider the possibility that there is a genetic linkage. Rafting is largely toward the oceanic crust, though radial rafting reflecting the Middle Arch structure has been suggested (Pilcher *et al.* 2014). The Smackover–Norphlet Rafts can be separated by diapirs or depositional troughs, making paleogeographic reconstructions very difficult, but essential to exploration well locations, as will be discussed in Section 3.3.4.

Further seaward, the uncertain zone between oceanic and continental crust is also accompanied by a basement step, though with less relief than observed on cross-section 1 (Figure 1.13). Salt appears to have crept onto unequivocal oceanic crust. Though the Louann Salt is now relatively close to its original position, the basinward translation necessitates modification of the term "autochthonous" salt to "**parautochthonous**" salt, following the nomenclature of Hudec *et al.* (2013a). This is also the area where a small segment of the Mississippi Fan–Atwater fold belt is present,

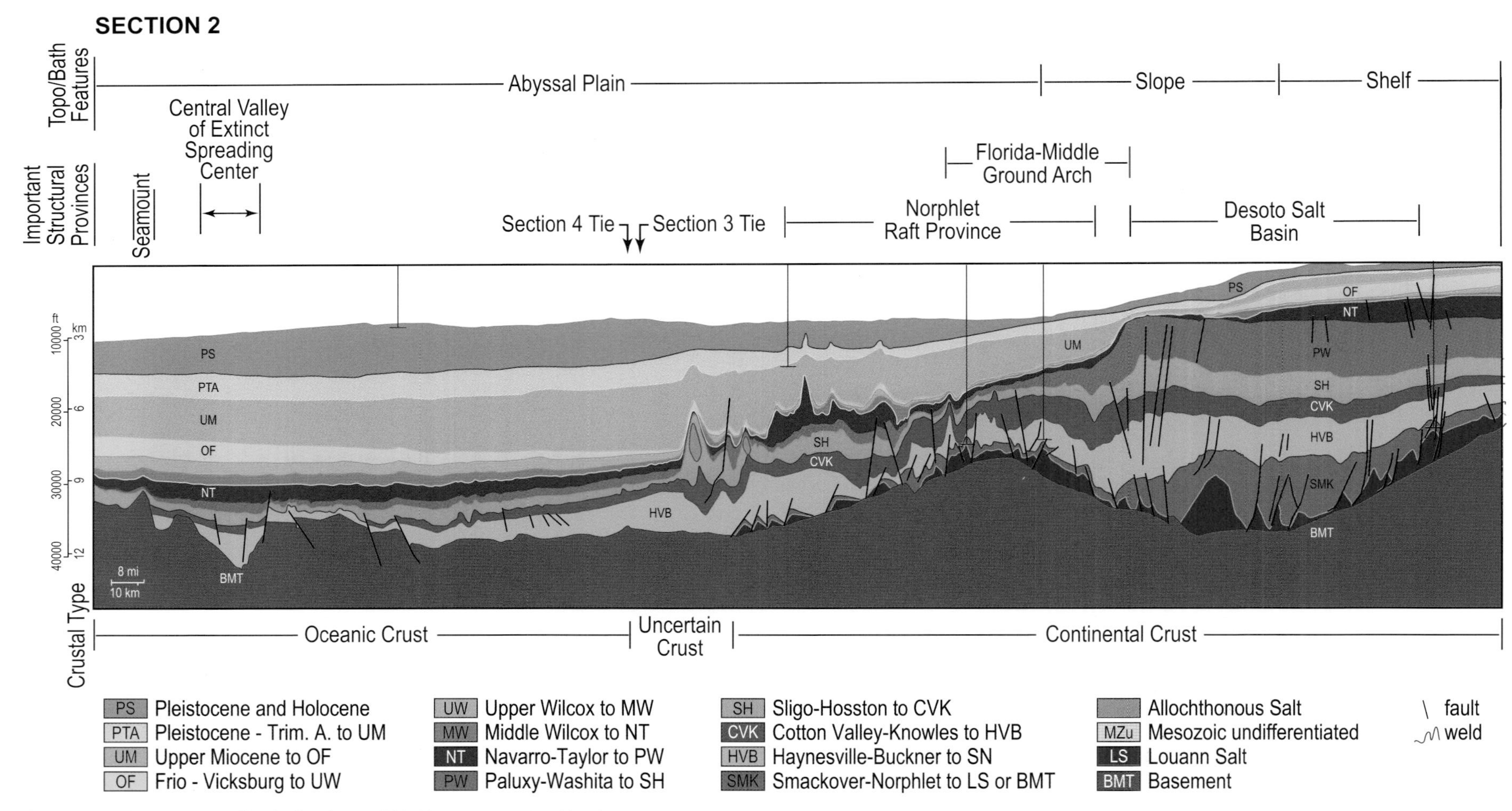

Figure 1.14 Cross-section 2: Florida shoreline to USA–Mexico international border.

with local thrust faults indicating some early crustal shortening.

The extinct spreading center is nicely developed toward the seaward end of the cross-section (Figure 1.14). Here is a large axial valley, about 20 km (12 miles) wide, with bounding basement structures (possible seamounts) and a dim to opaque infill interval below the Haynesville–Buckner seismic horizon. Normal faults dip toward the axial valley, similar to what has been previously described in the area (Snedden *et al.* 2014; Lin *et al.* 2019).

The total basin-fill is relatively thin, depressing the crust only to depths between 26,000 and 36,000 ft (8–11 km). The sedimentary interval is a bit deeper in the DeSoto salt basin at nearly 38,000 ft (12 km). Louann Salt is particularly thick here, exceeding 8000 ft (2.4 km) in the DeSoto salt basin, though this clearly reflects salt inflation.

Mesozoic platforms are well developed, particularly for the Jurassic Haynesville–Buckner (HVB), Cotton Valley–Bossier (CVB), and Cotton Valley–Knowles (CVK) at a position about 30 km (19 miles) seaward of the modern shelf edge. The Cretaceous Sligo–Hosston seems to be located in a similar position, indicating that the crustal boundary between thick and thin transitional crust has pinned the shelf margins due to changes in subsidence rates. Cenozoic shelf margins are all inboard of the Cretaceous and Jurassic platform margins. The Florida Escarpment is less pronounced here, instead a steep margin at Top Cretaceous (Top Navarro–Taylor) is observed, perhaps a byproduct of the Chicxulub impact event.

Growth faults are few in the Cenozoic interval and only Mesozoic growth along salt-detached faults is locally developed. Rafting is the dominant structural style, as discussed earlier.

Cenozoic deepwater reservoirs have been penetrated in portions of the area, but results to date have been disappointing in comparison to the Mississippi Canyon area, where giant discoveries (e.g., Thunderhorse Field) have been made. The lack of Cenozoic traps is one cause, though stratigraphic traps such as termination against the Cretaceous shelf margin have been considered, as will be discussed in Chapter 2.

1.5.3 Cross-Section 3: Onshore Texas to Onshore Florida

Cross-section 3 (Figure 1.15) is a transect from the onshore south Texas to the eastern basin margin and Florida onshore as described in cross-sections 1 and 2. The section in its central portion is located seaward of the Sigsbee Escarpment, where the salt canopy affects the sea floor. The abyssal plain section here illustrates trends in both Neogene and Paleogene strata, and thus is a veritable natural archive of the attendant sedimentary processes.

The variations between the western and eastern margins are notable. The sedimentary load on the west has depressed the crust to over 50,000 ft (15 km) near the present-day shelf margin offshore Texas (Figure 1.15). Gravity tectonics dominates the western margin, with numerous growth faults and multiple levels of fault detachment. The upper decollement is at a salt weld where Oligocene and Miocene age faults detach. The lower detachment surface for older Paleogene strata is founded upon the parautochthonous salt. The updip extension associated with this multi-level extension appears to be balanced, to some degree, by contraction within the Oligo-Miocene and Perdido fold belts, though local squeezing of salt can also occur (Radovich *et al.* 2007). Some faults nucleated at or near the Mesozoic platform margin or at the top of the Cretaceous. Faults also detach on the salt canopy.

The section crosses the Vicksburg Detachment, a well-known listric fault that nearly becomes horizontal as it slips along the Jackson Group Shales (Combes 1993; Feragen *et al.* 2007). The eastern portion of the cross-section shows limited salt stocks, which rise from the largely evacuated autochthonous Louann below, defining the eastern margin of the slope minibasins domain. Again, the Norphlet raft province is located just seaward of the Middle Ground Arch.

The Cenozoic sedimentary architecture is intimately convolved with the structural domains on the western margin. Shelf margins for each of the major Neogene and Paleogene units appear to be located at or just landward of the major growth and expansion of the various intervals. For some units like the Oligocene, much of the expanded interval represents slope deposition. Well penetrations indicate that much of the Oligocene interval is dominated by muddy lithologies and drilling in the Oligocene–Miocene interval has been challenging due to abnormal fluid pressures (P. Flemings, pers. comm.).

The Perdido fold belt is known to be linked to updip Oligo-Miocene extension (Trudgill *et al.* 1999; Gradmann *et al.* 2009; Radovich *et al.* 2011). These contractional folds have high relief and thickness due to the high sedimentation rates in the Cretaceous to Miocene interval that was shortened in the Neogene.

In the central portion of the cross-section (Figure 1.15), the abyssal plain, regional thickness trends provide a window on the source-to-sink processes of the Cenozoic basin-fill. Paleogene units (Middle and Upper Wilcox, Oligocene Frio–Vicksburg, and more condensed Upper Eocene Jackson Yegua Sparta) show clear eastward thinning, suggesting the Laramide source terranes were most important (Galloway 2008). The Lower Miocene shows a transition to eastward thickening in the Neogene stratigraphic interval, reflecting the rejuvenation of the Appalachians and rise of the Tennessee River as a major contributor to sand-prone fans in the Mississippi Canyon area (Galloway *et al.* 2011). The Pleistocene Mississippi Fan, the largest of the submarine fans to be formed, is apparent in the kilometer-scale interval above the Pleistocene Trim A (PTA, see Section 7.2) horizon that is banked against the Florida Escarpment. As will be discussed in Chapter 7, large Plio-Pleistocene channel–levee, mass transport, and lobate fans can be identified on high-resolution seismic data (Weimer 1990).

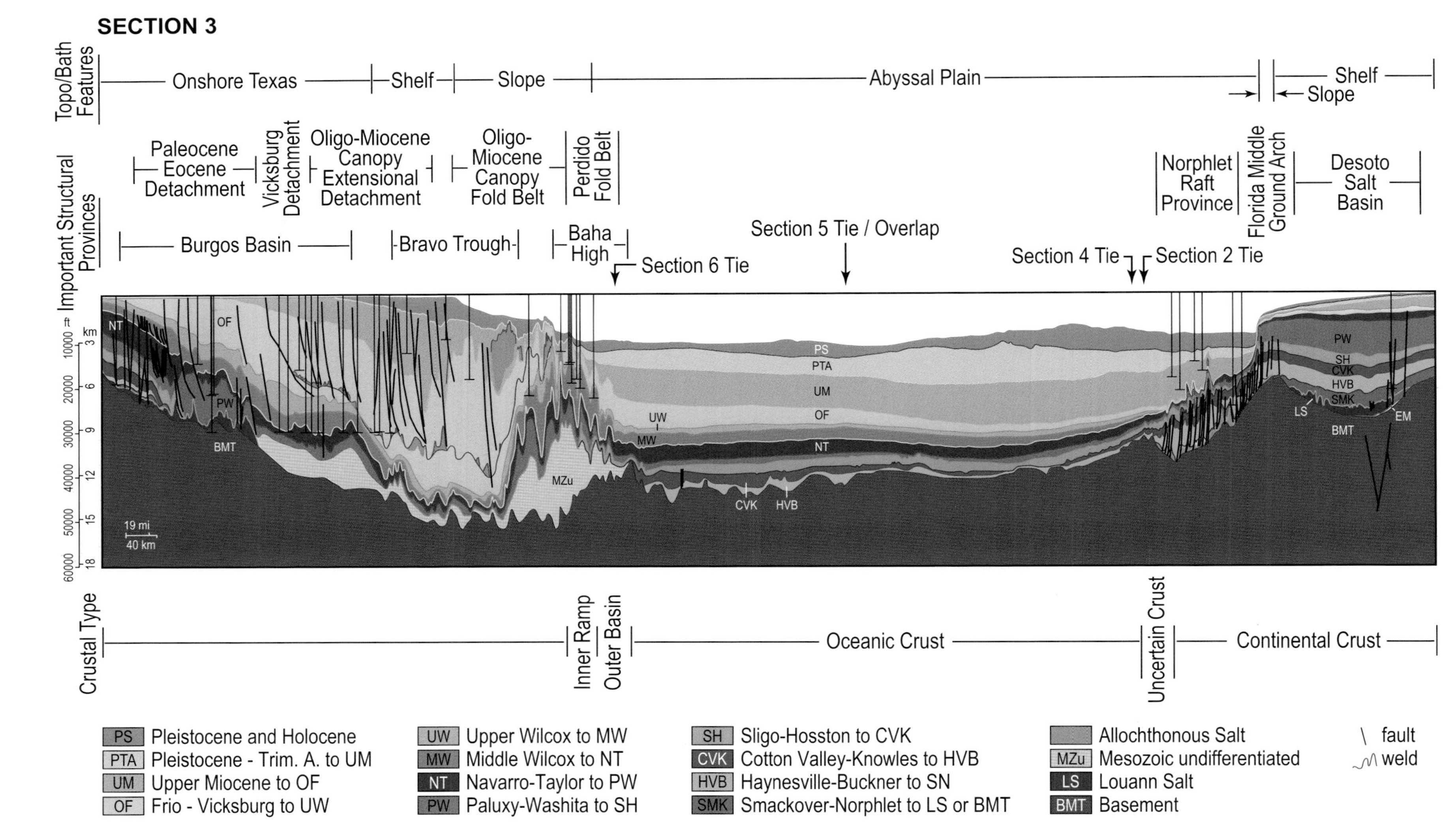

Figure 1.15 Cross-section 3: Onshore Texas to Onshore Florida.

1.5.4 Cross-Section 4: Black Warrior Basin to Yucatán Channel

Cross-section 4 (Figure 1.16) is a transect from the onshore Black Warrior basin to the USA–Mexico abyssal plain to the Cuban Platform, finally extending to the Yucatán Straits gateway to the Caribbean basin. The margin of the GoM basin can be defined at the hinge line between the Black Warrior basin, where Paleozoic sedimentary and basement rock are present near the surface, and the Mississippi salt basin to the south. The substantial thickening of the Jurassic strata in the Mississippi salt basin is clear evidence supporting placement of the GoM basin boundary here. Louann Salt also terminates near this margin and the Louann lapout and associated fault breakway zone are often used to demarcate the salt basin boundary (Ewing and Lopez 1991).

The Wiggins Arch basement structure borders the Mississippi salt basin to the south in Louisiana (Figure 1.16). Besides hosting a number of onshore discoveries, the Wiggins Arch acts as initiation point for successive downdip detachment zones starting with the Middle Cretaceous detachment zone. The crust is loaded to 50,000 ft (15.2 km), but as much as 40,000 ft (12.2 km) of that sedimentary interval is Cenozoic in age, a sign of the long-lived transport through the Mississippi River and its ancestors.

Isolated salt bodies and thick primary basins filled with Miocene to Cretaceous sediments give way to first extensive salt canopy just seaward of the modern shelf slope break. The Terrebone trough roho system, where extensional faults detach on one of the allochthonous salt bodies and/or welds, is denoted as the Oligo-Miocene salt weld (Hudec and Jackson 2011). Reconstructions of the Terrebone trough roho system show Early to Middle Miocene progradation expelled allochthonous salt seaward, toward the toe of the canopy, accompanied by considerable extension. By the Late Miocene, salt was largely expelled along the strike or dissolved, leaving the roho detachment, isolated **salt rollers**, and an extensive weld (McBride *et al.* 1998). Seaward rollover into an expulsion rollover near the Bay Marchand salt diapir (Schuster 1995) is not shown on this cross-section.

Further seaward are numerous supracanopy structures, including young secondary minibasins in Green Canyon and Atwater Canyon protraction blocks, where the section turns east–west (Figure 1.16). Below and at the seaward end of the salt canopy lies the Plio-Miocene Atwater fold belt, where deep salt diapirs (parautochthonous salt) occur along anticlinal axes. At this point, the cross-section turns to become more northwest–southeast trending across the abyssal plain and onward to Cuba.

Below the salt canopy on the modern slope of the USA sector is a relatively thick succession of Louann Salt that is conservatively estimated to be more than 5000 ft (1524 m) thick and to cover 220 km (136 miles) of lateral extent (Hudec *et al.* 2013a). The inner basin is the deepest portion of the Louann salt basin, where the greatest accumulation of evaporite-bearing interval is thought to have been deposited. Like elsewhere, there is substantial step up in acoustic basement, the inner ramp of Hudec *et al.* (2013a). Note that some portion of the relief is generated by the turn in the section at the Atwater fold belt.

The Cenozoic interval thins substantially toward Cuba and the Yucatán Straits to the south. Miocene strata alone thin from 8000 ft (2.4 km) to a few hundreds of feet (>30 m) as the interval lapouts onto the Cuba Platform margin. Mesozoic intervals are also thinner than known in the adjacent South Florida basin. The crystalline basement is as shallow as 12,000–16,000 ft (3.7–4.9 km) in the Yucatán Straits. These trends point to: (1) a distal position relative to major siliciclastic sources and linked river systems; and (2) the relatively recent joining of western and eastern Cuba microplates during the Eocene.

1.5.5 Cross-Section 5: Sabine Uplift to Sigsbee Escarpment

Cross-section 5 (Figure 1.17) extends from onshore Texas to deepwater GoM near the USA–Mexico international boundary. The Mexia–Talco fault zone is an extensional breakaway where salt thins to a zero edge (Hudec and Jackson 2011). The East Texas salt basin contains a series of generally north–south oriented diapirs and salt pillows toward the center of the basin where the original salt was thicker. Turtle structures formed by salt withdrawal into the adjacent diapirs is seen on nearby seismic lines (Jackson and Seni 1984). Note the over-thickened Albian and Aptian interval (Paluxy–Washita to Sligo–Hosston supersequences) located on the flanks of several salt domes. The intervening saddle is a remnant high that in some cases promoted reef development (Seni and Jackson 1983; Pashin *et al.* 2016).

Further seaward is a prominent basement structure called the Toledo Bend Flexure. It is notable as it marks the separation of the updip interior salt basins (East Texas, North Louisiana) and the central Louann basin proper (Hudec *et al.* 2013a). It is also thought to localize Mesozoic platform margins (Anderson 1979).

Strata south of the Toledo Bend Flexure dip rather steeply into the basin to the south, where Mesozoic horizons become difficult to trace basinward under a thick Cenozoic interval and allochthonous salt canopy (Figure 1.17). The Tiber well (KC 102 #1) penetrated the Top Cretaceous at 32,250 ft TVD-ss (true vertical depth subsea) (9830 m), which seismic mapping indicates is the regional level in the Keathley Canyon (KC) protraction block. In other wells, the Top Cretaceous may appear from first glance to be much shallower, but these are usually penetrations of salt-rafted carapace blocks, carried upward by differential salt movement, as described in Section 3.3.4.

Important transitions shown on the section include the rimmed platform margins, built up from the Jurassic to the end of the Albian, which give way seaward to several Cenozoic

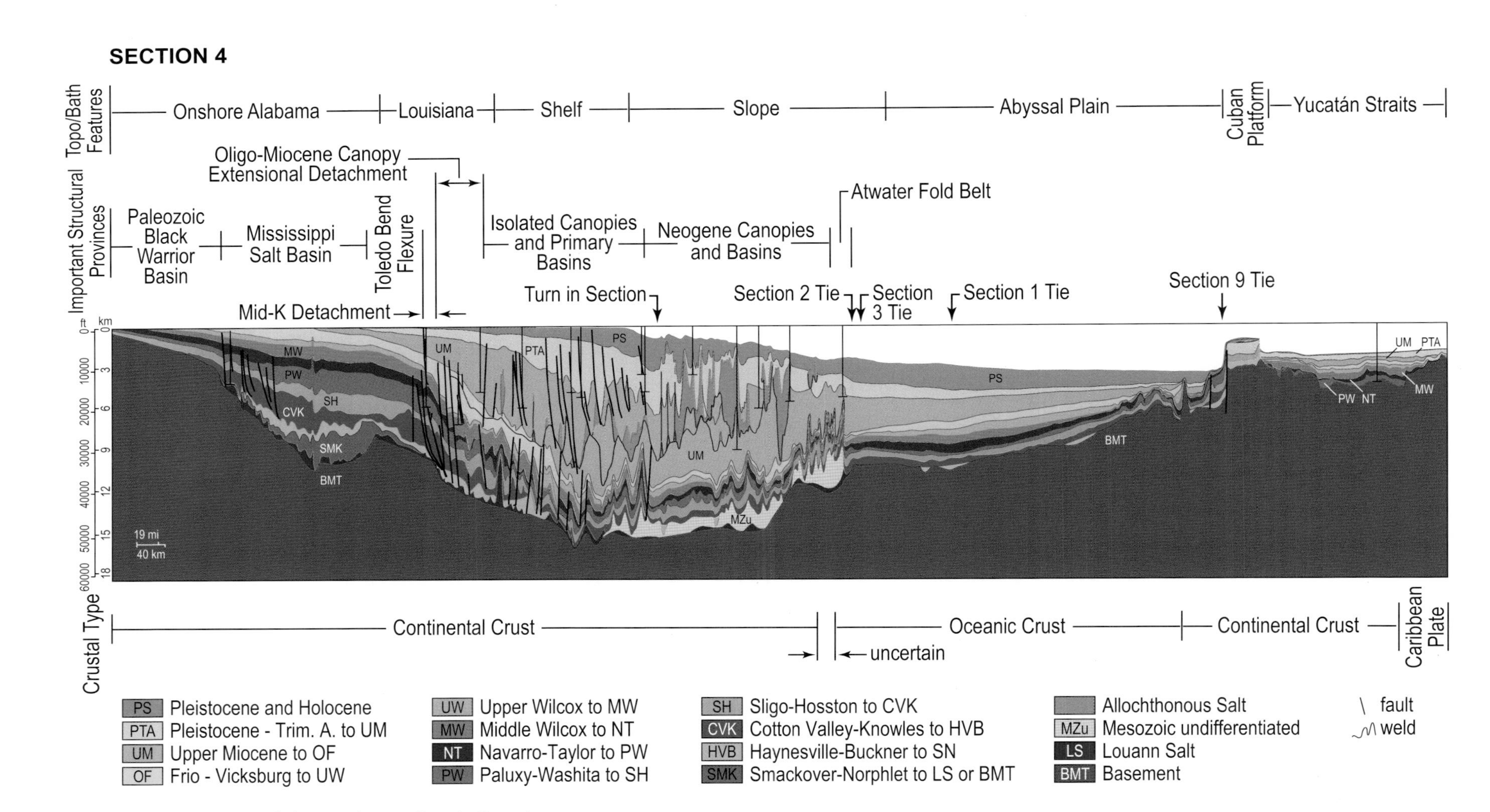

Figure 1.16 Cross-section 4: Black Warrior basin to Yucatán Channel.

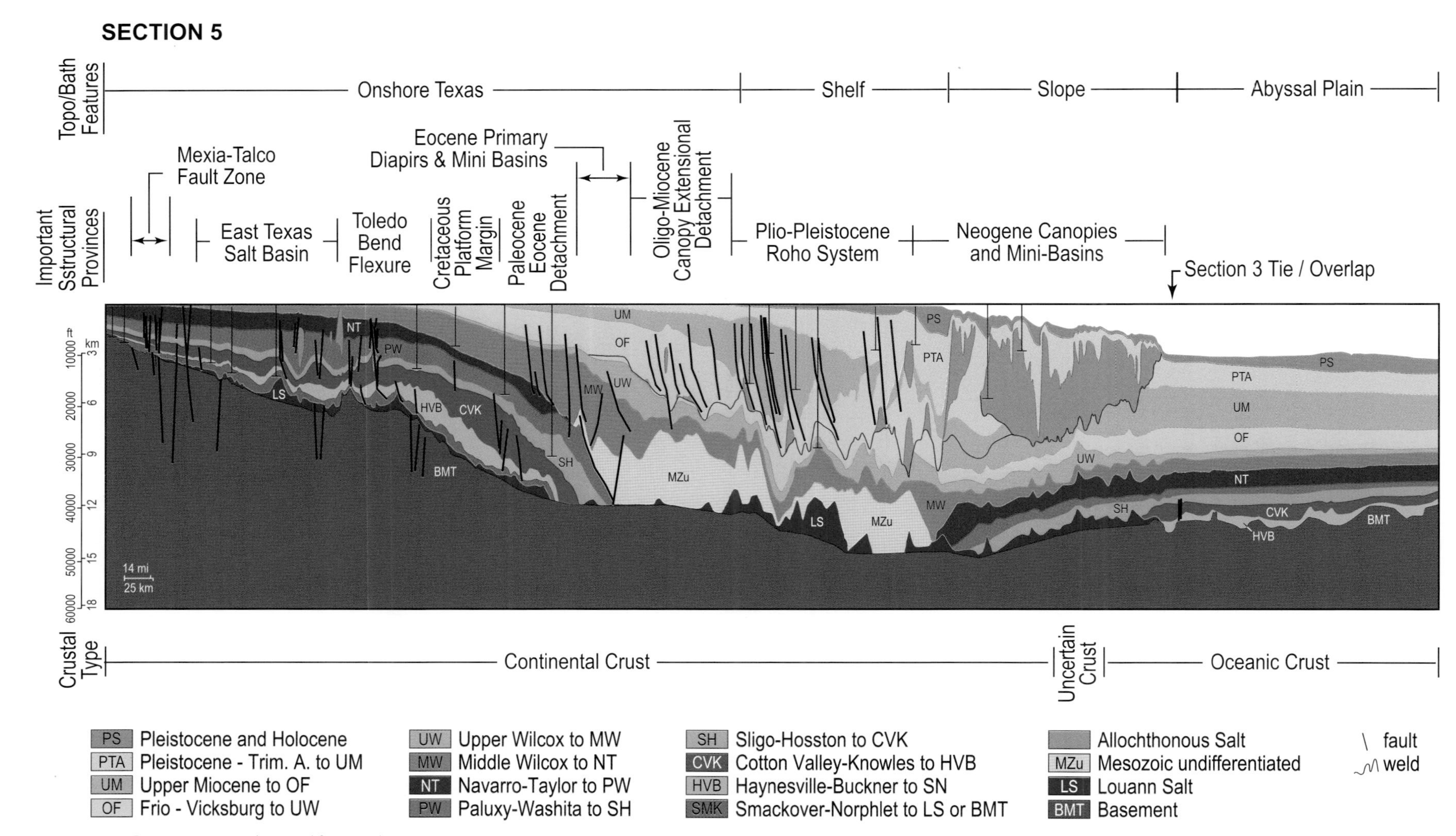

Figure 1.17 Cross-section 5: Sabine Uplift to Sigsbee Escarpment.

structural belts, including the Paleogene–Eocene **expansion zone**, the Oligocene–Miocene detachment zone, and the Pliocene–Pleistocene roho system on the present-day shelf (Peel *et al.* 1995). The present-day slope encompasses numerous Neogene canopies and secondary minibasins. These salt structures terminate at the Sigsbee Escarpment, where the salt canopy clearly impacts the sea floor morphology.

The section nicely illustrates the structure of the northern GoM basin depocenter located beneath the present continental shelf and slope. The Top Cretaceous is as deep as 40,000 ft (12.2 km) in places, loaded by Cenozoic siliciclastic deposition. The Cenozoic prism extends beneath the coastal plain and shelf, reaching its thickest point near the present continental margin. In many areas, the continental slope extends basinward to about the position of the transitional/oceanic crust boundary. Beneath this sediment prism, a large portion of the autochthonous Louann Salt has been expelled, forming a primary salt weld on the basal Jurassic unconformity that is a decollement zone for growth faults. Other detachments occur at salt welds, allochthonous salt canopies, or are rooted in decollements located within deep basinal mudstones of indeterminate age.

As with several previous sections, sedimentary architectures are influenced greatly by accommodation created by gravity tectonics. Shelf margins prograde progressively into the basin from Cretaceous to Neogene, reflecting the robust depositional systems extending from **source terrane** to basinal sink in this central GoM transect.

1.5.6 Cross-Section 6: San Marcos Arch to Sigsbee Escarpment

Cross-section 6 (Figure 1.18) starts at the San Marcos Arch, where Miocene uplift set up a steeply dipping basement surface, to the Perdido fold belt on the abyssal plain on the international border. Several levels of fault detachment are observed: (1) Paleo-Eocene detachment at or seaward of the Cretaceous margin; (2) Oligo-Miocene canopy detachment; and (3) the Corsair–Wanda fault zone.

In contrast to the central and northeastern GoM, this transect across the northwestern Gulf displays broad, complex Middle Cenozoic compressional domains, including the Perdido and Port Isabel (Oligo-Miocene canopy) fold belts. The Port Isabel fold belt is linked by a decollement to the Miocene Clemente-Thomas, Corsair, and Wanda fault zones of the Oligocene–Miocene canopy detachment province (Hall *et al.* 1993).

The Corsair–Wanda fault zone is particularly prominent on this section (Figure 1.18). Over 30,000 ft (9.1 km) of growth along the bounding fault is apparent on seismic sections, with much of it being Miocene in age. The fault detaches on the deep salt allochthon.

Like the Mississippi Fan fold belt, the Perdido fold belt is located near the original depositional limit of the basal (parautochthonous) Louann Salt (Fiduk *et al.* 1999). The isopachous Cretaceous to Early Cenozoic interval is considered prekinematic (deposited before deformation), while the synkinematic (during deformation) phase in the Miocene and younger interval shows lateral variations in thickness (Jackson and Hudec 2017). Additional contraction was accommodated by the compound salt canopy that has been injected up into the Oligocene and Miocene interval.

Interpretation of the remnant thickness of the autochthonous salt is challenging at the depths where it is present. Portions of the salt are deflated and welds likely remain in many unpenetrated structures. Amplitudes of folds in the Perdido trend suggest considerable salt thickness, but few wells penetrate deeply enough to verify this view.

1.5.7 Cross-Section 7: Quetzalcoatl Extensional Detachment, Northern Mexican Ridges to Chicxulub Crater

Cross-section 7 (Figure 1.19) is a west-to-east transect from the slope of eastern Mexico to the Yucatán Platform, the site of the Chicxulub impact event that ended the Mesozoic. The present-day physiography of a narrow shelf and steep slope reflects relatively recent Neogene tectonic activity associated with Pacific plate subduction (Padilla y Sánchez 2007; Witt *et al.* 2012). Associated loading and subsequent gravitational sliding in the Quetzalcoatl extension is linked with compression in the Mexican Ridges fold belt. Uplift in eastern Mexico associated with the Middle Miocene Chiapanecan orogeny resulted in deep incision and canyon formation along a narrow shelf and slope leading to delivery of large volumes of coarse-grained sediments to the basin floor (Ambrose *et al.* 2005).

While the fold and thrust belt shown on the middle of this section is commonly grouped with the Mexican Ridges to the south, newer seismic data suggests that this contractional belt may have also experienced the additional effects of salt being pushed from west to east (M. Hudec, pers. comm.). Though published evidence is currently lacking, there is a notable change in orientation of folds and faults south of this section (Figure 1.19) and a gap in the structure, implying some change in the tectonic forcing mechanism. Another observation is that the southern portion of the Mexican Ridges tends to show more expansion along the Quetzalcoatl faults versus the northern areas. This leads one to suspect pure gravity tectonics in the southern Mexican Ridges versus salt-involved compression in the north. Another difference in between this section and cross-section 8 to the south is the existence of a roho system detached at the Top Upper Miocene level, documented on better seismic data by CNH (2015b). Salt or a weld may be present at the Top Upper Miocene level, but is not shown in the CNH (2015b) compilation. The proximity to salt would be required in any case.

Cross-section 7 (Figure 1.19) continues across the relatively undeformed abyssal plain region of the south-central Gulf before passing across the Yucatán salt subprovince or subbasin

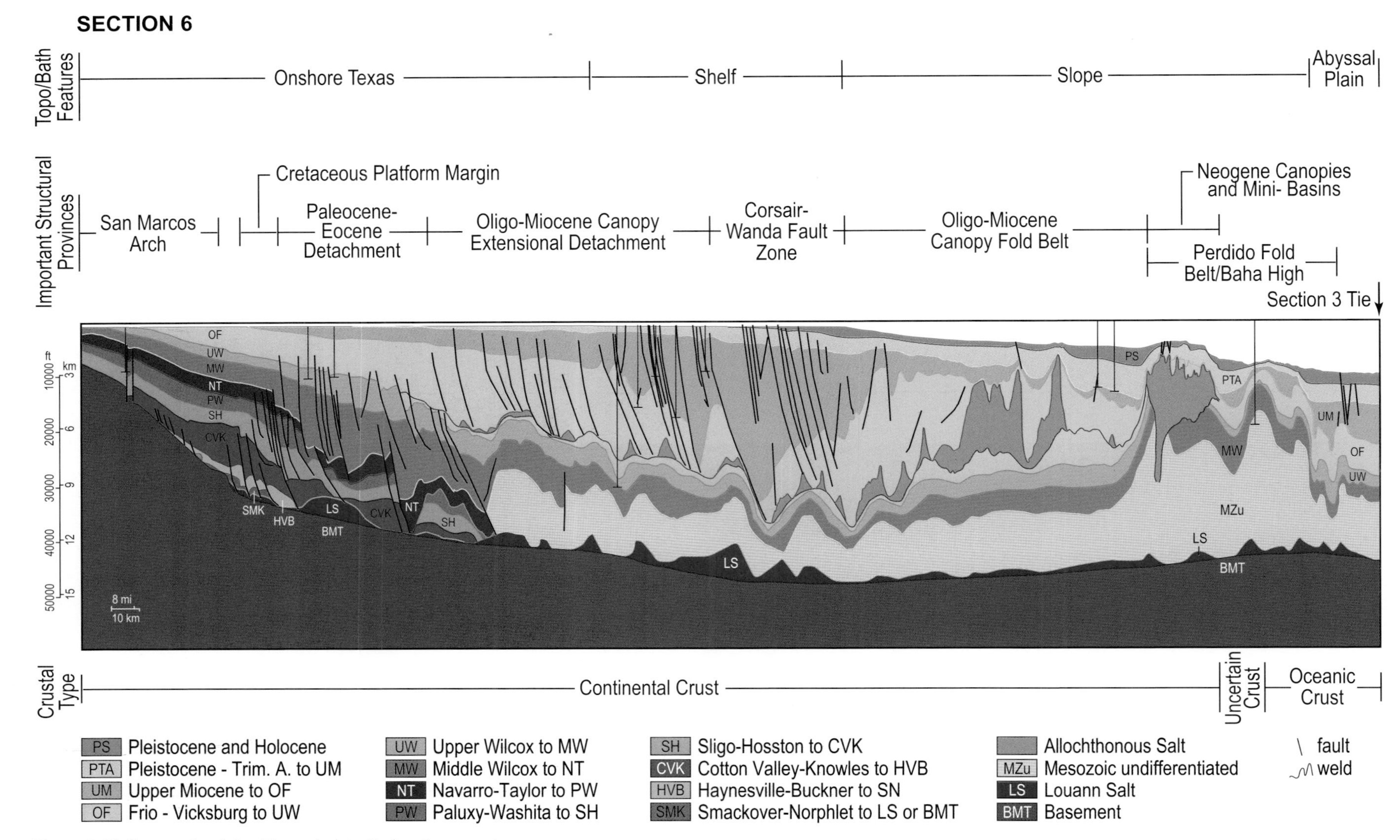

Figure 1.18 Cross-section 6: San Marcos Arch to Sigsbee Escarpment.

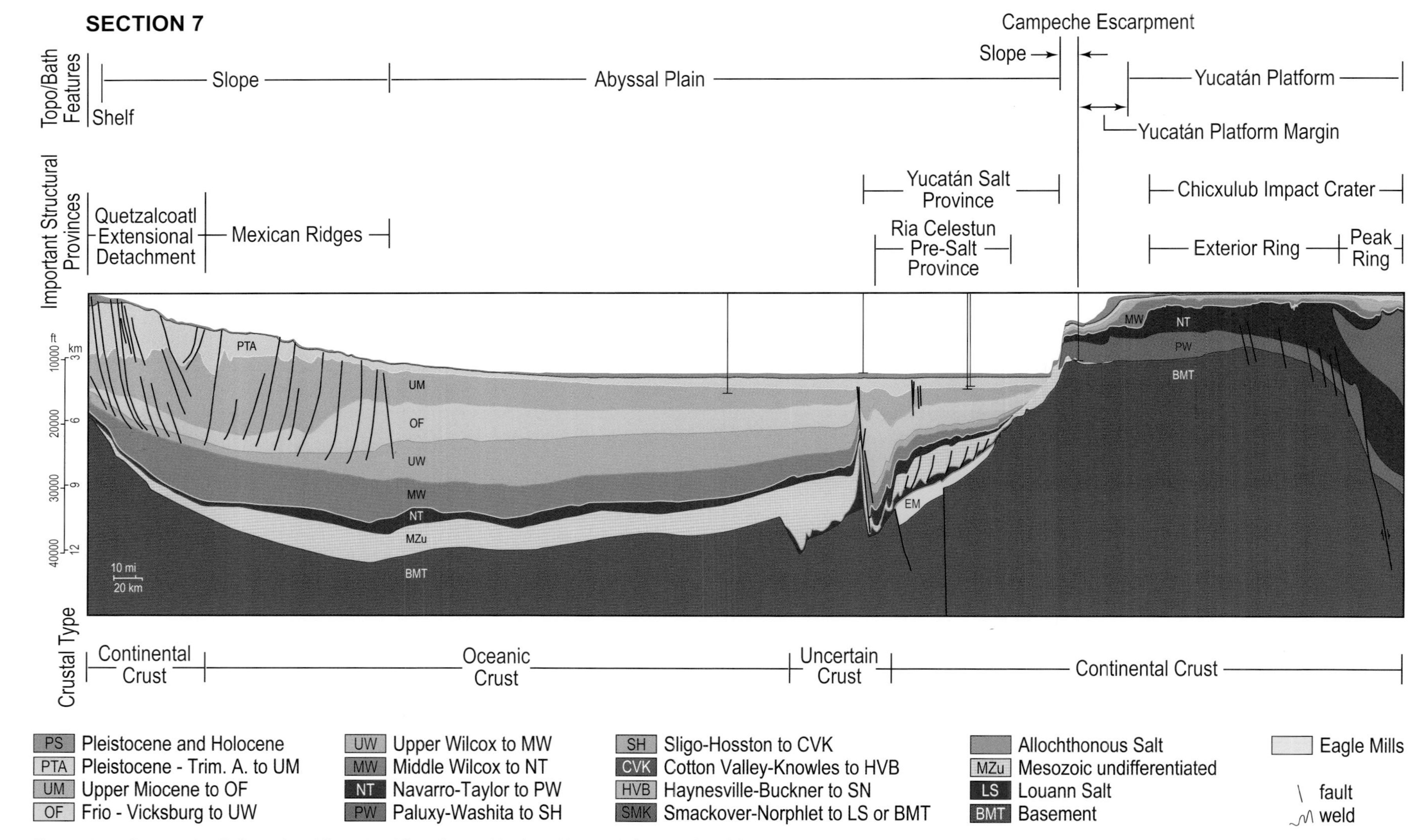

Figure 1.19 Cross-section 7: Quetzalcoatl Extensional Detachment, Northern Mexican Ridges to Chicxulub crater.

(terminology of Hudec and Norton 2018). Two tectonic styles are recognized: (1) a salt diapir complex with high-amplitude salt structures; and (2) lower-amplitude salt rollers. The former occurs near or at the transition from continental crust, thus uncertain crust. There is an interpreted basement step just seaward, where parautochthonous salt is observed to terminate.

The Yucatán salt roller domain has a remarkable similarity to the salt raft structures of the deepwater Norphlet salt raft exploration area of the northeastern GoM (Saunders *et al.* 2016; Hudec and Norton 2018). It has been suggested that this area is a conjugate to that Norphlet exploration arena, separated by sea floor spreading (Miranda Peralta *et al.* 2014; Steier and Mann 2019). However, there are no well penetrations in this area other than the shallow DSDP core sites (see Buffler *et al.* 1984).

The Yucatán salt roller domain also overlies a new, distinctly different structural province with possible pre-salt sedimentary fill. First noted by Williams-Rojas *et al.* (2012), this interval shows a wedge-shaped or rift-graben form, onlapping the Yucatán Platform margin (Hudec and Norton 2018; Rowan 2018). This interval may be analogous to the pre-salt Eagle Mills (Triassic to Middle Jurassic) of the eastern USA (Heffner 2013). High-amplitude, seaward dipping reflections (SDRs) appear at the base of the probable sedimentary interval, evoking global analogs of SDRs associated with initial stages of continental rifting (Norton *et al.* 2015). Alternatively, these may simply be layered volcanics (Hudec and Norton 2018), as commonly observed in the eastern USA pre-salt section (Heffner 2013). We informally refer to this area as the Ria Celestun pre-salt province, named after a local geographic feature. A similar pre-salt interval is noted on seismic sections in the Campeche subbasin to the southwest (Hudec and Norton 2018).

Cenozoic and Mesozoic stratigraphic units all taper and largely lapout against the steep Yucatán Platform margin. Continental crust basement rises from depths greater than 36,000 ft (11 km) to less than 10,000 ft (3048 m) over a short distance. Further inboard on the platform, basement abruptly drops and then rises to depths of less than 6000 ft (1829 m). This unusual basement architecture is a result of the Chicxulub impact, as documented by numerous studies (Denne *et al.* 2013; Sanford *et al.* 2016) and recent IODP coring at site M0077A (Morgan *et al.* 2016). Basement upwarp indicates the location of the peak ring, the deep crustal response to the bolide impact that ended the Cretaceous. Seaward of the peak ring, the Top Cretaceous reflection is relatively flat in the area of the exterior ring but drops 3–4 km on the platform margin. Clinoforming successions representing the post-impact crater fill are evident in this area.

As mentioned, few exploration wells are present in the eastern portion of the area, in spite of prominent salt-cored structural closures and prospective traps. Several factors may preclude any near-term drilling: (1) water depths greater than 12,000 ft (3.7 km); (2) distal thinning of Paleogene reservoirs like the Wilcox and parallel decreases in sand content away from siliciclastic source terranes. Neogene sandy intervals are present in DSDP core sites 87 and 91, but these are likely derived from southern Mexico rather than any local sources. The progressive sorting associated with the distal turbidity flows likely means a reduced grain size of any potential reservoirs.

1.5.8 Cross-Section 8: Mexican Ridges to US Abyssal Plain

Cross-section 8 (Figure 1.20) extends from the Quetzalcoatl extensional detachment to the southern end of the Mexican Ridges, across the Yucatán salt subbasin and continuing northward to the abyssal plain at the USA–Mexico border. As with cross-section 7 (Figure 1.19), both extensional faults and, further seaward, folds and low-angle thrusts of the Mexican Ridges are observed on this transect. However, the expansion along the Quetzalcoatl detachment faults is much greater, for reasons discussed earlier. Drilling of the Pemex Puskon #1 well documented the substantial growth along listric faults of the Quetzalcoatl extensional detachment zone (Alcocer 2012; Porres Luna 2018). The linked extensional–contractional system is thought to detach on overpressured Upper Eocene Shales, with a possible additional detachment surface in the Oligo-Miocene interval, similar to major multi-level detachments documented in the northern GoM (e.g., Radovich *et al.* 2007, 2011). The Upper Miocene interval in the extensional zone is generally thinner here than on cross-section 7, similar to observations made by CNH (2015b).

Large folds of the Mexican Ridges have wavelengths of 10–12 km (6–7 miles) and amplitudes of 300 m to 1 km (984–3280 ft) (Padilla y Sánchez 2007). The Mexican Ridges developed as a consequence of gravitational spreading processes, synchronous with growth faulting of the western onshore and continental shelf areas. Deformation occurred in several stages from Middle Miocene to the present day (Salomon-Mora *et al.* 2009). This deformation correlates with highly active **petroleum systems**, including migration and trapping of hydrocarbons, in turn forming direct hydrocarbon indicators, overpressured traps, gas chimneys, gas hydrates, and sea floor hydrocarbon seeps that are being investigated as part of new regional exploration efforts. Drilling has concentrated largely on the folds of the Mexican Ridges, with a few wells like Puskon #1 testing the updip extensional systems.

The Yucatán salt subbasin (high-amplitude salt diapir domain) in the middle of the section is entirely developed over uncertain or possible oceanic crust. Hudec and Norton (2018) hypothesized significant seaward translation after salt deposition due to the lack of a confining structural boundary, in contrast to the perched Campeche salt subbasin where the BAHA high is present. Shortening is evident in the shallow Cenozoic interval of the Yucatán salt subbasin, with the timing of deformation likely as Miocene or younger.

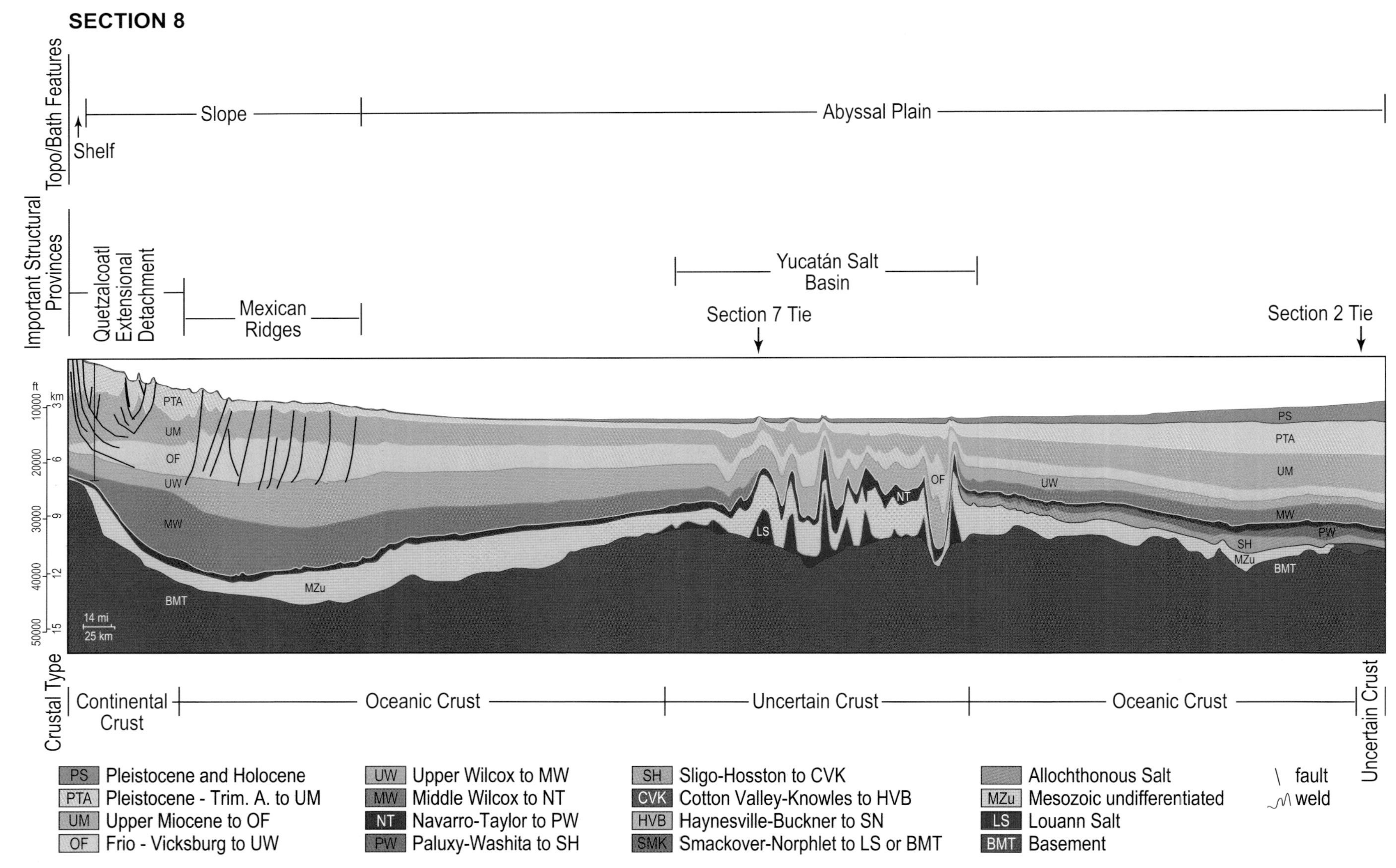

Figure 1.20 Cross-section 8: Mexican Ridges to US abyssal plain.

One interesting observation is the differential thickening and thinning trends of the Mesozoic and Cenozoic. Mesozoic strata thicken toward the salt diapir domain, while Cenozoic strata generally thin toward that area, trends noted both north and south of the subbasin. Local thickening into zones of salt evacuation is noted but does not change the inferred pattern. The Mesozoic thickening may signal development of an inner basin in Mexico, a conjugate to that in the northern GoM, where salt and overlying Mesozoic strata were deposited in greater magnitudes than elsewhere.

1.5.9 Cross-Section 9: Catemaco Fold Belt to Bahamas Platform

Cross-section 9 (Figure 1.21) is a basin-spanning strike transect from the Catemaco fold belt to the Bahamas Platform. The section includes a small segment of the Catemaco fold belt, unfortunately crossing oblique to the westerly verging folds. The Catemaco fold belt has been previously described as a linked extensional–contractional gravity-driven system with tectonic transport to the northwest (Mandujano-Velaquez and Keppie 2009). Northward salt evacuation in the Campeche salt subbasin (terminology of Hudec and Norton 2018) is thought to have occurred during the Middle Miocene Chiapanecan orogeny (Gutiérrez Paredes *et al.* 2017). However, new WAZ seismic data acquired in the Campeche salt basin suggests a longer duration of shortening, initiated in the late Paleogene and continuing today (Snyder and Ysaccis 2018). The nearby Veracruz basin was likely deformed during the Chiapanecan uplift, closely followed by uplift of the Anegada High and Los Tuxtlas volcanic massif (Jacobo Albarabn *et al.* 1992).

The section continues across the fringe of the Campeche salt subbasin, crossing into the Yucatán salt subbasin which is also observed on cross-sections 7 and 8 (Figures 1.19 and 1.20). Not obvious at this scale are the counter-regional fault systems that accommodated a thick Neogene interval, partly aided by salt expulsion rollovers (Gomez-Cabrera and Jackson 2009a, 2009b). CNH (2015a) notes in regional structural intervals that there is a major expansion of the Upper Miocene and Lower Pliocene in the adjacent Comalcalco basin, of up to 200–300 percent.

Like cross-section 7, cross-section 9 (Figure 1.21) carries onward across the Yucatán salt roller domain and underlying Ria Celestun pre-salt province, over the Yucatán Platform and Chicxulub exterior ring. Notable is the elevation of basement near the Chicxulub impact exterior ring <12,000 ft (<3.7 km), dropping several kilometers (to 20,000 ft; 6.1 km) on the rest of the platform. Basement appears to be quite shallow in the Florida Straits, less than 15,000 ft (<4.6 km) regionally and locally near the seabed, such as at Catoche Knoll, where DSDP cores 538 and adjacent cores 536 and 537 were retrieved (Buffler *et al.* 1984).

The cross-section continues to the Bahamas Banks or Platform where basement deepens, depressed under the thick Mesozoic succession (>11,000 ft; 3.4 km) largely Aptian to Albian carbonates (Ladd and Sheridan 1987; Epstein and Clark 2009). Evaporites are interpreted on the far eastern end of the line, making an appearance in the hypothesized seawater entry point for the GoM basin, as will be discussed in Section 4.4.

Exploration has primarily been concentrated in the Catemaco fold belt and adjacent areas (including key discoveries at Kunah #1 and other undeveloped resources), the Mexican Ridges, and adjacent Campeche salt subbasin. Limited drilling has been attempted elsewhere along the cross-section. The eastern end of the cross-section skirts the Cuban fold and thrust belt, where a handful of international companies have drilled wells without success in deepwater and older shallow-water wells near the Bahamas Banks (Epstein and Clark 2009; Melbana Energy 2017).

1.5.10 Cross-Section 10: US Abyssal Plain to South Florida Basin

Cross-section 10 (Figure 1.22) completes the circum-GoM tour, running from the international border to onshore Florida. The Florida Escarpment is particularly steep, coinciding with the interpreted continental to oceanic crustal boundary. The Mesozoic succession is relatively thin on oceanic crust (<7000 ft, 2134 m) compared to the South Florida basin, where it exceeds 12,000 ft (3.7 km) on the continental crust. Over the Sarasota Arch, a major basement-cored structure, the Mesozoic is as thin as 5000 ft (1524 m), documented by the nearby Charlotte Harbor-672 #1 and 622 #1 wells. Unlike cross-section 2, the Jurassic and Early Cretaceous platform margins are not observed, leading to alternative hypotheses of non-reefal development, or more likely K–Pg-related margin collapse (Denne and Blanchard 2013) and continued retrogradational failures of the margin well past the original platform margin position. Inboard well penetrations (e.g., Vernon Basin 654 #1) document only carbonate shelf and platform interior facies. Repeated margin failures and adjustments (Mullins *et al.* 1986) also reflect ocean current bottom erosion during a period of accelerated current flow in the Miocene and Pliocene, linked to progressive closure of the equatorial seaway and development of the Isthmus of Panama (Snedden *et al.* 2012).

As will be discussed in Section 4.5, the Glen Rose supersequence is particularly well developed in the South Florida basin, including extensive evaporites, mainly anhydrite of the Ferry Lake Sequence and stratigraphic equivalents (Punta Gorda Formation of Florida). The center of the basin is known to contain halite as well as anhydrite, indicating restricted conditions and hypersalinity during the Albian. The presence of evaporites in the Albian interval east of the Sarasota Arch is indicated by distinctive high-amplitude continuous seismic reflections.

The Cenozoic interval shows an opposite trend in thickness. On oceanic crust, 14,000 ft (4.3 km) of Cenozoic is present, much of it Neogene siliciclastic sedimentation

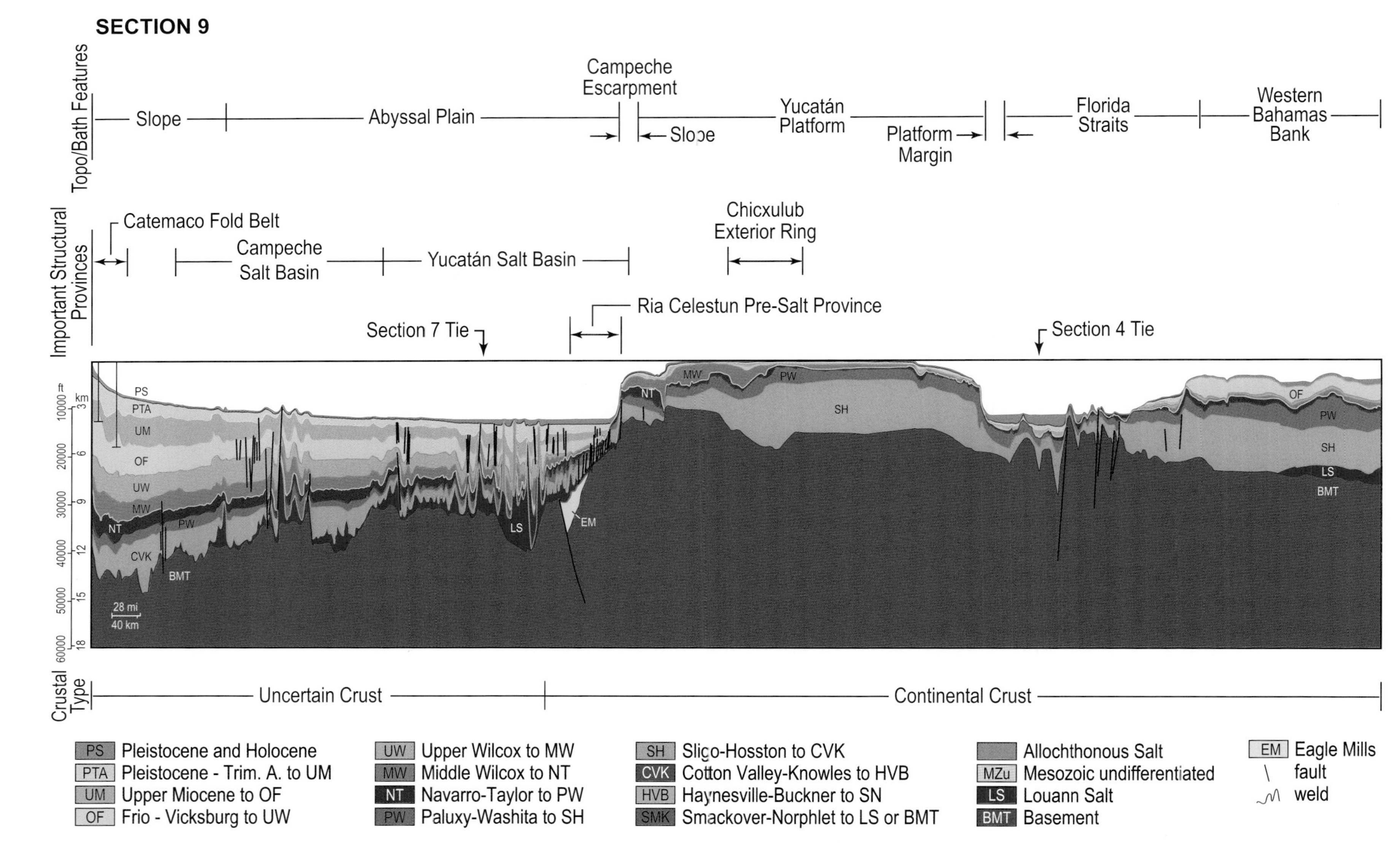

Figure 1.21 Cross-section 9: Catemaco fold belt to Bahamas Platform.

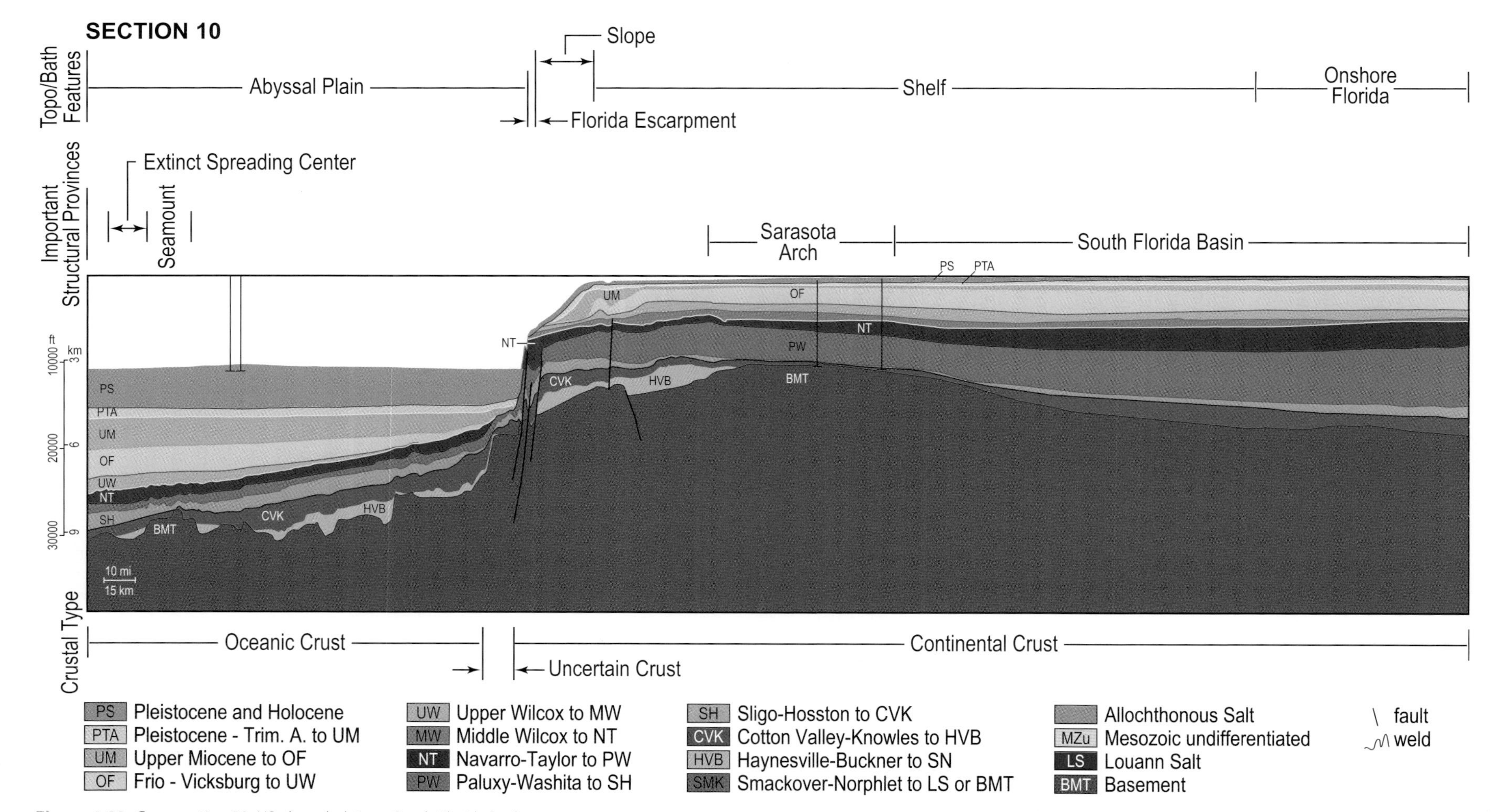

Figure 1.22 Cross-section 10: US abyssal plain to South Florida basin.

(9000 ft; 2744 m) linked to the Mississippi Fan and older systems like the paleo-Mississippi and paleo-Tennessee Rivers, which were sourced by the rejuvenated Appalachians. However, Cenozoic deposition on the platform is much less at 5000 ft (1524 m) and is dominated by carbonate sediments.

Due to the long-standing US drilling moratorium on the Florida shelf, no deepwater wells have been permitted; few wells have been drilled since 1985, all of these dry holes. Sizable onshore discoveries were made as recently as 1964 in the Sunniland trend (e.g., Felda Field) where carbonate reservoirs (grainstone banks and tidal shoals) are well documented (Mitchell-Tapping 1986, 2002). These small fields are largely sealed by extensive evaporites of the coeval Glen Rose interval. However, few wells have been drilled other than field infill wells, related to environmental concerns and other non-geologic factors.

1.5.11 Other Areas: Bravo Trough of Mexico

Between cross-sections 3 and 7 is a zone of major salt evacuation, only recently identified on new WAZ 3D seismic surveys (CNH 2015b; Hudec *et al.*, accepted). Depth to basement is in the range 45,000–52,000 ft (13–16 km), shallowing to 40,000 ft (12.2 km) on the adjacent BAHA high (Figure 1.4). Salt evacuation-related over-thickening of Oligocene sediments into this structural trough in the offshore portion of the Burgos basin is called the Bravo Trough (M. Hudec, pers. comm.). Some thickening of the Oligocene in this extensional zone was shown by Davison *et al.* (2015), but not to the scale of 8000 m (5 km) of expanded Upper Oligocene interval observed on new WAZ data in Mexico. A well drilled by Hess (Port Isabel 526 #1) in Bravo Trough penetrated nearly 17,000 ft (5.2 km) of sandstone-poor Oligocene interval before terminating. The thick Oligocene interval overlies a thin or absent Paleogene and Mesozoic interval, suggesting that a large salt body or diapir was present prior to Latest Eocene/Early Oligocene salt evacuation (Hudec, pers. comm.).

The lack of seismic reflectivity in the trough fill implies a shale-dominated interval. The US GoM interval with a similar seismic character is the basinal Oligocene (Frio–Vicksburg) interval of the Oligo-Miocene canopy detachment and contractional zones including the Port Isabel fold belt, as will be discussed in Section 6.5. Contributing rivers were likely mud-dominated, including volcanics altered to clays.

1.6 Temporal Reconstruction of Central GoM Line

Backstripping of regional cross-sections (Figure 1.23) reveals the dynamic interplay between deposition, wholesale mass transfer of salt, development of growth structures, and outbuilding of the Gulf margin that has characterized the basin's history (Diegel *et al.* 1995; Peel *et al.* 1995; McBride *et al.* 1998). Late Jurassic accumulation of up to 4 km of Louann Salt extended across the subsided, thinned transitional crust (Figure 1.23A). By the end of the Cretaceous, deposition had loaded and expelled much of the landward part of the autochthonous salt basinward, beneath the paleo-continental slope toe and northern basin floor (Figure 1.23B). Extension of the upper slope was accommodated by compressional deformation at the slope toe. A remnant layer of autochthonous salt provided the decollement horizon for basinward gravity spreading.

By the end of the Oligocene (Figure 1.23C), successive pulses of Paleogene deposition had prograded the continental margin over the Cretaceous slope, deflating the thick salt under-layer by intrusion of salt stock canopy complexes under the advancing continental slope and further inflation of the abyssal salt sheet. The Oligocene Frio growth fault zone migrated basinward with the prograding continental margin; here, detachment occurred within Upper Eocene mud as well as in the deeper salt. The resultant continental slope was a mix of sediment and near-surface salt bodies. Miocene–Pliocene deposition loaded the salt canopies, triggering passive diapirism and further gravity spreading, creating roho fault systems and isolated salt stocks separated by welds (Figure 1.23D). Thick secondary minibasin-fills separate the salt stocks. Loading also initiated extrusion of a salt sheet at the toe of the slope. Pleistocene deposition has filled updip minibasins and built the continental slope onto the distal salt sheet, where incompletely filled minibasins dominate present slope topography (Figure 1.23E).

1.7 Tectonostratigraphy, Chronostratigraphy, and Depositional Systems

With the focus of this book on the depositional history within the GoM basin, a brief description of various tectonostratigraphic and chronostratigraphic frameworks, stratigraphic terminology, and depositional classifications are necessary prerequisites. These are foundations for more detailed discussions of the Mesozoic and Cenozoic record in subsequent chapters. We also elaborate upon the evolving database of wells, seismic data, and reference papers in our research on the GoM.

1.8 Tectonostratigraphic Framework

Tectonics has a predominant role in creating the highland terranes that various fluvial systems tap for **terrigenous** source material, modifying routes from continental divides toward shorelines, creating accommodation in the receiving basins, forming bathymetric features that attract photic zone organisms that form carbonates, generating traps to allow hydrocarbon accumulations, and controlling burial that ultimately drives shale-prone source rock though time/temperature windows that generate oil and gas. The long-term structural history of the basin and its surrounding hinterland is the ultimate low-frequency spectrum upon which are superimposed high-frequency **eustatic sea-level changes**, climatic variations, and autocylic depositional processes.

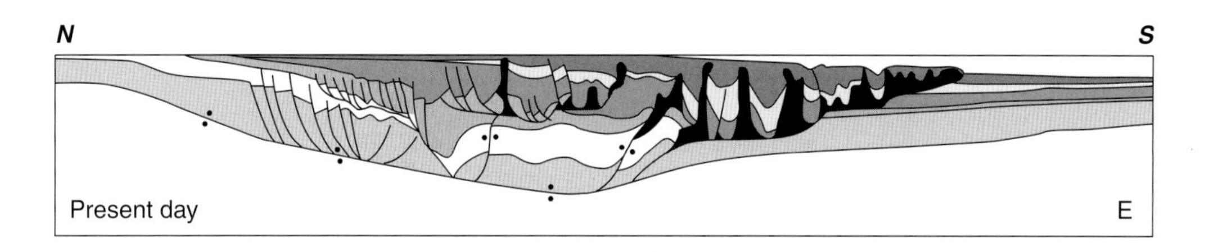

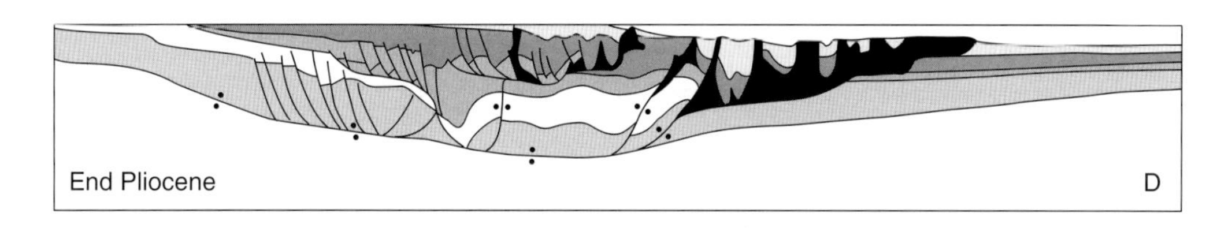

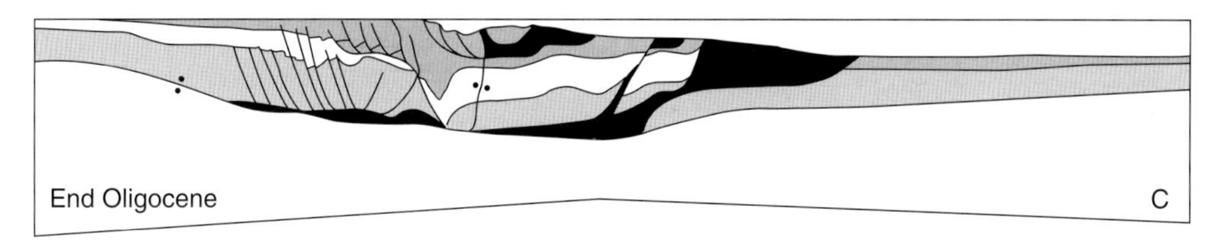

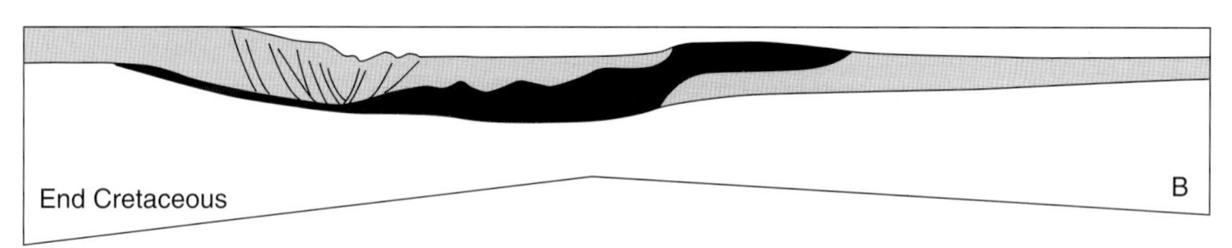

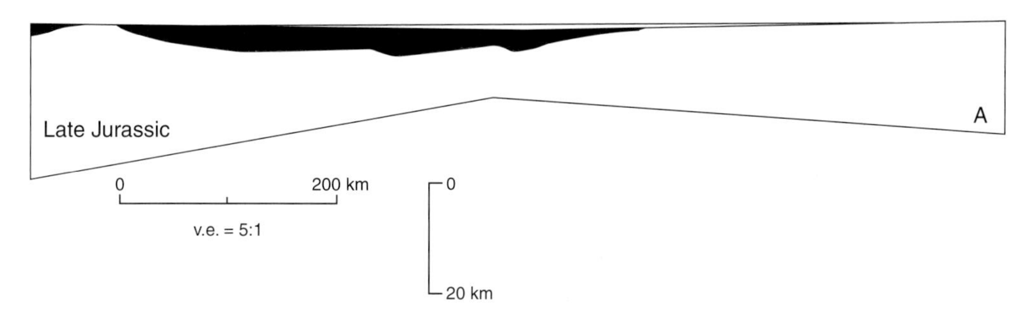

Figure 1.23 Sequential restoration of schematic central GoM section. Modified from Peel *et al.* (1995).

Thus, our over-arching stratigraphic framework, and the pathway we follow in this book from pre-basin history to the end of the Pleistocene, is a tectonostratigraphic scheme. Galloway (2009) first recognized this and subdivided the Cenozoic into four tectonostratigraphic phases:

1. Paleogene Laramide Phase
2. Middle Cenozoic Geothermal Phase
3. Basin and Range Phase (including Appalachian Rejuvenation)
4. Neogene Tectono-climatic Phase.

In spite of an equally long period of oil and gas exploration and scientific investigation, a similar tectonostratigraphic breakdown of the Mesozoic interval has not achieved consensus, in spite of considerable effort. Toward this end, we offer a new Mesozoic tectonostratigraphic classification, based on the same general principles as that of the Cenozoic framework (Figure 1.24):

1. Post-Orogenic Successor Basin-Fill and Rifting Phase
2. Middle Mesozoic Drift and Cooling Phase
3. Late Mesozoic Local Tectonic and Crustal Heating Phase.

These three phases cover the Marathon–Ouachita–Appalachian orogeny to end Cretaceous interval (299 Ma to 66 Ma) and naturally reflect plate tectonic forces that controlled tectonics, source terrane exposure, subsidence, accommodation, and even marine water entry to the nascent basin to form the Louann Salt body, the first basin-wide depositional

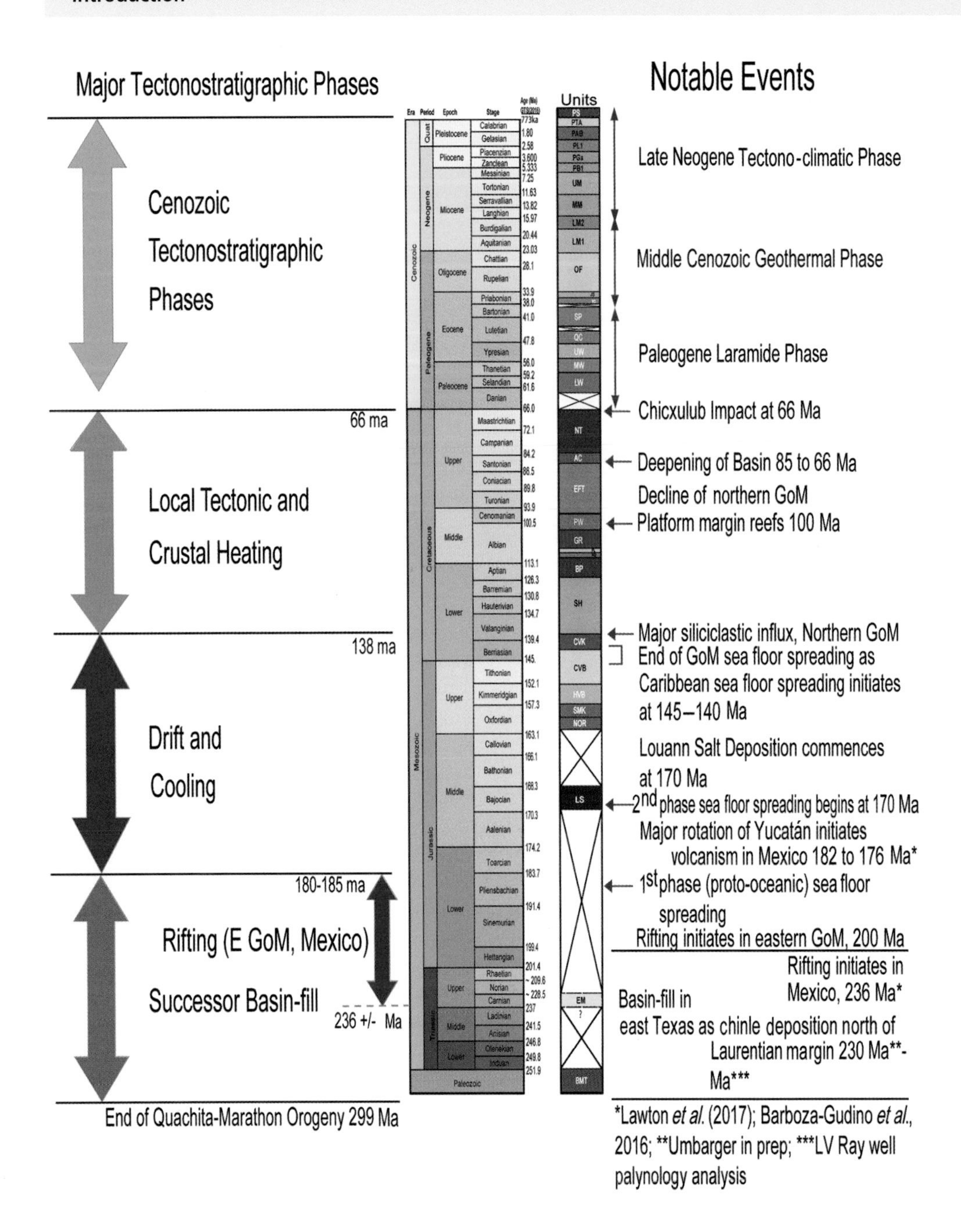

Figure 1.24 Major tectonostratigraphic phases, GoM basin and predecessors.

unit. Our tectonostratigraphic framework is based on new plate tectonic reconstructions, detrital zircon geochronology from deep wells, and analysis of new seismic reflection data in Mexico and the USA. Newly developed concepts depart from conventional GoM thinking both in terms of timing and kinematics, as will be described in detail in Chapter 3.

The stratigraphic terminology and chronostratigraphy that underpins unit-specific identification and correlation over regional to basin-scales is described in Chapters 2–8. Discussion of the Mesozoic and Cenozoic depositional systems classification and assumptions used in creating various depositional maps in this book immediately follows in this chapter.

1.9 Stratigraphic Terminology

Stratigraphic terminology used for naming depositional intervals in the greater GoM range from simple **lithostratigraphy** to **biostratigraphically** age-constrained **chronostratigraphy**. The differences between onshore and offshore nomenclature, reflecting the progressive shift from land to deepwater exploration, can be confusing. Some older formation names are clearly time-transgressive (e.g., Haynesville Shale; Figure 1.25) or facies-dependent (e.g., Ferry Lake Anhydrite, Gilmer Limestone). The southern GoM has similar issues and also suffers from a local lithostratigraphic nomenclature that is specific to each of six or seven geological provinces (e.g., Figure 1.26 for

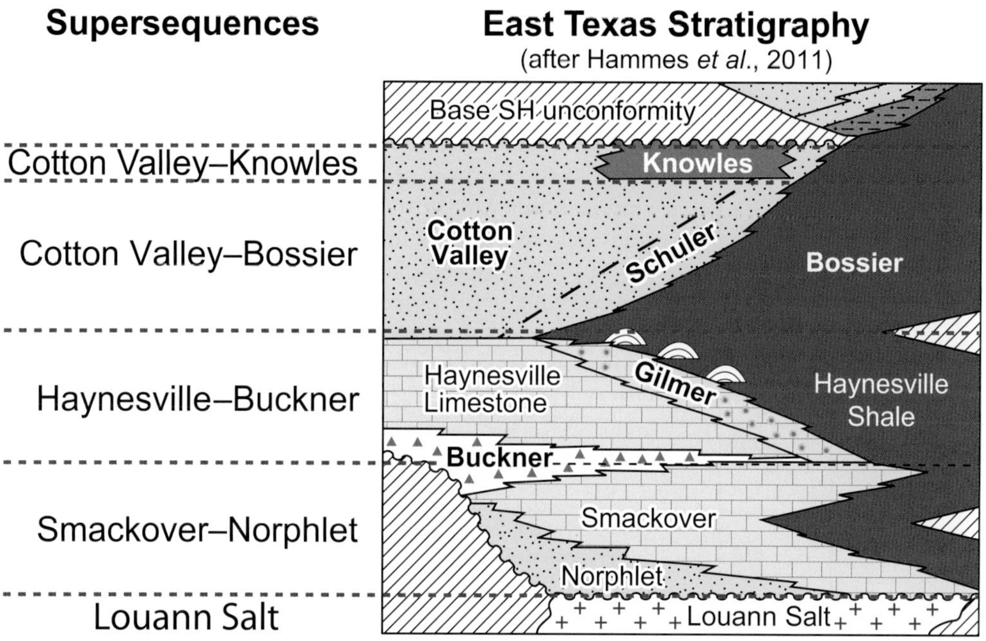

Figure 1.25 Early Mesozoic supersequences. Smackover–Norphlet supersequence. Lithostratigraphic units (e.g., Norphlet Formation, Smackover Formation) are often time-transgressive and essentially amount to paleo-environmental facies. Supersequences incorporate such units into chronostratigraphically significant regional- to basin-scale packages. Modified from Olson *et al.* (2015).

Tampico–Misantla province). Recent reports compiled and provided to the public by Mexico's National Hydrocarbon Commission has followed the same lithostratigraphic approach (CNH 2015a, 2015b, 2017b).

The ultimate goal of the stratigraphic framework in the GoM developed for the GBDS project and used in this book is to enable correlation from the Gulf coastal plain to the deep-water abyssal plain. The GoM exploration effort that began as early as the 1890s has generated a large volume of wells with available ditch (well) cuttings samples that are readily analyzed for **microfossil** and microfloral content. While early charts and zonations focused on benthic foraminifera, which had limitations due to paleo-environmental factors, modern well site biostratigraphy incorporates planktonic forams, calcareous nannofossils, and palynomorphs (Bolli *et al.* 1989; Styzen 1996; Olson *et al.* 2015; www.paleodata.com). Combined with the improved geologic timescales (Ogg *et al.* 2016), the resolution with the Neogene interval, for example, is fast approaching 100 ky or better (Snedden and Liu 2011). The structural complexity of the basin, illustrated by the 10 basin cross-sections (Section 1.5) also necessitates use of biostratigraphically age-constrained correlation surfaces.

Many companies and industry-support vendors have developed detailed chronostratigraphic classifications and biostratigraphic charts for the GoM. Key public domain charts include Styzen (1996), and those online at PDI (www.paleodata.com/chart), as well as the Mesozoic charts linked to Olson *et al.* (2015). Biostratigraphic data from wells drilled in federal waters is released to the public after 10 years or with lease relinquishment or termination. However, many of these BOEM "paleontology reports" are simple summaries of more detailed operator or vendor studies (Weber and Parker 2016). State surveys and universities have a limited number of biostratigraphic reports from wells drilled onshore or in state waters.

1.10 Mesozoic Chronostratigraphy, Northern GoM

Extensive exploration for northern GoM Mesozoic reservoirs actually preceded Cenozoic discoveries. Mesozoic hydrocarbon reserve additions reached a plateau around 1976, and interest shifted to the Cenozoic offshore. As a result, extensive use of microfossil datums was not well established for the Mesozoic prior to that shift in exploration focus, particularly in offshore parts of the basin. However, interest in Mesozoic stratigraphy has been rekindled as a function of two factors: (1) drilling of onshore unconventional plays including the Haynesville Shale gas play (Hammes *et al.* 2011; Wang *et al.* 2013) and Eagle Ford Formation oil and gas shale plays (Hentz and Ruppel 2011; Denne *et al.* 2014); and (2) improved seismic imaging below salt and the thick Cenozoic cover that often puts the Top Cretaceous at depths exceeding 30,000 ft (9.1 km) in slope and deepwater areas of the basin.

The Mesozoic stratigraphy used in this book is founded on microfossil datums that allow correlation from onshore to offshore areas (Figure 1.27; Olson *et al.* 2015). Similar to the Paleogene interval (e.g., Upper Wilcox), we have retained some older lithostratigraphic terms (Glen Rose, Austin Chalk), but each unit boundary is based upon age diagnostic information, including last appearance datums (LADs), first appearance datums (FADs), or, in some cases, faunal acmes (Olson *et al.*

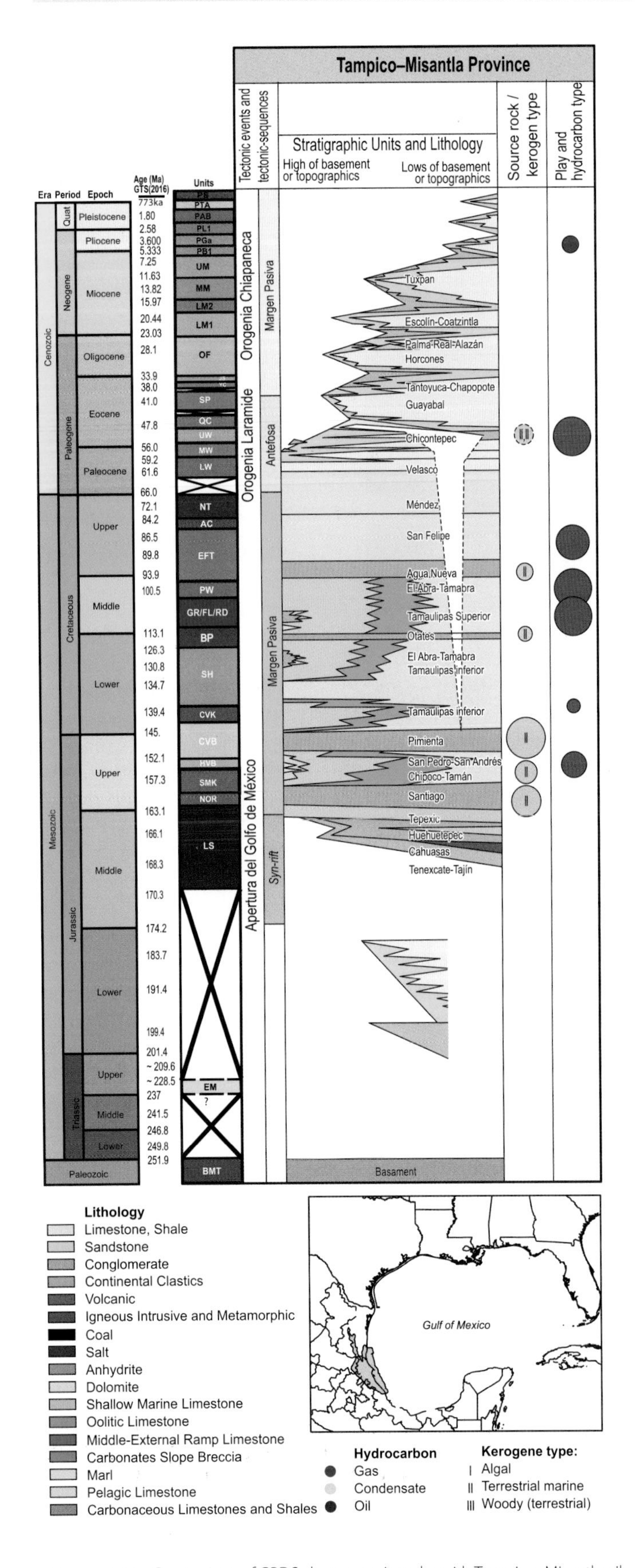

Figure 1.26 Comparison of GBDS chronostratigraphy with Tampico–Misantla oil and province stratigraphic column. Inset map shows Tampico–Misantla province.

2015). Our primary chronostratigraphic information comes from biostratigraphic sources and seismic stratal correlations. Our biostratigraphic data includes published and unpublished information from both onshore (outcrop and subsurface) and offshore sources (Scott 1984; Rogers 1987; Scott *et al.* 2002; Petty 2008; Denne *et al.* 2014). We detail our chronostratigraphic framework through a Mesozoic biostratigraphy table (full table available online at http://dx.doi.org/10.1190/INT-2014-0179.2). In compiling the table, we follow the global chronostratigraphic nomenclature proposed by SEPM Special Publication 60 (Hardenbol *et al.* 1998) and the chronostratigraphic designation system outlined by Snedden and Liu (2011). Additionally, we rely on the Mesozoic depositional architecture for the GoM previously outlined by Galloway (2008).

Practical considerations of basin-scale correlation and database size led us to establish chronostratigraphy at the supersequence level for much of the Mesozoic interval (Figure 1.27). **Supersequences** are longer-duration (5–10 million years) aggregates of sequences, with boundaries usually representing significant regional tectonic events (e.g., Top Paluxy–Washita supersequence). Because the underlying support for stratal correlation is biostratigraphy, we have designated 15 supersequences and a basement unit (BMT) in the GoM with two or three letters referencing lithostratigraphic names familiar to GoM workers (e.g., EFT for Eagle Ford–Tuscaloosa; Figure 1.27) for ease of use. These supersequences divide time-transgressive lithostratigraphic units (e.g., Smackover Formation, Norphlet Formation) into chronostratigraphically significant units (e.g., SN = Smackover–Norphlet; Figure 1.25). For additional details on the construction of the Mesozoic chronostratigraphy and examples of application, the reader is referred to Olson et al. (2015).

1.11 Mesozoic Chronostratigraphy, Southern GoM

Establishment of a chronostratigraphic system for the Mesozoic of Mexico has had to overcome several challenges. First, much of the Lower Mesozoic in accessible onshore outcrop sections is non-marine in origin, with fossil plants providing limited age control (Padilla y Sánchez and Jose 2016). Marine deposition is relatively rare in onshore localities until the Middle to Late Jurassic (Oloriz *et al.* 2003). Second, scarce ammonite macrofossils obtained in well cores have provided the primary age diagnostic information for Late Jurassic to Late Cretaceous offshore wells (Angeles-Aquino and Cantú-Chapa 2001; Cantú-Chapa 2009). This is in spite of excellent microfossil biostratigraphic zonations in the Cretaceous interval of northern Mexico (Longoria and Gamper 1977; Ice and McNulty 1980). Third, many of the detailed well reports with these age assignments remain proprietary (note at least four unpublished internal company reports were cited by Angeles-Aquino and Cantú-Chapa [2001]). An exception is the data-rich table included in the biostratigraphy of the Cretaceous–

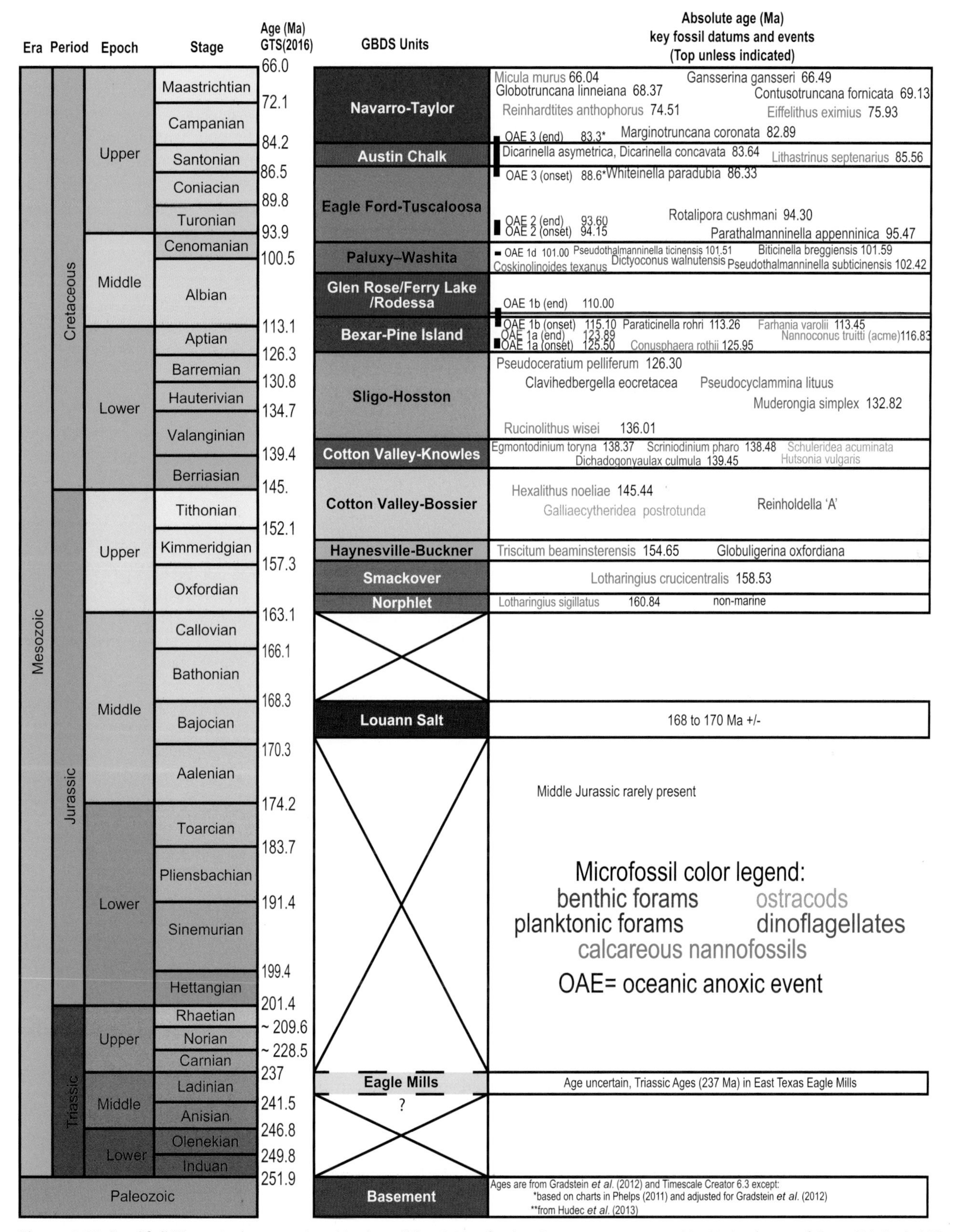

Figure 1.27 Simplified Mesozoic chronostratigraphic chart. Abbreviations for the 15 supersequences used in this book are as follows: EM, Eagle Mills; AC, Austin Chalk; BP, Bexar–Pine Island Shale; CVB, Cotton Valley–Bossier; CVK, Cotton Valley–Knowles; EFT, Eagle Ford–Tuscaloosa; FL, Ferry Lake Anhydrite; GR, Glen Rose; HVB, Haynesville–Buckner; LS, Louann Salt; NT, Navarro–Taylor; PW, Paluxy–Washita; RD, Rodessa; SH, Sligo–Hosston; SN, Smackover–Norphlet. The seismically defined basement unit is noted as BMT. OAEs are oceanic anoxic events and are noted in the GoM by Phelps (2011), with age dates from Gradstein *et al.* (2012), as well as other more recent publications (Elderbak *et al.* 2014; Lowery *et al.* 2014). Additional biostratigraphic datums are available for each supersequence and are detailed in the Mesozoic biostratigraphy table in Olson *et al.* (2015), supplementary material.

Paleocene boundary unit in offshore wells provided by Cantú-Chapa and Landeros-Flores (2001). Here, microfossils (primarily planktonic forams) allowed differentiation of Paleocene and Cretaceous intervals in key wells.

Recently, academic investigators have had some success using advanced absolute age dating techniques to provide sequence stratigraphic correlation points. Lehmann *et al.* (1999, 2000) used isotope chemostratigraphic results in work on the Lower Cretaceous outcrops of northeastern Mexico. U–Pb geochronology (see Box 1.2) based on first-cycle (volcanic) zircons obtained from Mexico outcrop intervals also provided important age constraints in certain Mesozoic intervals (Lawton *et al.* 2009; Lawton and Molina-Garza 2014). The summary stratigraphic chart of Martini and Ortega-Gutiérrez (2016) nicely illustrates the importance of first-cycle zircon U–Pb geochronology for better constraining onshore Jurassic stratigraphy and tectonostratigraphic evolution of the southern GoM. Unfortunately, the same approach is not, at present, being widely used on offshore samples.

Box 1.2 Detrital Zircon Analysis: Advanced Provenance Analysis

In recent years, detrital zircon geochronology has become the tool of choice for provenance analysis that supports detailed paleogeographic reconstructions. It has a number of advantages over previous approaches such as QFL (quartz–feldspar–lithic) ternary plotting from petrographic or compositional analyses that are particularly sensitive to diagenetic removal of framework

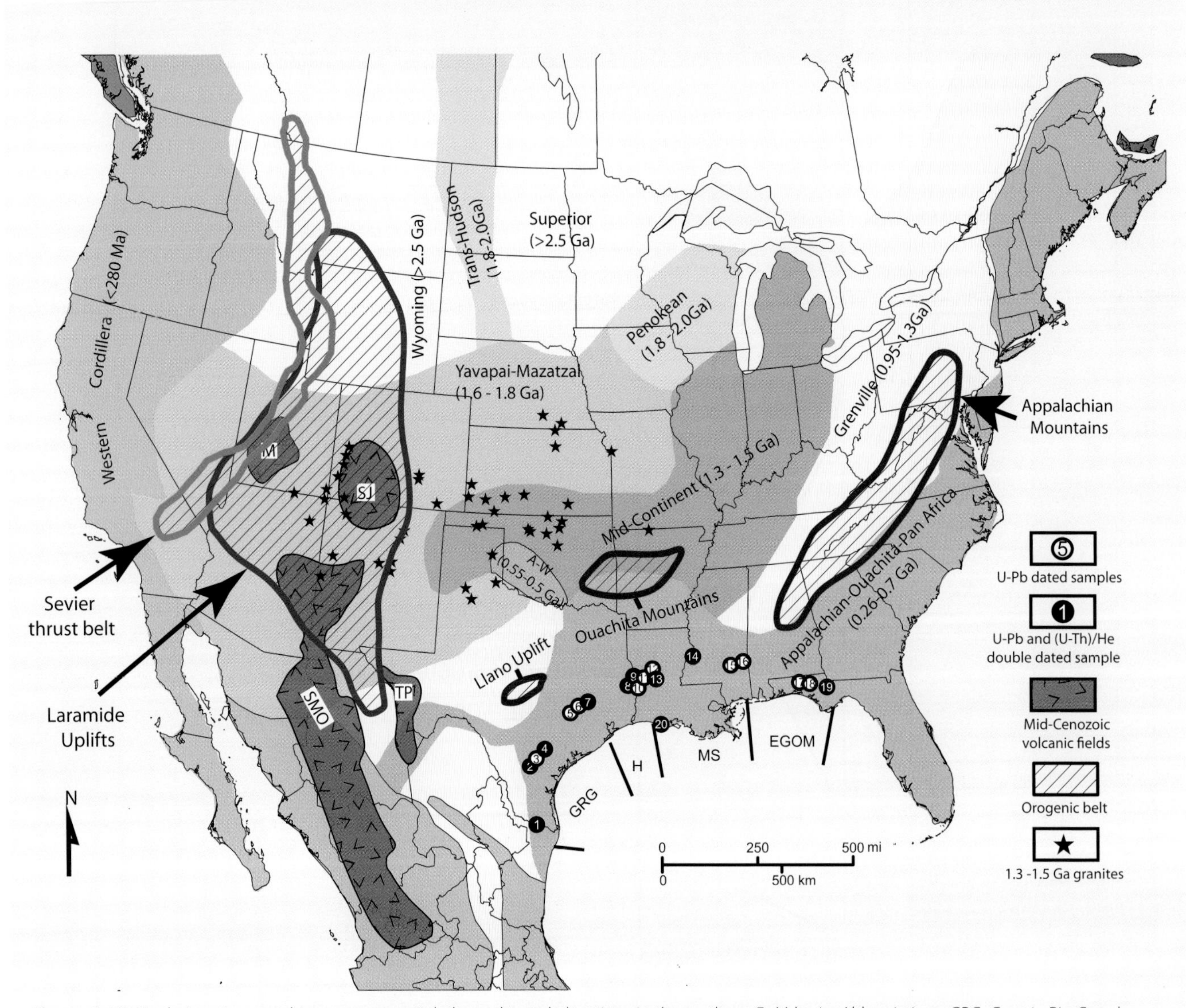

Figure 1.28 North America crustal terranes, orogenic belts, and sample locations in the northern GoM basin. Abbreviations: GRG, Greater Rio Grande Embayment; H, Houston Embayment; MS, Mississippi Embayment; EGoM, Eastern GoM Embayment; M, Marysvale volcanic field; SJ, San Juan volcanic field; TP, Trans-Pecos volcanic field; SMO, Sierra Madre Occidental volcanic field; A–W, Amarillo–Wichita. Modified from Xu *et al.* (2017).

Box 1.2 *(cont.)*

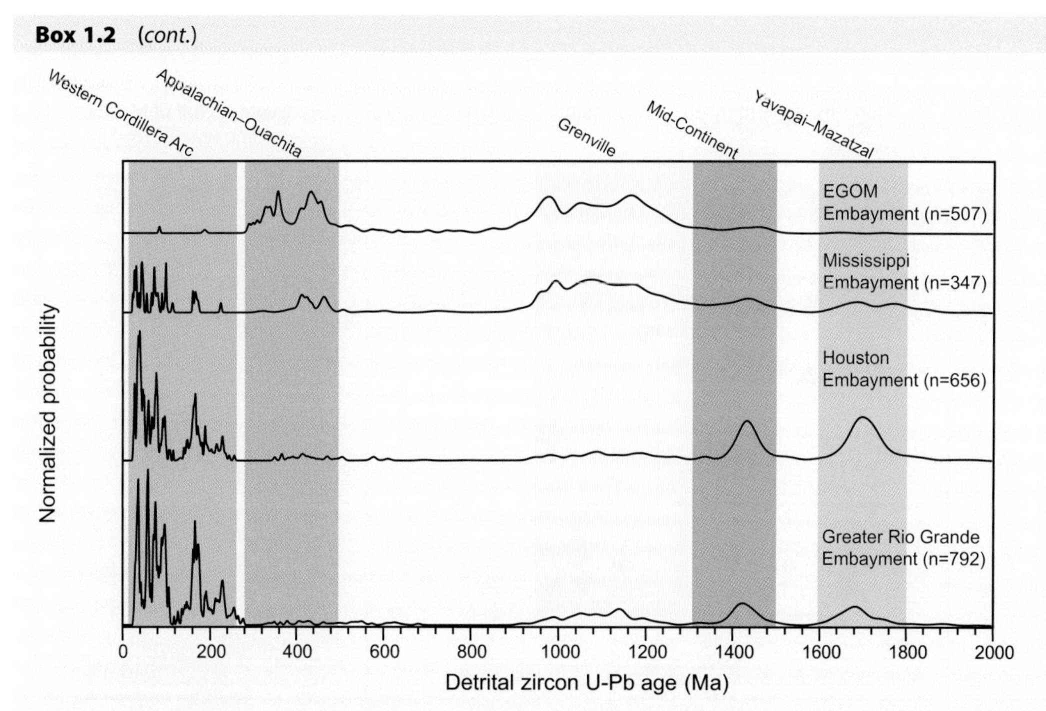

Figure 1.29 Comparison of U–Pb age spectra from detrital zircon grains in the Lower Miocene of Texas. Modified from Xu *et al.* (2017).

grains. Zircon is a heavy mineral, resistant to physical and chemical weathering, and very stable at surface to shallow crustal pressures and temperatures. The zircon uranium–lead (U–Pb) system has a high closure temperature (about 900°C), meaning that U and Pb do not escape the zircon crystal at lower temperatures. As zircon crystals initially have no Pb, the only Pb is from U isotopic decay. Ages derived from zircon U–Pb measurements thus provide the date of the original crystallization of the zircon crystal, assuming that has not been reset by exposures to temperatures over 900°C, rare in deep sedimentary burial without significant pressure–temperature metamorphism or igneous heating.

Because zircon crystallizes at high temperature and pressure, the U–Pb decay provides an age that can be matched to the timing of accretion of different basement terranes to North America (Figure 1.28; Blum and Pecha 2014; Xu *et al.* 2017). From U–Pb detrital zircon age spectra we can identify numerous source terranes such as the Western Cordilleran, Yavapai–Mazatzal, Wyoming, Trans-Hudson, Grenville, Mid-continent, and Appalachian terranes. Once enough zircons (typically 100–300 grains) have been collected and irradiated by laser ablation, a robust and diverse age spectra of the grains is a fingerprint of the contributing source terranes (Figure 1.29).

Zircon geochronology can be useful if there is some uncertainty about the stratigraphic age of a sample. Ages derived from U–Pb analyses of sedimentary rocks are logically considered as a maximum depositional age: young sedimentary intervals can incorporate older zircons but older sedimentary rock obviously cannot include zircons younger than its depositional age. The closest fit between true depositional age and depositional age from zircon geochronology is where first-cycle, volcanic airfall-derived zircons are abundant (Reiners *et al.* 2005). The strict criteria for determining maximum depositional age involves averaging the three youngest zircons that overlap in age at 2σ in a zircon population (Dickinson and Gehrels 2009; Gehrels and Pecha 2014).

One drawback to U–Pb ages derived from detrital zircon is the problem of recycling. Zircon can be liberated by exposure of basement, transported long distances to a new burial site, reburied and exhumed, still retaining the original U–Pb crystallization age. This can be a problem if a sandstone is potentially sourced from two different areas, but retains the signature of only the original source terrane, not the secondary site from which the rivers last drained.

To address this, a more advanced combined U–Pb and (U–Th)/He dating on single zircon grain or "double dating" approach is used to provide the age of cooling or exhumation (Rahl *et al.* 2003; Reiners *et al.* 2005; Xu *et al.* 2017). It makes use of the cooling temperature of zircon, which tells us when the source was uplifted. For example, in the case of the paleo-Greater Rio Grande River, zircons that crystallized 950–1300 Ma in the Grenville basement province were buried and later exhumed at three different sites (Great Plains, west Texas–New Mexico, and Llano area) during four different tectonic events ranging from pre-Cambrian to as recently as 40 Ma (Figure 1.30A). Using double dating one can determine which grains were recycled from the Colorado plateau and which came from the Llano area, for example. The same is true for the younger basement sources coming from the Rockies, in three uplifts ranging from 170 to 25 Ma (Xu *et al.* 2017; Figure 1.30B).

Box 1.2 (*cont.*)

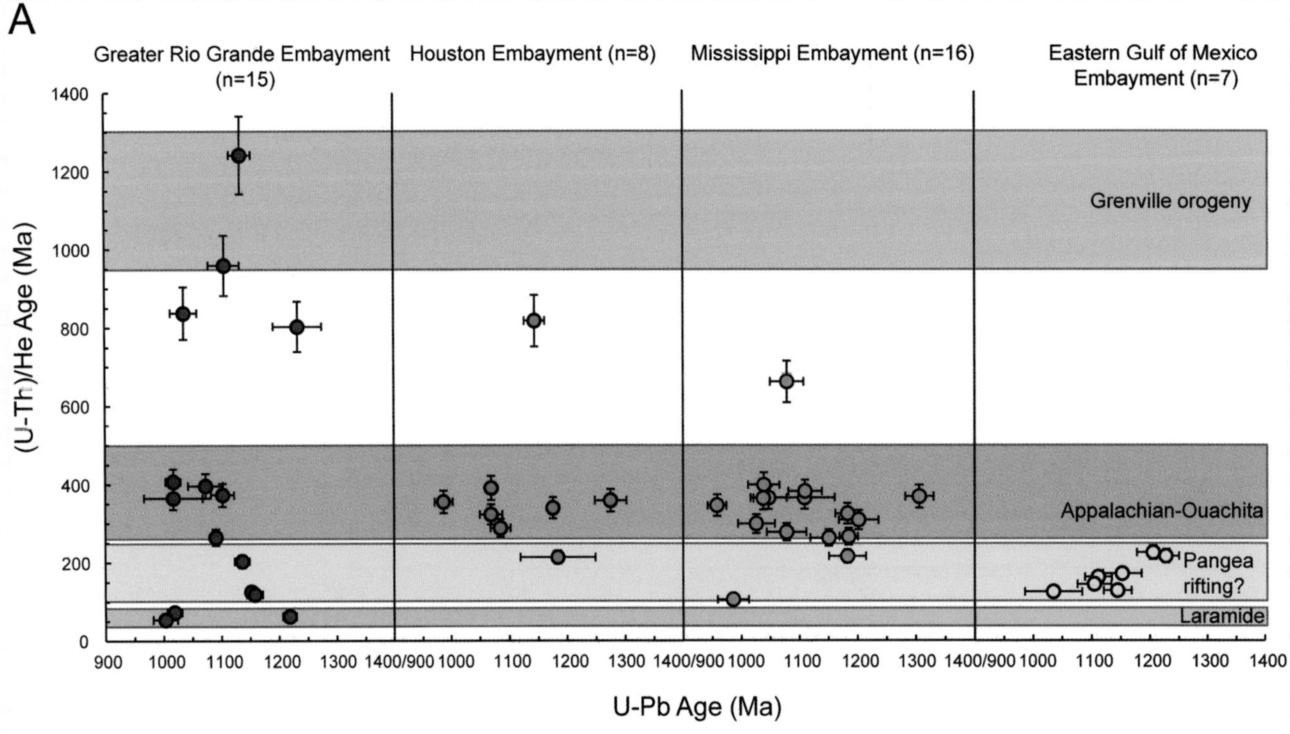

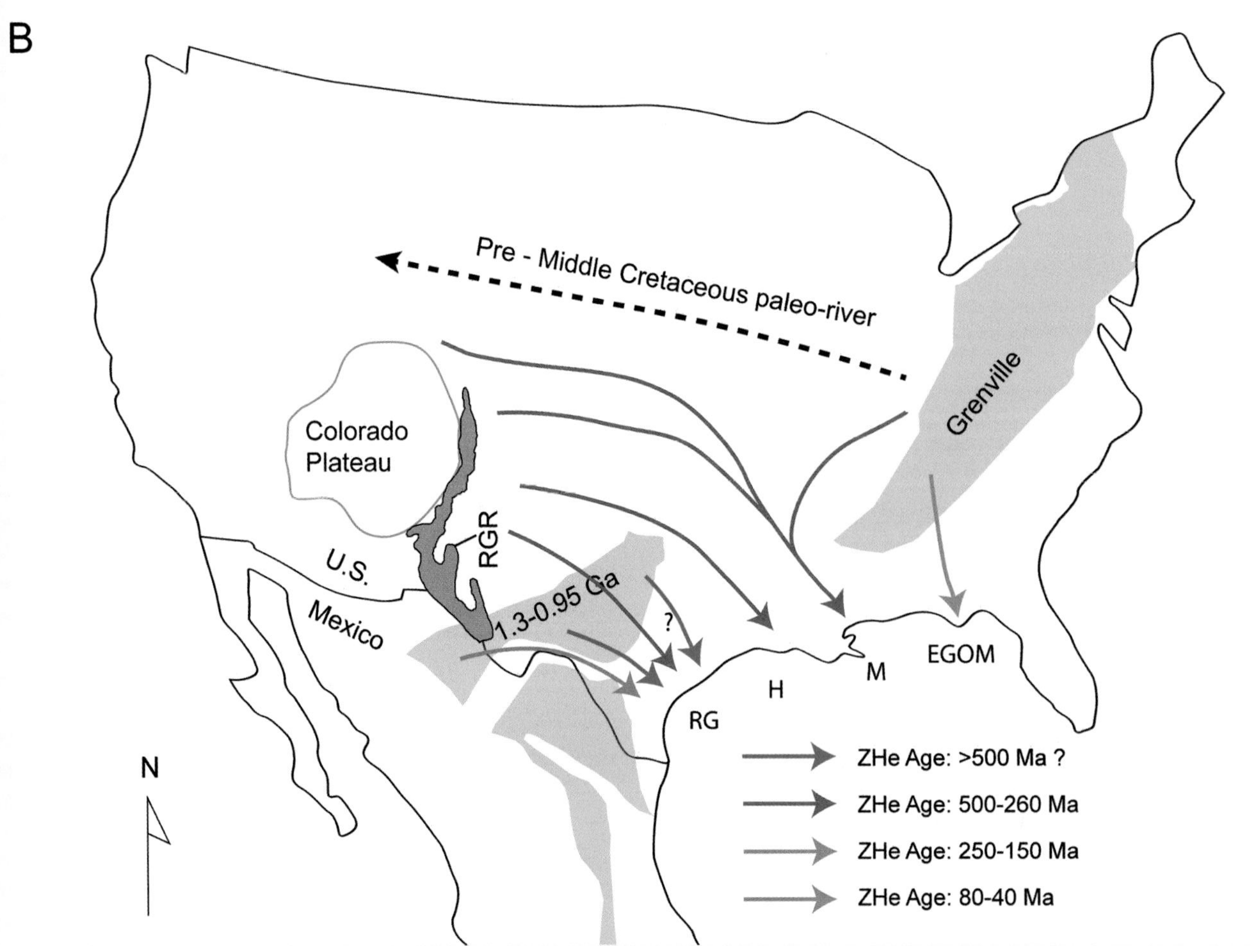

Figure 1.30 Detrital zircon recycling. (A) U–Pb–He ages of Grenville zircons in the Lower Miocene strata of the GoM basin. (B) Sediment routing of Grenville grains. RGR, Rio Grande rift. *n* = number of analyses. Color bars indicate different orogenic events. Modified from Xu *et al.* (2017).

1.12 Cenozoic Chronostratigraphy, Northern GoM

The northern Gulf basin stratigraphic framework, chronology, and nomenclature were established by the mid-twentieth century using conventional stratigraphic concepts. Alternating outcrops of sandy coastal to continental sediments and fossiliferous marine mudrocks provided an initial subdivision for Paleocene and Eocene strata. Early petroleum exploration revealed the subsurface stratigraphy beneath the coastal plain. The thick, repetitious, siliciclastic Cenozoic interval was initially subdivided using the marine shale tongues, and then widespread microfossil-bearing horizons were used to correlate and date the evolving stratigraphic framework. This concept of transgression-bounded genetic units was formalized in a seminal paper by Frazier (1974). Frazier argued that the Gulf Cenozoic fill recorded a succession of "depositional episodes" that deposited by a foundation of progradational marine and coastal facies that were, in turn, overlain and replaced landward by aggradational coastal plain and fluvial facies. This facies succession was capped by a relatively thin succession of transgressive or back-stepping coastal and marine shelf deposits. Importantly, much of the basin margin was sediment-starved at any moment of geologic time. Areas of starvation, bypass, and/or erosion most likely lay in the landward coastal plain and the offshore middle to outer shelf. Thus, the "Frazierian" genetic unit is bounded basinward by submarine starvation surfaces (condensed beds) created during and soon after transgressive retreat of coastal depositional systems. This surface would later come to be known as the *maximum flooding surface*. Such depositional episodes conform to the basic definition of a sequence as a contiguous suite of genetically related strata bounded in part by unconformities. If relative or eustatic sea-level fall punctuates the history of a depositional episode, the genetic unit will contain an internal subaerial unconformity within its landward strata. Fraser's model, in fact, was developed in and for the Quarternary stratigraphy of the Mississippi delta and coastal environs where eustatic sea level was a major factor.

Using the Frazierian depositional model, Galloway (1989a) defined the *genetic stratigraphic sequence* as a fundamental unit of GoM Cenozoic stratigraphy. A genetic sequence consists of all strata deposited during an episode of sediment influx and depositional offlap of the basin margin. It is bounded by a family of surfaces of marine non-deposition and/or erosion created during transgression, generalized as the maximum flooding surface. This pattern is readily recognized in the Paleogene interval, where transgressive marine shelf mudstone and glauconitic sandstone units extend to outcrop (Galloway 1989b). It also applies in Neogene strata, where prominent transgressive markers record **glacioeustatic** sea-level rise events (Galloway *et al.* 2000). Thus, genetic sequences typically correspond closely to widely used northern Gulf stratigraphic nomenclature.

The *depositional sequence* paradigm, which uses subaerial erosion surfaces as sequence boundaries, provides an alternative to the traditional Gulf basin lithostratigraphic framework and has been applied by several authors (Yurewicz *et al.* 1993; Mancini and Puckett 1995; Lawless *et al.* 1997), especially to Late Neogene strata that are strongly influenced by glacioeustasy (Weimer *et al.* 1998; Roesink *et al.* 2004). Depositional sequence models for carbonate and mixed successions, which are appropriate for the Mesozoic Gulf fill, are summarized and illustrated by Handford and Loucks (1993).

The synthesis of Gulf depositional history and physical stratigraphy as presented here largely utilizes the traditional Paleogene lithostratigraphic framework and the regional marine flooding horizons characterized by widely identified faunal markers within Neogene strata. Building upon the syntheses of Winker and Buffler (1988), Galloway (1989b), and Morton and Ayers (1992), Galloway *et al.* (2000) proposed a genetic stratigraphic framework that groups Cenozoic strata into a succession of 18 principal GoM depositional episodes (shortened to deposodes; Figure 1.31). Each episode records a long-term (ca. 2–12 Ma) cycle of sedimentary infilling, typically accompanied by shelf-margin offlap, along the divergent margin of the northern Gulf basin. Deposits of each episode are characterized by lithologic composition (predominantly sandstone and mudstone, with minor carbonate and evaporite), vertical stacking of lithofacies and parasequences, and relative stability of sediment dispersal systems and consequent paleogeography. Almost all of the depositional episodes terminated with a phase of deepening and/or basin margin transgression (Figure 1.31). Deposits of episodes are bounded by prominent, widely recognized, and well-documented stratigraphic surfaces. Bounding surfaces variously include marine starvation and condensed horizons, maximum flooding surfaces, marine erosional unconformities, and faunal gaps that are described and interpreted by multiple authors. They are widely recognized as fundamental stratigraphic building blocks of the basin-fill and are equivalent to the supersequences described for the Mesozoic. They constitute the physical stratigraphic equivalent of the chronostratigraphic deposode.

1.13 Cenozoic Chronostratigraphy, Southern GoM

Like the Cenozoic of the northern GoM, the chronostratigraphic framework of Mexico is based primarily on offshore well biostratigraphy, largely foraminifera of benthic and planktonic forms. Biostratigraphic work by Pemex and IMP has been occasionally incorporated into university theses and dissertations (Sánchez-Hernández 2013) or published papers (Vásquez *et al.* 2014; Gutiérrez Paredes *et al.* 2017). As an example, a data table of Gutiérrez Paredes *et al.* (2017) provides LADs and FADs of planktonic forams and calcareous nannofossils for the Upper Miocene to Lower Oligocene of 12 wells drilled in southern offshore Mexico. The majority of the fossil data conforms to the top unit boundaries of the Upper Miocene, Middle Miocene, and Oligocene Frio used here for the northern GoM. One important exception to the

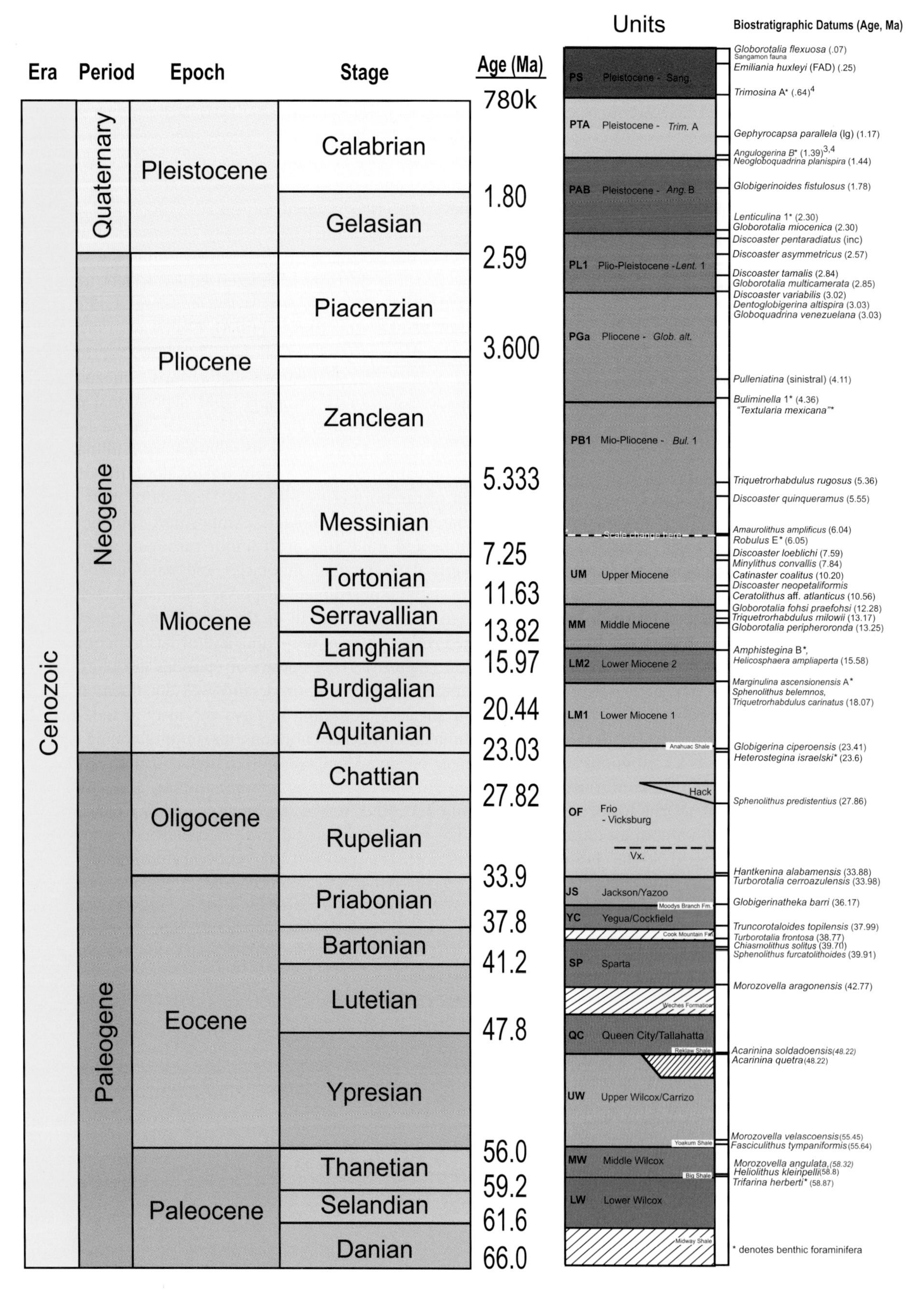

Figure 1.31 Cenozoic chronostratigraphic chart, including key biostratigraphic datums.

Neogene is the boundary of the Lower Miocene and Middle Miocene, which is lower within the GBDS stratigraphy discussed earlier. Gutiérrez Paredes *et al.* (2017) do point out, however, that the bounding Langhian stage is not well represented in the area, with only one well encountering that stage in cuttings. This is relevant to exploration, as Gutiérrez Paredes *et al.* (2017) show a large number of stratigraphic discontinuities in the Middle Miocene Serravallian interval that are accompanied by sandstone reservoir occurrence in the wells. That is consistent with the 9–16 Ma fast exhumation phase of the Chiapanecan orogeny in landward areas (Sanchez-Montes de Oca 1980; Witt *et al.* 2012).

The Paleogene chronostratigraphy is less well documented in public reports or published scientific papers. An exception is the detailed chronostratigraphic chart for the Chicontepec Canyon included in Vásquez *et al.* (2014). The biostratigraphic datums generally conform to global stage boundaries, but there are some notable departures that may reflect local conditions in this large-scale erosional canyon system and the repeated bypass of sands into the basin in the Eocene (Cossey *et al.* 2007).

Biostratigraphic charts provided in relatively rare university studies of Pemex wells provide some direct comparison to the northern GoM chronostratigraphy. For example, analysis of ditch (cuttings) samples by Gutiérrez-Puente (2006) in the Cupelado-10 well is shown as a range chart of various planktonic forams. As analysis was done using the standard micropaleontological scheme of Bolli *et al.* (1989), there is general equivalency of many biodatums in the Pliocene to Paleocene interval here, providing some level of comfort that age-constrained basin-wide correlations between the northern and southern GoM can be made.

1.14 Stratigraphic Framework of Cuba

Most structural and stratigraphic classifications consider Cuba as part of the greater Caribbean (Pardo 1975; Pindell and Kennan 2001). Our treatment of the area is therefore superficial, except where the stratigraphy of the adjacent GoM basin is concerned. Extensions of trends from the USA across the Florida Straits are relevant and the effects of various basin-wide events, such as the Chicxulub impact (K–Pg event) obviously are recorded in the rock record of Cuba. Additional discussion of the petroleum habitat of Cuba is included in Chapter 2. The subsections that follow focus on the Mesozoic and Cenozoic stratigraphic framework that is relevant for an understanding of the greater GoM basin.

1.14.1 Cuban Mesozoic Stratigraphic Framework

The Mesozoic stratigraphic framework of Cuba largely reflects the evolution of the GoM basin, as major differentiation of the GoM and Caribbean basins did not occur until the Late Cretaceous to Paleogene (Escalona and Yang 2013). While the proto Caribbean plate did form during the Late Triassic to Early Jurassic separation of North and South America, the stratigraphic intervals are remarkably similar (Figure 1.32). Initially, interpreted continental to shallow marine siliciclastics filled half-grabens (Escalona and Yang 2013), a pattern also observed in the northern and south GoM at this time. It is important to note that Sequence 1 of Escalona and Yang (2013) has not been penetrated in the offshore area to date, but its seismic character and geometry are suggestive of a synrift interval analogous to the Eagle Mills drilled in the northeastern GoM (Marton and Buffler 1999).

Late Jurassic rotation of the Mayan (Yucatán) block during GoM sea floor spreading also generated important tectonic elements in Cuba (Escalona and Yang 2013). Jurassic platform carbonates (Remedios district; Figure 1.32) and coeval distal slope or scarp facies of Oxfordian to Tithonian age show similarities in lithology with the limestones and carbonate mudrocks of the areas to the north (e.g., Smackover, Haynesville, Cotton Valley Formations).

This was followed by a period of relative tectonic quiescence in the Early Cretaceous, with progressive drowning of the proto Caribbean plate and deposition of deep marine carbonates (Sequence 2 of Escalona and Yang 2013). Shallow marine carbonates were restricted to the highest structural features (e.g., Upper Perros Formation of the Remedios district; Morena and Margarita Formations of the Placetas and Camajuani districts, respectively). Palenque Formation carbonates of the Remedios district are correlative to the Aptian to Albian interval of the GoM basin (e.g., Sligo–Hosston, Glen Rose, Paluxy–Washita supersequences; Figure 1.32). Ceno-Turonian equivalents of the Eagle Ford–Tuscaloosa and Austin Chalk supersequences (e.g., Purio Formation of the Remedios district) were deposited just prior to major plate collision in the Late Cretaceous, as described in Section 2.2 on plate tectonic reconstructions. DSDP core site 537, drilled to the north of Cuba, penetrated deep marine to shallow marine carbonates of Early Cretaceous age (Schlager *et al.* 1984).

DSDP cores to the north of Cuba also penetrated limestone breccia units with strong similarity to onshore Cuba deposits related to the Chicxulub impact event on nearby Yucatán. Sanford *et al.* (2016) described over 130 ft (40 m) of carbonate breccia in DSDP Leg 77 Sites 540 and 536 cores, linked to mass transport processes generated by the seismic wave that moved across the entire basin within minutes of the impact. The corresponding Cuba outcrops of the Penalver Formation and Cacarajícara Formations, also related to the impact event, are well documented (Tada *et al.* 2003; Cobiella-Reguera *et al.* 2015).

1.14.2 Cuban Cenozoic Stratigraphic Framework

The Late Cretaceous to Eocene strata collision between the greater Arc of the Caribbean and North American plates set up significant differences in stratigraphy between the two basins. The original Jurassic strata that were laterally continuous to the GoM basin were now subducted beneath the upper Caribbean plate in several stages, forming the Cuban fold and thrust

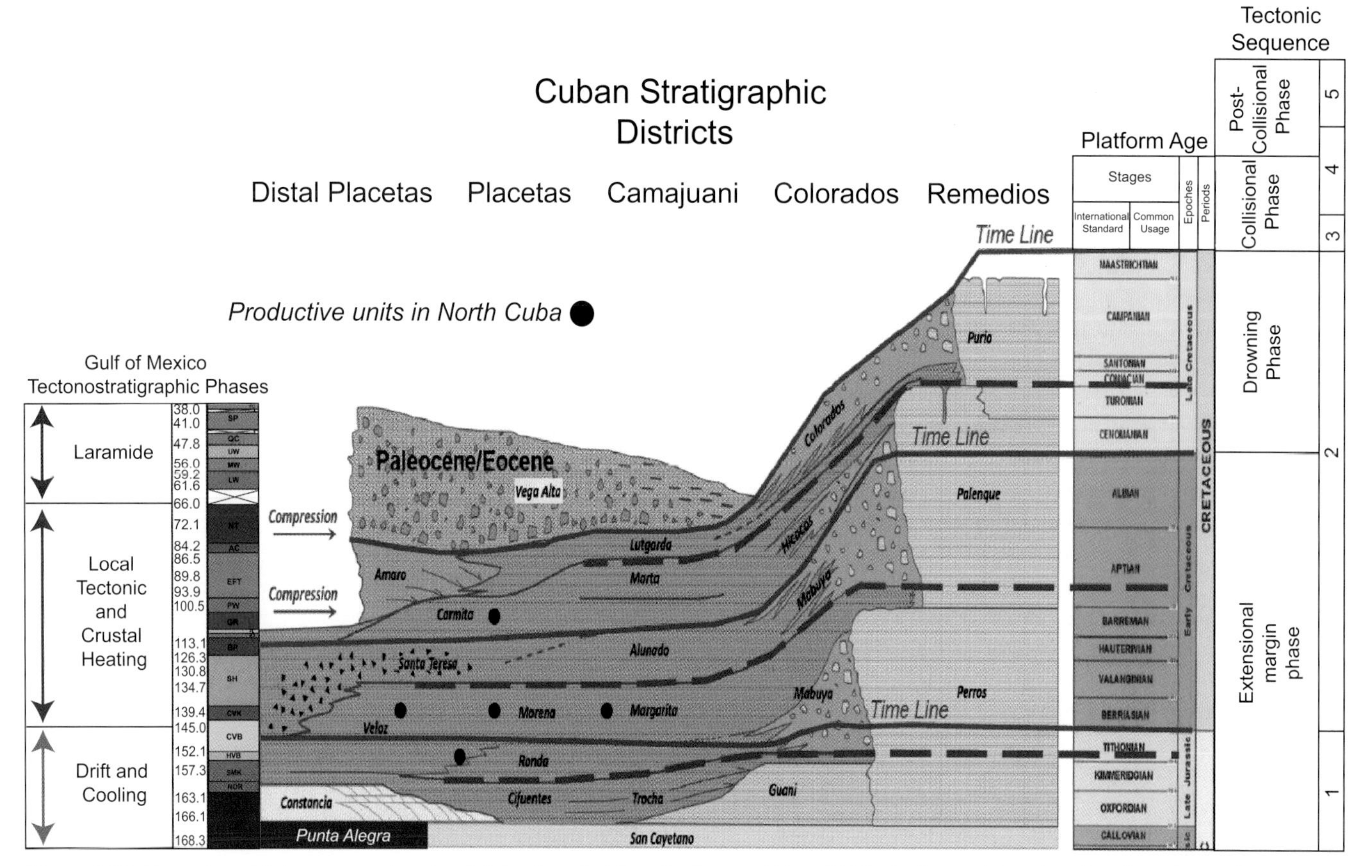

Figure 1.32 Mesozoic and Paleogene stratigraphy of Cuba. Compiled from Escalona and Yang (2013), Gordon *et al.* (1997), and Melbana Energy (2017)

system. Distal shales of Mesozoic age were thrust into the upper plate while more proximal carbonate facies are present in the lower plate, separated by a major mid-level detachment at the Eocene level. Escalona and Yang (2013) confine this collision phase to their Paleocene–Eocene age Sequence 3 and the Oligocene portion of their Sequence 4 (Figure 1.32).

Paleocene to Eocene foredeep sedimentation took place north of the thrust belt, as documented by deposition of the Vega Alta and Vega-Rosas Formations (Gordon *et al.* 1997; Melbana Energy 2017). Later back-thrusting within the upper plate further complicated the present-day structural architecture and has made unraveling the Cenozoic stratigraphy much more difficult (Escalona and Yang 2013). It is also important to note that the western half of Cuba merged with the eastern half during the west-to-east tectonic transport, so the pre-collision strata in western Cuba are linked more closely with the Yucatán (Mexico) stratigraphy (e.g., San Cayetano Formation; Haczewski 1976).

Post-collision Cenozoic strata are influenced by development of the Cayman trough and the Loop Current Gulf stream flowing through the Florida Straits. Large, deep sea erosional features (channels) and constructional sediment drifts of Miocene to Holocene age are present between Cuba and the Florida Escarpment, documenting vigorous bottom currents flowing from the northern Caribbean into the Straits of Florida and to the North Atlantic (Gordon *et al.* 1997). Post-collision strata constitute the Miocene portion of Sequence 4 and the entirety of Sequence 5 (Pliocene to end Pleistocene; Escalona and Yang 2013).

1.15 Depositional Systems Classification

Many classifications of past and present depositional environments exist. This is due to the tremendous amount of scientific effort that has gone into characterizing the various siliciclastic and carbonate settings in which sediments accumulate, to be buried and preserved in the rock record. For siliciclastic depositional systems, this book follows Galloway and Hobday (1996), and for carbonate systems the scheme discussed by Handford and Loucks (1993).

As work on depositional paleo-environments has continued since the original publication of these classifications, it is worthwhile to discuss updates and modifications to these schemes that are relevant for the greater GoM.

1.16 Update to Carbonate Depositional Systems in the GoM Basin

Advances in our understanding of carbonate depositional systems have also occurred as modern environments are newly

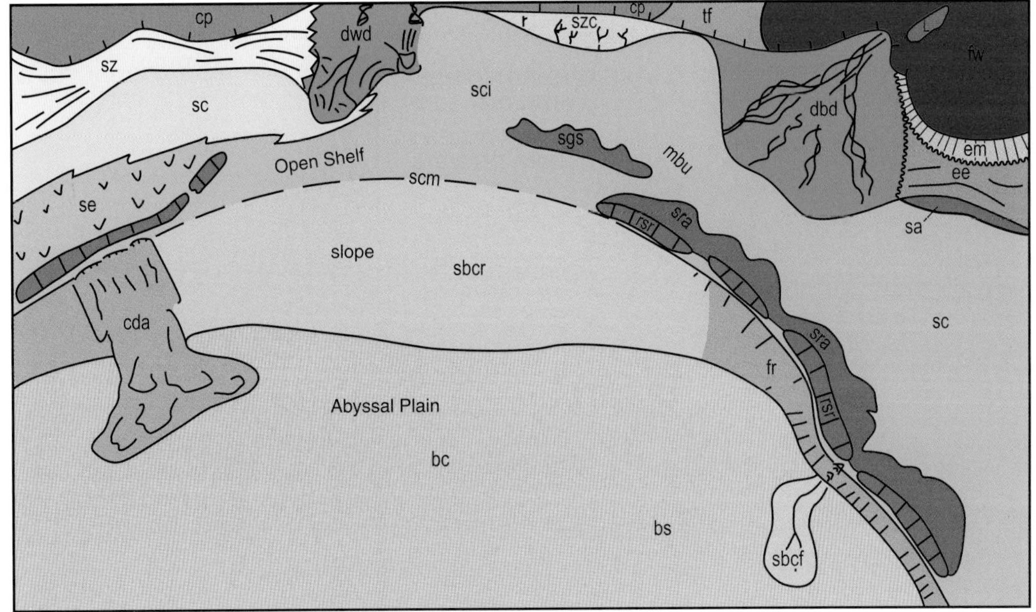

Figure 1.33 Schematic Mesozoic deposystems and classification.

investigated but also as better imaging and characterization of fossilized depositional systems has been carried out by industry and academia. These are particularly relevant to the GoM Mesozoic interval, as documented in well penetrations and numerous publications.

Unlike the siliciclastic-dominated Cenozoic interval of the GoM basin, the Mesozoic succession contains a large portion of carbonate facies, ranging from shallow tidal flat/sabkhas to rimmed shelf reefs to deepwater basin carbonates (Figure 1.33). The long time span of the Mesozoic also saw considerable evolution in different organisms, ranging from Jurassic **microbalites** to Cretaceous framework-building **caprinid rudistids** (Wilson 1975). The Mesozoic also chronicles the rise of massive rimmed shelf reef systems such as in the Aptian–Barremanian (Sligo) and Albian (Washita) and their decline after the Mid-Cretaceous.

Notable recent additions to the classification of Handford and Loucks (1993) include the shelf reef apron (abbreviated as sra), shelf grain shoal (sgs), inner and middle carbonate ramp (sci, scm), and others. For example, reef aprons are exceedingly common in modern systems (Vila-Concejo *et al.* 2013) and recognized in ancient Mesozoic systems as well (Adams 1985). These consist of grainy carbonates and debris transported locally from the rimmed shelf reef systems.

Detailed discussion of the characteristics of these carbonate deposystems and their characteristics in log and core is contained in the online poster titled "Gulf of Mexico Mesozoic Log Facies Interpretation" (www.cambridge.org/gomsb). Well log motifs, placed in a proper paleophysiographic context (e.g., coastal plain, shelf, slope, abyssal plain) define depositional environments for mapping purposes. Iteration with interval thickness, nearby well bores, and regional trends help constrain interpretations, as will be discussed in Section 1.17 and shown in the online resource titled "Gulf of Mexico Siliciclastic Log Facies Interpretation" (www.cambridge.org/gomsb).

At the heart of this book are the **paleogeographic** maps of the Mesozoic and Cenozoic stratigraphic units. One obvious way to validate paleogeographic maps of the embedded depositional systems is by comparison to modern analogs. While most biological components of a carbonate system have evolved since the Mesozoic ended 66 Ma, the physical processes of waves, currents, tides, winds, and sunlight that drive the areal distributions of carbonate systems have not changed.

The Great Barrier Reef (GBR) of Australia (southern sector; Figure 1.34) may be an appropriate analog for the Mesozoic carbonate systems of the Gulf for several reasons. First, the relative paleo-latitude of the Mesozoic (±20 degrees north of the equator) is comparable with the GBR southern sector (GBR-ss) at 22–24 degrees south of the present equator. Second, the GBR-ss is a mixed carbonate–siliciclastic system, with terrigenous input from multiple rivers (see Figure 1.34). In general, siliciclastics dominate landward areas, carbonates dominate seaward (outer shelf, slope, and deepwater) areas and mixing occurs between the two (Maxwell and Swinchatt 1970). A similar pattern is observed in at least four units of the Mesozoic that will be discussed (Paluxy–Washita, Sligo–Hosston, Cotton Valley–Knowles, and Cotton Valley–Bossier). Reciprocal sedimentation, where carbonates give way to sandstone moving paleo-landward, is well documented in the Mesozoic of the GoM, as it is in the GBR. Reefs can flourish in such a setting, as long as the mud content (and thus turbidity) of the input fluvial systems is low enough to permit photosynthesis. Third, the dimensions of key depositional elements are comparable. For example, the GBR extends over 2250 km of the

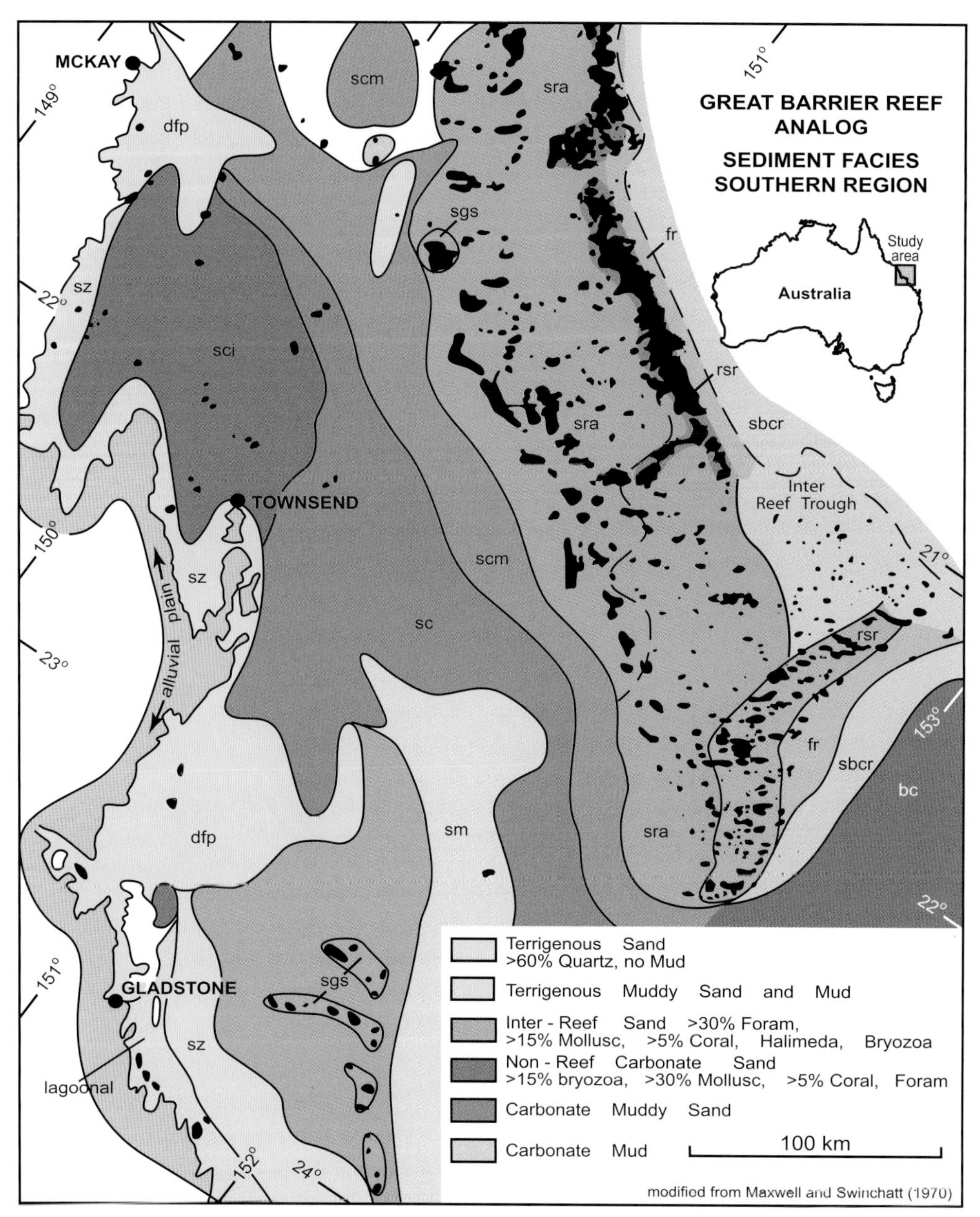

Figure 1.34 Great Barrier Reef analog for Mesozoic mixed carbonate and siliciclastic systems of the GoM. Modified from Maxwell and Swinchatt (1970). Depofacies classification from Galloway (2008). Abbreviations for depofacies used: scm, open shelf outer platform/ramp; sra, reef apron (landward); rsr, rimmed shelf margin (reef or grain shoal); sgs, shelf grain shoal; fr, forereef; sbcr, shelf-to-basin carbonate ramp or distal forereef; bc, carbonate-rich basin floor; sc, carbonate-dominated shelf; sci, open shelf, inner platform/ramp; sz, wave-dominated shore zone; dfp, fluvial-dominated platform delta; sm, mud-dominated shelf. Modified from Snedden *et al.* (2016b).

northeastern Australia margin (Harris and Kowalik 2005) versus the mapped extent of the Sligo rimmed shelf reef systems in the USA, which is at least 2500 km, with another 1000 km in Mexico if one considers the Yucatán margin.

One key difference between the GBR-ss and Mesozoic of the GoM may be the continuity of the rimmed shelf reef system itself. The GBR-ss is segmented at several scales, from small tidal passes that allow open exchange of oceanic and shelfal waters to larger interreef troughs (Figure 1.34) where the rimmed shelf reef is not developed. Most maps of the Sligo (Aptian–Barremanian) and Washita (Albian) systems show only a few tidal passes or interreef troughs (Goldhammer and Johnson 2001) breaking up the long extent of these systems. It may be that well control and 2D seismic line density is insufficient to resolve the tidal passes and other reentrants and thus greater continuity is incorrectly inferred. Even in the GBR, reefs extend along only 70 percent of the shelf edge (Harris and Kowalik 2005). Goldhammer and Johnson (2001) identified at least two interreef troughs or large tidal passes in the Mesozoic of onshore Texas.

The GBR-ss map also shows that the distinct seaward zonation of deposystems from landward to deepwater is mirrored in the paleogeographic maps from the Mesozoic. Shelf-to-basin carbonate ramp (sbcr) occurs in both the inter-reef trough and a distal equivalent of the forereef (fr). The ramp term is a bit of a misnomer, as rimmed shelf reef is not ramp-like but somewhere in the basin the bathymetry flattens out, but the log facies appear to be quite similar in both locations. Forereef is located seaward of the reef, and shelf reef apron (sra) is landward of the main reef. The shelf carbonate middle (scm), a generally carbonate mud-prone interval, is often positioned landward. Yet further landward is the shelf carbonate undifferentiated (sc). Shelf grain shoals (sgs) occur within this physiographic tract but are generally less continuous than the rimmed shelf reef. In the GBR-ss, these are highly variable in size and shape, and this is mirrored in the Mesozoic carbonate intervals. In the GBR, reefs are oriented relative to wind direction or prevailing currents (Harris and Kowalik 2005) and might be a control on grain shoal and patch reef development in the Mesozoic.

1.17 Update to Siliciclastic Systems in the GoM Basin

Classification of the Cenozoic siliciclastic depositional systems (Figure 1.35) follows Galloway and Hobday (1996). This approach emphasizes the process framework and nomenclature of physical geography and thus is specifically designed for creating paleogeographic maps delineating the landscapes and seascapes created during a depositional episode. The paleogeographic reconstructions that follow expand on and update previous syntheses of Galloway *et al.* (2000) and Galloway (2008).

This synthesis further benefits from a number of recent papers that have provided critical insights into global documentation of the processes, facies architecture, and geography of sediment transport systems and their constituent depositional and erosional elements. The interpretations and maps that follow are conditioned by their conclusions:

1. It has been long recognized that delta systems of large rivers are the major suppliers of sediment to the Gulf basin. The apex positions of large deltas are commonly localized by bedrock or long-lived alluvial valleys (Hartley *et al.* 2015). Thus the delta systems tend to geological longevity and reflect structurally defined basin margin topography. Along the Paleogene GoM margin, a number of specific uplifts bounded likely entry points for large rivers. Paleogene examples include the Tamaulipas Arch, Picachos Arch, Chittum Anticline, and Sabine Uplift. Beginning in the Oligocene, tilting uplift along the northern GoM

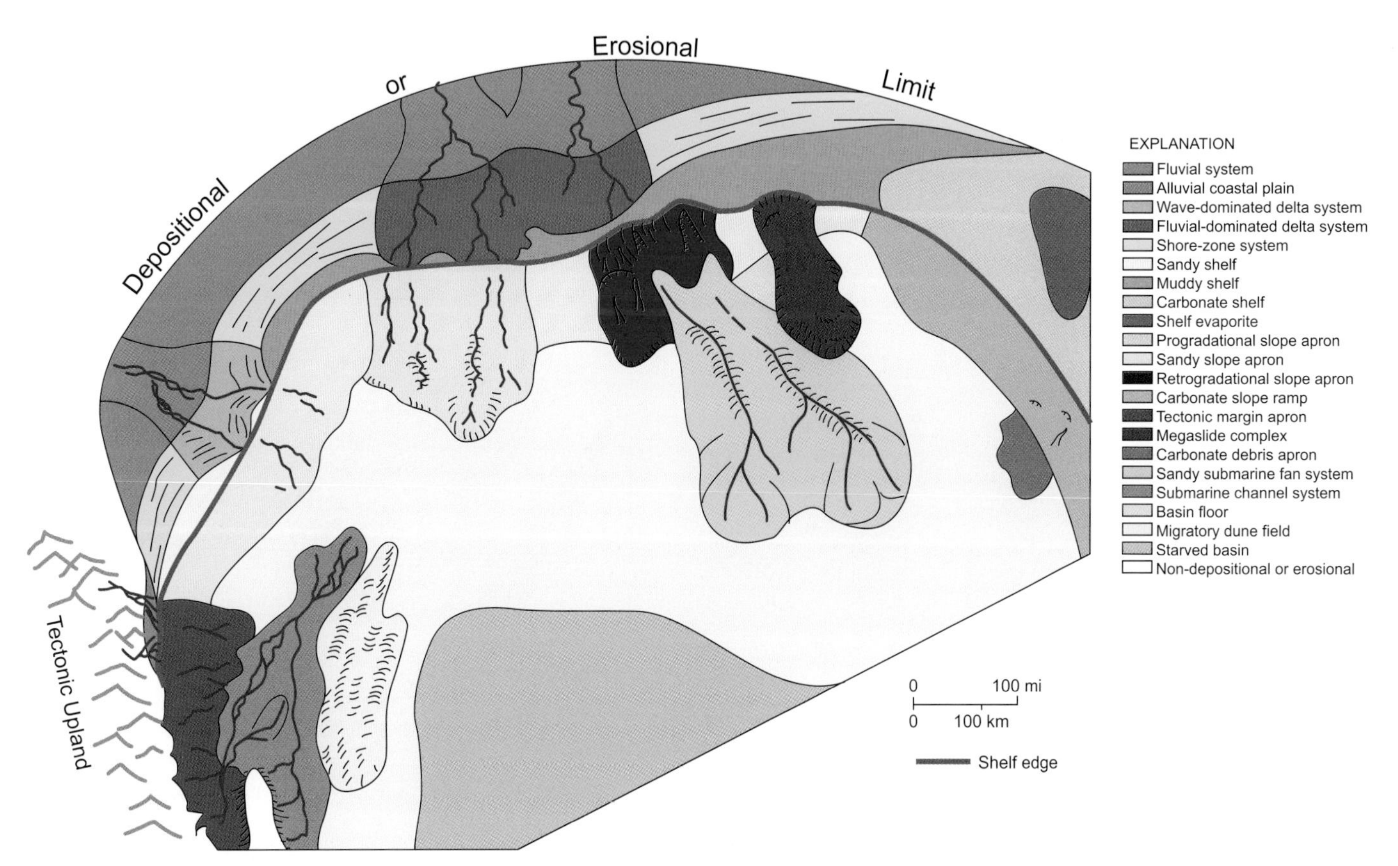

Figure 1.35 Depositional systems paleogeography typical of the Cenozoic deposodes of the Gulf basin. This figure provides a graphical explanation of the color scheme used in the paleogeographic maps.

Box 1.3 Submarine Fans, Ramps, and Aprons

Most sedimentologic literature has described, and continues to describe, sandy, deep marine facies using submarine fan models. Sequence stratigraphic systems tract models reinforced the application of fans as the primary sandy depositional elements of slope and basin settings, associating their origin with sea-level fall and lowstand. However, where regional datasets allow three-dimensional mapping of slope and basin facies, deepwater depositional systems display diverse geographies. Fan morphologies are only one of many areal patterns displayed. Reading and Richards (1994), using datasets from both Quaternary continental margins and ancient analogs, synthesized a suite of conceptual models that emphasized two major variables: (1) grain size of sediment supply, and (2) the geometry of the feeder system. They recognized that the pattern of sediment supply to the slope ranges from highly focused to widely dispersed along the length of the shelf edge. Based on the second variable, they differentiated point-sourced fans, arcuate-sourced ramps, and line-sourced aprons (Figure 1.36A–D). Ramps and aprons produce prisms of slope sediment whose along-strike breadth is subequal to or exceeds the run-out length of the depositing gravity flow dispersal system. Along-strike facies architecture is complex but repetitive. Differentiation of ramps and aprons is based on the degree to which slope feeders are dispersed along the strike.

Recognizing that the ramp model was associated with shelf-margin delta systems, Galloway (1998) suggested differentiation of slope/basin depositional systems into relatively focused, point-sourced fans and broadly sourced slope aprons (incorporating both aprons and ramps). Slope type was further differentiated based on the upslope depositional systems tract. Typical configurations for prograding slopes include arcuate delta-fed aprons and linear shelf-fed aprons. *Delta-fed aprons* are constructed where multilateral and/or sequential shelf-margin delta lobes cumulatively supply sand to the slope along a broad front that, over geologic time, can extend many tens of miles along the shelf margin. *Shelf-fed aprons* are typically linear, extending up to hundreds of miles along the strike. Retrograding slopes

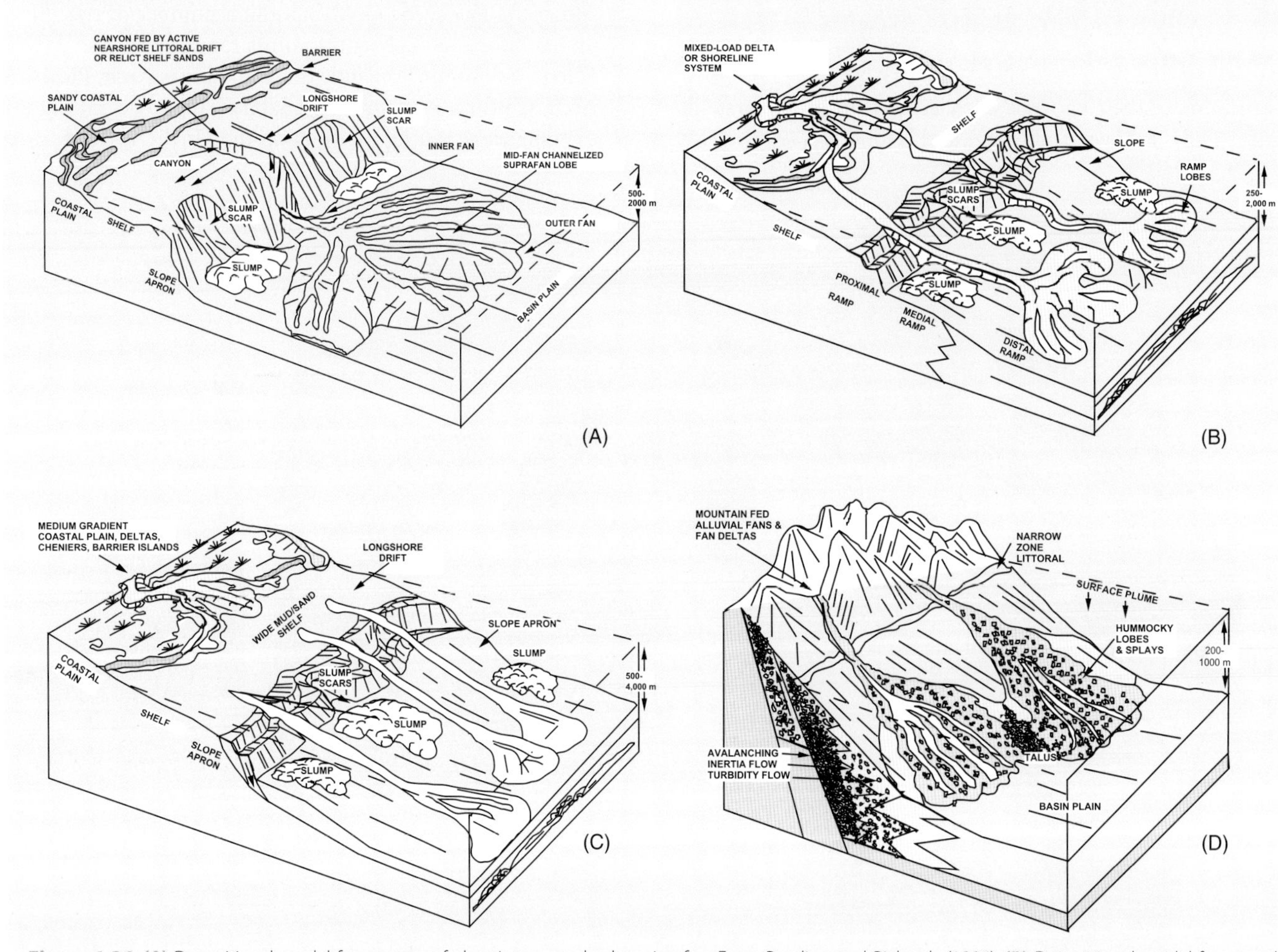

Figure 1.36 (A) Depositional model for a canyon-fed, point-sourced submarine fan. From Reading and Richards (1994). (B) Depositional model for a sandy delta-fed progradational apron. Note the direct connection between delta front and slope channels, which diverts sand directly from the sandy mouth bars and shoreface onto the upper slope. From Reading and Richards (1994) (C) Depositional model for a mud-rich progradational slope apron supplied by muddy deltaic or shelf systems. From Reading and Richards (1994). (D) Depositional model for a tectonic margin slope apron. A very narrow or absent shelf platform allows transfer of sediment load directly from uplands to the submarine slope. From Reading and Richards (1994).

Box 1.3 (*cont.*)

retreat by mass wasting and submarine erosion of the outer shelf and upper slope. Basinward, recycled upper slope and shelf-margin sediments deposit a *retrogradational slope apron*. Along tectonically active margins, adjacent upland sources shed sediment across an erosional terrane directly onto the subaqueous slope or across a narrow coastal zone of coalesced fans and fan deltas, depositing a typically coarse-grained *tectonic margin apron*.

Galloway (1998) models have been customized for mapping of common GoM slope/basin depositional systems. The paleogeographic maps relate all slopes to their updip depositional systems. Slope systems and their continental rise and abyssal plain extensions are differentiated into sandy submarine fans (Figure 1.36A), progradational delta and/or shelf-sourced aprons (Figure 1.36B,C), tectonic margin aprons (Figure 1.36D), and retrogradational aprons (Figure 1.37). Progradational slope aprons that front large shelf-margin delta systems and their adjacent progradational shore zones are commonly sandy. Progradational aprons are typically mud-dominated and front broad shelves or platform deltas that did not prograde onto the shelf margin. Retrogradational slope aprons formed where mass wasting and regrading recycled sediment from the upper slope and deposited an apron along the slope toe.

margin stabilized fluvial axes that cut across the uplifted basin rim (Galloway and Hobday 1996; Dooley *et al.* 2013).

2. A survey of modern world coastlines, which can be considered a snapshot of an instant of geologic time, shows that wave-dominated shores are more abundant than tide-dominated shores, and that both greatly exceed the fluvial-dominated shorelines (Nyberg and Howell 2016). Only the immediate distributary or river mouth preserves clear fluvial imprint. This has important implications. Detailed facies analysis of shoreline deposits of all types of deltas will reveal a dominance of marine features. Most of the delta front is being reworked most of the time by marine processes; shoreface successions displaying wave and tidal features will be abundant even within fluvial-dominated deltas. Differentiation of delta systems types, as done here, depends on interpretation and mapping of the entire suite of prodelta, delta front, and delta plain facies that comprise the deltaic depocenter. Our maps are drawn to emphasize the *maximum* extent of delta systems as defined by lithofacies distribution within the genetic sequence created by the deposode.
3. The transfer of sand from the shoreface to slope channels or canyons is highly constrained by the presence of an intervening shelf. Maximum shelf bypass distance is less than 5 km (Sweet and Blum 2016). For large-scale bypass of sand to the slope, the shoreface must extend essentially to the shelf margin, whether by progradation or by relative sea-level fall. Alternatively, submarine canyons must cut across the shelf to intercept the shoreface.
4. In consequence, high rates of shoreline progradation favor sand bypass to the slope and construction of sandy slope and basin depositional systems (Dixon *et al.* 2012; Gong *et al.* 2016). Our paleogeographic maps are drawn to emphasize the maximum progradational extent of deltas and shore zones, to highlight the regions within a genetic sequence where sand bypass is most favorable.
5. Sand-rich fluvial-dominated deltas and progradational sandy shorefaces also favor sand bypass to the slope (Dixon *et al.* 2012; Gong *et al.* 2016).
6. Particularly in climatic greenhouse times, large deltas are fully capable of prograding across transgressive shelves, bringing their sandy mouth bars and shoreface directly onto the shelf margin (Blum and Hattier-Womack 2009).
7. Local basin margin tectonics and morphology also play a major role in determining the timing and location of sand bypass to the slope and basin (Covault and Graham 2010). As will be shown, this is a dominant element for much of the Gulf margin of Mexico.
8. Geomorphic slope profiles of ocean basins include graded, tectonically over-steepened, stepped above grade, and ponded above grade continental slopes (Prather *et al.* 2017).

Several other generalizations apply to the mapping methodology and reconstruction of paleogeographies of the Cenozoic GoM:

1. The great majority of delta systems are either fluvial- or wave-dominated. Some tidal influence has been recognized in detailed facies analyses.
2. Across the northern Gulf margin, large delta systems have commonly prograded to the shelf margin, where they deposited distinctive assemblages of facies and intraformational structures common to shelf-margin deltas (Galloway and Hobday 1996).
3. Strike-fed shore zone systems are geographically, volumetrically, and economically important elements of the GoM basin-fill. They are well developed in several locations: delta system flanks, broad interdeltaic bights, and along coasts where numerous small streams flow from uplands to the adjacent coastline (Galloway and Hobday 1996; Figure 1.35). Gulf shore zone systems include wave-, mixed wave/tide-, and tide-dominated types.
4. Seascapes of the Cenozoic GoM contained diverse sediment transport pathways and depositional systems tracts, just as does the modern basin. In addition to submarine fans located at slope toes and commonly extending far across the continental rise and onto the abyssal plain, several different kinds of submarine slope

and basin paleogeographic systems are differentiated and mapped (Box 1.3). These include (1) slope aprons, characterized by line-sources along a broad length of the shelf margin; (2) sea floor channel systems; and (3) migratory submarine dune fields (Galloway 1998). Slope aprons can be further distinguished into progradational sediment prisms that construct offlapping continental margins and retrogradational aprons.

5. Using a global database, Prather *et al.* (2017) quantified average sand content deposited in continental margin depositional systems tracts. Shelves, which include coastal plain, delta, shore zone, and shelf depositional systems, average 27 percent sand content. The upper to middle slope decreases to 13 percent sand. In the lower slope and continental rise, which are characterized by decreasing declivity, sand content increases to 18 percent. Different slope profiles and sandiness of the fluvial–deltaic sediment input modify the site-specific percentages, but the pattern remains consistent; sand tends to bypass the upper slope, which is dominantly muddy, creating a bimodal pattern of vertical sand distribution within a prograding continental margin (Galloway and Hobday 1996).

Although volumetrically minor components of individual genetic supersequences, retrogradational slope systems display distinctive stratigraphic and structural architectures. Several create discrete petroleum plays. Structural and depositional elements of retrogradational margin aprons created by large-scale failure of the shelf margin are illustrated in Figure 1.37. Defining

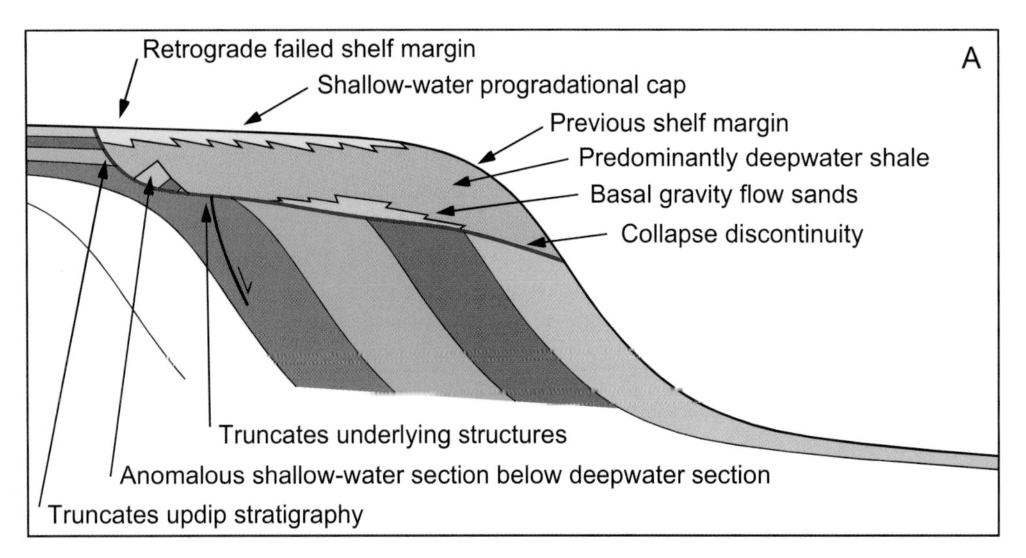

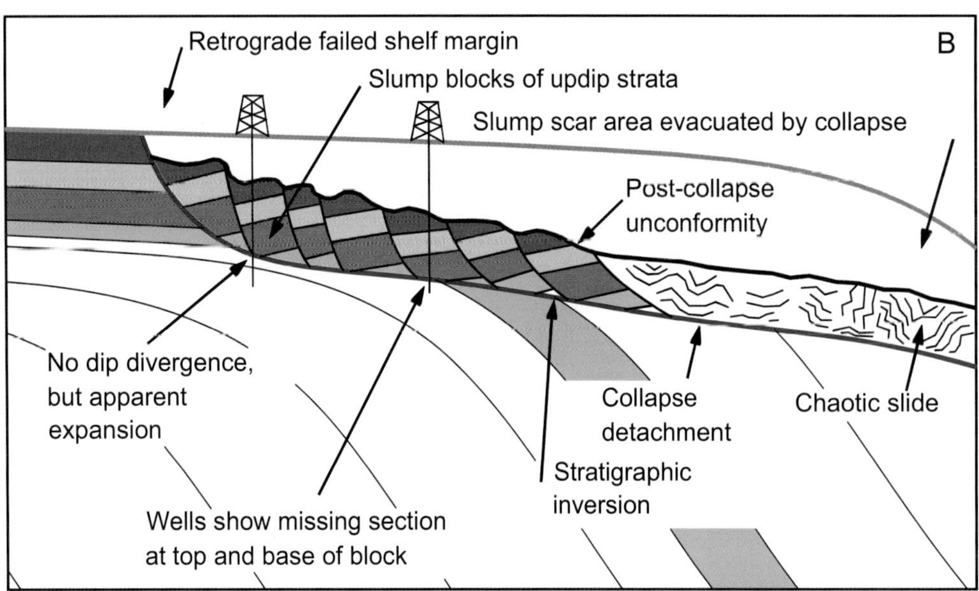

Figure 1.37 Structural and depositional architecture of failed retrogradational shelf margins. (A) Retrogradational wedge largely evacuated by mass wasting, creating a perched terrace upon which gravity flow sands, debris flows, and disconnected slump blocks may be deposited and preserved. (B) Retrogradational wedge within which slump blocks form a large part of the supra-discontinuity fill. In both, the position of the shelf edge was relocated landward from its original position at the top of the slope clinoform to the retrograded headwall position. From Edwards (2000).

elements include a basal erosional discontinuity, perched gravity flow and slump deposits, and a capping wedge of deepwater mudstone (Edwards 2000; Galloway 2005a).

1.18 Explanation of Paleogeographic Maps: Assumptions and Caveats

It is useful to consider the methods, assumptions, and caveats used to reconstruct the depositional history and paleogeography of the Mesozoic and Cenozoic intervals of the greater GoM basin. As mentioned in the discussion of the GBDS database (Section 1.9), wells and seismic data are the primary tools used in our reconstructions (see also Galloway *et al.* 2000). Wells are used for creation of lithofacies suites, and, where possible, are calibrated against published core cuttings information and tied to seismic data. Well log motifs (see "Gulf of Mexico Cenozoic Log Facies Interpretation" poster at www.cambridge.org/gomsb), stratigraphic unit thicknesses, and observed lateral trends in depositional facies guide thickness mapping (unit thickness maps) and structure mapping (unit top maps). Seismically derived thickness maps (isochore maps) and structure maps (structure contours) are also constructed where the density and quality of the 2D seismic grid permit.

These maps underlie and support the paleogeographic reconstructions for each stratigraphic unit. For example, unit thickness maps often help delineate and define depocenters. Depositional "thicks" (areas of prominent stratal thickening) at these depocenters often occur where sediment transported via extrabasinal fluvial systems (major pathways from highland source terranes) accumulate in large-scale deltas, which often act as important point sources for major submarine fans. Salt, where present, often enhances the thickness trends via salt evacuation. It should be noted that local over-thickening of units, for example in salt dome peripheral grabens, is averaged out by use of regional well control. Thinning onto salt highs or carapaces is dealt with in a similar fashion. Growth along extensional normal faults, however, usually can be related to a major sediment input point. By contrast, areas of low sedimentation are associated with development of thin carbonates, defined as condensed intervals *sensu stricto*.

In areas outside of the allochthonous salt canopy, seismic facies mapping adds confidence to the interpreted depositional environments. Seismic mounding, (seismic reflections showing double downlap) is often associated with major submarine fan development (Combellas-Bigott and Galloway 2006).

Identification of other structural and stratigraphic features also aids paleogeographic mapping. Depositional shelf margins often coincide with major fault detachments, as shown in several of the basin cross-sections described in Section 1.5. Submarine canyons are noted in several areas and units (e.g., Lavaca and Yoakum Canyons; Galloway and McGilvery 1995), which in turn are linked to submarine fan development in downdip areas (McDonnell *et al.* 2008).

Distinctive seismic architectures for carbonate systems are also noted and factored into mapping. Rimmed shelf reefs or platform margin reefs are particularly well developed for the Cretaceous stratigraphic units.

Together, the map suite defines location, areal extent, and total sediment volume associated with the major sand dispersal and carbonate development within the Gulf basin during each depositional interval.

There are a few caveats to consider when reviewing these paleogeographic maps. First, maps are reconstructions, back to the original position at the time of deposition, unless present-day position is indicated. Thus, plate reconstructions are used for Triassic, Jurassic, and Early Cretaceous depositional systems. Plate reconstructions of the GoM basin continue to evolve. For example, the timing of sea floor spreading described by Hudec *et al.* (2013a, 2013b) has already been modified (Norton, pers. comm.).

For some specific units, like the Smackover–Norphlet supersequence, post-depositional rafting has also been taken into account. Restoration back to the pre-rafted position has been carried out for the main Norphlet exploration area in the deepwater of the eastern GoM, following the kinematic model of Pilcher *et al.* (2014). If post-depositional rafting is not considered, the Norphlet map, for example, would depict a paleogeography that is far too broad relative to its original depositional geometry.

Finally, it is important to note that the GoM basin is the site of numerous ongoing studies, seismic surveys, and drilling campaigns that provide new information on the Mesozoic and Cenozoic on a yearly, if not monthly, basis. Use of detrital zircon U–Pb geochronology (Box 1.2) for provenance, for example, has had an especially large impact on reconstruction of ancient drainage systems (e.g., Snedden *et al.* 2018a). Our book, therefore, captures the state of the Gulf basin at the moment of publication and it is highly likely some of our interpretations will require future modification as new data becomes available.

1.19 Database

The greater GoM basin has long been known as a superb natural laboratory of sedimentary and structural processes. For example, our understanding of salt tectonics has advanced because of considerable work done in this basin and as featured in the work of Jackson and Hudec (2017), Hudec and Jackson (2011), and Rowan (1995). This is due in large part to quantity and quality of the information gathered in the course of oil industry studies of seismic data, testing of models by drilling wells, and supporting scientific studies of the basin.

Studies of the GoM date back many years. Since the publication of Amos Salvador's seminal DNAG volume J (Salvador 1991a), over 2500 papers have written on the GoM basin. Many of these are from industry workers providing their insights from seismic studies and well results. Another equally important source of information about the basin is the scholarly

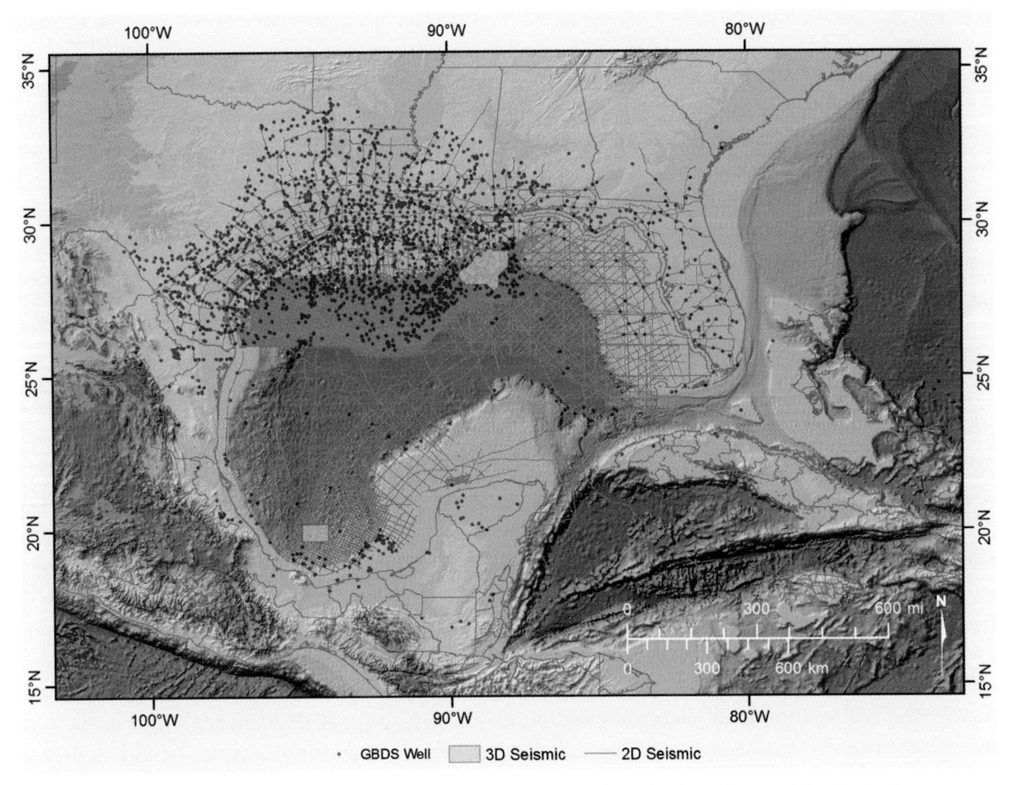

Figure 1.38 Well and seismic database of the Gulf Basin Depositional Synthesis project used in this book.

research at universities. At last count, over 40 students have written theses and dissertations on the GoM, ranging from near-surface sedimentary processes to the deep Louann Salt.

At the University of Texas and other national and international universities, faculty and research scientists have been leaders in the evolving understanding of the basin, often ahead of the industry interest. Dick Buffler, who for many years led the GBDS project, published early papers on DSDP core sites on the Mexico sector that now are being used for calibration of source rock and depositional systems in the Mexico deepwater rounds (e.g., Hessler *et al.* 2018). William Fisher and William Galloway's work on the onshore Wilcox (e.g., Fisher and McGowen 1967; Galloway and McGilvery 1995; Galloway *et al.* 2000) preceded drilling of the BAHA II well and opening of the Wilcox deepwater play. There are too many examples to cite within the limits of this introduction.

This book is founded upon a database built and maintained by the GBDS research project, Institute for Geophysics at the University of Texas at Austin, which enjoyed industry support for more than 20 years. This database includes over 2000 previously published papers, including many spatially referenced maps, but also well log and seismic data (Figure 1.38). The well data from the USA consists of released well data from federal and state waters; onshore US wells are courtesy of state surveys and third-party vendors like DrillingInfo™. Well data from Mexico are entirely public domain, largely university theses from National Autonomous University of Mexico and other Mexico universities.

Seismic data from federal waters was loaned to the GBDS project by seismic data companies, including ION Geoventures, TGS, Spectrum, MCG, and PGS. The data is mainly 2D seismic, with a few 3D surveys available to GBDS researchers and students.

Biostratigraphic data, so important to the stratigraphic age assignments, is mostly from BOEM data releases but also donations to the University of Texas at Austin. Other ancillary data (porosity, permeability, etc.) are provided on an individual basis via request to specific companies.

The ARCGIS database of Cenozoic and Mesozoic maps is the key derivative product from this 20+ years effort and the primary means of investigating the long and complex depositional history of the Gulf basin. The rest of this book sets forth to lay out the depositional framework, form the basin, and fill the basin with Mesozoic and Cenozoic sediments, the primary oil and gas reservoirs of this prolific hydrocarbon habitat.

Mesozoic Depositional Evolution

This part takes the reader from the pre-salt depositional history to the end of the Mesozoic in the GoM; from localized precursors through to nascent basin stage, to a well-evolved marine depositional basin. Chapter 2 opens with a new model for the post-Ouachita orogenic section, including Eagle Mills successor basin-fill. Chapter 3 describes the Middle Mesozoic continental Drift and Cooling Phase beginning with arid conditions and hypersalinity resulting in massive Louann Salt deposition, followed by the Norphlet eolian systems, ensuing transition to marine microbalites of the Smackover and development of platform margin reefs of the Kimmeridgian and Tithonian. In Mexico, prolific carbonate grainstones dominate shallow waters. Chapter 4 describes a phase of Late Mesozoic local crusting heating and regional tectonic uplifts, including the first major siliciclastic influxes into the basin with the Hosston and Ceno-Turonian (Tuscaloosa) depositional systems and the subsequent episode of enhanced organic richness represented by the Eagle Ford Shale and equivalents. Chapter 4 concludes with progressive basin deepening recorded by the Austin Chalk and Navarro–Taylor interval, fostering a carbonate-dominated land- and seascape just prior to the Chicxulub impact event that ended the Mesozoic.

Chapter 2

Post-Orogenic Successor Basin-Fill and Rifting Phase

2.1 Basin and Continental Framework

As briefly discussed in Section 1.8, the Gulf of Mexico (GoM) Mesozoic depositional history can be subdivided into a series of tectonostratigraphic phases (Figure 1.24). These phases reflect both the long-term tectonic evolution of the basin and its predecessors, as well as the shorter-term eustatic and climatic processes influencing sedimentation. While the Cenozoic phases have higher frequency (three phases over 66 million years), one can argue for three tectonostratigraphic phases over 170 million years or more since the suturing of Pangea and joining of Laurentia and Gondwana. The three phases that cover the post-Quachita–Marathon–Appalachian orogeny to end of Cretaceous (299 Ma to 66 Ma) are:

1. Post-Orogenic Successor Basin-fill and Rifting Phase
2. Middle Mesozoic Drift and Cooling Phase
3. Late Mesozoic Local Tectonic and Crustal Heating Phase.

We regard the first phase as a predecessor to formation of the GoM basin, but it is worthwhile to discuss this in detail as numerous tectonic and stratigraphic elements persisted into the Middle Mesozoic Drift and Cooling Phase, some extending into the Late Mesozoic Local Tectonic and Crustal Heating Phase. Galloway (2009) has argued that the basin initiated with deposition of the Louann Salt, the first stratigraphic unit that spans much of the area that is today known as the GoM. As discussed below, salt deposition was probably underway at 170 Ma, at the start of the Drift and Cooling Phase. New plate tectonic models suggest that accelerated opening of the Gulf began as an intrusive phase of oceanic crust generation below the accumulating mass of **evaporites** and later extrusive separation of salt bodies between the northern and southern GoM (Norton *et al.* 2018), a process also observed in other areas (Norton *et al.* 2015).

The end of the Late Mesozoic Local Tectonic and Cooling Phase, and the Mesozoic as a whole, was ushered in by the Chicxulub impact event at 66 Ma, which greatly altered the paleobathymetry and land surface of the GoM (Denne *et al.* 2013; Sanford *et al.* 2016). It also, to some degree, set up the basin configuration that the Cenozoic tectonostratigraphic phases modified by sediment input from the newly emerged Laramide highlands and rejuvenated Appalachian mountains (Galloway *et al.* 2011; Snedden *et al.* 2018a).

These three Mesozoic tectonostratigraphic phases naturally reflect the larger-scale geodynamics that controlled GoM basin opening and evolution. Plate tectonic forces drove the geodynamic systems that controlled subsidence and accommodation, uplift, and source terrane exposure, and even marine water entry to the nascent basin. Thus, our tectonostratigraphic scheme is based on current thinking regarding plate tectonics of the greater GoM basin since the breakup of Pangea, as described in detail in the following.

2.2 Plate Tectonic Reconstructions since 240 Ma

Geological views of the origin and evolution of the GoM are changing with evaluation of new deep-imaging seismic reflection and refraction data. The plate reconstructions of the GoM from 240 Ma to 140 Ma (Figure 2.1A–F) have evolved from research of the PLATES project at the University of Texas at Austin (www-udc.ig.utexas.edu/external/plates). The reconstructions are based on mapping of tectonic elements from multiple sources, including seismic, potential field, and geologic data. The main driver of GoM tectonics during this time is the motion between Yucatán and North America, driven by creation of oceanic crust. Deformation to the east in the Florida region separated the Yucatán motion from the central Atlantic, but we have no constraints on how this deformation was distributed. Another region of significant deformation was in Mexico. Since the earliest days of plate reconstructions it has been recognized that in a Pangea reconstruction the northwest part of South America has considerable overlap onto Mexico if both regions are mapped in their present-day geometries. We use a modified form of the megashear hypothesis (Anderson and Schmidt 1983) to move Mexico away from South America. We also recognize a single tectonic block in eastern Mexico that consists of a Permo-Triassic arc system that formed along the boundary between Pangea and the proto-Pacific Ocean (Norton *et al.* 2016).

The GoM basin opening was preceded by the Successor Basin-fill and Rift Phase associated with Pangea breakup (Figure 2.1) following the Marathon–Ouachita–Appalachian orogeny. By 240 Ma (Ladinian Stage of the Triassic), the Yucatán (Mayan block) had already joined with the North

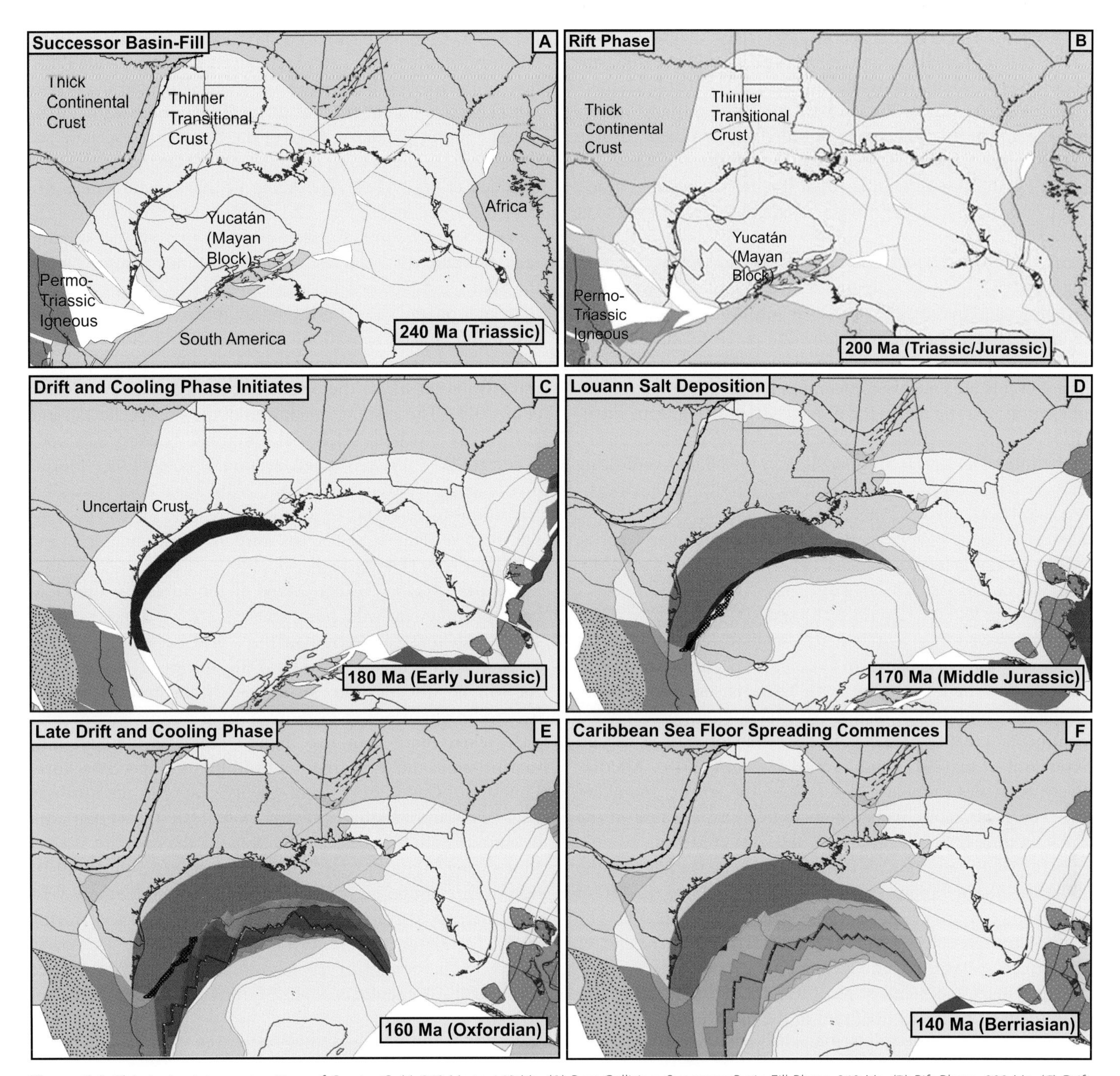

Figure 2.1 Plate tectonic reconstructions of Greater GoM, 240 Ma to 140 Ma. (A) Post-Collision, Successor Basin-Fill Phase, 240 Ma. (B) Rift Phase, 200 Ma. (C) Drift and Cooling Phase initiation, 180 Ma. (D) Louann Salt deposition, 170 Ma. (E) Late Drift and Cooling Phase, 160 Ma. (F) Caribbean sea floor spreading commences, 140 Ma. Reconstructions courtesy of Ian Norton and UT-Austin PLATES research project.

American plate, possibly bounded on the west by the Burgos lineament (Figure 2.1A). South America was located to the south and Africa to the east. Thick continental crust (orange) follows the Marathon–Ouachita belt from Mexico to Southern Arkansas, across Mississippi–Alabama to Georgia, where it reaches the Appalachians. Thinner transitional (light yellow) continental crust covered the future location of the GoM basin. Permian–Triassic igneous complexes are present in both Yucatán and onshore Mexico (Lawton *et al.* 2009; Xaio *et al.* 2017). Gaps in the plate model are shown in Mexico as later megashear motions are required to assemble that area.

Continental crust extension began as Pangea breakup commenced soon after, with continental rifting occurring in the eastern USA (e.g., South Georgia rift [SGR] system; Heffner 2013). Back-arc rifting initiated in Mexico around 236 Ma (Lawton *et al.* 2018), indicating the continued influence of Pacific margin subduction of the Farallon plate (Martini and Ortega-Gutiérrez 2016). The result was a series of right-lateral transtensional grabens filled with largely continental to marginal marine siliciclastics, as documented in onshore outcrops (Lawton *et al.* 2018). As will be discussed in Section 2.3.2, there is little evidence for rifting in east Texas and Louisiana, implying a more ductile lithospheric flexure response and ensuing deposition of the Eagle Mills here in a post-orogenic successor basin. Further, there is little room for pre-salt sedimentation on the northern Yucatán margin, a pattern that continues to at least 170–180 Ma.

By 200 Ma (Hettangian Stage of the Early Jurassic; Figure 2.1B), igneous activity initiated, manifested as surface lava flows and pyroclastics, as well as subsurface dykes and sills (Kidwell 1951). These may be linked to magnetic anomalies in both onshore Texas and offshore Mexico (Mickus *et al.* 2009). These events are roughly contemporaneous with emplacement of Central Atlantic magmatic province (CAMP) igneous intrusions, as well as igneous bodies in the SGR (Heffner 2013). Sometime later, seaward dipping reflections (SDRs), indicating massive basalt outflows, were emplaced in both the northeastern GoM and north of the Yucatán margin (cross-section 7, Figure 1.19). These underlie pre-salt sedimentary rocks, inferred from seismic character (Miranda Peralta *et al.* 2014; Curry *et al.* 2018).

Between 200 Ma (Figure 2.1B) and 170 Ma (Figure 2.1D) Yucatán moved in a southeasterly direction relative to North America. The crust between Yucatán and North America is colored dark blue in Figure 2.1C. There is some uncertainty about the composition of this crust, as it could be extended continental crust as predicted by the refraction models of Van Avendonk *et al.* (2015) and Eddy *et al.* (2014, 2018). Alternatively, as it lies outboard of the coastal magnetic anomalies (Mickus *et al.* 2009), it may be oceanic crust. Because of the thick sediments and salt overlying this area at present, seismic imaging (both reflection and refraction) is challenging and the differing interpretations are equally plausible. Final resolution will require more special-purpose data acquisition. For our tectonostratigraphic scheme, this marks the effective end of the Successor Basin-Fill and Rifting Phase, though there is only a fragmentary sedimentary record and a substantial age gap with the next phase, as discussed in the following.

Several important implications from these plate reconstructions are relevant to sediment routing of the pre-salt (Eagle Mills) depositional systems. Early in the continental stretching phase, the tight fit of Yucatán with North America limited space for sediment accommodation (Figure 2.1B,C). The east Texas/Louisiana/Arkansas area is the widest embayment north of the plate boundary where Eagle Mills deposition is well documented. The structural boundary could have controlled sediment routing, as discussed below. Later separation of the Yucatán (Mayan block) during and after the speculative first phase of sea floor spreading or simple rift extension (Figure 2.1C) may also mean that newly developed space was available for development of pre-salt deposition in the area north of Yucatán subbasin (see Section 1.5.7 and cross-section 7, Figure 1.19). The period of sediment accommodation in the pre-salt Yucatán subbasin may have been short, but accumulation rates must have been high, given kilometer-scale thickness of the interval between the possible SDRs and the base of salt in Mexico (Saunders *et al.* 2016; Hudec and Norton 2018).

At 170 Ma (Bajocian Stage of the Jurassic), we believe salt deposition commenced in the nascent Gulf basin (Figure 2.1D; Snedden *et al.* 2018c). Age dating of the Louann Salt is discussed further in Section 3.2.1. While a 170 Ma age for the Louann Salt is 7–8 million years earlier than previous estimates (e.g., Salvador 1987; Hudec *et al.* 2013a), this is a time when the South and North American plates are in closer proximity than later on and thus conditions are more conducive to basin restriction and evaporation. This is also a time of well-documented sea floor spreading, beginning initially as an intrusive event below the original Louann Salt body (Norton *et al.* 2016). The ensuing separation of the Campeche/Yucatán (Isthmian) salt bodies from the original Louann Salt body occurred as Yucatán rotated around a pole in the Florida Straits (e.g., Nguyen and Mann 2016). Sea floor spreading transitions to an extrusive process, increasing the gap between salt bodies. This initiates the Middle Mesozoic Drift and Cooling Phase (Figure 1.24).

At 160 Ma (Oxfordian Stage of the Jurassic), the GoM basin opening reached a point where there is a direct, progressively widening connection to the world ocean through the Florida Straits to the Atlantic–Tethyan seas (Figure 2.1E). As will be discussed in Section 3.3.1, the transition from hypersaline basin water to more normal marine salinity may have taken more than five million years, as the first fully marine fauna and flora are found within the Upper Smackover Limestone (Godo 2017). Later gravity sliding in the northern Gulf allowed Louann (northern GoM) salt to overlap with oceanic crust in the central US Gulf basin, the so-called Walker Ridge salient (Hudec *et al.* 2013a). However, seaward translation of the Campeche salt was probably limited by the BAHA high (hachured area in Figure 2.1D and E; Hudec and Norton 2018). Translation of various tectonic blocks along megashears in Mexico is thought to have continued assembly of Mexico south of the Tamaulipas Arch (Martini and Ortega-Gutiérrez 2016).

At 155 Ma to some time after 140 Ma, oceanic crust generation waned in the GoM, and sea floor spreading shifted to the Caribbean basin (Figure 2.1F). By 138 Ma, the Yucatán (Mayan) block has rotated into its present-day position, roughly coincident with a large influx of siliciclastics (Hosston and Travis Peak Formations) in the northern GoM (Galloway 2008). This ends the Middle Mesozoic Drift and Cooling Phase. In several areas of the basin, angular unconformities or substantial lacunas mark the base of the Sligo–Hosston supersequence (McFarlan 1977; Anderson 1979; Galloway 2008; Ewing 2010).

The now fixed continental and oceanic crustal blocks in the GoM are soon affected by a series of sub-regionally focused crustal heating and uplift events that continue episodically until the end of the Cretaceous (Figure 1.24). Local crustal heating and igneous intrusions occur in south Texas, Louisiana, and Mississippi along with formation of angular unconformities in east Texas and the Mississippi Embayment (Ewing 2009). Collectively, we refer to this timespan as the Local Tectonic and Crustal Heating Phase (Figure 1.24). This phase continued until the end of the Mesozoic, when the Chicxulub impact event dramatically changed the land- and seascape of the greater GoM basin, paving the way for the Cenozoic.

Superimposed on plate and local tectonics described above are first-order sea-level variations, reflecting changes in mid-ocean ridge volumes and subduction of water, among other factors (Conrad 2013; Haq 2014). Reexamination of Mesozoic sea-level variations in light of new chronostratigraphic information confirms prior work (Haq *et al.* 1987) that the Early Jurassic sea level began near present-day mean sea level, rising to a peak approximately 140 m above present-day mean sea level (pdmsl) in the Tithonian, and then stabilizing around 100–120 m above pdmsl until the Early Cretaceous (Haq 2017). A trough (80 m above pdmsl) in the Cretaceous worldwide sea-level curve in the Mid-Valanginian stage is followed by peaks in the Barremian and highest point (250 m above pdmsl) just above the Cenomanian–Turonian boundary (Haq 2014). As discussed in subsequent sections, the GoM Mesozoic record shows a variable response to these global sea-level changes, suggesting the stratigraphic record here is a convolved archive of tectonics and eustacy in a high sediment supply setting.

2.3 Tectonostratigraphic Models for Basin Precursor History

The general approach taken throughout this book is to present depositional models for the GoM that are supported by a preponderance of data currently available. The fragmentary sedimentary record that post-dates the breakup of Pangea but prior to formation of the basin and deposition of the Louann Salt is sufficiently unclear that we need to consider two alternative tectonostratigraphic models for the early Mesozoic (Figure 2.2A,B). The first is based on the more conventional model described in detail by Salvador (1987, 1991; Figure 2.2A) and the other is a newly developed concept that departs from the conventional model both in terms of timing and kinematics (Figure 2.2B). This alternative model (Figure 2.2B) is based on new data on plate reconstructions (Section 2.2), seismic reflection data (Section 2.3.2), and detrital zircon provenance work (Section 2.3.3).

Both models agree on the precursor to breakup, collision of Gondwana (South America, Africa, and Yucatán) with North America at the end of the Paleozoic to create Pangea (Figure 2.2). Deformation is recorded at the northern boundary to what will become the GoM basin, observed as a series of northward-directed thrust faults and foreland basin from the Marathon orogenic belt through the Ouachita Mountains to the Appalachians. Exposures of Pennsylvanian strata in the present-day Ouachita Mountains of Arkansas are an excellent

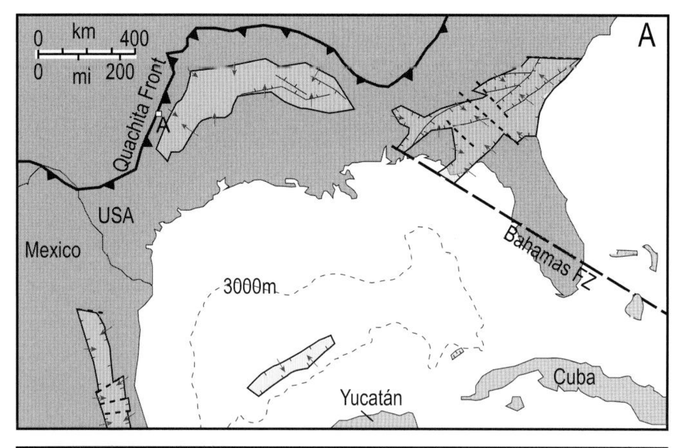

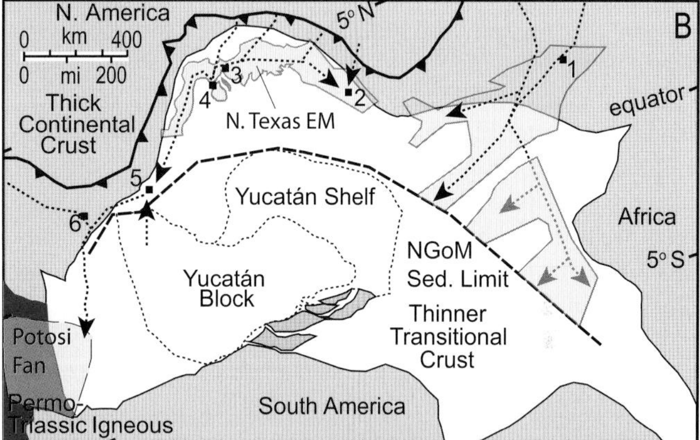

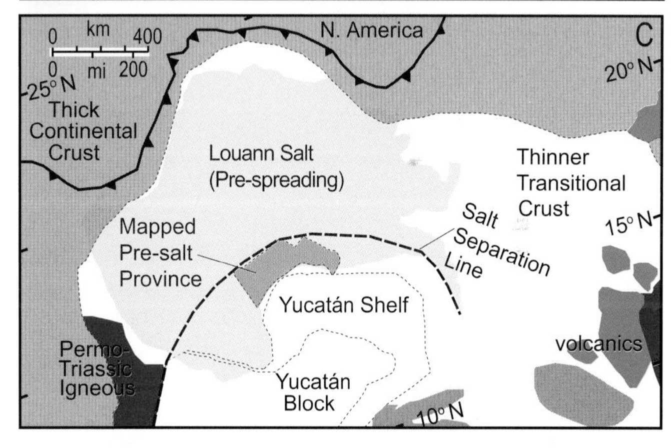

Figure 2.2 Alternative pre-salt models for GoM basin. (A) Conventional concept of Salvador (1991) showing pre-salt (Eagle Mills) rift system including the Ouachita–Marathon belt and localized sediment routing into adjacent grabens. Note that Salvador (1991) did not use a plate tectonic reconstruction. (B) New model for pre-salt, suggesting sedimentation in the central GoM fills a large-scale post-collision successor basin. Map restored to 240 Ma (courtesy I. Norton and UT PLATES project). Sediment routing trends based on key wells analyzed for detrital zircon geochronology as shown in Figure 2.6 are (1) Rizer #1; (2) McGee Unit 1; (3) McDonnell B3; (4) Exxon LV Ray Unit 1–2; (5) Superior McManus; (6) Amoco Stumberg. Potosi Fan outline from Dickinson *et al.* (2010). (C) New model restored to 170 Ma, showing interpreted Mexico pre-salt province (light green polygon) and overlying Louann Salt just prior to initiation of sea floor spreading. The Mexico pre-salt province lies between the Yucatán shelf margin and the salt separation line (terminology of Hudec *et al.* 2013a).

archive for reconstructing this tectonic episode (Gleason *et al.* 2007; Ewing 2016). In Texas, the Ouachita system is largely deeply buried below Mesozoic strata. A comprehensive analysis of wells penetrating the Paleozoic deformation is presented in the seminal work of Flawn *et al.* (1961). There are a few shallow outliers of this deformation, including the Sabine Island at the Texas–Louisiana border (see Section 3.4.2). Deep subthrust tests were drilled as recently as 1995 that tested the Ordovician Ellenburger below multiple thrust duplex structures. The Shell #1 Barrett well, drilled in Hill County (Well A in Figure 2.2A) tested a deep sub-Cretaceous structure called the Waco Uplift, which turned out to be largely Paleozoic metasediments (Rozendal and Erskine 1971; Vernon 1971; Nicholas and Waddell 1989). Pennsylvanian deformation transitioned to a period of subsidence in the Permian basin, accumulating as much as 14,000 feet (4270 m) of Permian strata in west Texas (Ewing 2016), though we logically exclude this interval from the GoM depositional fill.

What happens following the end of the Permian subsidence remains a matter of conjecture, due to the fragmentary stratigraphic record, with few early Mesozoic outcrops outside of Mexico or south of the Ouachita Mountains, and just a handful of wells drilled below the autochthonous Louann Salt in the onshore USA, particularly in the western part of the future GoM basin. General consensus among researchers suggests that the breakup of Pangea initiated in the early Triassic and separation of North and South America followed the Laurentian suture that can be traced from the Appalachians to the Ouachita–Marathon belt into Mexico. A particular scientific conundrum is the 90 million year hiatus between the Permian strata of west Texas (roughly 251 Ma) and the oldest ages of fully marine Upper Smackover strata (157–160 Ma) recorded in the Middle Mesozoic Drift and Cooling Phase of the Gulf basin (see also Section 3.3.1).

Understanding these pre-Louann or "pre-salt" depositional patterns is important for several reasons. First, recent pre-salt discoveries have opened new exploration frontiers in Brazil and Angola and added large hydrocarbon reserves (Arbouille *et al.* 2013). Second, newly acquired seismic data in deepwater Mexico has provided superb imaging of a newly identified pre-salt province off of northern Yucatán that has been considered for leasing by Commision de Nacional Hydrocarbons (CNH) (Saunders *et al.* 2016). Further, in 2017 Pemex announced plans for drilling a deep test (Yaaxtaab-1) of the pre-salt interval in the Bay of Campeche (CNH 2017a). Thus, the pre-salt of the northern GoM may be a depositional or tectonic analog for this new exploration frontier, as described in Section 1.5.7 and illustrated in Figure 1.20.

2.3.1 The Conventional GoM Early Mesozoic Rift Model

The presence of a post-Paleozoic, pre-Louann interval has been known in the northern GoM since the 1930s (Weeks 1938; Scott *et al.* 1961; Gawloski 1983; Salvador 1987, 1991b). Lithologies include red to greenish-gray shales and white sandstones and red dolomites (Woods and Addington 1973). Red bed successions, known as the Eagle Mills Formation (named after a well in Arkansas) have been encountered in a large number of oil and gas and even water wells (Salvador 1991b). The uncertainty of a Permian or Triassic age was resolved, in part, by the identification of a single leaf fossil (*Macrotaeniopteris magnifolia*) in the Humble #1 Royston, of Arkansas (Scott *et al.* 1961). The same leaf fossil is present in the Chinle Formation of Arizona and in the Newark Supergroup of Virginia. Later palynological analysis of the fossil algae *Coenobium plaesiodictyon* in a Cass County, Texas well confirmed a Triassic (Carnian) age for the Eagle Mills (Wood and Benson 2000). Fossil plants from red beds of the Eagle Mills equivalent La Boca Formation (Huizachal Group) in northern Mexico are less diagnostic, broadly indicating a Late Triassic to Early Jurassic age (Mixon 1963).

Linkage to the red beds of the Newark Supergroup is also appealing on the basis of lithology and tectonic process (Salvador 1991b). A model of the Eagle Mills red beds filling a series of discrete rift and graben systems during the Pangea breakup was adapted not just for the eastern GoM but the basin as a whole (Salvador 1991b; Figure 2.2A). This has evolved into what may be described as the conventional temporal model for the nascent GoM basin. In this chronostratigraphic scheme, rifting began soon after the end of the Permian (240 Ma), extending to about 205 Ma where there is a large (40 million year) hiatus until post-rift deposition of the Louann Salt beginning around 162 Ma (Salvador 1987, 1991b). The cause of the missing stratal interval is unclear, though it has been suggested that rifting was continuous until salt deposition, but shifted to the area under the present-day salt canopy where there are no well penetrations below autochthonous salt.

The Wood River Formation of the South Florida basin has yielded zircons with a maximum depositional age of 195–235 Ma from U–Pb analyses, but only partially covering the stratigraphic gap (Wiley 2017). However, the zircon sample counts of Wiley (2017) also tend to be low (often less than $n = 100$ per sample), raising questions about statistical significance of the results. The age of the Louann Salt may also be older than Callovian, as recent $^{87/86}Sr$ analysis has suggested an age approaching 170 Ma (see Section 3.2.1). Nonetheless, this gap of 90 million years or more remains puzzling. South Florida basin zircons show an affinity with Gondwana sources (Suwannee terrane), indicating proximity to the African continent, a pattern that continues into the Oxfordian (Lovell and Weislogel 2010; Lisi 2013; Wiley 2017).

North of the Ouachita–Marathon orogenic belt, outcrops of the Dockum Group stand in stark contrast to the entirely subsurface Eagle Mills of Texas. The Dockum Group and equivalent units of the Chinle outcrop in a belt from north Texas to Nevada, a distance of 2000 km, allowing detailed sedimentological analysis, paleocurrent measurements, and provenance work using detrital zircon (Mickus *et al.* 2009; Dickinson *et al.* 2010). Paleocurrents show fluvial transport to the northwest, likely coming from source terranes in the paleotopographic highs of the Ouachita orogenic belt on the south (Thomas 2011). Provenance analysis using U–Pb zircons shows expected Grenville sources from the Ouachita orogenic belt, with mixtures from other terranes (Dickinson and Gehrels 2008). Zircon analysis shows maximum depositional ages of 200–234 Ma (Umbarger, 2018), maintaining the enigmatic hiatus between the Louann Salt and the Triassic interval of the basin.

In Mexico, the Triassic to Middle Jurassic record, mainly archived in outcrop intervals, includes the Zacatecas, Nazas, and La Joya Formations of Mesa Central and Huizachal Group of the Sierra Madre Oriental (Barboza-Gudiño *et al.*, 2010). The Potosi submarine fan is believed to be connected to the El Alamar paleo-river of the Huizachal Group, influenced by the tectonics of the east Mexico Permo-Triassic continental arc (Stern and Dickinson 2010).

2.3.2 Alternative Model for Early Mesozoic Successor Basin-Fill and Rifting

Several new and even some older observations are inconsistent with the long-standing, conventional model of Salvador (1991). Closer examination of seismic data from Arkansas, Louisiana, and Texas in the area of Salvador's (1991) graben trend fails to show unequivocal evidence of a buried rift system (Figures 2.3 and 2.4). In north Texas, the Eagle Mills onlaps the deformed Top Paleozoic interval (see also Milliken 1988), suggesting depositional infilling of preexisting accommodation, not rift-grabens (Figure 2.5). Toplap stratal terminations against the base of the Louann Salt indicate a disconformable contact, not an angular unconformity. The Mexia–Talco fault zone, a breakaway fault zone at the landward termination of the salt, is noted just southward of the crest of the Waco Uplift (Figure 2.3).

Reviewing the original illustrations of the block-faulted Triassic strata in Arkansas used by Scott *et al.* (1961) and later repeated by Woods and Addington (1973), it is apparent that these models are not matched by seismic interpretations in the area. For example, a seismic profile across southwestern Arkansas depicted by Nicholas and Waddell (1989) shows the Eagle Mills is only separated from the underlying Pennsylvanian interval by a disconformity with no obvious rifting or erosional surface. Further, there is no system of half-grabens as seen in the SGR (e.g., Heffner 2013).

In east Texas, Exxon and Fina jointly drilled the LV Ray GU 1–2 well below allochthonous salt and encountered the Eagle Mills siliciclastic interval, reaching total depth (TD) at 18,498 ft (5640 m; White *et al.* 1999; Figure 2.4). Based on limited age information and long-distance correlation to the Mexico and the eastern USA Newark Supergroup, a rift model was proposed, albeit one with marine flooding of the axial portion of the rift system in east Texas to form a restricted marine evaporite unit within the Eagle Mills, called the Rosewood after a local field name (White *et al.* 1999). The syn-rift model offered by White *et al.* (1999) emphasizes a half-graben structural morphology, based on older 2D seismic data from the field area. Review of newly reprocessed regional 2D seismic lines that tie LV Ray GU 1–2 well and nearby Eagle Mills penetrations does not show a rift system half-graben morphology (Figure 2.4). The Eagle Mills (pre-salt) interval appears to drape an irregular deep (Paleozoic?) basement topography. An angular unconformity between the shallower Paluxy–Washita supersequence and overlying Eagle Ford–Tuscaloosa

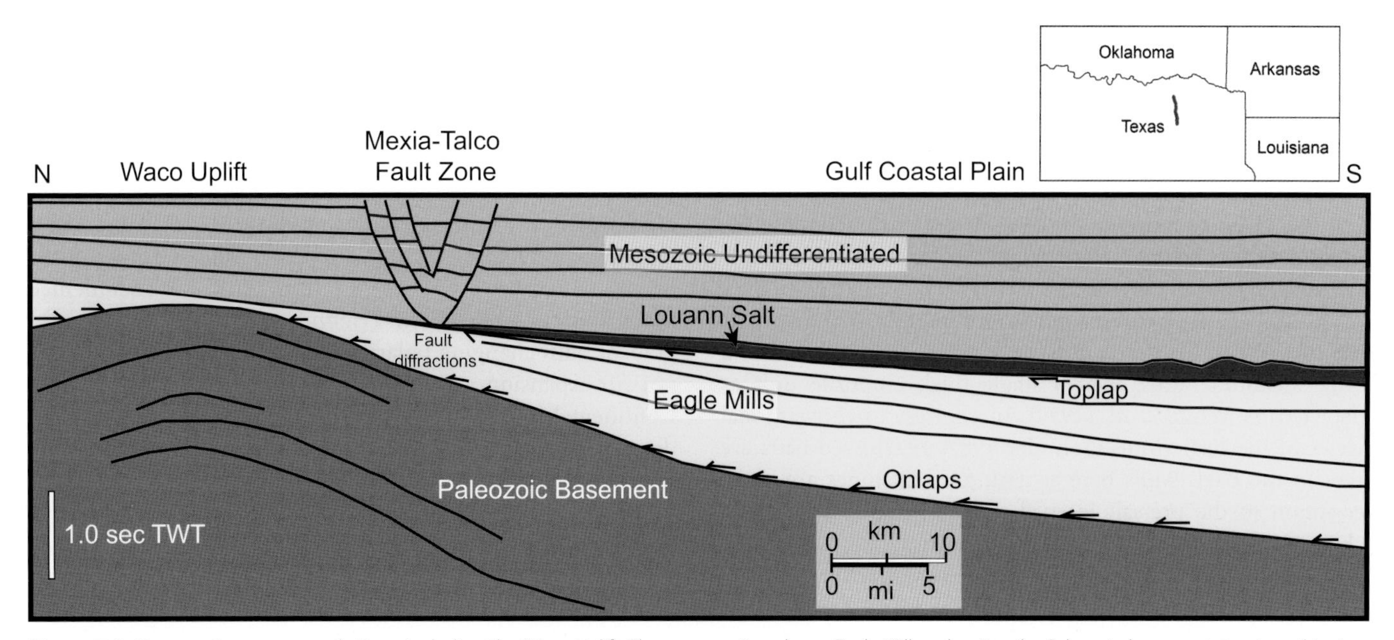

Figure 2.3 Cross-section across north Texas including the Waco Uplift. The cross-section shows Eagle Mills onlapping the Paleozoic basement structure, thinning onto the Waco Uplift, and is toplapped by the base Louann Salt.

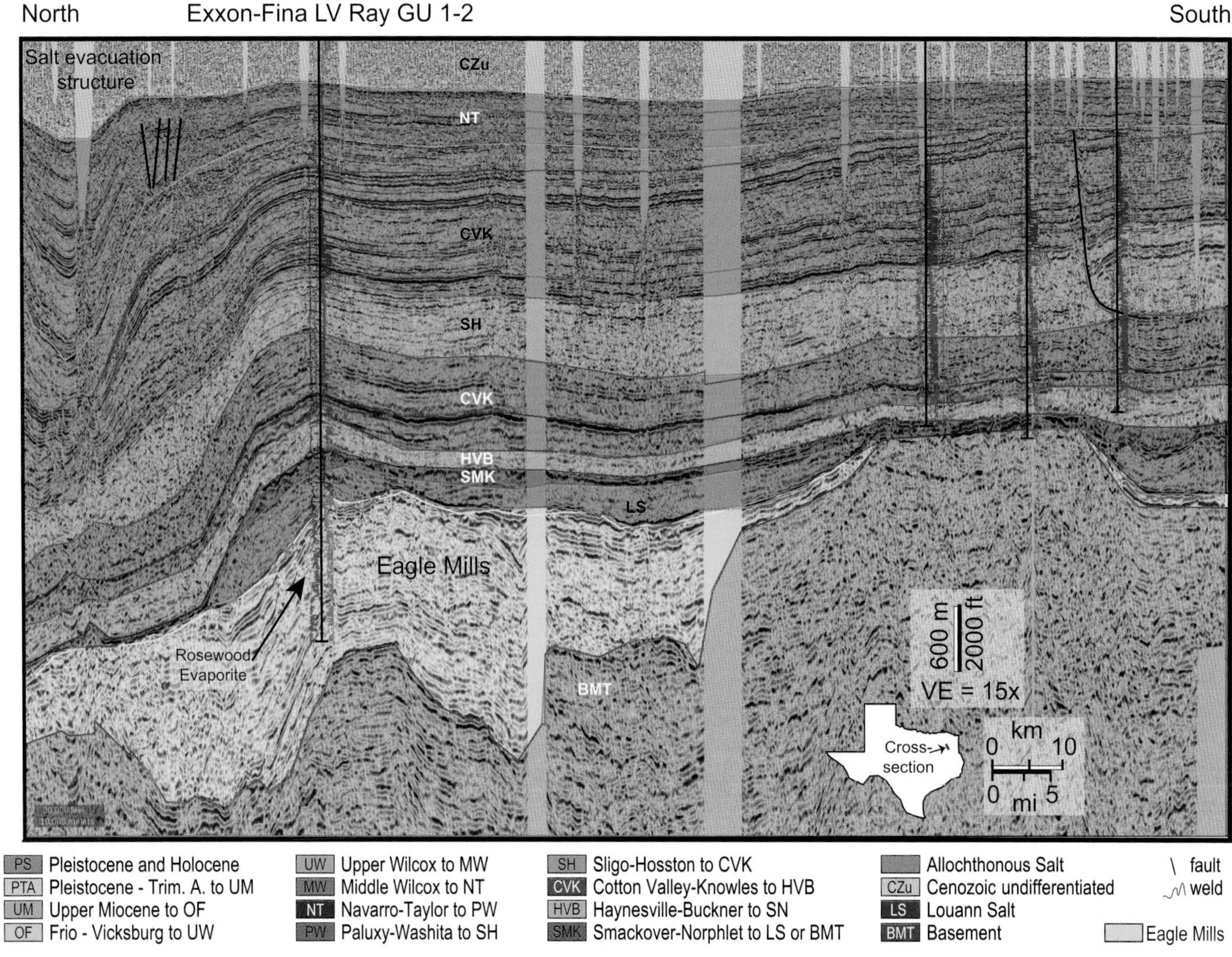

Figure 2.4 Seismic line interpretation across east Texas area including Exxon-Fina LV Ray well of Upshur, County, Texas Waco. Seismic line courtesy of ION.

supersequence coincides with the basement high point. Thus, part of the basement structure post-dates the Eagle Mills (Figure 2.4), probably related to salt evacuation. The Louann Salt has a roughly similar morphology to the Eagle Mills, thickening into lows and thinning onto basement highs. Faulting is not common in the deep interval, with later folding, rotation, and shallow salt evacuation more apparent. Little syntectonic thickening into the rare observed faults is evident.

Examination of the cuttings from the Rosewood evaporite interval reveals an alternation of thin-bedded anhydrites and gray siltstones rather than a single thick evaporite unit as depicted by White *et al.* (1999). In contrast to conventional views of the Eagle Mills (Salvador 1987, 1991b) red beds are rare in the Eagle Mills here as dark gray siltstones are more common to the pre-salt interval. Detrital zircon U–Pb geochronology from the Eagle Mills interval just below the Louann Salt in this well is discussed in Section 2.3.3.

Milliken (1988) mapped a large area of northeast Texas using a grid of older 2D seismic data. His Eagle Mills isopach map (Figure 2.5) shows a general basinward thickening, with local patterns following Paleozoic basement trends, for example, thinning over the Sabine Uplift and thickening in the east Texas salt basin. No expanded asymmetrical isopach thicks in half-grabens are evident. The erosional limit of the Eagle Mills on the west side of the successor basin follows but is well south of the Ouachita frontal thrust.

These observations of the deep Eagle Mills structure of the central GoM onshore contrast with the well-documented South Georgian rift system (Heffner 2013). In the SGR, half-grabens with changing polarity are the norm, with prominent syn-rift expansion and stratal rotation (Withjack *et al.* 2002). Continental red beds are common in the SGR, as seen in the deep Rizer #1 well of South Carolina (Goggin and Rine 2014; Rine 2014; Rine *et al.* 2014).

Salvador (1991b) and Thomas (2011) show a significant offset in the trend of the Iapetan rifted margin of southern Laurentia. The SGR-graben systems shift along a major transform (called the Bahamas fracture zone), in order to link with the Texas–La–Ark rift system to the west (Figure 2.3). This offset, later called the Florida Lineament, must be considered highly

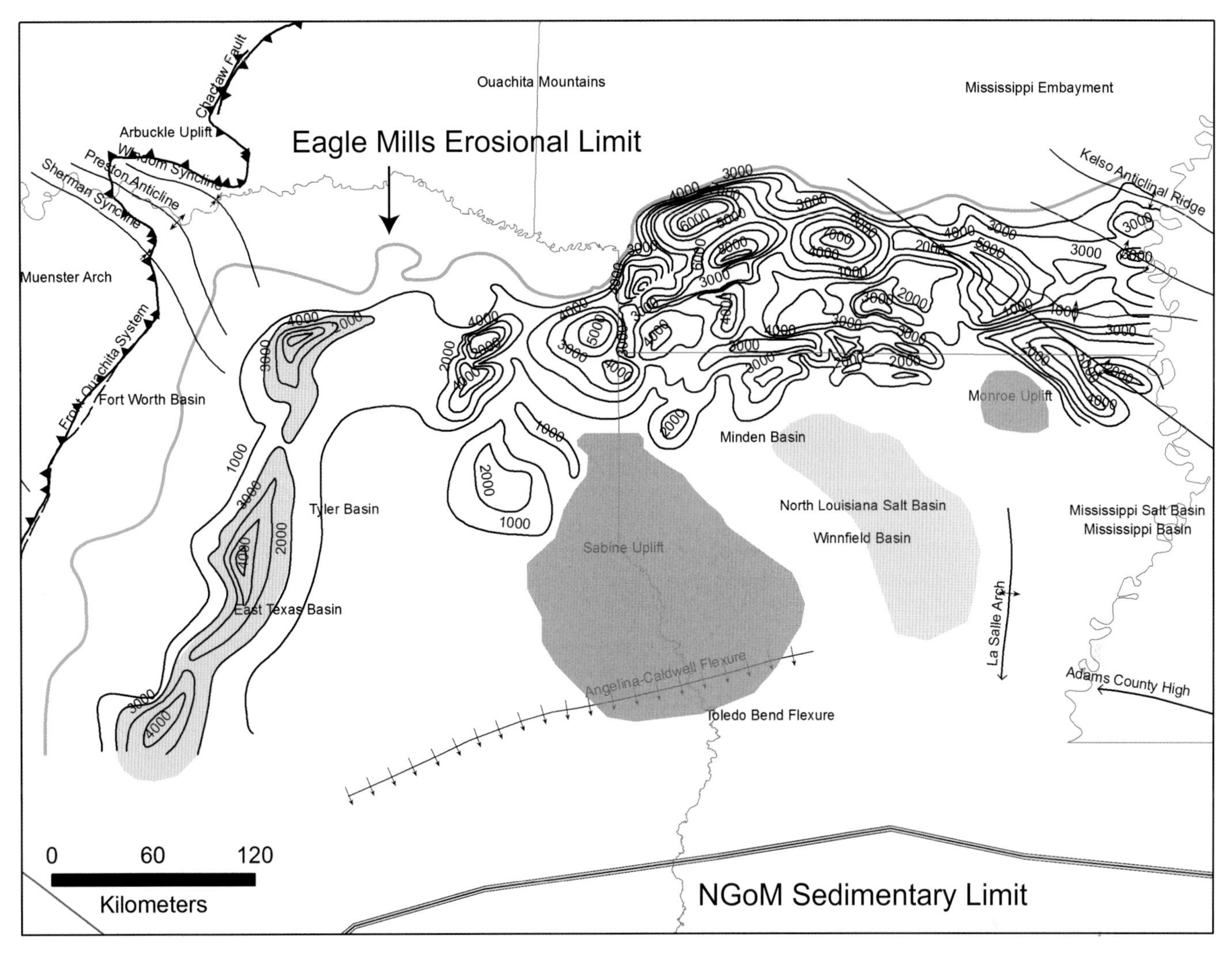

Figure 2.5 Eagle Mills isopach map based on well control and seismic mapping. Contours in feet. Modified from Milliken (1988).

interpretative as it is loosely based on alignment with onshore trends such as the Pickens–Gilbertown–Pollard fault system.

An alternative hypothesis, first described by Norton *et al.* (2018), is based on the concept that a significant portion of the Eagle Mills of the Texas–La–Ark area (Ouachita Embayment of Thomas 2011) is not filling a rift system of half-grabens but instead a successor basin or set of basins developed on the post-collision Ouachita structural surface. As discussed earlier, Eagle Mills strata of the Ouachita Embayment appear to dip monoclinally toward the south, presumably to a pinch-out or termination where the Yucatán block meets North America (e.g., Figure 2.3). This idea is also consistent with models suggesting a soft collision of Gondwana with Laurentia in the central GoM onshore, in contrast to the hard collision elsewhere (Heffner 2013). Further, it mirrors the observed accommodation pattern of the Dockum Group that appears to fill a low-relief depositional basin to the northwest of the uplifted Ouachita highlands source terrane (Riggs *et al.* 1996; Dickinson and Gehrels 2008). Dickinson *et al.* (2010) suggested that Triassic uplift of this central Texas pre-rift area was mainly a thermal precursor to the Jurassic opening of the GoM basin.

Several important implications emerge from the interpretation that the SGR system does not extend into the Louisiana–Arkansas–Texas area (Figure 2.2B). It is likely that the drainage systems of the SGR and central GoM onshore are not shared, axially or otherwise. This is confirmed by U–Pb age spectra (Figure 2.6; Section 2.3.3) showing at least two different source terranes and thus differing catchments and sediment routing patterns. It follows, then, that drainage catchments in Louisiana–Arkansas–Texas accessed a variety of source terranes and sediment routing was unconstrained by rift-graben topography. The influence of the North Atlantic basin opening (Withjack *et al.* 1998; Schlische 2003) apparently did not extend to areas west of the SGR. Thus, it is our conclusion that central Gulf onshore pre-salt deposition is more a reflection of the deposition upon the deformed Ouachita belt, as a post-orogenic successor basin-fill.

An alternative temporal sequence of events extending from the breakup of Pangea to the rift–drift transition is shown in

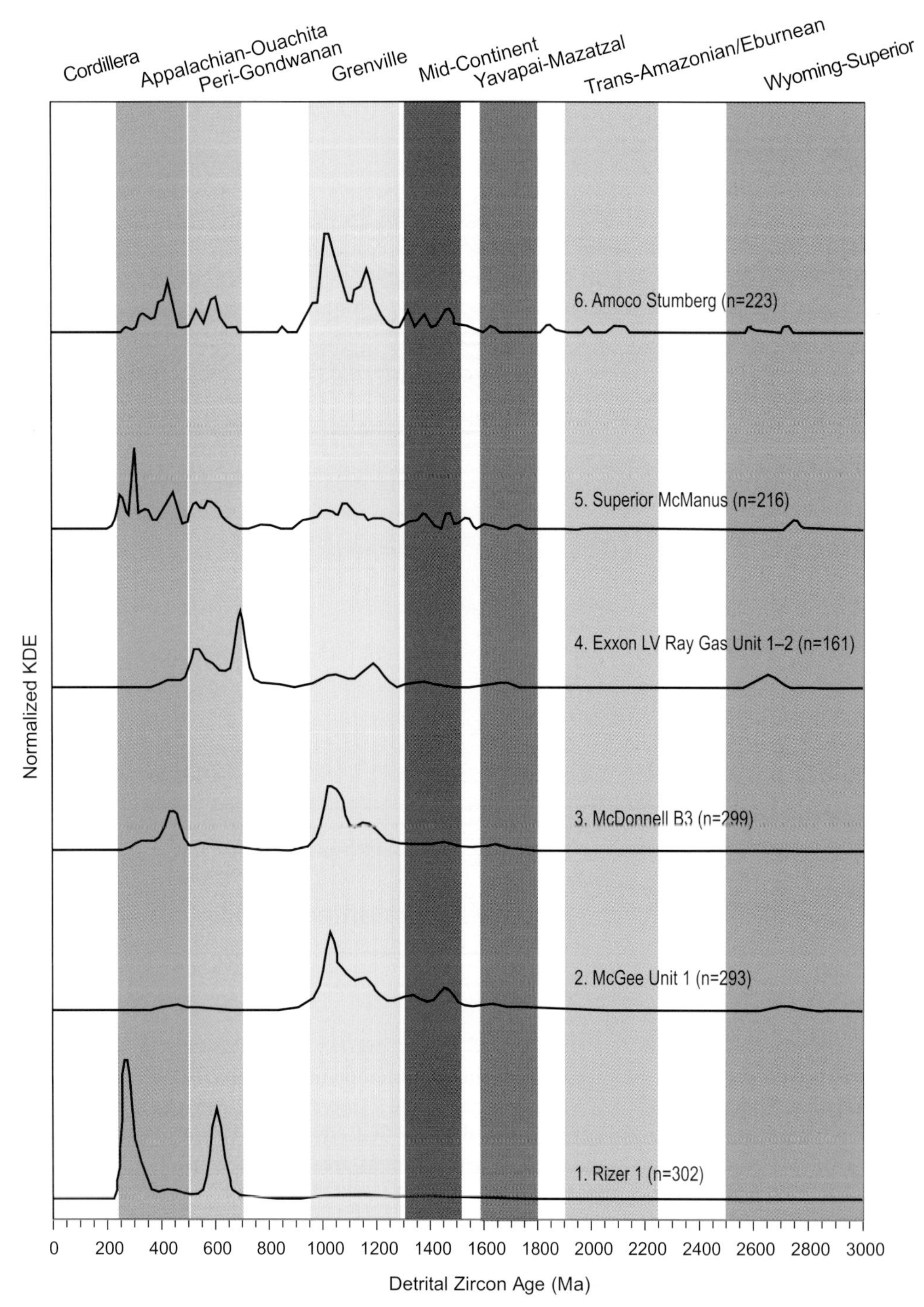

Figure 2.6 Comparison of detrital zircon age spectra from among Eagle Mills well penetrations in the northern GoM.

Figure 2.2B. The seismic observations from east Texas and Louisiana are more consistent with this alternative framework than the conventional view of GoM rifting as described by Salvador (1991b) and other authors. A revised model for the pre-salt (Eagle Mills) interval, founded on the plate tectonic evolution described in Section 2.2 (Figure 2.1) would follow this progression:

1. Pangea breakup and continental stretching results in development of successor basins in the central GoM, reflecting a more ductile lithospheric response than elsewhere in Mexico or the eastern GoM (Figure 2.2B). Like the roughly contemporaneous Dockum Group, Eagle Mills deposition in the central GoM onshore initiated with the Late Triassic uplift of a central Texas Ouachita source

terrane (Mickus *et al.* 2009; Dickinson *et al.* 2010), but other basement terranes were exposed. Eagle Mills sediments were routed to the drainage to the south, east, and west, bounded by the Yucatán block and magmatic material between Yucatán and North America (Figure 2.2B). The southern limit of central GoM Eagle Mills deposition is defined by the Gulf Coast Magnetic Anomaly, thought to indicate the Early Jurassic volcanic margin–rift axis (Mickus *et al.* 2009). Catchments of interior drainage systems extended across a variety of North American basement source terranes, as evidenced in the diverse U–Pb age spectra from pre-salt wells (Section 2.3.3). The continental divide was located at the uplifted Marathon–Ouachita belt; Dockum Group and Chinle equivalent fluvial systems fed sediment to the north and west. Thus, a large portion of the Eagle Mills of the Central GoM is successor basin sedimentation, westward of and separate from the SGR. Some early but local rifting initiated in northern Mexico, with deposition of the Cerro El Carrizalillo Member of the Plomosas Formation in a half-graben (Lawton *et al.* 2018). Maximum depositional age from U–Pb zircons here is 236 ± 1 Ma (Figure 1.24). This may also be related to back-arc effects from the Pacific plate, unrelated to GoM opening *sensu stricto* (Martini and Ortega-Gutiérrez 2016). Rifting in eastern North America soon commenced, including the SGR, and continued from 240 Ma to 200 Ma (Heffner 2013). Deposition of the Eagle Mills and Dockum Group proceeded through the Norian Stage in Texas and adjacent areas (Dickinson *et al.* 2010).

2. GoM continental stretching in a northwest–southeast direction caused southward migration of the Yucatán block (Figure 2.2C). This in turn created space and accommodation for basalt outflows (seaward dipping reflections) and later pre-salt sedimentary deposits in the offshore northeastern Gulf and Yucatán northern margin (see also Section 1.5.7, cross-section 7, Figure 1.19). In northern Mexico, back-arc rift-graben filling depositional units contain earliest dated volcanic U–Pb zircons of 191–193 Ma (Martini and Ortega-Gutiérrez 2016). In the northern GoM, rifting was limited to the area south of the present-day north-central GoM shelf that was later deeply buried below the Cenozoic strata (Van Avendonk *et al.* 2015).
3. Massive Central Atlantic Magmatic Province (CAMP) volcanism began, signaling termination of rifting in the Atlantic basin and initiation of sea floor spreading there at 200 Ma (Olsen 1997; Figure 2.1B).
4. By contrast, the main phase of sea floor spreading in the northern GoM commenced about 30 million years later than in the Atlantic, initiated around 170 Ma (Figure 2.1D). This is a clear indication of temporal separation of the plate tectonic processes and thus sedimentary processes in the GoM and North Atlantic basins.
5. Pre-salt deposition in northern Yucatán forms a seaward dipping wedge of continental deposition derived from erosion of exposed Yucatán basement (Figure 2.2C). Plate reconstructions suggest sedimentation here terminated around 170 Ma, coincident with initiation of major sea floor spreading.
6. Rifting persisted in northern Mexico onshore areas with ignimbrites in graben-fills dated as young as 176 ± 1 Ma (Lawton *et al.* 2018; Figure 1.24). Alternatively, syntectonic deposition here was related to transtensional motion as various Mexican basement blocks moved into position. Movement of the Mayan block allowed pre-salt deposition in the northern Yucatán margin, forming a seaward dipping wedge of continental deposition likely derived from erosion of exposed Yucatán basement.

2.3.3 Pre-salt (Eagle Mills) Sediment Routing

U–Pb zircon analyses, an advanced provenance technique (see Box 1.2), from northern GoM pre-salt wells reveals the complexity of sediment sourcing from various basement terranes, as well as differences between the central GoM onshore and the SGR (Figure 2.6). The typical SGR geochronological signature is that of Appalachian/Allegheny sources, as indicated by a peak at 250 Ma and a peri-Gondwana (Suwannee terrane) peak at 650 Ma (e.g., Rizer #1 well; Figure 2.6). Wells representing the central northern GoM onshore successor basin-fill (e.g., McDonnell B3 and McGee Unit 1; Figure 2.6) show prominent peaks in the 1000–1250 Ma range, indicative of Grenville basement sources (cf. Blum and Pecha 2014). By contrast, this Laurentian basement signature is absent in the Rizer #1 well. Pre-salt strata in the Exxon LV Ray Gas Unit 1–2 has a weak Grenville signature and a significant pan-African (peri-Gondwanan) set of peaks on the detrital zircon age spectra (Figure 2.6).

These age spectra support the view that the SGR and central northern GoM onshore pre-salt (Eagle Mills) interval was deposited by multiple drainage systems, some of which were probably small but steep rivers (Figure 2.2B). Even within the central northern GoM onshore pre-salt province, distinct differences over short distances indicate contributions of different tributaries whose catchment headwaters were anchored in different source terranes. On the southwestern margin is the Superior McManus #1 well (well 5 in Figure 2.2B) that shows contributions from multiple sources, but also confirms paleoflow to the southwest along the Yucatán/North America plate boundary (Figure 2.2B). Sediment drainage systems probably converged at the effective sedimentary boundary (northern GoM sedimentary limit of Figure 2.2B), defined by the volcanic margin coincident with the Gulf Coast Magnetic Anomaly (Mickus *et al.* 2009). This wide dispersion of pre-salt sediments is at odds with the conventional model of a more focused, half-graben controlled thickness trend that would be expected in a classic rift system.

The Amoco Stumberg well, located in Dimmit County, Texas near the Mexico border, has some similarities with late collisional succession formed prior to Eagle Mills deposition (Figure 2.6). Age spectra are comparable to the Pennsylvanian Age Haymond Formation of the Marathon basin (cf. Gleason *et al.* 2007), suggesting contributions from the north. Yet the Amoco Stumberg detrital zircon geochronology also has similarities to that of the Superior McManus age spectra, with aligned peaks of Appalachian, peri-Gondwanan, and Mid-Continent terranes. It is therefore possible that both wells also reflect a Yucatán basement block source, with prominent peri-Gondwanan peaks in the range of 400–600 Ma. This contrasts with earlier work suggesting that the Potosi fan, as sourced by the Yucatán block, was limited to drainage systems south of the Tehuantepec paleotransform (Ortega-Flores *et al.* 2014). The Amoco Stumberg and Superior McManus wells detrital zircon age spectra thus may point to a long-lived Yucatán source terrane feeding both Mexico and south Texas (Figure 2.2B). These fluvial networks may have served as tributaries for the El Alamar paleo-river (Barboza-Gudiño *et al.* 2010) that fed the Potosi submarine fan at the paleo-Pacific margin (B. Frederick, pers. comm.) More detailed statistical analyses are currently underway (Frederick *et al.* in review).

As explained in the alternative model discussion (Section 2.3.2), the pre-salt interval of the northern Yucatán subbasin is considerably younger than the Eagle Mills of the northern GoM, as accommodation for this deposition was probably not available until Yucatán had rotated sufficiently to create space between the Yucatán shelf and North America (Figure 2.2C). Prior emplacement of the possible SDRs, as noted on cross-section 7 (Figure 1.19), likely around 190–200 Ma, does help constrain the basal age of this younger pre-salt interval. This younger age of the Mexico pre-salt deposition, plus its location well away from higher heat flow continental crust, are both more favorable conditions for higher reservoir quality than commonly observed in the northern GoM.

Chapter 3

Middle Mesozoic Drift and Cooling Phase

3.1 Basin and Continental Framework

It has been argued that the Louann Salt is the first basin-wide Gulf of Mexico (GoM) depositional unit (Galloway 2008), as the pre-salt Eagle Mills and equivalents in Mexico are known to be more localized units. Deposition of the Louann Salt is also coincident with the initiation of sea floor spreading (albeit largely beneath the accumulating salt), ushering in the Middle Mesozoic Drift and Cooling Tectonostratigraphic Phase as we define it (Figure 2.1D). Subsequent deposition of the Norphlet and Smackover Formations shows the basin evolving from a harsh, arid climate and hypersaline water body to normal marine conditions. The ensuing rise of formidable Jurassic platform margin reefs continued throughout the rest of the phase to reach an acme in the Middle Cretaceous.

Economic prospectivity of the GoM basin is closely linked with the Louann Salt and its Campeche–Yucatán salt equivalent, as salt sets up traps and receiving basins, seals reservoirs by salt cutoff, and mitigates heat flow to the point that many older source rocks remain viable longer than would be otherwise expected. Thus, an extensive analysis of the origin, distribution, and movement of the original or "mother" Louann Salt is warranted.

3.2 Louann Salt Supersequence

The existence of regionally large and vertically thick salt deposits in the GoM basin (Figure 3.1) was first established by early twentieth-century drilling that discovered salt dome-associated oil fields near Beaumont, Texas (e.g., Spindletop discovery; Halbouty and Hardin 1956). In Mexico, salt was first penetrated in the Minas Viejas #1 well, a name that is often used for salt encountered in onshore areas of Mexico, whether it be allochthonous or autochthonous salt (Figure 3.2; Lopez Ramos 1982).

The presence of Jurassic-age salt in both the northern and southern GoM was established with early seismic surveys (Martin and Case 1975; Buffler *et al.* 1981). It was apparent that these salt bodies were once part of a continuous salt accumulation later separated by sea floor spreading (Salvador 1987, 1991a, 1991b). The northern salt body, which includes both autochthonous and allochthonous components, is now referred to as the Louann Salt and the southern portion is called the Isthmian salt basin (Cuenxa Salina del Istmo; Hudec *et al.* 2013a). Hudec *et al.* (2013a) consider the northern segment to encompass both peripheral "interior" salt basins (east Texas, northern Louisiana, and Mississippi) and the central Louann Salt body, separated by the Toledo Bend Flexure (Anderson 1979; cross-section 5; Figure 1.17).

The Isthmian salt basin is structurally separated into two subbasins, the Campeche and Yucatán salt basins (Hudec *et al.* 2013a). In the northern Mexico onshore areas, Jurassic evaporites are known from outcrop and subsurface drilling (Goldhammer and Johnson 2001; Lawton *et al.* 2001). These are believed to be analogous to the interior salt basins of the northern GoM and are separated from the central Louann Salt basin by the Tamaulipas Arch and its northern extension, the Salado-Burro Arch (Lawton *et al.* 2001).

The presence of evaporites in the present-day Chiapas region of Mexico is thought to represent the onshore extension of the Isthmian salt basin (Meneses-Rocha 2001). In a survey of Mexico salt, Castillon and Larrios (1963) noted that older onshore wells near the village of Minatitlan penetrated "salt" at depths ranging from 340 to 4360 ft. Nearly 40 different salt structures were identified by the 1960s, and many more have been identified on newer seismic data (Padilla y Sánchez and Jose 2016). These domes or diapir structures underlie Cenozoic strata in this oil-producing region (Padilla y Sánchez 2007). Wilson (1993) suggested that the Chiapas or Salina Basin salt (his name) was Oligocene in age, but this paper has been widely disregarded as the author attempted to date allochthonous salt by the age of surrounding strata. However, few of the wells that encountered salt near Minatitlan are publicly available.

Seismic sections published by Jennette *et al.* (2003) from the nearby Veracruz basin do not indicate the presence of autochthonous or allochthonous salt. Chiapas evaporites could be part of an interior (perched) salt basin akin to the interior (east Texas/Louisiana/Mississippi) salt basins. Tectonic reconstruction of the complex Chiapas area is evolving (Witt *et al.* 2012) and considerable uncertainty remains. Padilla y Sánchez's tectonic map of Mexico provides the most up-to-date map of salt distribution in Mexico (Padilla y Sánchez *et al.* 2013).

Elsewhere in the greater GoM basin, relatively pure halite has been noted in Cuba (Meyerhoff and Hatten 1968; Pardo 1975). Pardo (2009) argued for a Middle Jurassic age for what

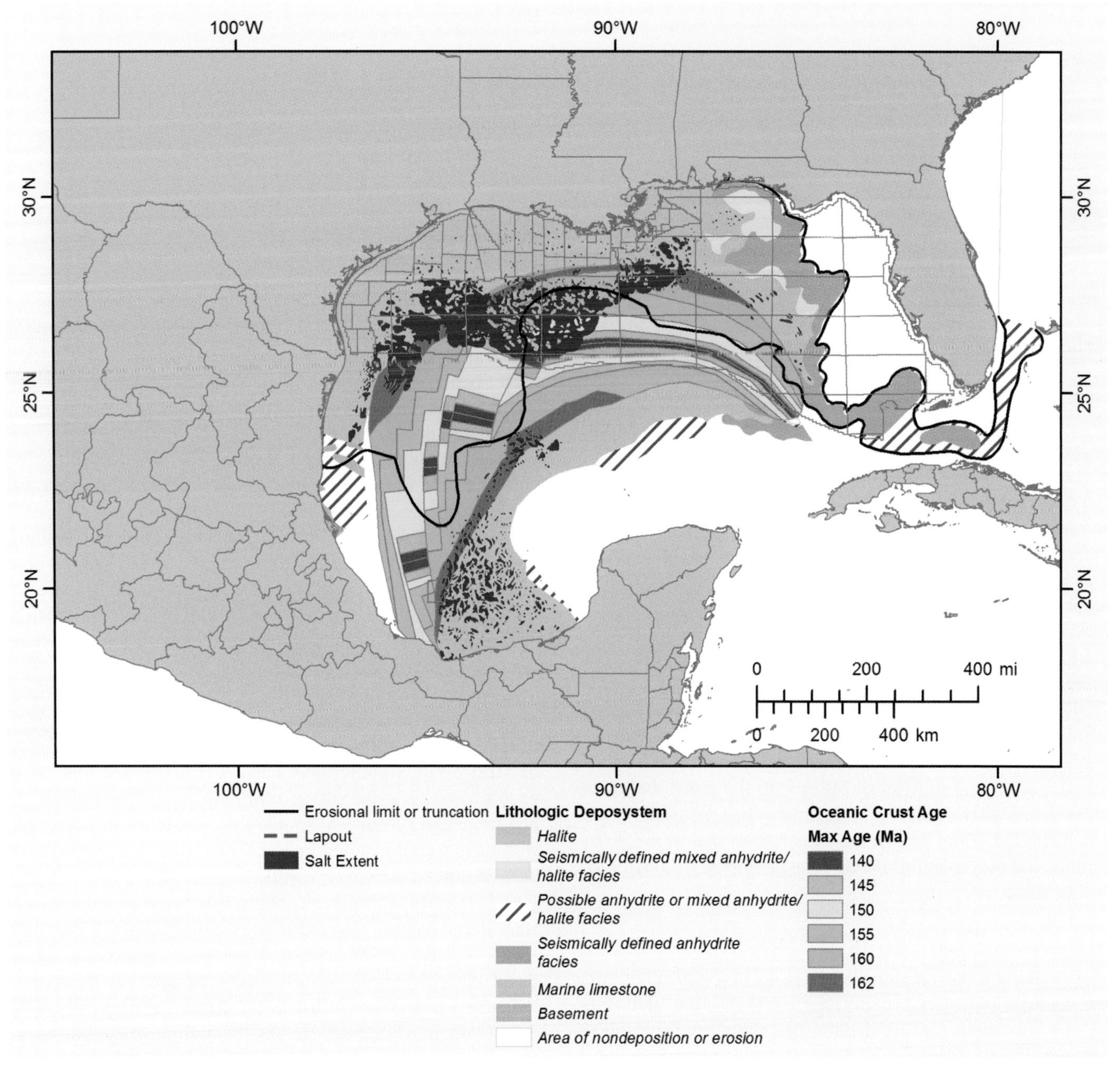

Figure 3.1 Louann Salt present-day diapirs and canopy (salt extent in red) with oceanic crust and original salt distribution inferred from seismic lithofacies indicated.

he called the Cunagua salt in the Kewanee #1 Collazo well based on unpublished analysis of spores in red shales within the salt. Salt has also been interpreted in the Bahamas (Schlager *et al.* 1984; Epstein and Clark 2009).

It should be noted that the present-day location of allochthonous salt provides only subtle clues to the original salt distribution, as salt has moved significant distances seaward and upward due to gravity gliding. Even large portions of the autochthonous salt have been deformed and folded, which Hudec *et al.* (2013a) refer to as "parautochthonous" salt.

3.2.1 Chronostratigraphy

The age of the Louann Salt has proven very difficult to determine, given its lack of faunal and floral content and nearly pure halite composition (Figure 3.2). Salvador (1987, 1991b) and later Hudec *et al.* (2013a) used 162–163 Ma for the age of the Louann Salt. However, this age is based on poorly age-constrained basalt xenoliths within salt diapirs (Stern *et al.* 2011) and limited marine biohorizons in distant Mexico outcrops (Cantú-Chapa 1998). Recently, questions have been

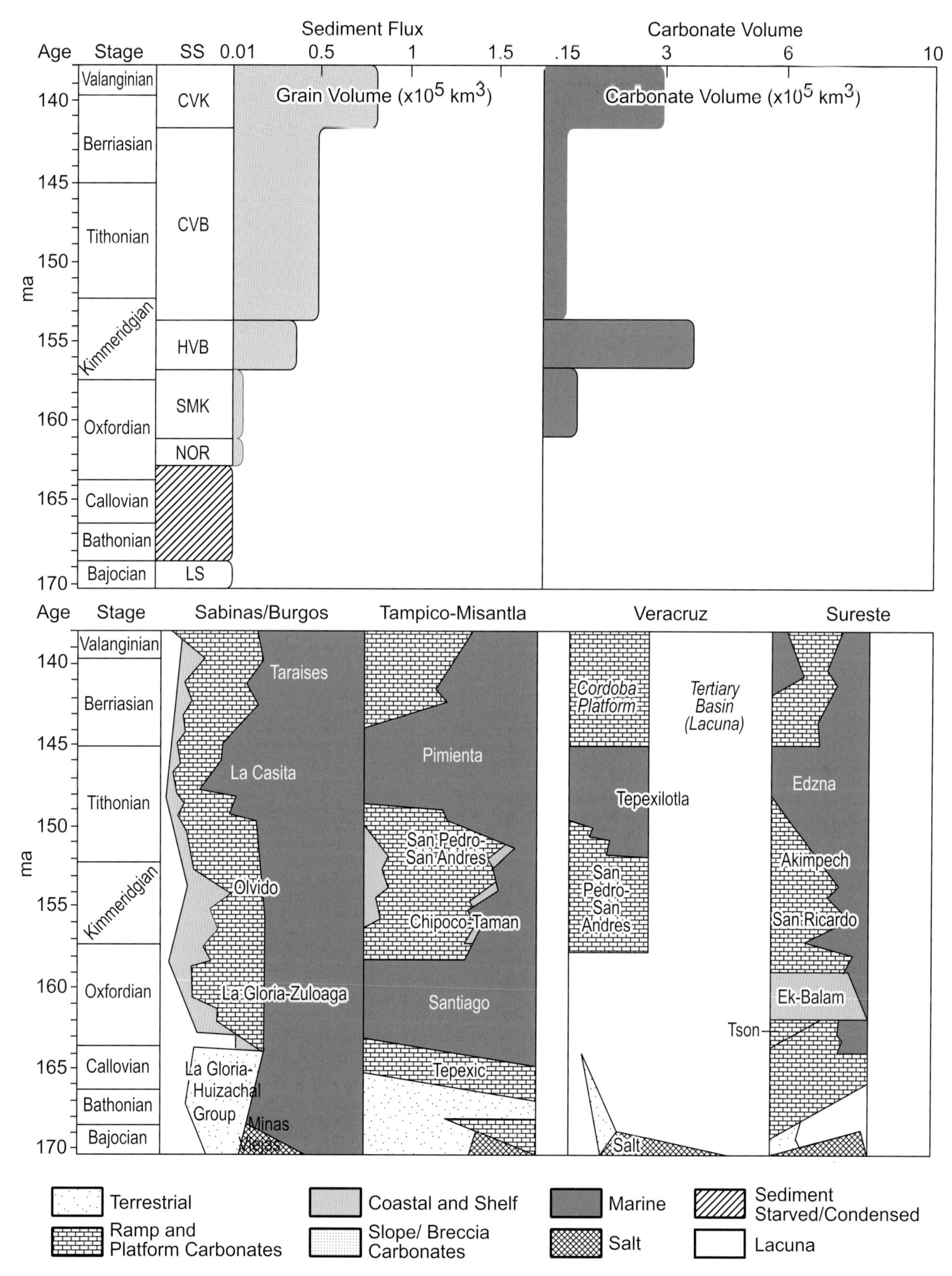

Figure 3.2 GoM Drift and Cooling Phase. Northern GoM: Upper, siliciclastic sediment flux and carbonate volume by supersequence. Lower, lithostratigraphy for four areas of the southern GoM.

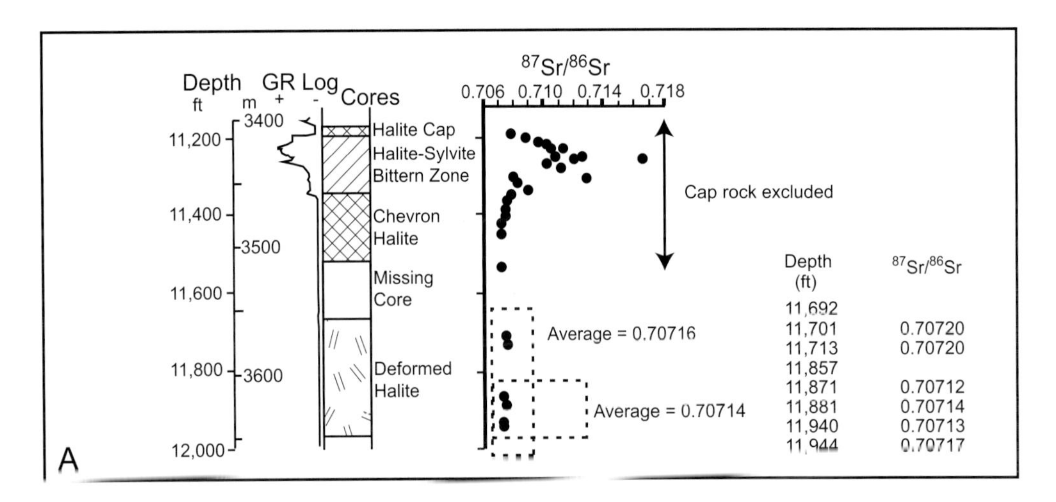

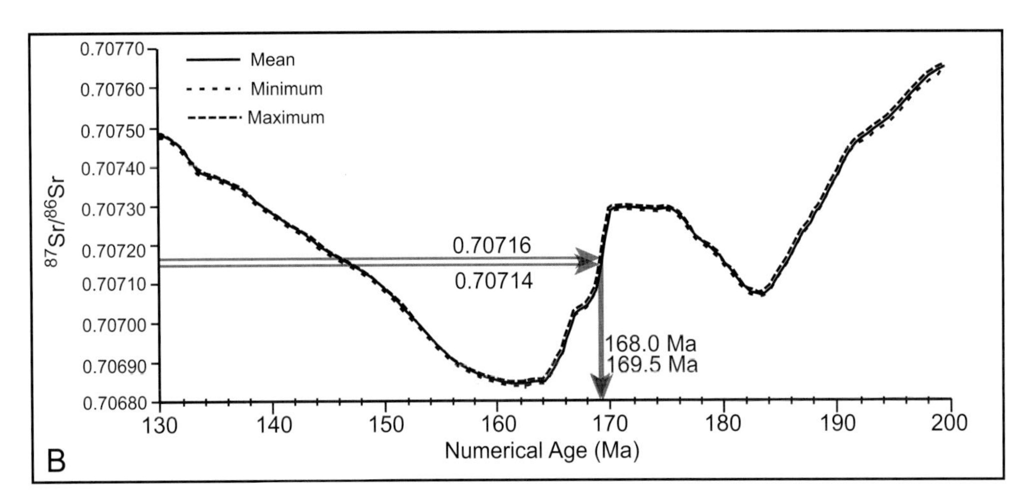

Figure 3.3 Louann Salt age analyses. (A) Humble Klepec #1 well $^{87/86}$Sr data of Land *et al.* (1995). (B) Average of four lowest samples $^{87/86}$Sr tied to the updated McArthur *et al.* (2001) seawater curve for 130–200 Ma.

raised about the affinities of ammonites used in the stratigraphic assignments (Martini and Ortega-Gutiérrez 2016).

Thus, non-biostratigraphic approaches must be considered. One way forward is to use strontium isotopic ratios from Louann halite and derive a proxy age by tying to the global ($^{87/86}$Sr) seawater curve (updated from McArthur *et al.* 2001; Figure 3.3).

Use of Sr isotopes for dating sediment is based on several assumptions: (1) measured material was precipitated in isotopic equilibrium with the open ocean, which may not be the case for some highly restricted portions of the Louann Salt basin and; (2) a significant terrestrial or detrital component within salt may skew results. Neither are a major concern here. The vast quantity of halite in the Louann Salt basin required continued access to seawater of the world ocean (F. Peel, pers. comm.). Second, the Louann Salt is remarkably pure halite and free of detrital material, excluding material incorporated during post-depositional salt movement.

Unpublished work of A. Pulham (pers. comm.) surveyed $^{87/86}$Sr dating reported for onshore salt domes of the interior salt basin. Strontium isotopes in the range of 0.70720–0.70690 were noted by Pulham in several Gulf Coast analyses. Earlier, Land *et al.* (1995) reported $^{87/86}$Sr from halite in Alabama onshore cores of the Louann Salt in the range of 0.70712–0.70727 (Figure 3.3A). In order to avoid possible dissolution contamination in the uppermost samples near the cap rock, we averaged the lowest sample results of Land *et al.* (1995) to derive an isotopic ratio of 0.70714–0.70716, well within the Gulf Coast range noted by A. Pulham (pers. comm.). Tying to the global seawater strontium isotopic curve yields a proxy age of 168–169.5 Ma (Bajocian stage or Middle Jurassic; Figure 3.3B). Further work is clearly needed, but there is some support for use of this age from a plate tectonic standpoint and depositional model standpoint, as will be discussed in Section 3.2.3.1.

3.2.2 Previous Work

Because of its scientific and economic importance, many models have been proposed for the origin and initial distribution of the Louann Salt (Salvador 1987; Land *et al.* 1995; Cantú-Chapa 1998; Padilla y Sánchez and Jose 2016; Peel *et al.* in prep.). At present, there is no consensus on: (1) the original source of the seawater that entered the nascent basin and evaporated to form the "mother" Louann Salt body; (2) the water depth of salt deposition; or (3) the thickness of the original salt deposit. In addition, several paleohydrologic process models are proposed to explain the development of such large salt deposits on a global basis (Jackson and Hudec 2017).

Two divergent schools of thought exist regarding the source of seawater that evaporated to form the Louann Salt. Salvador (1987, 1991b) proposed that Late Jurassic salt formed in shallow hypersaline bodies with seawater entering from the Pacific through Mexico. Land *et al.* (1995) showed a similar salt map distribution and seawater influx point but added an outlet to the Atlantic or Tethyan seaway. Cantú-Chapa (1998) and Padilla y Sánchez and Jose (2016) also argued for a Pacific source based on marine deposits of roughly similar age as the Louann Salt in southern Mexico.

Recently, Peel *et al.* (unpublished report) have suggested that a "chain of basins" connected to the Jurassic marine Tethyan (paleo-Atlantic) seaway allowed progressive deposition of non-NaCl compounds prior to entry into the greater Louann Salt basin. Halite is the dominant mineral phase of the Louann Salt. Ignoring post-depositional inclusions found in the salt, the autochthonous Louann Salt is remarkably pure, nearly 100 percent NaCl (Fredrich *et al.* 2007). Chemical analyses from several locations, mostly interior basins, show that the Louann is deficient in $CaSO_4$ and we would extrapolate this to the larger Louann deep basin. It should be noted that some seismic velocity studies (e.g., Cornelius and Castagna 2018) suggest that the allochthonous salt is considerably less pure due to the presence of various inclusions incorporated during salt movement, as well as destruction of the original layered fabric of the salt.

Peel's new model argues that seawater incoming from the Tethyan Ocean was concentrated to the point of sulfate deposition in a series of predecessor rift basins prior to reaching the Gulf. We mapped and interpreted the Louann Salt in small structural basins bounded by basement horsts in the eastern GoM and adjacent Florida Straits, lending this model some degree of plausibility (Figure 3.4). Peel estimated that given notional evaporation rates of 14 m/yr and an original salt thickness of 3–5 km, it is possible that the entire Louann Salt body was formed in a relatively short period of time (5000–250,000 years).

Of course, this rather short duration estimate relies upon an assumption about the original thickness of autochthonous salt, prior to salt evacuation and movement. Original deep salt thicknesses of 2–4 km were estimated by Salvador (1987). Hudec *et al.* (2013a) suggest a similar thickness was developed in the "inner basin" of the present-day deepwater prior to salt movement upward and seaward to form the allochthonous salt canopy (see Section 1.4.2.2). Unfortunately, wells that do fully penetrate the autochthonous salt are located in the interior basins of Texas, Louisiana, and Mississippi, or on the Sabine Arch/Toledo Bend Flexure, where the salt is significantly thinner than in the Louann central basin. The thickest penetration of deep salt outside of that area is in the Cheyenne well (LL 398 #1), which encountered nearly 2 km of salt. However, the Louann here is present in a significant salt structure so thicknesses probably reflect post-depositional salt inflation. In the Isthmian salt basin of Mexico, parautochthonous halite is penetrated in the Ek-Balam field area, where at least 200 ft (61 m) of salt is shown in logs and lithologic descriptions (Cantú-Chapa 2009). Seismic sections document at least 4 km of salt in this area but much of that is allochthonous salt and highly inflated (CNH 2015a).

3.2.3 Louann Salt Supersequence Paleogeographic Reconstruction

Reconstruction of the original salt basin as it appeared prior to sea floor spreading has been difficult to carry out due to the subsequent allochthonous salt emplacement that obscures stratal relationships at depth. The following sections describe (1) preferred plate tectonic configuration for salt deposition; and (2) the methodology and results of seismic mapping focused on identification of Louann facies transitions from the deep basin to the onshore lapout points. Generation of such a paleogeographic map allows testing of several existing hypotheses regarding the origin of the GoM salt, a critical component of this prolific petroleum system. In addition, understanding the original distribution of the Louann Salt can be a guide to its present-day location in frontier areas such as the Mexico deepwater.

3.2.3.1 Plate Tectonic Reconstructions for Original Salt Distribution

Plate tectonic restoration for the original salt distribution was considered at 170 Ma (versus younger ages of 162–163 Ma suggested by Hudec *et al.* 2013a) based on chronostratigraphic evidence described in Section 3.2.1. Another reason for favoring an older age for the original salt is that plate reconstructions (Section 2.2) show a more limited ocean gateway between South America and Yucatán earlier in the Mesozoic geohistory (Figure 2.1D versus Figure 2.1E). Evaporation as a result of limited influx of marine water might be enhanced by a more restricted area between continental blocks.

A large number of wells penetrate parautochthonous salt in the northern GoM. Most of these are located in onshore areas above continental crust. Some have actually passed through salt into underlying Eagle Mills or basement rock. However, an increasing number of wells have partially penetrated salt in the deepwater realm of Mississippi Canyon and Destin Dome protraction blocks. In these areas, the Jurassic interval rafted apart after the younger Smackover carbonate deposition. Thus, restoration of well locations to a pre-rafted position was required for the paleogeographic map. Salt rafting is discussed as part of Section 3.3.4 (Norphlet Sandstone).

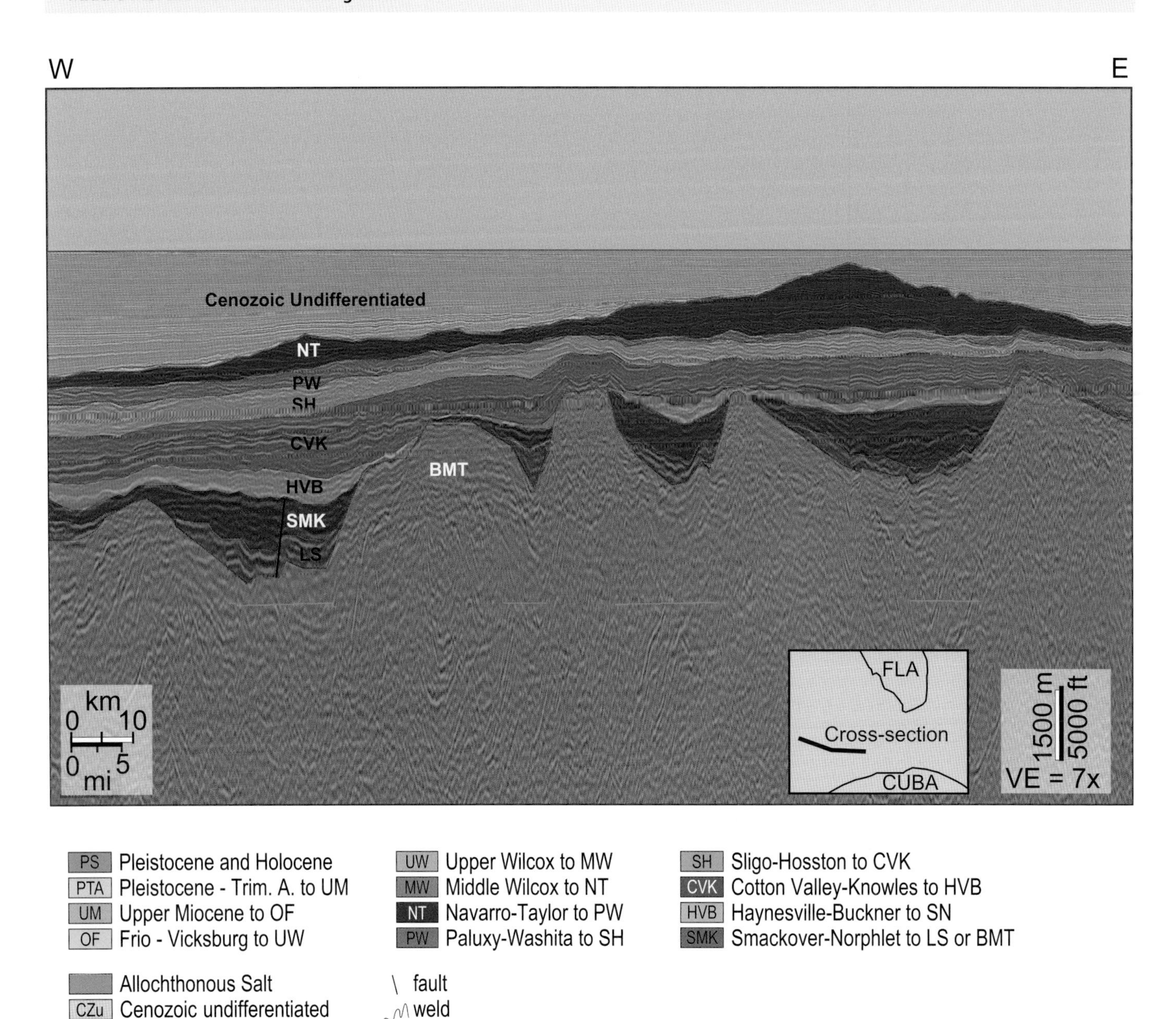

Figure 3.4 Seismically based cross-section through small structural basins of the Florida Straits.Seismic line courtesy of ION.

3.2.4 Louann Salt Seismic Facies

Our paleogeographic reconstruction for the Louann Salt also relies heavily upon seismic facies analysis that differentiates the Louann into two end member lithofacies, halite and anhydrite, with discrete zones of mixed halite and anhydrite lithofacies. Observations of structural character also help discriminate halite and anhydrite, as halite is far more ductile than anhydrite (Jackson and Hudec 2017).

3.2.4.1 Louann Anhydrite Lithofacies

The high-amplitude continuous seismic character of the inferred anhydrite seismic lithofacies is distinctive, as is its structural response to sediment loading (Figure 3.5). There are few observed instances of ductile flowage, diapirism, flowage, or piercement. Basal stratal detachment or rafting is generally absent. The unit generally lapouts on the Middle Ground Arch or other structural highs. Local basement highs generally perch above the regional level of this unit.

This anhydrite lithofacies is calibrated against the Sake well (DC 726 #2) which penetrated a nearly pure anhydrite interval of about 200 ft (61 m) thickness from 17,890 to 18,100 ft (5454–5518 m) measured depth (MD); Figure 3.6). In cuttings, the anhydrite is white, soft, non-translucent, and with reddish inclusions. Petrographic thin sections of cuttings show finely crystalline anhydrite minerals with high birefringence.

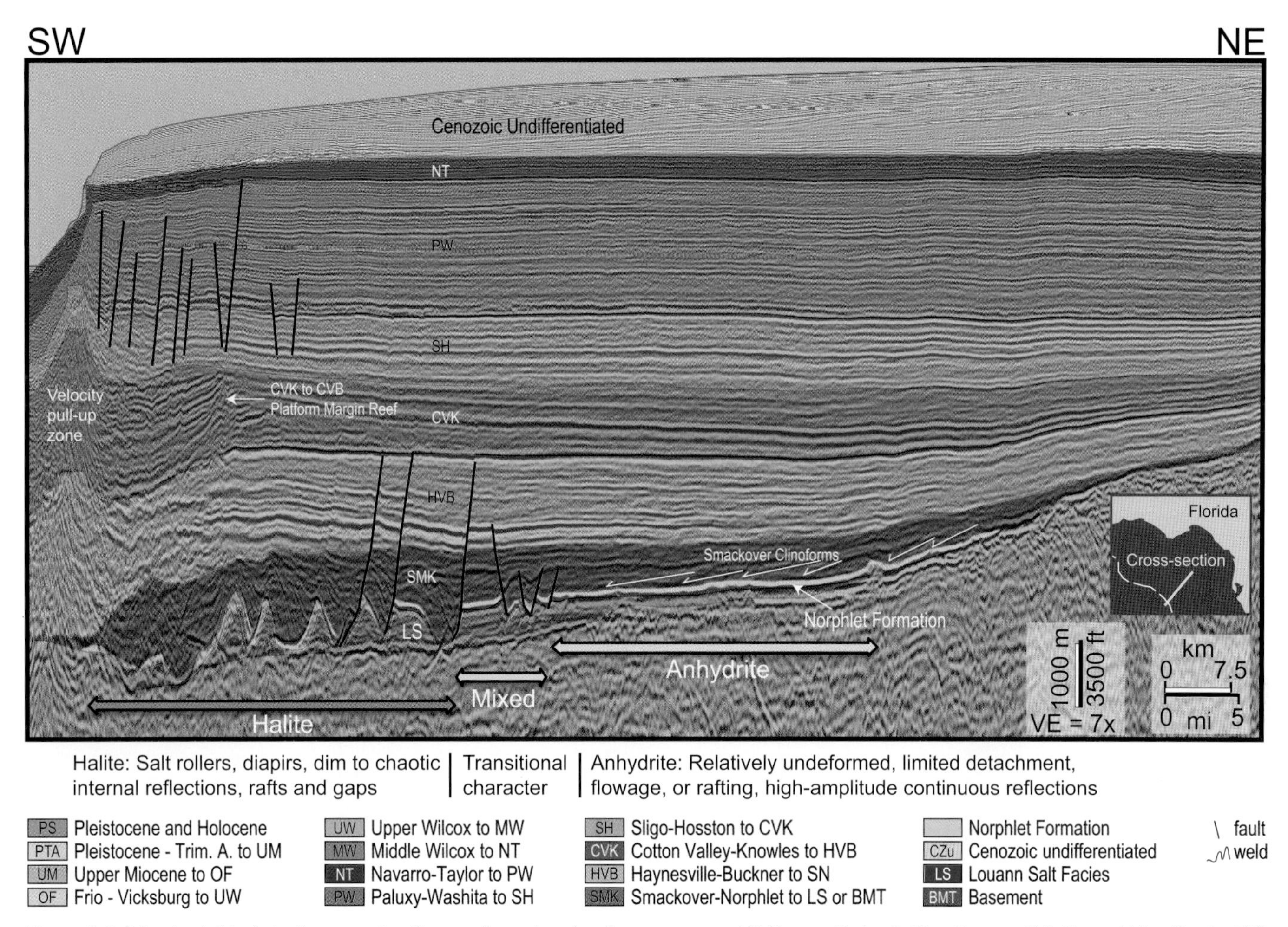

Figure 3.5 Salt seismic lithofacies interpretation. See text for explanation. Supersequences: NT, Navarro–Taylor; S, Sligo–Hosston; CVB, Cotton Valley–Bossier; HVB, Haynesville–Buckner; SMK, Smackover; NOR, Norphlet; LS, Louann Salt; BMT, basement. Seismic line courtesy of Spectrum.

Overlying Norphlet red beds and underlying gray siltstones of the pre-salt Eagle Mills equivalent interval are distinctively different in both cuttings and log response (Figure 3.6A). Synthetic seismograms show that even at 18,000 ft, the anhydrite interval is seismically resolvable due to its high impedance contrast and 200 ft (61 m) thickness (Figure 3.6B). Reflection seismic lines tied to the Sake #2 well show a series of high-amplitude continuous reflections that can be mapped laterally over a large area.

Based on seismic lithofacies mapping and seismic correlation, we interpret this anhydrite unit as a sabkha deposit that formed landward of the Louann Salt in both interior (perched) salt basins and the main Louann Salt basin (Figure 3.7).

We also interpret a similar facies transition to occur in the southern GoM during the Late Jurassic. This is based on seismic observations in offshore Yucatán. New regional 2D seismic lines extending across the Isthmian salt basins (Campeche and Yucatán) show a high-amplitude seismic reflection package present in structural terraces attached to the Yucatán Platform (see cross-sections 7 and 9; Figures 1.19 and 1.21). Anhydrite-bearing intervals are not continuous around the platform margin, as in some areas the strata lapout against the steep Yucatán Platform margin (Figure 3.7). Based on limited seismic data, we hypothesize two areas where anhydrite is present on terraces attached to the restored eastern and western margin of the Yucatán Platform margin.

It should be noted that in this area and also onshore areas further north, the vertical transition from Louann Salt to Norphlet siliciclastic is marked by a similar distinctive anhydrite unit informally called the Pine Hill Anhydrite (Tew *et al.* 1991). The Geolex (see Lexicon of Geological Names, https://ngmdb.usgs.gov/Geolex/search) states that the Pine Hill is white, finely crystalline with reddish inclusions and maximum thickness of 210 ft (64 m), closely matching that observed in the Sake #2 well some 300 miles (500 km) to the southeast. The Pine Hill in the southwest Alabama type locality is thought to have been deposited in an isolated evaporitic area (sabkha) landward of the Brevard Anticline (Oxley and Minihan 1969; Raymond *et al.* 1988) where average thicknesses of 7–40 ft (2.1–12.2 m) and a maximum thickness of 100 ft (30.5 m) are locally encountered above the Louann Salt. Thus, we regard this anhydrite unit as both a lateral facies

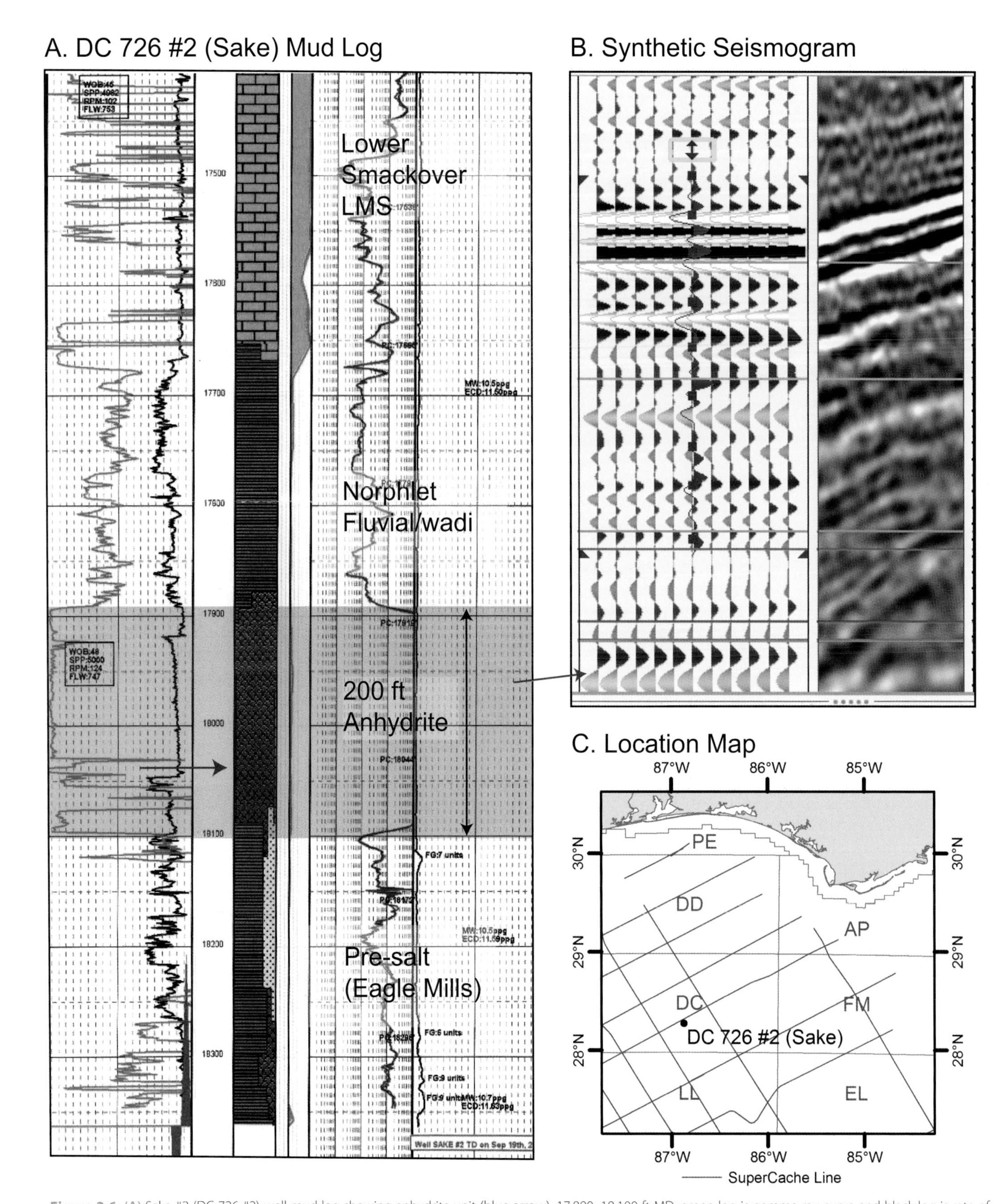

Figure 3.6 (A) Sake #2 (DC 726 #2) well mud log showing anhydrite unit (blue arrow), 17,890–18,100 ft MD; green log is gamma ray curve and black log is rate of penetration curve. (B) Sake (DC 726 #2) synthetic seismogram showing normal polarity extracted wavelet tied to PSTM Deep East Survey. (C) Location map.

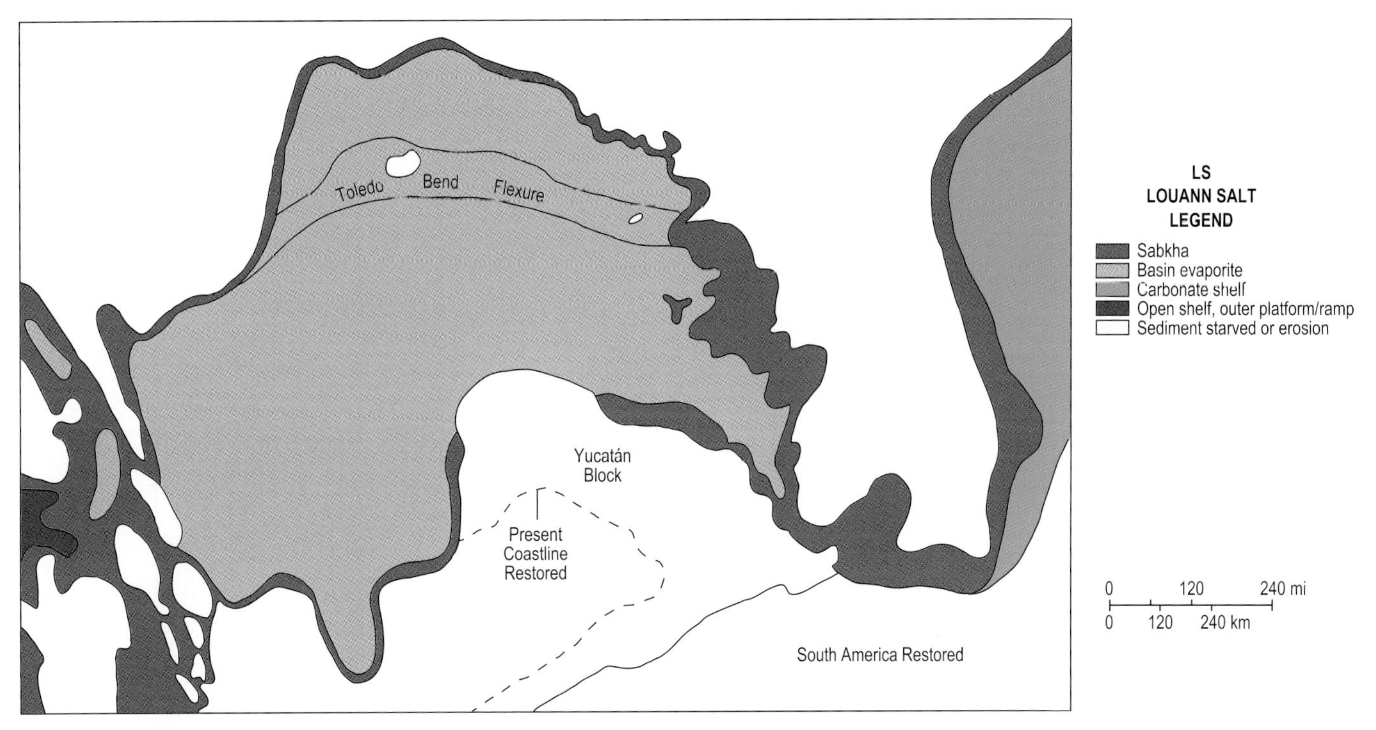

Figure 3.7 Louann Salt supersequence paleogeographic reconstruction.

equivalent of Louann halite and a product of a restricted sabkha that prograded over the Louann on its basin margin.

Jurassic non-halite evaporites are also known from both outcrops and older Pemex wells drilled in the Monterey Trough (including the La Popa basin) and Sabinas basin (Lopez Ramos 1982; Cantú-Chapa 1998; Lawton *et al.* 2001). Much of the exposed outcrop section is gypsum, the hydrous form of anhydrite, though halite is thought to be present in the subsurface over small areas of the La Popa basin center, based on structural character and limited well penetrations (Lopez Ramos 1982; Lawton *et al.* 2001). In our paleogeographic reconstruction (Figure 3.7), we show halite confined to the La Popa basin centers. The dominant facies in these interior basins is depicted as anhydrite, in order to match the dominance of gypsum in outcrop intervals (Lawton *et al.* 2001). A similar map pattern is hypothesized for the Ferry Lake Anhydrite interval of the Glen Rose (GR) supersequence, where halite deposits are restricted to the center of the South Florida basin (Section 4.5.3).

3.2.4.2 Halite Lithofacies

On a large scale, the halite lithofacies has both distinctive structural and seismic characteristics (Figure 3.5). In the eastern Gulf, where the parautochthonous salt is not obscured by a thick allochthonous salt canopy, there is abundant evidence of ductile flowage, including salt rollers, diapirs, piercement structures, and rafts (cross-sections 1–3; Figures 1.13–1.15). Basal detachment of listric faults in salt is common. The top of salt is often seismically distinct as a high-amplitude reflection but internally the halite seismic facies can vary from dim to chaotic.

The ductility of salt is well-documented and a variety of salt structures are found in the GoM, ranging from salt-cored folds to salt domes, allochthonous salt canopies, and more (as discussed in Section 1.4). Although present-day parautochthonous salt is quite discontinuous, it is logical to assume original halite must have been continuous over most of the area in order to provide a source of halite to the numerous salt structures and canopies. The Louann Salt supersequence paleogeographic reconstruction reflects that assumption. Hudec *et al.* (2013a, 2013b) proposed that basinward salt gliding allowed halite to encroach upon the oceanic crust formed after salt deposition in the Walker Ridge salient.

3.2.4.3 Mixed Halite–Anhydrite Lithofacies

A mixture of halite and anhydrite can be inferred from structural character in several locations transitional between anhydrite and halite (Figure 3.5). Some faulting and detachment is observed, but salt rollers seldom form. The unit is weakly deformed with some reversal of dip from regional but little vertical flowage. Faulting is more planar, with limited listric curvature at the base of the interval. This structural response to sediment loading suggests an admixture of ductile and more brittle lithologies that could be interbedded halite, anhydrite, or even carbonates. Present-day thickness is more representative of original thickness, adjusting for compaction.

The facies is penetrated in a number of wells in the Destin Dome area. Seismic character is transitional from the high-amplitude reflections of the anhydrite to the dim character of the halite (below the Norphlet–salt contact). The transitional seismic facies character may also reflect sub-equal amounts of salt and anhydrite, and elsewhere the Louann stratigraphic interval is dominated by one or the other. For simplicity, we

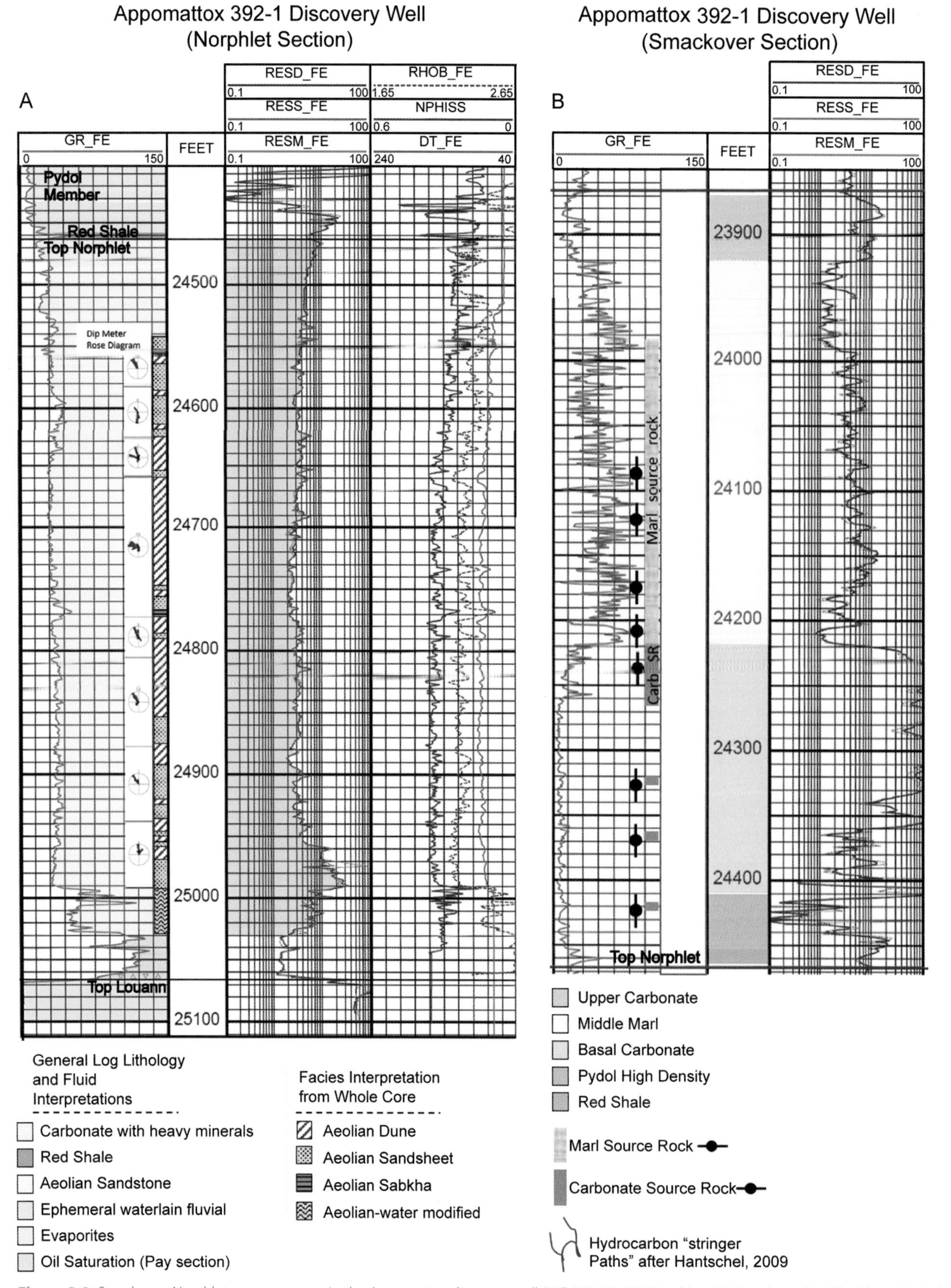

Figure 3.8 Smackover–Norphlet supersequence in the Appomattox discovery well (MC 392 #1). (A) Norphlet. (B) Smackover.Modified from Godo (2017).

have grouped this mixed facies with the anhydrite lithofacies in the paleogeographic map (Figure 3.7).

3.2.4.4 Summary

The paleogeographic reconstruction of the Louann Salt (Figure 3.7), based on both seismic observations and plate tectonic reconstruction, leads one to question the proposed marine connection between the Louann basin and the paleo-Pacific Ocean as suggested by Cantú-Chapa (1998) and later Padilla y Sánchez and Jose (2016). Evidence for a Pacific Ocean influx is based on Reineckeid (Pacific affinity) ammonites in marine outcrops south of Cuidad Valles (Cantú-Chapa 1998). However, the Pacific affinities of these ammonites has recently been questioned, with suggestions that these are more similar to Tethyan fauna or are cosmopolitan (Martini and Ortega-Gutiérrez 2016).

The mapped configuration of narrow passages and dominance of anhydrite (and near-surface gypsum) in Mexico also argue against a Pacific seawater source (Figure 3.7). Further, Goldhammer and Johnson (2001) suggested some local connections between the larger Louann Salt basin and the Mexican interior basins through a marine passageway crossing the Tamaulipas–Chiapas transform fault zone and associated segmented series of basement highs. Clearly, the narrow connections between Mexico and the greater Louann Salt basin may have limited hydrologic connectivity in the opposite direction. At the very least, Pacific seawater must find its way a considerable distance from the ocean gateway suggested by Cantú-Chapa (1998) and Padilla y Sánchez and Jose (2016).

In any case, present-day locations of the earliest Middle Mesozoic marine outcrops as cited by Cantú-Chapa (1998) and later Padilla y Sánchez and Jose (2016) are suspect, given the extensive post-depositional structural dislocation as Yucatán rotated into its present-day position with sea floor spreading (Martini and Ortega-Gutiérrez 2016; see also Section 2.2). The other, more interpretive reason is that the segmented series of basement-controlled small basins in the Florida Straits (Figure 3.4) may represent a vestige of the "chain of basins" needed to explain how halite of the deep Louann basin achieved near purity, as suggested by Peel *et al.* (unpublished report).

Finally, the long-term (first-order) architecture of carbonate depositional systems in Mexico is more reflective of Atlantic Ocean influence than the Pacific. The cyclic patterns of carbonate platform progradation, aggradation, and drowning in Mexico are better explained in light of Atlantic Ocean eustatic controls versus a tectonic modulator related to Pacific plate motion (Martini and Ortega-Gutiérrez 2016). The architecture of carbonate systems is discussed in more detail in subsequent sections and chapters.

3.3 Smackover–Norphlet Supersequence

The Smackover–Norphlet supersequence is one of the most important reservoir units both onshore and offshore in the eastern GoM. The Smackover Formation has enjoyed a long onshore exploration history, with discoveries peaking in the late 1970s (Lore *et al.* 2001). Subsequent attempts to extend Smackover production into offshore areas failed, with numerous dry holes in the Destin Dome protraction block in the early 1980s. Interest in the Norphlet Sandstone, which generally played second fiddle to the Smackover in onshore areas, was enhanced with the 1979 discovery of gas in the deep (>20,000 ft) Norphlet of the Mobile Bay area, the Mobil-operated Mary Ann Field (Marzano *et al.* 1988; Mankiewicz *et al.* 2009; Ajdukiewicz *et al.* 2010). However, the seaward extent of the eolian reservoir was not well known and concerns over the economic viability of high-pressure/high-temperature (HPHT) gas in the deepwater limited further exploration.

In 2003, Shell and its partners stepped basinward over 100 miles to discover oil in the Norphlet of DeSoto Canyon with the Shiloh well (DC 269 #1; Godo 2017). It is likely that Shell's basin modeling suggested that the location of Jurassic prospects on transitional crust (with lower overall heat flow) and colder deepwater sea bottom temperatures (since the Jurassic) indicated that Oxfordian source rocks would be within the oil window. Although Shell eventually relinquished the Shiloh discovery as subeconomic, information from this well and subsequent drilling and geologic analysis clearly set up the large (<700 MMBOE) discovery called Appomattox (MC 392 #1; Figure 3.8A) that followed (Godo 2017).

Many of the Norphlet discoveries in the DeSoto Canyon and eastern Mississippi Canyon areas are structural traps set up by rafting apart of the interval above salt, with significant gaps where there is no Norphlet or Smackover (Pilcher *et al.* 2014; Figures 1.13 and 3.9; Table 3.1). Raft gaps are generally filled in by younger Jurassic and Early Cretaceous sediments. The structures are complex and diverse, ranging from four-way anticlinal closures to three-way fault-dependent closures (Godo 2017). Imaging is challenging because of the depth and velocity structure, which mean it is difficult to precisely pick the top of the reservoir and top of the salt (Herron 2014). Note also that reverse faulting is observed on seismic data and we have observed repeated intervals in several wells, including the Appomattox sidetrack well (MC 392 #1 ST02BP01).

The paleogeographic maps for the Norphlet and Smackover are among the most complex ever attempted in the GoM. Reconstructions include restoring the wells to the pre-sea floor spreading positions and then accounting for rafting on salt (see discussion in Sections 3.3.3 and 3.3.4). In addition, Norphlet and Smackover data from Cuba and Mexico was included in order to provide the basin-scale perspective. Finally, consideration of dryland depositional systems (eolian and fluvial wadi) and calibration against well logs and available paleo-wind information is a prerequisite for understanding this supersequence.

3.3.1 Chronostratigraphy

The Oxfordian (Jurassic) Smackover–Norphlet supersequence is the first major sedimentary unit to be deposited on the Louann Salt (Figure 3.2). We include the variably developed Pine Hill Anhydrite within the Louann Salt supersequence.

Table 3.1 Rafting estimates, deep-water Norphlet play, based on seismic interpretation.

Operator	Block	Prospect Name	Well Number	Interpreted Direction of rafting	Estimated amount of rafting (miles)	Estimated amount of rafting (km)
Shell	DC269	Shiloh	1	WSW	14	22
Shell	DC268	Antietam	1	WSW	13	22
Shell	DC353	Vicksburg	1	SW	15	24
Shell	MC392	Appomattox	1 ST2 BP1	SW	19	31
Shell	LL399	Cheyenne	1	Not rafted	0	0
Shell	DC486	Fredericksburg	1	SW	16	25
BHP	DC726	Sake	2	Not rafted	0	0
Anadarko	DC535/491	Raptor	1, 1 ST1	SW	9	14
Marathon	DC757	Madagascar	1	SSW	15	24
Shell	DC843	Swordfish	1	SSW	16	26
Shell	DC529	Petersburg	1	SW	17	28
Statoil	DC231	Perseus	1	SW	1	2
Shell	MC525	Rydberg Deep	2	SW	17	28
Murphy	DC178	Titan	1	WSW	15	23
Chevron	MC607	Ballymore	1	SW	28	45
Shell	DC348	Appomattox NE	3 ST1BP0	SW	19	31
Shell	DC348	Appomattox NW	3 ST0BP0	SW	19	31
Shell	DC398	Gettysburg West	1 ST0BP0	SW	14	23
Shell	MC475	Leesburg	1 ST0BP0	SW	30	48
Shell	MC566	Fort Sumter	2 ST0BP0	SW	24	38
Chevron	MC607	Ballymore	Not yet released	SW	28	45
				Average	15	24
				Maximum	30	48
				Minimum	9	14

Abbreviations: DC, DeSoto Canyon; MC, Mississippi Canyon; LL, Lloyd Ridge; ST, sidetrack; BP, bypass

Given its lack of marine-fossil datums, the Norphlet is poorly dated, generally considered to be older than the Middle to Upper Oxfordian Smackover and its equivalent in Mexico, the Zuloaga Limestone (Myczynski *et al.* 1998; Figure 3.2). Cantú-Chapa (2009) suggested a Lower Oxfordian age for the Ek-Balam Formation, a possible correlative of the Norphlet, based on the presence of fragments of the ammonite *Ochetoceras* sp. in a core from Balam Field well #101. Onshore in northern Mexico, red beds of the La Joya Formation (Huizachal Group) are thought to be Middle to Late Jurassic, based on fossil plants (Mixon 1963; Figure 3.2). A small hiatus may be present at the top of the Norphlet (Brand 2016), but determining age relationships is particularly challenging in this non-marine setting.

In the northern GoM, the Smackover is thought to encompass middle and upper portions of the Oxfordian Stage (Oxfordian-7 maximum flooding surface, 157.3 Ma of geological timescale (GTS) 2016 of Ogg *et al.* 2016), while the Norphlet occupies the lower portion of the Oxfordian. The base of the Oxfordian (Oxfordian-1 sequence boundary) and possibly the Norphlet itself is age-designated as 163.5 Ma (GTS 2016 of Ogg *et al.* 2016), with an uncertain hiatus between it and the underlying Louann Salt which initiated at 170 Ma, discussed earlier. The base of Smackover or Top Norphlet is equally problematic as the first marine fossils in the Smackover do not appear until near the top of the unit, near the Top Oxfordian biohorizon (Godo 2017; Figure 3.8B).

Other complications in age designation for this supersequence include significant changes in the timescales (e.g., Ogg *et al.* 2016 versus Gradstein *et al.* 2012) commonly used in plate reconstructions and the time-transgressive nature of the Smackover and Norphlet lithofacies (Figure 1.25). The paleogeographic maps included here depict carbonates being deposited in during both Smackover and Norphlet, consistent with Figure 1.25. Carbonate and sandstone grain volumes, measured entirely within the subsurface of the northern GoM, are a small fraction of volumes within the overlying Haynesville–Buckner supersequence (Figure 3.2). This likely reflects progressive opening of the Gulf basin with sea floor spreading and other factors.

3.3.2 Previous Work

Much has been published about the onshore Smackover, primarily from subsurface samples and a limited amount of work on outcrop intervals of Mexico and Cuba. The first Norphlet schematic reconstructions showed a vast area of lowlands, shallow lakes, and evaporitic basins, with eolian sand seas (ergs) limited to the areas just west of the Appalachians (Salvador 1991b). Marine sediments were not believed to be present until late Oxfordian time, represented by the Smackover Limestone and its equivalent in Mexico, the Zuloaga. However, our work and that of others (Goldhammer and Johnson 2001) has suggested that the Norphlet and Smackover "lithofacies" are time-transgressive (Figure 2.2) and a shallow marine seaway was likely present in the area between the Yucatán (Mayan) block and the North American continental plate. Such a pattern is typical of many of the Mesozoic supersequences in the GoM.

With discoveries of oil and gas in the Smackover across the onshore northern GoM, much more was learned about this unit through drilling. The Smackover of onshore areas is often divided into two units: (1) a lower argillaceous, often organic-rich zone informally called the "Lower Brown Dense" (Sassen *et al.* 1987); and (2) an upper zone of shallow-water carbonates (Moore 1984). Initial core studies suggested the Upper Smackover contained dolomitized grainstone banks developed across a wide carbonate ramp (Moore 1984). Later investigations revealed that some of these organic build-ups were microbial build-ups, with both primary and secondary porosity (Mancini *et al.* 2004). With new 3D seismic surveys, it was shown that the best Smackover fields were developed over basement highs, an example being the Little Cedar Creek Field of Alabama (Haddad and Mancini 2013).

In deepwater areas of the Mississippi Canyon and DeSoto Canyon protraction blocks, Godo (2017) has identified three components of the Smackover (Figure 3.8B): (1) basal transgressive Smackover carbonates (100–300 ft; 33–100 m thick) with an unusual iron mineral-rich transition zone to the underlying Norphlet eolian sandstones (see also Brand 2016); (2) Middle Smackover mudstones (200 ft; 66 m) representing distal clinoform deposits; and (3) Upper Smackover Limestones (~100 ft; 30.5 m) formed in a shallow, wave-influenced environment of near-normal marine salinity.

Although the Smackover is quite a prolific reservoir onshore, attempts to extend production into offshore areas have been unsuccessful. The lack of commercially viable porosity and permeability in the Smackover of deepwater areas is probably due to pervasive cementation, but also a lack of grainy carbonates versus more micritic limestones (Godo 2017).

By contrast, exploration has shown that porosity preservation and diagenetic enhancement of reservoir properties does occur in the Norphlet Sandstone of offshore areas. The Norphlet Sandstone is best known from onshore subsurface case examples and field studies in the Mobile Bay area (Marzano *et al.* 1988). Onshore mapping of the unit from Arkansas to Alabama helped define a large eolian sand sea (erg) and coeval fluvial wadi systems that drained the western Appalachian highlands (Pepper 1982; Mancini *et al.* 1985). The Norphlet thins to the west, reflecting distance from the Appalachian highlands, and may not include eolian sediments.

In the Mobile Bay area, a series of linear dunes, aligned with northerly paleo-winds, were thought to be located near the paleo-shoreline for the Norphlet (Marzano *et al.* 1988). The best porosity and permeability are found in eolian sandstones that contain chlorite rim cements, the chlorite acting as a sink for excess silica that normally ends up as quartz cement in these deeply buried sandstones (Ajdukiewicz *et al.* 2010). The linear dunes are quite thick in Mobile Bay and actually "sink" into the underlying Louann Salt, forming a rugose top and base of reservoir surface that sets up gas compartments with variable column heights (Mankiewicz *et al.* 2009).

In Mexico, the Smackover–Norphlet supersequence is known from outcrops and the subsurface (Figure 3.2). The Smackover–Norphlet supersequence of the northern GoM is roughly equivalent to the Zuloaga Limestone and La Gloria Formation sandstones present in northern Mexico outcrops (Angeles-Aquino and Cantú-Chapa 2001). Offshore, in the Campeche area, these units are thought to be coeval with the Ek-Balam Group reservoirs of the Ek-Balam Field (Mitra *et al.* 2007). Limited biostratigraphic analysis suggests the sandstones of the Ek-Balam Group are Oxfordian in age (Cantú-Chapa 2009). This and the comparable eolian paleoenvironment (Roca-Ramisa and Arnabar 1994) argues for inclusion in the Norphlet paleogeographic map.

3.3.3 Plate Tectonic Reconstruction

Plate reconstructions for the Smackover and Norphlet depict a smaller basin than the present-day GoM, a narrow connection to the world ocean between the Florida block and the Yucatán block, and a tectonically complex connection across Chiapas to the south (Figure 2.1E). Opinions vary on the location of the western half of Cuba, which could have been attached to Yucatán as a separated segment later gathered up and joined to the eastern Cuban segment during the Eocene collision of the Great Arc of the Caribbean with Florida and the Bahamas. It is possible that sandstones of the San Cayetano Formation are also age-equivalent with the Norphlet of the northern GoM (Haczewski 1976). The large exposed land surface on Yucatán could have served as a sand source during the dry climatic regime of the Oxfordian in the basin.

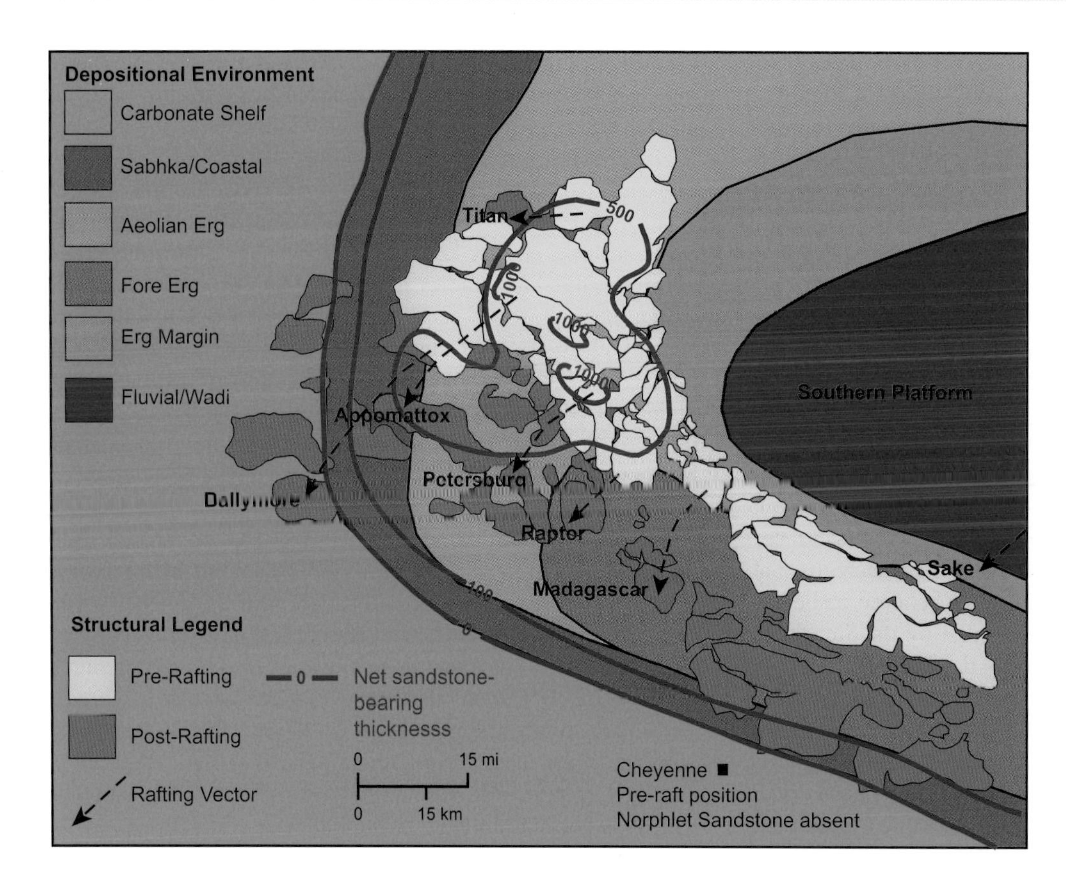

Figure 3.9 Radial translation model of rafts from Pilcher *et al.* (2014) as shown on paleogeographic restoration adjusted for post-Smackover rafting and sea floor spreading. Selected Norphlet prospects/discoveries are shown with both original and present-day locations. Arrows indicate rafting vectors. Raft shapes and locations modified from Pilcher *et al.* (2014). Restored net sandstone-bearing interval thickness contours shown in red.

3.3.4 Restoration for Raft Tectonics

Outboard of the Florida Escarpment, centered on western DeSoto Canyon and eastern Mississippi Canyon protraction blocks, lies an area of salt-detached raft blocks (Figure 3.9; Table 3.1). These blocks separated and moved basinward by gravity gliding on autochthonous Louann Salt in the Late Jurassic to Early Cretaceous (Pilcher *et al.* 2014). The raft blocks differ from earlier recognized carapaces as drilling has determined that the rafts include a pre-kinematic interval of the Smackover and Norphlet versus the condensed Mesozoic strata normally found in carapace blocks (Hart *et al.* 2004). The rafted blocks are the primary targets of the Norphlet exploration play, forming large four-way anticlines and three-way fault-dependent closures. Drilling results are a mixed bag, ranging from large discoveries (e.g., Appomattox, Ballymore), subcommercial finds (Shiloh, Gettysburg West), and outright failures (multiple wells).

Our observations of the raft timing and direction of gravity gliding are roughly comparable to the recent tectonic restoration put forward by Pilcher *et al.* (2014; Table 3.1). First, consensus is that the Smackover and Norphlet units are pre-kinematic units, as the thicknesses determined by BOEM-released wells fit a regional trend and the Norphlet can be correlated across the area (Figure 3.10). Second, the synkinematic interval that contemporaneously filled the expanding gaps between gravity gliding blocks includes the Haynesville–Buckner (HVB), Cotton Valley–Bossier (CVB), and Cotton Valley–Knowles (CVK) supersequences. Third, we agree that faults detach on salt and that rafting continues downdip until welding out. Fourth, we believe that the paleo-slope was generally to the south and west.

Our reconstruction does differ in some respects and offers a possible explanation of the driver for rafting and its relationship to sea floor spreading to the south of the zone of detachment. Pilcher *et al.* (2014) suggest a radial divergence of rafts, based on 3D seismic mapping of raft blocks and fitting the blocks to an arc around the structural nose of the Middle Ground Arch (Figure 3.9). No genetic cause of the rafting was offered, other than to cite analogs in the Kwanza basin of Angola.

It is intriguing to consider the similar timing of rafting and sea floor spreading as determined by stratigraphic correlations from this area to the area of newly created oceanic crust (Snedden *et al.* 2013, 2014). The synkinematic units of the HVB, CVB, and CVK were clearly deposited following sea floor spreading, when cooled oceanic crust has already subsided, creating a seaward paleo-slope for gravity gliding of the raft blocks.

We would also suggest that the rates of gravity gliding of the rafted blocks might be similar to the estimated half rates of sea floor spreading, about 20–22mm/year (Christeson *et al.*

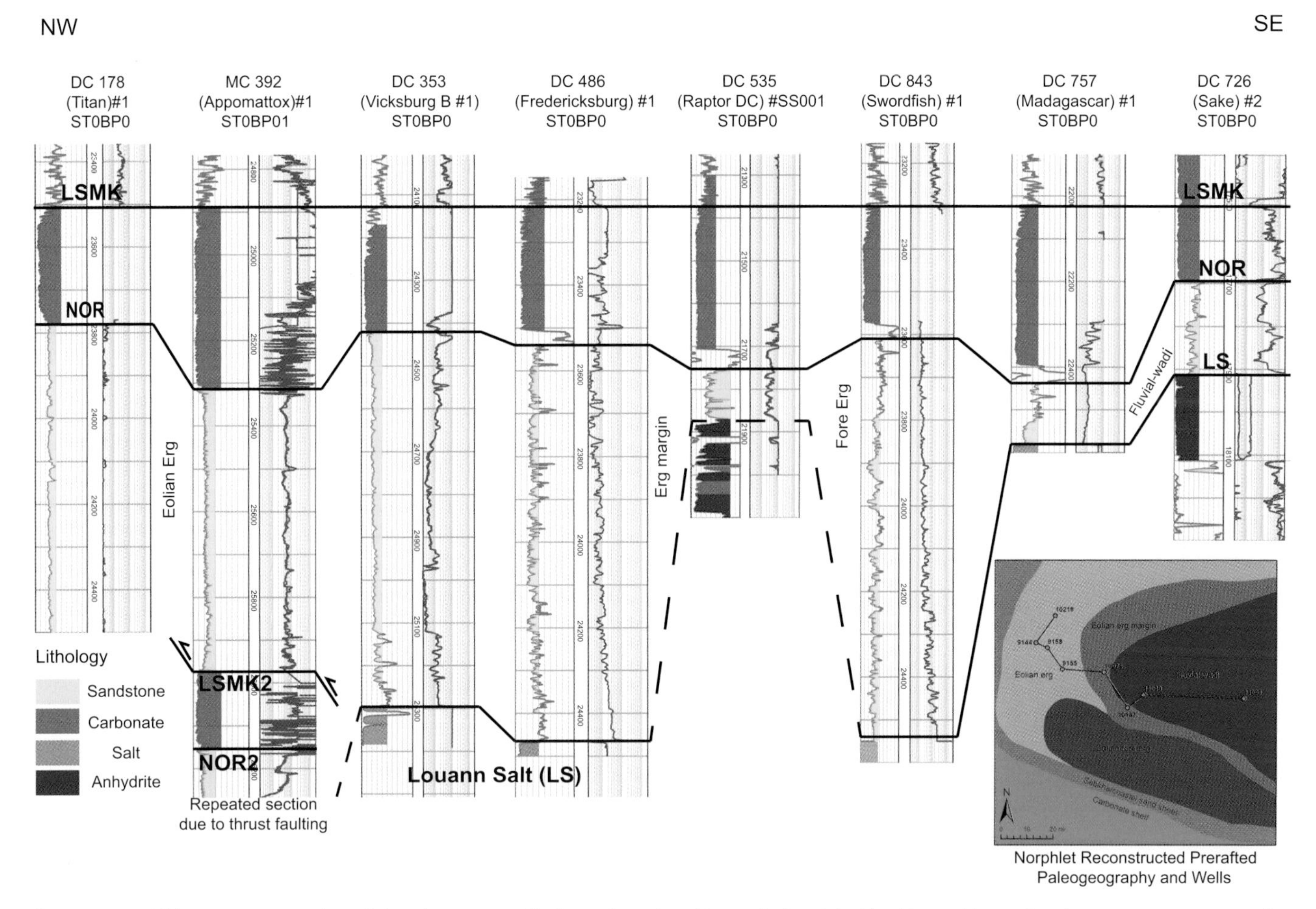

Figure 3.10 Well log cross-section through key deepwater wells showing stratigraphic continuity of the Norphlet and Lower Smackover across numerous raft blocks. Inset map shows cross-section on interpreted Norphlet paleogeography discussed in the text.

2014). This is a relatively slow rate of sea floor spreading, typical of mid-ocean ridge systems with a more diffuse, unfocused magma supply (Snedden *et al.* 2014). We are also drawn to the idea of low-rate gliding of raft blocks to explain how the pre-kinematic interval of the Smackover and Norphlet maintains its structural and stratigraphic "integrity" over displacement distances of 25–50 km. Well log correlations (Figure 3.10) and anecdotal information in Godo *et al.* (2011) and Godo (2017) do not indicate any significant overturning or stratigraphic inversion. The slow rate of displacement may have preserved depositional relationships across the dryland deposystem depicted in our paleogeographic reconstruction. There are thrust faults (e.g., indicated by repeated intervals in the Appomattox sidetrack well [MC 392 #1 ST02BP01]) but these can be explained by oblique gliding of blocks across the offshore nose of the Middle Ground Arch (M. Hudec, pers. comm.).

The reconstruction in Figure 3.9 depicts a general northeast–southwest gravity gliding of rafts, based on vectors suggested in physical models of salt flow across basement arches (Smith 2015). Physical models successfully replicate thrust-faulting parallel to the Middle Ground Arch basement trend and extension perpendicular to the arch trend. Areas of convergence and divergent flow salt oblique to the arch trend fit patterns observed in well penetrations.

Our estimates of the total amount of displacement for individual wells are remarkably similar to those of Pilcher *et al.* (2014), but differ somewhat in the direction of rafting as we interpret most of the raft blocks to have migrated southwest and south, toward cooled, dense oceanic crust. Estimated displacements range from 14 to 48 km (9–30 miles; Table 3.1). Based on our reconstruction, for example, the Ballymore (MC 607 #1) discovery structure rafted some 45 km (28 miles). Originally close to the Appomattox structure, the present-day gap with Appomattox is now 22 km (14 miles). Some rotation of pre-kinematic blocks may also have occurred during rafting, which likely affected Norphlet paleo-wind reconstruction so present-day cross-bed dip azimuths from image logs will need to be corrected for raft rotation.

3.3.5 Norphlet Deposystems: A Look into Ancient Dryland Deposition

Analysis of the Norphlet Sandstone requires consideration of new paleo-environmental deposystems for the GoM basin, which generally has few preserved ancient dryland sedimentary deposits, and certainly few potential reservoirs. Modern large eolian systems are globally rare in occurrence, confined to select locations within the "horse latitudes" between 30°N and 30°S, where precipitation is limited and dry and hot conditions prevail (Parrish and Peterson 1988). Our plate tectonic reconstruction places the main northern Oxfordian GoM dryland system in a similar geographic position just north of 20°N.

Modern dryland systems, which include eolian and arid, ephemeral fluvial systems such as wadis, have been the subject of considerable scientific study (Glennie 1972; Kocurek and Havholm 1993; Hernandez Calvento *et al.* 2017), which facilitates investigation of ancient systems as with other GoM units. We interpret subsurface well log response in the context of core calibrated facies and paleophysiographic position. For example, eolian depositional facies in Mobile Bay Norphlet wells have been described in considerable detail (Marzano *et al.* 1988). The deepwater Norphlet well cores from Shiloh, Vicksburg, and Vicksburg B are also discussed by Douglas (2010). Godo (2017) detailed Norphlet Sandstone lithology and facies in the Appomattox Field area (Figure 3.8A).

In most exploration areas, the Norphlet reservoir is represented by a single seismic reflection at best, limiting the ability to do seismic facies analysis (Herron 2014). Nonetheless, depositional processes can be inferred from the log response, the paleogeographic map position, and the location relative to structural trends like the Middle Ground Arch. The Gulf of Mexico Mesozoic Log Facies Interpretation Poster (www.cambridge.org/gomsb) shows subdivisions of the dryland deposystem classification. Figure 3.11 is a schematic representation of the Norphlet deposystem as adapted from Douglas (2010). Finally, trends in hydrocarbon seep or reservoir oil composition (e.g., sulfur content) that provide indications of the presence of the Oxfordian source rocks were used to guide paleogeographic mapping in areas with limited Jurassic penetrations (see also Section 9.6.1).

3.3.5.1 Eolian Erg

The most important Norphlet reservoir deposystem in both onshore and offshore areas is the eolian erg (Figure 3.11). Loose sand was presumably transported by winds from exposed highlands and alluvial plain into a large, arid sand sea or *erg*, which may have, at its maximum, covered over 58,000 km^2. This is comparable to the some of the largest modern ergs, including the Namib Desert that nearly exceeds 81,000 km^2 in desert surface area.

Sandstones formed in this deposystem generally have the highest overall net to gross ratio, porosity, and permeability, consistent from onshore to deepwater well penetrations (Ajdukiewicz *et al.* 2010; Godo 2017). Avalanche-dominated grain flow facies have the highest overall permeability (Douglas 2010). Repeatedly stacked sets of thick, high-angle (up to 30 degrees) cross-beds are common in Norphlet cores of this paleo-environment (Mancini *et al.* 1985). Giant cross-bed sets enable high confidence dip analysis from dipmeter, FMS, or other tools, as will be discussed in the paleogeographic reconstructions. Both straight-crested and sinuous-crested bedforms have been inferred from the dipmeter/FMS data (Godo 2017).

Lower-angle sets likely resulting from grain fall deposits in the lee of large dunes are also common (Marzano *et al.* 1988). Other features indicate soft-sediment deformation common to wind-formed dune systems (Douglas 2010). Finer-grained beds suggest deposition by wind ripples and wet and dry interdunes (Godo 2017). Porosities in excess of 20 percent are present in eolian erg sandstones (Douglas 2010), but in some offshore wells solid hydrocarbons can be found in the pore systems of the reservoir rocks (Godo *et al.* 2011). Dry dune facies exhibit the best reservoir properties as this is usually the topographically highest part of the dune (Godo 2017).

The expression of this deposystem on well logs is relatively consistent. Gamma-ray logs show higher readings than the overlying Smackover Limestone but are consistently lower than shales and other mixed sandstone/shale intervals. Resistivity trends reflect hydrocarbon content but generally are lower than the highly resistive Smackover. Blocky log motifs are common (Figure 3.10).

In the deepwater exploration area of the DeSoto and Mississippi Canyon protraction blocks, sedimentological analyses point to development of large barchan and barchaniod dunes as the primary eolian bedforms (Douglas 2010). Paleoflow was interpreted as trending largely to the northeast, with subsidiary orientations toward the west (Figure 3.11).

At Mobile Bay, this paleo-environment features a series of elongate, north–south-aligned linear sand dunes (Marzano *et al.* 1988). Dunes are thick enough to cause the deformation of the contact with the underlying Louann Salt (Mankiewicz *et al.* 2009). Thickness in Mobile Bay can exceed 1000 ft (300 m), with similar thicknesses in deepwater.

Offshore isochore map trends of dunes are somewhat obscured by raft tectonics, but mapping shows a series of depositional maximums near the reconstructed position of the Appomattox well at the center of the erg (Figure 3.9). Subsidence of eolian sandstones into the underlying Louann Salt apparently has caused post-depositional inverting of the Norphlet and local formation of small turtle structures (Godo 2017).

3.3.5.2 Eolian–Erg Margin

The landward margin of an erg system often shows a transition to the fluvial wadi systems draining the continental highlands (Figure 3.11). The style of bedform can also change from large barchaniod to smaller barchan-type dunes (Glennie 1972). Wet interdune and even temporary lakes are sometimes developed, as the dune bedforms cause damming of ephemeral river flow during infrequent rain storms (Krapf *et al.* 2003). This results in interbedding of eolian sands and fluvial muds.

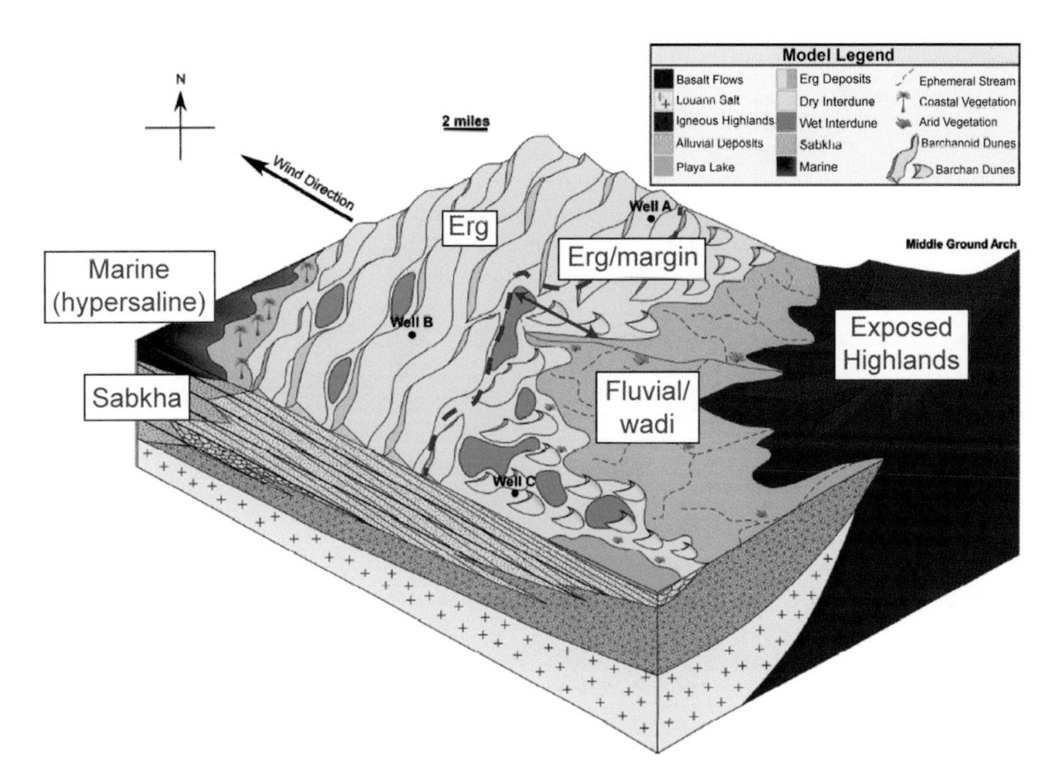

Figure 3.11 Dryland systems of the Norphlet Sandstone in the northeastern GoM. Middle Ground Arch highlands are in close proximity to fluvial wadi (alluvial) deposits shedding off of the highlands. Distal alluvial deposits interfinger with the eolian system. Well A = Shiloh (DC 269 #1); Well B = Vicksburg (DC 353 #1); Well C = Fredericksburg (DC 486 #31). Modified from Douglas (2010).

The erg margin, as a result, has a distinct serrate log motif of interbedded sandstones and silty shales. In the Norphlet, this is evidenced by wells drilling on the margin of the Norphlet erg system, though eolian erg can also be partially developed within a given single well. Overall reservoir thickness is much reduced in comparison to the erg deposits. An example well is Raptor (DC535 #1), which penetrated a relatively thin (118 ft; 36 m) Norphlet sand interval (Figure 3.10) and generally poor reservoir quality. The unit is transitional into the underlying Pine Hill Anhydrite as observed in the basal portion of the Fredericksburg well (Well C of Douglas 2010; Figure 3.11) and Vicksburg B well (Figure 3.10).

3.3.5.3 Fluvial Wadi

Fine sand in the Norphlet eolian erg deposystem presumably was moved by wind from adjacent eroding highlands, but paleotransport of coarser material may have occurred in ephemeral streams, so-called "wadis" of arid dryland systems (Glennie 1972; Figure 3.12). While several authors have suggested alluvial fans were present immediately adjacent to the eroding Appalachian highlands in onshore wells, no penetration of this facies in offshore areas has been cited. Coarse but well-rounded clasts reported in the Norphlet (e.g., Mancini *et al.* 1985) could have been transported in flash floods typical of wadi systems (Krapf *et al.* 2003). While local alluvial fans may be present, the irregular distribution challenges log and seismic-based mapping and a broad category called fluvial wadi is used here.

Log response across the fluvial wadi deposystem shows fining-upward motifs (e.g., Sake well DC 726 #2 in Figure 3.10), with scattered dips on dipmeter logs that point to smaller bedforms and common planar flat bedding. Fluvial wadi reservoirs of the Norphlet are usually thinner than eolian erg sandstones, typically 200 ft (61 m) or less. Examples include the entire Norphlet interval of the Madagascar and Sake #2 wells (Figure 3.10). Cuttings from the Sake #2 well show a dominance of reddish-colored siltstones and shales, with rare sandstones (Figure 3.6A).

3.3.5.4 Coastal Sand-Sheet/Sabkha

The seaward margin of modern erg systems usually shows a transition into marginal marine environments like evaporitic sabkhas and wave-reworked eolian deposits (Fryberger *et al.* 1983; Figure 3.11). Collectively, we refer to these as the coastal sand-sheet/sabkha deposystems of the Norphlet. Like many wave-dominated shoreline systems, planar flat bedding dominates due to the intense, near-bed shear stress associated with shoaling waves. Planar (flat to low-dip bedding) is also present in the upper part of the Denkman zone of the onshore Norphlet, overlying the large cross-bed sets of the Norphlet eolian erg deposits (Mancini *et al.* 1985). The width of the sabkha/coastal zone can be substantial, sometimes three or four GoM lease blocks wide.

The uppermost portion of the Norphlet has commonly been interpreted as marine-influenced, transitioning into the Smackover marine limestones (Pepper 1982). This unit has

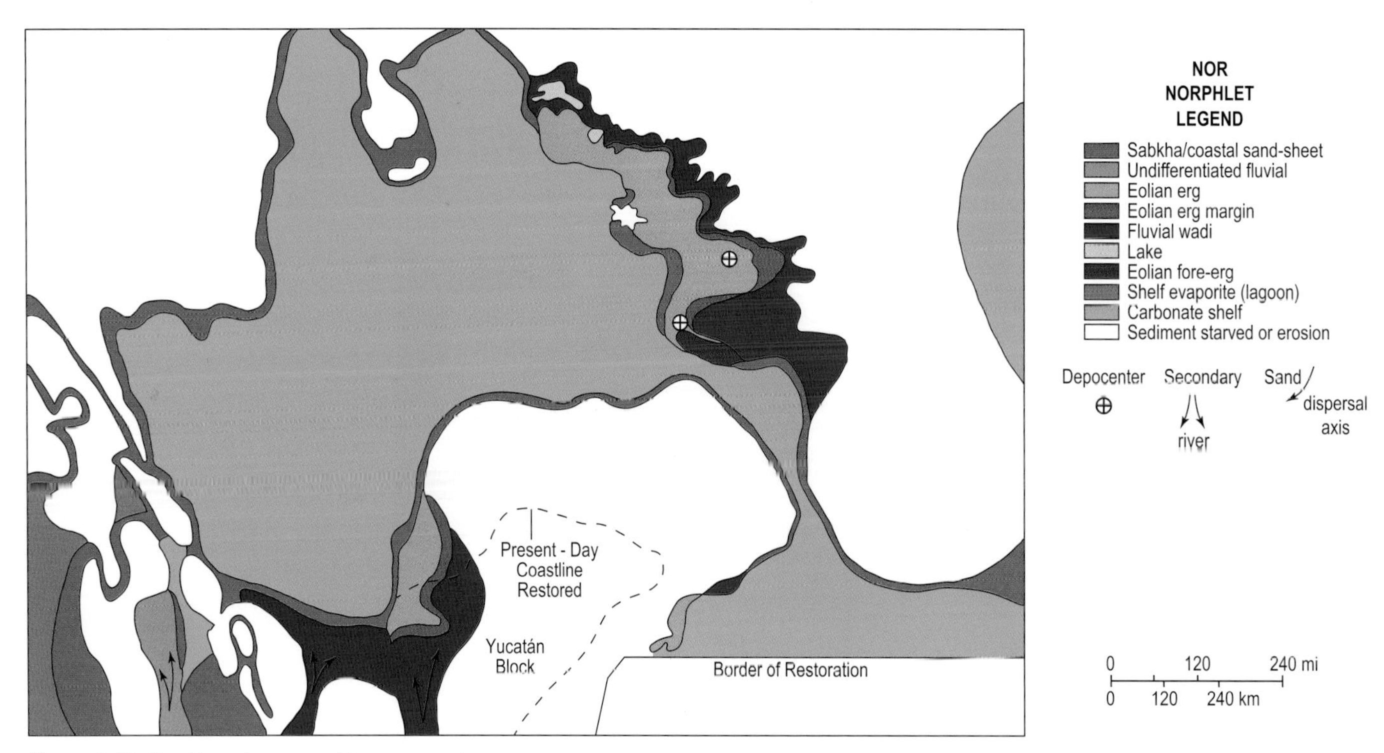

Figure 3.12 Norphlet paleogeographic reconstruction.

finer grain size and contains a higher percentage of carbonate and/or quartz cements and thus lower reservoir quality than the underlying eolian erg sandstones (Mancini *et al.* 1985). Sedimentary structures suggest marine reworking in local areas that may have modified the original orientation of the dune systems and more east–west-oriented paleo-wind directions are noted in some wells (Marzano *et al.* 1988). Log motifs are usually blocky to slightly coarsening upward (based on upward-decreasing gamma-ray) as exemplified by the MO 950 #1 well, which contains a relatively thin (<100 ft; 30 m) Norphlet, transitional to the Pine Hill Anhydrite, a sabkha evaporite deposit that itself is transitional to the underlying Louann Salt. In the deepwater wells penetrating the Norphlet, we have interpreted this unit to be locally developed between the eolian erg deposits and the Smackover Limestones or the variably developed iron-rich (mainly hematite and siderite) high-gamma ray zone (Brand 2016). Godo *et al.* (2011) describe a zone interpreted as mixed coastal sand-sheet and waterlain sabkha facies, which we include here vertically and laterally west of the erg center.

3.3.5.5 Lake

In spite of the arid climatic conditions prevailing during Norphlet deposition, lacustrine mudrocks and carbonates have been noted in both deepwater and onshore locations (Godo *et al.* 2011; Figure 3.11). Episodic flash flooding may have caused formation of temporary lakes in low-drainage areas, within and marginal to the erg. The presence of temporary lake systems such as this may explain the occurrence of large but irregular areas where Norphlet sand is not present, replaced by carbonates (Marzano *et al.* 1988). Onshore, observed log and lithology patterns are dominated by fining-upward cycles of sandstone and carbonates, transitioning into highly resistive dolomites. Net sandstone content is generally lower than erg or erg margin. Ostracods are present, as reported in BOEM biostratigraphy reports from offshore wells. From a spatial standpoint, the location of such deposits within or near the eolian erg margin is consistent with a lake system, rather than a marine incursion or coastal lagoon, the latter an alternative suggested by Marzano *et al.* (1988).

3.3.5.6 Eolian Fore-Erg

Eolian sand seas (ergs) expand and contract as a function of climatic, tectonic, or eustatic effects (Herries 1993). During erg expansion, the eolian systems advance with the dominant wind and interact with the margin of paleo-environments located downwind. The fore-erg represents the leading edge of erg progradation and is distinct from erg margin (transitional to fluvial wadi). The erg expands during dry periods while ephemeral fluvial systems transgress the erg during wetter climatic episodes (Krapf *et al.* 2003). Ergs may also migrate into marginal and marine settings, dependent on the shoreline orientation relative to the dominant wind direction. Climate-driven cyclicity, the so-called drying-upward cycle, is most apparent in this setting (Pettigrew *et al.* 2017) and such patterns have been described in the Norphlet of offshore wells (Godo *et al.* 2011).

Because of the interfingering of eolian and non-eolian facies, we would expect serrate log motifs superimposed on fining-upward log trends, a pattern we note in at least one well, as described below. We would also anticipate the development of internal barriers and baffles due to the interbedding of layers with widely varying reservoir properties.

As will be discussed in the paleogeographic map (Section 3.3.6), we believe that the Norphlet of the deepwater play fairway advanced to the southeast along the southern margin of the Middle Ground Arch, into the marine seaway separating the Yucatán and Florida blocks (Figure 3.12). Based on our reconstructions, the only wells in this "downwind" area are Swordfish (DC 843 #1) and possibly Madagascar (DC 757 #1). This trend does not bode well for extension of the play to the southeast.

3.3.6 Norphlet Paleogeographic Reconstruction

The Norphlet reconstruction (Figure 3.12) depicts a shallow-water basin surrounded by exposed Paleozoic and older continental crust. Connection of the basin to the Atlantic Ocean may have been limited to a narrow passage between the Yucatán and Florida blocks, though alternative Pacific-sourced models have been proposed (Cantú-Chapa 1998; Padilla y Sánchez and Jose 2016; see Section 3.2.4.4).

The position of the western Cuban block is hypothetical, shown as attached to the eastern position of Yucatán (Figure 3.12). As mentioned, it is possible to reinterpret the large cross-bedded sandstones of the San Cayetano Formation (roughly time-equivalent to the Norphlet of the northern GoM) as eolian. Cross-bed orientations measured by Haczewski (1976) would then rotate with our new plate tectonic model to show paleoflow from the north and northwest (rather than the northeast prior to reconstruction). The vast exposed surface of the Yucatán block probably provided the supply of loose detritus needed to nourish this presumed erg system in western Cuba (as well as the Ek-Balam area of the Salina del Istmo).

The most prominent siliciclastic deposystem in the Norphlet of the GoM is a large erg (eolian sand sea) system located on the eastern margin of the basin, stretching from the state of Mississippi to the DeSoto Canyon deepwater protraction block (Figure 3.12). As described above, the northern extent of this erg system is well-defined by onshore drilling and in maps published by Mancini *et al.* (1985) and Marzano *et al.* (1988). The southeastern extent is conjectural, based on trends in log facies in wells that appear to contain more serrate patterns and isochore thinning trends typical of the "fore-erg" facies.

The northern GoM erg system transitions paleo-landward to erg–margin and fluvial wadi systems draining the Laurentian and Gondwanan highlands, as suggested by Lovell (2013). Seaward, the erg is bordered by a coastal sand-sheet/sabkha deposystem on the margin of the shallow marine basin.

A smaller eolian depositional system, a possible erg, is located in a reentrant on the west side of Yucatán, a position defined by the Norphlet penetrations in the Ek-Balam and adjacent fields (Roca-Ramisa and Arnabar 1994; Angeles-Aquino and Cantú-Chapa 2001; Figures 3.12 and 3.13). Relatively little information has been published on the Norphlet in Mexico, onshore or offshore. In the Bay of Campeche, Oxfordian-age sandstones of the Bacab Formation are roughly time-equivalent to the Norphlet of the northern GoM (Cantú-Chapa 2009). Like the Norphlet of the northern GoM, Bacab Sandstone exhibits large-scale cross-beds (Figure 3.13B) and have been interpreted as eolian dune (Spanish "duna") deposits (Roca-Ramisa and Arnabar 1994). Published dipmeter information shows consistent dips of 25 degrees or more in multimeter thick sets with westerly to northerly wind orientations (Figure 3.13C), but post-depositional tectonic rotation complicates interpretations (Rios Lopez 1996).

The Norphlet, not to be confused with sandstones of the younger Smackover, tapers westward and has been penetrated in few Texas wells. One rare core of the Norphlet in the Cities Service Peeler Ranch #1 well shows about 5 ft (1.5 m) of sandstone, interbedded with anhydrite and dolomite, with disrupted algal laminations, mud chips, and some ripples and bioturbation (D. Budd, pers. comm.). Sedimentary structures are consistent with deposition in the sabkha/coastal sand-sheet facies as we define it here. Presumably, this deposystem extended around the GoM basin; however, there may have been some "rocky" coastlines with limited preserved shoreline deposition.

3.3.7 Paleo-wind Interpretation

Cross-bed orientations interpreted from dipmeter/FMS data indicate highly variable orientations across the northern GoM (Hunt 2013; Lisi 2013; Hunt *et al.* 2017; Figure 3.14). Parrish and Peterson's (1988) continental-scale model suggests paleo-winds feeding this erg system varied around the basin, perhaps on a seasonal basis (Figure 3.14). Lisi (2013) reported that prevailing summer winds were from the west-southwest and winter winds from the north. These bimodal wind directions may explain why the linear dunes of Mobile Bay tend to be oriented largely north–south (see Section 9.6 and Figure 9.4), aligned with the winter wind orientation, while some bedding in coastal facies wells show some westerly wind directions (Hunt *et al.* 2017). Moore *et al.* (1992) developed a numerical simulation of Jurassic paleo-winds showing strong easterly prevailing winds from the Tethyan realm.

Based on limited 2D seismic mapping, Hunt *et al.* (2017) interpreted paleo-transport in nearby fluvial wadi systems as structurally controlled, a radial pattern coming off of the Middle Ground Arch toward the west or into the DeSoto salt basin from the south (Hunt *et al.* 2017; Figure 3.14). However, one key difference between our reconstruction and that of Hunt *et al.* (2017) and Lovell and Weislogel (2010) is the larger map extent of our mapped fluvial wadi deposystem across the Middle Ground Arch (Figure 3.12). We are able to correlate a Norphlet seismic reflection over a large part of the Middle Ground Arch to a termination more or less coincident with the Norphlet truncation line as suggested by Gohrbant (2002). It should be noted that the simplified paleogeography of Hunt

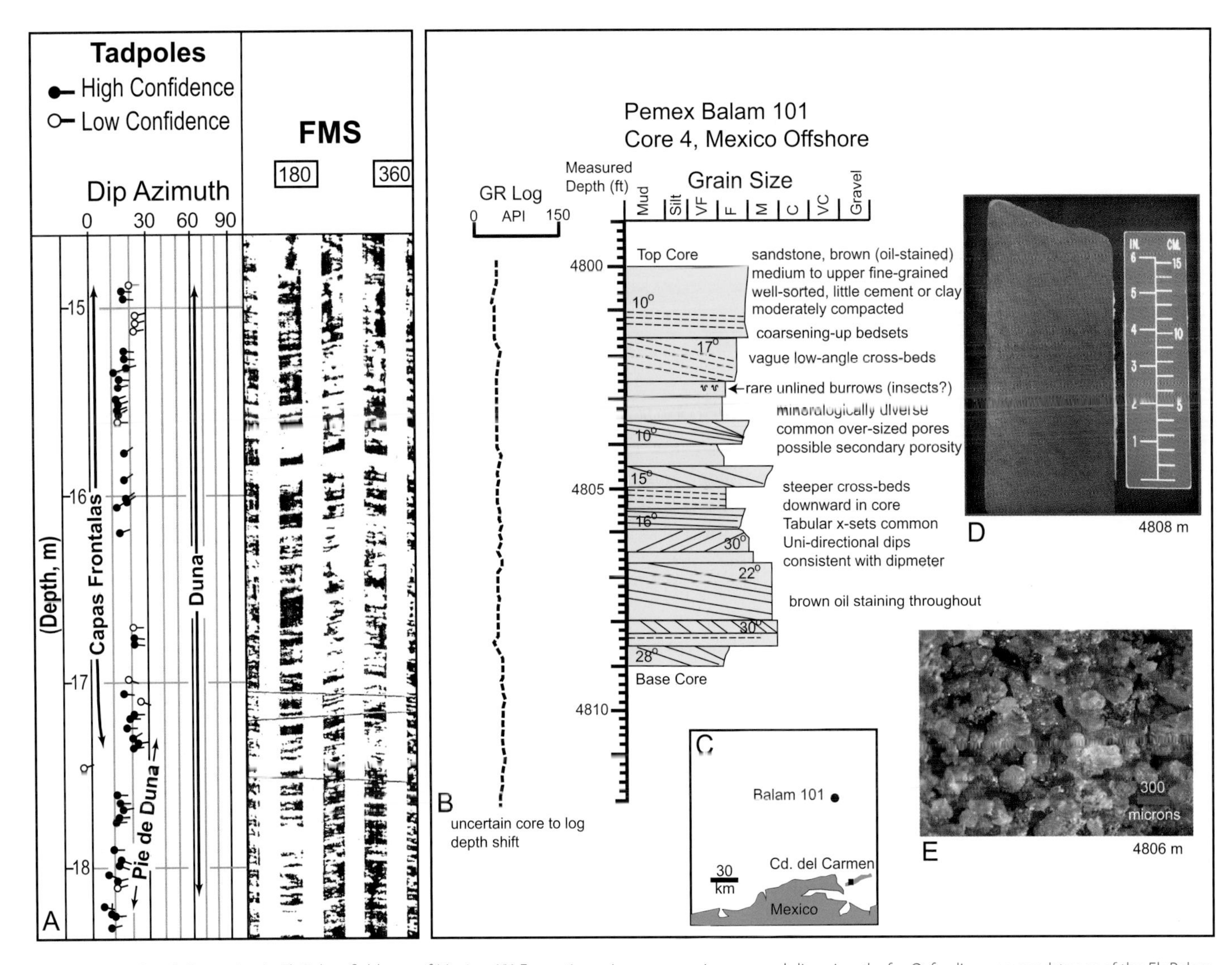

Figure 3.13 Bacab Formation in Ek-Balam field area of Mexico. (A) Formation micro-scanner images and dip azimuths for Oxfordian-age sandstones of the Ek-Balam Field, Mexico. Spanish duna = dune in English. Depths are disguised, but are in the 4000–5000 m range. Modified from Roca-Ramisa and Arnabar (1994). (B) Core description of Bacab Formation in Balam 101 well of Mexico. (C) Location of Balam 101 well. (D) Core photograph of Bacab Formation in Balam 101 well, 4808.3 m MD. (E) Magnified view of Bacab Formation Sandstone, Balam 101 4806 m MD.

et al. (2017) does not depict a fluvial wadi system large enough to support the large Norphlet erg system we have mapped (over 58,000 km^2). Sand supply is a critical element of eolian erg development (Herries 1993).

Information on eolian cross-bed orientation in the deep-water exploration wells of DeSoto Canyon and eastern Mississippi Canyon protraction areas is limited due to data restrictions. Dipmeter rose diagrams for eolian dunes vary from a dip azimuth of 180–290 degrees in the middle portion of the Appomattox well (Godo 2017) to an orientation of nearly 330 degrees in the lower portion of the same well. Douglas (2010) reported a northwesterly trend of grain flow cross-bed dips from a 50 ft (15 m) interval of Well B (Vicksburg-DC 353 #1). It is uncertain whether these dip azimuths have been corrected for post-depositional tectonics including raft rotation as discussed above. In addition, syndepositional modification of dips could have also occurred as large dune sand bodies subsided into the ductile underlying Louann Salt, as observed in Mobile Bay. Taking these dip azimuths at face value, however, points to a local trend of eolian paleotransport to the northwest.

We would expect some variation in wind direction due to the influence of the topographically prominent Middle Ground Arch (Figure 3.14). The northerly winter winds of Parrish and Peterson's (1988) model may have reoriented around the arch to flow more to the south (Figure 3.14). Interaction of these with Tethyan westerly winds south of the Middle Ground Arch is also possible, producing a net convergence of eolian transport in the center of the erg near the Appomattox discovery, where we mapped multiple depositional thicks (Figure 3.9). Smaller-scale structural reentrants may have locally altered dune orientations. Lisi (2013)

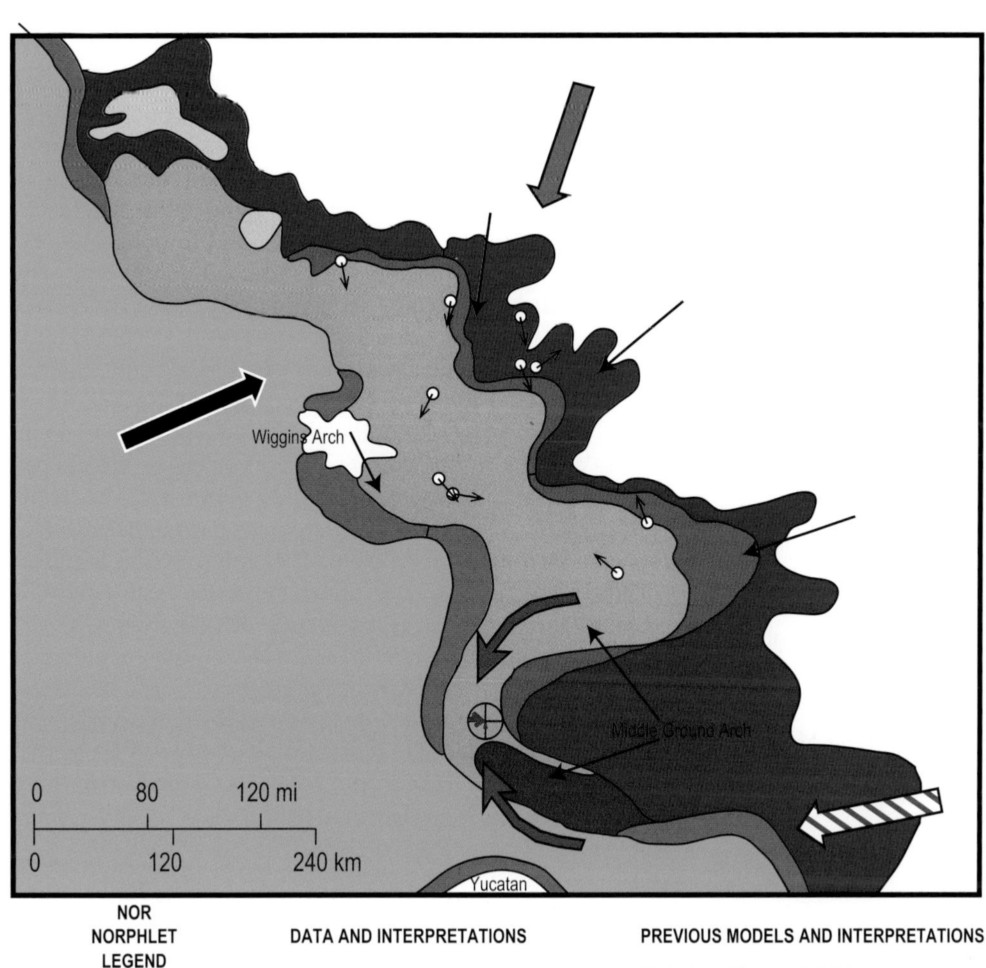

Figure 3.14 Interpreted paleo-winds driving Norphlet eolian sedimentation trends in the Eastern GoM, superimposed on the Norphlet paleogeographic and isochore reconstruction. Paleo-wind vectors partly from Parrish and Peterson (1988), Lisi (2013), Lovell (2013), and Hunt *et al.* (2017).

suggested that the paleo-Apalachicola Embayment may have channeled winter winds and forced a more west-northwest orientation as interpreted from dipmeter data in Hunt (2013).

This paleo-wind model does explain the trend of decreasing net sandstone south of the Middle Ground Arch as documented in well penetrations such as Madagascar (DC 757 #1) and Swordfish (DC 843 #1; Figure 3.10). Alternatively, this area may have been part of or adjacent to the initial "fore-erg" or downwind area of the central erg prior to a shift to more Tethyan-dominated southeasterly wind flows that were preserved prior to the Smackover transgression. Reworking by the Tethyan winds have robbed the fore-erg of loose sand that subsequently accumulated in the thickest part of the Norphlet near Appomattox.

Overall, our paleo-wind reconstruction (Figure 3.14) is consistent with provenance information published by Lovell (2013), Lisi (2013), and Hunt (2013) hypothesizing sand supply from the north (Laurentia) and east (Gondwana). Convergence of the two eolian wind systems may explain why the isochore thick or depocenter appears to be located on Destin Dome and DeSoto Canyon protraction block (Figure 3.9). Sources of loose sand would include the fluvial wadi systems flowing radially from the Middle Ground Arch, as suggested by Hunt *et al.* (2017).

These fluvial wadi deposystems may have also transported more volcaniclastics than Norphlet wells have documented in the Conecuh Embayment. Heatherington and Mueller (2003) analyzed rhyolites in Florida that apparently overlap in age with the Norphlet. Evidence of volcanics in the Raptor well (DC 535 #1) in a reconstructed position closer to the arch are consistent with that interpretation. Erosion and transport of igneous-sourced sands is quite possible.

In the Ek-Balam field area of Mexico, dipmeter interpretation suggests prevailing paleo-winds were from the west (Ramisa-Roca and Arnabar 1994; Figure 3.13A) and north (Rios Lopez 1996). These winds may have reworked fluvial wadi deposits derived from the exposed Yucatán Platform area into north–south-oriented transverse dunes, concentrating the eolian deposits in a relatively small embayment west of the Yucatán Platform where the Ek-Balam field complex is located (Figure 3.12). Preliminary detrital zircon provenance work indicates that the Mayan (Yucatán) block is the primary original source terrane for the Oxfordian sandstones in the Balam 101 well (Snedden and Stockli 2019). Petrographic work indicates quartz content in the Oxfordian sandstones is quite high (>90%; Rios Lopez 1996), likely due to derivation from this older granitic basement terrane at Yucatán.

3.3.8 Smackover Paleogeographic Reconstruction

The Smackover (middle to upper Oxfordian) reconstruction (Figure 3.15) shows a roughly similar basin configuration to the Norphlet, with a few notable exceptions. After marine transgression, the large eolian erg of the Norphlet is replaced by a wide, low-angle ramp transitioning from shore zone clastics, inner ramp carbonates, and shelf to outer-platform carbonates and mudstones. A bedload-dominated river system, which apparently fed a wide siliciclastic shore zone system, is present along the present-day Mississippi–Arkansas state boundary. Here, a gross sand-prone interval of >600 ft (183 m) is present, suggestive of a major fluvial input, likely a river draining the paleo-Appalachian highlands. We refer to this fluvial system as the paleo-Mississippi, as it is roughly oriented along the present-day Mississippi Embayment. We differentiate this river from paleo-fluvial systems operating to the southeast, south of the Southern Platform (see Figure 1.3), for which we use the term "Grenville" river as these sandstones are largely derived from Grenville basement source terrane. Weislogel *et al.* (2016) noted a marked decrease in content of Grenville-age zircon (1.0 Ga) and increase in Gondwanan-age zircon (525–680 Ma) in the underlying Norphlet of the Conecuh Embayment, in contrast to the Norphlet north of the Southern Platform/Middle Ground Arch, which is dominated by Grenville affinity zircons.

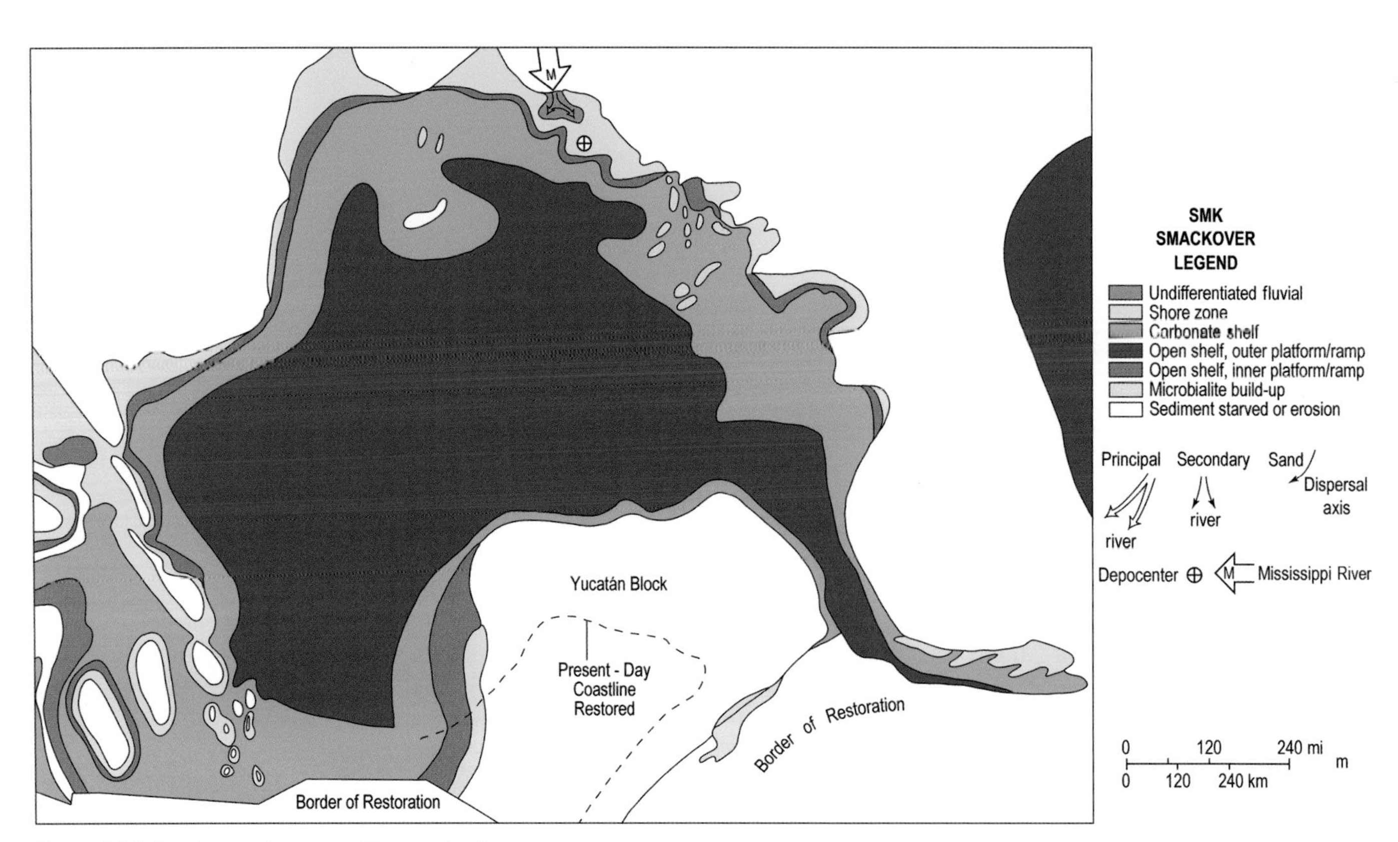

Figure 3.15 Smackover paleogeographic reconstruction.

Presence of a major point source for siliciclastics through this paleo-Mississippi system is further confirmed by oil geochemistry from Oxfordian source rocks. Reservoir oils and seeps yield with a low sulfur content, an indication of siliciclastic influence (see Section 9.8.1). A possible secondary river input coming from the Ouachita Mountains may have been present in northeast Texas, though well control is limited. Small pockets of shore zone siliciclastics are distributed along the Smackover seaway, which is dominated by carbonates.

Important Smackover carbonate deposystems include a wide belt of shelf grain shoals extending from Texas to Arkansas, inner-shelf carbonates rimming basement highs in Mexico, and microbialite build-ups in the eastern GoM onshore. High-energy oolite-dominated shoal water complexes are common in the Upper Smackover, including the main reservoirs of the large Jay Field of Florida (Figure 3.16). Even on the generally low-resolution spontaneous potential (SP) curve, the upward-shoaling pattern common to shelf grain shoal deposits is apparent in the Upper Smackover Limestone.

Microbial build-ups are also important carbonate reservoirs in the Smackover, often nucleating on basement highs in Mississippi and Alabama (Mancini *et al.* 2004). Microbialites and thrombolites represent carbonate deposition generated by microbial activity under favorable marine conditions (Kennard and James 1986). Microbialite build-ups are important producing reservoirs in southwest Alabama, found in fields such as Appleton (Figure 3.17), often positioned upon Paleozoic basement highs (Mancini *et al.* 2006).

Further offshore, seismic-scale clinoforms are identifiable in the basal Smackover, apparently downlapping upon the Norphlet Sandstone (Figure 3.5). Progradation of the Lower Smackover likely occurred after initial marine flooding of the Norphlet eolian erg and margins. The low-angle clinoforms imply that water depths were not large (<100 m) at this point. This is evident from a visual comparison of these features to large-scale, higher-angle platform margin clinoforms of later Jurassic supersequences (CVK, CVB, HVB; Figure 3.5).

In Mexico, the Smackover-equivalent Zuloaga Limestone (Figure 3.2) contains common *Gryphaea* sp. clam shells (Sierra de Minas Viejas area), tidal laminites (Sierra de Bunuelos), and ooid grainstone and packstones (Sierra de Enfrente). The Zuloaga Formation is thought to have formed on a broad carbonate ramp extending south from the emergent Coahuila Peninsula (Finneran 1984). However, the earlier reconstructions of Finneran (1984) and Salvador (1991b) may have failed to take into account the large amount of post-Smackover lateral displacement of Mexican basement blocks required to move Yucatán into its present position, essentially linking facies which were never in close proximity. One trend that is evident in both Mexico and the USA is the observed backstep of the Smackover depositional systems upward, which is thought to reflect progressive introduction of marine seawater from the east (Martini and Ortega-Gutiérrez 2016).

Shoreline siliciclastics are identified in Cuba (Barros 1987) and possibly rimming the southern rim of the Florida block (Salvador 1991b). In the Bay of Campeche, the upper portion of the Oxfordian Ek-Balam Group contains shallow-water carbonates (here designated as carbonate shelf), although local anhydrites are present in the adjacent inner-shelf, suggesting local hypersalinity during extremely arid periods (Angeles-Aquino and Cantú-Chapa 2001).

Connections to the Atlantic Ocean may have become more extensive during Smackover (late Oxfordian) time. The southeastern connection, partially blocked by fluvial and eolian sedimentation during Norphlet (early Oxfordian) time, may have allowed more open marine conditions to prevail in the latest Oxfordian.

In fact, the evolution of the GoM basin from hypersaline to a normal marine salinity is recorded within the Smackover succession. Godo (2017) noted that the basal Smackover Limestone in the deepwater areas of Mississippi Canyon and DeSoto Canyon contains only algae adapted to hypersaline conditions. Normal marine forms appear only in the Upper Smackover. This suggests that flooded basin developed over the Louann Salt deposits evolved slowly from hypersaline to normal marine conditions, perhaps over a 10 million year transition.

3.4 Haynesville–Buckner Supersequence

The Haynesville–Buckner (HVB) supersequence, as we define it, is stratigraphically positioned within the Middle Mesozoic Drift to Cooling Phase. It is marked by much thicker intervals of siliciclastics and carbonates than the Smackover–Norphlet supersequence, suggestive of higher crustal subsidence rates and an open, normal marine basin (Figure 3.18). Compacted thicknesses often exceed 4000 ft (1220 m) in the onshore of East Texas, North Louisiana, Mississippi, and DeSoto salt basins. By contrast, maximum interval thickness within the Smackover and Norphlet rarely exceeds 1000 ft (305 m).

The HVB also represents the initiation and development of a prominent carbonate platform margin in the northeastern GoM, locally referred to as the Gilmer Limestone (Figure 3.19). Seismic lines in the DeSoto Canyon to Viosca Knoll protraction blocks show strata with pronounced, rather steep clinoforms transitioning paleo-landward to seismically dim signatures (Figure 3.19). Such seismic geometries are characteristic of carbonate shelf margins (Phelps 2011), described here as the platform margin deposystem, which includes rimmed shelf reef and forereef deposition (see online poster, The Gulf of Mexico Mesozoic Log Facies Interpretation Poster at www.cambridge.org/gomsb). This transition from the Smackover microbialite-dominated ramp carbonates to a Gilmer Platform margin reefal framework represents a significant change in the basin history. Development of the first platform margin reefs in the GoM at this time may be due to three factors: (1) bathymetric deepening of the Gulf basin, with higher subsidence rates; (2) continued influx of marine waters from the Atlantic (Tethyan) Ocean, which presumably provided nutrients that promoted reef growth; and (3) alignment with the maximum fetch of the newly opened basin, with reefs providing a bulwark against wave action. The HVB platform

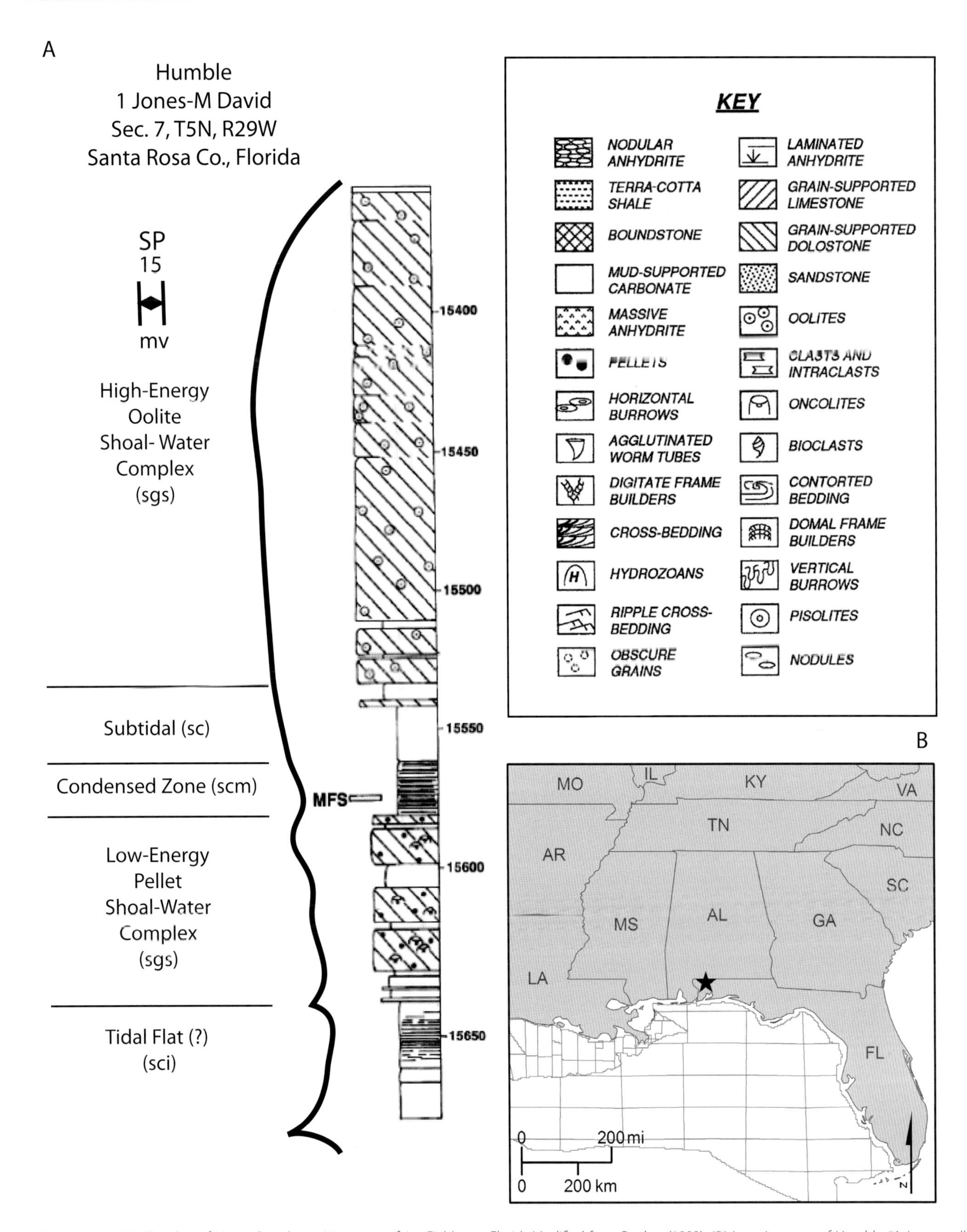

Figure 3.16 (A) Core log of Upper Smackover Limestone of Jay Field area, Florida.Modified from Prather (1992). (B) Location map of Humble #1 Jones well.

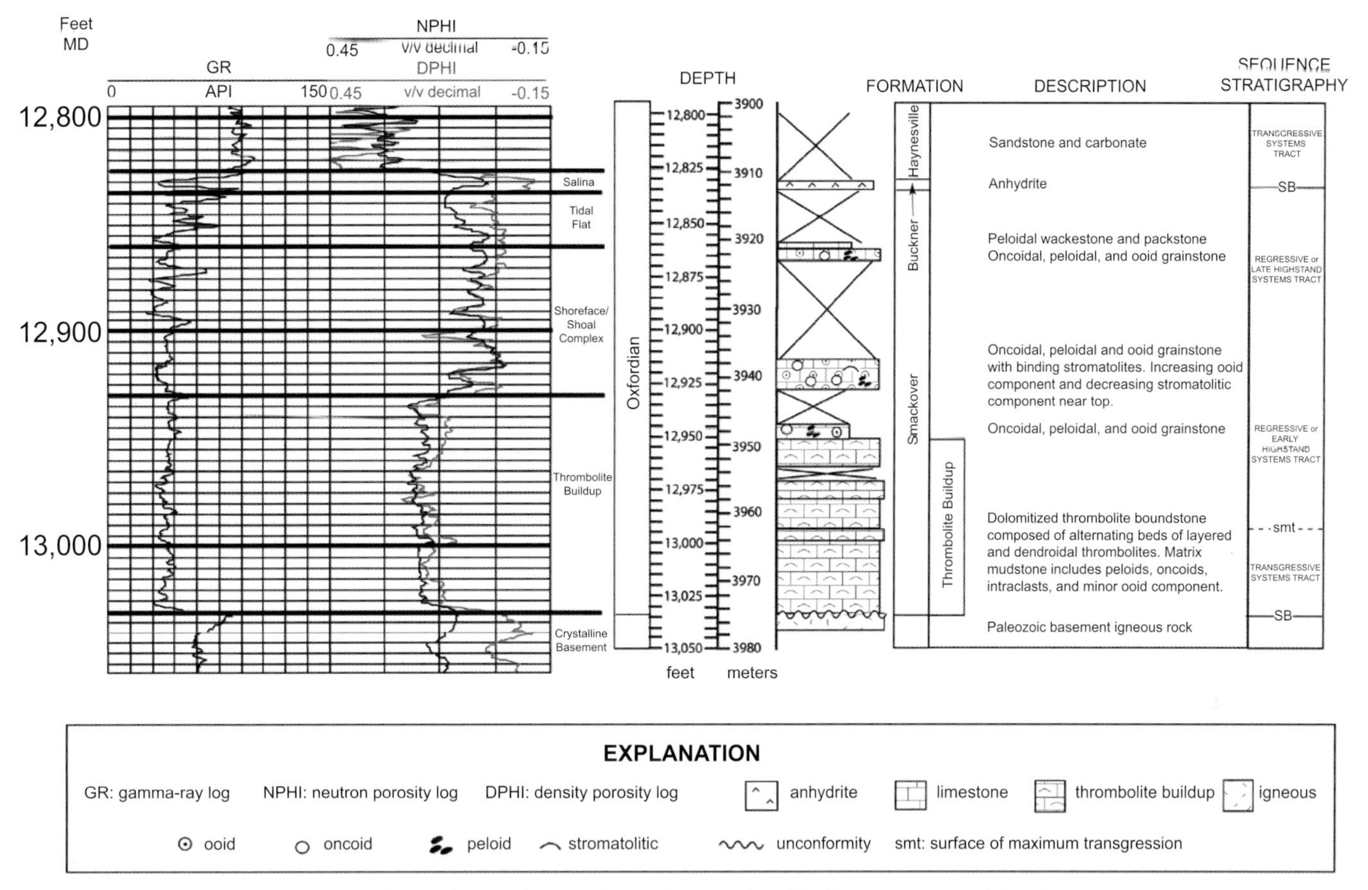

Figure 3.17 Thrombolite (microbialite) build-up in the Smackover of the Appleton Field.Modified from Mancini *et al.* (2004).

margin clinoforms are relatively low-angle compared to the overlying Cotton Valley–Bossier, reflecting a long-term trend in reef development (Figure 3.19).

Late Jurassic reefs are well-documented in Europe and the Middle East, beginning as early as the Oxfordian (Wilson 1975), so it is possible that conditions for reef development in the GoM basin were not suitable until the Kimmeridgian for the reasons stated above, but also due to a change in climatic conditions (less arid) and continued evolution of basin water toward more normal marine salinity. The local development of Buckner Anhydrites within the HVB supersequence suggests temporary climatic regressions back toward aridity or restriction leading to hypersalinity.

The HVB supersequence also contains significant volumes of sandstone, likely sourced from the Appalachian Mountains (Essex *et al.* 2016). Wells drilled onshore and on the eastern shelf area of the GoM penetrated 1000 ft (305 m) or more of sandstone (Petty 2008). However, apparently very little of the fluvial to deltaic sand made its way past the shelf edge into the slope and basin, likely due to reef blocking (see Box 3.1).

3.4.1 Chronostratigraphy

Deposition of the HVB supersequence extends from 156.86 Ma to 153.86 Ma, Oxfordian-8 sequence boundary to Kimmeridgian-4 sequence boundary (Olson *et al.* 2015). This is a duration of three million years (timescale of Gradstein *et al.* 2012), relatively short in comparison to the underlying Smackover–Norphlet supersequence (approximately five million years) and overlying CVB supersequence (12 million years). At least two key fossil horizons can be identified within the HVB, and we expect more to be defined as drilling information is released.

3.4.2 Previous Work

Much has been recently published about the HVB, primarily from onshore areas. The Sabine Platform area of east Texas and western Louisiana has received the most attention, as Haynesville Shale gas play enjoyed a brief but robust period of exploration centered on the Texas–Louisiana border counties/parishes of Panola and DeSoto (Hammes *et al.* 2011; Wang *et al.* 2013). The HVB supersequence here is estimated to contain significant onshore gas resources, with the Sabine Platform area alone estimated to harbor at least 60 TCFG of undiscovered reserves (Dubiel *et al.* 2012). However, interest in the Haynesville Shale gas play waned as natural gas prices dropped in 2012–2013, and a recovery has yet to emerge at the time of this writing.

Offshore, the HVB supersequence has yet to produce significant volumes (Petty 2008). Many of the Norphlet

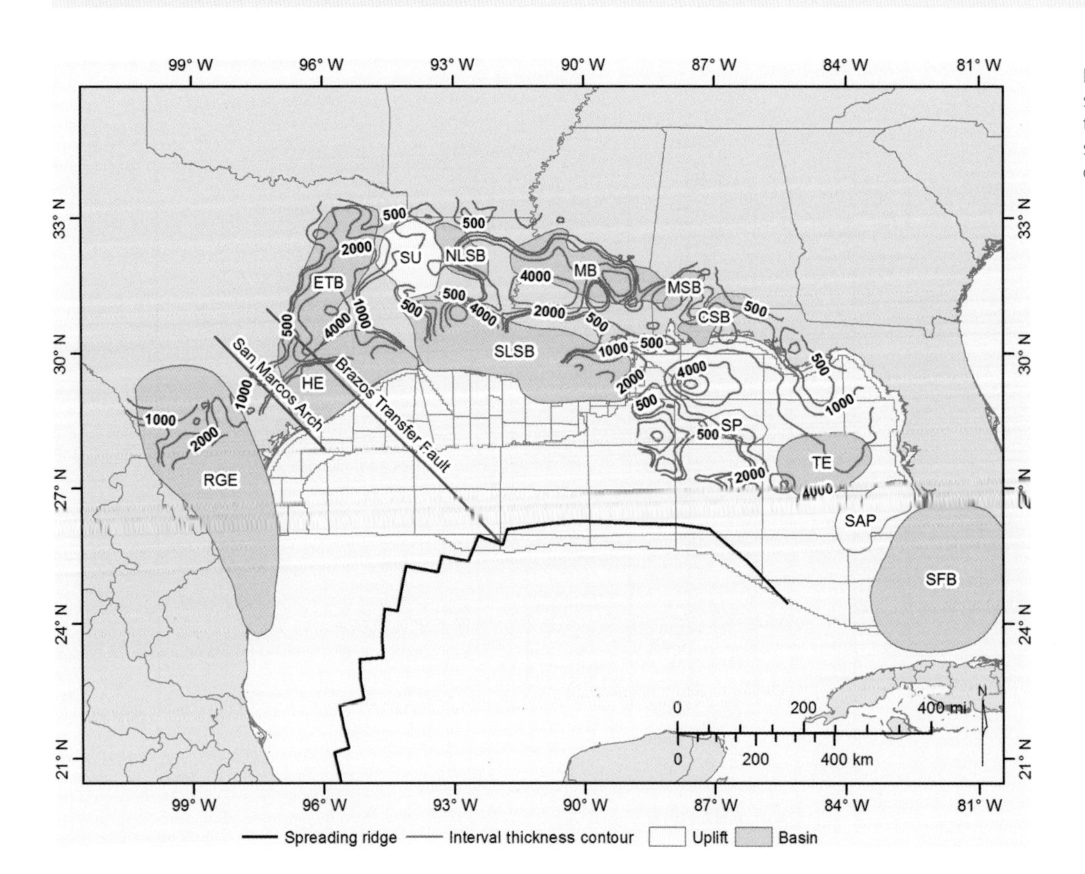

Figure 3.18 Haynesville–Buckner supersequence isochore map with key tectonic features. Abbreviations for structural features are explained in the caption to Figure 1.3.

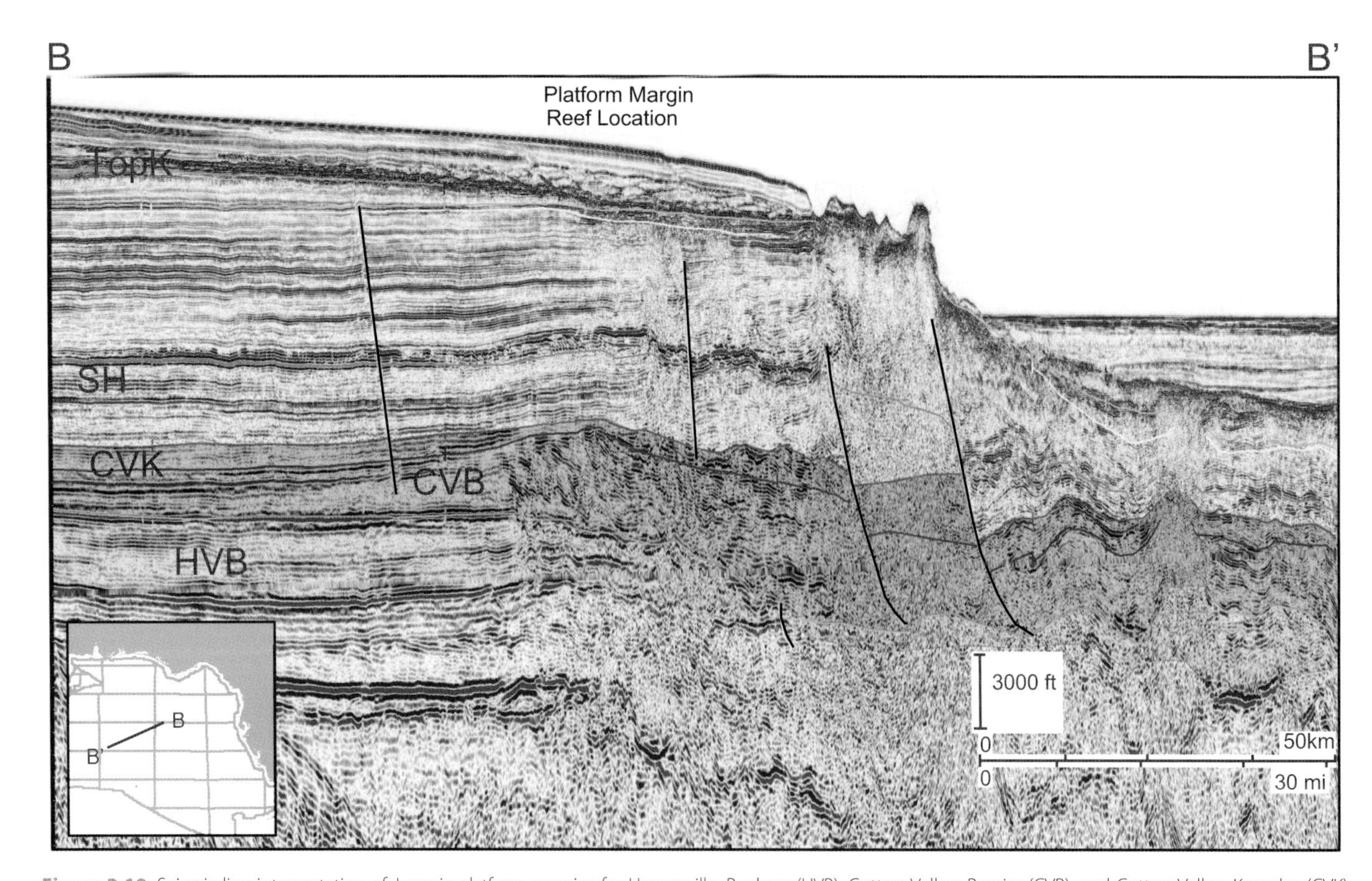

Figure 3.19 Seismic line interpretation of Jurassic platform margins for Haynesville–Buckner (HVB), Cotton Valley–Bossier (CVB), and Cotton Valley–Knowles (CVK) supersequences. Modified from Cunningham *et al.* (2016).Seismic line courtesy of Spectrum.

discoveries in the deepwater DeSoto Canyon and eastern Mississippi Canyon areas are set up by rafting apart of the Jurassic intervals over salt, with significant gaps where there is no Norphlet or Smackover present (Pilcher *et al.* 2014). Haynesville and younger units often fill in the gaps between rafted blocks and can act as seal rocks, given the generally low sandstone content, and low porosity and permeability.

However, even in this limited, data-rich area some differences in interpretation have emerged. Anderson (1979) used older well data to define broad east–west deposystem trends that showed little relation to local tectonic features like the Sabine Island and Strickland highs, which later drilling indicated were an influence on deposition (Cicero *et al.* 2010; Hammes *et al.* 2016).

From detailed core description and well correlation, Hammes *et al.* (2011) determined that the Haynesville Shale was deposited in deep, partly euxinic and anoxic basin surrounded by carbonate shelves. Earlier, Hammes *et al.* (2016) had suggested this same area was a lagoon flanked by shallow grainstone shoals and the basement-cored Sabine Island complex (terminology of Nicholas and Waddell 1989).

In the eastern GoM, Dobson (1990) used older industry seismic and well control to map the Haynesville in the Tampa Embayment, the shallow shelfal area, where she recognized a carbonate shelf margin bounding updip deltaic siliciclastics. MacRae (1994) also observed shelf-margin reefs, though located further landward than Dobson (1990). Petty (2008) constructed a paleogeographic map of the Haynesville across the Destin Dome and Viosca Knoll protraction blocks and adjacent shallow-water areas, showing shore zone and delta plain deposits.

The HVB supersequence of the northern GoM includes evaporites, known as the Buckner Formation (Forgotson and Forgotson 1976), that occupies a similar stratigraphic position and lithology with the upper portion of the Olvido Formation in northern Mexico outcrops (Humphrey and Diaz 2003; Figure 3.2). Offshore, in the Campeche area, Angeles-Aquino and Cantú-Chapa (2001) used well-based biostratigraphic datums to define a Kimmeridgian-age unit of dolomitic limestone, siltstones, and bentonitic shales that is called the Akimpech Formation.

Oolitic carbonate banks and shoals are the primary reservoir facies of the coeval Kimmeridgian (Jurassic Superior) interval of Mexico (Figure 3.20; Treviño García 2012). Grain shoals were preferentially developed over elevated bathymetry related to salt pillows and fault blocks (Chernikoff *et al.* 2006). As described further in Section 9.7, enhanced porosity development occurs in the oolite bank crest facies (Treviño García 2012). In this way, the Kimmeridgian ooid banks of Mexico are analogous to the well-known grainstone shoals of the Arabian Shield and Jura Mountains of Europe (Wilson 1975).

In Cuba, Barros (1987) used mainly outcrops to develop a succession of paleogeographic maps showing largely a coeval carbonate-dominated shelf to outer-platform margin.

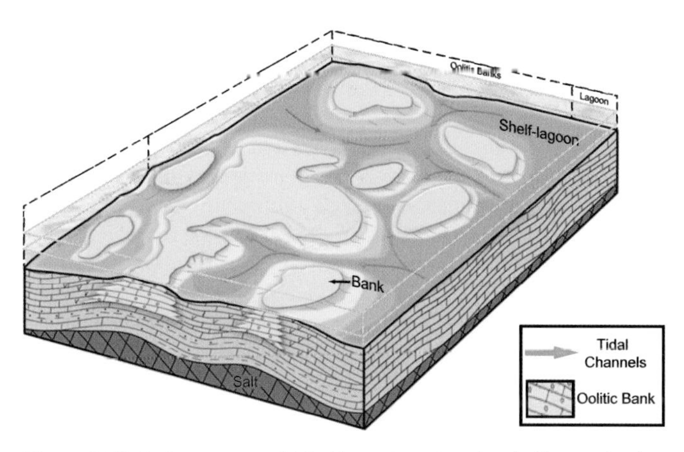

Figure 3.20 Sedimentary model for Upper Jurassic oolite shelf grain shoals in Pilar de Akal contractional zone of Mexico.Modified from Treviño García (2012).

Box 3.1 What is Reef Blocking? Limits on Sandstone Entry into the Deep Basin

The rise of carbonate platform margins (rimmed shelf reefs and forereefs) after the Oxfordian may explain why siliciclastics derived from updip fluvial and deltaic systems are generally rare in coeval deepwater strata. A present-day analog in northeast Australia, where deposition is mixed carbonate–siliciclastic at the Great Barrier Reef (GBR) margin, allows us to gain insight into the processes that contribute to siliciclastic deposition in the deepwater in a reef-rimmed basin. Puga-Bernabéu *et al.* (2013) analyzed high-resolution multibeam bathymetry and side-scan sonar data to characterize submarine canyons and interpret depositional processes at the GBR margin.

Two types of submarine canyons are observed at the modern GBR margin: shelf-incised canyons and slope-confined canyons (Puga-Bernabéu *et al.* 2013). The heads of the shelf-incised canyons are incised into the shelf break at water depths of 60–80 m (197–262 ft) and are either reef-blocked, partially reef-blocked, or shelf-connected (Figure 3.21). Puga-Bernabéu *et al.* (2013) noted deeper

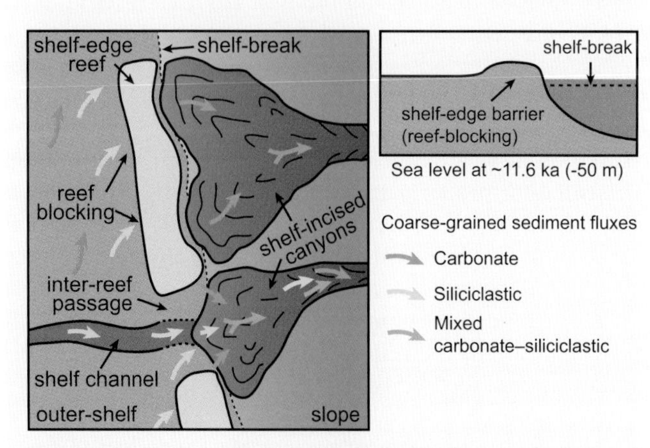

Figure 3.21 Example of modern Great Barrier Reef control on slope and basin sediment fluxes. Modified from Puga-Bernabéu *et al.* (2013).

Box 3.1 *(cont.)*

oceanographic waters on the slope of slope-confined canyons. During the growth of an extensive platform margin reef, the reef location and morphology control the amount and type of sediment supplied to the basin. Puga-Bernabéu *et al.* (2013) concluded from examination of slope drop cores that submarine canyons entirely blocked by reefs receive only the carbonate reef material transported into the canyon. However, partially reef-blocked submarine canyons can receive mixed carbonate and siliciclastic input. Canyons that link up with and connect with shelf channels are often dominated by siliciclastic input and basinal deposits can have a large component of sand (Figure 3.21). This pattern of reef blocking by large platform margin reefs continued from the Kimmeridgian Haynesville until the demise of the Albian platforms in the northern GoM. Thus, relatively rare basin entry points are the only areas where significant slope and basinal sandstone accumulations would be expected.

3.4.3 Haynesville–Buckner Paleogeographic Reconstruction

Like the Norphlet and Smackover, the HVB supersequence paleogeographic reconstruction is complex. Reconstruction includes restoring the wells to the pre-sea floor spreading positions and then accounting for salt rafting. In addition, HVB-equivalent stratigraphic data from Cuba and Mexico (onshore and offshore) are included in order to provide a basin-scale perspective.

3.4.3.1 Plate Tectonic Reconstruction

Plate reconstruction for the three-million-year duration of the HVB supersequence follows the same process as with the underlying Smackover–Norphlet supersequence (Section 3.3). Current models suggest that the main phase of sea floor spreading ended as late as the Berriasian (139–138 Ma) in the central Gulf (Hudec *et al.* 2013a; Snedden *et al.* 2013; Figure 2.1). As mentioned, sea floor spreading drove the counterclockwise rotation of the Yucatán block away from North America (Kneller and Johnson 2011; Christeson *et al.* 2014; Figure 2.1).

3.4.3.2 Restoration for Raft Tectonics

As described with the Smackover–Norphlet supersequence, restoration of well locations back to their pre-rafted position is essential for proper paleogeographic mapping. The process used for the HVB is effectively the same, with the exception that the unit was deposited *during* rafting, as observed with the raft gap infill patterns, so that only a partial restoration of the raft gap is needed. Determining what percentage of total raft distance to use in the restoration requires some type of calibration, as described below.

Our approach is to use three wells in the DeSoto Canyon area that penetrate platform margin reef, and shelf-to-basin ramp deposystems for such a calibration; Appomattox (MC 392 #1 ST2BP1), Fredericksburg (DC486 #1), and the Raptor original hole (DC 535 #1), respectively (Figure 3.22). The HVB platform margin reef extent was mapped on 2D seismic data from onshore and offshore shelfal areas into the raft province itself. Variations in percentage of total rafting from 0 to 100 percent were considered to ascertain which estimate adequately returned the near-reef (forereef) depofacies wells to the correct position for a continuous reef trend. The results of this method suggest that about 80 percent of the total rafting occurred after deposition of the HVB, comparing favorably with the percentage of post-HVB temporal duration (82 percent) between the end of the start and end of rafting, a time span of 32 million years. This match suggests that the rafting process was more or less continuous, with an almost "creep-like" rate rather than episodic translation downslope toward cooled and subsided oceanic crust. This value of 80 percent was applied to all rafted wells in the restoration to pre-rafted positions shown in the HVB paleogeographic map.

3.4.4 Discussion

The paleogeographic map for the HVB (Figure 3.23), restored to a pre-spreading, pre-rafted state shows a smaller basin than present-day GoM, but somewhat larger than that for the Smackover–Norphlet supersequence (Figure 3.15). The connection to the world ocean through the paleo-Florida Straits is narrower than present day, but local changes in sill depth and

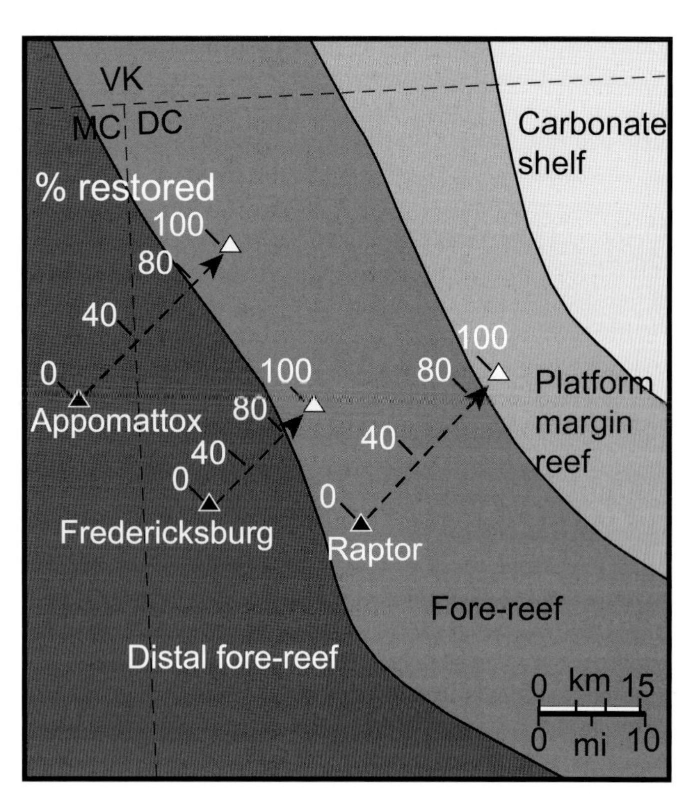

Figure 3.22 Rafting reconstruction using Appomattox, Fredericksburg, and Raptor OH wells to restore HVB near-reef depofacies into the correct paleogeographic position, prior to rafting. Percentages indicate the amount of rafting used in restoration. Black triangles indicate original present-day locations, white triangles show pre-rafted positions.

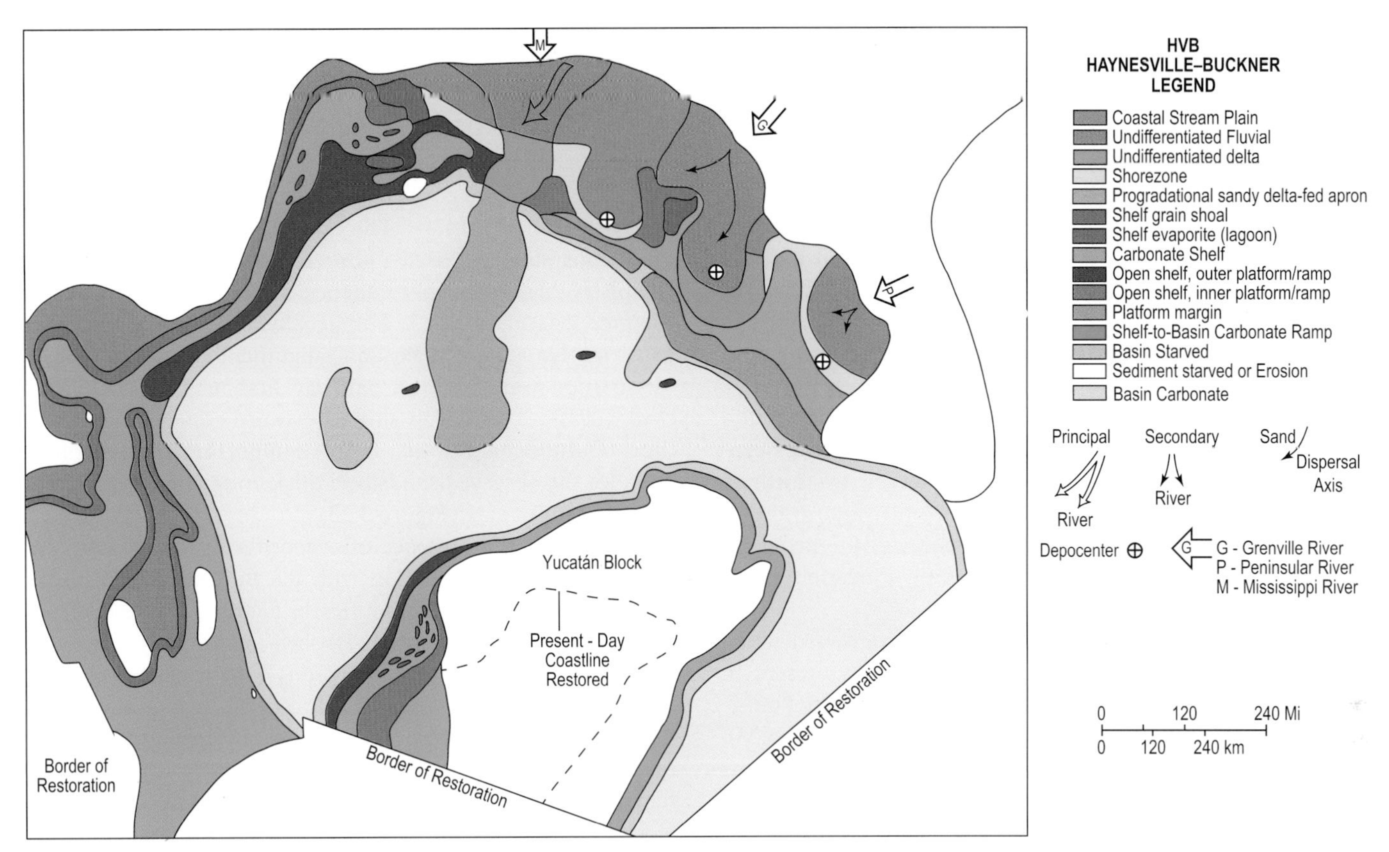

Figure 3.23 Haynesville–Buckner supersequence paleogeographic reconstruction.

the effect on bottom water circulation probably affected source rock deposition as well.

The eastern GoM is dominated by a platform margin reef system mapped with 2D seismic data as extending from the Vernon protraction block to onshore eastern Louisiana. This represents initiation of the framework reef system that continued into the Paluxy–Washita (Albian) supersequence of the Cretaceous. The eastern termination is thought to be on the margin of the Sarasota Arch; however, seismic data in this area is relatively poor quality due to attenuation below the Top Cretaceous seismic surface. The western extent is constrained by seismic and well control. In onshore areas, the platform margin reef system is called the "Gilmer Reef," a lithostratigraphic designation (Forgotson and Forgotson 1976). Deposition of the Buckner Anhydrite is thought to be in an evaporitic depression behind the reef system (Moore 1984).

An expanded zone of distal forereef (shelf-to-basin carbonate ramp) is illustrated in the present-day DeSoto and Mississippi Canyon protractions blocks and is worth explanation (Figure 3.23). The Appomattox well (MC392 #1) contains 620 ft (189 m) true vertical thickness (TVT) of HVB strata with a gamma-ray pattern more typical of forereef than distal forereef or carbonate-rich basin floor. The nearby Fredericksburg well (DC 486 #1) contains less overall HVB interval (500 ft; 152 m TVT), but a similar log motif and carbonate lithology on the mud log. However, these wells are 30–32 km (19–20 miles) from the present-day location of the HVB shelf edge, so must have been rafted basinward at least that same distance. This implies that these wells were, prior to rafting, in close proximity to Raptor (DC 535 #1) and Shiloh (DC 269 #1) wells, which contain platform margin reef facies based on log motif. The log response suggests limited porosity in the HVB at these wells, which improves top seal quality of the underlying Smackover and Norphlet reservoirs in the raft blocks.

Seismic and well-based mapping of the HVB supersequence shows two important depocenters located updip of the aforementioned carbonate platform margin (Figure 3.23). In southern Alabama, about 1000 ft (330 m) of sandstone-prone fluvial sediment was deposited behind the platform margin. A similarly thick coastal plain siliciclastic depocenter is centered in southeastern Mississippi, again behind the coeval platform margin carbonate deposystem. We refer to the former depocenter as linked to the paleo-Mississippi system and the latter depocenter as related to the Grenville River. Essex *et al.* (2016) report detrital zircon populations in southern Alabama Conecuh Embayment wells, with a dominance of Laurentian Grenville Province and Appalachian-sourced zircons, mixed with Gondwanan Suwannee terrane grains. A smaller depocenter in Florida is appropriately termed the Peninsular River as it is in close proximity to the emergent Peninsular Arch.

The western drainage system, the paleo-Mississippi system, has a fluvial-dominated delta and flanking shore zone systems.

A shelf-margin delta system is hypothesized, based partly on the published maps of Cicero *et al.* (2010), and is depicted as extending to the western termination of the rimmed shelf reef system.

These three depocenters represent a significant volume of siliciclastics trapped behind this platform margin reef system. The concept of "reef blocking," based on modern analogs like the GBR of Australia, is an important concept for the Mesozoic of the northern GoM (see Box 3.1). Simply put, framework reef systems like this are effective barriers to the source-to-sink transfer of quartz-rich sediment into slope and deepwater paleo-environments. Interreef passages, where slope channels can connect to shelf channels and deltas, are relatively small in size and number, limiting deep-water sand systems. This pattern is observed to continue until the Ceno-Turonian, when local tectonics and drainage catchment expansion allowed siliciclastics to prograde beyond the shelf edge for the first time (Snedden *et al.* 2016b).

Regional mapping suggests that further westward along the paleo-shoreline, a mud-prone shelf basin was present in the Sabine Platform area of east Texas and western Louisiana, the center of the Haynesville Shale gas play (Figure 3.23). The notable "gap" (white polygon indicating non-deposition) on the state line is the Sabine Island complex, an important basement-cored high that acted somewhat like a buttress or margin to the mud-prone basin. This reconstruction argues for a shallow shelf basin where organic matter was preserved, versus a deep marine basin as implied by Hammes *et al.* (2011). The actual bathymetric shelf margin may be to the south, on trend with Sabine Island, as suggested by Hammes *et al.* (2016). However, we do depart from Hammes *et al.* (2016) in our interpretation of the paleo-environment of the "Gray Sand" of Louisiana (Terryville Field) as shore zone to shelfal siliciclastics rather than submarine fans. Core descriptions and photographs in Judice and Mazzullo (1982) contain either equivocal paleo-environmental indicators (cross-beds, ripples) or diverse trace fossil assemblages more consistent with shelfal deposition (see the Williamette 31-1 well).

Further west toward the Texas and Louisiana border, the HVB supersequence was deposited on a carbonate ramp, similar to that of the underlying Smackover (Ahr 1973). Platform margin build-ups are replaced by local grainstone banks and shoals. The separating siliciclastic wedge is discussed below. The Sabine Uplift does not become structurally prominent until the Late Cretaceous, but the east Texas salt basin played a role in influencing the location of these grainstone shoals and banks. The term "Gilmer" is also used to refer to this belt of high-energy carbonate shoals on the west flank of the still-submerged but structurally stable Sabine Uplift (Ahr 1981). The local variability of the grainstone banks was similar to modern tidal shoals of the Bahamas. Regressive cycles (relative sea-level lowstands) contributed to development of chalky intragranular porosity in the ooid grainstones (Ahr 1981). This ramp morphology continues across Texas into Mexico, where the large, exposed structural platforms are rimmed by inner-shelf carbonates.

A prominent shelf-to-basin prograding, delta-fed sandy apron is shown as extending across the central Gulf deep-water. This deposystem is hypothetical, being deeper than all current well penetrations. The existence and general trend of this deposystem is based upon seep and reservoir oil geochemistry of HVB supersequence samples, as discussed in Chapter 9 (see Box 9.1). The geochemical indicators point to a marly source rock and alignment with the paleo-Mississippi source-to-sink pathway first mapped by Cicero *et al.* (2010). This apron shifts westward somewhat in CVB time (Cunningham *et al.* 2016). Supporting evidence also includes the observed sandstone thickening aligned with this sediment input axis (Figure 3.23).

In Mexico, the presence of evaporites in the Olvido Formation implies a correlation with the Buckner Anhydrite of the northern GoM (Salvador 1991b; Figure 3.2). Reconstruction of the Kimmeridgian (HVB) shelf in the Campeche basin shows extensive sabkhas and lagoons with restricted circulation (inner-platform/ramp), landward of shelf grain shoals (Figure 3.23). However, uncertainties in reconstruction of Mexico, particularly the Sureste and onshore Chiapas area, remain. In the La Popa basin of northern Mexico, carbonate blocks encased in diapiric evaporites, analogous to northern GoM carapaces (cf. Figure 1.9) provide some limited insights to the early Mesozoic of the basin. It is estimated that the carbonate blocks have moved as much as 3 km upward from their original position, but contain a relatively narrow macrofaunal range of early to middle Kimmeridgian (Vega and Lawton 2011). This helped to differentiate the Olvido evaporites (gypsum in outcrop) from the older halite and gypsum beds of the Minas Viejas (Figure 3.2; Vega and Lawton 2011). Vega and Lawton (2011) speculated that evaporite deposition may have continued uninterrupted from the Callovian to Kimmeridgian stages. However, the degree of isolation and separation of this area from the greater Louann basin is unknown and difficult to reconstruct in light of complex tectonics and the fragmentary nature of salt diapiric carapace blocks in the La Popa basin.

In central Cuba, HVB-equivalent strata include the Artemisa (Rosario belt), Cifuentas (Placentas belt), and Guasasa (Sierra Organos belt) Formations, all interpreted as shallow marine inner- to outer-platform carbonates, part of the long-lived passive margin that developed on the south edge of the North American plate (Schenk 2008). Given uncertainties in plate tectonic restorations of western Cuba, it is excluded from the paleogeographic map.

3.5 Cotton Valley–Bossier Supersequence

The CVB supersequence is an important unit as it includes, near its base, one of most prolific source rocks for the GoM basin. Tithonian-centered source rocks extend from the

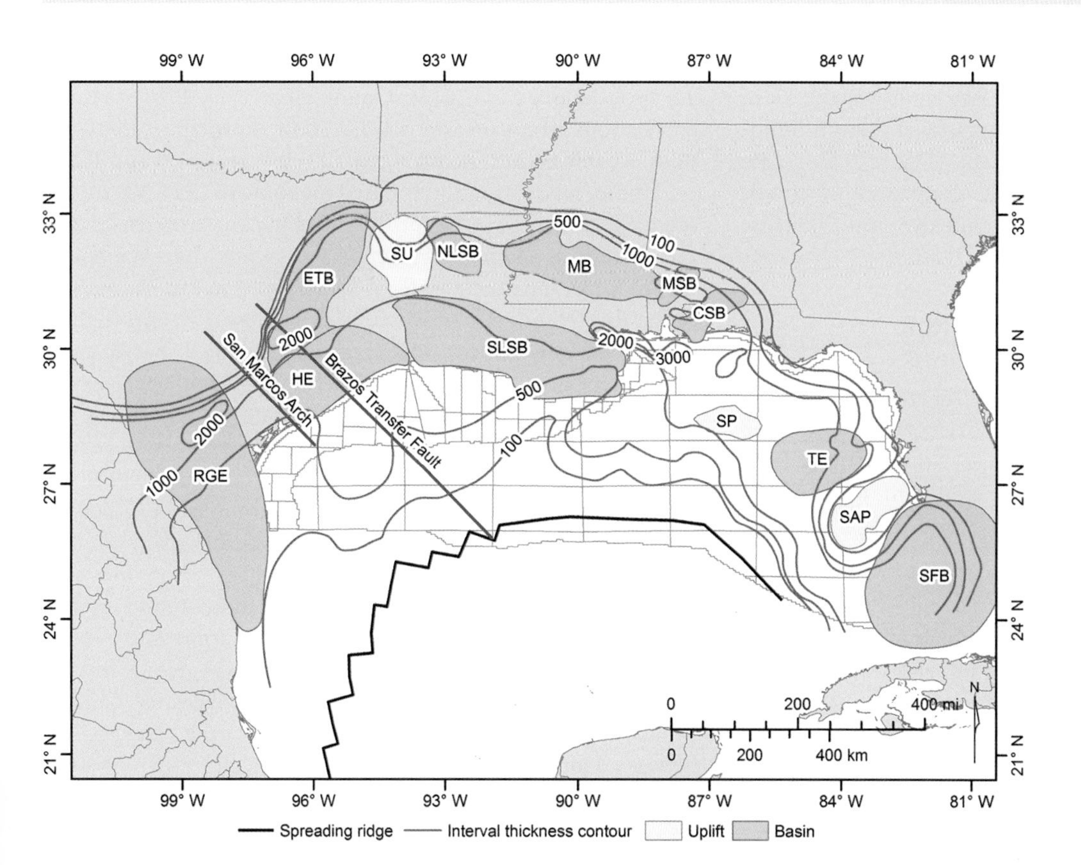

Figure 3.24 Cotton Valley–Bossier isochore map with key tectonic elements. Abbreviations for structural features are explained in the caption to Figure 1.3.

uppermost Haynesville to Lower Bossier (Hammes *et al.* 2011; Cunningham *et al.* 2016). Like the HVB supersequence, locally large-magnitude thicknesses (>1000 ft; 330 m) reflect increased crustal subsidence with continued sea floor spreading as is typical with this Drift and Cooling Tectonostratigraphic Phase (Figure 2.1). Trends in thickness reflect structural features; e.g., thicks in the South Florida basin and various salt basins and thinning on the Sarasota Platform and Southern Platform/Florida Middle Ground Arch, for example (Figure 3.24). Regionally smaller thickness maxima in the Mississippi–Alabama–Florida area are major depocenters usually related to fluvial input points or salt deflation zones. Pronounced thinning in the Garden Banks–Green Canyon–Walker Ridge–Keathley Canyon protraction blocks probably reflects attenuation of the CVB on salt flowing seaward toward new oceanic crust over the so-called Walker Ridge salient of Hudec *et al.* (2013a) and downlapping onto new oceanic crust on its flanks. The sharp boundary between thin and thick CVB offshore across the Brazos transfer fault may be a function of accommodation created by deeper crust/acoustic basement and perhaps thinner and less structured salt during the Tithonian interpreted to the west versus east of the fault (Hudec *et al.* 2013b). Other thickness variations can be related to local salt tectonics and faulting. The overall lack of detail in the isochore map (Figure 3.24) in deepwater areas reflects limits on well control and deep seismic imaging below the salt canopy.

3.5.1 Chronostratigraphy

The CVB supersequence extends from the Kim4_mfs (153.76 Ma) until Be5_sb (141.9 Ma), an approximately 12-million-year duration (Olson *et al.* 2015). This low-order chronostratigraphic unit is characterized by 19 separate biohorizons in offshore wells.

3.5.2 Previous Work

The CVB has been the subject of numerous papers as it encompasses both conventional (Klein and Chaivre 2002) and unconventional reservoirs (Cicero *et al.* 2010; Hammes and Frébourg 2012). This interval is recognized as the primary source of thermogenic hydrocarbons around the GoM basin (Jacques and Clegg 2002; Cunningham *et al.* 2016). In Mexico, coeval strata are variously named as the Edzna (Cantú-Chapa and Ortuño-Maldonado 2003; Figure 3.2), Pimienta Formation (Magoon *et al.* 2001), or Tepexilotla and La Casita Formations (Oloriz *et al.* 2003), depending on location. The presence of coeval shallow-water carbonates in Cuban outcrop localities was noted by Barros (1987), and this suggests stratigraphic continuity with the Florida Platform.

Platform margin reefs in the CVB supersequence are not well-documented in the literature, in contrast to well-publicized counterparts in the Kimmeridgian–Tithonian interval along the Tethyan seaway (Wilson 1975; Kiessling *et al.* 1999; Leinfelder *et al.* 2002). These massive (600–900+ m thickness) European platform margin reefs may be correlative to the roughly coeval interval in the Viosca Knoll and the Main Pass area of the eastern Gulf, where a platform margin build-up of 3000 ft (914 m) scale is present (Figure 3.19). Late Jurassic reefs of the Tethyan realm are dominated by corals, even where siliciclastics are present (Leinfelder *et al.* 2002), as in the eastern GoM.

3.5.3 Paleogeographic Reconstruction

Paleogeographic reconstruction of the CVB supersequence follows the same approach described for the Smackover–Norphlet and HVB supersequences, discussed earlier. The CVB strata fill gaps between Smackover–Norphlet rafts, so a correction has been applied to place pertinent wells in the original paleogeographic position.

3.5.4 Discussion

Plate reconstructions show the GoM basin steadily opening as the Yucatán Platform block is nearing the end of its rotation (Figure 3.25). However, the narrow and restricted early ocean basin would still have been subject to factors enhancing density stratification such as input of metal-rich hydrothermal fluids from the spreading ridge and input of brines due to dissolution of Louann Salt exposed at the sea floor during downslope salt advance (Hudec *et al.* 2013b; Cunningham *et al.* 2016). Seismic correlations in northeastern GoM show downlap of the CVB onto oceanic crust (Snedden *et al.* 2013; Figures 1.6 and 1.21).

Updip, seismic, and well-based mapping of the CVB supersequence thickness shows pronounced thickness increases that are denoted as sedimentary depocenters (cross-hair symbol in Figure 3.25). One depocenter is located on the eastern margin of the Rio Grande Embayment, where a fluvial system feeding shore zone deposits in south Texas is suggested, based upon well penetrations reported by Ewing (2010). Another depocenter, in this case, shelfal to ramp carbonates, is located in the western half of the East Texas salt basin. Two other depocenters, dominated by platform margin (reefal) carbonates, are mapped in eastern onshore Louisiana and the western border of the Destin Dome protraction block (Figure 3.25).

The CVB supersequence includes both shoreline and fluvial-dominated delta siliciclastics, generally found updip, and carbonates, located at the platform margin and within the basin (Figure 3.25). The platform margin clinoforms of the HVB continue into the CVB interval, suggestive of continued carbonate platform growth near the passageway between Florida and Yucatán (Figures 3.5 and 3.19). Seismic mapping shows westward expansion of this platform margin

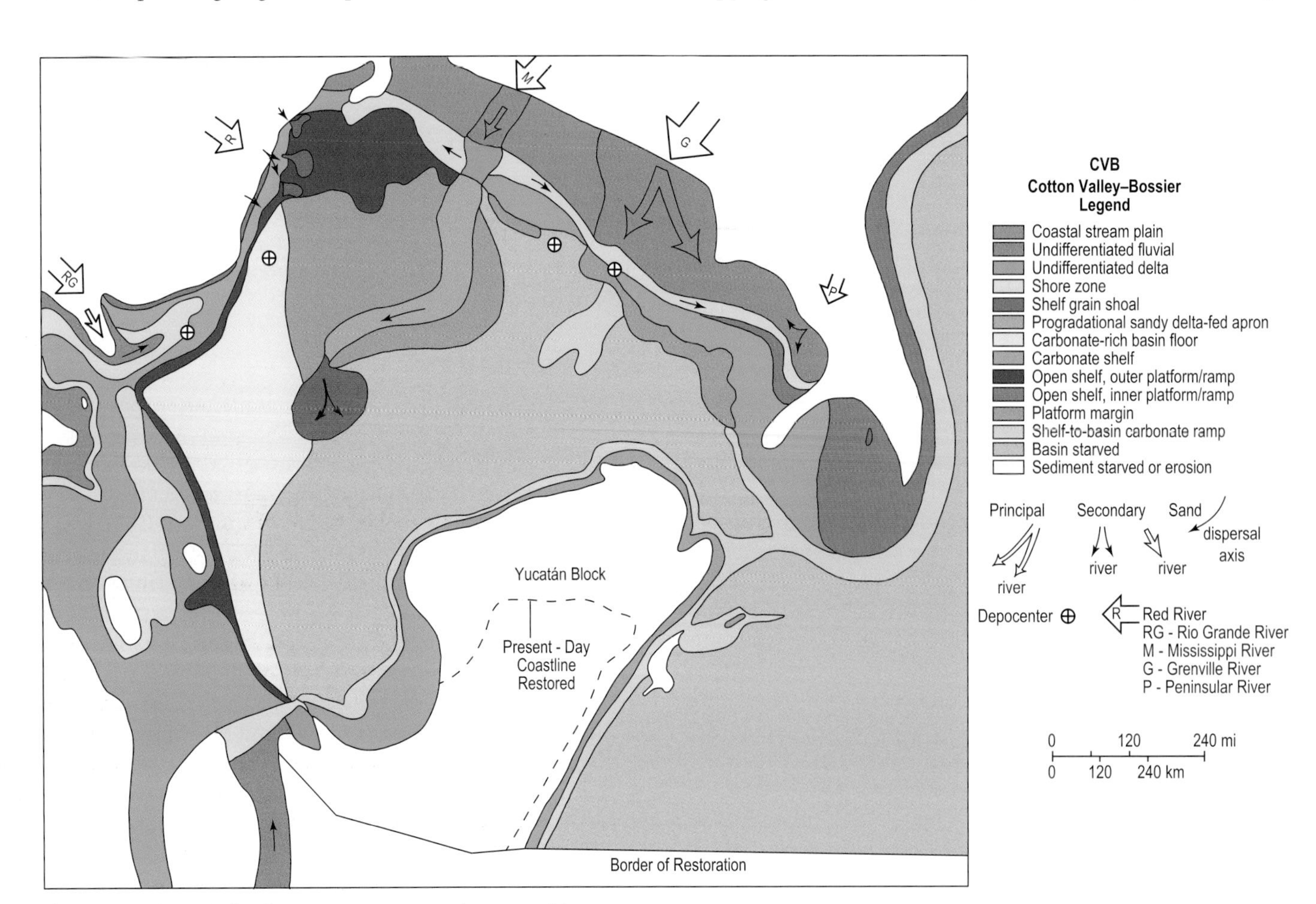

Figure 3.25 Cotton Valley–Bossier supersequence paleogeographic map.

facies deposystem further westward than observed for the equivalent deposystem in the HVB. However, like the HVB, the CVB platform margin reef was confined to the eastern area of the basin, from eastern Louisiana to Viosca Knoll–Main Pass, and terminating southeastward in the Vernon protraction block. This area is also associated with mapped sedimentary thicks (and thus depocenters) for the carbonates within the CVB. Examination of the seismic datasets suggests little or no reef development further westward, which fits with paleogeographic maps of the western GoM by Goldhammer and Johnson (2001) and Cicero *et al.* (2010). Cicero *et al.*'s (2010) reconstructions show a pronounced continental shelf break separating the Haynesville Shale basin from the deeper GoM, but do not assign a carbonate reef system there. Goldhammer and Johnson (2001) suggest the CVB interval of the western GoM was characterized by a gentle ramp extending eastward. The position of the CVB platform deposystem may reflect two important processes: (1) alignment of the reef system perpendicular to the maximum fetch of the basin; and (2) proximity to the narrow seaway between the Yucatán (Mayan) block and the Florida Platform, where marine waters presumably brought nutrients and faunal elements.

Siliciclastics in the CVB supersequence are shown as being derived from multiple highland terranes or longer river systems (Figure 3.25). Major siliciclastic terranes include the exposed Appalachians, which source both the paleo-Mississippi River and the Grenville fluvial system centered on present-day Alabama and Western Florida. Minor clastic input in north Texas (paleo-Red River or Ouachita source terrane) and Florida (Peninsular River) are also depicted in the CVB paleogeography.

However, very little of the siliciclastics appear to have reached the deepwater, with one notable exception. The deepwater well penetrations of the CVB encountered either carbonates (shelf-to-basin carbonate ramp) or starved basin, organic-rich shales. A widespread but thin package of carbonates is recognized in the Mississippi Canyon area, probably derived by downslope transport from the prominent carbonate platform margin.

The lone exception is shown as an elongate progradational apron extending from onshore areas to the deepwater of the northwestern GoM. This distribution is inferred from characteristic hydrocarbon signatures of reservoir oils and sea floor seeps in this area, signaling a marly or siliciclastic source bed. Details of this methodology are described in Box 9.1, and Tithonian-centered source rocks are further discussed in Section 9.8.1. The slope to basin apron system is possibly linked to a deepwater fan or channel complex and Cicero *et al.* (2010) indicate potential sediment input from the paleo-Mississippi River system. Of course, it is possible that the deepwater areas mainly received mud and clay, and sandstone content might be quite limited.

As mentioned earlier, the CVB supersequence contains the important Tithonian hydrocarbon source rock at its base. Preservation of organic matter, derived either from siliciclastic or carbonate sources, is promoted when anoxia is present, as during periods of sluggish oceanic circulation, particularly in bottom waters. A more detailed discussion of the Tithonian source rock genesis and characteristics is provided in Section 9.8.1, as well as complementary discussions in Boxes 9.1 and 9.2.

3.6 Cotton Valley–Knowles Supersequence

The CVK supersequence is the last unit deposited in the Jurassic, and one that continues into the Lower Cretaceous (Figure 3.2). In the northern GoM, it is best known from subsurface penetrations and fields in east Texas and Louisiana (Figure 3.26). Platform margin reefs are well documented, including those locally called the Knowles, Calvin, and Winn Limestones (Loucks *et al.* 2017b). Trends in sandstone thickness reflect substantial fluvial input from the Grenville River of Mississippi and Alabama and the paleo-Red River of east Texas and Louisiana (Figure 3.26). The paleo-Red River system is likely part of a smaller drainage basin probably connected to the Ouachita tectonic front. Wave action clearly acted to distribute sand along the depositional strike, forming the large Terryville and Calvin shore zone systems. In Mexico, the coeval interval is dominated by carbonate deposition of the Taraises and updip equivalents (Figure 3.2). Sub-equal volumes of sandstone and carbonate are present in the subsurface of the northern GoM.

3.6.1 Chronostratigraphy

The CVK supersequence extends from the Be5_sb (141.9 Ma) to the Va2_ssb (138.2 Ma), an approximately 3.7-million-year duration. This low-order chronostratigraphic unit is characterized by nine separate biohorizons in offshore wells, including the last appearance datum (LAD) of the ostracod *Schuleridea acuminata*. A supersequence boundary marks the top of the CVK unit in the Valanginian (Va2_sb).

3.6.2 Previous Work

The CVK supersequence includes sandstones (e.g., Terryville of Coleman and Coleman 1981) and carbonates in the Knowles Limestone of Cregg and Ahr (1983), Finneran *et al.* (1984), Petty (2008), and Loucks *et al.* (2017b). As with the underlying HVB and CVB supersequences, sandstones are largely confined to updip areas while carbonates dominate downdip areas, including the paleo-slope and basin. A number of local lithostratigraphic names are used for units within the CVK supersequence, such as Upper and Lower Terryville (sandstones), Massive Sandstones, B Lime (carbonates), Dirgin Member (limestones), B Sand, and Schuler Formation (sandstones) (Bailey 1983; Dyman and Condon 2006a). Local facies changes probably reflect paleo-environmental transitions from alluvial (Schuler) to shore zone (Terryville) to shelf (Knowles), shelf-margin (Calvin, Winn; see Loucks *et al.* 2017b) and deep basin (Upper Bossier).

The Knowles Limestone of east Texas includes a wide diversity of algae, foraminifera, sponges, corals, stromatoporiods, and other Tethyan fauna from shelf to basin. Tintinnids can be correlated to the Berriasian to Valanginian equivalents

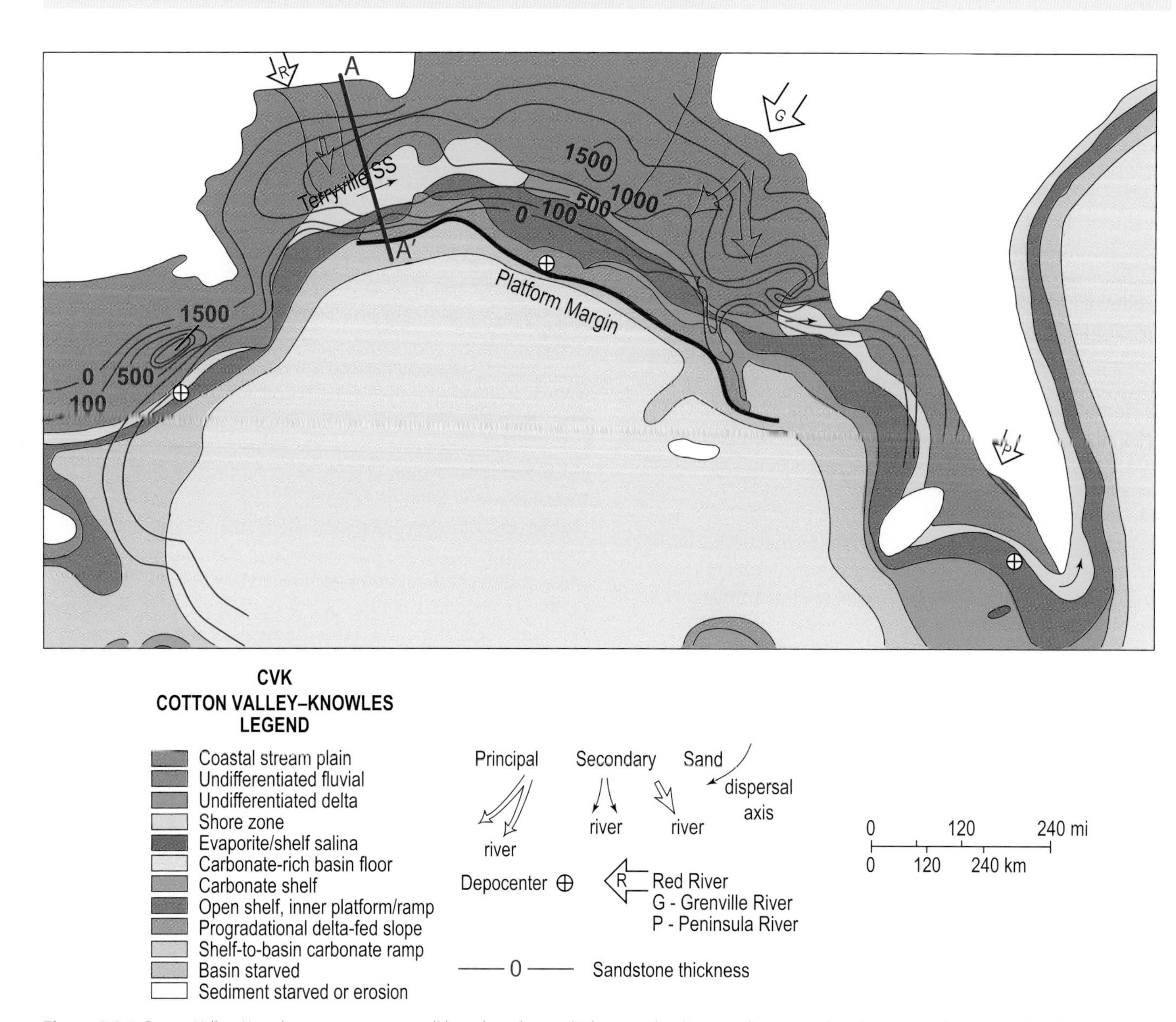

Figure 3.26 Cotton Valley–Knowles supersequence well-based sandstone thickness with relevant paleogeographic elements and interpreted sediment routing pathways.

in the Mediterranean, indicating an open marine connection to the Atlantic (Scott 1984). In the platform margin itself, CVK Limestone contains a variety of reef-building and reef-associated organisms, including corals, stromatoporoids, and *Lithocodium* sp., a binding cyanobacterial coating (Loucks *et al.* 2017b; Figure 3.27). Corals and stromatoporoids play a particularly prominent role as reef builders (Crevello and Harris 1984; Finneran *et al.* 1984). Bindstones also dominate the platform margin reef, while calcareous packstones and grainstones are common in the back-reef apron and in local banks and ramp build-ups (Cregg and Ahr 1983). In contrast to the Sligo and Washita platform margin reefs, rudists are rare (Loucks *et al.* 2017b). Slope facies cores yield packstones and rudstones including shallow-water high-energy flat ooids, broken corals, and stromatoporoids.

Loucks *et al.* (2017b) reconstructed the temporal sequence of platform margin reef evolution from Knowles to Calvin to Winn limestones (Figure 3.27). This section further demonstrates how effective "reef blocking" (Box 3.1) was at preventing Terryville and Calvin Sandstone from bypassing the coeval platform margin. Petty (2008) also noted that carbonate platforms developed over siliciclastic wedges, a pattern also observed in the overlying Sligo–Hosston supersequence.

3.6.3 Paleogeographic Map Reconstruction

Sea floor spreading was essentially complete, as Yucatán had rotated into its present position. Salt rafting is limited at this time, so no adjustment has been made.

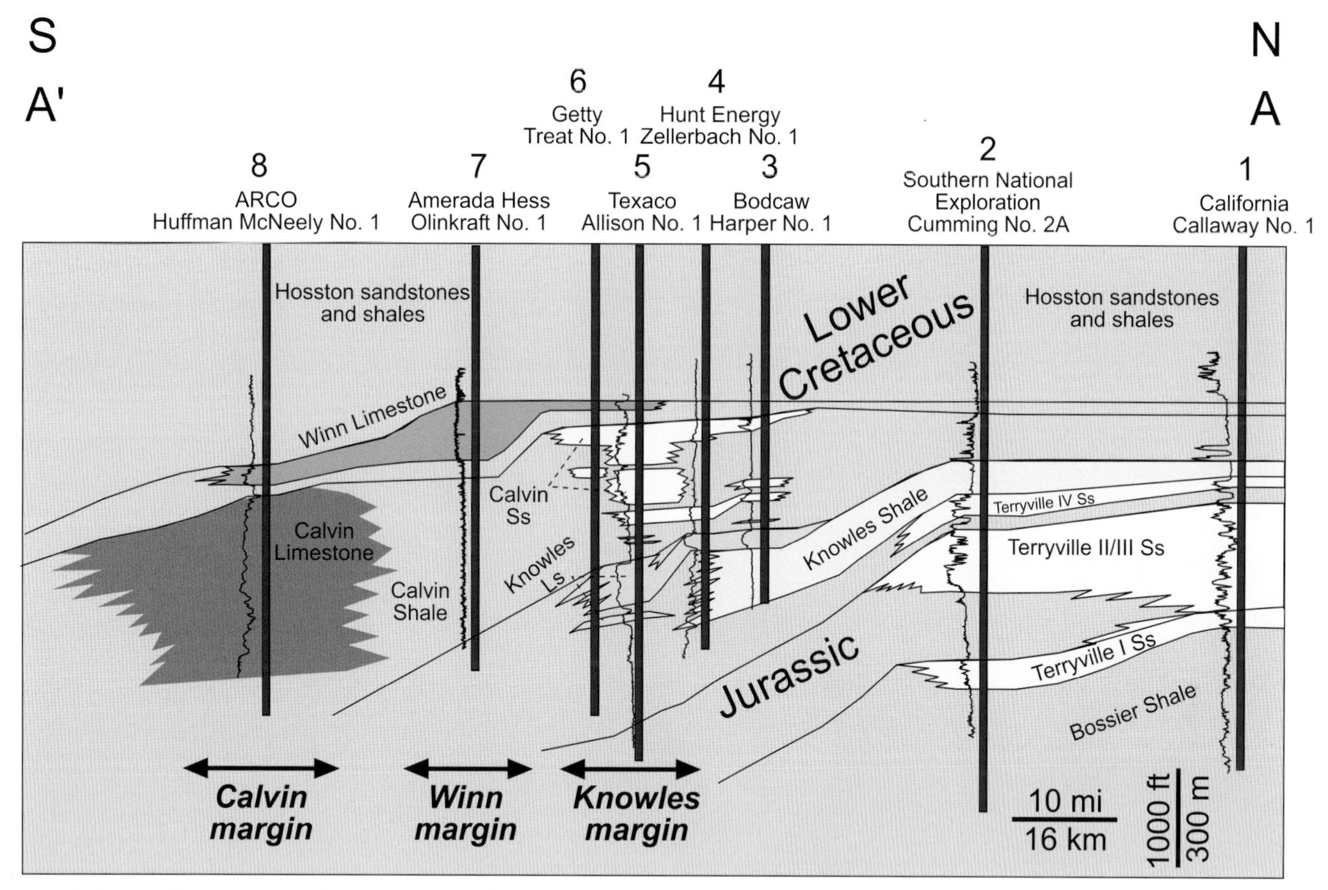

Figure 3.27 Stratigraphic cross-section showing the regional lithostratigraphy that includes: Calvin and Winn shelf-margin trends. Correlations are based on wireline logs. Modified from Loucks *et al.* (2017b).

3.6.4 Discussion

Platform margin reefs have been identified on regional 2D seismic sections, comparable to published trends of Anderson (1979), Finneran *et al.* (1984), and Loucks *et al.* (2017b). However, CVK platform margin reefs are relatively narrow (<10,000 ft/3048 m; Figure 3.28) in comparison to the overlying Sligo and Washita (Stuart City) platform margin reefs, as will be discussed later in this book.

Trends in gross CVK supersequence thickness, developed from well and seismic data, show two prominent depocenters (Figure 3.26). One depocenter is positioned on an area that evolves into the Maverick basin later in the Cretaceous. This elongate trough is largely filled by shore zone siliciclastics. The second depocenter is at the Knowles carbonate platform margin of central Louisiana, at the area of enhanced subsidence and accommodation at the transition from thick to thin transitional crust. Several other platform margins for younger supersequences occur in the immediate vicinity (Yurewicz *et al.* 1993).

Apparent thinning of the CVK occurs over the basement-cored highs like the Southern Platform, the Pensacola Arch–Monroe Uplift, and Adams County High, as defined by Ewing and Lopez (1991). In deeper waters offshore, the CVK is difficult to identify due to depth and imaging problems, but is recognized in carapace structures such as at Norton (GB 754 #1), suggesting greater extent than mapped here.

As mentioned, sandstones are concentrated along the border of Texas and Louisiana, where Coleman and Coleman (1981) recognized a thick succession of shore zone clastics called the Terryville Sandstone (Figure 3.26). This extends eastward into the Mississippi salt basin, reflecting early salt movement and influence on deposition. Much of the siliciclastic input is trapped in extensive strike-fed, shore zone systems of Florida, east Texas/western Louisiana, and south Texas, with few or no siliciclastics entering the deepwater. Petty's (2008) maps depict a similar pattern, with shore zone clastics giving way downdip to mud-rich inner carbonate platform (the carbonate shelf deposystem).

As with other supersequences discussed thus far, concomitant increases in carbonate thickness are noted as sandstones thin paleoseaward (Figure 3.26). The thickest carbonate intervals are associated with platform margin (reef) deposystems extending from Louisiana to the DeSoto Canyon protraction block (Figure 3.28). The platform margin reef trend has been traced to the Texas–Louisiana border based on well control, but is not identified further westward (Loucks *et al.* 2017b).

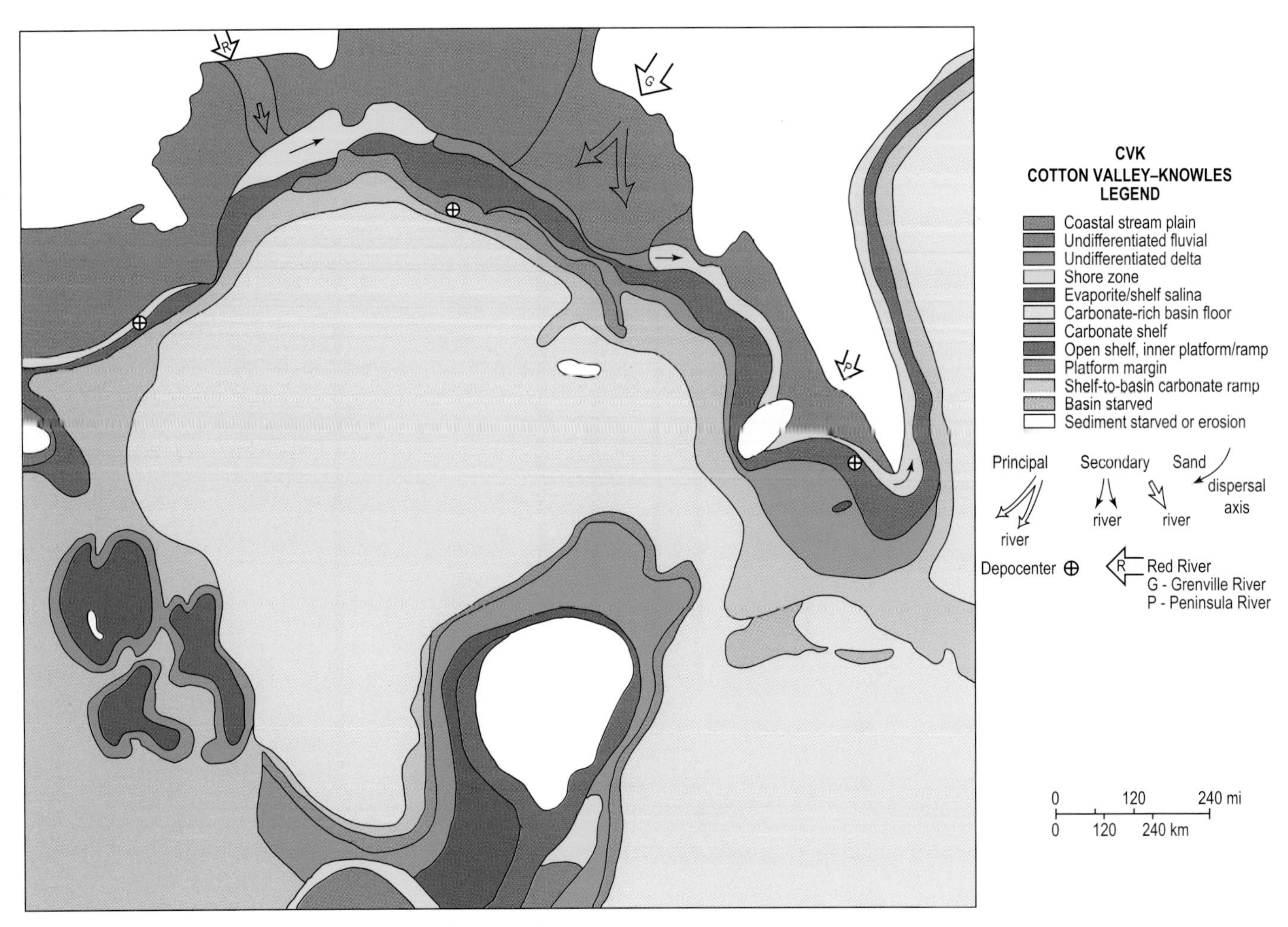

Figure 3.28 Cotton Valley–Knowles supersequence paleogeographic map.

A prominent protrusion of CVK carbonates into slope paleoenvironments is noted in the Mississippi Canyon area (Figure 3.28).

Platform margin (reef) deposystems, as discussed above, show a westward expansion from the eastern focused systems of the underlying CVB. Outside of this area, a broad carbonate ramp extending across Texas and Mexico (Goldhammer and Johnson 2001) is mirrored by similar ramp profiles in Florida and into Cuba (Dobson 1990; Petty 2008). Yucatán may have had a pronounced shelf margin, based on Salvador's (1991b) reconstruction. No CVK strata are present across inland Yucatán, based on well control provided by Ward *et al.* (1995).

Chapter 4

Late Mesozoic Local Tectonic and Crustal Heating Phase

4.1 Basin and Continental Framework

Deposition in the remainder of the Mesozoic Era may be broadly viewed as influenced by local tectonism and volcanism that commenced some time after the end of sea floor spreading in the Gulf of Mexico (GoM) (Figure 2.1). A significant volume of siliciclastics in Florida can be linked with uplift of the Peninsular/Ocala Arch at the start of this phase (Snedden *et al.* 2016a; Figure 4.1). While numerous unconformities punctuate the Cretaceous interval, erosional vacuities are regional but not basin-scale (Ewing 2009). A broad array of region-specific domal uplifts develop or are reactivated in the Late Mesozoic (e.g., Sabine Uplift, Rusk Uplift; Ewing 2009). Crustal heating events are also spatially limited, as volcanism occurs in small centers in southwest Texas, north Texas, Louisiana, and Mississippi (Byerly 1991). In some restricted areas such as the Maverick basin of south Texas, nutrient input from volcanic ash falls may have acted as an "iron fertilizer," driving enhanced pelagic sediment accumulation and increased primary organic productivity of the Eagle Ford Shale (Frébourg *et al.* 2016; Alnahwi *et al.* 2018).

The volume of siliciclastic sediment influx that begins this tectonostratigraphic phase is the largest of the Mesozoic Era in the northern GoM (Figure 4.1). Thus, understanding the distribution of reservoirs, sources, and seals in this initial Sligo–Hosston supersequence requires investigation of regional to local controls, from specific drainage networks linked to nearby uplifts, small silled basins and incised valleys enriching organic matter for conventional and unconventional source rocks, and local carbonate development. This tectonostratigraphic phase was abruptly and catastrophically terminated by the Chicxulub impact event at 66 Ma, one of the most important global events.

4.2 Sligo–Hosston Supersequence

The Middle Cretaceous rise of rudistid pelecypods to become a major component of platform margin reefs worldwide (Kiessling *et al.* 1999) began in the GoM with the Sligo–Hosston (SH) supersequence. Caprinid and requeinids rudists evolved from simple pelecypods and combined with corals, sponges, algae, and microbial binders to create reef mounds and buildups that formed formidable isolated platforms and continental shelf margins. Globally, reef expansion was favored by its location in a wide equatorial region from 40°N to 20°S along the Tethyan seaway (Wilson 1975). As discussed below, it did this in the northern GoM basin in spite of significant siliciclastic input from the Appalachians and other North American source terranes.

The SH supersequence is sandstone-dominated at its base (Hosston, Travis Peak, and Sycamore Formations) and carbonate-dominated at its top (Sligo Formation). This reflects a long-term marine onlap or encroachment and lateral transition from marine carbonate deposition paleo-seaward to marginal and non-marine sandstone accumulation paleo-landward. This is superbly illustrated in a core and outcrop-based schematic cross-section from south Texas (Phelps *et al.* 2014; Figure 4.2). The Sligo limestone-dominated upper half and Hosston (Sycamore) sandstone-dominated lower half of the supersequence may be defined as sequence sets, as each contains discrete embedded sequences. A prominent microbial–coral–rudist reef complex marks the Sligo platform margin, as documented by well control and seismic studies (Bebout 1977; Scott 1990; Yurewicz *et al.* 1993; Fritz *et al.* 2000; Phelps *et al.* 2014).

The top of Sligo seismic reflection is one of the most prominent correlation surfaces in the Gulf, facilitating basin-wide mapping (Figure 4.3). The major basins, arches, and platforms are clearly evident in the trend of structural highs and lows. The significant dip change at the Sligo shelf margin likely indicates the change from thick to thin transitional crust, where many of the Mesozoic carbonate platform margins tended to build up. In the deep basin, the Bravo Trough and Veracruz Trough reach depths of over 40,000 ft (12.2 km).

Because of the potential future importance of the Hosston sandstone-dominated sequence sets to frontier deepwater exploration in the northeast GoM, separate paleogeographic maps, and discussions of those maps, were prepared for the Hosston and Sligo portions of the SH supersequence.

4.2.1 Chronostratigraphy

The SH supersequence spans the Va2 ssb (138.2 Ma) to Ap3_sb (126.01 Ma), a duration of 12.19 million years. Ten biohorizons are included in this chronostratigraphic unit. The basal boundary of the SH also roughly marks the termination of sea floor spreading in the GoM (Snedden *et al.* 2013), though this remains less well constrained due to a limited

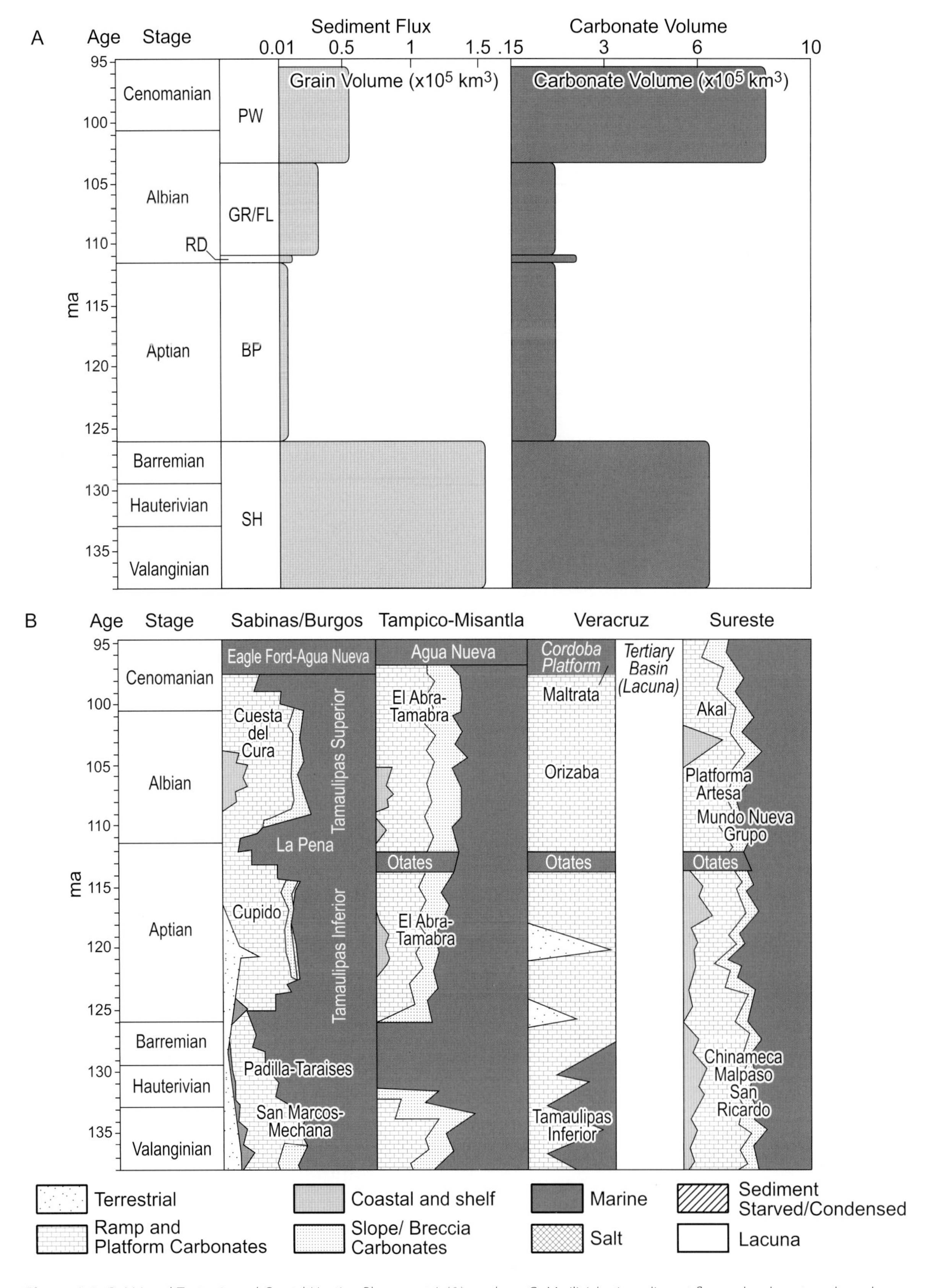

Figure 4.1 GoM Local Tectonic and Crustal Heating Phase, part I: (A) northern GoM siliciclastic sediment flux and carbonate volume by supersequence. (B) Lithostratigraphy for four areas of the southern GoM.

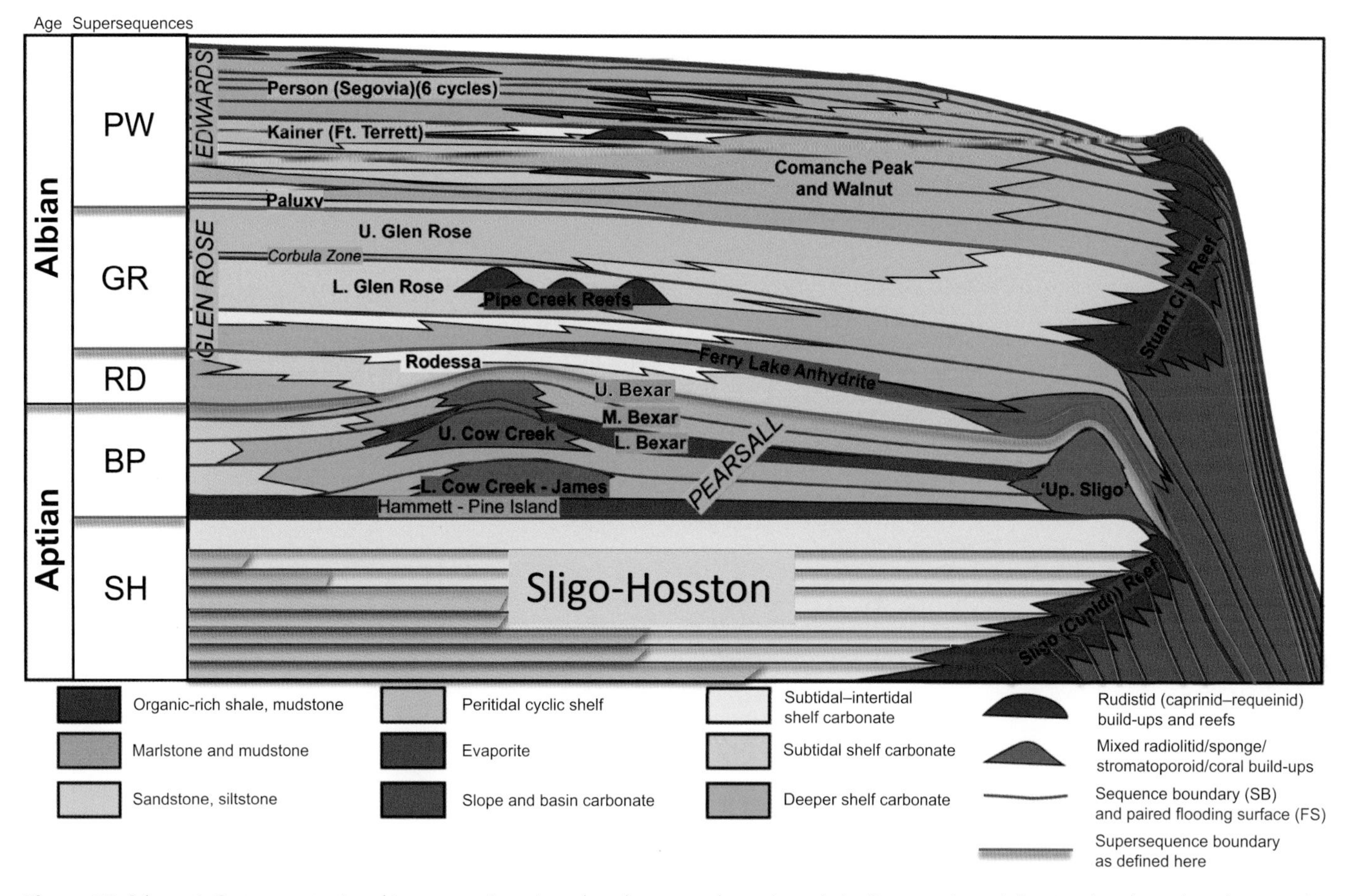

Figure 4.2 Schematic Cretaceous stratigraphic cross-section oriented northwest–southeast through the Texas continental, showing the relationship of stratigraphic units, sequences, and depositional episodes. Modified from Ewing (2016) and Phelps *et al.* (2014).

number of Jurassic magnetic anomalies. This event may have been accompanied by a structural reorganization of the GoM basin, most acute in the eastern GoM with possible uplift or rejuvenation of the Peninsular Arch and Ocala Uplift in Florida.

Because of the common lithostratigraphic confusion about the Sligo and Hosston, we have elected to define biostratigraphically, for the first time, the Hosston sequence set of offshore areas. The type biostratigraphic interval includes biohorizons from several wells in the Destin Dome and DeSoto Canyon offshore protraction blocks. The top of the Hauterivian Stage is based on the identification of the nannoplankton *Cruciellipsis cuvillieri*, identified by Olson *et al.* (2015) as clearly within the Hosston interval of two wells. Biodatums within the Barremian Sligo and Berriasian Cotton Valley–Knowles (CVK) also help constrain the top and base of the Hosston.

As discussed in Section 3.1, the lowest point of global sea level in the Jurassic is within the Valanginian stage (Haq 2017), represented in the GoM basin by the Hosston sequence set. The occurrence of a large volume of siliciclastic sediment generated during this same interval of time appears to be more than coincidental (Figure 4.1). It is, however, difficult to separate the influence of global sea-level lowering from basin-specific tectonic factors, including possible structural reorganization following the end of GoM sea floor spreading, uplift of the Peninsular and Ocala Arches, etc.

4.2.2 Previous Work

Barremian to Aptian carbonate units are usually referred to as the Pettet (onshore east Texas and Louisiana) and Sligo Formations (McFarlan and Menes 1991).

The Valanginian–Hauterivian siliciclastics have a variety of lithostratigraphic names, including Travis Peak (onshore Texas) and Hosston Formation (east of Texas to Florida). Less is known about both units in offshore areas, as a limited number of wells penetrate the interval (Petty 1999). However, it is noted that lithostratigraphic names are often applied without reference to the age of the unit penetrated in both onshore and offshore areas. For example, any sandstone encountered in the Sligo interval is typically referred to as "Hosston" regardless of whether it is Valanginian–Hauterivian. This does make chronostratigraphically based mapping a challenge in some areas.

The Valanginian–Hauterivian-age Travis Peak and Hosston Formations of onshore Texas, southern Arkansas, Louisiana, and Mississippi represent basinward-thickening wedges of siliciclastic material derived from fluvial–deltaic depocenters

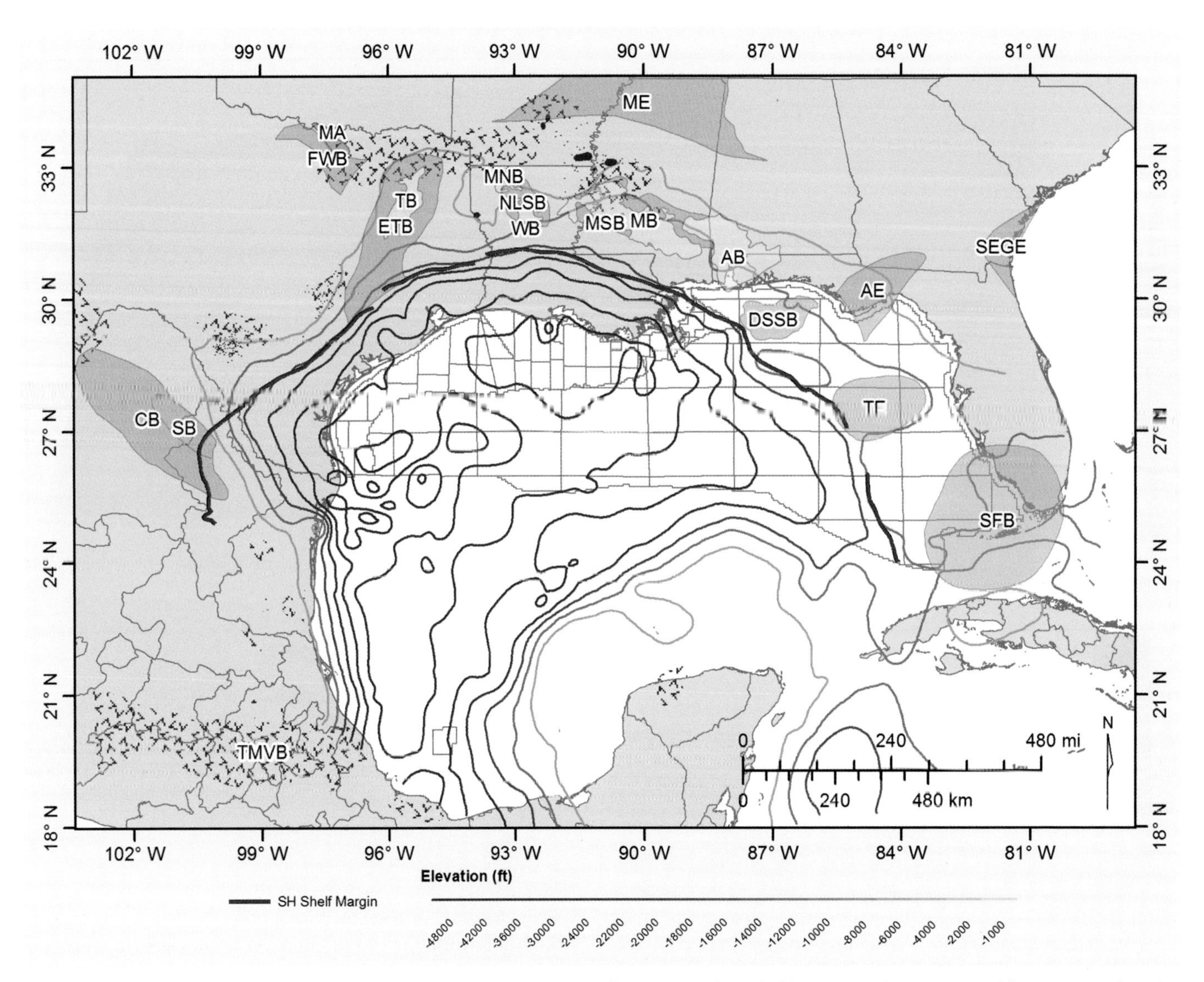

Figure 4.3 Structure map, Top Sligo–Hosston supersequence. Important tectonic features are indicated. Abbreviations of major structural features are explained in Figure 1.3.

in east Texas and western Mississippi, apparently sourced by the ancestral Red River and ancestral Mississippi River. Dyman and Condon (2006b) report four discrete stratigraphic intervals in the Travis Peak, based on the findings of Saucier (1985): (1) a basal interval of mixed sandstone and shale, representing delta-fringe deposits; (2) two thick middle intervals of stacked, aggradational braided stream sandstones, transitioning from low-sinuosity fluvial to high-sinuosity fluvial and floodplain deposits; and (3) an upper interval of sandstones and mudstones, representing coastal plain/paralic/marine deposits.

Ewing (2010) identified a sandstone-rich basal Hosston in south Texas, including a prograding base, aggrading alluvial plain, and a transgressive sandstone top, conformably overlain by a thick, continuous shale. Ewing (2010) argues the basal Hosston is a lowstand systems tract deposit overlying a major sequence boundary (at the contact with the underlying CVK supersequence). With the deltaic and shoreline systems lowstand deposits identified, Ewing (2010) speculated on the potential for a downdip lowstand fan exploration target.

The Hosston seismic interval can be mapped in the deepwater areas and traced back to the slope of the Florida Escarpment, Destin Dome area, and other areas (Bovay 2015). The acoustically dim (low continuity, low amplitude [LCLA]) seismic reflection package can be followed into the deepwater of DeSoto Canyon and Mississippi Canyon protraction blocks, southwest of the paleo-shelf margin (Figure 4.4). This LCLA seismic facies is hypothesized to be more sandstone-rich than the high-amplitude, high-continuity seismic reflection packages that in other Mesozoic intervals of the Gulf basin are typically associated with carbonate-rich intervals (Cunningham *et al.* 2016; Snedden *et al.* 2016a). Further support comes from 2D seismic data to ties with key eastern Gulf wells Shiloh (DC 269 #1), Appomattox (MC 392 #1), and Vicksburg B (DC 353 #1; Bovay 2015).

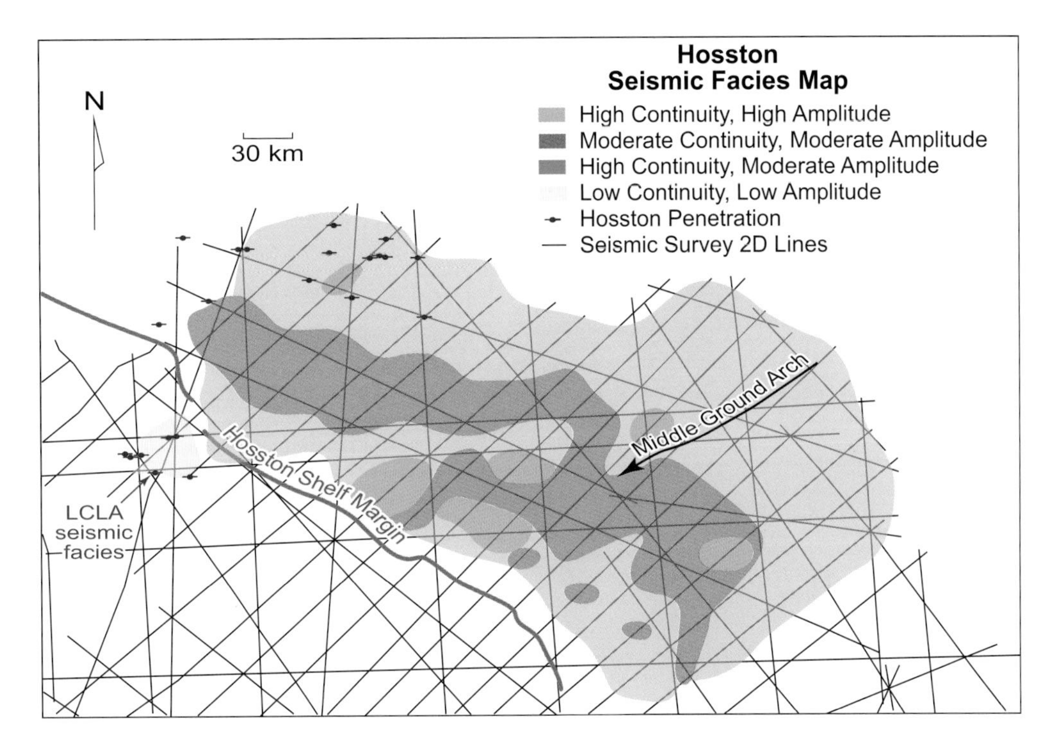

Figure 4.4 Hosston seismic facies map. Modified from Snedden *et al.* (2016a).

4.2.3 Hosston Sequence Set

The Hosston sequence set can also be defined separately from the Sligo sequence set using seismic data tied to well-based biostratigraphy (Snedden *et al.* 2016a). Mapping on the modern Florida shelf identified a seismic surface between the Hosston and overlying Sligo (Bovay 2015). It is defined here as the deepest seismic trough between the continuous, high-impedance Sligo carbonates and the sometimes discontinuous, weak, low-impedance Hosston siliciclastics underneath. Previously, workers had suggested a more gradational contact, due to expected interfingering of Sligo carbonates and Hosston siliciclastics (Galloway 2008). However, the presence of a sand-prone lowstand systems tract in Texas suggests otherwise (Ewing 2010). In areas near the platform margin, a set of low-angle clinoforms can be observed below the Top Hosston sequence boundary. These downlap onto a maximum flooding surface at the base of the Hosston.

4.2.4 Hosston Sequence Set Paleogeographic Map Reconstruction

The prominent platform margin reef system of the overlying Sligo sequence set likely had a predecessor of Hauterivian age, which may have evolved from the more limited carbonate build-up called the Winn Limestone (Loucks *et al.* 2017b; Figure 3.27). The lower Valanginian platform margin reef system, known as the Calvin, is considered part of the CVK supersequence (see Section 3.6). The lateral extent of these two systems is not well documented in the supersequence due to limited drilling in comparison to the shallower Sligo platform margin, but likely was a sufficiently large barrier to preventing coeval land-derived sandstones from reaching the slope and deepwater.

Interpretation of well logs and seismic data, as well as regional source-to-sink reconstructions in the eastern GoM, suggests the Hosston may have found its way past the obstacle of this coeval platform margin reef. Two possible basin entry points for siliciclastics during the Valanginian–Hauterivian can be identified (Figure 4.5). The northern entry point is located in proximity to an extensive fluvial bed load system likely draining the southern and western Appalachians, which as discussed earlier (Section 3.3.8) we refer to as the Grenville River system. Seismic observations of the platform margin deposystem show continuity across Main Pass and Destin Dome protraction blocks but an absence of characteristic seismic geometries across a portion of the DeSoto Canyon block. The presence of a coeval deltaic system nearby supports this hypothesis. Following the model of Puga-Bernabéu *et al.* (2013; see Box 3.1), these feeders may have been sourced from an interreef passage that connected to shelf channels and the updip deltaic system. While it is possible that these features are carbonate-dominated, the proximity to a sand-prone alluvial system suggests otherwise.

A second basin entry point can be interpreted in the area of the Vernon protraction block (Figure 4.5). Platform margin seismic observations straddle the block, but within the block no obvious reef signatures are apparent. This interreef passage is depicted as linking to a pathway from the Peninsular Arch across the Florida shelf just north of the Sarasota Arch. Preliminary seismic observations point to a potential deepwater sandy apron in the vicinity of Vernon, Lloyd Ridge, Elbow, and small portions of the Henderson and DeSoto Canyon

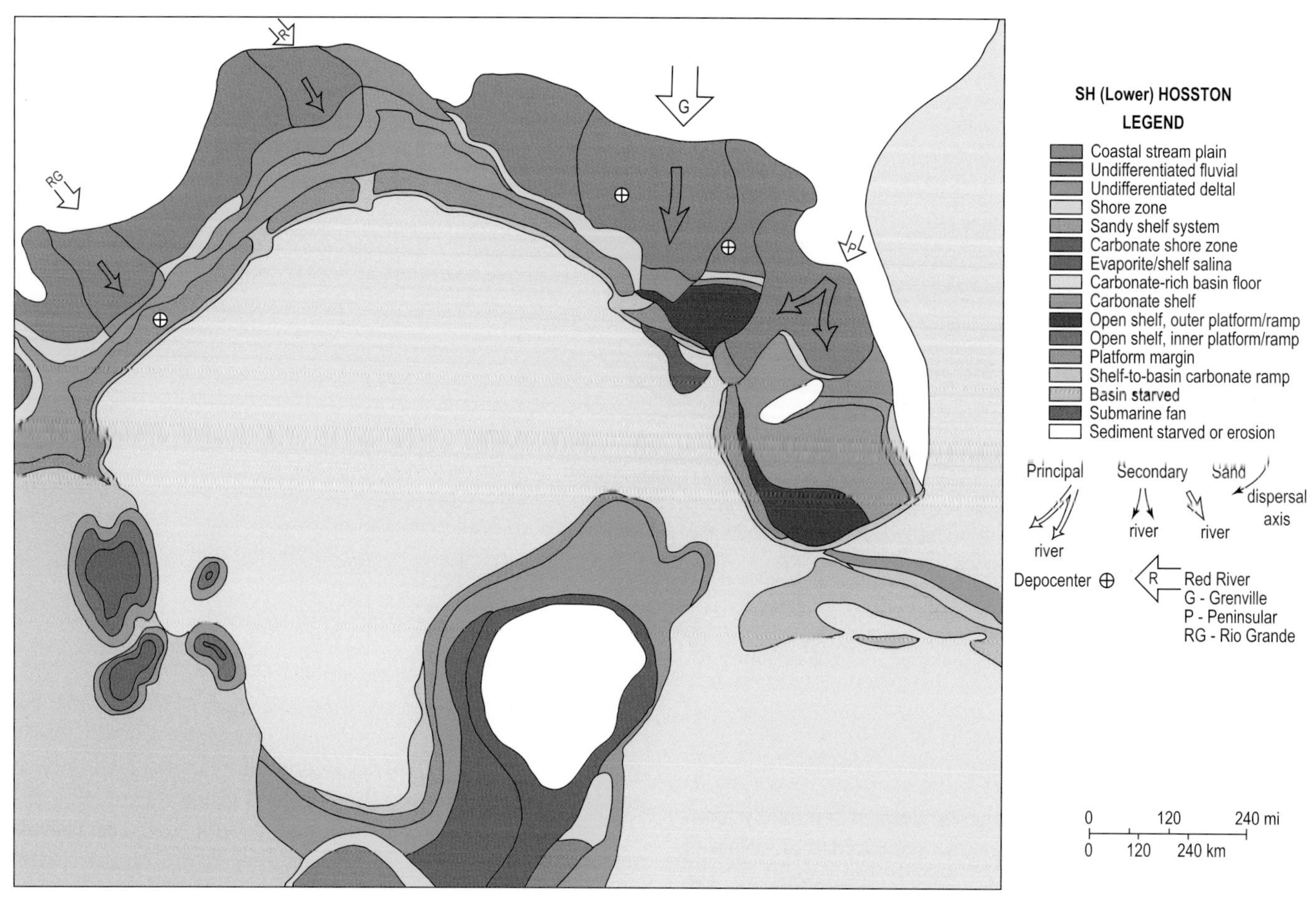

Figure 4.5 Hosston sequence set paleogeographic map.

protraction blocks (Snedden *et al.* 2016a). Alternatively, siliciclastics may have been derived from the northern entry point discussed above.

Onshore, the Travis Peak fluvial-deltaic system can be found in the surface and subsurface of east Texas. However, there is no evidence suggesting that this siliciclastic fairway extends into the deepwater, trapped behind the platform margin system documented here and by Yurewicz *et al.* (1997).

4.2.5 Provenance of the Hosston Sequence Set

Once basin entry points are identified for the Hosston, the potential run-out length of deepwater sandstones in submarine fans can be estimated from consideration of the source-to-sink scaling relationships (Snedden *et al.* 2016a; see Box 4.1). Potential siliciclastic source terranes can also be evaluated using detrital zircon U–Pb dating (see Box 1.2), which indicates basement source terrane and thus confirms the paleo-river catchment length, a necessary component of source-to-sink scaling predictions. To refine the Hosston paleogeographic reconstruction, sandstone samples taken from two intervals in an onshore Florida well, the Stanolind-Sun Perpetual Forest #1 well, drilled in Dixie County, Florida and processed for U–Pb detrital zircon analysis (Figure 4.6). Comparison of the age spectra of the upper half (4180–4700 ft measured depth [MD]) and lower half (4700–5230 ft MD) of the Hosston interval of the Stanolind-Sun Perpetual Forest #1 well are virtually identical, showing dominant basement source terrane age populations of 493–699 Ma, usually associated with the Gondwanan Suwannee terrane (Lovell and Weislogel 2010; Weislogel *et al.* 2015). A secondary peak at 1992–2240 Ma is likely related to sediment derived from the Trans-Amazonian basement terrane (Figure 4.6). Grenville basement sources (peaking at 1 BY), typical of the Appalachians (Weislogel *et al.* 2015), are largely absent in the lower half core sample and show a minor peak in the upper half core sample.

Comparison to detrital zircons from a nearby Paleozoic basement well penetration analyzed by Lisi (2013) shows a remarkably similar age spectra (Figure 4.6). This suggests local derivation of the coarse sandstones in this well from similar-age Paleozoic basement rock known to underpin the Ocala and Peninsular Arches of Florida (Winston 1976; Heatherington and Mueller 2003). The lack of Appalachian (Grenville-age) source material implies a separate sediment system than commonly recognized further to the north (Weislogel *et al.* 2015).

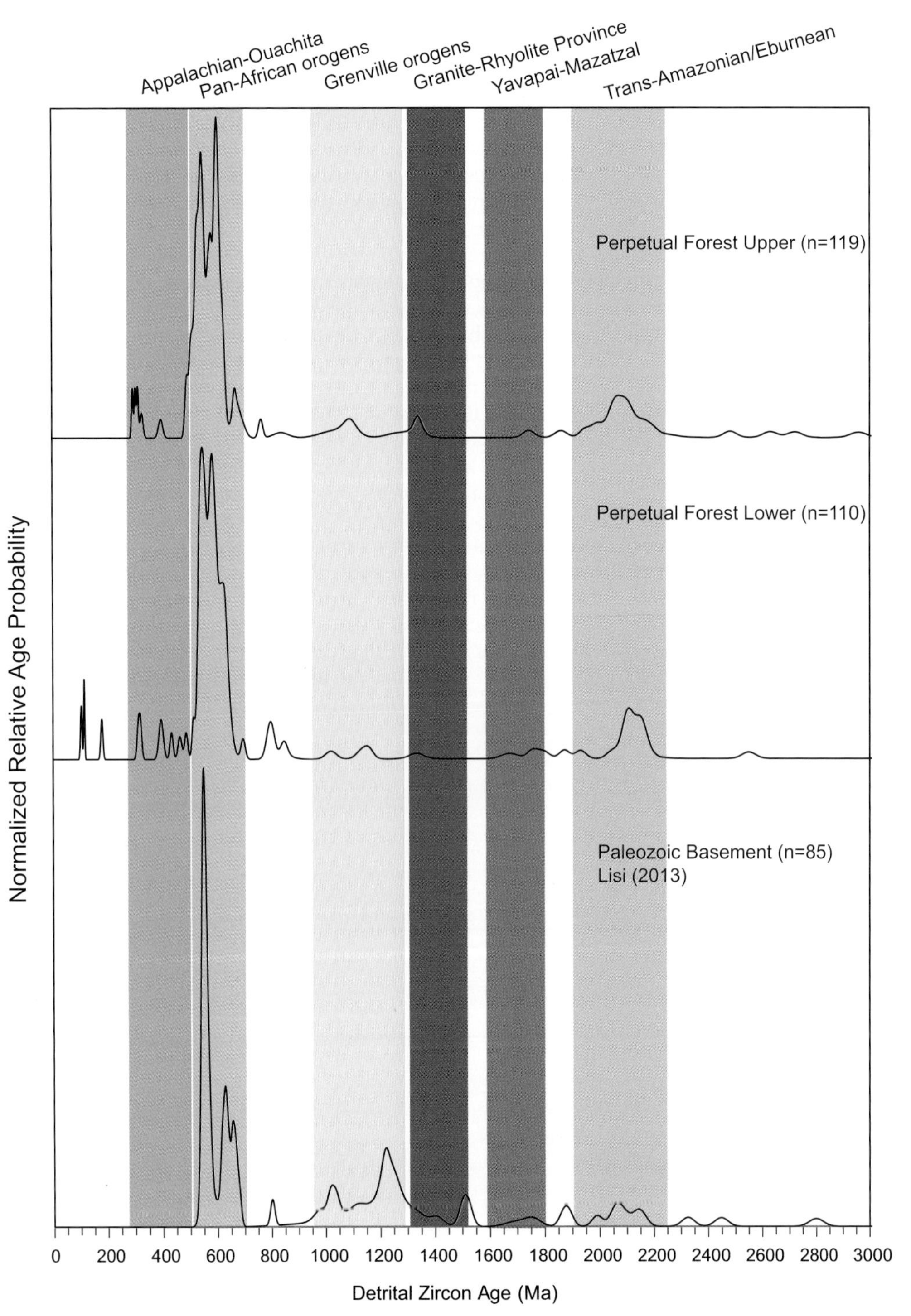

Figure 4.6 Detrital zircon results, Perpetual Forest well as compared to Paleozoic basement sample of Lisi (2013). Modified from Snedden *et al.* (2016a).

Box 4.1 What are Source-to-Sink Analyses and Empirical Scaling Relationships?

Source-to-sink analysis is a broad and rapidly evolving scientific approach to paleogeographic reconstructions, but one that also has practical applications relevant to the global search for hydrocarbon resources (Sømme *et al.*, 2009a; Helland-Hansen *et al.* 2016; Walsh *et al.* 2016). Quantification of the scales of modern and Pleistocene systems suggests linkages within and between segments of sediment dispersal systems that terrigenous clastics follow from highland source terranes toward the basinal sinks. This makes it possible to predict the unknown geomorphological dimensions of one segment from empirical measurements of another (Sømme *et al.* 2009a; Bhattacharya *et al.* 2016). For example, deepwater depositional systems can

Box 4.1 *(cont.)*

be linked in many cases to the rivers that carry the sediment, and thus should scale with fluvial system properties (Blum and Hattier-Womack 2009). The approach has utility for interpretation of ancient processes as well as being a tool for subsurface exploration, particularly in areas where seismic reflection resolution is of poor quality, including areas of poor illumination due to thick salt canopy cover (e.g., Meyer *et al.* 2007).

A first application of the source-to-sink empirical scaling approach for an ancient, deepwater subsurface system was in the hydrocarbon-bearing Maastrichtian–Danian Ormen Lange deepwater system of offshore Norway (Sømme *et al.* 2009b). This test of a small depositional system suggested great promise for - first-order prediction of reservoir dimensions such as submarine fan length and width. Snedden *et al.* (2018a) validated the methodology for larger, continental-scale systems in the Cenozoic of the greater GoM. Nyberg *et al.* (2018) reanalyzed and refined the scaling relationships from an expanded submodern dataset.

Key morphological dimensions measured in these studies include catchment area and length (or longest river), backwater length, delta length, delta area, shelf length, and submarine fan length and width (Snedden *et al.* 2018a). In most applications, reconstruction of catchments is the most difficult due to the complexity of tectonics and preservation of the drainage basin. However, the advent of detrital zircon provenance analysis (see Box 1.2) has, along with other techniques, greatly enhanced catchment reconstructions.

For estimation of longest river lengths, we focused on the prominent trunk river systems that are linked to the basinal sinks. Examples from the GoM Mesozoic (Tuscaloosa) and Cenozoic (Wilcox) are discussed further by Blum and Pecha (2014), Blum *et al.* (2017), and Milliken *et al.* (2018). Submarine fan length and width in the validation dataset is a direct function of the robust Cenozoic northern GoM well and seismic database. We use fan run-out length (distance from shelf margin to fan termination) due to ease of recognition of the shelf edge versus the continental toe of the slope, given the challenges in its identification.

A key relationship between source-to-sink segments is between longest river length (which itself scales with catchment length) and submarine fan run-out length (Figure 4.7). Most GoM Cenozoic submarine fan run-out lengths fall in the range of 0.1–0.5 times the longest interpreted river lengths. This includes depositional bodies of all scales within our database, including at least one (of the Upper Miocene Rio Bravo) that has been reinterpreted as a deepwater current-modified sand body (Snedden *et al.* 2012). However, a significant number of deepwater bodies, all of which are classified as submarine aprons, fall outside of that range (points 2–7 of Figure 4.7). These and other outliers for the Lower Miocene and a few other units may indicate some uncertainty in defining sand body dimensions in these poorly organized, deepwater systems (e.g., point 8 of Figure 4.7; see also Box 1.3, which explains the concept of a depositional apron). There is a weaker but still strongly positive correlation ($R^2 = 0.4$) between fan width and run-out length (Figure 4.8), reflecting some uncertainty in delineating fan width due to the laterally extensive salt canopy.

Testing of these submarine fan dimensions is ongoing, but has shown some success, particularly when calibrated against mapping of seismically defined depocenters aligned with known basin entry points (Snedden *et al.* 2018b). Short but steep systems in

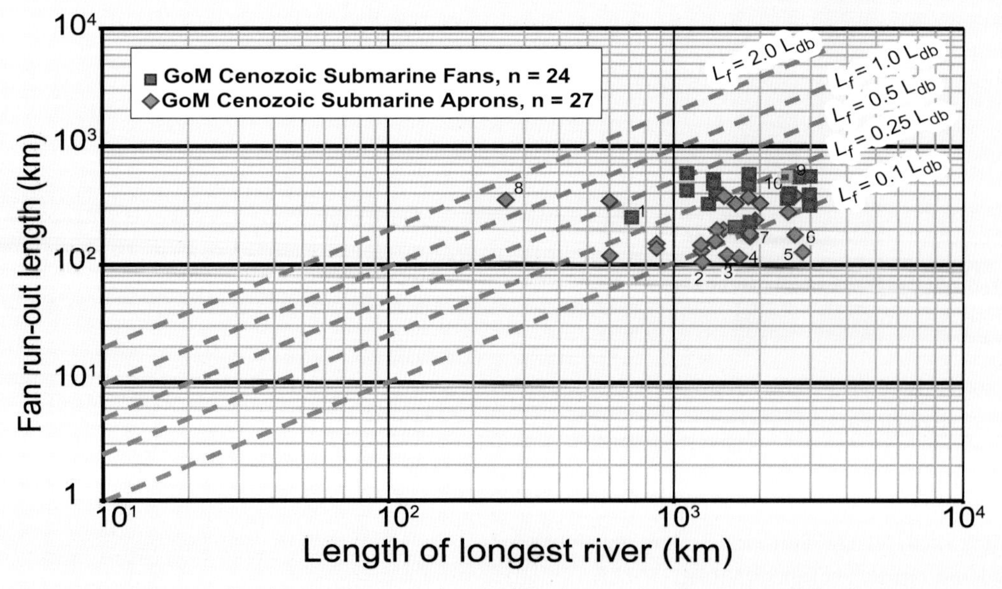

Figure 4.7 Longest river length versus fan run-out length (defined as slope length + fan length) for all GoM Cenozoic deepwater systems including fans (red squares) and aprons (green diamonds). Indicated deepwater system points (refer to Figure 1.31 for deposode names): (1) Upper Wilcox Mississippi Fan; (2) Lower Miocene 1 Rio Grande apron; (3) Oligocene Frio Houston-Brazos apron; (4) Yegua Cockfield-Jackson apron; (5) Oligocene Frio Mississippi apron; (6) Lower Miocene 1 Mississippi apron; (7) Lower Miocene 1 Red River apron; (8) Oligocene Rio Bravo north apron; (9) Lower Wilcox Mississippi system measured at centerline of catchment; (10) Lower Wilcox Mississippi system plotted at minimum possible river length as a test of sensitivity, discussed in the text. Abbreviations used: L_f, fan run-out length; L_{db} drainage basin catchment length. Modified from Snedden *et al.* (2018a).

Box 4.1 *(cont.)*

areas of high rainfall are known to produce larger fans than expected, but most systems scale to the size of the catchment.

A further extension of these empirical scaling relationships can be carried out where the catchment is poorly constrained but single story point bar thickness can be measured in outcrop or from subsurface logs (Xu *et al.* 2016b). Channel belt sand body thickness scales to bankfull discharge and is a reliable first-order proxy for the drainage basin area, a proxy that is more robust if climatic regimes can be independently constrained (Milliken *et al.* 2018).

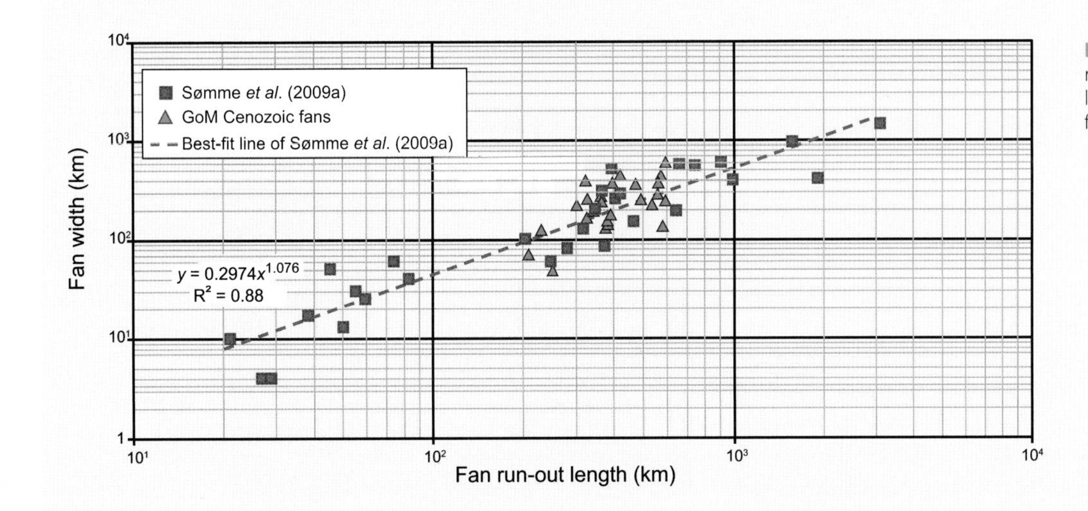

Figure 4.8 Empirical scaling relationships for fan run-out length versus fan width. Modified from Snedden *et al.* (2018a).

4.2.6 Hosston Source-to-Sink Predictive Scaling Relationships

In order to further constrain the Hosston paleogeographic map, empirical scaling relationships based on the submodern data of Sømme *et al.* (2009b) were employed (Box 4.1) to evaluate sediment routing pathways. These relationships also help to place our seismic observations into a better context for the purpose of paleogeographic reconstruction.

A critical relationship note in Sømme *et al.*'s (2009b) compilation of modern and submodern data is between longest river length, a proxy for fluvial catchment size, and submarine fan length. Snedden *et al.* (2018a) validated this approach for Cenozoic depositional systems of the northern GoM. Determining ancient river courses and thus the length is obviously challenging for Mesozoic systems like the Hosston, but regional data point to three potential river sources to the eastern GoM, as discussed below.

Average river length of Hosston source-to-sink systems were estimated for Florida-sourced systems (here called the Peninsular River) and Appalachian-sourced systems (here called the Grenville River), as listed in Table 4.1. Five potential catchment (drainage basin) lengths were considered for these and the modern Chattahoochee–Apalachicola catchment.

For the Peninsular River catchment, average distance from the interpreted river headwaters to shelf edge is approximately 424 km (263 miles). The average length of 424 km would scale with a submarine fan approximating 10–50 percent of the length of the river system (Sømme *et al.* 2009b), predicting a sand body of 42–212 km (26–131 miles) in length from the coeval platform margin (Table 4.1).

Average river length of the Appalachian-sourced Hosston source-to-sink system (here called the Grenville River system), defined as the length from the headwaters to the shelf edge and calculated for five potential pathways, is 887 km (550 miles; Table 4.1). This average river length is much shorter than the same pathway drawn following the exact path of the modern Chattahoochee–Apalachicola fluvial system (1107 km; Table 4.1). An average river length of 1349 km (836 miles) derived from the schematic map of Blum and Pecha (2014) showing an Early Cretaceous drainage divide, which represents the longest pathway that could be considered for this system (Table 4.1).

Together, these empirical scaling relationships would suggest submarine geobodies (fans and aprons) in the range 42–674 km (26–418 miles; Table 4.1). The mapped fan of the Appalachian-sourced system, using seismic data alone, is approximately 70 km (43 miles) long. This is shorter than most of the empirical predictions but may reflect the complexity of salt-influenced depocenters and enhanced accommodation that limited fan run-out distance here. Like the paleo-Suwannee River, there is potential for the longest river length of the paleo-Chattahoochee–Apalachicola River to be much shorter.

The river channel that extends to the Lower Cretaceous drainage divide (Blum and Pecha 2014), yields a fan length of 135–674 km (84–418 miles). This is nearly twice as long as seismic facies mapping would suggest is the fan extent (Snedden *et al.* 2016a; Figure 4.4). This interpretation is

Table 4.1 Source river and predicted submarine geobody run-out lengths for Hosston source-to-sink systems

Catchment (drainage basin)	Potential routes measured	River channel length (km)	Expected geobody length range (km) using 0.1–0.5 × drainage basin length
Peninsular (Florida) catchment	5	424	42–212
Grenville catchment	5	887	89–443
Modern Chattahoochee–Apalachicola catchment	5	1107	111–553
Distance to Appalachian drainage divide of Blum and Pecha (2014)	5	1349	135–674

Note: Potential routes used to calculate geobody length (submarine fan or apron) are based on Sømme *et al.*'s (2009b) relationship: geobody lengths are 10–50 percent of the estimated river length. River length includes the straight distance from the river mouth to the shelf break, in addition to the path over land. Paleo-Suwannee River paths are approximated from approximate center of Ocala Arch. Paleo-Chattahoochee–Apalachicola river paths are interpreted by considering the approximate location of the modern headwaters. Distance to hypothesized drainage divide is based on Blum and Pecha (2014)

considered least likely, given the uncertainty of the Cretaceous drainage divide reconstruction.

Seismic data indicates the mapped apron of the Peninsular system is about 200 km (124 miles) in length at the long end of the predicted run-out length for the distribution, but within the predicted range. Certainly there is potential for the longest river length to be much less. Another caveat is the tendency for submarine apron lengths (defined in Box 4.1) to show more scatter in plots against longest river length (Snedden *et al.* 2018a; Figure 4.7).

The effect of salt rafting also needs to be considered. Salt rafting of Mesozoic strata during the Jurassic and Lower Cretaceous (Pilcher *et al.* 2014) separated numerous fault blocks in this area, requiring restoration of wells back to their original position. Hosston Sandstone was penetrated by the Appomattox (MC 379 #1) well that, when restored some 23 km (14 miles) to the east (an average rafting distance from Pilcher *et al.* 2014), would demarcate a minimum fan length of 90 km, about 10 percent of the average paleo-Chattahoochee–Apalachicola river length.

In summary, seismic interpretations and empirically based predictions discussed above were utilized to revise and refine the Hosston paleogeographic map (Figure 4.5). Individual polygons for the paleogeographic features were edited according to the seismic interpretations, resulting in a relatively short sandy fan in Mississippi Canyon/DeSoto Canyon area, shown in the Hosston paleogeography as a progradational sandy delta-fed apron derived from the Grenville drainage network (Figure 4.5). The apron system fed by the Peninsular River sediment routing system linked to the Peninsular Arch is depicted as a strike-trending slope apron, consistent with seismic mapping and estimates from consideration of source-to-sink scaling relationships (Figures 4.7 and 4.8).

This section illustrates that paleogeographic reconstructions can evolve as new technologies are employed for provenance analysis (Box 1.2), as well as novel uses of empirical scaling relationships derived from modern and submodern datasets (Box 4.1). This point will be emphasized in consideration of the Eagle Ford–Tuscaloosa supersequence (Section 4.8) and several Cenozoic units, including the Wilcox (Section 5.2) and Lower Miocene (Section 6.6).

4.2.7 Sligo Sequence Set Paleogeographic Map Reconstruction

Thickness trends with the Sligo sequence sets (upper portion of the SH supersequence), as determined from both seismic and well data, mirror Mesozoic structural trends, with depositional thinning around the LaSalle Arch, Angelina-Caldwell Flexure (so-called Gulf Flexure of Dooley *et al.* [2013]), and lapouts onto the Sarasota Arch. Prominent carbonate-prone depocenters were noted in south Texas, a precursor to the Maverick basin, and, surprisingly, directly a carbonate-dominated depocenter seaward of the San Marcos Arch (Figure 4.9). Both depocenters are at or just landward of the platform margin deposystem.

Sandstones are relatively rare in the Sligo sequence set on the western side of the basin, limited to a small fluvial system (paleo-Red River) in far northeast Texas that may be linked to the Ouachita Mountains, though provenance information is lacking. An equally small system in south Texas may be a predecessor of the Rio Grande River. By contrast, siliciclastics are prominent in southern Alabama (Grenville River) and northern Florida (Peninsular River), bounded to the south by the Sligo platform margin (Figure 4.9).

It is no coincidence that the sandstones are partitioned in such a fashion; in spite of the structural reorganization and uplift of the Peninsular Arch and Ocala Arch that generated a large amount of terrigenous siliciclastics, reef blocking has effectively limited basinward transport of siliciclastics at this time (Snedden *et al.* 2016b). This transition of siliciclastics to carbonates within the Sligo sequence sets is well-documented in south Texas as well (Phelps *et al.* 2014; Figure 4.2).

As illustrated in the paleogeographic reconstruction (Figure 4.9), the Sligo platform margin (reef) system stretched across the entire northern GoM, as interpreted here and

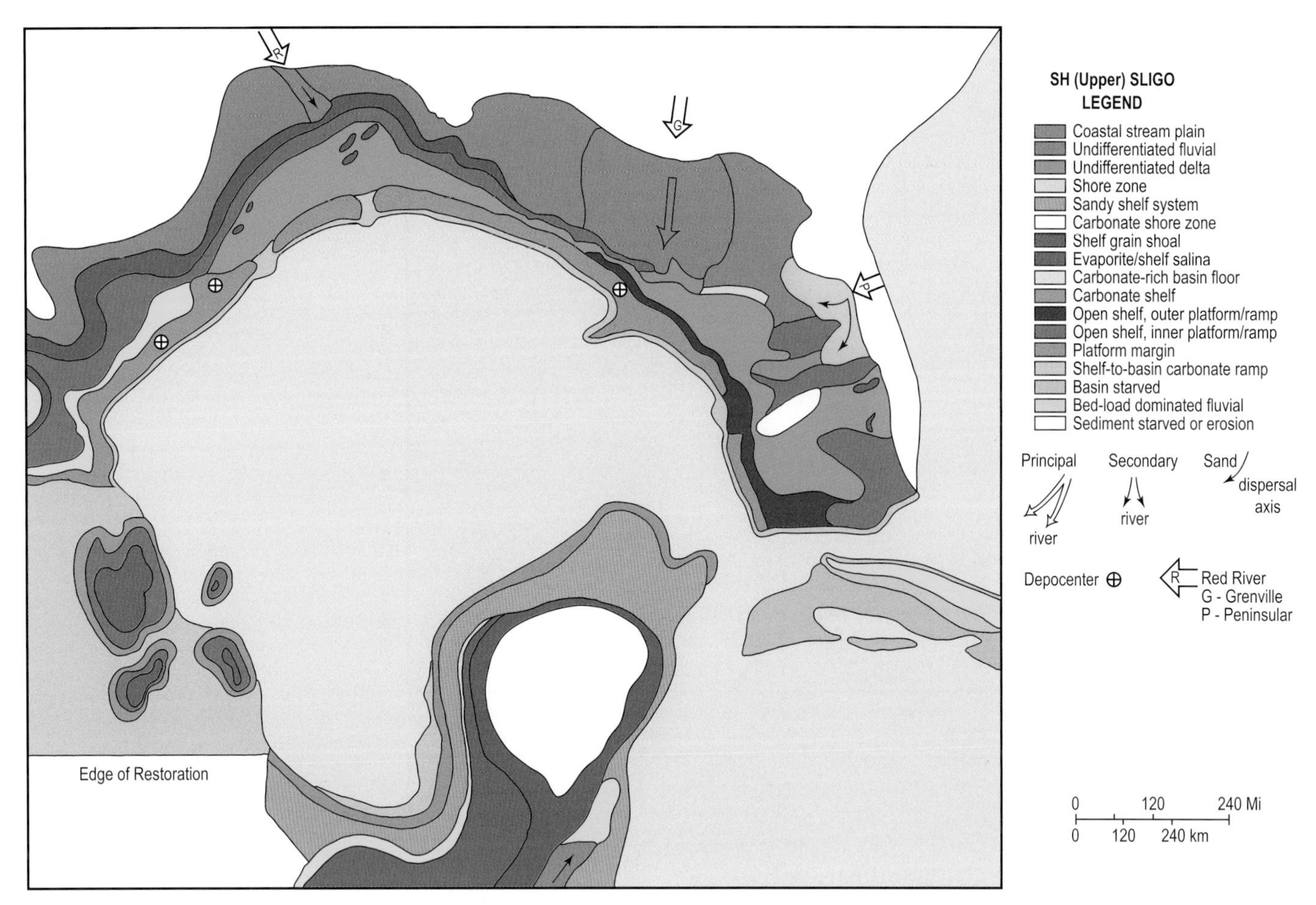

Figure 4.9 Sligo (Sligo upper sequence set) paleogeographic map.

suggested in earlier paleogeographic maps of Goldhammer and Johnson (2001) and Randazzo (1997). In Mexico, the Cupido Formation contains reefal carbonates, but these appear in large, semi-circular isolated carbonate platforms (Figure 4.9). The estimated extent of the Sligo rimmed shelf reef is on par with the largest known reef system in modern times, the Great Barrier Reef (GBR) of Australia (Figure 1.34). From seismic observations, previous publications, and analogy to the GBR, we interpret reentrants or tidal passes to occur along the reef trend, but it is likely that more were present than illustrated in this paleogeographic map.

Like the modern GBR, siliciclastics are delivered by bed-load-dominated rivers, which in turn feed deltas that are mostly restricted to updip areas, well landward of the rimmed shelf reef. In mixed siliciclastic–carbonate systems with well-developed rimmed shelf reefs, entry into basinal areas is often quite limited due to "reef blocking," where siliciclastics are trapped on the shelf (see Box 3.1; Puga-Bernabéu *et al.* 2013). While carbonate turbidities may originate at the shelf edge, siliciclastic turbidity flows form where incised canyons connect to shelf channels (shelf breaching), typically through interreef passages (Figure 3.21).

One mappable deepwater extension of the Sligo age-equivalent interval is shown in the Mississippi Canyon protraction block. This may be dominated by carbonates as it is positioned near a semi-continuous portion of the Sligo platform margin. Wells like EW922 #1 show Sligo-age carbonates penetrated (albeit salt overturned) over 270 km from the coeval shelf edge.

The Sligo paleogeographic map demonstrates the existence of a nearly continuous reef system across the northern GoM, indicating that the laterally restricted platform margin deposystem of the CVK, limited to the eastern GoM, Louisiana, and east Texas had spread across most of the northern GoM continental margin (Figure 4.9). This must have occurred during deposition of the Valanginian–Hauterivian Hosston–Travis Peak interval. Following termination of sea floor spreading around 145–138 Ma, the newly created and stable structural margins may have fostered reef growth or reef-building organisms needed to adapt to the newly expanded GoM basin and attendant changes in wave fetch, bottom currents, surface flow, and oxygen levels. How quickly this occurred is a matter of conjecture, but it is clear from the paleogeographic map of the Hosston that Valanginian-age

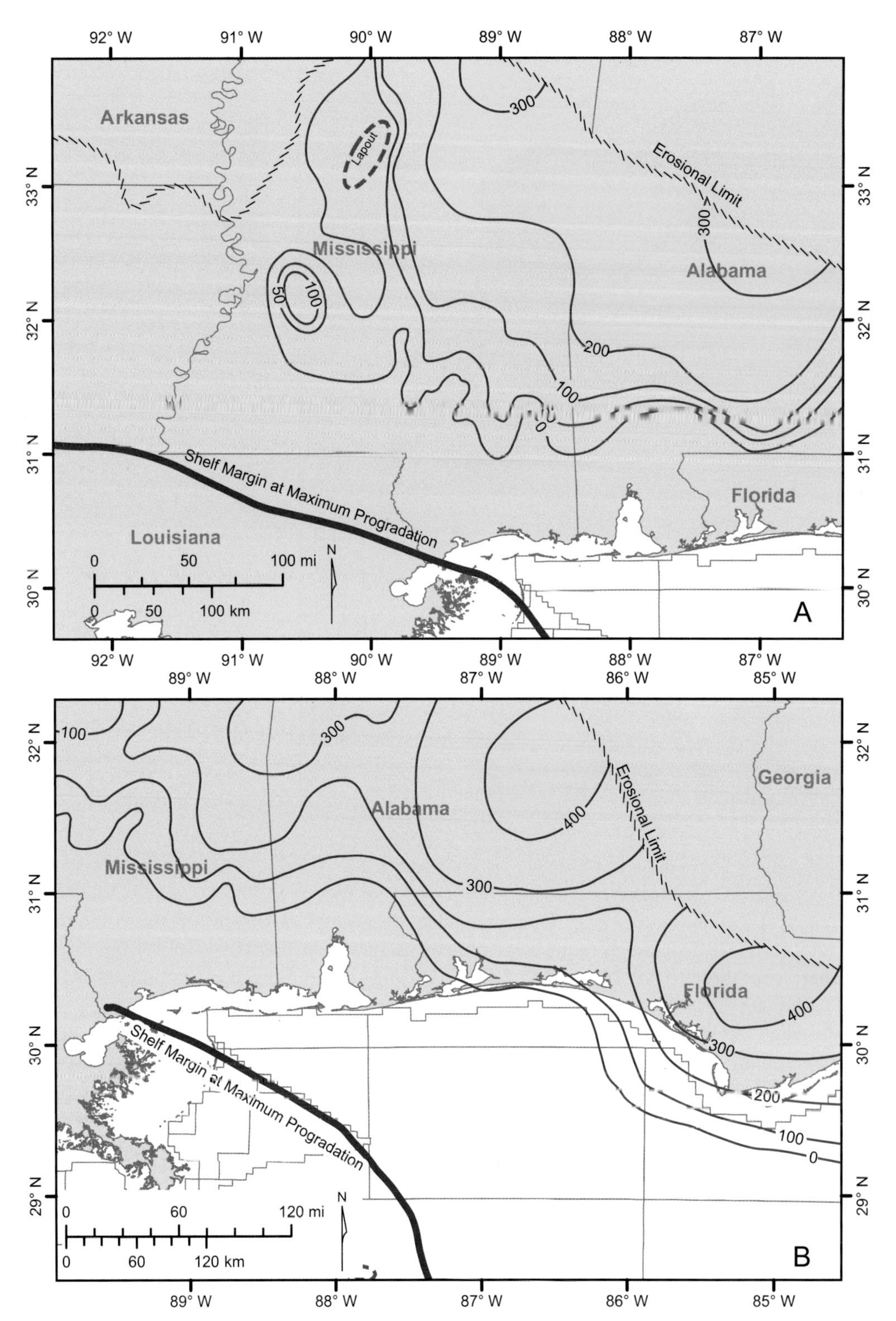

Figure 4.10 Sandstone-bearing interval thickness of (A) Bexar–Pine Island supersequence and (B) Rodessa supersequence.

siliciclastic transport had to pass through a limited number of interreef passages or reentrants in order to deliver sands to the deepwater GoM. This is an important issue, given large salt-related expulsion rollovers (see definition in Section 1.3.7) that could contain Hosston-equivalent sandstones. This is further discussed in Section 9.15.

4.3 Bexar–Pine Island Supersequence

The Bexar–Pine Island (BP) supersequence, also known locally as the Pearsall Shale, is a shale-prone set of sequences that include discrete shoaling-upward carbonate units such as the James Limestone (east Texas) or Cow Creek Limestone (south

Texas). In the subsurface, it extends all the way to Florida, where it is known as the West Felda Shale. In Mexico, the BP is roughly equivalent to a basal portion of the Lower Tamaulipas Formation (Tamaulipas Inferior; Lehmann *et al.* 1999; Figure 4.1). In Cuba, platform carbonates of the basal Palenque in the Remedios area are roughly time-equivalent (Melbana Energy, 2017; Figure 1.32).

The Bexar–Pine Island supersequence was largely deposited during a global rise and subsequent highstand in relative sea level (Haq 2014). Locally, this is manifested as a shale-dominated supersequence and reduced carbonate volume in comparison to the underlying SH supersequence (Figure 4.1). Carbonate platform margins construction paused and a distally steepened ramp developed across much of the GoM (Phelps *et al.* 2014; Figure 4.2). Both carbonates and siliciclastics are present and are thought to be contemporaneous. The timespan of BP deposition extends across oceanic anoxic events (OAEs) 1a and 1b, which likely also impacted the productivity of the carbonate factory (Weissert *et al.* 1998; Phelps *et al.* 2014; see Box 4.2). Tethyan carbonate platforms in central Europe all diminished or drowned at around the same times as here in the northern GoM (Phelps *et al.* 2014).

Sandstones derived from Appalachian source terrane are present in onshore Mississippi and Alabama (Figure 4.10A). This continues into the overlying Rodessa supersequence (Figure 4.10B), suggesting a long-lived drainage system routing sediment from the Appalachians, which we earlier named the Grenville River (Section 3.3.8). However, the siliciclastic sediment influx is much reduced in comparison to the underlying SH supersequence (Figure 4.1).

Box 4.2 What Are Oceanic Anoxic Events?

Oceanic anoxic events (OAEs) are geologically brief (<1 million years) episodes of oxygen-depleted conditions in the global ocean that resulted from profound perturbations in the carbon cycle. They were originally defined as intervals of globally synchronous black shale deposition (Schlanger and Jenkyns 1976), but subsequent work has shown that individual black shales are often diachronous (e.g., Tsikos *et al.* 2004), and OAEs are best defined by their positive carbon isotope excursion (Jenkyns 2010).

Oceanic anoxic events are driven by a net increase in nutrients in the global ocean, possibly related to large igneous province volcanism (Leckie *et al.* 2002), either by direct injection of nutrients to the ocean or through enhanced continental weathering due to strengthening of the hydrologic cycle caused by global warming (Pogge von Strandmann *et al.* 2013). The resulting enhanced productivity results in anoxia as blooming phytoplankton die and decay. In many cases, regionally widespread organic preservation and burial occurs, and it is estimated that source rocks deposited during Cretaceous and Jurassic OAEs could be responsible for up to 50 percent of the global hydrocarbon endowment (Klemme and Ulmishek 1991).

Oceanic anoxic events often, but not always, occur during globally high sea-level states (Leckie *et al.* 2002). For example, the Cenomanian–Turonian OAE2 took place during a major peak in global sea level (Haq 2014). Both the paleo-deep ocean and adjacent continental shelves, including the GoM, are impacted by these OAEs (Núñez-Usche *et al.* 2016; Lowery *et al.* 2017). Eustatic sea-level rise can bring existing oxygen minimum zones onto slopes and even into shelf-depth waters. In the GoM, this is documented in the early Turonian when deep, oxygen-depleted water masses migrated onto the shelf, resulting in anoxia on the Mexican carbonate platforms and dysoxia in other areas (Lowery *et al.* 2017). The deeper shelf, in the productive area of the Eagle Ford trend and across most of Mexico, was bathed in these low-oxygen waters throughout this time interval.

Organic-rich source bed deposition is not the only result of OAEs. Significant biological turnover during an OAE is well documented (Leckie *et al.* 2002). Lowery *et al.* (2017), in a study of a complete cored interval of the OAE2 in Louisiana, noted a major increase in the planktonic foraminiferal population, a decrease in agglutinated relative to calcareous forams, and significant changes in planktonic type. Radially elongated chambers in Cretaceous foraminifera deposited during OAEs were noted by Coccioni *et al.* (2006). Super-anoxic events have been suggested as causes for mass extinctions at the Permo-Triassic boundary (Grice *et al.* 2005). Oceanic anoxic events thus represent important biohorizons, and this is especially true of the GoM Mesozoic interval (Olson *et al.* 2015).

Oceanic anoxic events also have detrimental impacts on carbonate production and reefal development. On the Comanche Platform of Texas, OAEs are often associated with reef termination, as seen at the top of the SH supersequence (OAE1a), midway through the Glen Rose (GR) supersequence (OAE1b), and at the top of the Georgetown Formation (OAE1d; Phelps *et al.* 2014, 2015; Figure 4.11). Anoxia associated with OAE1a–b is known from shale samples taken from deepwater wells in the northern GoM (Lowery *et al.* 2017). In Florida, the detrimental effects of OAE1b anoxia greatly affected the Sunniland carbonate interval and ultimately resulted in deposition of an organic-rich shale, the primary source of the Upper Sunniland oil (Liu 2015). Recovery of the carbonate factory commenced slowly as dysoxia diminished, initially confined to nearshore areas but eventually recovering to form caprinid rudist patch reefs and ooid grain shoals on the shelf. This pattern of carbonate termination (crisis), slow recovery, and eventual return to equilibrium is also observed on the Comanche Platform (Phelps *et al.* 2015), as associated with OAEs and their aftermath.

Oceanic anoxic events have also been recognized in Mexico (Núñez-Useche *et al.* 2016; Figure 4.11). OAE1a has been isotopically constrained to the lower part of the La Pena Formation in northeast Mexico (Figure 4.1; Bralower *et al.* 1999). OAE2 is known from outcrops of the Agua Nueva and Morelos Formation (Núñez-Useche *et al.* 2014) and further west in the Ojinaga Formation in the Chihuahua Trough (Frush and Eicher 1975). However, relative sea-level rise may have played a greater role in platform reef demise than anoxia in Mexico, as many of the central Mexican carbonate platforms were drowned after OAE2, by the early Turonian highstand (Núñez-Useche *et al.* 2014, 2016; Lowery *et al.* 2017).

Box 4.2 *(cont.)*

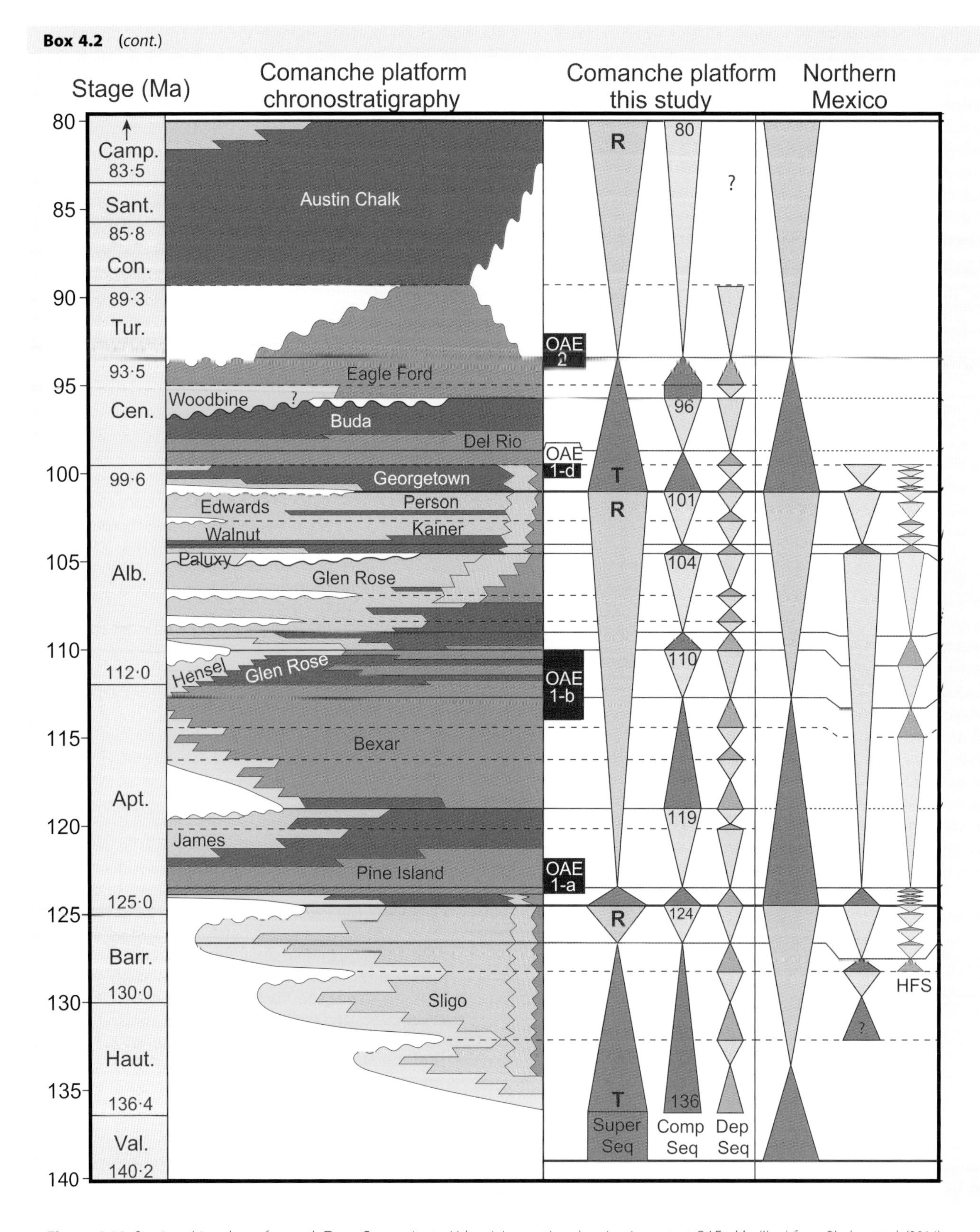

Figure 4.11 Stratigraphic column for south Texas Campanian to Valanginian section showing important OAEs. Modified from Phelps *et al.* (2014).

4.3.1 Chronostratigraphy

The BP supersequence extends from Aptian-3 sb to Albian-1_sb, approximately 15 million years (111.12–126.01 Ma). This includes two major OAEs (OAE1a and OAE1b; Bralower *et al.* 1994; Phelps *et al.* 2014; Figure 4.11). At least six depositional cycles, likely sequences, can be identified from detailed well log correlations (Hull and Loucks 2010).

4.3.2 Previous Work

Loucks (1977) provides core-based descriptions of the BP carbonate shoal facies, including echinoid–mollusk skeletal grainstones to patch reefs dominated by coralgal–stromatoporid–rudist boundstones. Lagoonal skeletal packstones and wackestones developed between shoals with updip tidal mudflats dominated by terrigenous mudrocks. In the subsurface of south Texas, the Pearsall Formation includes three members, in ascending order, the Pine Island, Cow Creek, and Bexar (Loucks 1977). The middle carbonate-rich Cow Creek Limestone is known in east Texas and Louisiana as the James Limestone, an important hydrocarbon-producing horizon (Kosters *et al.* 1989; Webster *et al.* 2008) discussed in Section 9.11. Considerably less work has been done on the coeval units of the northeastern GoM (e.g., the Lehigh Acres Formation of Florida), which are less hydrocarbon-productive.

4.3.3 Paleogeographic Map Discussion

In central Texas, basal Hensel sands are thought to be derived from erosion of the Llano Uplift (Hull and Loucks 2010; Figure 4.12). These shore zone sandstones are restricted to exposures west of Fredericksburg, Texas, suggesting more western sources. Shore zone and wave-dominated deltaic sandstones are also locally present on the eastern margin of the Texas basin, possibly derived from the Ouachita Mountains. However, the largest sand-prone depositional system in the BP supersequence is the western Appalachian (Grenville River) derived siliciclastic system of onshore Mississippi and Alabama (Figure 4.12). Sandstone reservoirs formed in fluvial–deltaic systems nicknamed the Hazlehurst, Collin, and Richton lobes (Reese 1976; Figure 4.10A).

Elsewhere in the GoM basin, carbonates dominate the BP supersequence (Figure 4.12). The Cow Creek Limestone includes the largest size and thickness of grain-dominated shoal water carbonates (shelf grain shoal deposystem) formed on the flank of Uvalde Embayment. Other carbonate bodies are present in the Bexar Shale interval (Hull and Loucks 2010). Carbonate shoals also occur in east Texas and in the South Florida basin. Petty (1999) schematically depicted an abnormally large bank of "interior grainstones" extending from DeSoto Canyon to the Florida Middle Ground protraction

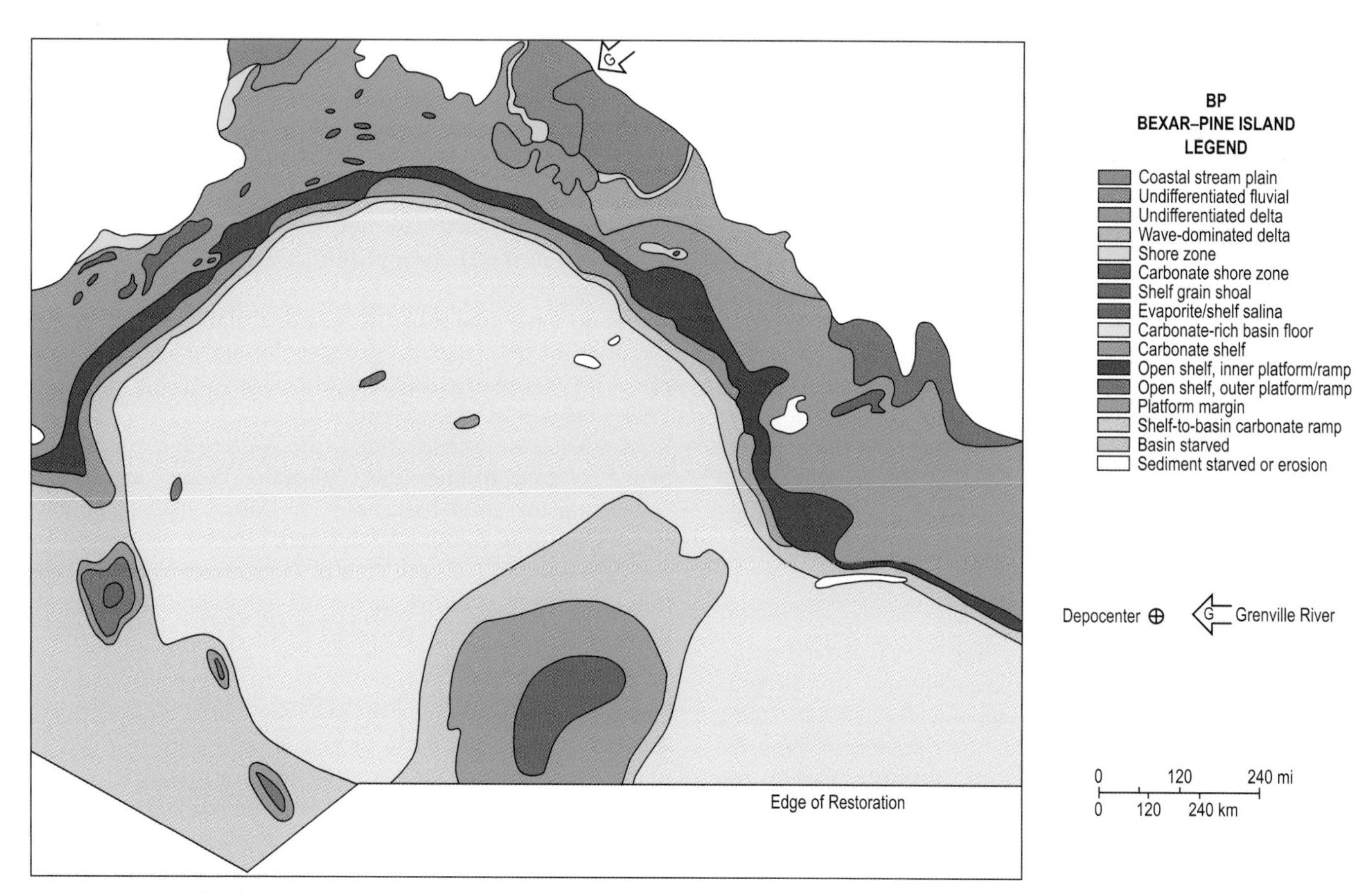

Figure 4.12 Bexar–Pine Island supersequence paleogeographic map.

blocks, which we reinterpret as a reef apron, the landward part of the platform margin deposystem. However, examination of logs in offshore wells FM 455 #1 and FM 252 #1 does not indicate the characteristic log signature of carbonate grain shoals seen elsewhere in the BP supersequence.

4.4 Rodessa Supersequence

The Rodessa (RD) supersequence has a short timespan of deposition and is also areally more limited than almost all of the chronostratigraphic units in the GoM. Regional stratigraphic correlation identified the Rodessa as a grainy carbonate- or sandstone-bearing unit below the Ferry Lake or "Massive" Anhydrite, but above shales of the BP supersequence. The Rodessa represents the earliest platform margin reef development coming out of the long-term retrogradation and OAEs of the underlying BP supersequence (Figure 4.2).

The Rodessa extends in the subsurface from east Texas to Florida, where it is lithostratigraphically equivalent to the Able and Twelve Mile members of the Lehigh Acres Formation (Randazzo 1997). In south Texas, the Rodessa is not generally distinguished due to lithostratigraphic similarities with the lower portion of the GR (Figure 4.11). The RD supersequence is also not well-recognized beyond the Albian platform margin, where the top RD seismic horizon converges with other reflections by downlap in the deep basin.

In Mexico, the RD may be roughly equivalent to the Otates member of the Lower Tamaulipas Formation (Lehmann *et al.* 2000). The Otates and coeval La Pena are shale-dominated in most Mexican basins (Figure 4.1).

4.4.1 Chronostratigraphy

The RD supersequence extends from Al1_sb (111.12 Ma) to Al3_sb (110.73 Ma), an approximately 0.4-million-year duration. In spite of the short duration, at least four carbonate shoaling-upward cycles are documented in surface exposures and subsurface wells (Kosters *et al.* 1989; Phelps *et al.* 2014, 2015). This suggests development of multiple parasequences or high-frequency sequences (*sensu stricto* Mitchum and Van Wagoner 1991) during a major carbonate growth and aggradation phase. RD supersequence carbonate volume in the northern GoM actually exceeds that of the underlying BP and overlying GR supersequences, deposited over much longer timeframes (Figure 4.1).

4.4.2 Previous Work

Most work has been done on the well-known skeletal grainstone shoals (shelf grain shoal deposystem) that are reservoirs for oil and gas in the East Texas salt basin (Kosters *et al.* 1989). Little thought has been given to the question of how the Rodessa transitions seaward to the platform margin reef system that existed since the Aptian–Barremanian (Sligo) and continued into the late Albian (Washita). The lack of regional seismic data, until recently, inhibited consideration of the larger framework, as was done for the Comanche shelf and platform margin of central Texas (Phelps *et al.* 2014). Locklin (1985) presented a paleogeographic map for northeast Texas, which showed the transition from shore zone sandstones, carbonate shelf with common shelf grain shoals to a basinal carbonate mud-dominated paleo-environment. No platform margin was illustrated by Locklin (1985), but presumably a relict or palimpsest reef must have existed as discussed in the following section.

4.4.3 Paleogeographic Map Reconstruction

It is logical to posit that a platform margin reef system, so prominent in the underlying Aptian–Barremanian succession (SH supersequence) and one that reached its acme in the overlying Albian Washita, must have existed during RD deposition (Figure 4.13). A barrier reef system probably allowed development of shoals and local patch reefs as envisioned by Locklin (1985), as without it wave conditions would have inhibited shelf carbonate bank development. The dominance of skeletal material in the Rodessa argues for a reef framework in the vicinity. This is a similar relationship as observed on the Comanche shelf of central Texas, where Upper GR sand shoals transition downdip to the Stuart City Platform margin reef (Figure 4.2; Phelps *et al.* 2014). As discussed earlier, rimmed shelf reefs are present from Hauterivian to the late Albian, except for the retrogradational phases (early Aptian, early Albian, and early Cenomanian) when distally steepened ramps were developed (Phelps *et al.* 2014). At least three wells studied here confirm the presence of a platform margin reef development in the RD supersequence, with the best development in east Texas and Louisiana (Figure 4.14).

In Mexico, we illustrate a hypothetical platform margin reef at the Yucatán Platform, but there is little well control here and seismic facies are not well-developed on this steep margin. Isolated carbonate platforms, so prevalent in the Barremanian (SH supersequence) of onshore Mexico to the west, appear to have diminished in size and abundance, with the exception of the Tuxpan (Tampico–Misantla) and Coahuila platforms that persisted into the late Albian (Lehmann *et al.* 1999; Phelps *et al.* 2014; Figure 4.13).

A possible early connection to the Western Interior Seaway may have gone through the Chihuahua Trough in Mexico, connecting the GoM basin with the Mural shelf of Sonora (C. Kerans, pers. comm.). Albian reefal carbonates present in outcrops of the El Calosa area of Northwestern Mexico are thought to be correlative to the GR Limestone and Pearsall Formation of the GoM (Warzeski 1987).

4.5 The Glen Rose Supersequence

The GR supersequence can be broadly split into two depositional sequence sets: (1) vertically and areally restricted carbonates of the Lower GR; and (2) extensive, open-platform carbonates of the Upper GR (Phelps *et al.* 2014; Figure 4.2). The early phase of the GR supersequence is commonly marked by widespread deposition of evaporites. These evaporites are

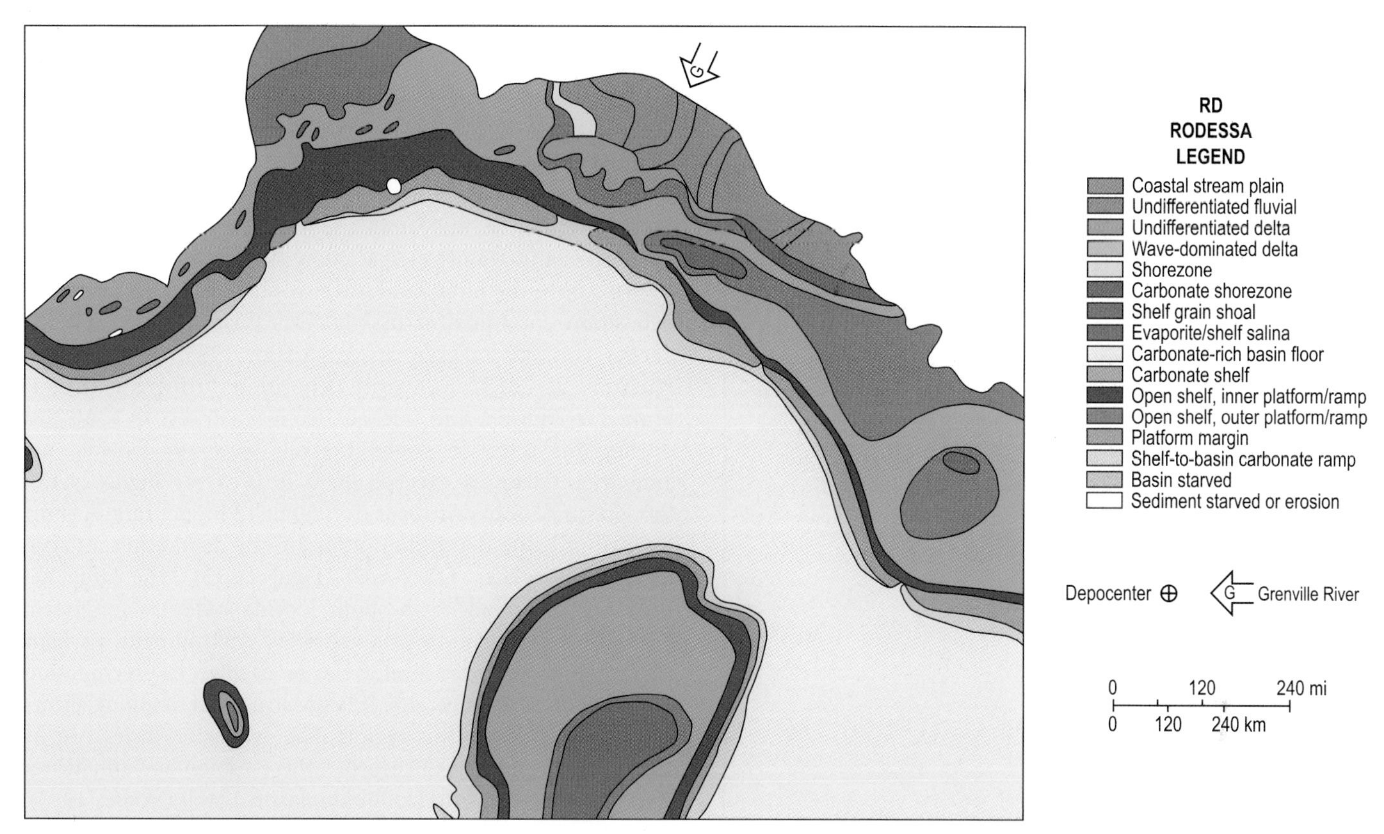

Figure 4.13 Rodessa supersequence paleogeographic map.

generally mapped as the Ferry Lake Anhydrite, a lithostratigraphic unit developed along the northern Gulf Coast, and the equivalent Punta Gorda Anhydrite in Florida. Thus, this evaporite unit extends in the subsurface from the flank of the San Marcos Arch in east Texas to south Florida (Figure 4.15). It is interbedded with shallow-water carbonate mud and carbonate grainstones/packstones formed in shoals and sediment drifts (Mitchell-Tapping 1986). Evaporitic facies of the Ferry Lake indicate a hypersaline, restricted shelf environment, and suggests the inception and maturation of a basin-wide barrier reef system that impeded the exchange of waters on the shelf with normal marine waters offshore. The GR Formation itself is named for outcrops near the town of Glen Rose, Texas where dinosaur tracks are a popular geologic and tourist attraction (Ewing 2016).

In Mexico, carbonate platforms and margins expanded, with deposition of the Akal Formation of the Sureste Basin, the Orizaba Formation of the Cordoba Platform in the Veracruz basin, the El Abra and Tamabra of Tampico–Misantla, and Cuesta del Cura of northern Mexico onshore Sabinas basin (Figure 4.1). Platform interior basins in Mexico were also characterized by widespread evaporites (McFarlan and Menes, 1991).

4.5.1 Chronostratigraphy

The GR supersequence (including the Ferry Lake Anhydrite) spans the Al3_sb (110.73 Ma) to Al9_sb (103.13 Ma), and covers 7.6 million years. Equivalent lithostratigraphic units include the Mooringsport of Louisiana, and in Florida the Rattlesnake Hammock, Lake Trafford, and Sunniland Formations (Petty 1995). It is the second of three supercycles in the Early Cretaceous corresponding to carbonate platform growth and demise, between the Valanginian–Aptian SH and the Albian–Cenomanian Paluxy–Washita (PW) supersequences. The GR supersequence was terminated with a drowning event at the Al9_sb and the deposition of the Paluxy Sand, marking onset of the PW supersequence.

4.5.2 Previous Work

For the GR carbonates, we integrate our interpretations with depositional trends first identified by Barros (1987), Salvador (1991b), Yurewicz *et al.* (1993), Ward *et al.* (1995), Goldhammer and Johnson (2001), Pollastro *et al.* (2001), and Hentz and Ruppel (2011). For the Ferry Lake evaporites, a notable contribution is the work of Loucks and Longman (1987), who documented evaporite–carbonate cycles that dominate the Ferry Lake lithofacies in Henderson County, Texas (Figure 4.16). Cyclic sedimentation is also well documented in this portion of the GR in areas to the southwest, where evaporite deposition does not occur (Phelps *et al.* 2014; Figure 4.11). In these areas, high-frequency cycles are manifested as numerous depositional sequences, with shallow-water facies successions of subtidal through supratidal deposits described in shallow-water environments (Cleaves 1977). In coeval middle shelf environments, subtidal, shoaling-upward

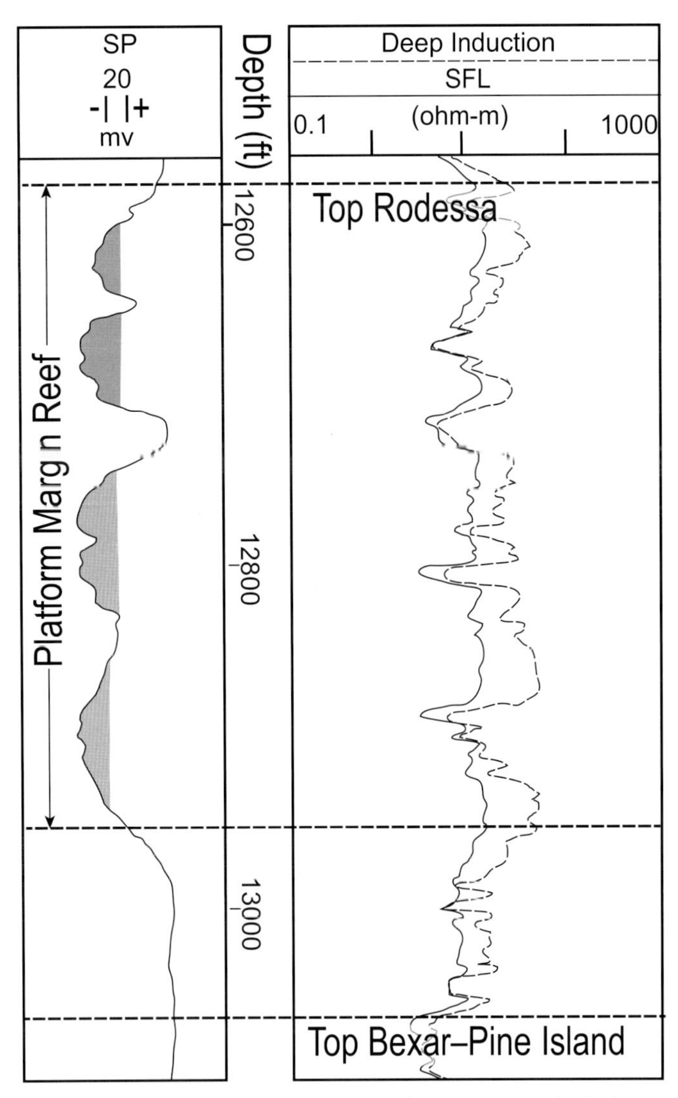

Figure 4.14 Example of platform margin reef development in the Rodessa supersequence of east Texas. Log from Damson #1 Carter, Tyler County, Texas. Porous and permeable zones shaded on SP log. Abbreviations: SP, spontaneous potential; mv, millivolts; SFL, shallow focused resistivity log.

deposits of terrigenous mud, carbonate debris, and sandstone are common in south Texas (Bay 1985).

The Albian expansion of platform margin reefs commenced in the GR of Texas with the evolution and proliferation of the caprinid rudistids, large, slender, and asymmetric pelecypods. Caprinids appear to have adapted well to both restricted back-reef/lagoonal and open marine platform margin conditions with enhanced wave activity (Wilson 1975). Rudists met their demise at the end of the Cretaceous, coincident with the Chicxulub impact, and were replaced in Cenozoic reefs by forerunners to modern corals (Kiessling *et al.* 1999).

Glen Rose slope deposits (forereef to toe of slope) observed in cores are dominated by skeletal debris, peloids, breccia clasts, and planktonic forams. These grade seaward to basinal calcareous micrites (Phelps *et al.* 2014).

Above the Ferry Lake and coeval Punta Gorda Anhydrite of Florida, a widespread, normal marine limestone was deposited across the entire Gulf shelf, mapped variously as the Lower GR (Texas), Mooringsport (Louisiana–Alabama), and Sunniland (Florida). The Mooringsport of east Texas and Sunniland of south Florida are both hydrocarbon-productive but volumetrically limited. In the Sunniland trend, reservoirs are small patch reefs and grainstone/packstone banks (Mitchell-Tapping 1986). Above the Sunniland interval, shallow-water carbonate deposition continued across the Gulf coastal plain, and the barrier reef system grew apace (Adams 1985; Perkins 1985; Phelps *et al.* 2015). Anhydrite deposition during this time is limited to Florida, and thickens from northwest to southeast during this time. In some intervals in south Florida, net anhydrite thickness exceeds 2000 ft (610 m; Figure 4.15), though single anhydrite beds over 50 ft (15 m) are rare. A short interval of halite deposition occurs in the depocenter, as seen in the Collier Co. #12-2 well (Figure 4.15). The large net thickness of anhydrite in south Florida indicates long-term but episodic deposition in a restricted environment, perhaps also modulated by sea-level cycles or changes in precipitation on Milankovitch timescales. Seismic structural mapping shows a depression across this region. The evaporitic facies can be traced to the Bahamas Platform, where Albian-age anhydrites were penetrated by the Doubloon Saxon 1 well (Walles 1993).

Individual beds and bedsets of anhydrite are correlative on well logs across long distances (Pittman 1989; Petty 1995). Thicker anhydrite-dominated intervals are often evident on seismic surveys as bright parallel reflections, though the high-impedance carbonates often interbedded with the anhydrite do contribute to the increased reflectivity.

4.5.3 Paleogeographic Map Reconstruction

The paleogeographic map depicts the younger part of the GR supersequence, when anhydrite deposition is limited to south Florida and much of the northern GoM shelf is an open carbonate shelf environment (Figure 4.17). Most of the shelf is composed of shallow-water carbonates with grain shoals and patch reefs that make up the primary hydrocarbon reservoir on the GR shelf, with small exploration plays both in east Texas (Mooringsport) and south Florida (Sunniland). Sunniland patch reef build-ups are largely caprinid rudistid bafflestones, floatstones, and rudstones (Liu 2015) associated with ooid shoal facies (Loucks and Crump 1985). At least 14 Sunniland ramp interior reef/shoal complexes can be identified across Lee, Collier, Hendry, and Dade counties of south Florida (Liu 2015; Figure 4.18). These are the primary reservoirs of the Sunniland petroleum reservoir field trend, as discussed in Section 9.13.

Like the overlying PW supersequence, there is limited deepwater bypass of terrigenous material, likely due to the barrier reef system ("reef blocking"; see Box 3.1). One possible exception to this trend is a small delta system in southern Mississippi/Louisiana which was identified from seismic data

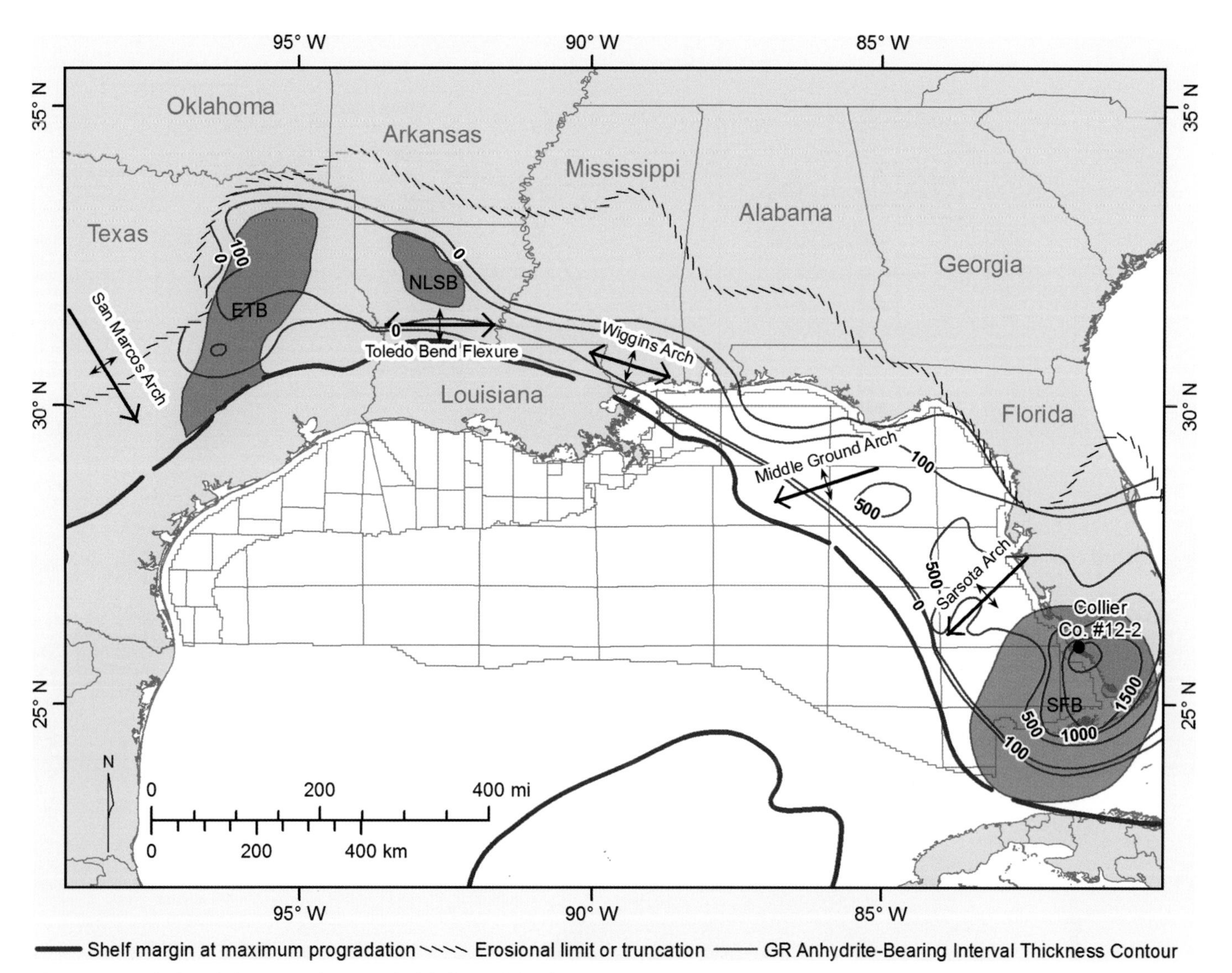

Figure 4.15 Anhydrite thickness map for Ferry Lake lithofacies of the Glen Rose supersequence. Net anhydrite thickness is mapped from well log interpretation of anhydrite occurrence. This is defined as intervals of very high gamma-ray and very low resistivity, controlled where possible (about half of the wells) with lithologic log data from cuttings. Location of halite penetration in center of South Florida basin by Bass Collier #12-2 well of Florida is indicated.

(Figure 4.17). The Upper GR here exhibits seismic-scale clinoforms, which occupy a downdip gap in the platform margin reef system. Unfortunately, no well penetrations occur in this area, and so it is difficult at present to determine whether this small delta is sandstone- or mudstone-prone. This is also in proximity to large salt structures (Section 1.3.7) located downdip in eastern Mississippi Canyon that may be related to progradation of a siliciclastic unit driving salt evacuation. The presence of similar features in the same region during the GR supersequence suggests a persistent fluvial source in this area, connected to an interreef passage or tidal pass. This feature mapped in the GR bolsters the interpretation of a persistent Lower Cretaceous (SH to GR) deepwater entry point.

In Mexico, GR carbonates formed large, isolated carbonate platforms with interior evaporitic lagoons, rising above a deep carbonate ramp (Goldhammer and Johnson 2001; Figure 4.17). Deeper shelf sites are composed of foraminiferal mudstones and wackestones representing a continuation of the Tamaulipas Formation from the SH supersequence. Locally restricted carbonate platforms are known from deposition of the evaporitic Acatita Formation (Lehmann *et al.* 1999). The deepwater GoM is shown as a starved basin center with pelagic carbonates transitioning to a carbonate slope/forereef depositional system near the shelf margin (Figure 4.17).

The GR rimmed shelf reef system extends from south Texas to south Florida, based on seismic mapping of the platform margin across the northern GoM. Like the Sligo reef, the GR rimmed shelf reef system is similar in scale to the GBR of Australia (Figure 1.34). Also like the Sligo, we interpret the occurrence of tidal passages through the reef based on seismic observations and previous publications; because of the low resolution of seismic data versus the scale of interreef passages, it is likely that there are more passages than illustrated here.

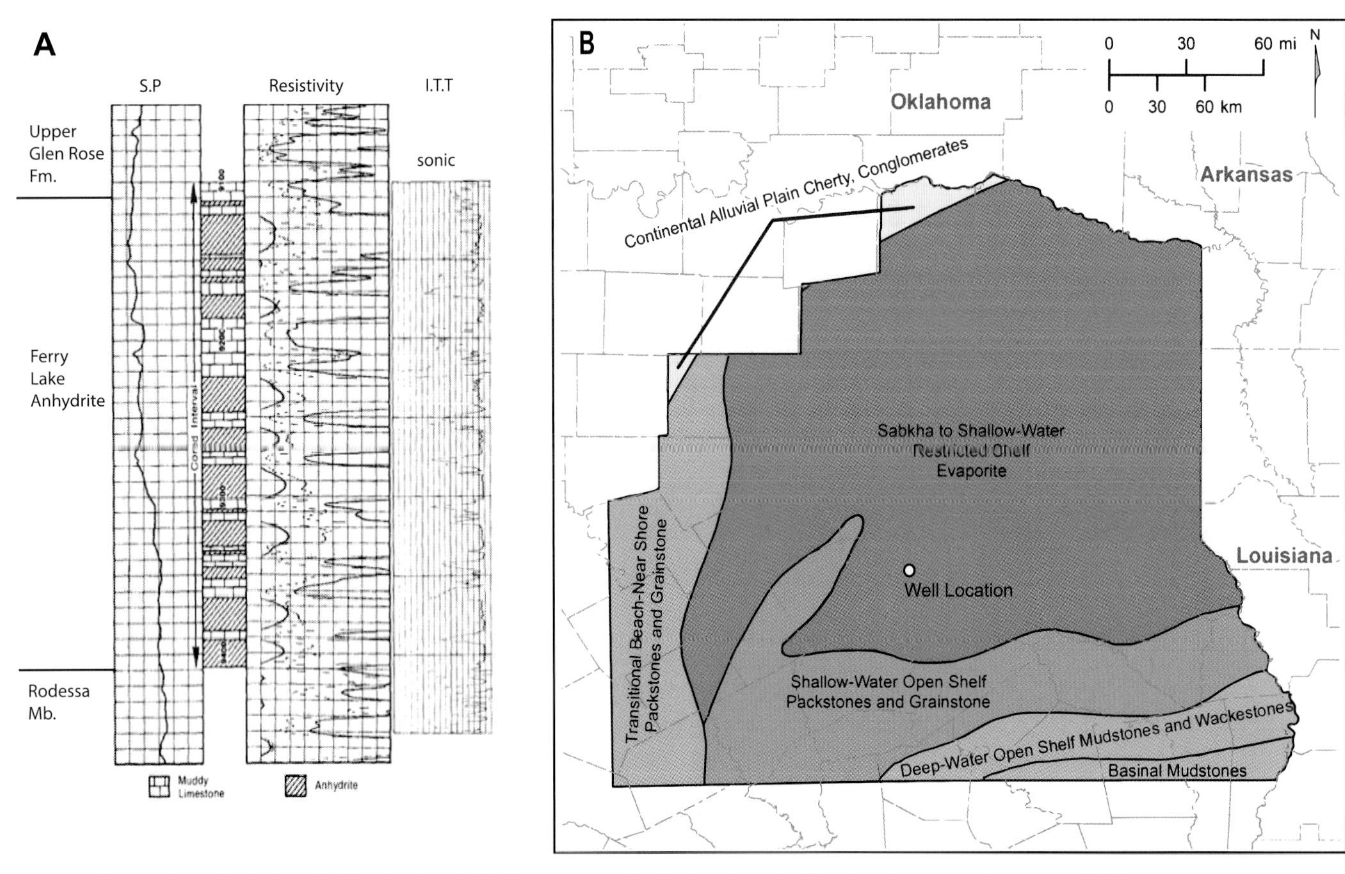

Figure 4.16 Ferry Lake lithofacies. (A) Cities Service #1 J.B. Kitchens well. (B) Well location. Modified from Loucks and Longman (1987).

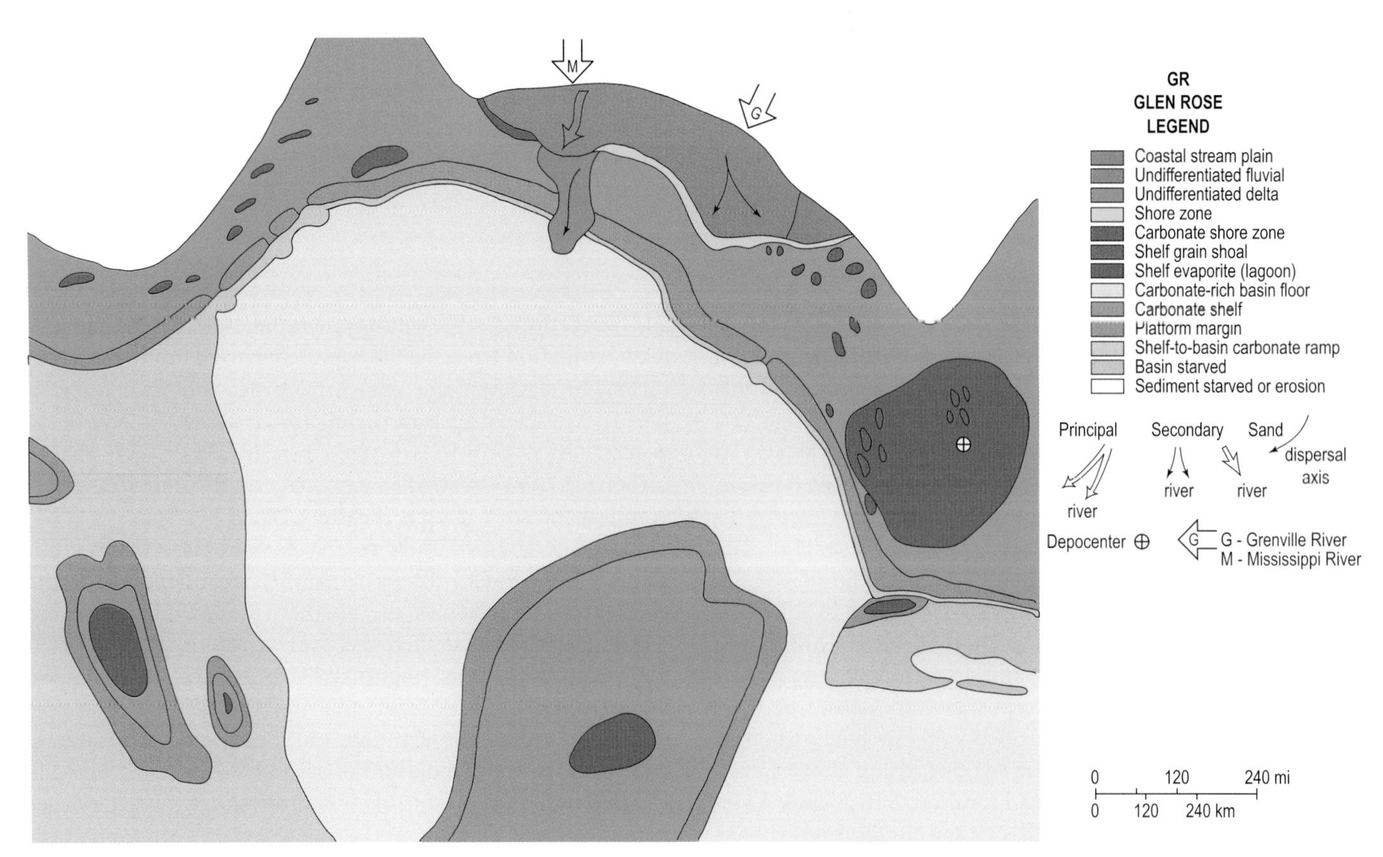

Figure 4.17 Glen Rose supersequence paleogeographic map.

The correlation between cyclic deposition of anhydrite facies and low-amplitude sea level changes in the Albian (e.g., see charts in Snedden and Liu 2011; Haq 2014) leads one to conclude that relative sea-level changes and restriction behind the barrier reef system promoted anhydrite deposition in the GR supersequence. However, platform margin reef systems are normally segmented by interreef passages that allow free exchange of marine seawater between shelf and basin (Maxwell and Swinchatt 1970; Figure 1.34). Thus, development of hypersaline conditions behind the platform margin may have involved tectonic obstructions to limit marine seawater exchange. The Wiggins Arch, for example, is sub-parallel to the contemporaneous platform margin and may have acted in just such a fashion (Figure 4.15). Another structural trend potentially playing a role is the Toledo Bend Flexure. The San Marcos Arch, which trends perpendicular to these structural features, may have acted as a lateral barrier (Figure 4.15).

Comparison of the Ferry Lake Anhydrite thickness map (Figure 4.15) and GR paleogeographic reconstruction (Figure 4.17) suggest the possibility that lagoonal restriction east of the San Marcos Arch was probably due to fewer interreef passages and a more continuous platform margin reef. Restricted circulation must have been particularly acute during the early GR time during the massive and widespread deposition of the widespread Ferry Lake/Punta Gorda Anhydrite facies. This episode of evaporite deposition is the most significant since the end of Louann Salt deposition (see Section 3.2).

It is notable that the GR is the only carbonate supersequence in the Early Cretaceous to contain such thick anhydrite deposits, and therefore is suggestive of a unique confluence of environmental and paleogeographic circumstances with: (1) enhanced evaporation/reduced precipitation; (2) a more restrictive reef system with fewer interreef passages; and (3) tectonic obstructions preventing communication with the

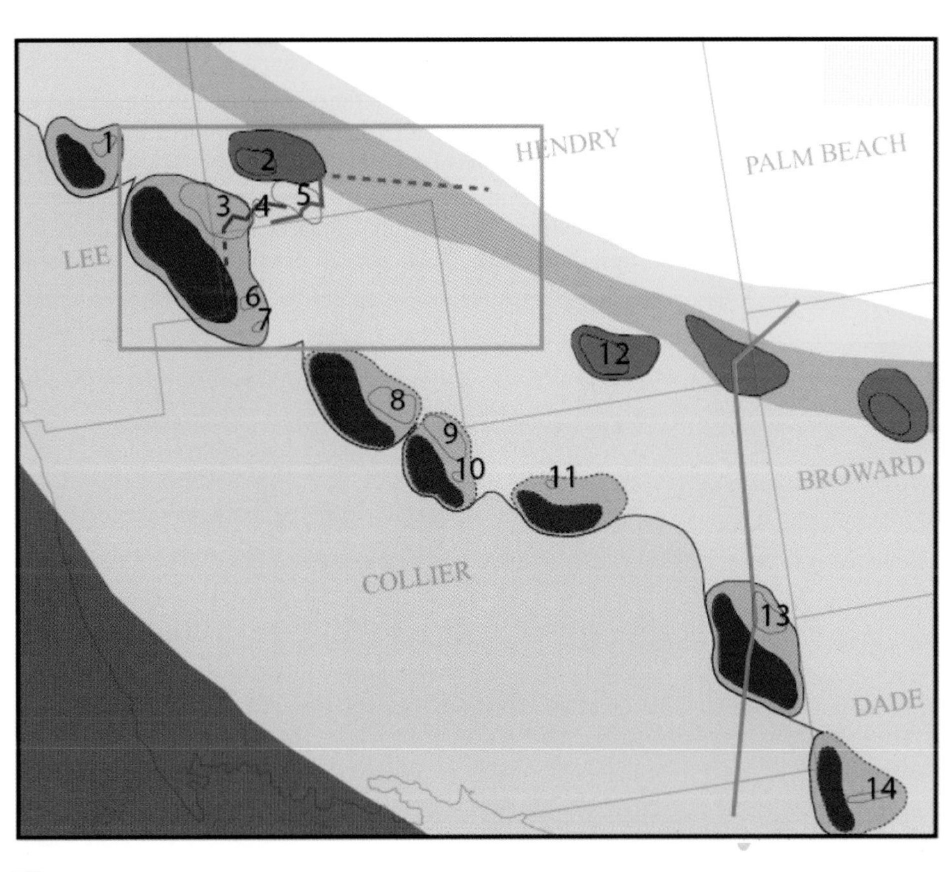

Figure 4.18 Facies and interpreted depositional environments of the Sunniland ramp interior carbonates in the South Florida basin during the active growth stage of these caprinid reef build-ups. Sunniland play fields are indicated by number keyed to the map. Modified from Liu (2015).

open ocean. Restricted conditions were most acute during the early portion of the GR supersequence, and only present on the Florida Platform for most of the Upper GR. Thus, it is possible that the number of interreef passages increased as the GR carbonate system matured. It is also significant that there is no Ferry Lake Anhydrite present west of the San Marcos Arch, except for a small deposit in the Maverick basin, known locally as the McKnight Formation. The lack of evaporites here suggests the area south of the San Marcos Arch was better connected to the open ocean, perhaps due to smaller reef buildups or more interreef passages.

4.6 Paluxy–Washita Supersequence

In the GoM, platform margin reefs reached an acme in extent, thickness, and influence on basin deposition during deposition of the PW supersequence (Yurewicz *et al.* 1993; Phelps *et al.* 2014; Figure 4.19). Platform margin reefs expanded across the northern GoM but also may have developed on the margin of Yucatán, though there is comparably less well control there. In Texas, the term Stuart City trend is reserved for the Albian (including the GR) shelf margin. Well and seismic studies indicate this shelf-margin rimming reef tract is at least 1000 km (620 miles) long by 5–10 km (3–6 miles) wide (Wilson 1975; Phelps *et al.* 2014). As documented by abundant well control, carbonate-bearing interval thicknesses consistently averaged 2000 ft (610 m) or more along the platform margin in the northern GoM and in areas landward of the reef (Figure 4.20).

Major platform margin reef systems also flourished in Mexico as evidenced by the development of the Golden Lane, Valles, Actopan, El Doctor, and other carbonate platforms (Goldhammer and Johnson 2001; Figure 4.1). However, there are some notable differences with the northern GoM platform margin reefs. A number of isolated carbonate platforms formed seaward of the Mexican Albian reef margins, often nucleating upon basement structures and salt diapirs (Gutteridge *et al.* 2019). Slopes along isolated Mexico platforms were quite steep (5–30°; Janson *et al.* 2011; Figure 4.21) versus Stuart City forereef slopes of less than 2° (Phelps *et al.* 2014). Breccia-dominated debris aprons often formed on the steep Mexican forereef slopes. Siliciclastics are less prominent updip of the Mexico bank margins where platform interior evaporitic facies are more common (Figure 4.1). Platform margin reefs continued longer in Mexico, persisting until the end of the Cenomanian, versus an abrupt decline of the Stuart City Platform margin reef system at the end of the Albian in the northern GoM (Wilson 1975). But a key difference for petroleum accumulation is the post-depositional uplift and structural rotation at Golden Lane that put the Tamabra forereef reservoirs in a favorable trap configuration (Janson *et al.* 2011). The Stuart City trend did not experience such tectonism. Only local closures develop in the Stuart City due to carbonate buildups, differential compaction, or extensional faulting at the shelf

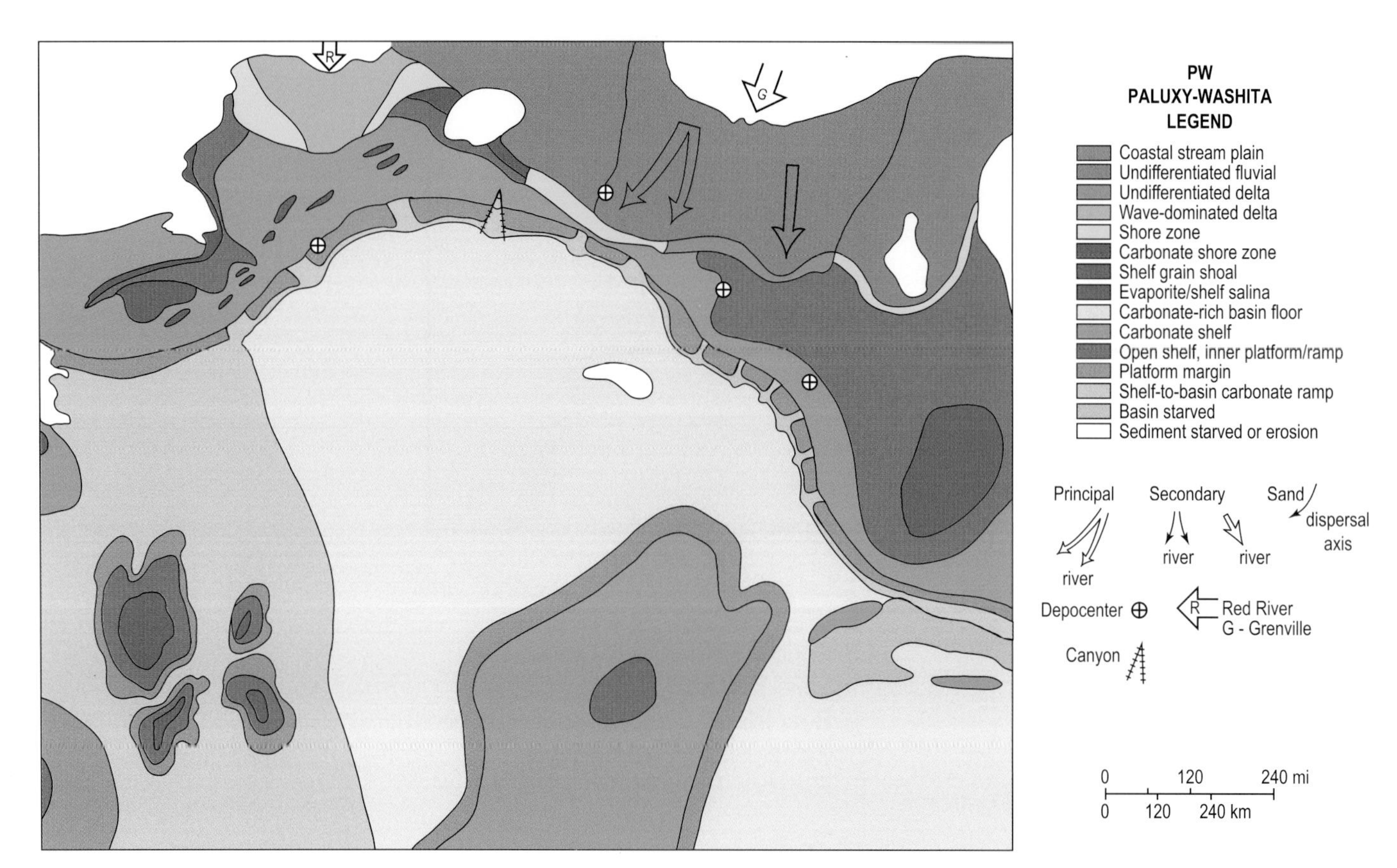

Figure 4.19 Paluxy–Washita supersequence paleogeographic map.

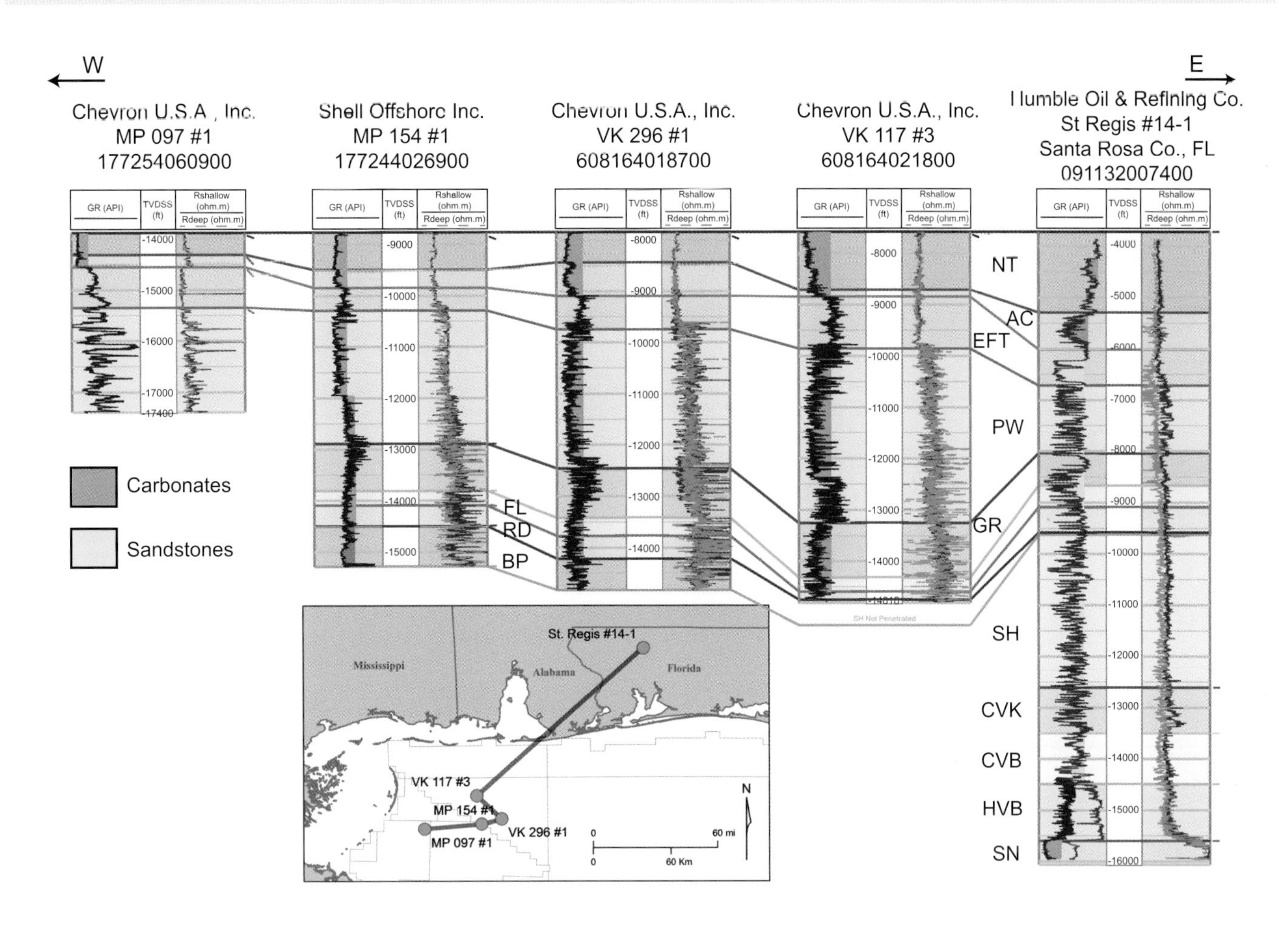

Figure 4.20 Mesozoic cross-section, Central GoM Mesozoic units, Oxfordian to Top Cretaceous. Abbreviations: AC, Austin Chalk; BP, Bexar–Pine Island Shale; CVB, Cotton Valley–Bossier; CVK, Cotton Valley–Knowles; EFT, Eagle Ford–Tuscaloosa; FL, Ferry Lake Anhydrite; GR, Glen Rose; HVB, Haynesville–Buckner; NT, Navarro–Taylor; PW, Paluxy–Washita; RD, Rodessa; SH, Sligo–Hosston; SN, Smackover–Norphlet.

margin. Pervasive marine cementation is also detrimental to reservoir quality in the Stuart City (and Sligo; Achauer 1977) versus meteoric diagenesis and even karst and cavern formation at the Golden Lane and Tuxpan Platforms (Janson 2004; Janson *et al.* 2004).

In the northern GoM, coeval siliciclastics, formed largely in fluvial to nearshore and deltaic paleo-environments, were limited to areas north of the nearly continuous reef margin (Figure 4.20). Siliciclastics bypassed the margin in the USA in rare interreef passages in the Main Pass area (Figure 4.19). Siliciclastic volumes are substantial (Figure 4.1), as seen in well-known Paluxy and Dantzler Sandstone of the central and eastern GoM (Figure 4.20).

4.6.1 Chronostratigraphy

The PW supersequence extends from Al9_sb (103.19 Ma) to Ce3_ssb (96.24 Ma), a duration of about seven million years. Seventeen biohorizons characterize this Late Albian to Early Cenomanian succession. The interpreted chronostratigraphy of this supersequence is also supported by use of oxygen and carbon isotope excursions indicating ties to the global perturbation of the carbon cycle associated with OAE1d at the Albian–Cenomanian boundary (Figure 4.11; Box 4.2). Termination of the Albian platform margin may be coincident with the onset of OAE1d, just prior to the Cenomanian (Phelps *et al.* 2015).

4.6.2 Previous Work

The sedimentological character, distribution, and macrofaunal content of the Washita Group and its equivalents (Stuart City, Edwards, Georgetown; Figure 4.2) has been well-documented by Goldhammer and Johnson (2001), Yurewicz *et al.* (1993), Pitman (2014), Barros (1987), and Phelps *et al.* (2014). In Texas, platform margin lithofacies are dominated by sponge–microbial–coral boundstones, with caprinid rudistids in the reef wall (Phelps *et al.* 2014). In contrast to earlier work (e.g., Wilson 1975; Bebout *et al.* 1977), new work suggests that caprinid rudists are far more common in the shelf reef apron located landward of the platform margin, mainly as rudstones and grainstones (Phelps *et al.* 2014). Here, local subaerial exposure is indicated by carbonate cement textures.

The PW supersequence platform margin can be traced from onshore Texas to offshore areas of the eastern GoM. In

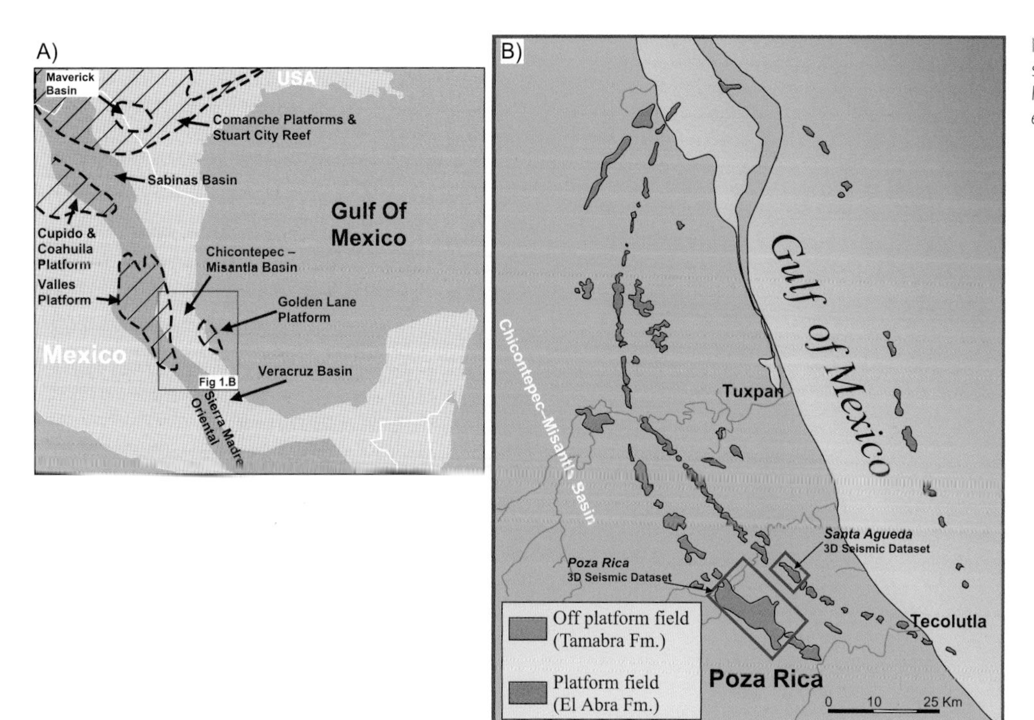

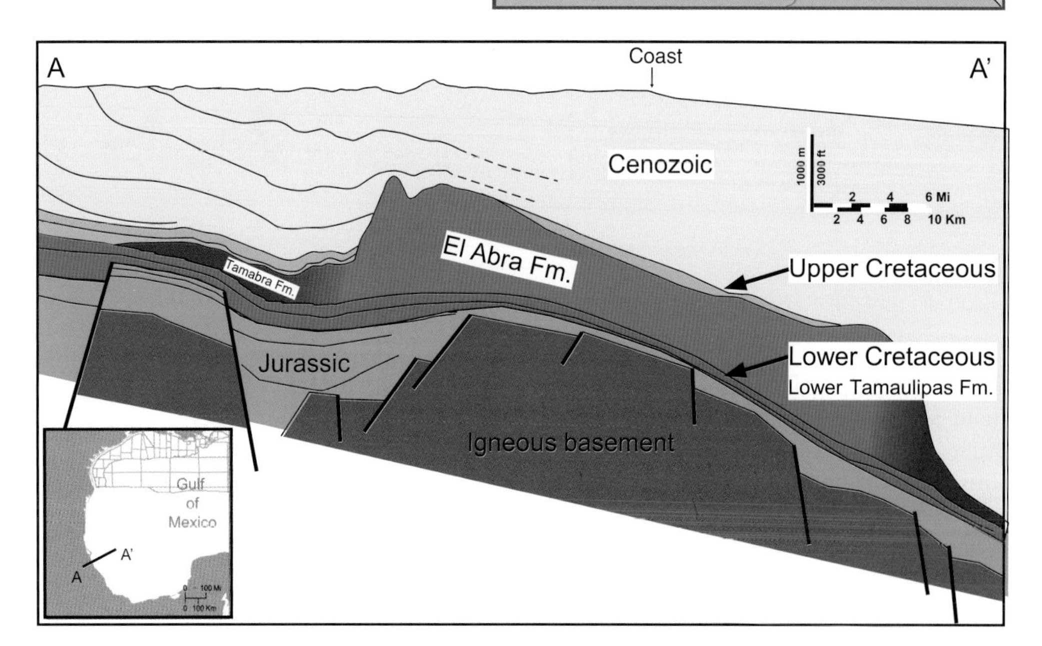

Figure 4.21 Regional schematic cross-section of the Golden Lane Platform. Modified from Coogan *et al.* (1972), Chen *et al.* (2001), and Janson *et al.* (2011).

the Main Pass/Viosca Knoll area of the eastern Gulf, the name Andrew Limestone is used to refer to the undifferentiated Washita–Fredericksburg interval (Petty 1999). At least three separate carbonate platform margins can be identified on logs, though these are more amalgamated on 2D seismic data, particularly due to structural complexity imposed by salt tectonics. As in Texas, stromatoporiods and coralline red algae are major faunal components of the platform margin reef in the eastern GoM, with rudists locally abundant (Petty 1999).

The Golden Lane Platform and Poza Roca trend have been investigated thoroughly due to their economic importance (Chen *et al.* 2001; Janson *et al.* 2004, 2011). The El Abra Formation was deposited in a large-scale, isolated carbonate platform with steep flanks (Figure 4.21). The coeval slope deposits are referred to as the Tamabra Formation. Both formations contain oil reservoirs, some of the most prolific in Mexico, as described in Section 9.14. Post-depositional tectonics resulted in the Tamabra slope deposits being rotated into a position structurally higher than the coeval shallow-water platform (Chen *et al.* 2001; Janson *et al.* 2011).

Ward *et al.* (1995) presented well data from Mexico Yucatán onshore areas showing that the long-emergent Yucatán

(Mayan) block was finally covered by more than 1000 ft (328 m) of carbonate deposition in the Albian. Contemporaneous source rocks are developed in areas adjacent to the Atlantic Ocean (Florida Straits and Cuba), as documented by Pollastro *et al.* (2001) and Moretti *et al.* (2003).

4.6.3 Paleogeographic Map Reconstruction

The PW platform margin reef trend is inferred to extend across the entire GoM basin. Stratal lapouts are mapped on the Ocala Arch, northern Louisiana, and the San Marcos Arch (Figure 4.19). Carbonate depocenters are found in central Texas (Stuart City trend), Main Pass/Viosca Knoll protraction blocks, and further east in the Elbow protraction block. A siliciclastic-dominated depocenter, with thicknesses of 1000 ft (610 m) or more, is present in the Mississippi salt basin of Mississippi and Alabama, probably related to the Grenville River coming out of the long-emergent Appalachians. The effectiveness of the Albian reef blocking (Box 3.1) at preventing bypass to the deepwater basin is apparent in the limited thickness of sandstones penetrated in deepwater areas (Figure 4.20). In the deepwater of the GoM, PW-age units are present in a number of carapaces and deep tests (e.g., GB 754#1 [Norton prospect]; EW 922 #1 [Wrigley], and AC 557#1 [BAHA II]) but are entirely carbonates. Alternatively, it is possible that Albian sand-prone sediment gravity flows may have entered the basin through the Main Pass entry point (Figure 4.19) but avoided the salt-influenced structural highs targeted for drilling.

Onshore, the PW supersequence is dominated by carbonates of the Washita Group and its stratigraphic equivalents (Stuart City, Edwards, etc.). Lesser amounts of sandstones are found in two stratigraphic intervals, the lower called the Paluxy Formation and an informal upper sandstone interval called the Dantzler. However, there is some question as to the age of the Dantzler sandstone and it may actually be part of the basal Tuscaloosa (Chasteen 1983). In any case, the presence of thick Paluxy sandstones signals the continuing influence of fluvial systems draining the southern Appalachians in the Albian, with a major sandstone depocenter in the Mississippi salt basin. Sandstones thin laterally into Texas, possibly indicating waning influence of the Quachita source terrane. However, mapping the unit into the Viosca Knoll to the Main Pass area suggests some systems in the latter part of the PW timeframe may have managed to find a pathway into the GoM through the PW platform margin reef system.

4.7 Summary of Post-Oxfordian Mesozoic Deposition

Following deposition of the Oxfordian Smackover–Norphlet supersequence, carbonates dominated in the GoM, coincident with evolution of an extensive platform margin reef system in the northern Gulf and large isolated carbonate platforms in Mexico (Figure 4.1). Over 60 million years, platform margin reefs evolved from a limited extent in the eastern Gulf during deposition of the Jurassic Haynesville–Buckner (HVB) supersequence (Figure 4.22A) to a Gulf-wide shelf margin by the Mid-Cretaceous PW supersequence (Figure 4.22 B). Prominent shelf margins of the HVB and Cotton Valley–Bossier (CVB) supersequences were restricted to areas near inflow from the Atlantic (world) Ocean, aligned with the maximum fetch of the early basin. Reef expansion to the west in SH was accompanied by the rise of caprinid rudistids and associated faunal elements, culminating in the northern GoM with the Washita Group (Stuart City) reefs (Figure 4.22B). In Mexico, isolated carbonate reef platforms expanded in the Albian (Figure 4.22B) and continued into the Cenomanian within the tectonostratigraphic phase.

Siliciclastics are generally subordinate to the carbonate systems during the Drift and Cooling Tectonostratigraphic Phase. Siliciclastic input from multiple rivers steadily increased

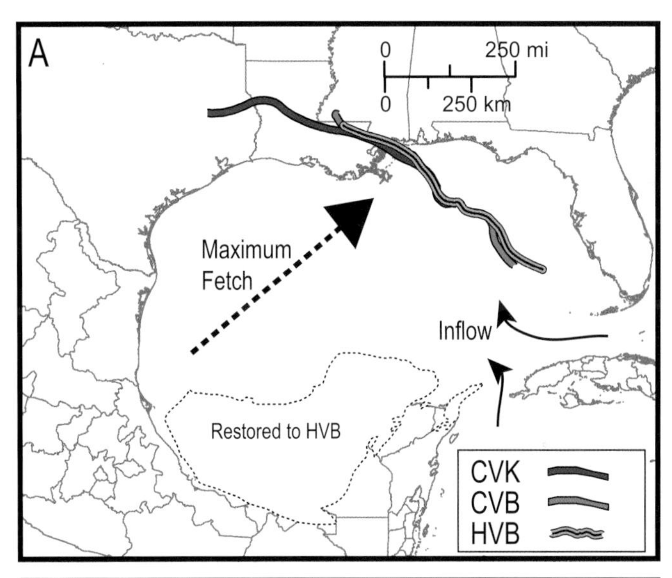

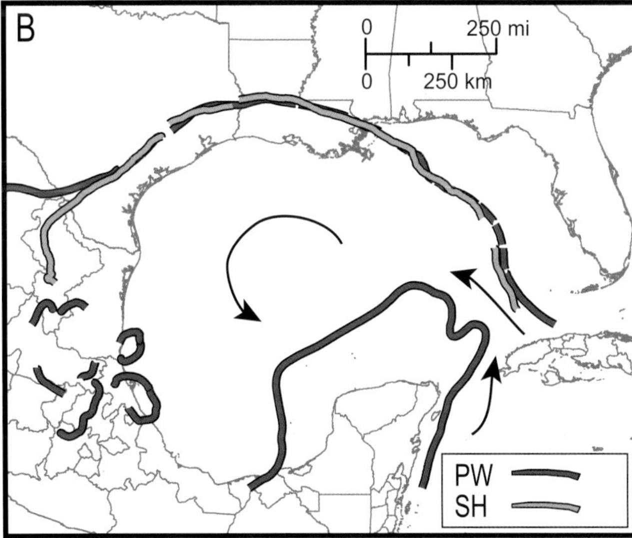

Figure 4.22 Platform margin reef distribution during the Drift and Cooling Phase. (A) Jurassic: HVB, CVB, and CVK supersequences. (B) SH and PW supersequences.

from the Oxfordian into the latest Jurassic, followed by a huge influx in the earliest Cretaceous (Hosston–Travis Peak; Figure 4.1). However, this also parallels expansion of Lower Cretaceous carbonate platform margin reefs, effectively trapping siliciclastics on the shelf and preventing sands from reaching the deep basin. After a major retrogradation in the Aptian (BP supersequence), platform margin construction accelerated in the Albian, again acting to limit siliciclastic bypass to the basin. This pattern of reef blocking is abruptly terminated in the Ceno-Turonian, as described in Section 4.8.

The base of the Eagle Ford–Tuscaloosa supersequence represents the demise of Albian (PW) platform margin reef systems in the northern GoM (Phelps *et al.* 2015). In Mexico, platform margin reefs persisted longer, eventually drowning in the Turonian (Núñez-Useche *et al.* 2016). In Cuba, reefs may have continued briefly into the Cenomanian (Figure 1.32), though information is limited here. What follows the Albian stage is a voluminous increase in siliciclastics of Ceno-Turonian-age sandstones into the northern GoM (Figure 4.23), and bypass of the relict Albian (PW) platforms (Figure 4.24) into the deep northern GoM (Snedden *et al.* 2016b). Carbonate volumes declined by nearly an order of magnitude (Figure 4.23).

Thus, the long-term pattern of reef blocking of siliciclastics changed dramatically in the Upper Cretaceous (Cenomanian–Turonian). This transition is so abrupt that some authors (e.g., Cox and Van Arsdale 2002) have suggested a major tectonostratigraphic boundary is present between the PW and Eagle Ford–Tuscaloosa supersequences, the so-called Middle Cretaceous unconformity (MCU). The MCU was thought to extend into the deep basin, but subsequent work has demonstrated that the MCU was miscorrelated with the Top Cretaceous disconformity (Dohmen 2002; Denne *et al.* 2013). While the basal Tuscaloosa Sandstone is present in a number of incised valleys in onshore Louisiana (Ambrose *et al.* 2009, 2015; Woolf 2012), it is not a major, basin-wide tectonostratigraphic boundary. The transition from a carbonate-dominated interval to one with common siliciclastics represents major expansion of fluvial catchments and a major drainage reorganization in North America as indicated by detrital zircon provenance results (Blum and Pecha 2014). Continued sea-level rise throughout the Late Cretaceous (cf. Miller *et al.* 2005) also challenged platform margin reef systems to keep up (Phelps *et al.* 2014). A direct connection between the northern GoM and the Western Interior Seaway was probably established by the Ceno-Turonian global highstand in sea level (Haq 2014), changing water circulation patterns in a substantive fashion.

The Cordilleran orogenic belt in Mexico, which includes the Mexican fold and thrust belt and foreland basins to the east, was tectonically active in separate areas spanning the Turonian to Campanian and Campanian to Maastrichtian stages (as well as the Cenozoic Ypresian to Lutetian stages; Fitz-Díaz *et al.* 2018). Subduction of the Farallon slab in the Late Cretaceous to Eocene is thought to be the main driver here, and probably had effects in the northern GoM. Our discussion of the Late Mesozoic interval in Mexico is focused on areas east of the fold and thrust belt, where the bulk of the hydrocarbon discoveries are located.

4.8 Eagle Ford–Tuscaloosa Supersequence

The Eagle Ford–Tuscaloosa (EFT) supersequence north of the San Marcos Arch has at its base a sandstone-dominated sequence set, called the Woodbine Sandstone in the east Texas basin and Tuscaloosa Sandstone east of the Sabine Uplift (Figure 4.23). The basal contact is typically unconformable due to fluvial incision here but is also in disconformable contact with Albian carbonates in areas away from these paleovalleys, such as in southwest Texas (Donovan *et al.* 2012).

To the south of the San Marcos Arch, the EFT supersequence is largely devoid of sandstones and is dominated by calcareous mudstones of the world-class Eagle Ford unconventional source rock play of south Texas. The Eagle Ford also extends to Mexico, where the equivalent Agua Nueva play is in its early stages of development. Future deepwater exploration potential remains in some of the largest undrilled prospects in the Mississippi Canyon protraction block that will likely rely upon the Tuscaloosa Sandstone as the main reservoir. The younger Tuscaloosa Marine Shale of Louisiana also has some potential as a future unconventional play, though recent drilling results have been disappointing (Lowery *et al.* 2017), as discussed in Section 9.15.1.

4.8.1 Chronostratigraphy

The EFT supersequence extends from Ce3_ssb (96.24 Ma) to Sa1_sb (85.99 Ma), a duration of 10.3 million years (timescale of Gradstein *et al.* 2012). The Eagle Ford also contains numerous volcanic ash beds and first-cycle zircons that broadly confirm the above age range of the coeval Boquillas Formation that outcrops in southwest Texas (Pierce 2014). However, the Eagle Ford Group of subsurface wells is particularly rich in fauna and flora, with over 30 biohorizons noted within this Cenomanian to Late Turonian succession (Denne *et al.* 2014; Olson *et al.* 2015). The Eagle Ford Shale is also known to include a global OAE (OAE2, 93.6–94.3 Ma; Lowery *et al.* 2014; Eldrett *et al.* 2015; Figure 4.11). The reader is referred to Olson *et al.* (2015) and Alnahwi *et al.* (2018) for details on biostratigraphic and chronostratigraphic datums used to define this supersequence.

The base of the Turonian occurs in the lower portion of the Tuscaloosa Marine Shale of Louisiana, which in turn is overlain by the regressive sandstones of the Upper Tuscaloosa, equivalent to part of the upper Eagle Ford (roughly equivalent to the Langtry Formation of southwest Texas; Donovan *et al.* 2012). In updip intervals, notably in the Tuscaloosa outcrop belt, the Tuscaloosa Marine Shale is absent and sandstones of Cenomanian–Turonian age, entirely equivalent to the Eagle Ford, are present (Lowery *et al.* 2017).

As discussed earlier in Section 3.1, Cretaceous sea level reached a global peak at the Ceno-Turonian boundary (Haq 2014). This may explain a propensity for organic enrichment

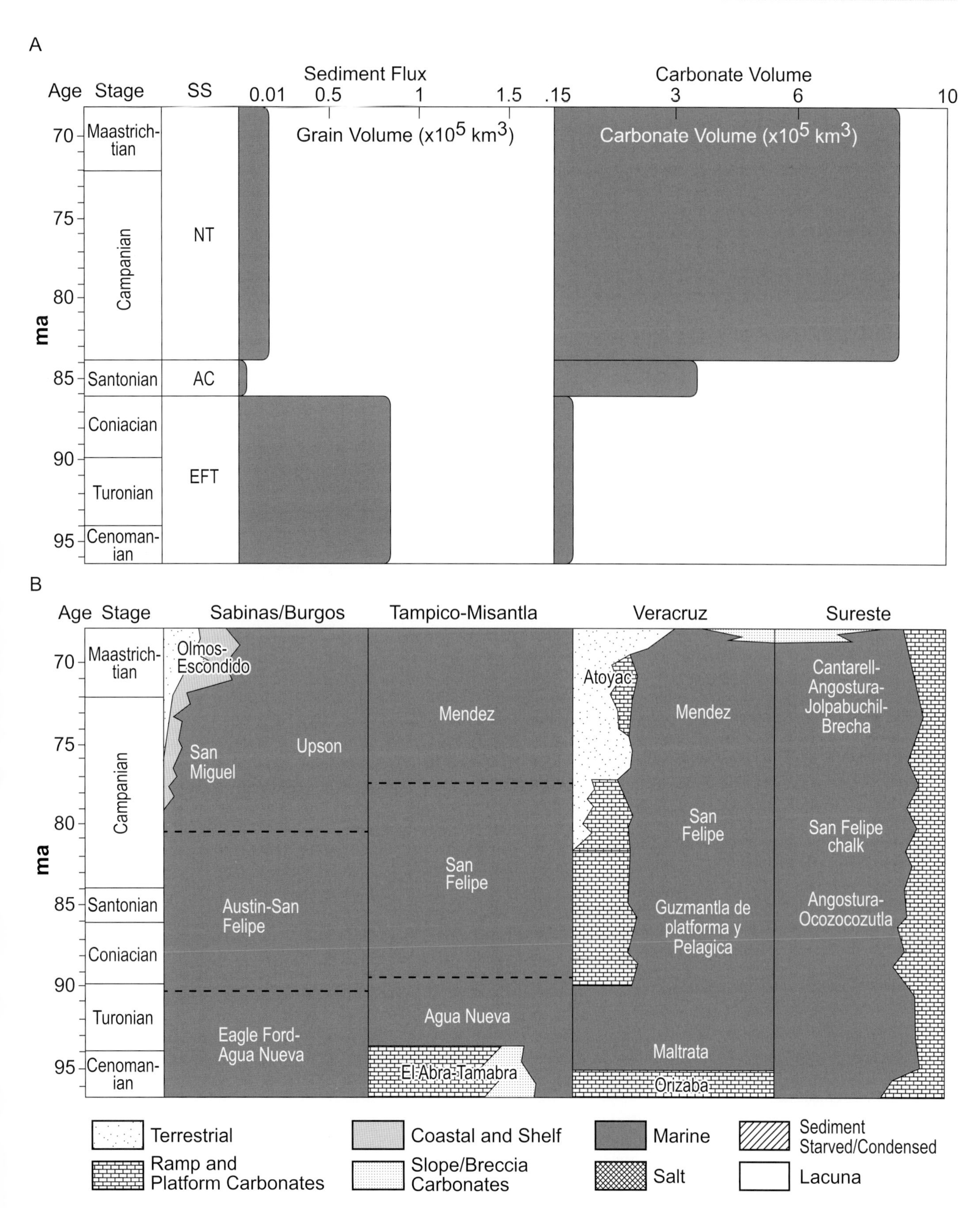

Figure 4.23 GoM Local Tectonic and Crustal Heating Phase, part 2. (A) Northern GoM siliciclastic sediment flux and carbonate volume by supersequence. (B) Lithostratigraphy for four areas of the southern GoM.

and source rock development observed in many localities, as oxygen minimum zones expanded during OAEs (Lowery *et al.* 2017). The GoM stratigraphic record does not align particularly well with this global sea-level highstand, as work described in Section 9.15.1 shows that the highest organic content with the Eagle Ford occurs somewhat earlier than this peak (Phelps *et al.* 2015). Furthermore, the second largest volume of siliciclastics generated in the Mesozoic occurs in this general timeframe, as represented by the Tuscaloosa and Woodbine Formations of the northern GoM (Figure 4.23). This is, of course, consistent with our general view that the GoM Mesozoic (and Cenozoic) stratigraphic record is a convolved archive of tectonic, eustatic, and sediment supply factors, interacting over multiple long- and short-term cycles.

4.8.2 Previous Work

Cenomanian- to Turonian-age sandstones (Woodbine, Lower Tuscaloosa, and equivalents; Figure 4.23) have been of great interest in the Gulf Coast since discovery of the supergiant east Texas field in 1930 (Alexander 1951). The Lower Woodbine Group west of the Sabine Uplift includes fluvial channels, deltas, strandplains, shelf, and slope sandstones (Foss 1979; Phillips 1987; Ambrose *et al.* 2009), but is absent in south Texas due to non-deposition and erosion (Adams and Carr 2010). East of the Sabine Uplift, the Lower Tuscaloosa Sandstone is roughly coeval and includes fluvial, deltaic, and slope deposits (Barrell 1997; Woolf 2012). Sandstones in Louisiana may extend into the Turonian that in Texas is dominated by shales of the Eagle Ford Formation (Adams and Carr 2010).

Numerous unconformities punctuate the late Albian to Coniacian interval (Ewing 2009), implying local tectonic influence during a period of high relative sea level (cf. Snedden and Liu 2011). The base of this supersequence coincides with what was thought to be a Gulf-wide unconformity known as the MCU. Later work has indicated that this unconformity is only locally developed in the Mississippi Embayment, where a substantial lacuna separates Upper and Lower Cretaceous strata (Ewing 2009). Cox and Van Arsdale (2002) speculated that the present-day Mississippi Embayment was uplifted during westward passage of the area over the Bermuda hotspot in the Mid-Cretaceous. An alternative hypothesis is uplift related to changes in subduction of the Farallon slab (Liu 2015).

Whatever the driver, uplift of the Mississippi Embayment is the likely cause for deep incision of Tuscaloosa-age valleys as mapped by Woolf (2012). These, in turn, may have provided both a partial sediment source and pathway for delivery of siliciclastics to the deep GoM basin. This idea is further explored later in discussion of the basin-scale paleogeography.

As mentioned, the MCU of onshore areas was initially miscorrelated with the Top Cretaceous or GBDS Top Navarro–Taylor (NT) supersequence (Faust 1984). Drilling of the Showboat well (AT 336 #1) revealed that the basin-wide, high-amplitude seismic reflections were actually at the Cretaceous–Paleogene boundary (Dohmen 2002). However, the importance of the base EFT or base Tuscaloosa supersequence boundary for bypassing quartz-rich sands to the basin was stressed by Horn (2012).

Much work has been done on the Eagle Ford Formation of south Texas, primarily driven by economic interest in the unconventional shale play itself that is detailed in Section 9.15.1. Recently, Alnahwi *et al.* (2018) studied long subsurface cores and described the major lithofacies of the Eagle Ford and their transitions in a dip transect from the Maverick basin to the south of the underlying Stuart City Platform margin in south Texas. The low-relief relict shelf margin appears to have partitioned the Lower Eagle Ford, in particular, into an updip restricted shelf basin and downdip open marine basin. Lithologic and geochemical parameters vary between these settings, with highest organic content within the restricted shelf basin updip of the margin.

In Mexico, the EFT supersequence includes the Agua Nueva and basal San Felipe Formations (Burgos–Tampico–Misantla basins) and Maltrata (Veracruz) formations that are known for their source potential (Prost and Aranda-Garcia 2001; Ortuno-Arzate, *et al.* 2003; Blanco *et al.* 2011; Figure 4.23). These are carbonate–mudstone-dominated intervals with little sandstone content (Padilla y Sánchez and Jose 2016).

Our basin-scale reconstructions build upon earlier regional-to-basin maps of the Ceno-Turonian, including the work of Salvador (1991a), Goldhammer and Johnson (2001), Seni and Jackson (1983), Randazzo (1997), Braunstein *et al.* (1949), and Jordan *et al.* (1949). However, seismic mapping, calibrated by and integrated with northern GoM wells, was most critical in this reconstruction.

4.8.3 Paleogeographic Reconstruction

Two major depocenters are closely associated with the expanded Tuscaloosa interval of central Louisiana, where Mid-Cretaceous detachment occurs seaward of the relict Albian (PW supersequence) platform margin (Figure 4.25). Here, the EFT supersequence thickens to over 2000 ft (610 m) near or at the shelf margin. The EFT supersequence deposition buried the relict Albian (PW) platform margin with a thick shelf-margin deltaic succession (see Figures 4.24 and 4.25). This is the first time in the Mesozoic that a siliciclastic system was able to prograde the shelf margin seaward of the underlying relict carbonate margin.

However, the most important features of the paleogeographic reconstruction are several areas of significant bypass of the relict Albian margin. The first, which we refer to as the Keathley Canyon depositional axis (KC axis), is a prominent sandstone-dominated submarine fan that extends from the platform margin depocenters in southern Louisiana to the ultra-deepwater blocks of Alaminos Canyon (Figure 4.26). A second interpreted bypass of the margin passes though Viosca Knoll and Main Pass protraction blocks onto the Mississippi Canyon protraction block. We refer to this as the Mississippi Canyon depositional axis (MC axis) (Figure 4.27).

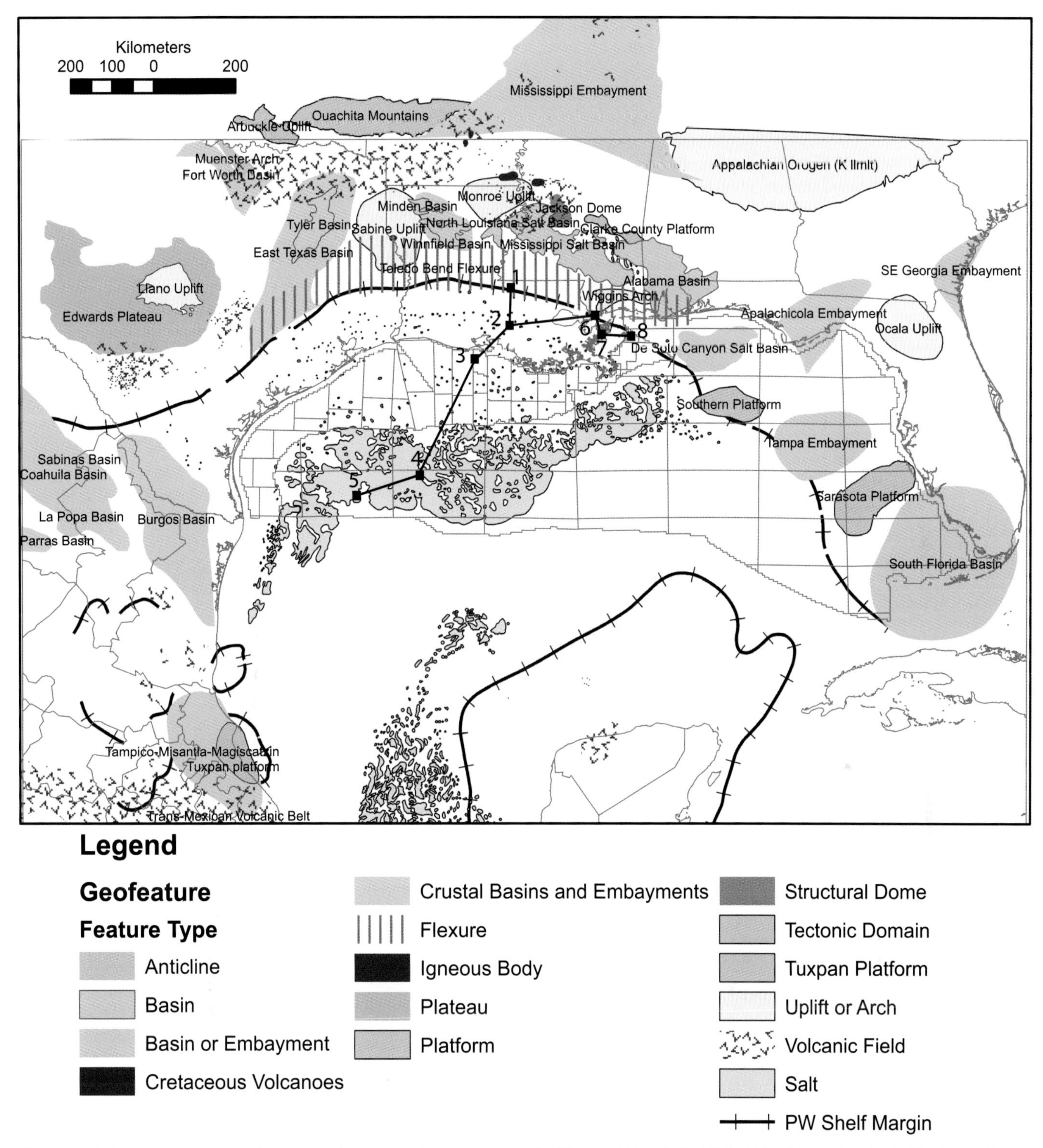

Figure 4.24 Important structural and depositional features and cross-section wells of Figures 4.26 and 4.27. Wells: (1) Shell #1 Hughes; (2) Highlander #1; (3) Davy Jones II; (4) Tiber #1; (5) BAHA II; (6) Chevron #3 SL 664; (7) Arco Biloxi Marshlands #P-2; and (8) Chevron Main Pass 253 #6. Modified from Snedden *et al.* (2016b). PW, Paluxy–Washita supersequence.

For sediment routing to the KC axis, the Tuscaloosa fluvial–deltaic system is interpreted to be predominant over the Woodbine system and most likely the primary source of Ceno-Turonian-age sandstones transported into the deepwater of Keathley and Alaminos Canyon protraction areas (Figure 4.26). The sand-rich fan fairway of the KC axis extends from the Tuscaloosa shelf edge system over 500–600 km (310–372 miles) to the Tiber (KC 102 #1) and BAHA II wells.

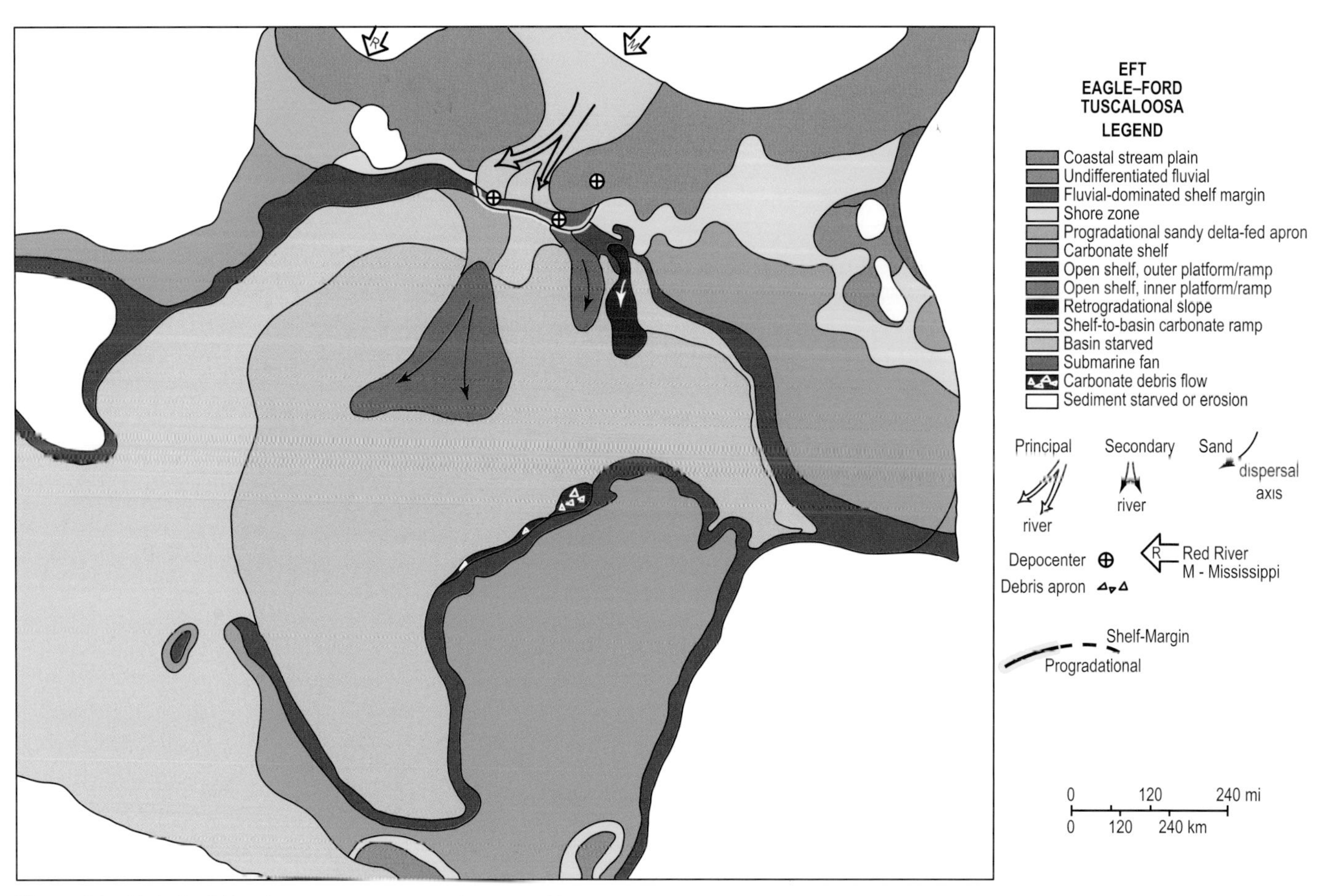

Figure 4.25 Eagle Ford–Tuscaloosa supersequence paleogeographic map.

Width of the sand-rich fan as shown in Figure 4.25 should be considered a minimum, based on existing but limited well control. Future seismic mapping and drilling is likely to increase basin lateral extent, more resembling deepwater distributive systems.

Key offshore wells penetrating the deep Tuscaloosa along the KC axis include: (1) South Marsh Island 234 #14 (Davy Jones II prospect), which encountered the Ceno-Turonian Sandstone at approximately 29,000 ft (8840 m) MD and; (2) The Freeport McMoRan Jeanerette Minerals #1 (Highlander #1 prospect), with Tuscalooosa at 28,000 ft (8537 m) MD (Figure 4.26). Paleoecological analyses from these two wells have not been released, but the location of Highlander #1 at 72 km (45 miles) and Davy Jones II at 160 km (100 miles) from the coeval shelf margin, respectively, suggest upper slope to lower slope paleo-environments. Of course, it is well known that slope systems sometimes contain transported shallow-water fauna (Armentrout and Clement 1990).

The Woodbine fluvial–deltaic system is separated from the Tuscaloosa system by the Sabine Uplift and does not appear to be supplying large volumes of sediment to the KC axis. The paleogeographic reconstruction (Figure 4.25) and previous work (Ambrose *et al.* 2009) shows the Woodbine (east Texas) shoreline and slope apron probably terminated well inboard of the modern-day coastline. Moreover, the Woodbine system is not a likely siliciclastic source for the deepwater, as the mapped downdip limit is located much farther north than for the Tuscaloosa. Regional mapping indicated that the coeval shoreline is interpreted to be located more than 500 km (310 miles) updip of the BAHA II or Tiber #1 wells (Snedden *et al.* 2016b) and it is unlikely that the extensive well control south of this point (targeting Eagle Ford and deeper intervals) would miss such a prominent sand fairway.

The MC axis is depicted as a set of smaller delta-fed bypass zones, though the extent of these is less well documented due to limited deep control outside of the carapace, rafted blocks, or overturned penetrations (Figure 4.27). Mesozoic strata in carapace and rafted blocks tend to be thinner, often stratigraphically condensed, and finer-grained than Mesozoic strata in the original depositional position (Fiduk *et al.* 2014). In addition, deepwater sediment gravity flows would tend to avoid emerging salt highs in these carapace/rafted blocks.

The eastern GoM deepwater area has a large number of salt-related growth structures that may have been local deposystems for sand-rich sedimentary gravity flows in the MC axis (McDonnell *et al.* 2010; Bovay 2015; Snedden *et al.* 2016b; Figure 4.28). Large Mesozoic-age expulsion rollovers (see Section 1.3.7) have been identified in the Mississippi Canyon

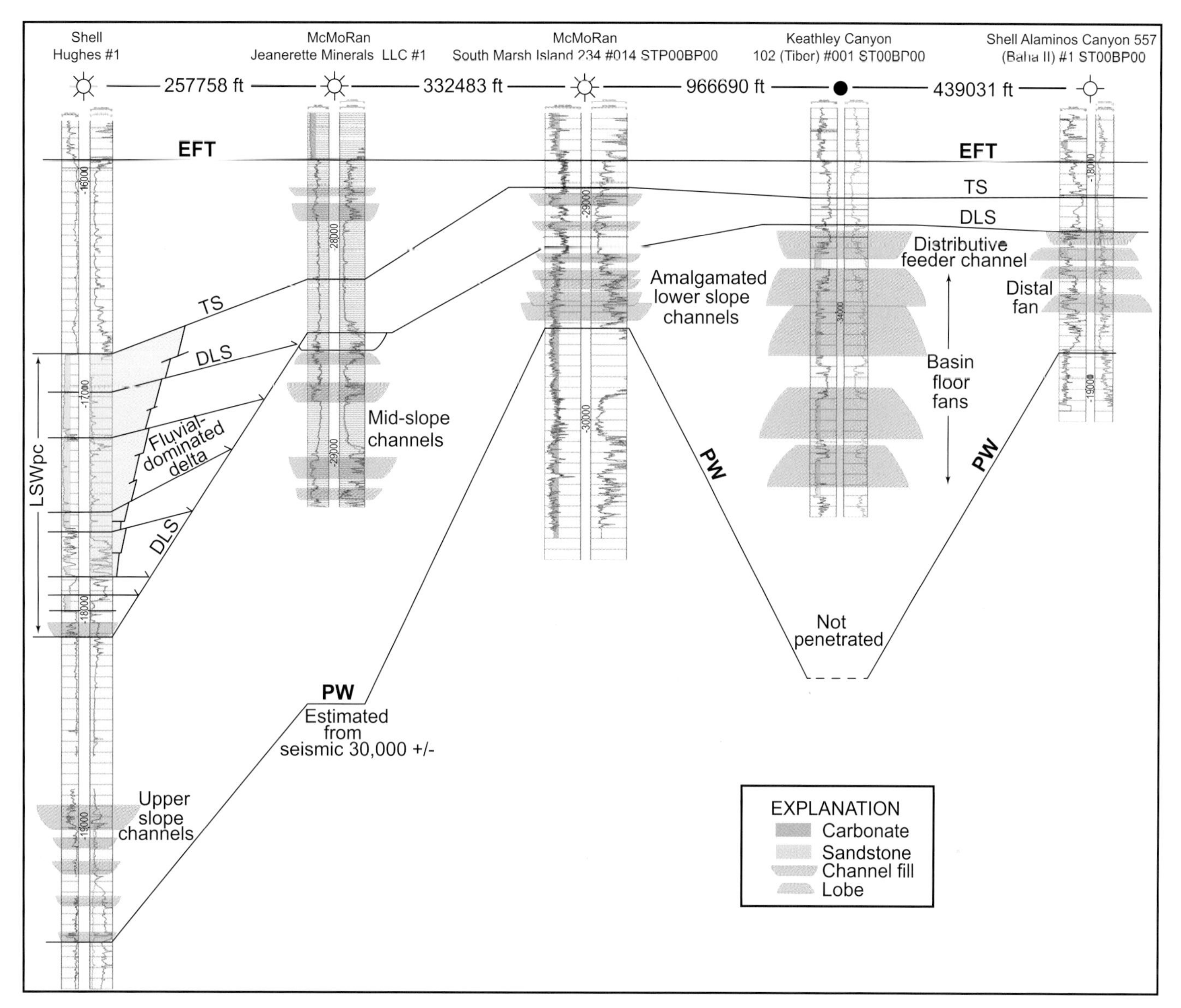

Figure 4.26 Stratigraphic well cross-section, Keathley Canyon axis of Tuscaloosa Sandstone. See Figure 4.24 for cross-section location. Modified from Snedden *et al.* (2016b).

area, some of the largest undrilled structures in the northern GoM (Figure 4.29). Individual structures have thicknesses of 4000–10,000 ft (1220–3050 m) and structural closures of 2000–3000 ft (610–915 m) are not uncommon (Harding *et al.* 2016). Key thick intervals within the expulsion rollovers are Upper Cretaceous to Lower Cretaceous. The Upper Jurassic raft province (see Figure 1.14) is on the eastern margin of this area of expulsion rollovers. However, none of the Norphlet wells drilled to date have penetrated the expulsion rollovers in optimal locations where this Cretaceous interval is presumed to be at its thickest.

To address the uncertainty of reservoir content, burial history and trap timing of these expulsion rollovers, Harding *et al.* (2016) used proprietary processed and merged 3D wide azimuth (WAZ) surveys (examples discussed in Section 9.25) to correlate Mesozoic surfaces and create structure and isopach maps over a 4000 square-mile area of Mississippi Canyon within the MC axis. From this work, allochthonous salt was determined to have been progressively deflated from north to south, generating elongate progradational structures aligned with similar-trending sediment feeders (Figure 4.28). Cretaceous salt walls presumably funneled sediment gravity flows to the south, before reaching salt ridge backstops that have subsequently been deflated in the Miocene (Harding *et al.* 2016).

The best-documented example is what is called the Galapagos minibasin wedge (Figure 4.29). The expulsion rollover geometry results in a series of wedges with the maximum thicknesses progressively offset and younging from north-northeast to south-southwest. These expulsion rollovers

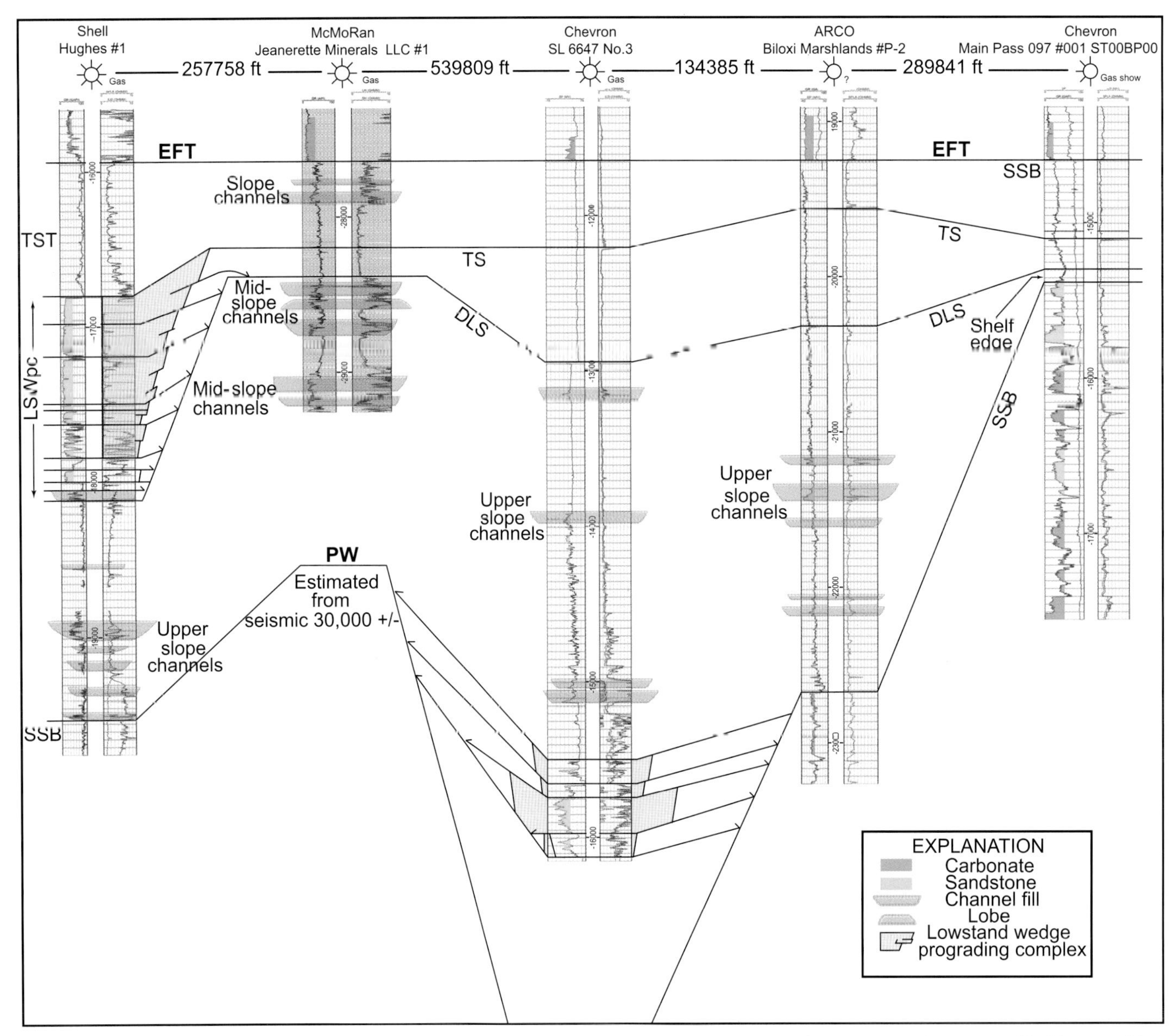

Figure 4.27 Stratigraphic well cross-section, Mississippi Canyon axis of Tuscaloosa Sandstone. See Figure 4.24 for cross-section location. Modified from Snedden *et al.* (2016b).

prograde south-southwest until reaching a backstop at a buried salt ridge (Harding *et al.* 2016). This expulsion rollover trend is in a separate structural trend from the Appomattox discovery wells (MC 392 block) that targeted the Norphlet Sandstone reservoir (see Sections 3.3 and 9.6).

Mapping these features in Mississippi Canyon indicates a distinct set of depocenter orientations (northwest–southeast and northeast–southwest), suggestive of sedimentary progradation, though at a scale far greater than sedimentary clinoforms (McDonnell *et al.* 2010). It is possible to align the expulsion rollovers trending northwest–southeast with the MC axis input point as shown on the EFT paleogeographic map (Figure 4.25), suggesting Tuscaloosa sandstones may be present within these large untested structures. Northeast–southwest trending structures imply an eastern source, possibly the Paluxy Sandstone of the PW supersequence or the Hosston Sandstone of the SH supersequence (see Section 4.2). However, the Hosston Sandstone scenario is less likely, given the short rivers draining from the Ocala and Peninsular arch areas of Florida (see Section 4.2.6; Snedden *et al.* 2016b).

Nonetheless, the presence of a large volume of sand contained in Tuscaloosa shelf-edge delta systems (Figure 4.25) suggests a high probability of sandstones in these Mesozoic expulsion rollovers. Another line of evidence for sand-prone Tuscaloosa feeder channels to these structures is evident in a core from the Arco P-2 Biloxi Marshlands well, exhibiting

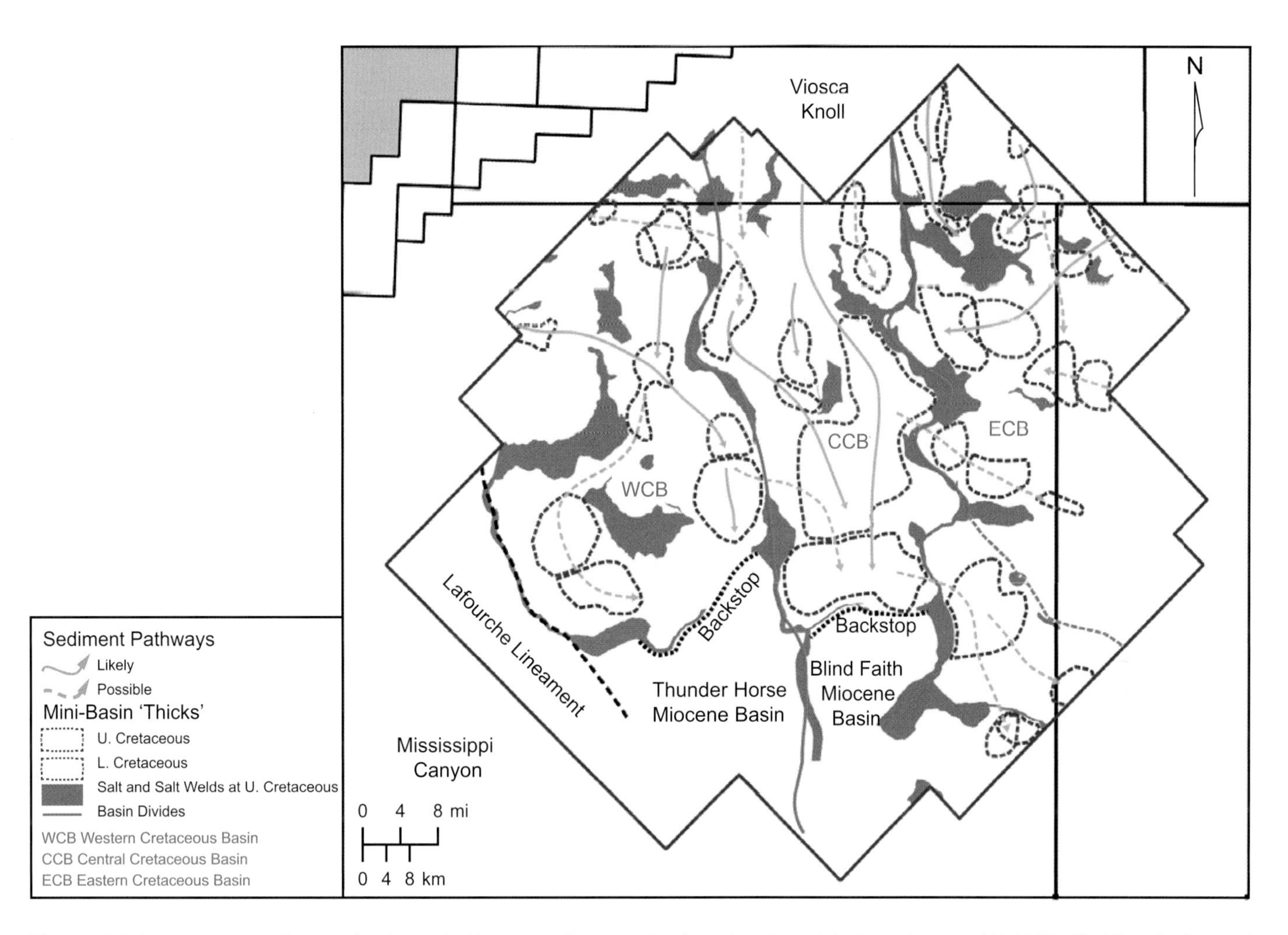

Figure 4.28 Seismic mapping of potential pathways for Mesozoic sediment routing through various salt basins and structural highs. Modified from Harding *et al.* (2016).

structures compatible with deposition from high-density turbidites (Figure 4.30). Horn (2012) suggested that regional data indicate slope channel systems, a general interpretation with which we concur, though it is more likely that these slope channels lead into MC rather than KC, where the Tiber (KC 102 #1) well is located. Horn's (2012) model is also at odds with the prior sedimentological interpretation of Woolf (2012), who suggested a shallow marine, wave and bottom current-influenced paleo-environment for the sandstones in the Biloxi Marshlands P-2 well.

Because of the importance of the Biloxi Marshlands P-2 well for reconstructing the sediment routing pathway for the basal EFT (Tuscaloosa) sandstones, it is worthwhile to examine the cored succession in some detail (Figure 4.30). Interpretation of sedimentary structures and bedset successions in the cores of the Biloxi Marshlands P-2 does not support the wave and current genesis of the sandstones as proposed by Woolf (2012). Sedimentary structures include flame and load structures, convolute bedding, and sheared bedding, and micro-scoured contacts. Bedding successions vary from Bouma sequence Ta–Tb turbidites (Figure 4.30) to nearly complete Bouma sequences (Tabcd). Taken as a whole, these sedimentary features point to sedimentary gravity flows, turbidity flows, and hybrid flows (Talling 2013) as the primary transport mechanisms. This is consistent with the paleo-slope setting inferred from our paleogeographic reconstruction for the EFT supersequence.

Detrital zircons collected from Cenomanian-age outcrops of the Gulf coastal plain and reported by Blum and Pecha (2014) largely confirm the EFT paleogeography (Figure 4.25). Major siliciclastic input is probably derived from the paleo-Mississippi River (Tuscaloosa) system linked to the Appalachian–Grenville source terrane, with lesser contributions from the Quachita Mountains (Woodbine system) and Ocala and Peninsular Arches of Florida. The lack of detrital zircon ages younger than 275 Ma suggest little or no contribution from the western USA at this time (Blum and Pecha 2014). In fact, Woodbine detrital zircon age spectra are similar to Canadian McMurray Formation outcrops, an observation that led Blum and Pecha (2014) to posit that the Appalachian–Ouachita Cordillera formed a Mid-Cretaceous drainage divide between the Boreal Sea and the GoM.

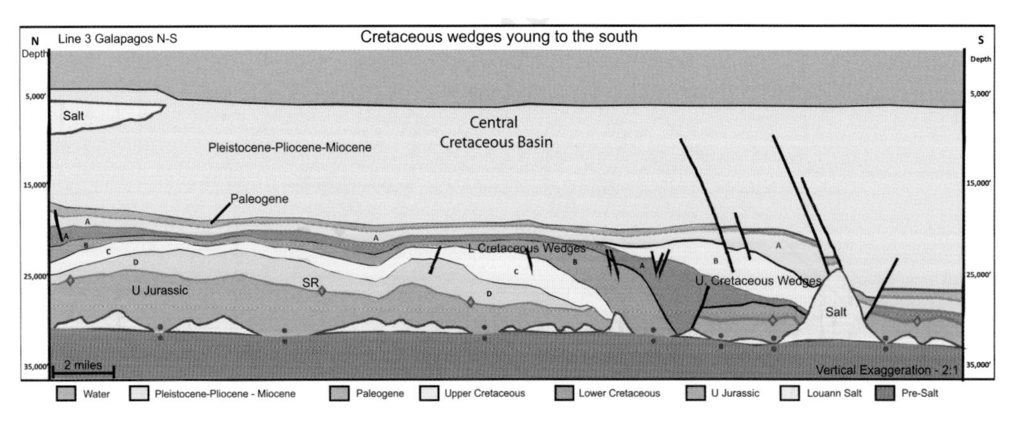

Figure 4.30 Interpreted Mesozoic expulsion rollovers in Mississippi Canyon. Modified from Harding *et al.* (2016).

The shore zone system shown as extending east of the Tuscaloosa depocenters probably served to trap the bulk of the siliciclastics coming from the waning source terrane of the Appalachians. Appalachian influence is not felt again until structural rejuvenation during the Miocene (see Chapter 7).

The timing of Tuscaloosa deposition into shelf margin expanded growth fault basins relative to the deepwater transport is uncertain, but may mirror the pattern observed in the Paleogene Wilcox. The growth fault depocenters of the south Texas Wilcox were likely filled in Upper Wilcox time, after deposition of the Lower and Middle Wilcox in the deepwater (see Chapter 5). A similar temporal model may explain the relationship between the onshore and offshore Tuscaloosa, with the deepwater Tuscaloosa being somewhat older than onshore Tuscaloosa.

Carbonates and calcareous shales within the EFT supersequence are mainly found within the Eagle Ford Formation, located above the Woodbine intervals of east Texas where present. East of the Sabine Uplift, shales of the Tuscaloosa Marine Shale overlie the Tuscaloosa Sandstone. Thus, the paleogeographic map shows shales are concentrated in areas east and west of the main Tuscaloosa sand transport fairway (Figure 4.25). Reduced deposition (often 100 ft/31 m or less in thickness) across the northern GoM is a departure from the underlying PW supersequence carbonate trends, where platform margin thicknesses often exceed 1000 ft (305 m). Thin carbonates (and thicker calcareous shales) are also present in other offshore wells, many of which are carapace or raft penetrations.

An exception to this is in the area of the Maverick basin of Texas, where thick calcareous shales and shaley carbonates are present in the Eagle Ford Formation and its equivalents (Figure 4.31). The Middle and Upper Eagle Ford are particularly carbonate-rich, while the Lower Eagle Ford is described most often as a calcareous shale (Hentz and Ruppel 2011). The Eagle Ford thins to as little as 10 m over preexisting structures, such as salt domes and the PW supersequence platform margin, and transitions seaward to basinal shales (Figure 4.31; Hammes *et al.* 2016).

Age-equivalent shale units in the Boquillas Formation of west Texas and equivalent Mexican formations in the Chihuahua Trough are also quite thick. Frush and Eicher (1975) measured 700–900 ft (213–275 m) of similar lithofacies in the Big Bend region, which is outside of the focus area of this book.

Of course, the Eagle Ford Formation is a major unconventional play in Texas, producing over 1.0 BBO as of 2014 (Stoneburner 2015). The Eagle Ford source rock characteristics are described in detail in Section 9.15.1.

A broad carbonate ramp prevailed over most of Mexico, extending to the Yucatán Platform. Minor fluvial sources from Chiapas and even Guatemala are present, but contribute little to the carbonate-dominated shelves that existed at this time. The Cuba Platform was a broad area of non-deposition (Figure 4.25).

4.9 Austin Chalk Supersequence

The Austin Chalk (AC) supersequence is a relatively thick (up to 1000 ft/310 m) carbonate-dominated succession that extends across the northern Gulf and southward into Mexico. Deposited just after the peak Cretaceous global sea-level highstand (Haq 2014), the unit is known for its relatively normal marine fauna and coccolith-dominated chalks and marls (Lundquist 2000). Chalks and carbonates of similar age are referred to as the Tokio (Louisiana), Eutaw (Mississippi/Alabama), and Pine Key (Florida). In Mexico, time-equivalent units are called the San Felipe (Figure 4.23).

4.9.1 Chronostratigraphy

Olson *et al.* (2015) estimate the AC supersequence to have had a duration of about two million years, extending from 85.99 Ma to 83.64 Ma, Santonian-1 sequence boundary to Sa3-600 flooding surface (ages from Gradstein *et al.* 2012; terminology of Snedden and Liu 2011). Nine biohorizons characterize this short-duration but relatively thick and fossil-rich unit. A comprehensive reference on AC biostratigraphy and paleoecology can be found in Lundquist (2000).

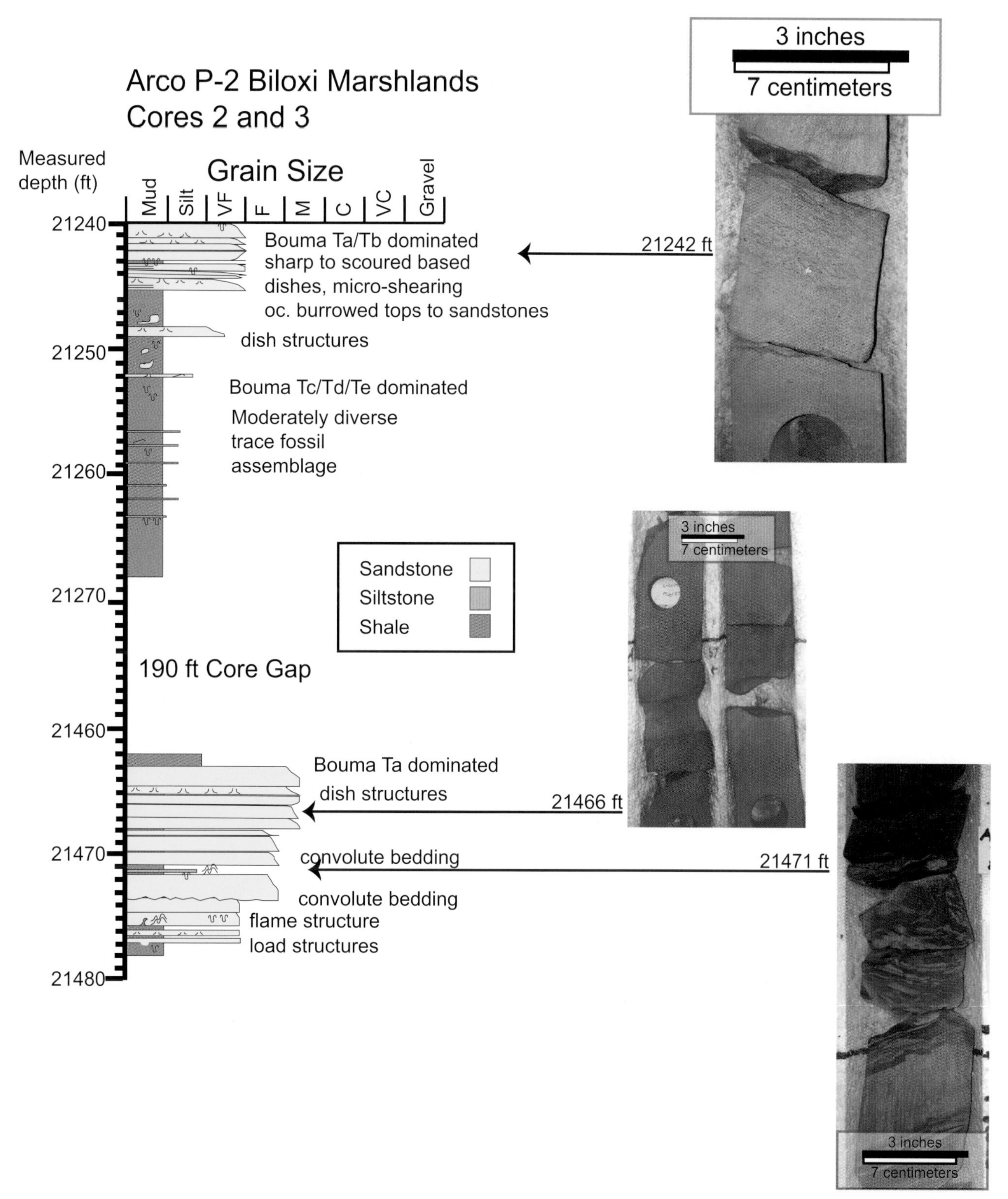

Figure 4.30 Sedimentary structures and bedset successions of Tuscaloosa Sandstone cores from Biloxi Marshlands P-2 well (for location see Figure 4.24).

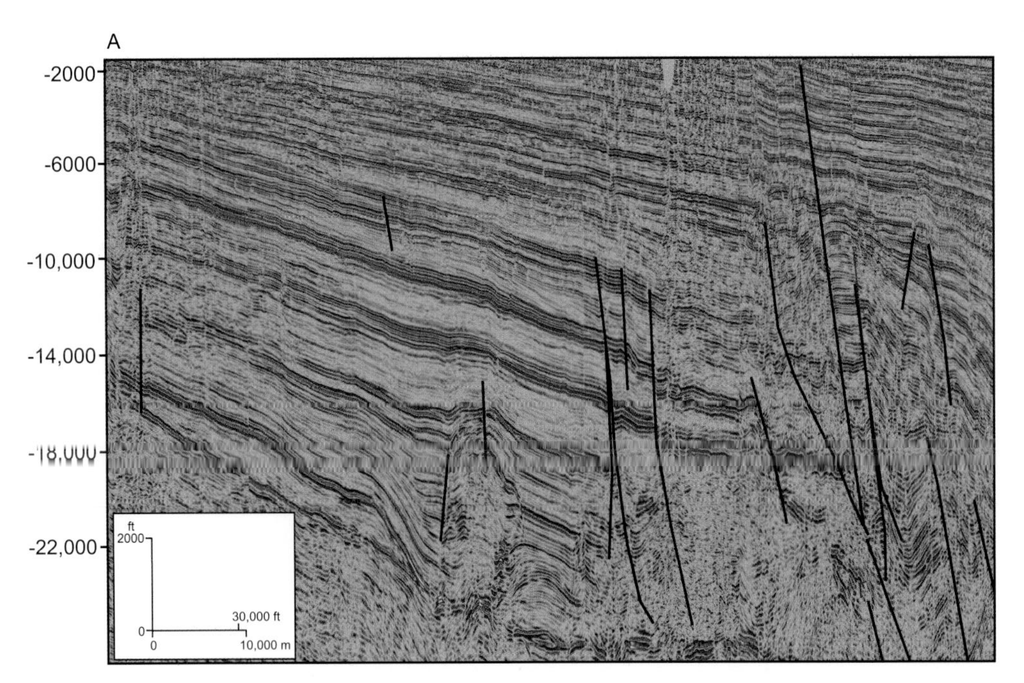

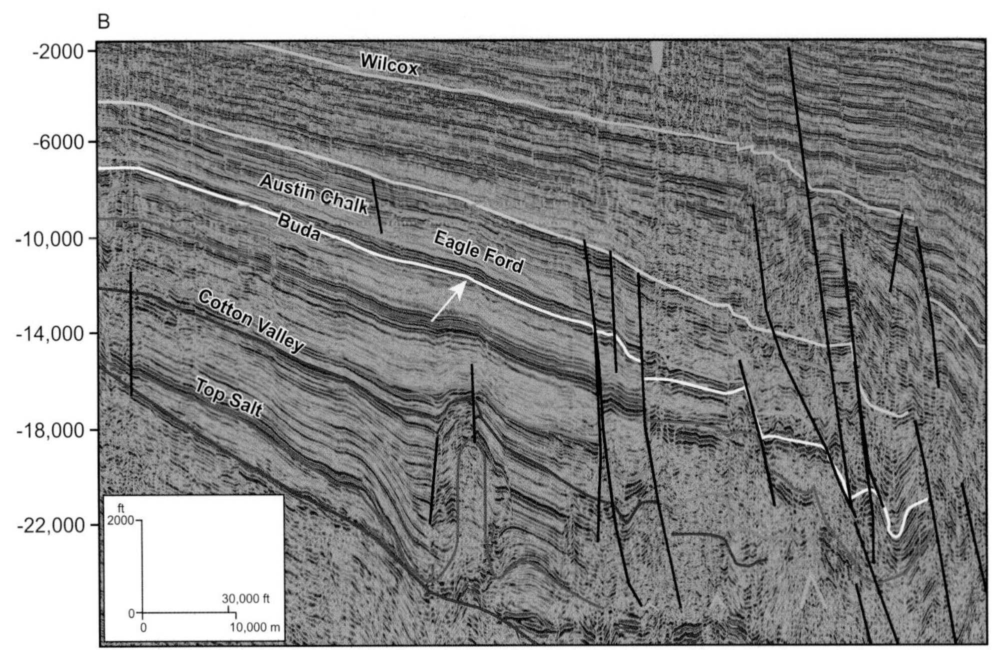

Figure 4.31 Seismic dip line across Atascosa to San Patricio Counties of south Texas. (A) Uninterpreted line. (B) Interpretations showing Eagle Ford horizon in light pink; Austin Chalk in orange, and Buda in white. Thinning of the Eagle Ford is observed over a salt dome (white arrow). Seismic line courtesy of ION. Modified from Hammes *et al.* (2016) with seismic interpretation provided by the senior author.

4.9.2 Previous Work

Early outcrop-based work in Texas recognized a white chalky unit that was lithologically distinct from calcareous claystones of the Taylor above and Eagle Ford Shale below (Lundquist 2000). Young (1977, 1985) used macro- and microfauna to subdivide the Austin Chalk Group into seven formations: Sprinkle (Big House/Pflugerville), Burditt, Dessau, Jonah, Vinson, and Atco Formations (Figure 4.32). Based on planktonic to benthic ratios, diversity, equitability, the distribution of planktonic morphotypes, and the depth associations of suites of benthic species, Lundquist (2000, 2015) interpreted the basal Atco Formation and Upper Dessau to have formed in the deepest water conditions, though no fixed value was estimated, described as "very deep to moderately deep." Our work and that of others suggests shelf-depth paleowater depths, in the range of 100–200 m. Simple reconstructions depict a continuous seaway extending from the Boreal to Tethyan Oceans that originated in the Ceno-Turonian (Eagle Ford interval) and continued into the Coniacian–Santonian (AC) timeframe (Lundquist 2000).

Earlier paleogeographic maps by Salvador (1991a) and Randazzo (1997), based on outcrops and limited well control, depict a wide marine seaway with locally deep bathymetry where chalks tended to form. Longman *et al.* (1998) later suggested that chalk deposition was associated with the influence of the Tethyan water masses extending from Texas up through Kansas and South Dakota.

Volcanic activity in southwest Texas, centered on the Uvalde area, occurred during this time (Byerly 1991; Ewing 2009). The Middle and Upper AC interval contains a significant number of volcanic ash beds (Hovorka and Nance 1994), as does the Eagle Ford immediately below.

Dravis (1981) investigated AC cores and outcrops to describe chalks formed in two broad depositional settings: (1) a shallow marine platform updip and near the San Marcos Arch with structurally positive features; and (2) a relatively deeper-water, basinal setting in downdip regions. A gradual transition from shelf to basin was suggested. Shelfal chalks are light-colored, with abundant and diverse macrofauna that coexisted with planktonic microfossils and nannofossils. Local oyster bioherms are present in outcrops. *Thalassinoides* burrows are common, pointing to shallow, well-oxygenated conditions. Deepwater chalks, by contrast, are darker and macrofossil-poor except for interbedded carbonate debris flows, and dominated by *Chondrites* burrows often associated with stressed, oxygen-poor conditions (Dravis 1981).

In the eastern Gulf, Iannello (2001) identified unusual seismic anomalies on the Florida Escarpment portion of the Middle Ground Arch. She interpreted these as dissolution collapse features in the AC and attributed these to diagenetic alteration caused by subsurface groundwater flow set up by recharge from onshore areas.

Hovorka and Nance (1994) observed large scours in the walls of tunnels dug in the AC of east Texas. They attributed formation of these meter-to-decimeter deep and 30 m average width scours to long-lived deepwater currents, trending northwest–southeast, parallel to the trend of the seaway connecting Texas to the Western Interior US. This observation is consistent with Durham and Hall's (1991) work on the "Waco Channel," an incision with 300 ft (91 m) estimated relief from outcrops of the Austin Chalk Group near Waco, Texas (Figure 4.32). The "channel" (more like a canyon-scale feature) apparently cut out all but the ATCO and lower portions of the Vinson Formations.

In Mexico, limestone/dolostone intervals can exceed thicknesses of 2000 ft (610 m) in the San Felipe formation (Padilla y Sánchez and Jose 2016; Figure 4.23). Deposition in the Veracruz and Sureste basins of Mexico is dominated by carbonate platforms founded on large structural platforms. These transition to chalks locally. In Cuba, the AC supersequence equivalent is largely missing, possibly due to erosion associated with the K–Pg impact event and also the collision of the Cuban block with the northern GoM passive margin in the Paleogene (Schenk 2008).

4.9.3 Paleogeographic Map Reconstruction

The reconstructed AC paleogeographic map (Figure 4.33) depicts a largely carbonate-dominated basin, with limited siliciclastic input from the north, northeast, and a small system in eastern Florida. Chalk deposition was widespread, reflecting increased water depth due to elevated sea level, warm equable climates, and limited fine terrigenous sediment input from the Ouachitas, Appalachians, or Rockies (Lundquist 2015). We

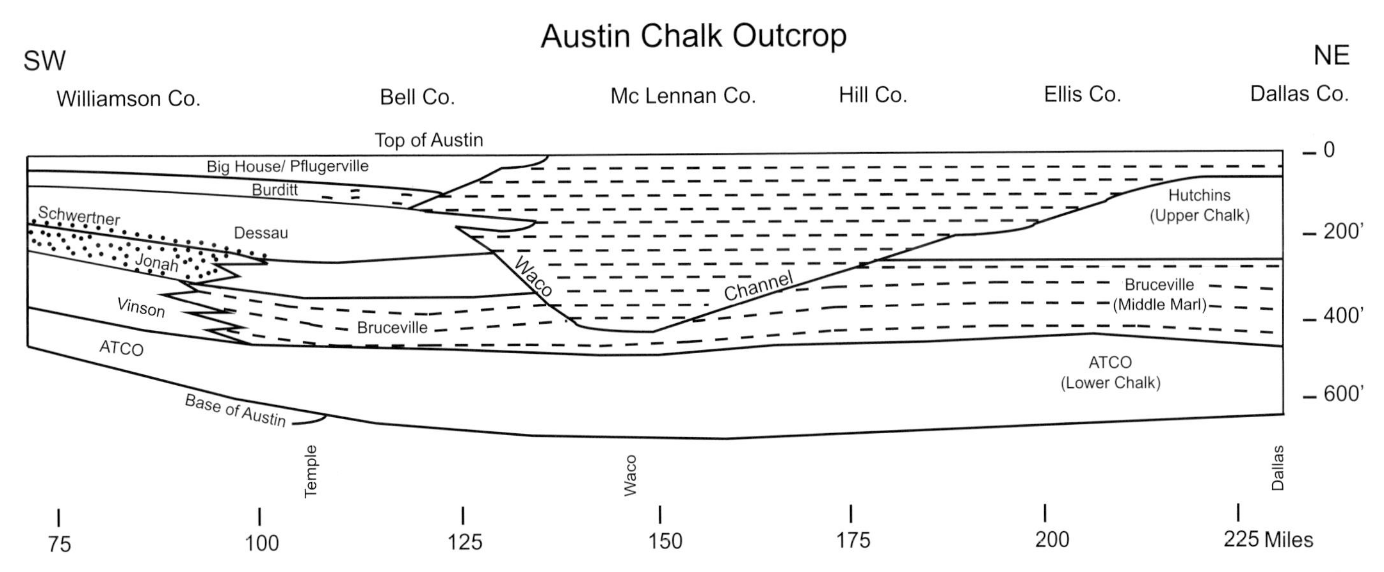

Figure 4.32 Stratigraphic cross-sections of the Austin Chalk supersequence as seen in outcrop correlations across the Waco channel northeast to Dallas. Modified from Durham and Hall (1991).

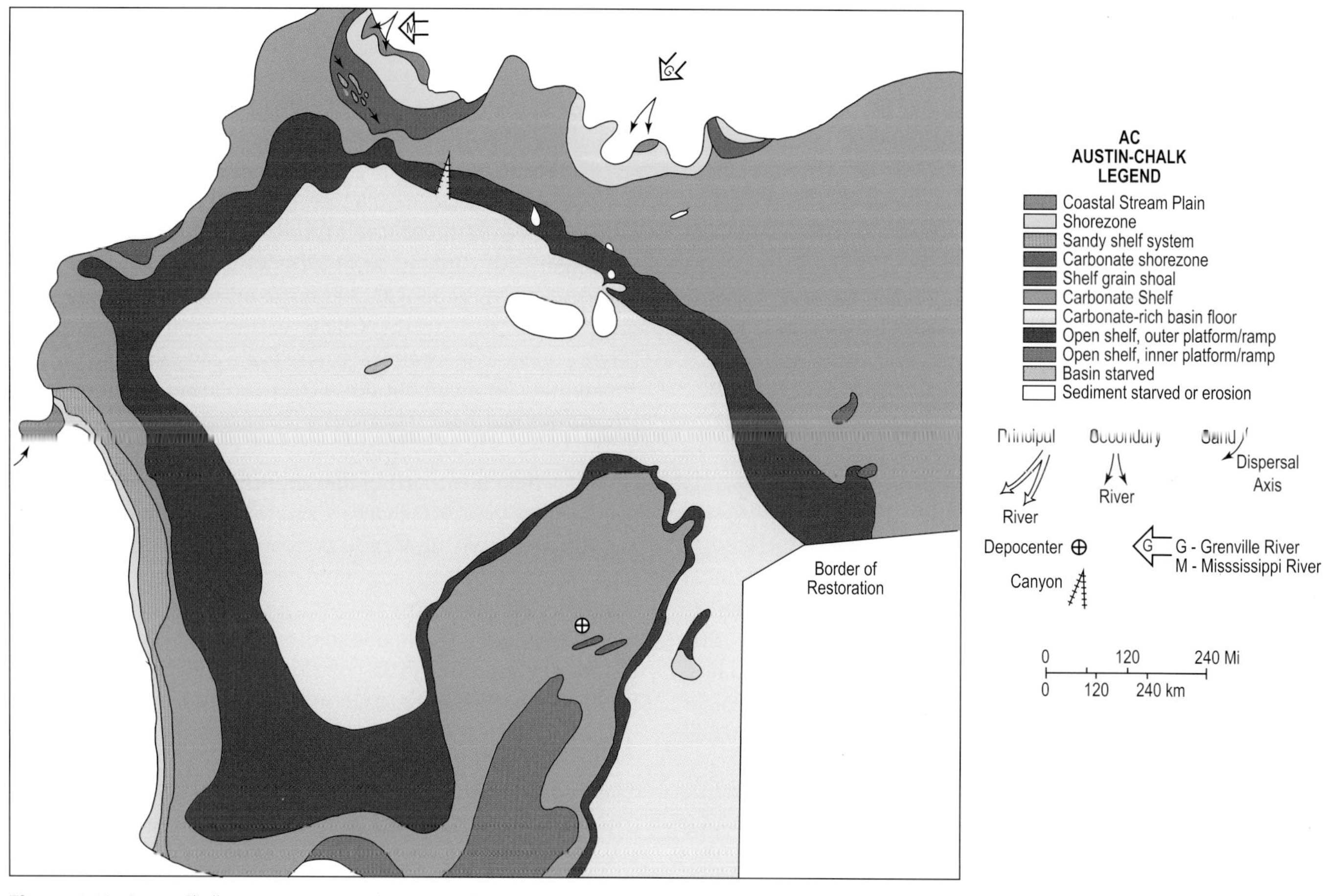

Figure 4.33 Austin Chalk supersequence paleogeographic map.

interpret carbonate shelf to outer shelf to basinal paleoenvironments extending across the entire basin. This reconstruction is consistent with Texas core/outcrop work of Dravis (1981) suggesting a ramp-like transition from shallow-water chalks to basinal chalks.

The likely connection with the Western Interior Seaway, which extended from the Boreal Sea to the GoM, was probably through east Texas, between the emerging San Marcos Arch and Sabine Uplift structural features. Evidence of vigorous deep-current movement in the form of scours and channels (Durham and Hall 1991; Hovorka and Nance 1994) supports this view.

Strong current flow may also explain the presence of shoreline-detached, shelf sand bodies about 270 km to the east, along the Texas–Louisiana border (Figure 4.34). A sandstone body with a gross thickness of 100 ft (31 m) is present in the Upper Austin Chalk (Energy Reserve #1 Simon Herold well). While the presence of other shelf sand bodies is conjectural, work on modern shelf sand ridges indicates that these features tend to occur in groups or "fields" of ridges, oriented parallel or slightly oblique to the prevailing storm-current flow (Snedden *et al.* 2011). Strong bottom currents may have also eroded areas of the deepwater of northern Mississippi Canyon and DeSoto Canyon (Figure 4.33).

An alternative connection to the Western Interior Seaway is through west Texas and adjacent to the Chihuahua Trough, a passage south of the Llano Uplift. However, this does not explain the intra-formational scours and channels (Durham and Hall 1991; Hovorka and Nance 1994) nor development of shelf sand ridges as noted above.

The paleogeographic reconstruction indicates limited sandstone influx from the north (paleo-Mississippi River) and northeast (Grenville River), reflecting small catchments draining Quachita and Appalachian source terranes. It is possible that a portion of the Ocala–Peninsular Arch in Florida remained emergent and supplied minor terrigenous material to the marine seaway extending across Florida.

Seismic mapping and well control indicates a prominent depositional thick onshore Yucatán Peninsula reflective of substantial carbonate deposition (Figure 4.33). Structural mapping suggest a rather rugose Top AC surface reflecting local salt tectonics. There is a coincidence of salt-related turtle structures (Pitman 2014) in the East Texas salt basin and thicknesses of the AC supersequence that can locally exceed

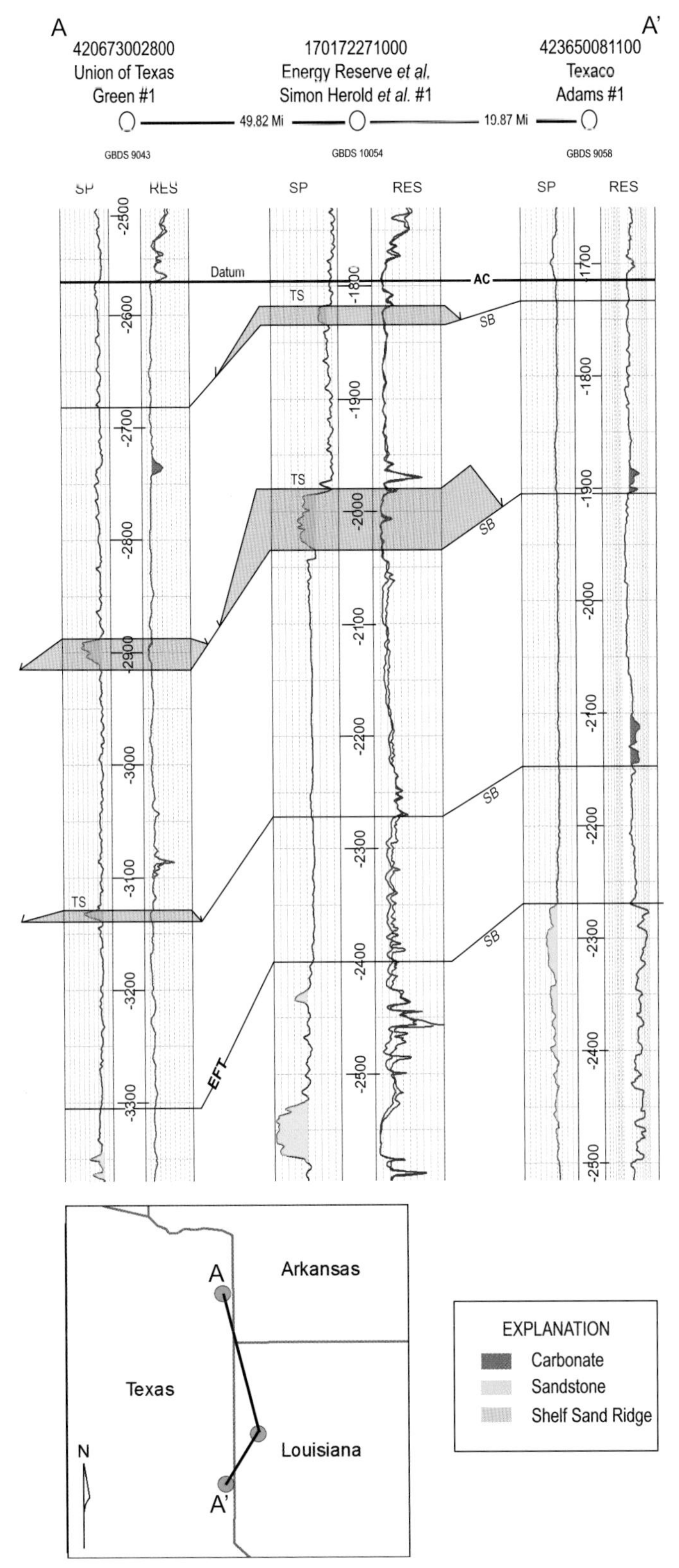

Figure 4.34 Well log cross-section through Austin Chalk shelf sand ridges. Inset map shows cross-section location.

1000 ft (305 m). In general, the AC supersequence expands, stepping off of the PW supersequence shelf edge. It should be noted that in the offshore areas, the largest seismic thickness maximums are uncalibrated by well control in deep structural lows.

In Mexico (Figure 4.33), we depict a narrow shore zone following the trend of the emergent Sierra Madre tectonic belt, assuming continuity with the overlying NT supersequence, as will be discussed in the next section. The Yucatán Platform is shown as a shallow carbonate shelf, hosting a few shelf grain shoals as penetrated by two Pemex wells (Y4 and Y5). Ward *et al.* (1995) used Pemex data from these wells to report *Textularia* and *Dicyclina*-rich grainstones and packstones in the Austin Chalk equivalent of these wells.

In contrast to the NT supersequence, discussed in Section 4.10, most of the AC supersequence shows limited influence of Laramide source terranes, the exception being this narrow Sierra Madre-influenced shore zone in Mexico. The global high sea-level state of the Santonian stage (Miller *et al.* 2005; Haq 2014) probably played a role in this regard.

4.10 Navarro–Taylor Supersequence

The NT supersequence is the last unit deposited in the GoM prior to the Chicxulub impact event that ended the Mesozoic Era. As such, it provides a look at the landscape and seascape around the GoM just prior to this deep impact. The known composition of the Cretaceous–Paleogene (K–Pg) boundary deposit is also better understood by examining this unit in detail. The K–Pg boundary deposit is discussed in a separate section (Section 4.11) due to its scientific and economic importance.

4.10.1 Chronostratigraphy

Olson *et al.* (2015) determined from industry biostratigraphic data that the NT supersequence temporally extends from 83.64 Ma to 68.20 Ma (timescale of Gradstein *et al.* 2012) or from Sa3_600fs to Ma5_sb (nomenclature of Snedden and Liu 2011). The upper boundary is the event horizon associated with the Chicxulub impact and its associated physical processes of landform and bathymetric modification. However, many workers place the Top Cretaceous at 66.0 Ma (Renne *et al.* 2013; Ogg *et al.* 2016; Lowery *et al.* 2018). Last appearance datums (LADs) in GoM wells do not exactly coincide with 66 Ma due to impact-related erosion and sediment reworking (e.g., Bralower *et al.* 1998), but from a practical standpoint we consider these to be the same. The timespan of 15.4 million years is the longest duration of any of our supersequences.

4.10.2 Previous Work

Snedden and Kersey (1982) did detailed core descriptions and well log correlation in Webb County, Texas to assemble a depositional model of a fluvial to mixed wave-and-fluvial–deltaic system that deposited the "updip" Olmos Sandstone, part of the NT supersequence. Stratigraphic correlations indicate the Olmos Sandstone grades seaward to a storm-dominated shelf sand-sheet (so-called "downdip Olmos"). Tyler and Ambrose (1986) primarily relied upon log motifs, isopach trends, and a few cores to develop a more regional model of constructive and destructive (shore zone) deltas.

McGowen and Lopez (1983) used outcrop and subsurface data from a few counties in northeast Texas to construct a depositional model of shore zone to shelf sandstones and local deltaic systems. Tidal influence was noted in some areas. Shelf sandstones were thought to have been influenced by strong tidal currents.

Further along the strike to the southwest is the Serbin Field of Bastrop and Lee Counties, Texas where storm-dominated muddy shelf sandstone reservoirs with a patchy distribution are mapped. Abundant mollusk fragments in tempestites (storm deposits) are common and, together with the *Cruziana* ichnofacies, these suggest a shallow marine environment not far from a siliciclastic shore zone (Ogiesoba *et al.* 2018).

In Mexico, coeval intervals are the shale-prone Upson (Sabinas/Burgos basins), Mendez (Tampico–Misantla and Veracruz basins), and shales below the K–Pg boundary **breccia** of Sureste (Figure 4.23).

Chalk and chalky marls are particularly common in the Campanian–Maastrichtian strata of the GoM. In northern Louisiana, four chalk-dominated lithostratigraphic units are commonly recognized: Ozan, Annona, Marlbrook, and Saratoga (Crane 1965). Core-based sedimentological analyses suggest deposition in relatively deep marine waters, below storm wave base, away from terrigeneous input, but well-oxygenated as suggested by common benthic and planktonic foraminifera and diverse types of burrow structures (Loucks *et al.* 2017a).

Based on detrital zircon U-Pb geochronology, Potter-McIntyre *et al.* (2018) suggested the initiation of the south-flowing drainage that evolved into the Paleogene Mississippi River began in the Maastrichtian in the Illinois basin. However, the abundance of deep marine chalks in the southern end of the embayment (Loucks *et al.* 2017a) indicates a lag in siliciclastics reaching a large portion of the Gulf basin.

4.10.3 Paleogeographic Map Reconstruction

As with the underlying AC supersequence, the NT supersequence base map includes western Cuba in its position attached to Yucatán, prior to the Eocene movement to its present-day location (Figure 4.35). The GoM at this time was largely a marine carbonate-dominated basin, with siliciclastic deposition limited to the north and west sides of the area (Figure 4.35). Shelf margins were largely inherited from the relict Albian (PW) margin rimming the basin. Siliciclastics in northeast Texas, representing the Nacatoch Sandstone (McGowen and Lopez 1983), were likely derived from the Ouachitas (paleo-Red River system). In south Texas, the Maastrichtian to Campanian age Escondido–Olmos–San Miguel Sandstone intervals indicate the first major influx of sandstones yielded from source terranes in the Sierra Madres. Associated fluvial to shallow marine Difunta Group siliciclastics are found in the Sabinas, La Popa, and adjacent basins (Lawton *et al.* 2001), reflecting deformation associated with the coeval Hildalgoan orogeny. A siliciclastic shore zone–shelf depositional system extends from the Sabinas basin south along the foreland trough to Chiapas. The Petrerillos Formation (Delgado Member) sandstones, at or immediately below the K–Pg surface, are thought to be shallow marine in origin, grading seaward to marls (Schulte *et al.* 2012).

The NT paleogeography depicts the GoM just prior to the Chicxulub impact event. The distribution of paleo-environments may help explain the composition of the K–Pg breccia unit encountered in offshore well penetrations. The K–Pg in the entirety of the basin is largely composed of carbonates (Denne *et al.* 2013). NT supersequence carbonate-prone paleo-environments, by our mapping, represents about 96 percent of the 2930 km^2 area present at the time of the K–Pg impact event (Figure 4.36). Fluvial to shallow marine sandstones amount to just 115 km^2 of the area mapped, excluding areas of erosion. In deepwater wells, the dominant lithology of the K–Pg is pelagic carbonates (Denne *et al.* 2013) derived from just 33 percent of the depositional area mapped, suggesting that reconfiguration of the deep marine seascape was substantial and attenuation of seismic waves and tsunamis on the shallow shelf did limit, to some degree, the generation of debris flows there. These map-based observations support the large carbonate volume estimates for the NT supersequence (Figure 4.23). Surprisingly, the Late Cretaceous carbonate volumes for the northern GoM exceed that of the Albian acme of carbonate platform margins here. This is due to the Late Cretaceous relative sea-level highstand that expanded carbonate deposition. A connection to the Western Interior Seaway likely still existed, but the seaway was much reduced in size, as the connection to the Boreal Sea was severed by the Late Campanian.

One area that differs with previous published paleogeographic reconstructions is the Nacatoch siliciclastic system of northeast Texas (Figure 4.35). McGowen and Lopez (1983) interpret a tide-influenced shore zone and shelf system, including detached, offshore sand bars. While we generally agree with the shore zone and shelf sand trend, we believe that the shelf sandsheet is a transgressive remnant of a broader deltaic platform.

Wells drilled further to the south of the McGowen and Lopez (1983) study area have penetrated an unusually thick, sand-prone succession called the "Ozan" Sandstone in this well. The Pan Am #1 Lutcher Moore Lumber well, drilled in Newton County, Texas, encountered a 360 ft (109 m) thick sandstone interval above the AC (Figure 4.37). Lithology logs confirm the presence of sandstone. At least five other wells penetrated this same sandstone, though log quality does vary. The log motif looks similar to what Galloway (2008) refers to as a wave-dominated shelf margin delta (online poster, The Gulf of Mexico Cenozoic Log Facies Interpretation Poster, www.cambridge.org/gomsb). The over-thickening of the delta (versus normal 50 ft/15 m thick deltas) is likely due to higher accommodation at the paleo-shelf edge located here. Further, we interpret the shelf-edge delta to have been deposited at a

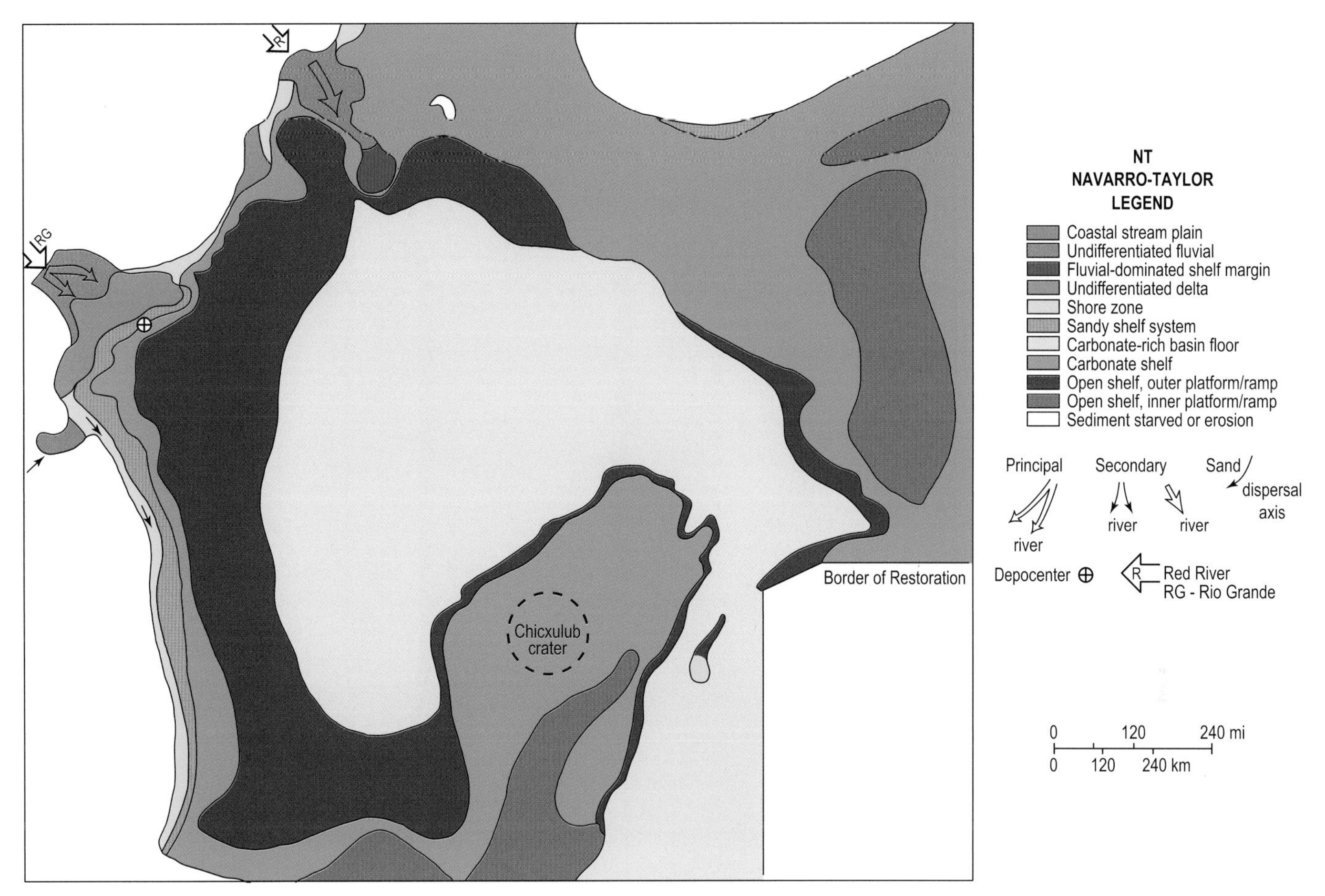

Figure 4.35 Navarro–Taylor supersequence paleogeographic map.

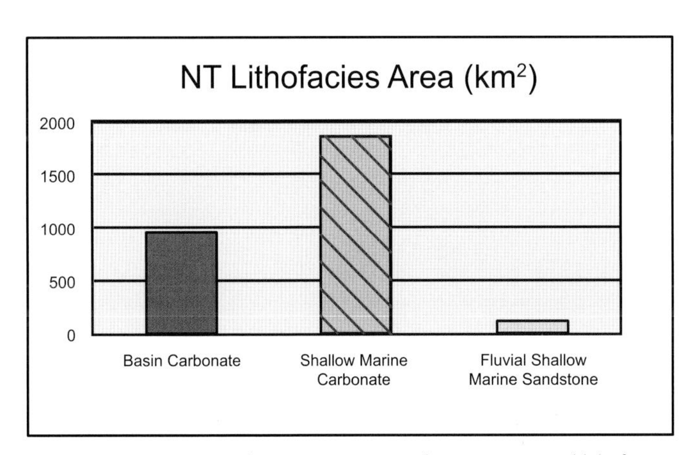

Figure 4.36 Navarro–Taylor supersequence paleo-environmental lithofacies distributions just prior to the Cretaceous–Paleogene (K–Pg) boundary impact event.

relative sea-level lowstand and transgressed, leaving a sandy retreat path back toward the shelf sand-sheet described by McGowen and Lopez (1983). Similar transgressive retreat paths and sand distribution patterns are observed in Quaternary systems of the GoM (Suter and Berryhill 1985). The possibility of slope and basin-floor fans connected to this shelf-edge delta system should be considered, as relatively few wells have drilled much below the Top Cretaceous in the offshore areas. Elsewhere in onshore areas, the Ozan Formation (Taylor Group) is generally a mix of shallow-water chalk, shale, and fine-grained sandstone.

Comparing the NT paleogeography to structural trends suggests that the western Quachita Mountains may have played a role in sourcing this sandstone protrusion into the basin. As discussed below, a relatively flat platform extending across the area between the Houston Embayment and the Toledo basin may have provided the progradational ramp for the Ozan delta to prograde into a shelf edge position (Figure 4.35).

Beyond the NT shelf margin, inherited largely from the Albian shelf edge, depths and structural complexity increase substantially. Estimation of depth to Top Cretaceous (NT) in the salt canopy area should be considered with caution, given subsalt illumination and imaging challenges, velocity variations, and limited well control outside of salt carapaces and salt-cored folds. Recent WAZ seismic surveys imaging below the salt canopy indicate the Top Cretaceous is generally in the range of 32,000–41,000 ft (9754–12,497 m; Michael Hudec, pers. comm. 2016).

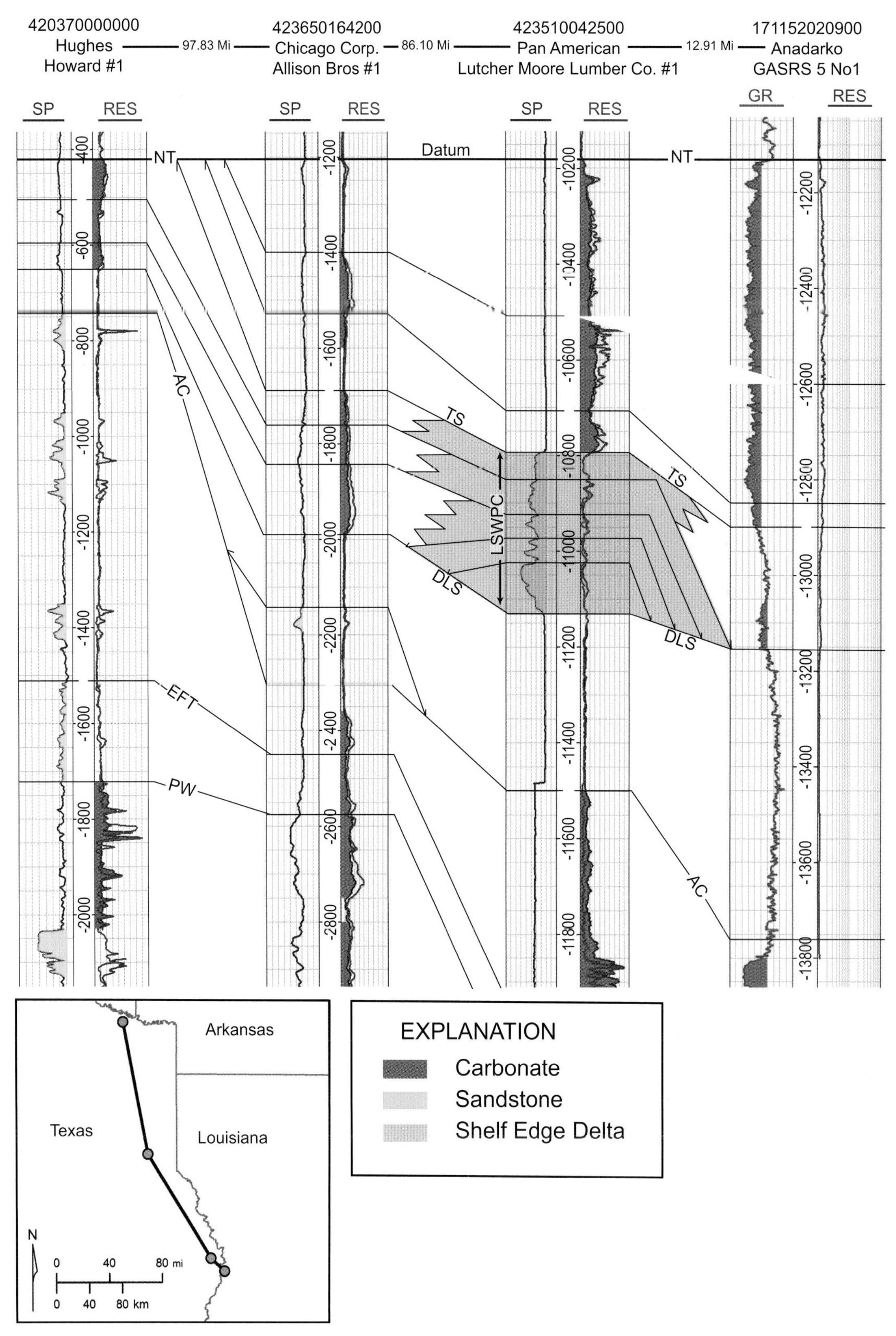

Figure 4.37 Well log cross-section in northeast Texas showing Ozan Sandstone of the Navarro–Taylor supersequence.

A major depocenter in south Texas (>3000 ft; 915 m) can be related to the Laramide-/Hidalgoan-sourced siliciclastics of the Difunta Group and US equivalents in the Escondido, Olmos, and San Miguel Formations (Figures 4.23 and 4.35). A smaller depocenter (not indicated) is associated with Nacatoch/Ozan deposition, presumably sourced from the Ouachita source terrane.

As mentioned earlier, a substantial amount of carbonate deposition accumulated in the Chicxulub target zone on the Yucatán Peninsula prior to impact and may have contributed to the predominant carbonate lithology of the K–Pg boundary deposit (Figure 4.35). Much of the Bay of Campeche area was covered by marine carbonate paleo-environments, with significant anhydrite deposited in local sabkhas. Ward *et al.* (1995) noted up to 50 percent evaporite in the Mesozoic interval of Pemex wells drilled in Yucatán. Siliciclastics are limited to the Hidalgoan tectonic front of Mexico and areas near the Chiapas region, beyond the mapped area (Witt *et al.* 2012).

Structural features such as basement highs do show some influence on thickening and thinning trends in the NT, though some are counter-intuitive. The Sabine Uplift, which did seem to influence thinning of earlier Cretaceous intervals, was submerged and less prominent (Figure 4.35). All the salt-hosting basins, Tyler of east Texas, north Louisiana, Mississippi, and DeSoto, are associated with thickening of the interval. Sediment thicks centered on Grand Isle, Louisiana and other areas of the eastern GoM probably reflect some influence of salt, such as salt flank grabens. In western Mexico, the emerging platforms (e.g., Tuxpan) are carbonate-dominated, with siliciclastics confined to the Laramide-front shorelines. The eastern Gulf, including Florida, was also a site of significant carbonate deposition (Figure 4.35).

4.11 Cretaceous–Paleogene (K–Pg) Boundary Unit

The K–Pg boundary deposit in the GoM is the product of a short-duration event, linked conclusively to the bolide impact event at the Chicxulub crater in Yucatán, Mexico (Denne *et al.* 2013; Morgan *et al.* 2016; Sanford *et al.* 2016). Specifically, the base of this unit marks the end of the Cretaceous Era and its time span is entirely within the Paleogene (Molina *et al.* 2006).

A history of the search for a cause behind the end Cretaceous mass extinctions, including an extraterrestrial bolide, the identification of an impact crater candidate, and the collection of core information from the Chicxulub crater peak is described in Box 4.3. The sections that follow focus on the processes and products that followed from the impact event, and how this impact event reshaped the GoM basin and paved the way for the Cenozoic Era.

Box 4.3 The Chicxulub Impact Event: A History of Scientific Research

One of the greatest scientific breakthroughs in our understanding of the Mesozoic of the GoM started with a puzzle: Why did so many genera, both on land and at sea, terminate at the end of the Cretaceous? While the extinction of non-avian dinosaurs at this time is well known, the disappearance of 76 percent of all species on Earth is also well documented (Schulte *et al.* 2010; Lowery *et al.* 2018). A father-and-son team, Luis Alvarez (physicist) and Walter Alvarez (geologist), and analytical support staff investigated two European sites where the K–Pg boundary is well exposed: Gubbio, Italy and Stevns Klint, Denmark (Alvarez *et al.* 1980). Elemental analyses of samples from thin clay layers, where extinction of Cretaceous fauna was documented by previous biostratigraphic work, showed a dramatic increase in the amount of iridium, a platinum group element, about 30 and 160 times above background levels, respectively at the two sites (Figure 4.38). Concurrent work in Spain produced the same result (Smit and Hertogen 1980). Iridium is depleted in the Earth's crust relative to its cosmic abundance, such as in meteorites. Later work has since demonstrated similar iridium spikes at the K–Pg boundary at over 350 sites globally (Schulte *et al.* 2010; Figure 4.39). Other indications of an impact in these sites are shocked quartz and spherules, small (<1 mm) rounded balls, the condensed products of molten and vaporized rock (e.g., Bohor *et al.* 1984).

Alvarez *et al.* (1980) and Smit and Hertogen (1980) theorized that the cause of the K–Pg mass extinction was a bolide (impacting object) of at least 10 km (6 miles) diameter striking the Earth and causing a massive dust cloud that blocked out the Sun and stopped photosynthesis, attacking the food chain at its

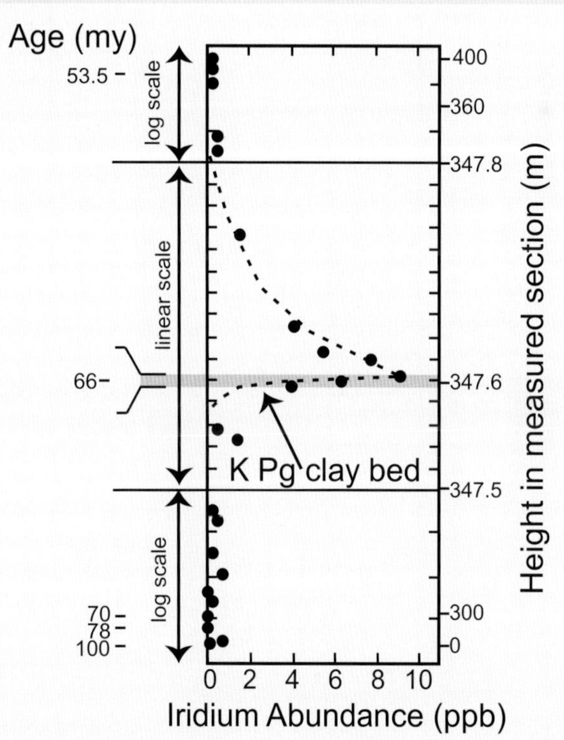

Figure 4.38 Iridium spike at the K–Pg boundary near Gubbio, Italy. Modified from Alvarez *et al.* (1980).

Box 4.3 *(cont.)*

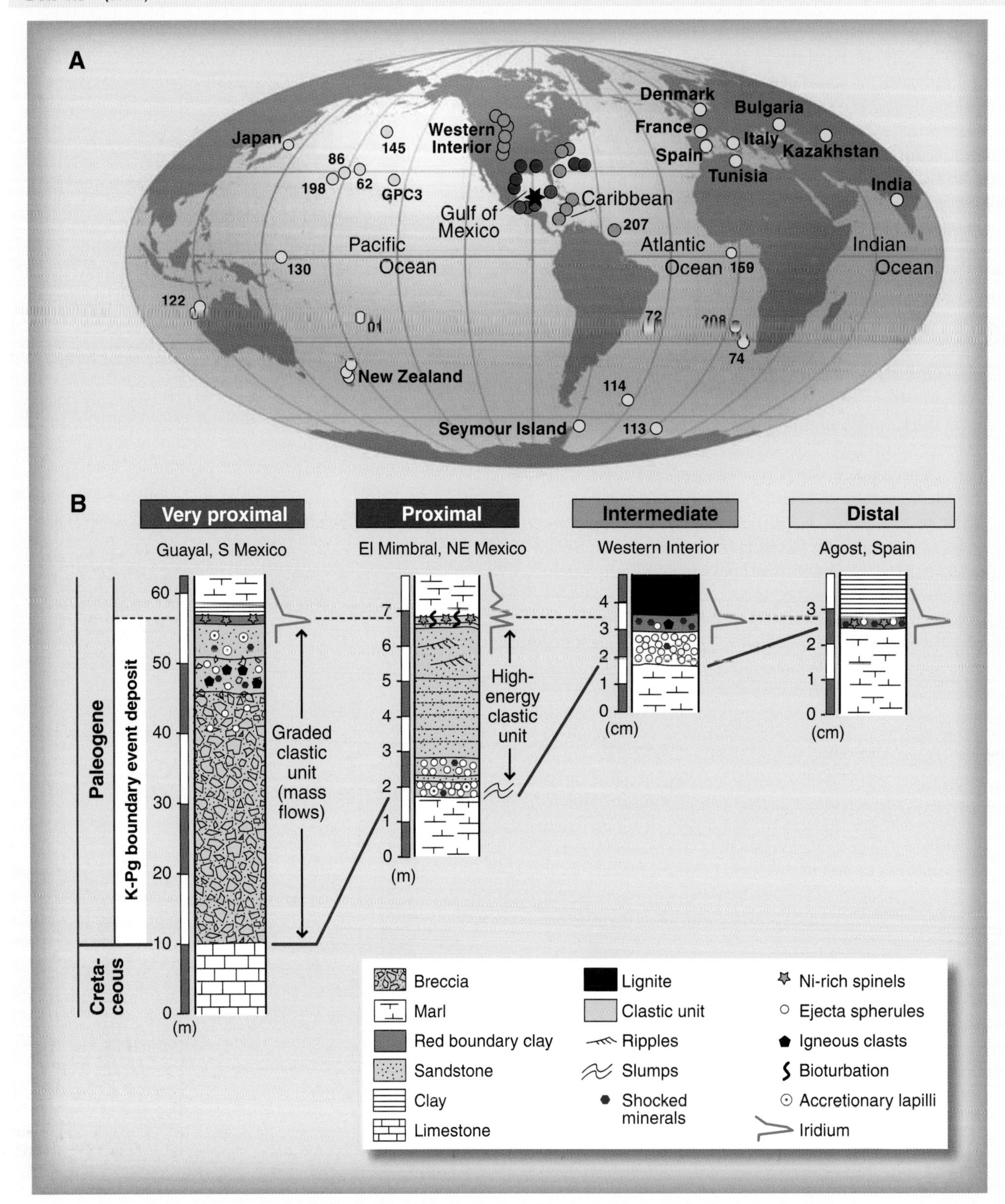

Figure 4.39 Global trend of outcrops of K–Pg boundary deposits. Modified from Schulte *et al.* (2010).

Box 4.3 (*cont.*)

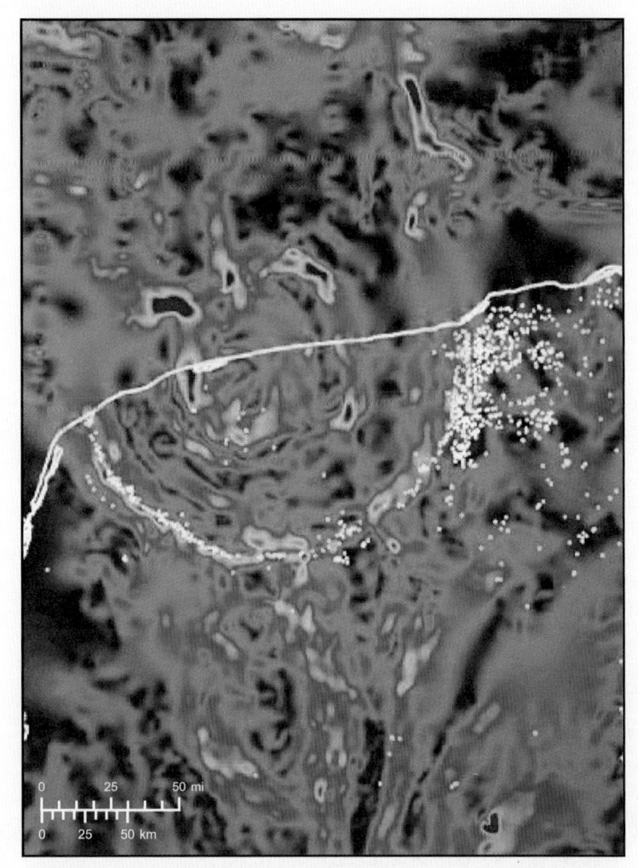

Figure 4.40 Gravity map over the Chicxulub crater. From Wikipedia (https://en.wikipedia.org/wiki/Chicxulub_crater). Also see Gulick *et al.* (2013).

base. This hypothesis was initially roundly criticized (e.g., Officer and Drake 1985), and a central question focused on the obvious: Where is the impact crater?

During earlier collection of airborne magnetic data for the purpose of identifying oil- and gas-bearing structures in Mexico, Pemex geologists and consultants identified a circular structure that extended from offshore to onshore northern Yucatán, centered on the small town of Chicxulub. The structure was at least 100 km (62 miles) in diameter, clearly suggestive of a bolide impact (Penfield and Carmargo-Zanoguera 1981; Hildebrand *et al.* 1991; Figure 4.40). Unsuccessful Pemex exploration wells targeting the magnetic anomaly provided a database of well information that allowed Hildebrand *et al.* (1991) to identify impact melt rock, shocked quartz, and other evidence of an extraterrestrial origin of the structure, and date the crater to the K–Pg Era. Ward *et al.* (1995) also noted a high content of anhydrite in Pemex wells in the area. Anhydrite is an evaporite mineral containing sulfur, which if widely distributed following impact could have had a deleterious effect on global fauna and flora (Artemieva *et al.* 2017).

New seismic data, both seismic reflection and refraction, was collected across the offshore areas in 1996 and 2005 by geologists

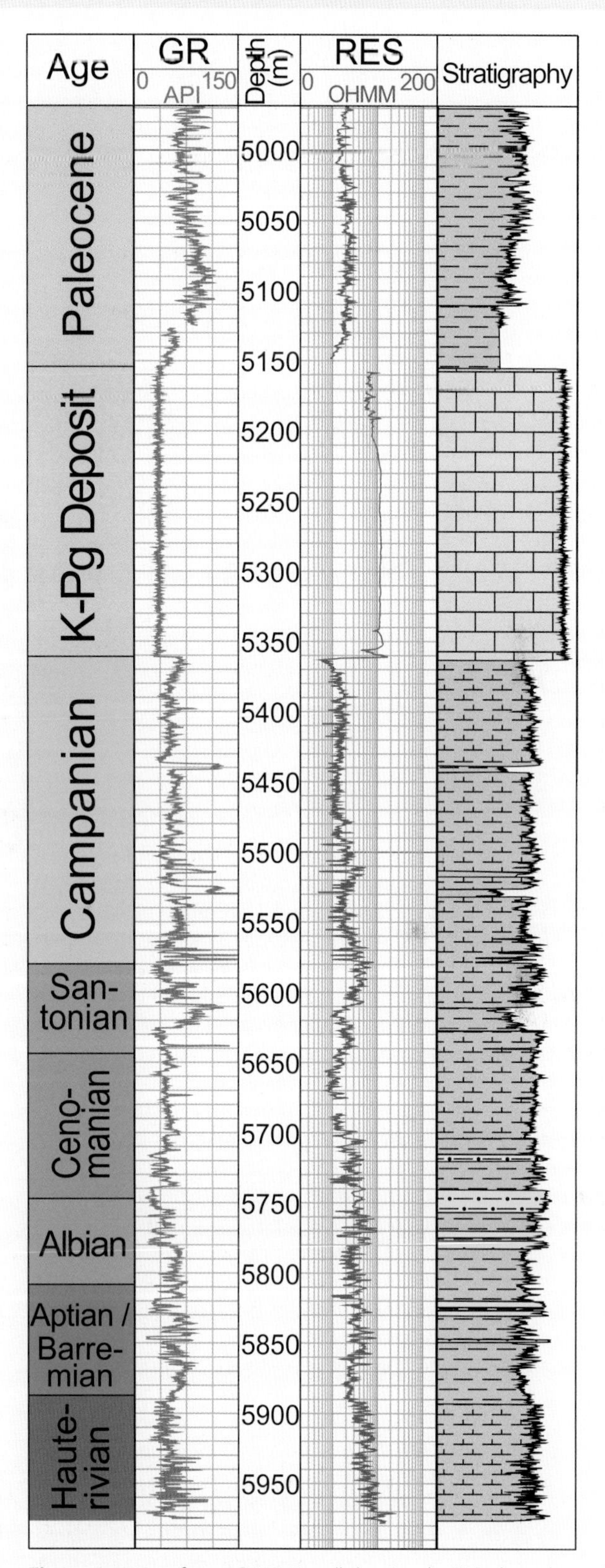

Figure 4.41 Log from AC 557 #1 well showing the K–Pg boundary unit. Modified from Denne *et al.* (2013).

Box 4.3 *(cont.)*

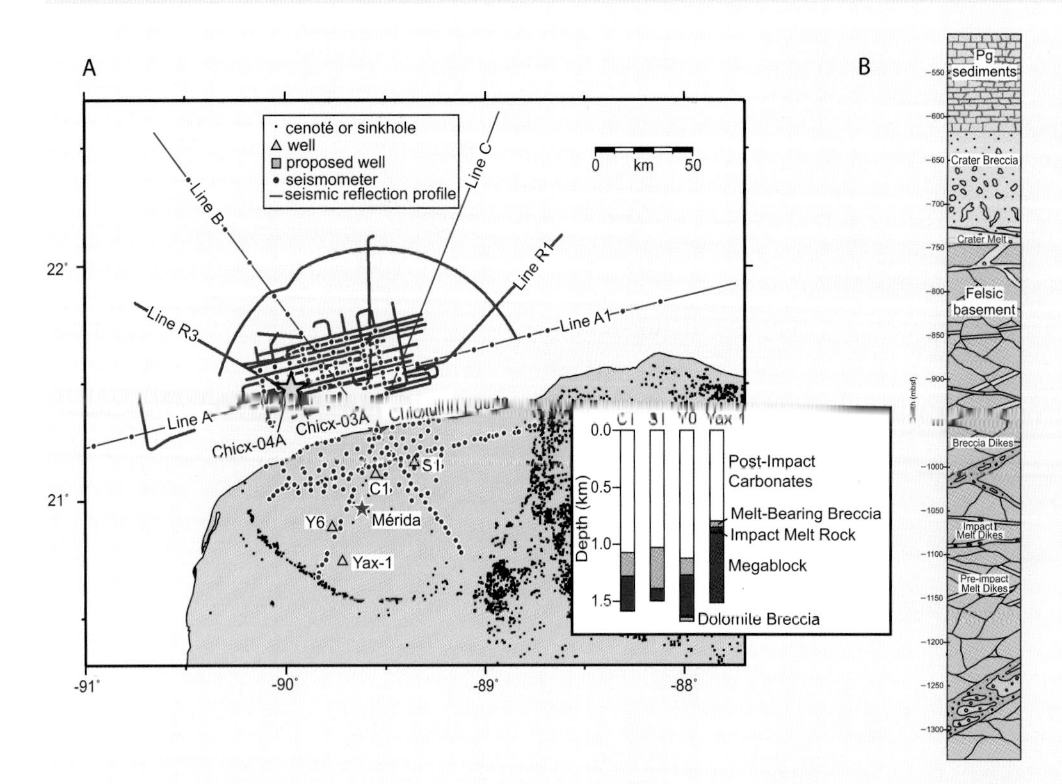

Figure 4.42 IODP Expedition 364. **(A)** Map of Chicxulub crater and site of coring site M0077A. Modified from Gulick *et al.* (2013) **(B)** Core schematic interpretation for M0077A. Modified from Morgan *et al.* (2016).

with the University of Texas at Austin, Imperial College, London, and UNAM, Mexico (Gulick *et al.* 2008). Two new conclusions emerged from this new data: (1) the final Chicxulub crater diameter was probably around 180 km (112 miles); and (2) the effect of the impact was noted as deep as 35 km (22 miles), indicating that the impact had deformed the Moho, the contact between the mantle and crystalline crust (Christeson *et al.* 2001). The onshore outer-rim of faults of the crater concentrated groundwater flow, creating a ring of sinkholes, called cenotes, which are well documented as important sources of water for the Mayan civilization.

The oil and gas industry had long noted in offshore areas a prominent seismic reflection package of Cretaceous age, initially and erroneously called the MCU (Mid-Cretaceous unconformity) because of the erosional truncation and hiatus below the K–Pg horizon. This was later confirmed to be at the Top Cretaceous (Dohmen 2002). In Mexico, a fractured limestone breccia unit at the K–Pg boundary is the major reservoir unit in the Akal–Reforma trend including the Supergiant Cantarell Field (Grajales-Nishimura *et al.* 2000; Cantú-Chapa and Landeros-Flores 2001). Carbonate breccias are also noted in Cuban outcrops, with thicknesses exceeding 400 m (1312 ft; Tada *et al.* 2003).

Some time later, explorationists began to recognize some key characteristics of the K–Pg boundary deposit in the deepwater areas of the northern Gulf, where the unit was much thicker (regionally to >1000 m in salt-related minibasins) than in onshore or shelf-depth waters (Denne *et al.* 2013). The large impedance contrast at Top Cretaceous was determined to be due to the high compressional velocity of the deposit, matching mud log observations of a hard limestone breccia with generally low porosity (Scott *et al.* 2014). On logs, the K–Pg deposit was typically a blocky, low gamma-ray log motif, sometimes capped by slight increases in gamma-ray at its very top (Scott *et al.* 2016; Figure 4.41). Well-based biostratigraphic analyses show a mixture of redeposited fossils of late to early Maastrichtian and Campanian age (Denne *et al.* 2013). This matched the Cretaceous–Tertiary boundary "cocktail" biostratigraphy of Deep Sea Drilling Project (DSDP) sites in the southeastern GoM that Bralower *et al.* (1998) described as characteristic of the Chicxulub impact event.

The most recent scientific investigation is the joint IODP–IDCP Expedition 364, which drilled to, and obtained core samples from, the peak ring, the central portion of the crater at a site offshore of Merida, Mexico (Figure 4.42A). The Chicxulub crater represents the only preserved peak ring crater on Earth; thus cores of the basement rock provided important constraints for modeling how peak rings formed on other planets (Morgan *et al.* 2016). The core was obtained using a lift boat, a rig normally employed by the oil and gas industry for drilling in shallow coastal waters like offshore Yucatán Platform. Cores retrieved from 505–1335 m depth included 130 m of suevite (melt-bearing impact breccia), melt rock and shocked granitic basement cut with dikes of impact melt rock (Figure 4.42B; Morgan *et al.* 2016).

Logs and seismic data indicate that the granitic basement rock has unusually low densities and seismic velocities; the density of the felsic basement varies between 2.10 and 2.55 g/cm^3, with a mean of 2.41 g/cm^3, and P wave velocities vary between 3.5 and 4.5 km/s, with a mean of 4.1 km/s. These values are unusually low for felsic basement, which typically has densities of >2.6 g/cm^3 and seismic velocities of >5.5 km/s (Christeson *et al.* 2018).

Box 4.3 *(cont.)*

Other signs of impact-induced deformation in the core include pervasive fracturing and formation of cataclastites and ultracataclastites (Morgan *et al.* 2016). Petrographic analyses revealed the presence of distorted twin lamellae in plagioclase feldspars, and quartz grains with planar deformation features (Riller *et al.* 2018).

This core calibration of the seismic reflection and refraction data allows reconstruction of the impact and transient crater formation processes (Morgan *et al.* 2016). In as little as five minutes following impact, it is estimated that the transient crater went from 30 km deep to a peak of 20 km high and then back down to its current depth of 740 m below sea floor (Riller *et al.* 2018). By comparison, the Marianas Trench is only 11 km deep and Mount Everest is only 9 km high. The observed shock metamorphic features suggest that the peak ring rocks were subjected to impact pressures of ~10–35 gigapascals (Morgan *et al.* 2016). Impact melt, which is formed at high shock pressures, is also an important component of the peak ring.

Estimating the size of the bolide and the energy released at impact continues with this new core calibration. Earlier estimates suggest a bolide of 17.5 km diameter, striking the Earth at a 20° angle and a speed of 20 km/s. This likely generated a force of 4.2–12 × 1023 joules (Hildebrand *et al.* 1991), equivalent to a magnitude 11 earthquake (Day and Maslin 2005). No human has experienced an earthquake of this size.

Another important insight from the new scientific research at Chicxulub is the rapid recovery of life within the crater following impact. The picture established from analysis of foraminifera, calcareous nannoplankton, trace fossils, and elemental abundance data indicates that life reappeared soon after impact and in fact a high-productivity ecosystem was established within 30,000 years, possibly as soon as a thousand years (Lowery *et al.* 2018).

Given its proximity to the Chicxulub crater, the deepwater GoM is therefore a regional sink and thus a natural repository of information on the processes and products outside of the crater. There is also increasing recognition that this seminal tectonostratigraphic event, one that ended the Mesozoic, also dramatically reshaped the GoM basin, its seascape, and its adjacent landscape, paving the way for the Cenozoic (Denne and Blanchard 2013; Sanford *et al.* 2016; Scott *et al.* 2016).

4.11.1 Chronostratigraphy

This short-duration unit begins immediately after Ma4_700fs (66.76 Ma) chronostratigraphic horizon. Renne *et al.* (2013) assign the impact to 66.0 Ma. Mass transport deposits were all emplaced within hours to days, and the settling of the fine particulate material from the atmosphere and through the water column is estimated to have only taken a few years (Sanford *et al.* 2016).

4.11.2 Previous Work

While the land record of this event has been chronicled in numerous papers (e.g., Yancey 1996; Tada *et al.* 2003; Schulte *et al.* 2012), only recently have workers investigated the products of this impact in the deep GoM basin (Bralower *et al.* 1998; Denne and Blanchard 2013; Denne *et al.* 2013; Scott *et al.* 2014; Sanford *et al.* 2016).

4.11.3 Impact-Related Processes and Products

As discussed in Box 4.3, the Chicxulub impact likely generated a magnitude 11 earthquake at ground zero. The present-day Chicxulub crater is centered on the town of Chicxulub near the coast of the modern Yucatán Peninsula, which did not become emergent until the Eocene. Known processes that result from large earthquakes in such a setting include seismic shaking (ground roll) and formation of tsunamis (Sanford *et al.* 2016). At the impact site, the formation of suevites (i.e., melt-bearing impact breccias) and melt rock was contemporaneous with the injection of sedimentary material and vaporized impactor into the atmosphere, as well as sediment gravity flows directed seaward (Morgan *et al.* 2016). However, the timing of the arrival of all these events must have varied with distance from the crater as a function of the differential velocities of each.

Sanford *et al.* (2016) estimated the velocities for seismic and mega-tsunami waves and applied this thinking to reconstruct the timing of failure and depositional events within the first few hours of impact. This temporal reconstruction also explains the products of the event documented from well, core, and seismic data around the entire basin (Figure 4.43). The sequence of events is illustrated in the context of a schematic cross-section extending from the impact site to the Florida Platform (Figure 4.44).

Seismic shaking or ground roll is the first and primary initiator of both failure and sediment transport following impact and formation of the transient crater (Figure 4.44A, B). With velocities roughly an order of magnitude faster than the expected mega-tsunami, the seismic wave probably reached the Yucatán Platform margin within two minutes, given known distances from the crater (Figure 4.44C; Sanford *et al.* 2016). This undoubtedly caused slope failure. Present-day seismic lines across the Yucatán margin in this area (cross-sections 7 and 9; Figures 1.19 and 1.21) document over-steepened margins and presence of multiple terraces, which are overlain and essentially "healed" by post-impact Cenozoic carbonate deposition. The structure of the exterior ring clearly relates to impact-related processes (Figure 4.40). Seismogenic debris flows would have initiated from margin failure at Yucatán and even as far north as the Florida Platform (Figure 4.44D).

At the crater site, suevite deposits overlie impact melt rock, which in turn overlies shocked basement rock on the crater's peak ring (Morgan *et al.* 2016). These are roughly contemporaneous with the K–Pg boundary unit as a whole. Outside of the crater, the largest volume of sedimentary product are muddy

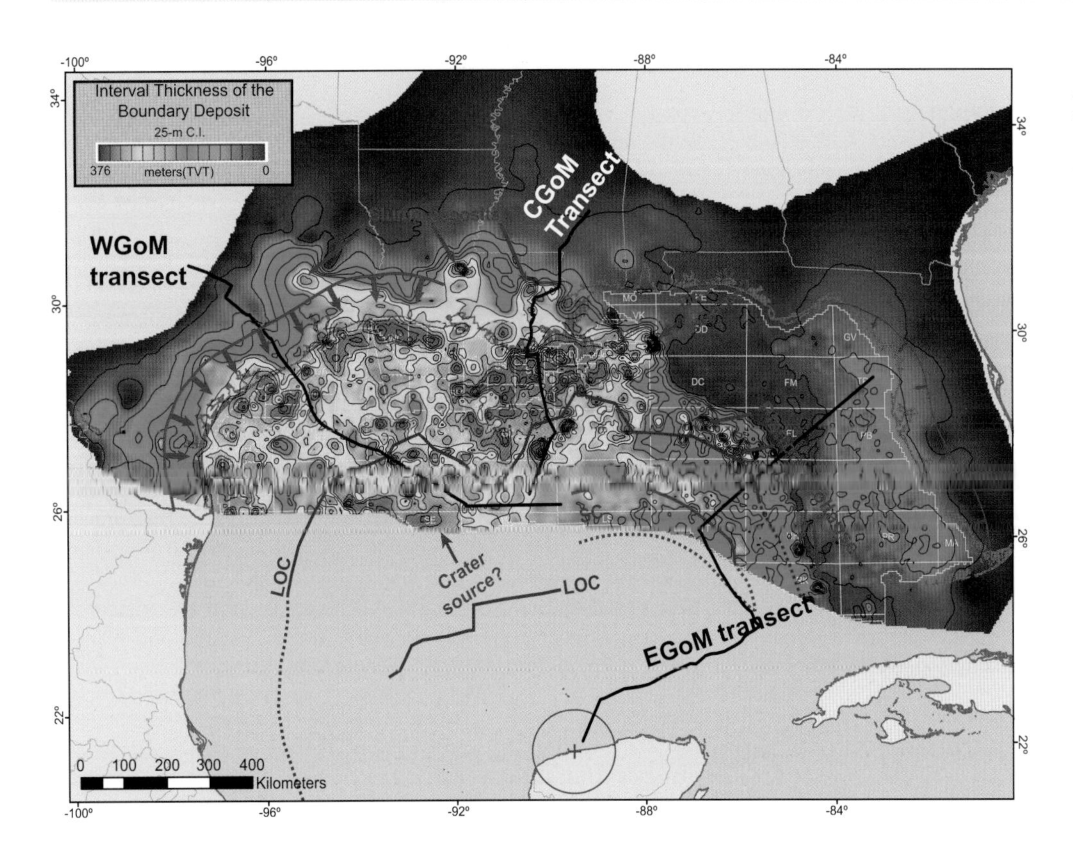

Figure 4.43 Thickness of K–Pg based on 2D seismic reflection interpretation. Cross-section (black line) is shown in Figure 4.44. Modified from Sanford *et al.* (2016).

debrites, documented in DSDP sites 536 and 540 in the Florida Straits that resulted from platform margin collapse and remobilization as debris flows that entrained carbonate blocks of various sizes, along with a matrix of carbonate mud likely present at the sea floor. In the DSDP core sites 536 and 540, debrites can be described as greenish-gray pebbly mudstones with irregular clasts floating in a mud-rich carbonate matrix (Sanford *et al.* 2016).

Tsunami waves from the point of impact likely followed within 30 minutes, but not before the seismic wave passed across the Gulf basin and reached the Florida Platform margin, causing platform margin collapse at this time (Figure 4.44E). Seismic lines across the Florida Platform show over-steepened margins there as well, but continuing Cenozoic retrogradational failure (Mullins *et al.* 1986) or even bottom current erosion (Snedden *et al.* 2012) may have subsequently modified that escarpment. Likely failure of the Florida margin within four minutes of impact is recorded in large, irregular blocks formed in Elbow and Lund protraction blocks of the northern GoM (Snedden *et al.* 2014; Figure 4.45).

By this point, much of the Chicxulub crater present-day structure was in place, including the peak ring, central basin, inner and outer ring, and exterior ring faults (Figure 4.44E). Muddy debris flows presumably left a record across the entire Florida Straits, though coring is limited to DSDP sites, and Cuban deepwater well logs are not publicly available. It is possible that Florida-sourced debris flows, directed back toward the deep basin, may have resulted from the seismic wave reaching the Florida shoreline in the first five minutes of impact (Figure 4.44E). Presumably these added to the muddy debrite record in the US side of the basin.

The order of magnitude slower velocity of the mega-tsunami propagating from the crater suggests arrival at the Yucatán margin and DSDP core sites much later, 30 minutes or more after impact (Figure 4.44F). The 40 m thick debrite interval at sites 536 and 540 is capped by a 10 m thick interval of turbidites (Sanford *et al.* 2016). The turbidity flow deposits include normal to reverse-graded beds of carbonate grainstones and pebble-rich zones (Sanford *et al.* 2016). Multiple bedsets point to a series of turbidity flows, perhaps coming from several point sources (Yucatán, Florida, and possibly Cuba). Turbidity flows probably initiated at the Florida margin as the mega-tsunami reached that margin within 45 minutes (Figure 4.44G). Bounce-back of the mega-tsunami probably generated some thin turbidite deposition on the Florida shelf and adjacent basin, all within an hour of impact (Figure 4.44H)

The final subaqueous product is likely to have been the settling out of carbonate mud over several months to years following the impact (Artemieva and Morgan 2009). This may have included a fine fraction of silt related to crater airfall ejecta that settled into the global oceans. Assuming limited post-depositional reworking, the iridium-enriched dust layer is part of the mudstone cap (Sanford *et al.* 2016).

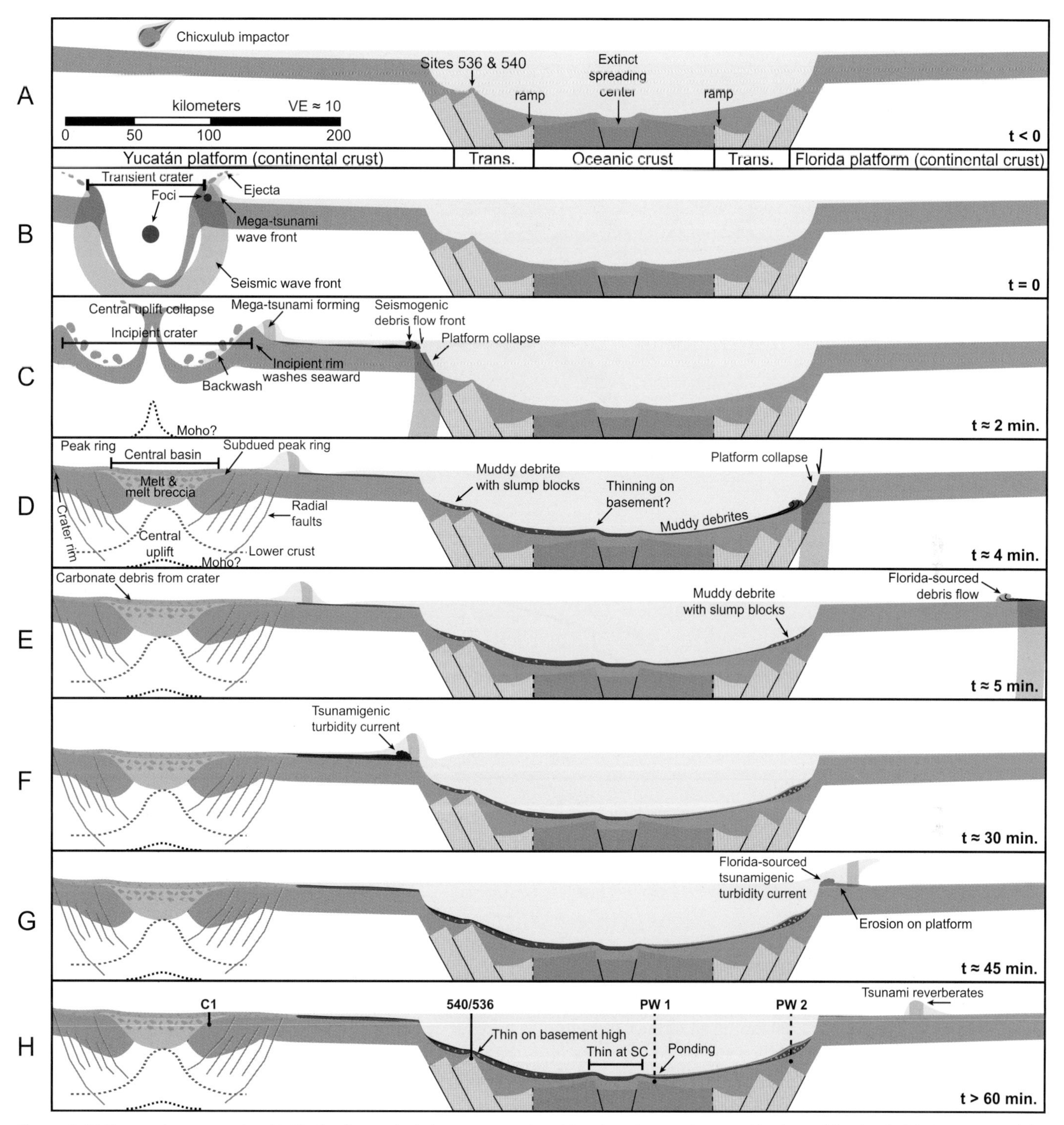

Figure 4.44 Temporal reconstruction (A–H) of sedimentological processes and resulting deposits occurring over 60 minutes following bolide impact to end the Mesozoic. Location of the cross-section is shown in Figure 4.43. Modified from Sanford *et al.* (2016).

Sanford *et al.* (2016) proposed a trimodal model of post-impact processes related to the differential velocity of the seismic ground wave versus tsunami waves, and suspension settling of carbonate mud and airfall ejecta. However, this pattern would not be expected to develop globally, due to distances involved and landscape versus seascape settings. In the next section, observations about the K–Pg boundary unit in other portions of the basin are considered in the context of the Sanford *et al.* (2016) model of Chicxulub-related impact processes. This includes seismic data interpretations and observations from well penetrations around the basin.

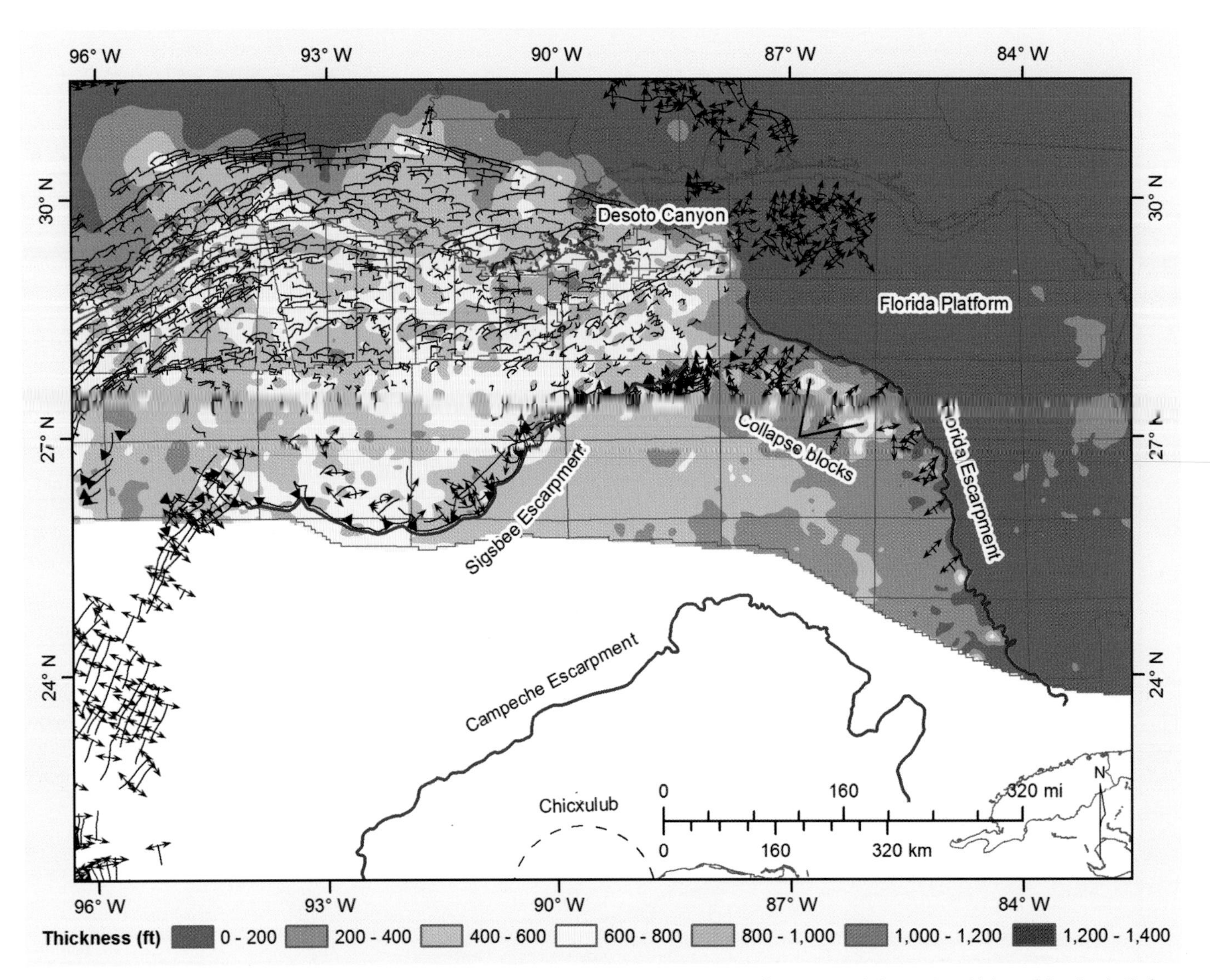

Figure 4.45 K–Pg boundary unit thickness in the eastern GoM. Area of large intact blocks likely derived from seismic shaking induced failure of the Florida Platform margin indicated by arrows. Modified from Snedden *et al.* (2014).

4.11.4 Observations of the K–Pg Boundary Deposit around the GoM

In northern Louisiana, about 1200 km north of the Chicxulub crater (Figure 4.46), Justiss Oil drilled and cored the K–Pg in the Justiss #1 La Central IPNH well in the Olla Field, LaSalle Parish, Louisiana. Shellhouse (2017) made a detailed description of the cored succession that included contacts between the K–Pg boundary unit and underlying chalk of the Navarro Group NT supersequence (Olson *et al.* 2015) and the overlying Midway Group shales (Figure 4.47; Shellhouse 2017). The cored succession can be divided into three gross units: (1) a lower light-colored, tan, marly chalk; (2) an upper, darker-colored, gray, marly chalk; and (3) a silty dark shale with occasional siderite concretions. The three zones are separated by clasts and pebble-bearing zones (Figure 4.47), suggestive of a hiatus or unconformity (Shellhouse 2017).

The lower zone (core depth of 4582.3–4540.75 ft/ 1396.7–1384.0 m; Figure 4.47) is rich in trace fossils including *Thalassinoides* and *Chondrites* burrows and shell fragments. A scour structure is evident in the core. The lower zone is bounded at its top by a clast-bearing zone (Shellhouse 2017). In thin section, the high fossil content, largely calcareous foram tests, and broken macrofossils, is evident. Insoluble residue content is about 25 percent (dry weight; Kinsland *et al.* 2017). Log gamma-ray and density logs reflect this difference with the middle zone. The pebble zone separating the two marly chalk zones is a fossiliferous pebble zone similar to that below the contact.

The middle zone (4540.75–4495.5 ft/1384.0–1370.2 m; Figure 4.47A) of darker-colored marly chalk has a higher content of clay than below (Shellhouse 2017). Insoluble residue content is 45 percent (Kinsland *et al.* 2017). It is equally fossiliferous but more pebble and clast-rich than the

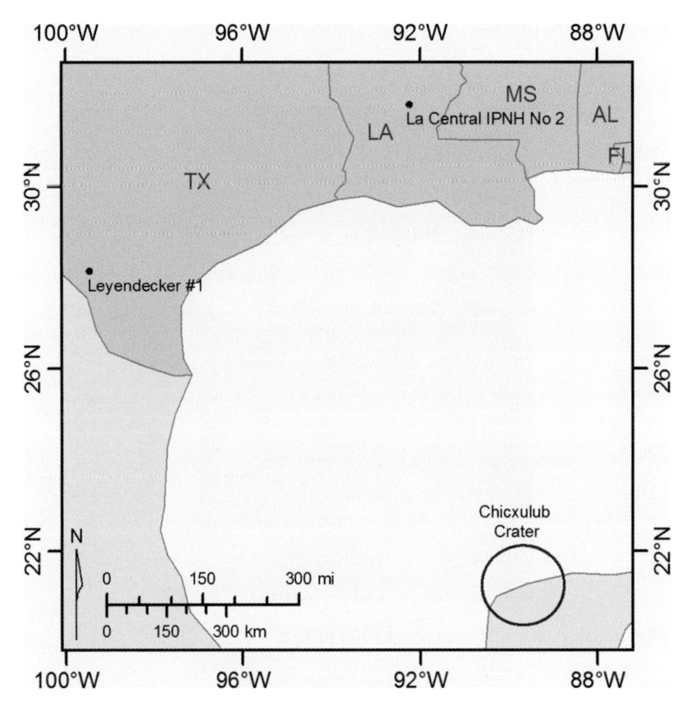

Figure 4.46 Map showing location of Justiss Oil #1 La Central IPNH No 2 well and south Texas Leyendecker #1 well that cored the K–Pg boundary unit. Modified from Kinsland and Snedden (2016).

underlying unit, excluding the bounding pebble-zones. *Thalassinoides* trace fossils are also common as well as *Helminthopsis* burrows. The unit is distinguished from the underlying chalk by the darker color, indicating a higher carbonate mud content as well as a lack of any primary sedimentary structures like scours and horizontal laminations. In thin section, fossil content and high clay content is apparent.

The upper contact of the darker-colored chalk with the overlying Midway Shale is another pebble-bearing zone at 4496.5–4495.5 ft (1370.5–1370.2 m; Figure 4.47B). However, this zone contains an unusual diversity of clast types and sizes. Fossiliferous claystone clasts are 1–7 mm, rarely up to 4 cm in size, in a black–dark brown mudstone–claystone matrix. Several clasts contain inclusions, and alteration rim color changes in clasts are common. There are several laminations dipping ~60° from horizontal. The contact with the overlying Midway Group dark gray shales is sharp to erosional (Figure 4.47C).

Thin sections from the upper contact include fossiliferous pebbles and an unusual clast and overlying matrix containing round calcite-filled structures of 0.2 mm size, interpreted to represent spherules of 0.2 mm diameter (Figure 4.48; Shellhouse 2017). Spherules are interpreted as condensed products of molten and vaporized rock from the crater that were dispersed to the upper atmosphere and distributed globally (see Box 4.3). The contact contains up to 20 percent spherules.

The overlying dark gray to black shale interval (4495.5–4463 ft; 1370.2–1360 m) is dominated by mudstone with occasional siltstones and bands of red–brown siderite. The shales are nearly devoid of shell fragments, trace fossils, or bioturbation, with only the rare *Helminthopsis* burrow. Thin sections show micro-laminated mudstone (98 percent) with about 2 percent silt grains, but silt content does change to small calcareous grains (15 percent of rock) downward toward the contact. Samples from shale without reddish bands yield insoluble residues >90 percent, as is expected of a shale (Kinsland *et al.* 2017). Samples in the reddish banded intervals of the core are high in iron, as indicated from XRF (X-ray fluorescence) data, probably from siderite.

4.11.4.1 Sedimentary Process Interpretation

Earlier 3D seismic studies had noted large, dune-like bedforms at the top of the K–Pg boundary deposit in a nearby location that are thought to indicate tsunami wave passage (Egedahl *et al.* 2012; Strong 2013; Strong and Kinsland 2014). This argues for impact-related processes even at the distal (>1000 km) location of the Justiss core (Figure 4.46).

As discussed earlier, the seismic ground roll or wave that moved outward from the impact crater affected onshore Florida and should also have reached northern Louisiana, where our paleogeographic reconstruction shows a broad carbonate shelf accumulating chalks (Figure 4.35). Seismic shaking could have caused failure and basinward sliding of the upper (darker-colored) chalk unit from an updip location to the present well location. This more nearshore location could have been more prone to advection of clay-sized material from muddy rivers, thus explaining the higher overall clay content and darker color of the lower chalk unit (Shellhouse 2017). Thus, the base of the K–Pg boundary deposit is picked at the lower clast-bearing zone separating the lower and upper chalks (Figure 4.47A), the latter thought to be in place prior to impact. The upper chalk is thus interpreted as a mass failure deposit, and similar to mud log descriptions of the K–Pg boundary deposit in shelfal locations (Denne *et al.* 2013; Scott *et al.* 2014). Thus, the end of the Cretaceous at this location is a structural contact between *in-situ* chalks below and a mass transport deposit, the overlying darker-colored chalks (Figure 4.47).

The upper pebble-bearing zone is the best candidate for the contact between the K–Pg boundary deposit and the Midway Shale (Figure 4.47B). The concentration of calcite-replaced spherules, irregular clasts, and sharp contact with the Midway supports this view (Figure 4.47).

A final supporting line of evidence for picking the top of the K–Pg boundary deposit, not included in the thesis of Shellhouse (2017), is apparent in XRF-based elemental analysis over this contact (Figure 4.47C). A prominent spike in zinc and zirconium and minor increases in other elements over background levels occurs just below the upper pebble-bearing zone, a light-orange-colored bed at 4495.50 ft (1371 m). Zinc and zirconium are known to be proxies for platinum group elements like iridium. This suggests accumulation of airfall iridium and a final period of wave reworking prior to deposition of the Midway Shale group.

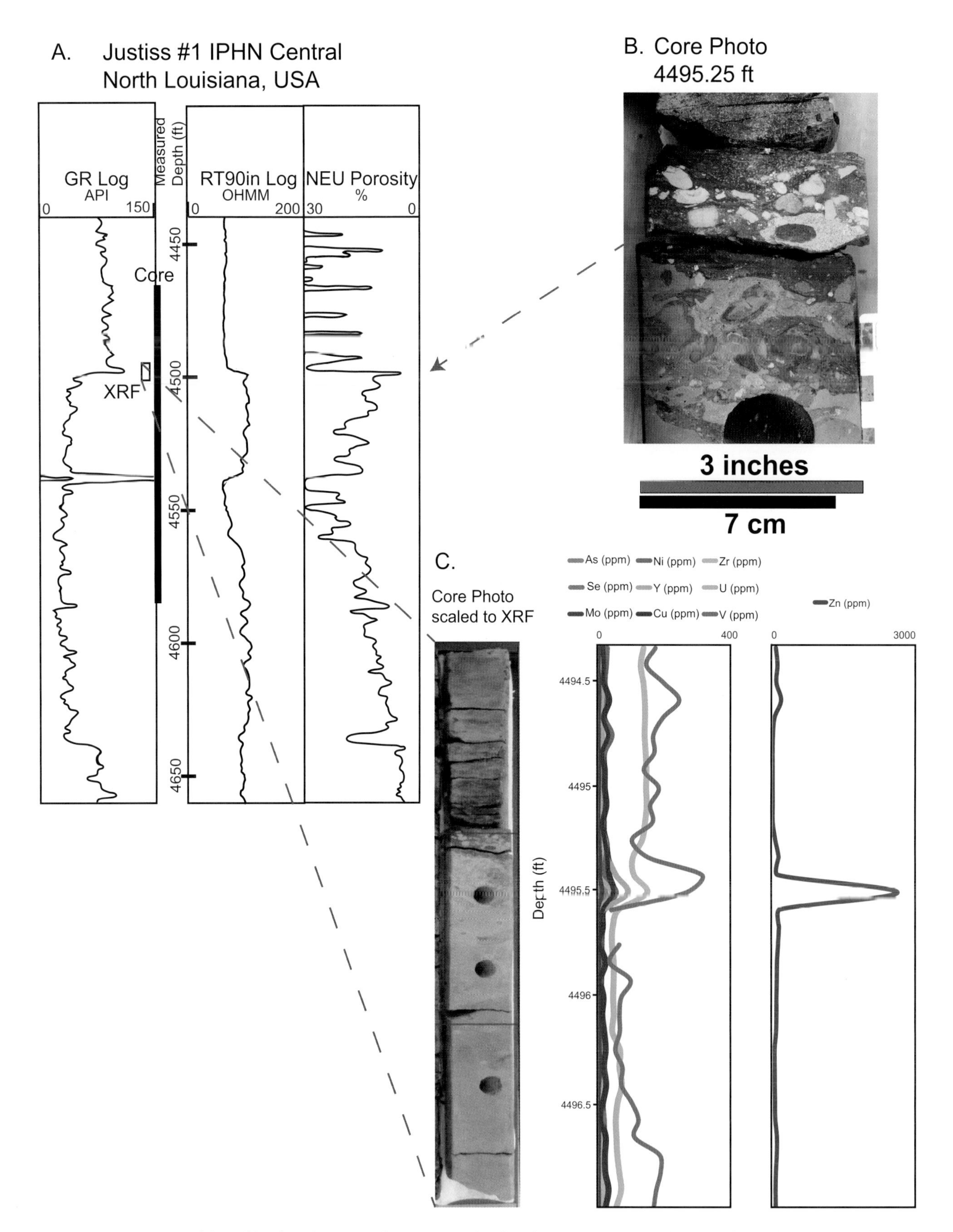

Figure 4.47 (A) Portion of the well-log from the Justiss Oil #1 IPHN Central well. High gamma-ray zone at about 4540 ft is the likely boundary between underlying *in-situ* chalk deposits and overlying chalks redistributed by impact-related processes. Upper contact between Midway Shale is highlighted in (B) and (C). Note the depth shift of 3 ft (1 m) between the core depth and the log depth. (B) Core photo at the Midway/K–Pg contact. (C) Photo of core and corresponding X-ray fluorescence measurements showing elemental variations across the K–Pg boundary with overlying Midway Shale.

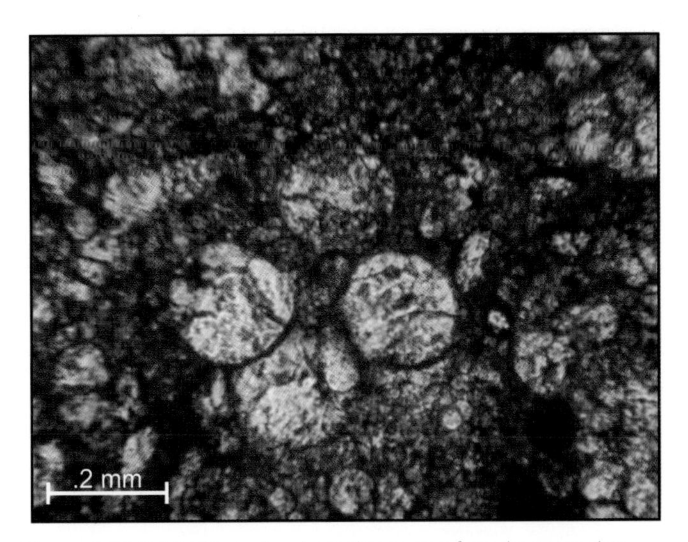

Figure 4.48 Calcite-filled spherules in thin section from the upper clast-bearing zone, in plane-polarized light. Each spherule is ~0.2 mm in diameter. Modified from Shellhouse (2017).

Another surprise in our work is the lack of organic content in the overlying Midway Shale, in spite of the rather dark color that is normally associated with organics. One possibility is that the dark color reflects a high inertinite content related to combustion of organics like vegetation and trees in adjacent land areas, later washed into this area (Kinsland *et al.* 2017).

In another location at a similar distance (1260 km; 780 miles) from the Chicxulub impact crater, the Murexco A-1 Leyendecker well was drilled in the Tom Walsh Field of Webb Co. Texas, penetrating the Late Cretaceous Escondido Sandstone (Figure 4.46). From 5280 ft to about 5295 ft core depth, the A-1 Leyendecker well core contains a 15 ft (4.5 m) thick fine-grained quartzose sandstone with low-angle, hummocky cross-stratification and a sharp basal contact (Figure 4.49). The upper contact is a gradational succession grading from buff sandstones to burrowed admixtures of sandstone, siltstones, and shales (Figure 4.49).

Like the Justiss #1 La Central IPNH well, the paleogeographic positions of this well is interpreted to be on the NT shelf, within the middle to inner neritic zone. However the Leyendecker well is proximal to a wave-dominated delta system of the Escondido Formation versus open shelf setting of the Justiss well with no nearby siliciclastic input (Figure 4.35).

The possibility of tsunami-generated bottom currents was likely enhanced in the Rio Grande Embayment. This may explain the sharp, erosional base to the Leyendecker Sandstone (Figure 4.49). Hummocky cross-bedding in the cored interval of the Escondido Sandstone indicates combined wave and current flow. The gradational top to the interval implies decelerating flow as the tsunami waned (Figure 4.49). Pervasive bioturbation in silty shales above the sandstone implies enhanced food supply that occurred as organics were swept from updip coastal plains into the ocean. Seismic and well log correlation shows the K–Pg boundary deposit to be laterally continuous over the Tom Walsh Field (Figure 4.50). Production from the K–Pg boundary deposit supports this view of lateral continuity over at least 1.6 km (1 mile), as the reservoir shows a natural decline in both oil and gas, with a cumulative production of 122 MMCFG and 2550 BO.

The Justiss and Leyendecker cores provide rich insights into the processes that occurred at a considerable distance from the impact. If seismic shaking affected the basin at such a distal location, then it is logical that areas closer to the impact would be dominated by these impact-related processes.

4.11.5 Seismic-Based K–Pg Unit Thickness Map

Mapped distribution of the K–Pg boundary deposit in the northern GoM strongly suggests that sediment was sourced from virtually every shallow-water province north of the border, supporting the model of Sanford *et al.* (2016) that the majority of sediment flowed southward rather than directly sourced from the crater and transported to the north (Figure 4.43). A notable thick extending across southern Keathley Canyon, southeastern Alaminos Canyon, and Sigsbee Escarpment protraction blocks is probably linked to the Mexico-derived K–Pg deposits.

Isochore map trends are thus interpreted to indicate K–Pg sediment routing (Sanford *et al.* 2016; Figure 4.43). In the western Gulf, the Texas and Louisiana shelf appears to have acted as a significant line source for sediment in shelf-proximal deep waters, while the Florida coastline also appears to have been a minor line source. On the Louisiana and Mississippi shelf, isolated thicks represent slump deposits on the platform margin.

In the eastern Gulf, the Florida coast appears to have been a minor sediment line source to the Florida Platform, which itself was an area of minimal deposition as a result of its elevation. The platform itself, however, acted as a substantial sediment source, though isolated thicks representing slump blocks suggest that the primary mechanism of sediment redistribution in this province was platform collapse and glide block translation, likely induced seismically (Figure 4.45; Snedden *et al.* 2014). We interpret the bulk of the mass transport blocks off Florida to be of carbonate lithology, though these have not been penetrated by wells. In the distal deep waters of the eastern Gulf, ponding of sediment suggests that the extinct spreading center acted as a southern barrier to transport.

The central-northwestern area of the salt province is the area of greatest sediment accumulation, probably sourced by collapse and excavation of DeSoto Canyon as per the model espoused by Denne and Blanchard (2013). Erosional truncation is evident on seismic data in this area.

In the deepwater areas of the northern GoM, it is clear that the thickness of the deposit within the salt canopy varies substantially, with local thicks generally corresponding to syn-depositional salt lows. A local thickening (485 ft; 148 m) around the Tiber well (KC 102 #1) is a subsalt trend that reflects actual regional NT thickness. The nearest well is Salida (GB 989 #1), which is a partial penetration of the NT. Other nearby wells are carapace or salt-encased block penetrations

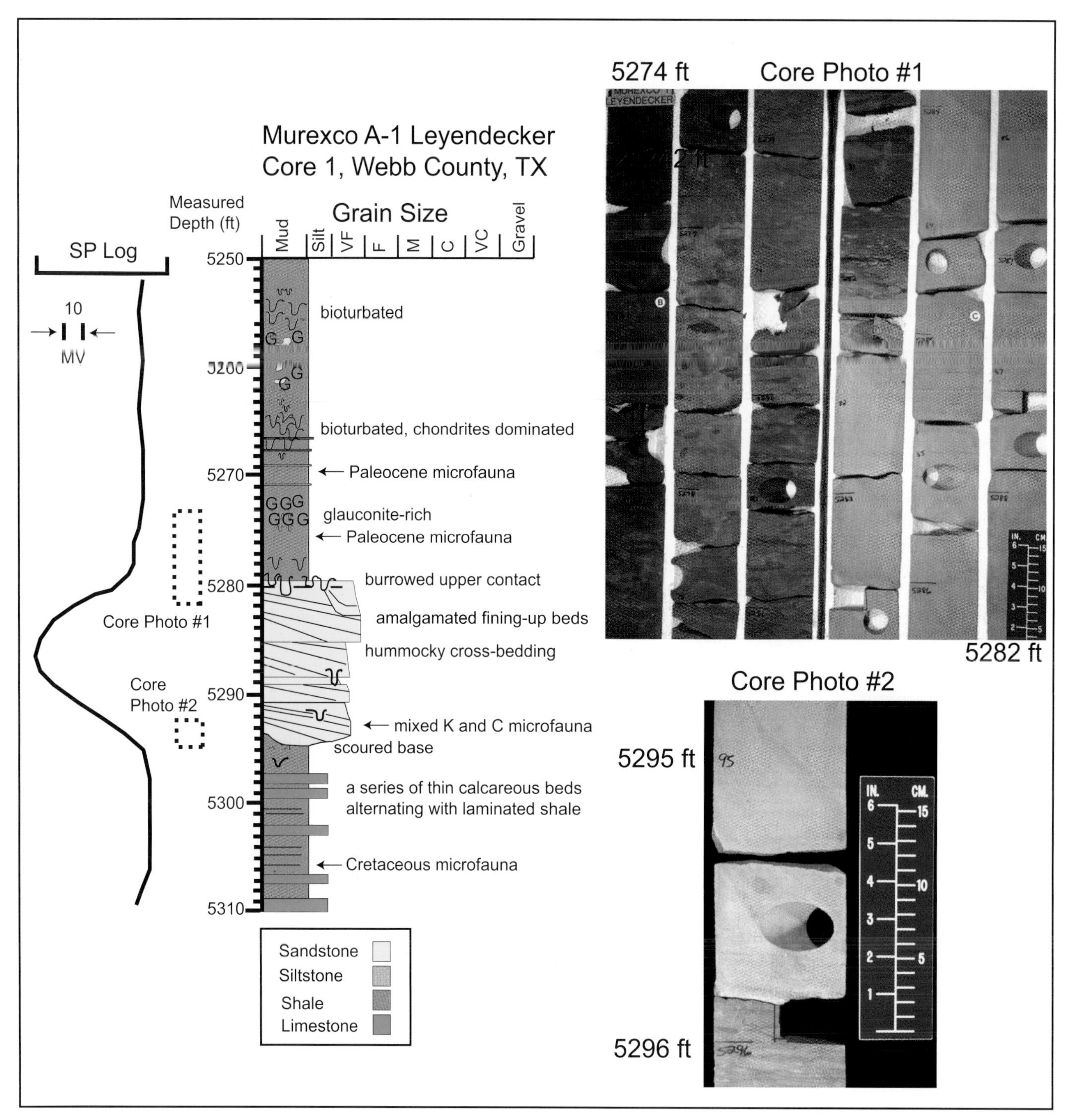

Figure 4.49 K–Pg (Escondido) Sandstone core description and core photos from Murexco A-1 well core, Webb County, Texas. Well location is shown in Figure 4.46.

that were excluded from the well seismic-based isochore mapping algorithm.

The salt province contains the thickest of deposits, on average, suggesting that this deep basinal domain was a regional sink for sediment generated throughout the Gulf. Local thicks are observed in welds and other areas of salt deflation that may have been Cretaceous lows, synclines, and grabens (Denne *et al.* 2013; Scott *et al.* 2014). The temporal model of Scott *et al.* (2014) notes the importance of allochthonous salt-flank grabens in trapping thick (1000 ft/300 m) K–Pg boundary deposits, thus creating a less rugose seascape prior to Wilcox deposition (Figure 4.51).

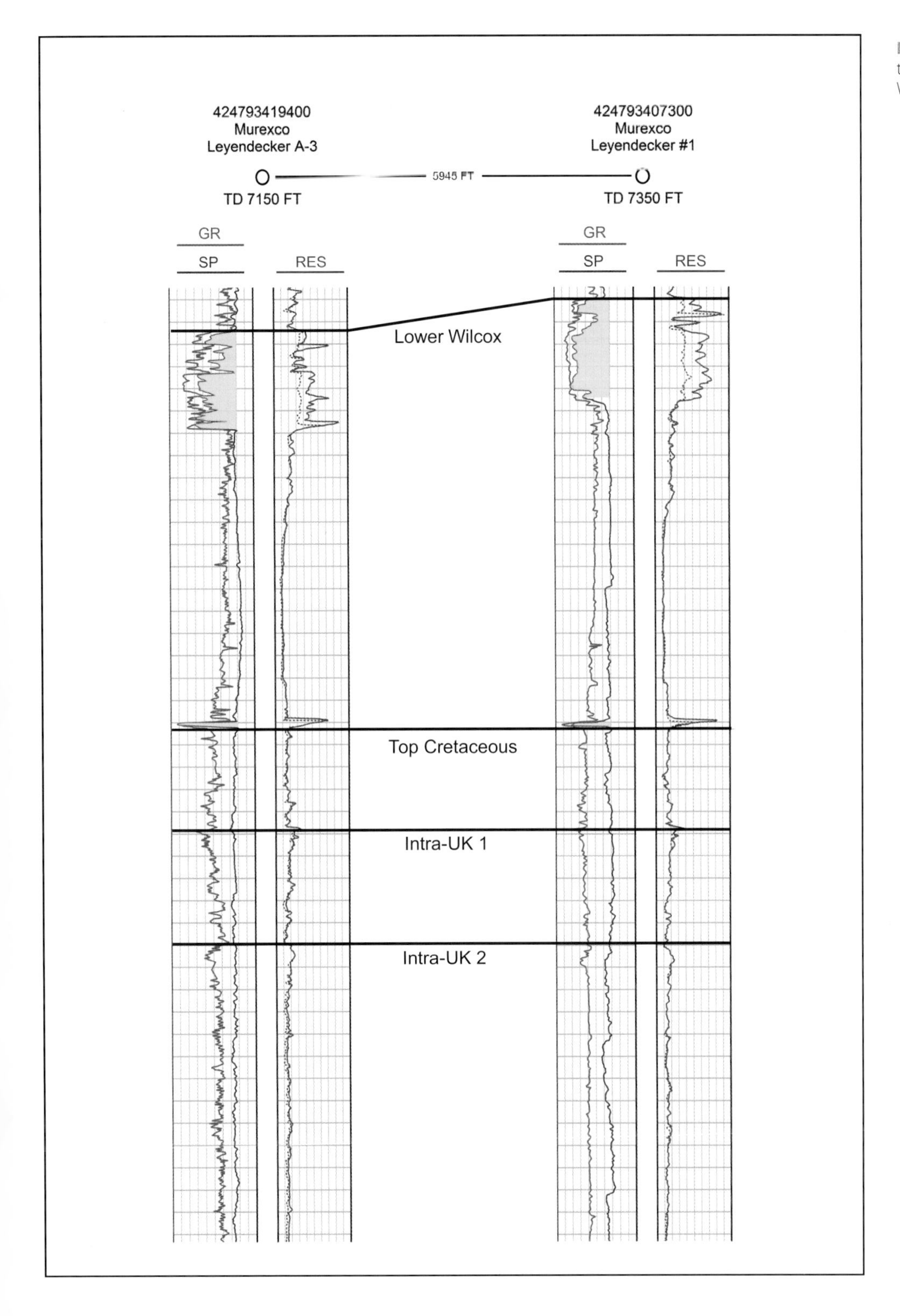

Figure 4.50 Well Log cross-section through the K–Pg unit in the Tom Walsh Field of Webb County, Texas.

4.11.6 K–Pg Boundary Deposit in Mexico

In Mexico, considerable work on the K–Pg boundary deposit has been undertaken as the breccia is a very important producing reservoir in several large fields of the Reforma–Akal trend of southeast Mexico (Figure 4.52). The K–Pg is the main reservoir of the Cantarell complex of four fields discovered initially in the 1970s (Alka, Nohoch, Chac, and Kutz) and the underlying Sihil thrust block first tested by Cantarell 418C in 1999 (Aquino Lopez and Gonzalez 2001; Mitra *et al.* 2005). The K–Pg boundary deposit reservoir accounts for over

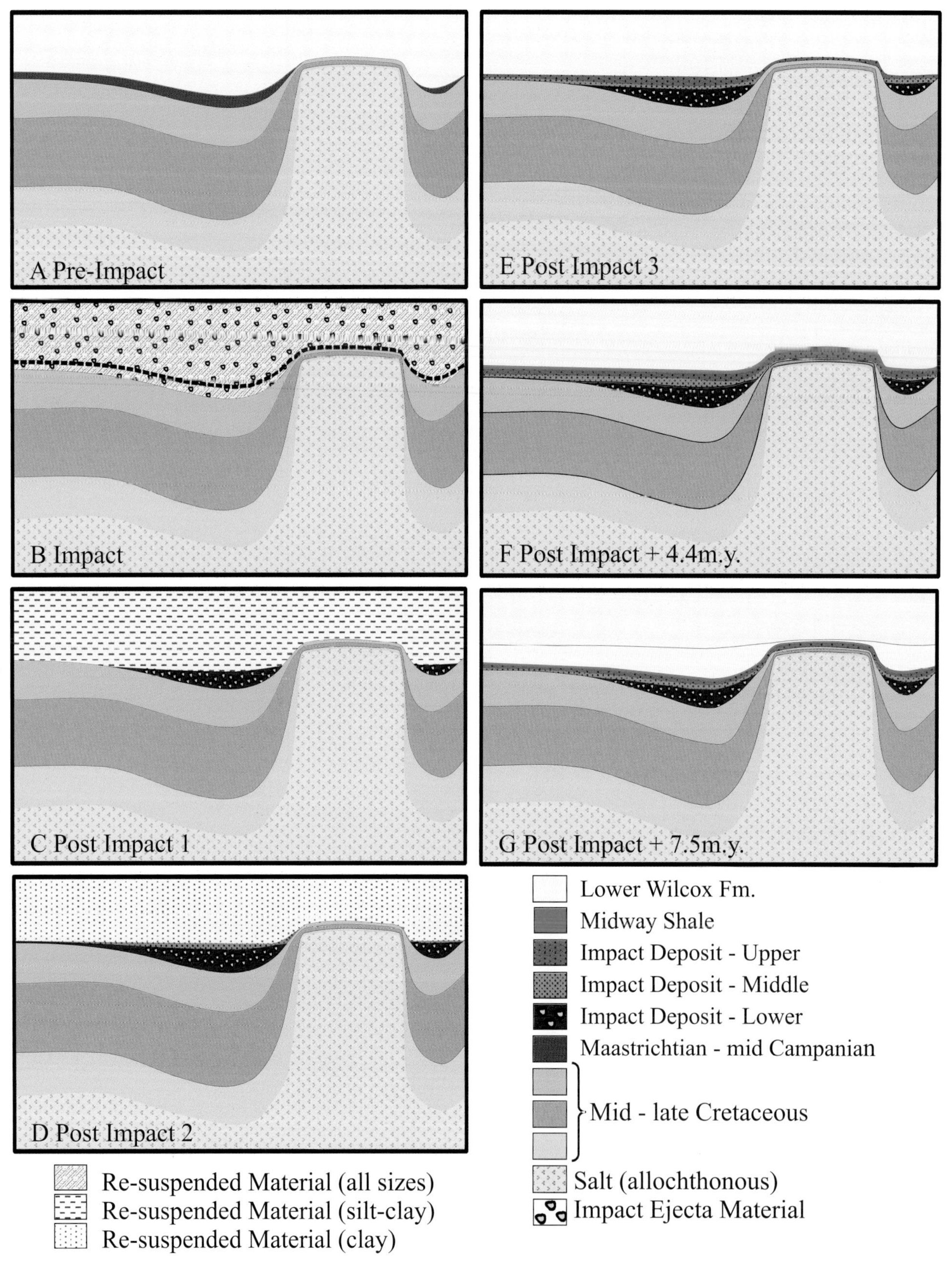

Figure 4.51 Temporal model for infilling pre-impact, salt-influenced structural lows during emplacement of the K–Pg boundary deposit. Modified from Scott *et al.* (2016).

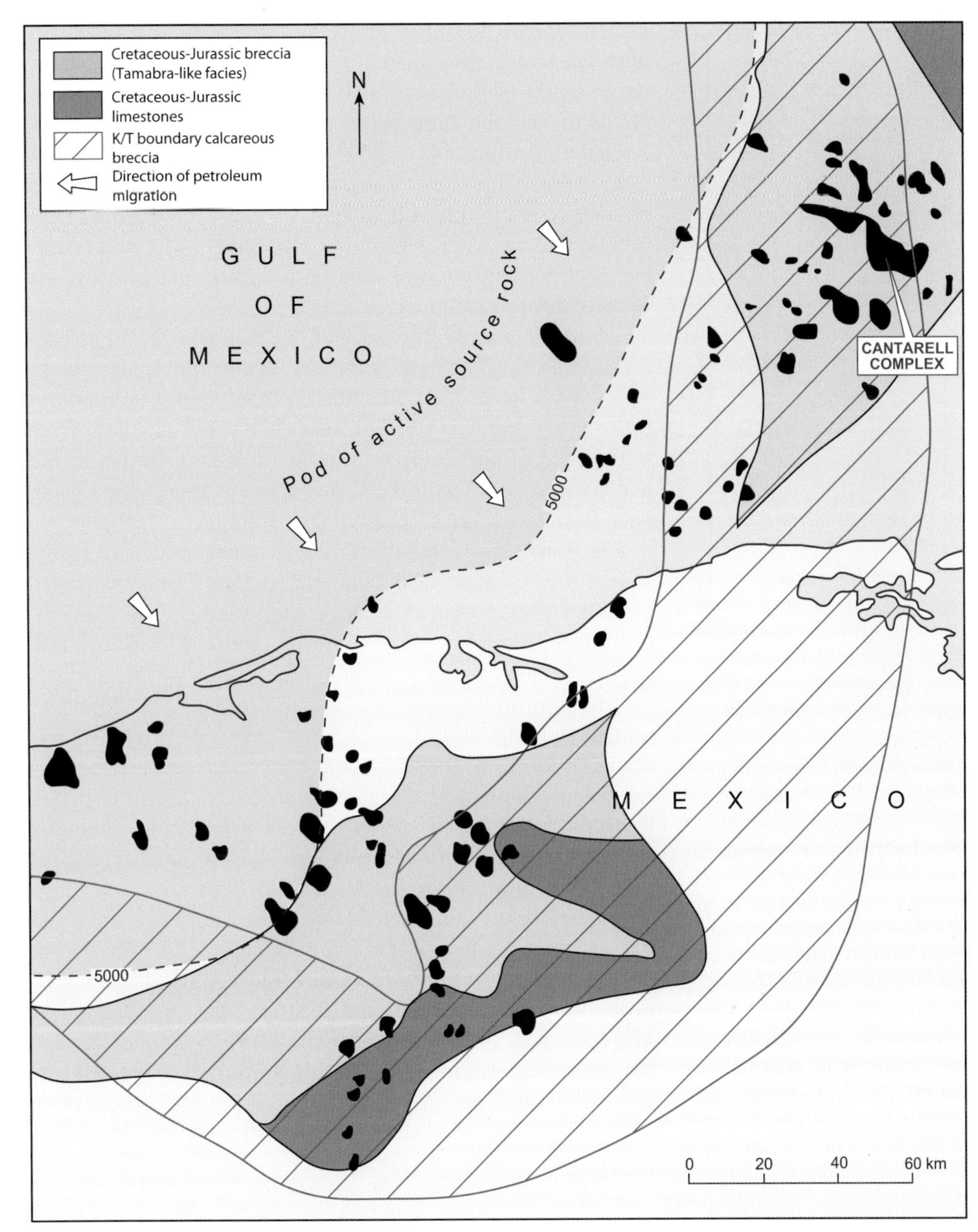

Figure 4.52 Interpreted trend of the K–Pg boundary deposit in Mexico. Modified from Magoon *et al.* (2001).

70 percent of the ultimate recovery from these fields. K–Pg deposits are also an important reservoir in Ek-Balam (Mitra *et al.* 2007) and Ixtoc Fields (Murillo-Muneton *et al.* 2002). The linkage of this breccia unit in these fields to the Chicxulub impact event is based on proximity to the crater but also the presence of shocked quartz, altered glass, and feldspars with planar deformation features (Grajales-Nishimura *et al.* 2000).

In spite of its importance as an oil reservoir and abundant well penetrations, a number of disagreements exist among Pemex geologists concerning the age and dominant depositional process. In addition, the suggested map distribution raises other issues, given the insights on the impact-related sedimentological processes described in Section 4.11.3.

The K–Pg breccia reservoir of southeast Mexico offshore areas was originally assigned by Cantú-Chapa and Landeros-Flores (2001) to the Cretaceous Cantarell Formation on the basis of *Globotruncana* sp., a Cretaceous foraminifera not actually found in the type well Chac-2, where the K–Pg boundary deposit has a thickness of nearly 300 m. However, this conclusion ignored earlier work on onshore outcrops where a breccia unit with similar lithology and thickness was shown to lie below an iridium anomaly and interpreted ejecta material, clearly connected to the end-Cretaceous Chicxulub

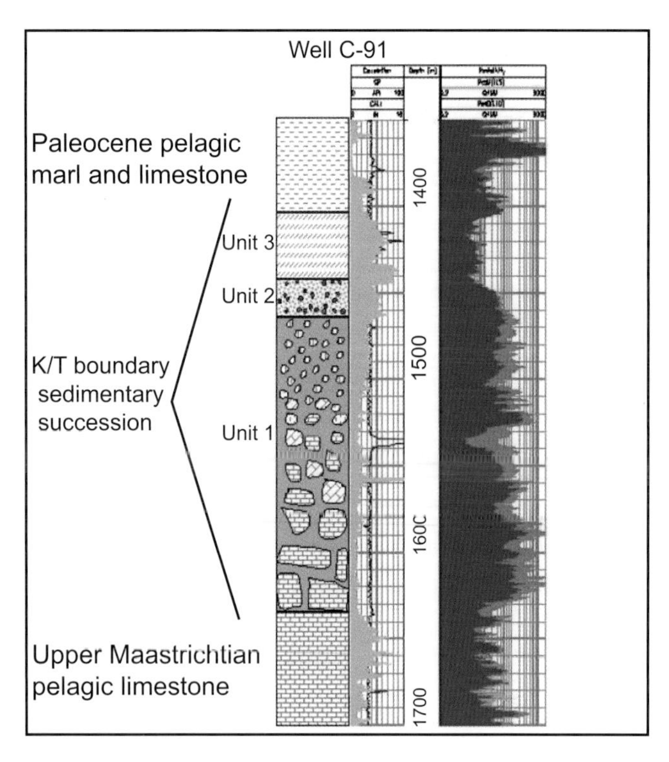

Figure 4.53 Representative stratigraphic column showing the major lithofacies units of the K–Pg reservoir in the Cantarell oil field of Mexico. Modified from Murillo-Muneton *et al.* (2002).

impact event (Grajales-Nishimura *et al.* 2000). Subsequent work on the breccia reservoir at Cantarell by Murillo-Muneton *et al.* (2002) supported the findings of Grajales-Nishimura *et al.* (2000) with detailed sedimentological analysis of the K–Pg boundary deposit in cores (Figure 4.53). Further, Cantú-Chapa and Landeros-Flores (2001) also failed to recognize the "K–T boundary cocktail" assemblage of reworked Maastrichtian–Campanian microfauna that Bralower *et al.* (1998) had previously identified from study of the DSDP and Ocean Drilling Program (ODP) sites across the GoM and Caribbean.

Murillo-Muneton *et al.* (2002) used the Cantarell C-91 well to subdivide the K–Pg boundary deposit into three units below definitive Paleocene pelagic marl and limestones (Figure 4.53). Unit 1 is a basal coarse carbonate breccia, with poor sorting and clast sizes up to 30 cm size. Lithoclasts appear to be derived from shallow marine paleo-environments, platform margin to sabkha/tidal flats, with little evidence of deepwater facies, in contrast to descriptions of the K–Pg in the northern GoM deepwater (see Section 4.11.4). This and possible Albian fossils suggested derivation from the adjacent Yucatán Platform (Murillo-Muneton *et al.* 2002). A large percentage of the interval was dolomitized during burial diagenesis and fractured, possibly associated with the Middle Miocene contractional event (Grajales-Nishimura *et al.* 2000). Evaporite molds, possibly anhydrite, were noted. The breccia is probably a debris flow product of massive failure of the Yucatán margin following impact (Grajales-Nishimura *et al.* 2000), consistent with the similar units suggested for the northern GoM.

Unit 2 is a normally graded interval with few carbonate lithoclast sizes greater than 2 cm (Figure 4.53), but the shallow-water origin of the clasts is still evident. Feldspar and quartz grains in this unit show shock metamorphism in thin section (Grajales-Nishimura *et al.* 2000; Murillo-Muneton *et al.* 2002). Thus, some ballistic ejecta derived from the Chicxulub itself is clearly present in this unit, in contrast to the underlying coarse breccia unit (Grajales-Nishimura *et al.* 2000). The unit is only partially dolomitized and thus porosity and permeability are reduced relative to unit 1.

Unit 3 is a shale interval with what is referred to as bentonite beds, possible basal seal rocks for the underlying breccia reservoirs (Figure 4.53). It is unclear from Pemex descriptions if this is a volcanic ash bed or, more likely, the clay layer from settling out of post-impact atmospheric fallout. Dolomite and feldspar are found with the shale as well as more ejecta products of impact glass (presumably spherules) and shocked quartz. Current-generated bedding is noted, similar to DSDP core sites 536 and 540, suggesting some tsunami-related bottom current reworking, or, as suggested by the tsunamigenic turbidity flows, paralleling the model of Sanford *et al.* (2016).

Mapping by Pemex geologists shows the K–Pg boundary unit to thin significantly over a relatively short distance, less than 500 km from the Chicxulub crater across the Campeche salt basin (Figure 4.52; Magoon *et al.* 2001). This interpreted thinning trend, based largely on well penetrations targeting structural highs, contrasts with the K–Pg boundary deposit map distribution in the northern GoM. This raises questions as to whether the salt-created lows of the Campeche salt basin may contain a substantial K–Pg boundary deposit thickness, as observed in the northern GoM (Figure 4.51). These structural lows are largely undrilled at the present time. Magoon *et al.*'s (2001) map of the K–Pg boundary deposit (Figure 4.52) proposed a radial distribution of K–Pg deposits from the crater center, but also argued for the breccia units to be largely ballistic (airfall) products of the impact, which is unlikely for unit 1, given the large lithoclast size (Figure 4.53).

4.11.7 K–Pg Boundary Deposit in Cuba

In Cuba, the K–Pg boundary deposit has mainly been investigated in outcrop intervals (Figure 4.54), but rarely mentioned in subsurface studies. Unlike Mexico, it is not considered an important reservoir, though a USGS assessment of undiscovered Cuban hydrocarbon resources does consider the potential for traps at Top Cretaceous level in the North Cuba basin (Schenk 2008).

The best-documented outcrops of the K–Pg boundary deposit in Cuba are found in the Penalver Formation of north-central areas and the Cacarajícara Formation of western Cuba (Tada *et al.* 2003; Scott *et al.* 2014; Cobiella-Reguera *et al.* 2015; Figure 4.54). Because of the complex tectonic history, including major plate movement as recently as the

Eocene, reconstruction of the original positions of these outcrops is challenging (Tada *et al.* 2003). The Penalver Formation is thought to rest upon the original Cretaceous Cuba volcanic arc, whereas the Cacarajícara Formation lies above the North American Mesozoic paleo-margin (Cobiella-Reguera *et al.* 2015).

The Penalver Formation at its type locality near Cuba is over 590 ft (180 m) thick and includes two gross sedimentological packages: a heterogeneous breccia-bearing lower unit, interpreted as a sediment gravity flow deposit, and an upper unit, grading from calcarenites to calcilutites, thought to be formed by reworking and sediment resuspension associated with tsunami waves (Figure 4.54A). In the lower unit, large mudstone lithoclasts from the underlying Cretaceous Via Blanca Formation and obvious shallow marine bioclasts point to local derivation from the Cuba Platform. Quartz grains with planar deformation features, an indication of shock metamorphism and crater airfall derivation, are only found in the upper unit. The mixed assemblage of Aptian to late Maastrichtian microfossils in the upper unit (Cobiella-Reguera *et al.* 2015) is comparable to the K–T boundary cocktail of Bralower *et al.* (1998). Overall, the Penalver Formation here shows upward-fining trends in both silicate grains and carbonate lithics, mirroring the K–Pg log motif pattern in Mexico (e.g., Figure 4.53 versus Figure 4.54).

The Cacarajícara Formation of western Cuba (Rosario belt) is considerably thicker (>700 m) than the Penalver Formation, reflecting proximity to the impact (Figure 4.54B). A basin floor setting for the Cacarajícara is envisioned, versus the slope setting for the Penalver (Tada *et al.* 2003). Ceno-Turonian stratigraphic units unconformably underlie the K–Pg boundary deposit. An absence of Paleocene microfossils partially constrains its age, but it is likely similar to other K–Pg boundary deposits.

Tada *et al.* (2003) recognize three members to the Cacarajícara: a lower breccia, middle grainstone, and upper "lime" mudstone (Figure 4.54B). The breccia member contains cobbles and pebbles of shallow to deep marine origin with no obvious grading. A 25 m boulder of Aptian to Albian-age chert was noted, suggestive of mass slope failure. The middle grainstone member also contains lithoclasts, but clearly fines upward. Shallow marine rudists and forams are common, pointing to derivation from a shallow carbonate platform. The upper "lime" member is mainly massive to faintly laminated carbonate mudstones and wackestones (Tada *et al.* 2003). Shocked quartz grains are found throughout the three members and spherules are present in the basal breccia member.

A debris flow origin of the lower breccia member has been suggested (Kiyokawa *et al.* 2002), while the shocked quartz content suggests some ballistic contributions (Tada *et al.* 2003). A variety of mechanisms for the middle grainstone member have been proposed, ranging from high-density turbidity flows (Kiyokawa *et al.* 2002) to tsunami-related processes (Tada *et al.* 2003). The origin of the upper "lime" mudstone member is also thought to tie to tsunami-related processes.

Tada *et al.* (2003) present a process model for the K–Pg boundary deposit in Cuba, based on the Penalver Formation outcrops, that is grossly similar to the GoM basin-wide model

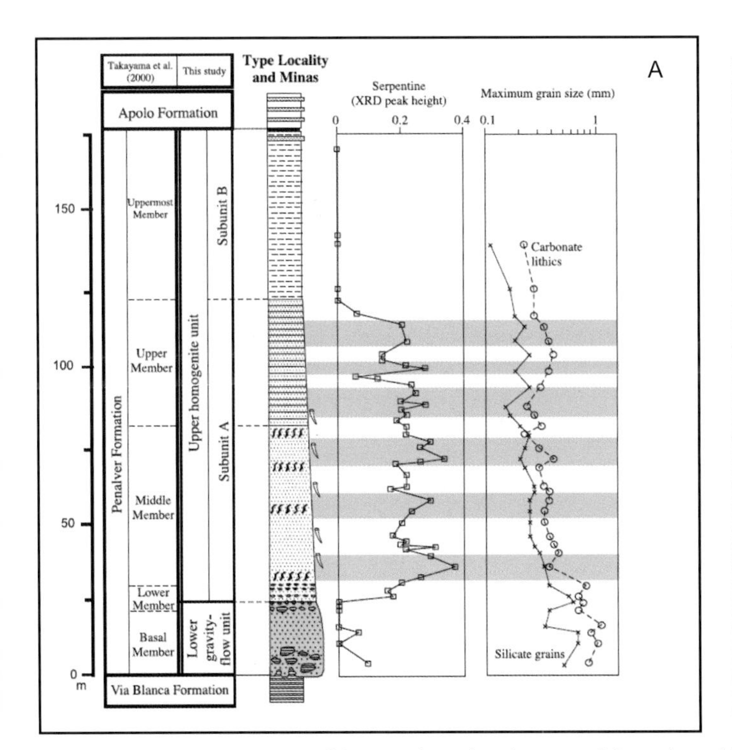

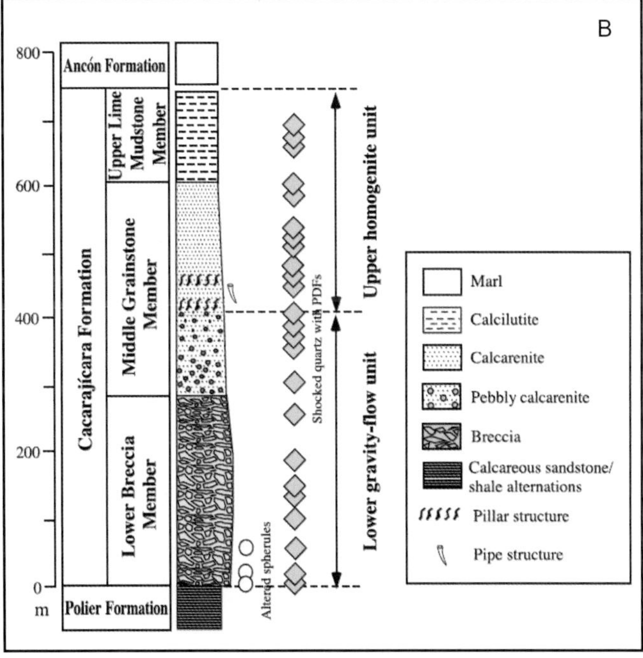

Figure 4.54 Columnar sections of the K–Pg boundary deposit in Cuba and possible correlations based on stratigraphic subdivisions described in the text. (A) Penalver Formation at its type locality. (B) Cacarajícara Formation along the San Diego River. From Tada *et al.* (2003).

proposed by Sanford *et al.* (2016). Cuban and Yucatán Platform margin failure associated with seismic shaking and ground roll generated lithoclast-rich debris flows, followed by tsunami-genic turbidity flows and waves. Though not explicitly discussed by Tada *et al.* (2003), clearly airfall (ballistic) material from the crater (e.g., shocked quartz) contributed to deposition of the K–Pg boundary deposit in Cuba.

Other outcrops of the K–Pg boundary deposit in Cuba are thinner (e.g., Moncana, Amaro, and Santa Clara Formations), but data from these generally support the above observations and interpretations, and the reader is referred to Alegret *et al.* (2005) and Cobiella-Reguera *et al.* (2015) for a detailed discussion. The future potential of the K–Pg as a reservoir in Cuba should be considered, though exploration to date has not targeted this reservoir as a major objective (Escalona and Yang 2013).

4.11.8 Landscape and Seascape at the End of the Mesozoic

The end of the Cretaceous and termination of the K–Pg deposition set the stage for the next major episode of deposition, as will be described in Chapter 6. Through its history, allochthonous salt movement in the GoM has greatly modified the paleo-seascape that influenced shallow and deepwater depositional processes. Reconstructing the Late Cretaceous canopy is challenging due to both present-day imaging but also the extensive post-K–Pg Cenozoic salt deflation and inflation. Even rafting of Mesozoic blocks (Fiduk *et al.* 2014; see Section 1.3.8) complicates reconstruction of the pre-impact seascape.

Nonetheless, Scott *et al.* (2016) have attempted a temporal model for K–Pg boundary unit deposition in relation to the extant Cretaceous canopy (Figure 4.51). The rugose sea floor that included salt diapirs and minibasins probably existed at the end of the Cretaceous prior to the impact (pre-impact and impact). Massive platform failures and generation of debris flows from the seismic shaking undoubtedly served to both remove unconsolidated sediments near the sea floor, but also to deposit the thickest part of the K–Pg unit in the deepest structural lows (Figure 4.51C). Seismic observations of K–Pg boundary deposits thickening into paleo-lows of salt welds, as well as thinning and onlap onto parautochthonous salt structures, supports this model (Scott *et al.* 2016). Sanford *et al.* (2016) noted a similar pattern in other areas of the basin.

This process of filling salt-structured lows on the contemporaneous sea floor could have smoothed out the sea floor over a large area (Scott *et al.* 2016). This more uniform bathymetry may have existed well into the Paleogene, prior to the next major depositional phase in the northern GoM, the Lower Wilcox. Salt movement would likely have recommenced with this next period of sediment loading. This fits with the observed continuity and isopachous nature of the Wilcox, at least in the outboard trend of Alaminos Canyon, Keathley Canyon, and southern Walker Ridge protraction blocks (Meyer *et al.* 2007). However, the inboard Wilcox (e.g., Green Canyon and Garden Banks protraction blocks) clearly varies in thickness laterally, as a function of primary (subcanopy) basin influences on sedimentation (Moore and Hinton 2013). This implies that the K–Pg boundary deposit was unable to completely fill primary basin structural lows, or those basins continued to act to funnel and rout sediment gravity flows further south.

4.12 Middle and Late Mesozoic Summary

The reconstructed paleogeographies discussed in Chapters 3 and 4 provide a foundation to consider the larger stratigraphic controls and continental-scale source-to-sink pathways for the Middle and Late Mesozoic timeframe in the GoM. This is relevant to the current and future potential of reservoirs embedded within this section. The earlier successor basin-fill and syn-rift sedimentary picture is more uncertain, limiting generalizations that can be made for that interval formed prior to GoM basin opening. However, well and seismic data, combined with advanced scientific methods (e.g., detrital zircon provenance work), has greatly illuminated the subsequent tectonostratigraphic phases: (1) Drift and Cooling; and (2) Local Tectonics and Crustal Heating.

Prior to the Ceno-Turonian (basal Tuscaloosa Sandstone) shelf margin progradation, siliciclastics seldom extended beyond the shoreline or outer shelf, trapped behind large carbonate platforms. This pattern of contemporaneously depositing fluvial to marginal marine sandstones updip and shelf to basinal carbonates downdip began in the Oxfordian with Smackover deposition (Figure 4.55). This is not reciprocal alternation of sandstones and carbonates, though relative sea-level changes likely promoted carbonate deposition during transgressive and highstand phases and sandstones in lowstand phases.

Our paleogeographic mapping indicates that the dominant pattern for the Middle and Late Mesozoic is one in which updip siliciclastics deposited in alluvial to marginal marine paleo-environments transition to, and are often trapped behind, prominent carbonate platform margins. This "reef blocking" mechanism (see Box 3.1) was quite effective at preventing quartz sand from being transported into the deep basin, except at discrete basin entry points created by structural or depositional processes (e.g., interreef passages). The Hosston Sandstone (Valanginian–Hauterivian) is the lone exception of an Early Cretaceous siliciclastic interval able to reach at least the slope, though in very restricted areas. There is little evidence to date in well penetrations that large volumes of terrigenous siliciclastics of the HVB, CVB, CVK, or PW supersequences bypassed the platform margin.

By the Late Mesozoic (EFT supersequence), this pattern changed as, for the first time, siliciclastics surmounted the preexisting or relict carbonate platform margin and moved large volumes of siliciclastics into the basin (Figure 4.55). The best-documented example is bypass of Tuscaloosa sandstones into the central and eastern GoM, as discussed in Section 4.8 and by Snedden *et al.* (2016b). Subsequent global sea-level rise (Haq 2014) limited further bypass, as

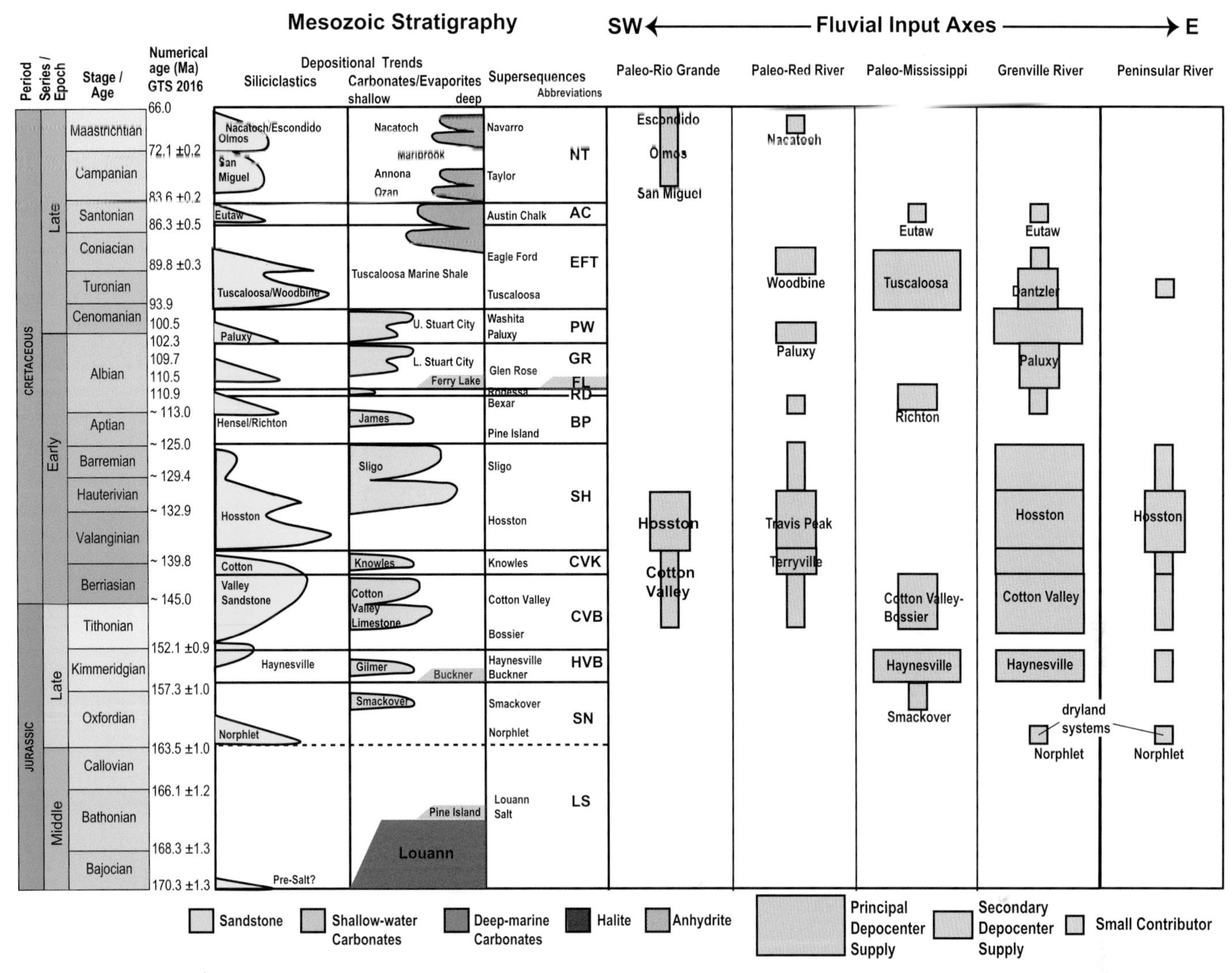

Figure 4.55 Summary chart of Mesozoic stratigraphy, depositional systems and source-to-sink sand routing for northern GoM.

Campanian–Maastrichtian sandstones were largely confined to shelf and shoreline zones. For future exploration play opportunities, work should focus on finding these potential basin entry points for other Mesozoic units, as these sandstones are distinctly better reservoirs than carbonate reservoirs formed in or transported into the deepwater.

The long-duration depositional cycles of the Middle and Late Mesozoic clearly fit the concept of supersequences (Mitchum and Van Wagoner 1991; Snedden and Liu 2011). Assemblages of highstand sequences (often carbonate platform-dominated), lowstand sequences (sandstone-dominated at entry points), and transgressive sequences (often shale-dominated) built the sedimentary architecture of the basin.

These paleogeographic reconstructions, supplemented with new provenance information, continue to shed light on the contributions of various North American source terranes and linked drainage networks (Figure 4.55). We recognize possible forerunners of the modern Rio Grande, Red River, and Mississippi Rivers (with added-prefix "paleo" for interpreted Mesozoic rivers in Figure 4.55) as well as Grenville and Peninsular Rivers first defined here.

In the early part of the Mesozoic, the Grenville and Peninsular Rivers transported siliciclastics found in the SN, HVB, CVB, CVK, and SH supersequences. The Peninsular River (and its Gondwanan/Suwannee source terrane) waned after deposition of the Hosston Sandstone (Figure 4.56). However, the Grenville River (Appalachian source terrane) likely continued as an important sediment routing pathway into the Ceno-Turonian (EFT supersequence).

The paleo-Mississippi River, likely sourced from the western Appalachians and areas to the west, contributed to the HVB and CVB supersequences but expanded with the Tuscaloosa fluvial system in the Ceno-Turonian (Figure 4.56).

Our mapping of fluvial drainage systems suggests that the paleo-Red River, predominantly derived from the older Ouachitas terrane, supplied significant grain volumes to local systems like the Terryville (CVK supersequence), Travis Peak (SH supersequence), Paluxy (PW supersequence), and

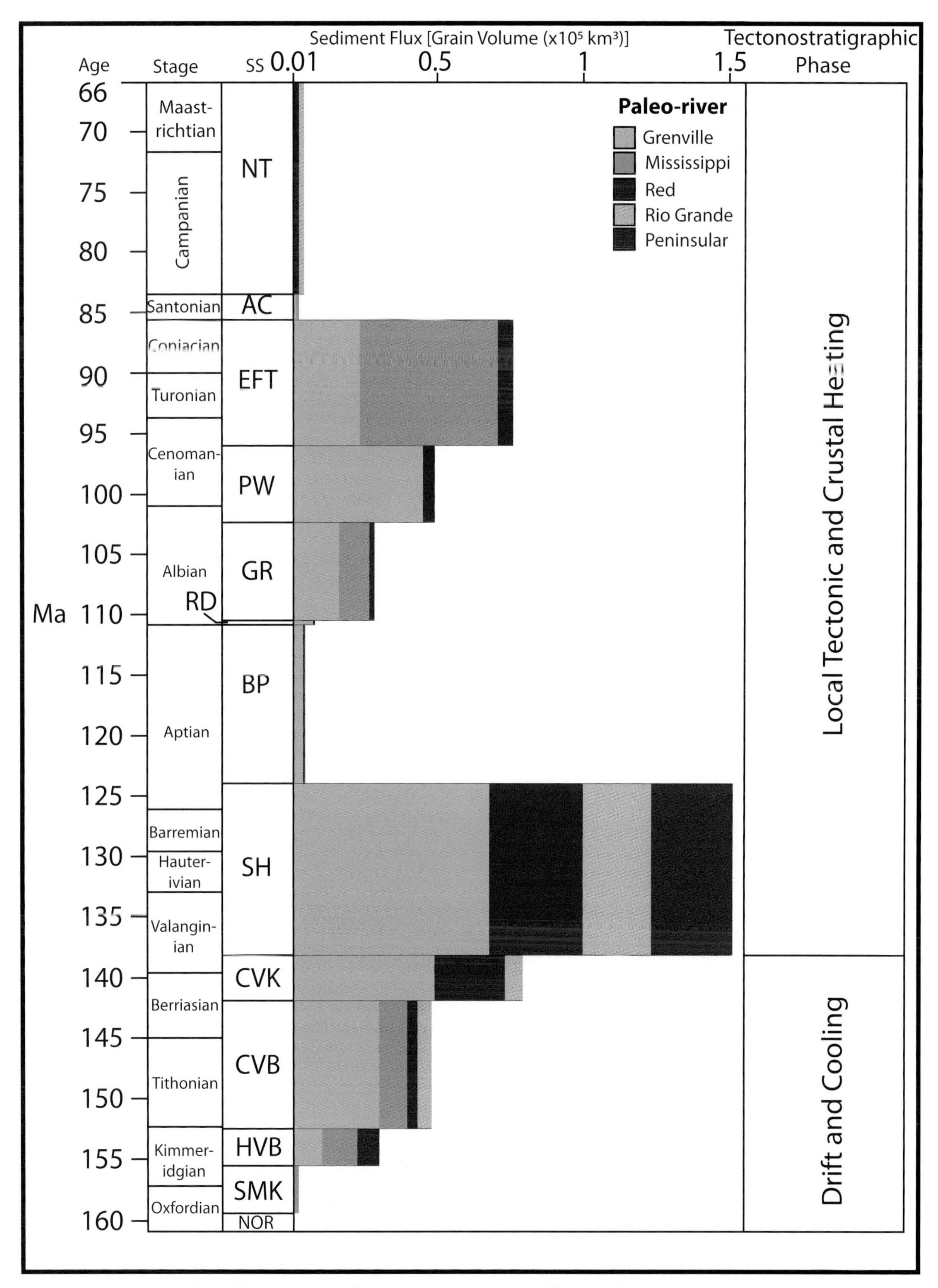

Figure 4.56 Siliciclastic sediment flux (grain volume) for Mesozoic supersequences, differentiated by interpreted paleo-river system and tectonostratigraphic phase.

Woodbine (EFT supersequence). The Tuscaloosa Sandstone depositional system, part of the EFT supersequence, is the first to support transport of siliciclastics into the deepwater and evidence suggests a structurally controlled routing from Appalachian source terranes (Blum and Pecha 2014; Snedden *et al.* 2016b).

By comparison to the above, the paleo-Rio Grande River was far less important to fluvial transport than the other paleo-rivers considered. Siliciclastic grain volumes peak in the SH supersequence with lesser contributions in the CVB, CVK, and NT supersequences (Figure 4.56). One reason is that the Rocky Mountains of North America do not appear to be a major sediment source terrane until Laramide tectonics are underway in the early Cenozoic (Figure 4.55). Progradation and retrogradation of the Escondido, Olmos, and San Miguel Formations in shoreline and shelf systems is probably linked to Late Cretaceous Cordilleran tectonics in the Sierra Madre Mountains of Mexico. The final contribution of the Quachita Mountains, worn down by erosion since the Paleozoic, is represented by the Nacatoch Sandstone of east Texas and north Louisiana (Figure 4.55). We expect future investigations to further refine the source-to-sink sediment routing history proposed here.

In Mexico, deepwater Middle to Late Mesozoic sandstones have not been extensively penetrated or produced. Oxfordian-age siliciclastic reservoirs are thought to be largely contained in the area around the Cantarell Field, where Bacab Formation sandstones produce oil at Ek and Balam fields (Mitra *et al.* 2007). Oxfordian sandstones are penetrated in wells near Tampico–Misantla (Horbury *et al.* 2003), but production information is generally lacking. In general, the grain volume in northern GoM Oxfordian dryland systems is relatively small compared to other units (Figure 4.56).

Part III Cenozoic Depositional Evolution

This part describes the depositional evolution of the Cenozoic GoM basin and the peripheral tectonic basins of its western margin that merge into and overprint its original structure. Chapters 5, 6, and 7 discuss each of the three phases that comprise the final 65 million years: the Paleogene Laramide Phase, the Middle Cenozoic Geothermal Phase, and the Neogene Tectono-climatic Phase. Chapter 8 summarizes the evolving patterns of sediment sources and inputs, depositional systems paleogeography, and evolution of the continental margin.

Chapter 5

Cenozoic Depositional History 1
Paleogene Laramide Phase

5.1 Cenozoic Introduction and Overview

The Cenozoic of the GoM displays a continuous record of basin infilling by terrigenous clastic sediment. Two broad patterns emerged within the first few million years of the Paleogene. Across the northern Gulf, sediment supply by large continental rivers initiated continental margin offlap that continued throughout the remaining Cenozoic. In the western Gulf, supply was through numerous local drainages, and margin bypass rather than offlap prevailed until the Late Neogene. Beginning with Paleocene Laramide deformation, ongoing plate convergence along the Pacific margin of Mexico created a series of regional uplifts and basins that impinged on the flank of the Gulf and provided sources and sinks for lithologically diverse sediment.

5.1.1 Foundations of Modern Understanding of Gulf Basin Depositional History

This chapter and the two chapters to follow build on earlier syntheses of the Gulf of Mexico (GoM) and its stratigraphic record. The history of petroleum exploration in the Cenozoic interval of the GoM basin extends back to the early twentieth century. First the coastal plain, then the continental shelf, and ultimately the continental slope all became theaters of active drilling. As a consequence, one of the challenges of the basin is the sheer volume of subsurface geologic data that exists. Beginning in the late 1970s, regional correlation projects tackled the task of creating suites of subsurface sections, establishing a reasonably standardized nomenclature for principal stratigraphic units and markers. Notable reference correlation sections for the northern Gulf margin include those of Dodge and Posey (1981), Bebout and Gutiérrez (1982, 1983), Morton *et al.* (1985), Shideler (1986), Reed *et al.* (1987), Morton and Jirik (1989), and Galloway *et al.* (1994).

Grover Murray's *Geology of the Atlantic and Gulf Coastal Province of North America* (1961) was the first modern overview of northern Gulf basin geology. As the name implies, data and analysis were limited to the outcrop and subsurface geology of the modern coastal plain. Winker (1982, 1984) published the first comprehensive maps and discussion of GoM shelf margins, including their recognition criteria, types, structural styles, evolution, and relation to deltaic axes and continental sediment sources. The Geological Society of America publication, *The Gulf of Mexico Basin* (Salvador 1991a), as volume J of the GSA series *The Geology of North America*, was the milestone compilation of all facets of the geology, origin, evolution, and resources of the entire GoM basin. Chapter 11 (Galloway *et al.* 1991) reviewed the Cenozoic evolution of the Gulf. The publication and its related maps and correlation charts encompassed the entire basin, including Mexico and the deep marine basin. Using maps created in the early years of development of a basin-wide geo-database, Galloway *et al.* (2000) published descriptive historical synthesis of the Cenozoic interval of the basin. This synthesis was expanded and updated by Galloway (2008). The framework and maps that follow reflect the evolution of these syntheses, utilizing an additional decade of well and subsurface data.

5.1.2 Cenozoic Basin Framework and Tectonostratigraphic Phases

Throughout the Cenozoic, the GoM basin included five distinct geologic/depositional provinces: the Florida Platform, northern divergent margin, western margin, Sureste (Campeche), and the Yucatán Platform.

The Florida and Yucatán Platforms share many common features. Both are largely composed of aggradational carbonate and evaporite strata. They are bounded by carbonate ramps terminating in high-relief slope-toe scarps (Florida scarp and Campeche scarp) that approximate the position of the foundered Cretaceous fringing reef complex. They have shed minor amounts of carbonate sediment onto the abyssal plain.

The northern margin extends from the northeast corner of the Gulf, westward to the Burgos basin in Tamaulipas state. This expanse constitutes the classic divergent margin of the Gulf discussed in most English-language literature. Throughout the Cenozoic, it was dominated by deposition of terrigenous clastic sediment supplied by multiple continental rivers. By global standards, sediment supply and accumulation rates were high. As a result, Cenozoic offlap of the continental margin sedimentary prism (slope, shelf, and coastal plain) filled nearly one-third of the inherited Mesozoic basin volume. High rates of supply and a history of multiple, large delta systems favored large-scale shelf margin bypass. About one-third of the total Cenozoic sediment volume was transported to and accumulated on the continental slope and the basin floor. The combination of widespread salt substrate, load-induced overpressure, and high shelf-to-basin relief resulted in pervasive, basin-scale gravity tectonics.

In contrast, the western basin margin was dominated by tectonic subsidence, uplift, and tilting driven by the ongoing convergence of the North American plate with the Cocos, Rivera, and Pacific plates. A suite of relatively small (in comparison to the greater GoM basin) tectonic basins overprinted the Mesozoic margin of the Gulf basin. Adjacent uplifts provided local sediment sources, commonly draining directly from uplands onto the submerged basin margin slope. Erosion, cannibalization, and tectonic over-steepening and failure added complexity to the stratigraphic record of the western Gulf. Development and preservation of fluvial, coastal plain, and shore zone systems was limited. Rather, most sediment was bypassed to the slope and structural basin floor. From there, it commonly found avenues to spill from the local tectonic downwarp onto the abyssal plain of the open GoM. In contrast to the northern GoM, margin offlap and gravity tectonic structures are prominent only in young Neogene basin-fill.

The Sureste (Campeche) province similarly experienced a complex tectonic history. In the Neogene, a confluence of short but significant rivers progradationally filled the inland basins, building an offlapping platform into the Gulf of Campeche and onto deep salt. Plate convergence and gravity tectonics combined to create complex structural provinces.

Much like the Mesozoic fill, the Cenozoic fill of the Gulf can be subdivided into three phases. Unlike Mesozoic phases, they do not reflect processes of basin opening and evolution. Rather, they record a combination of impacts driven by convergence along the western North American plate margin, by intra-plate tectonism, by evolving patterns of continental climate and consequent drainage basin evolution and runoff, and finally by global climate change and resultant glacioeustasy. We have dubbed them for discussion the Paleogene Laramide Phase, the Middle Cenozoic Geothermal Phase, and the Neogene Tectono-climatic Phase. The depositional history of each phase is the subject of a chapter, beginning here with the Paleogene Laramide Phase.

5.2 Paleogene Laramide Tectonostratigraphic Phase

Following the Chicxulub cataclysm and subsequent erosional and depositional modification of the GoM, sedimentation resumed the languid Late Mesozoic pace. A broad, shallow shelf extended to the landward limits of the northern basin. The morphologic shelf margin was largely a gentle rollover capping the sediment-draped, foundered Mid-Cretaceous reef-rimmed platform edge. Carbonate/evaporite platforms dominated Florida and Yucatán. The central Gulf was a deep, sediment-starved abyssal plain. The basin connected to the open Atlantic through the Florida Straits, at the south end of the platform, and the Suwannee channel, which separated the Florida Platform from the continental land mass.

The Western Interior of the North American plate was not, however, tectonically quiescent. Laramide deformation progressed from north to south, forming the Front Range and other mountains of the US Western Interior and the Sierra Madre Orientale of Mexico. Between ca. 75 Ma and 45 Ma, these uplands were deeply eroded due to uplift by thrust faults (Cather *et al.* 2012). Intermontane basins and the Great Plains remained near sea level, providing ample accommodation space for the initial surge of sediment. However, during Early Paleocene many of these basins were largely filled, and drainage systems integrated to form trunk rivers that flowed eastward toward the GoM (Galloway *et al.* 2011). In Mexico, Laramide compression extended to and overprinted the Mesozoic margin of the Gulf basin, creating uplifts and adjacent foreland troughs, enhanced Gulf-ward tilt, and reactivating basement structures (Alzaga-Ruiz *et al.* 2009a). A curvilinear series of foreland troughs and outboard marginal bulges extended south to north along the western periphery of the Gulf basin. These tectonic elements include the Parras–La Popa, Salinas, Chicontepec, and Veracruz basins, the Tamaulipas and Picachos Arches, and Tuxpan Platform.

Paleoclimate across the GoM was largely arid (Scotese 2017). The humid tropical northern margin and its adjacent coastal plain and continent were the notable exceptions.

5.2.1 Chronostratigraphy and Depositional Episodes

The Cenozoic Era, and the Laramide tectonostratigraphic phase, began with deposition of relatively thin and/or mud-rich strata across the entire northern and western Gulf margins (Figure 5.1). The Early Paleocene record consists of disconformities and muddy sediment of the Midway and Velasco Groups. However, by Middle Paleocene time, sand-rich accumulation dominated.

The northern Gulf record includes five principal deposodes (Figure 5.1A): (1) Lower Wilcox; (2) Middle Wilcox; (3) Upper Wilcox; (4) Queen City; and (5) and Sparta. The foreland troughs along the western margin of the basin record a major depositional episode of coarse clastic filling encompassing the Middle Paleocene–Early Eocene (Figure 5.1B; Salvador and Quezada-Muneton 1989; Lawton *et al.* 2001). In the La Popa basin, it is bounded above and below by marine mudstone. In the Chicontepec and Veracruz basins, the same deposode is recorded by the Chicontepec sandstone and conglomerate. It too is bounded by marine mudstone units, the Velasco and Guayabal Formations. The Sureste province contains only thin deep marine mudstone of the Nanchital Formation. Following Middle Eocene interruption of coarse sediment supply and consequent transgression/foundering, the remaining Laramide Phase strata record renewed coarse sediment accumulation; the Carroza Formation in La Popa, and the Tantoyuca-Chapopote in the Chicontepec and Veracruz Troughs.

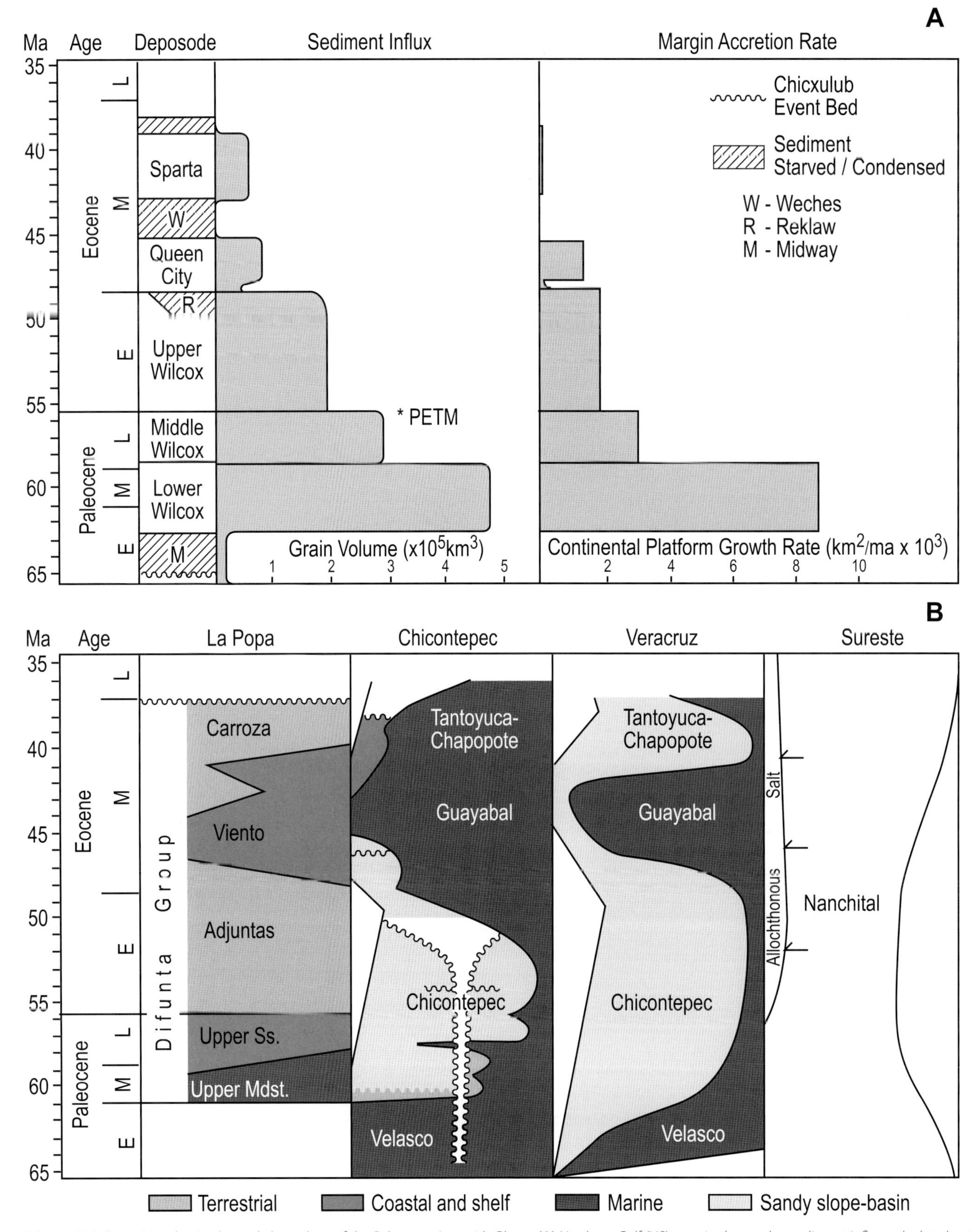

Figure 5.1 Depositional episodes and chronology of the Paleogene Laramide Phase. **(A)** Northern Gulf (US) margin deposodes, sediment influx, calculated as total grain volume, and continental margin accretion rates. **(B)** Stratigraphic units of the peripheral tectonic basins of the Mexican Gulf margin. Patterns distinguish terrestrial, coastal and shelf, deep marine, and sandy, deep marine facies associations. Expanding continental, shallow marine, or sandy marine facies indicate regressive episodes due to increased sediment influx or to uplift to the basin margin and consequent forced regression. The Sureste area includes the Macuspana, Comalcalco, and Ithmus basins.

5.2.2 Previous Work

Early Paleogene units have been topics of some of the foundational studies on depositional history and paleogeographic reconstruction of terrigenous clastic sediments and their relationship to occurrence of energy resources. Echols and Malkin (1948) used subsurface data to recognize and document the record of a large deltaic system in the Wilcox Group of Louisiana. The depositional systems paradigm was developed at the Texas Bureau of Economic Geology (BEG) by Fisher and McGowen (1967) in the course of their interpretive synthesis of the Wilcox Group across the Texas coastal plain. This seminal paper introduced the concept that traditionally defined stratigraphic units consist of three-dimensional, genetically related bodies of sediment, which they termed "depositional systems," that can be recognized, interpreted, and mapped. Using the newly developing studies of modern sedimentary environments (especially in the northern GoM coastal plain and shelf), they differentiated fluvial- and wave-dominated delta systems, shore zone systems, and coastal plain fluvial systems as dominant components of the Wilcox Group. Galloway (1968) extended the BEG synthesis of the Paleocene Wilcox eastward across the Louisiana–Mississippi coastal plain.

Subsequent studies examined relationships between Wilcox depositional systems and their rich endowment of petroleum (Edwards 1981) and lignite-attendant coal-bed methane (Ayers and Lewis 1985). Updates of the original Fisher and McGowen analysis incorporated deep well control and seismic data to evaluate potential geopressure/geothermal energy resources (Bebout *et al.* 1982: Winker *et al.* 1983). Ramos and Galloway (1990) documented the importance of tidal processes and facies within the Queen City Formation, expanding the spectrum of coastal types beyond the previously recognized fluvial- and wave-dominated analogs. Xue (1997) further updated Paleocene Wilcox studies, using a genetic supersequence framework to differentiate and map individual Wilcox genetic units.

Comparable regional syntheses of Laramide Phase units of Mexico remain to be done. In part this reflects the subdivision of the western GoM into several relatively small and structurally overprinted foreland basins. Outcrops are limited, faults and unconformities truncate many units, and basin-fills are dominated by marine turbidite and basinal mud facies. Drilling, especially of deeply buried and offshore units, remains sparse, and data availability has been limited. Representative modern overviews of the depositional framework of the onshore La Popa and Chicontepec basins include Lawton *et al.* (2001) and Vásquez *et al.* (2014), respectively. Gonzales and Medrano (2014) combined well and seismic data to map turbidite channel–lobe elements in the Paleogene fill of the Gulf margin Veracruz basin.

5.3 Middle Paleocene Lower Wilcox Deposode

Rapid influx of sediment into the northern GoM began about two million years into the Early Paleocene (Figure 5.1A). Regional sediment influx abruptly increased from negligible to one of the highest rates recorded in the basin history. Progradation across the foundered relict Cretaceous shelf platform was rapid. Preserved clinoforms show that the subjacent Early Paleocene shelf was as much as 1000 ft deep at its marginal rollover (Figure 5.2). Upon reaching the relict margin, the combination of increased depth, requisite deposition of a thick continental slope sediment prism, and relatively steep basinward slope gradient initiated the first of the many curvilinear belts of extensional growth faults that comprise the Wilcox fault zone (cross-sections 4–6; Figures 1.16–1.18). See Box 5.1 for discussion of shelf recognition criteria. The Lower Wilcox supersequence is typically 4000–6000 ft (1200–1800 m) thick in the fault-expanded interval beyond the relict margin. During the ~4-million-year duration of the deposode, total continental margin offlap from the Burgos basin of Tamaulipas to the east-central Gulf

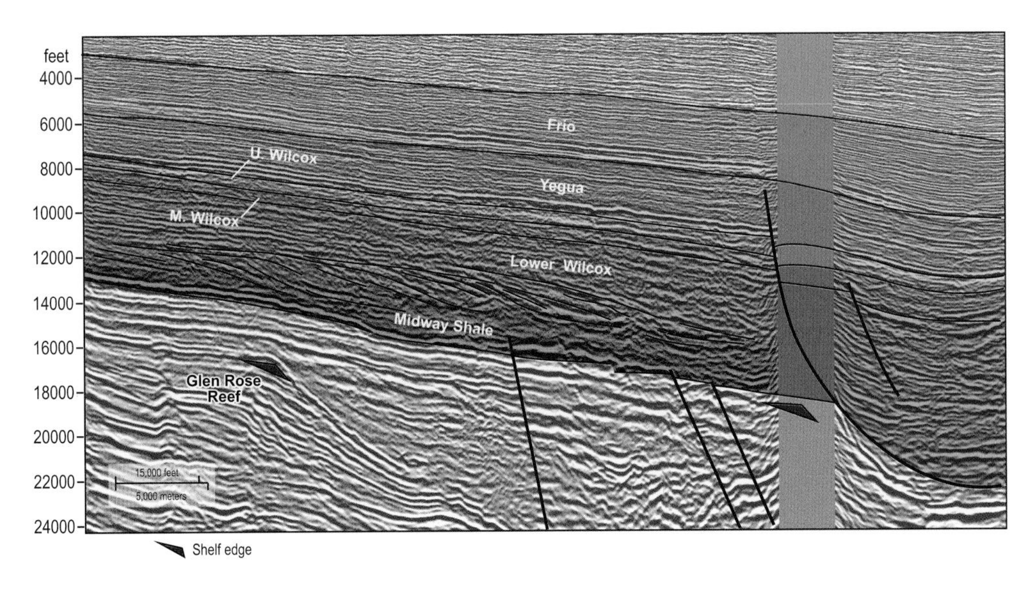

Figure 5.2 Seismic line interpretation illustrating Lower Wilcox clinoforms produced by progradation across the foundered Upper Cretaceous platform. Initial growth faulting occurred where the delta prograded over the subjacent shelf edge. Maximum clinoform relief exceeds 1000 ft (305 m). The lower clinoforms and horizontal clinoform toes lie within the lithologically defined Midway Shale, which consists of the subjacent shelf and prodelta deposits of the Lower Wilcox prograding deltaic and shore zone systems. For location, see Figure 5.3. Seismic line courtesy of ION.

margin of Louisiana ranged from 24 to 72 km (15–45 miles). Continental platform area (consisting of coastal plain, shore zone, and shelf) grew at a rate of more than 5000 km^2 (17 miles^2) per million years (Figure 5.1A). Total grain volume of the Lower Wilcox supersequence in the northern GoM is nearly 500,000 km^3 (120,000 miles^3; Figure 5.1).

5.3.1 Paleogeography

Lower Wilcox paleogeography of the northern GoM basin was dominated by two principal fluvial–deltaic systems, the Rockdale delta in east Texas and the Holly Springs delta in Louisiana (Figure 5.3). The Rockdale system was first described and named by Fisher and McGowen (1967) in their seminal Wilcox paper. Subsequently, Galloway (1968) mapped and described the Holly Springs delta system. Xue and Galloway (1993) updated and elaborated the depositional history of the Texas coastal plain Wilcox.

The Lower Wilcox depositional systems array established a recurrent paleogeographic pattern typical of subsequent northern GoM deposodes. Principal geographic elements present in each of the Cenozoic deposodes include (Figure 5.4): (1) multiple, broad, coastal alluvial aprons deposited by large, extrabasinal rivers; (2) intervening coastal plains deposited by numerous, basin-fringe and intrabasinal rivers; (3) broadly arcuate depositional headlands consisting of the composite succession of individual prograded delta lobes, their reworked margins, and bounding destructional facies; and (4) broadly concave coastal interdeltaic bights. The extrabasinal rivers provide the bulk of sediment supply to the basin margin. The headland morphology created by progradation of the deltaic coastline focuses wave energy, amplifying wave energy flux, reworking, and longshore transport. Longshore drift transports sediment along the shoreface and inner shelf toward and into the interdeltaic bights. Depending on the balance between longshore sediment flux and rate of sediment supply

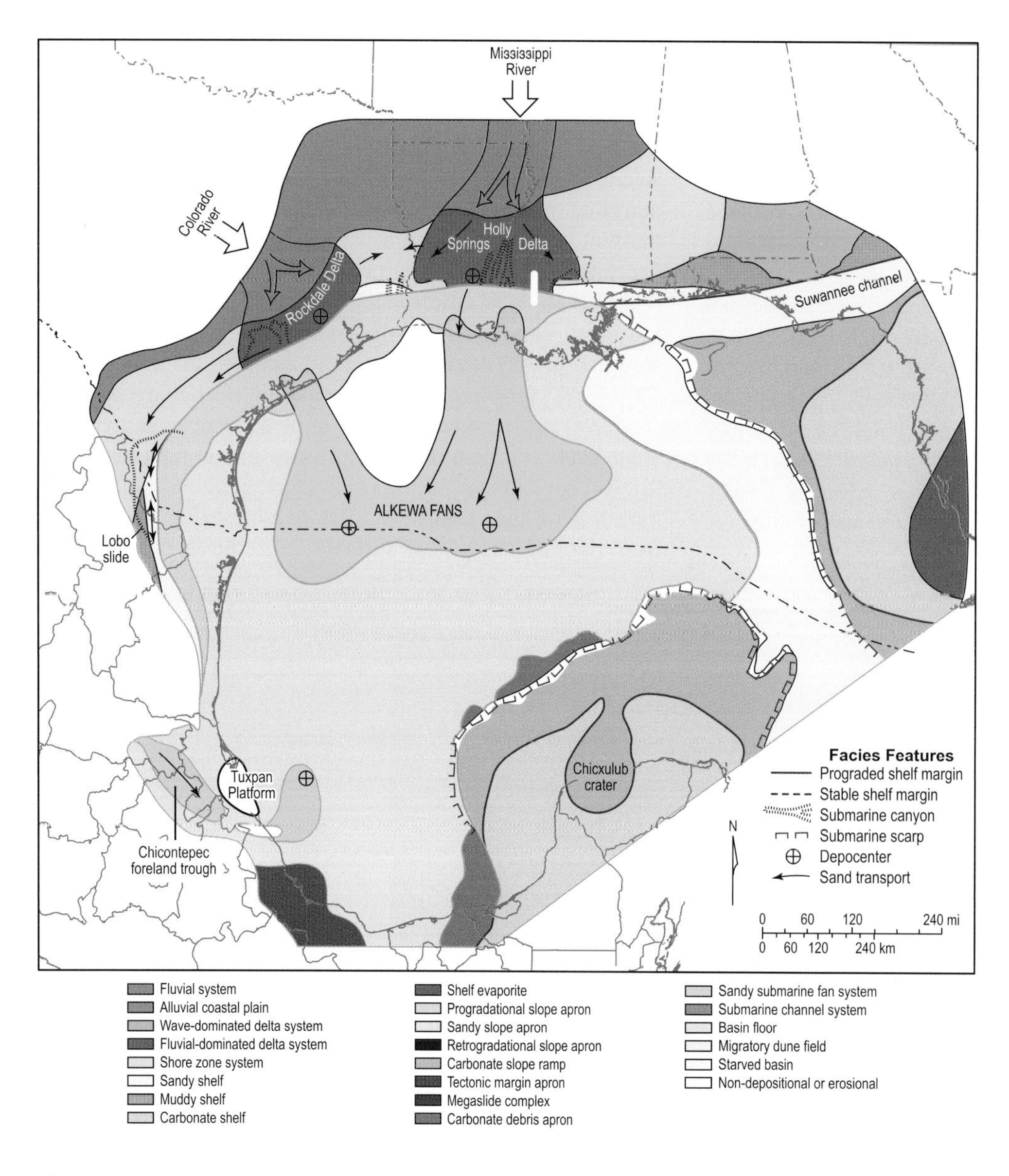

Figure 5.3 Paleogeographic map of the Paleocene Lower Wilcox deposode. Depositional system outlines and shelf edge reflect their positions at maximum progradation. White box locates seismic line shown in Figure 5.2.

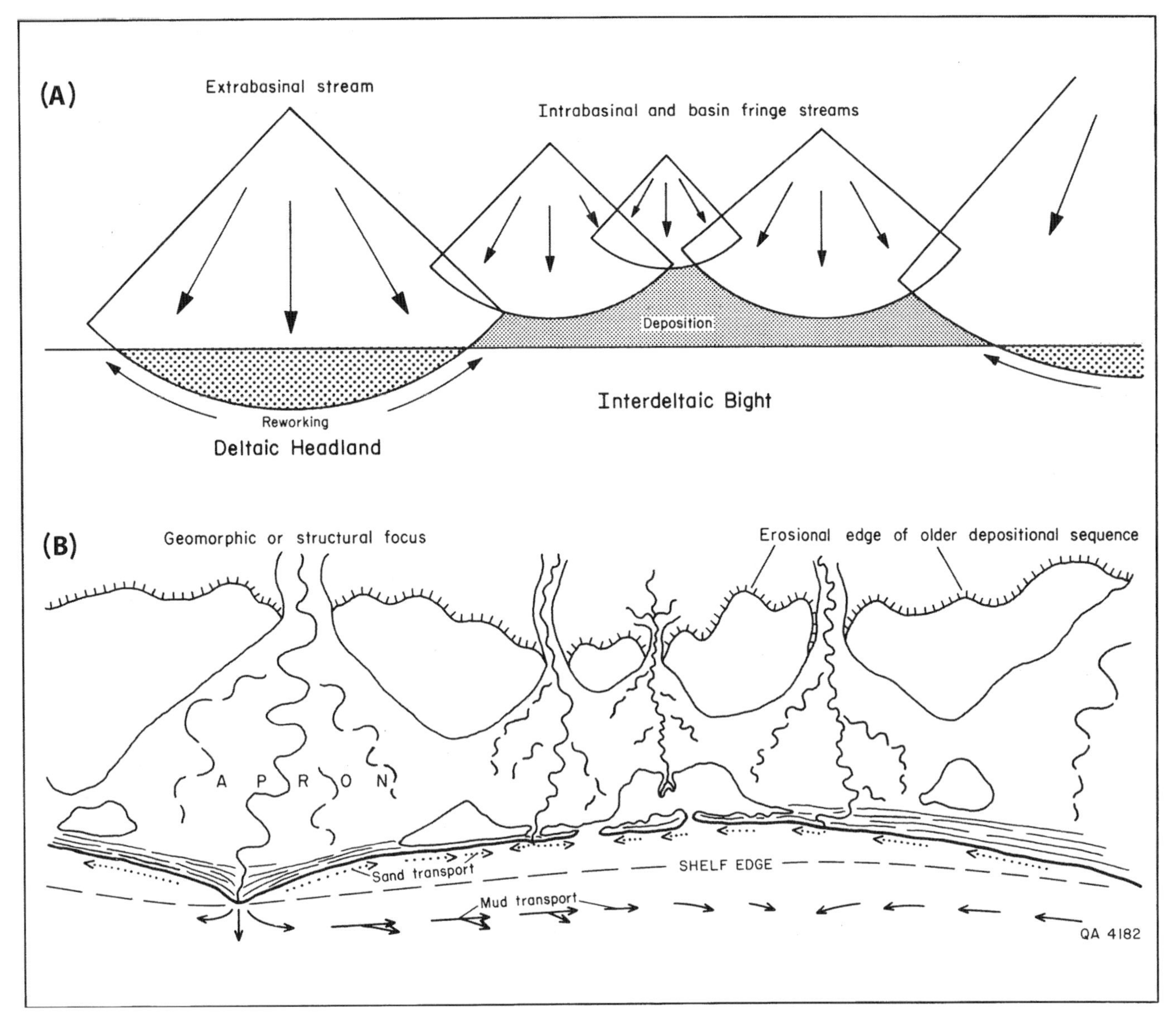

Figure 5.4 Schematic **(A)** and artistic rendering **(B)** of the headland/bight geomorphology typical of both the modern and Cenozoic GoM coastlines of the northern Gulf. Multiple large, extrabasinal rivers dominate sediment input to the basin, constructing convex progradational deltaic headlands. Between headlands, concave coastal bights extend across toes of the smaller alluvial aprons of intrabasinal and basin-fringing streams. Longshore transport of sand and mud from reworked deltaic headlands augments direct sediment input from the secondary streams. From Galloway *et al.* (1986).

through the smaller intrabasinal and basin-fringe rivers, the bight coastline forms a strandplain or spit/barrier island lagoon with associated lagoon and bays. Specific morphologies and areal extents of coastal elements commonly vary both along strike and through the history of a depositional episode.

Wilcox paleogeography reflects a recurrent theme found in the northern GoM supersequences. Both delta systems are coincident with the two principal Lower Wilcox depocenters (Figure 5.5A). Of the two deltas, the Rockdale was the largest in terms of sediment supply and depositional volume. Deltaic depocenters are clearly outlined by the sand thickness contours (Figure 5.5B). Updip, the Simsboro sand records the large, sandy fluvial system (the first ancestral Texas Colorado drainage system) that fed the delta (Galloway *et al.* 2011). Simsboro sand bodies radiate from an apex in northeast Texas (Ayers and Lewis 1985). The Rockdale delta was fluvial-dominated, but extensive reworking of sand southwest along shore nourished development of a broad, sandy, strike-oriented shore zone and shelf systems that aggraded more than 1000 ft (300 m) of wave and storm reworked sand-rich facies (Figure 5.3). Facies of the south end of the shelf and shore zone were displaced early after their deposition by the Lobo slide. Sandy shelf deposits filled the extensional gaps created by displacement of the slide sheet (Shultz 2010). Detailed core studies suggest tidal influence, especially in delta abandonment and shore zone facies (Zhang *et al.* 2016).

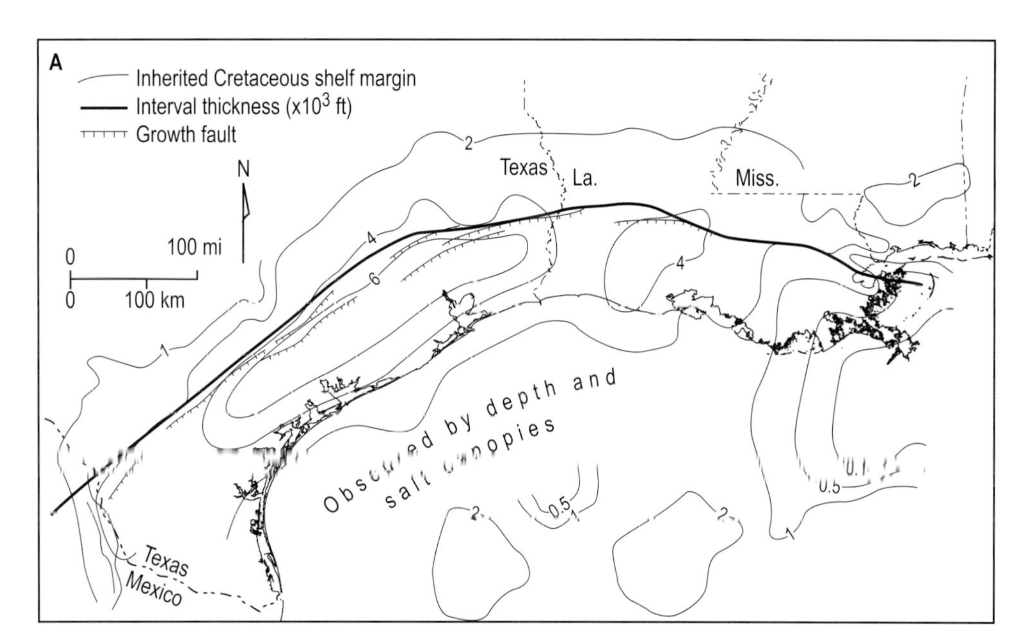

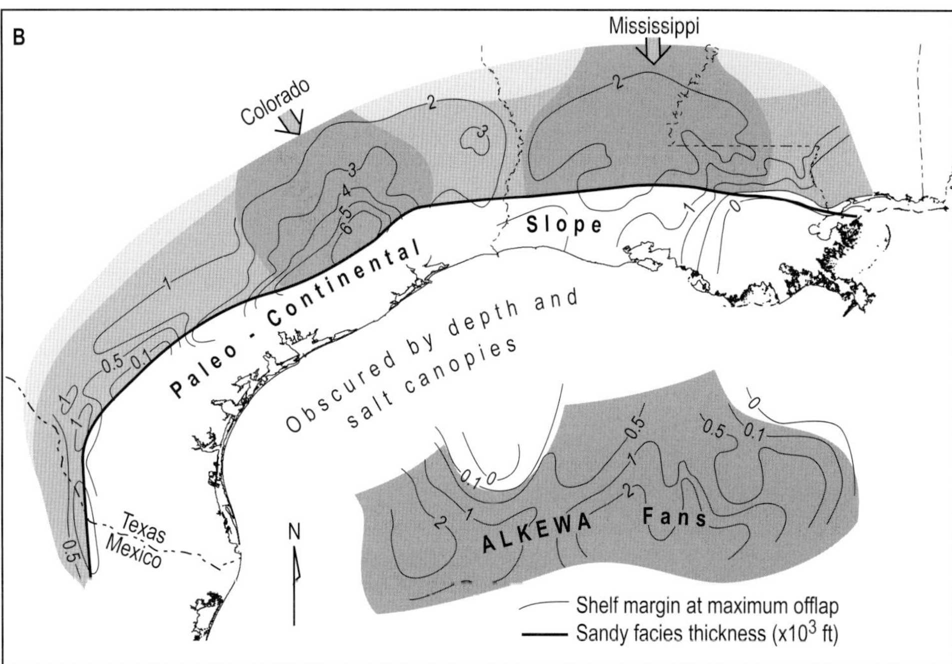

Figure 5.5 **(A)** Thickness of the Lower Wilcox genetic supersequence in the northern GoM basin. **(B)** Gross thickness of sandy facies. The dominant Colorado fluvial–deltaic axis is outlined by the 6000 ft contour and 3000 ft contour on the interval and sand maps, respectively. The Mississippi axis centers within the 4000 ft and 2000 ft contours. Basinal secondary depocenters containing up to 2000 ft of gross sand outline the ALKEWA fans.

The Holly Springs delta, though covering a comparable area, deposited less total sediment and sand (Figure 5.5) than the Rockdale. Convergence of contours northward into the mouth of the Mississippi River Embayment shows this delta to have been fed by an ancestral Mississippi River drainage system. The coastal bight between the deltas accumulated a broad strandplain. Reworked sediment from the east margin of the Holly Springs delta combined with sediment from the small rivers draining the southern Appalachians to construct a thin strandplain succession that extended eastward into the mouth of the Suwannee channel. Nourished by two continental rivers, these greenhouse-climate era deltas and their flanking shore zones repeatedly prograded onto the shelf margin (Zhang *et al.* 2016).

Lower Wilcox fluvial, delta, and shore zone systems all contain abundant, commercial lignite deposits. Lignite beds extend from the outcrop belt into the mid-depth subsurface (Kaiser *et al.* 1980; Ayers and Lewis 1985). The wet tropical climate and abundant surface water produced shallow water tables that favored swamp and marsh plant growth, burial, and preservation.

Paleocene Wilcox margins display an array of submarine canyons and slumps at a scale and abundance not replicated again in the GoM until the Late Neogene (Galloway *et al.* 1991). As described in Chapter 4, large-scale failures were generated by the Chicxulub meteorite impact across the width and breadth of the Gulf margin (Murillo-Muneton *et al.* 2002; Denne and Blanchard 2013; Cossey and Bitter 2016; Umbarger and Snedden 2016). These significantly modified the morphology of the Early Paleocene continental slope and appear to have localized substrate instabilities that, in turn, nucleated subsequent failures as Lower Wilcox deposition loaded the margin (Galloway *et al.* 1991). In the updip Rio Grande Embayment, south Texas, and the adjacent Burgos basin, Laramide foreland deformation uplifted and tilted inboard Gulf basin strata, triggering syndepositional gravity slides, most notably the regional Lobo megaslide (Figure 5.3).

Wilcox canyons display a range of evolutionary maturities. Some are largely intact retrogradational slump scars (for example, the pre-Lavaca canyons); others, such as the Lavaca canyon, display erosionally modified slump morphologies. The most mature canyons are highly elongate and excavated tens of miles across transgressive shelf platforms (i.e., Middle Wilcox Yoakum Canyon; Figure 5.6). Reconstructed Wilcox canyon depths ranged from a few hundred feet to more than 3000 ft (900 m) in the largest mapped example (Galloway *et al.* 1991; White and Snedden 2016). Lower Wilcox canyons cluster geographically in three areas (Figure 5.3): (1) on the western flank of the Rockdale delta; (2) between the Rockdale and Holly Springs deltas; and (3) at the front of the central Holly Springs delta.

The observed unique abundance of Cenozoic submarine canyons within the Paleocene Wilcox genetic supersequences poses an interesting question. Galloway *et al.* (1991) proposed a confluence of several factors favoring canyon excavation during these first few million years of clastic margin offlap. (1) The Paleogene Wilcox margin was a belt of rapid depositional loading onto an inherited Mesozoic margin. (2) This slope, created under conditions of slow accumulation of carbonate-rich fine sediment, was depositionally out-of-grade for rapidly prograding delta-fed aprons. The slope had been further destabilized and modified by post-Chicxulub impact events. (3) Finally, and uniquely to the Paleocene Laramide Phase, the proximity of dynamic compressional foreland tectonism along the western GoM meant that seismic events were common. Seismic activity is a well-recognized trigger of continental slope failure.

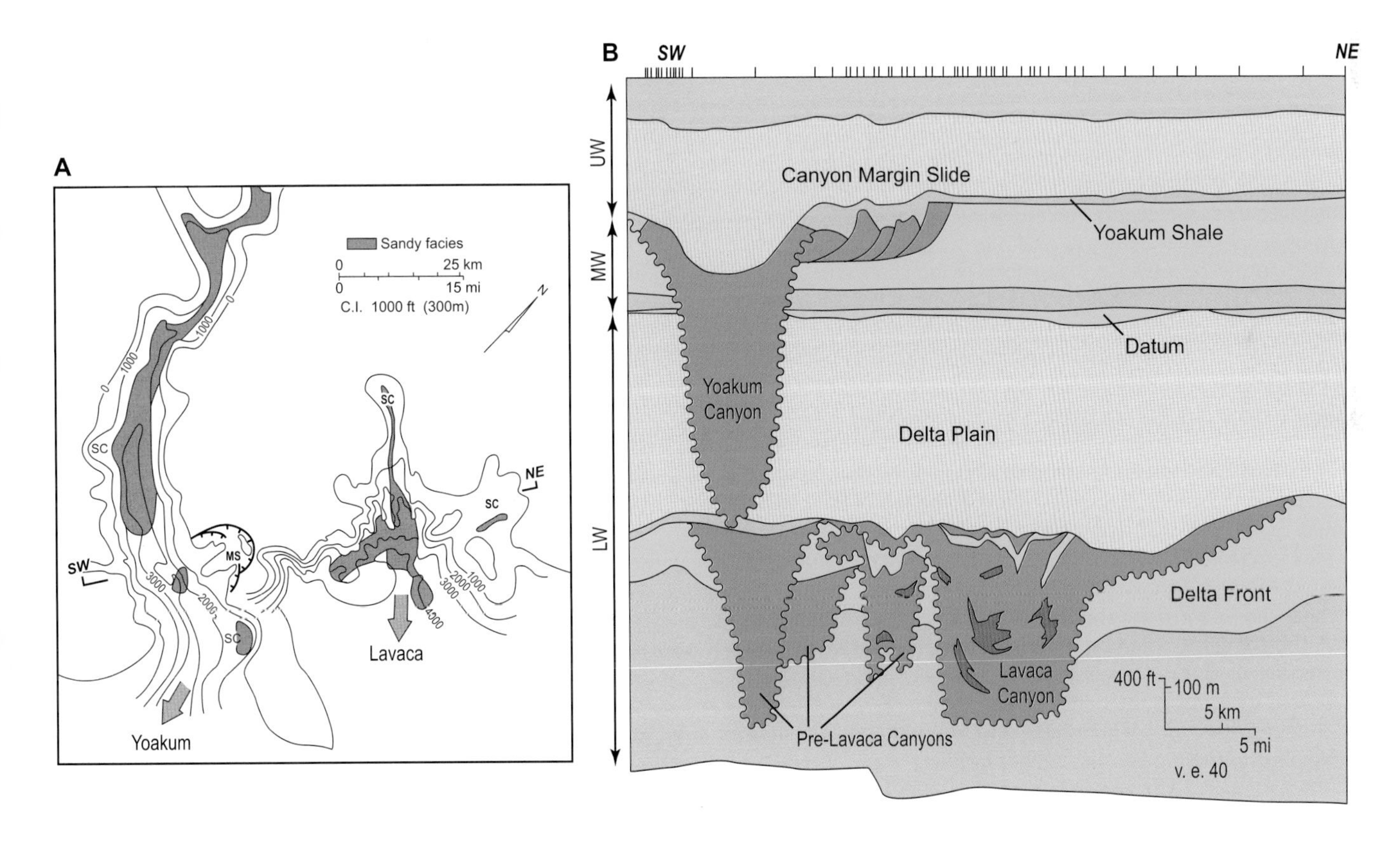

Figure 5.6 **(A)** Decompacted isopach maps of the Lavaca (Lower Wilcox) and Yoakum (Middle Wilcox–Upper Wilcox boundary) canyon fills. Contours approximate original canyon paleobathymetries. Gray shading shows distribution of local sand facies within the mud-dominated canyon fill. SC: arcuate, retrogradational, perched, canyon-wall slump scars. MS: incompletely evacuated canyon-wall slump. **(B)** Stratigraphic setting and cross-sectional morphology of basal Lower Wilcox canyons and the Yoakum Canyon, which excavated underlying Middle Wilcox and Lower Wilcox sequences during a transgressive flooding event that terminated the Paleocene Middle Wilcox deposode. The top Lower Wilcox flooding surface is the datum. Very closely spaced wells and seismic data reveal the multiple rotated blocks of the incompletely evacuated slump on the northeast flank of the Yoakum Canyon. The preservation of intact, angular, upturned slump blocks required submergence below the effective wave base at the time of slumping into the canyon. Deposition of the Yoakum Shale buried and preserved the irregular topography created by the unconsolidated slump blocks. **(C)** Seismic line interpretation of the Lower Wilcox Lavaca and Yoakum Canyons. Mass wasting extends into Upper Cretaceous strata, revealing the persistence and scale of Early Paleocene margin instability. H1–H5 seismic horizons define mass wasting events. (A) and (B) modified from Galloway *et al.* (1991); (C) modified from White and Snedden (2016; in prep.). Seismic line courtesy of ION.

C

DEPTH

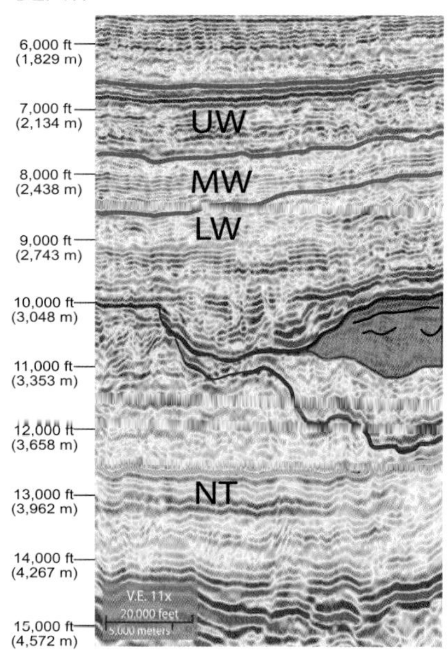

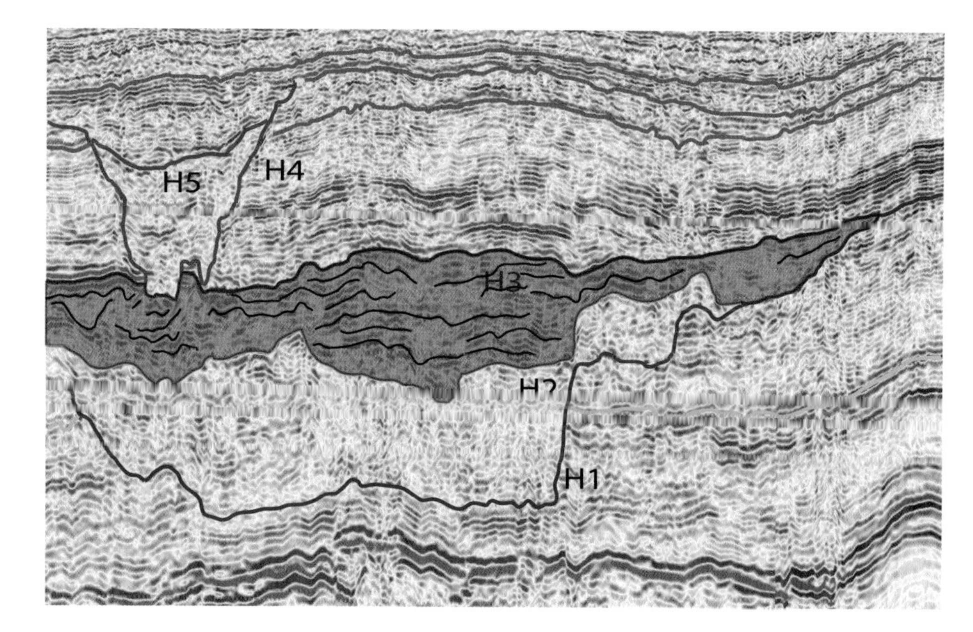

Figure 5.6 *(cont.)*

Box 5.1 Shelf Edge Recognition Criteria

In structurally simple sedimentary basins, the shelf edge is readily identified by its preserved clinoform stratal geometry. Relatively flat, gently dipping shelf strata roll over into steeper dips of the slope, reflecting the depositional topography of the continental margin. With few exceptions, simple clinoform geometry is rarely seen on the northern GoM margin. Progradation of the shelf edge required construction of an offlapping foundation consisting of many thousands of feet of muddy continental slope strata. This prograding slope sediment prism was typically underlain by thick autochthonous and/or allochthonous salt. High sediment loading rates created instabilities with resultant slumping and sliding, and, at depth, compaction disequilibrium and overpressurization of thick marine mud. Syndepositional deformation by gravity spreading and salt tectonics created a dynamic array of structures that overprinted and obscured depositional architecture. Consequently, several additional criteria for paleo-shelf edge recognition must be applied. Winker compiled, listed, and applied these criteria in his 1984 compilation of GoM shelf margins. Galloway (2005a) updated Winker's work across the northern Gulf margin.

Useful aids for identification and mapping of structurally complex northern paleo-shelf edges include:

1. Paleobathymetry. Microfaunal assemblages change from shelf (neritic) to slope (bathyal) associations across the shelf edge.
2. Change from traction transport (shelf) to gravity mass transport (slope) processes. Sand transport on the shelf is dominated by marine currents, creating a familiar array of bedforms and consequent sedimentary structures such as cross-stratification. Sediment transport direction and resultant facies trends are strike-oriented, reflecting the prevailing longshore and along-shelf transport. On the continental slope, gravity-driven transport dominates erosion and deposition. Transport and facies trends are dip-oriented.
3. Change from progradational to aggradational facies successions. Progradation of shoreface and shelf deposits typically creates upward-coarsening facies sequences that are relatively continuous and thus correlative over distances of miles. Upper slope facies are characterized by abrupt facies boundaries both vertically and laterally. Continuity, especially along strike, is poor.
4. Increased rate of interval thickening. Rapid increase of slope depth basinward creates accommodation space for thick stratal units.
5. Presence of submarine gorges, canyons, and slump scars. The upper slope is an unstable setting characterized by mass wasting, failure, and erosional scour. Resultant discontinuities, affecting hundreds of meters of strata, disrupt facies continuity and further limit correlation of markers beds or surfaces.
6. Regional increase in dip. If structural deformation is moderate, regionally averaged stratal dip increases basinward of the shelf margin.
7. Maximum displacement rates across growth faults. Extensional stress is focused at the upwardly convex inflection created by the continental shelf edge. Normal faulting typically shows greatest growth rates across the upper slope-to-shelf transition. Note, however, that growth faults rarely create significant sea floor topography.

While no single criterion is perfect or everywhere applicable, confluence of several typically defines the position of the shelf margin. Criteria 1, 4, 5, 6, and 7 delimit a general position of the continental margin. Criteria 2 and 3 have the potential for high-resolution definition of a paleo-shelf edge, but their use requires multiple well log suites and/or core data.

In 2001, drilling of the BAHA II well into the Perdido fold belt made an unexpected discovery: a thick interval of Wilcox sand containing (non-commercial) oil. Subsequent nearby discoveries at Trident and Great White confirmed a major new play in basinal Wilcox reservoirs. The Cascade discovery expanded the play 440 km (275 miles) to the east, into the central Gulf. Additional exploratory drilling confirmed a robust interval of sand-rich basinal Wilcox strata that extends southward beneath the modern abyssal plain and landward beneath the Neogene salt canopy. These basinal sands lie 300–550 km (200–350 miles) from the Paleocene Wilcox shelf edge. Released logs from well penetrations outline two large abyssal plain fans that together comprise the ALKEWA fan system (named for the Alaminos Canyon, Keathley Canyon, and Walker Ridge protraction areas in which it lies). Depocenters for each fan exceed 2000 ft (600 m) in total thickness and are sand-rich (Figure 5.5). Logged and cored intervals display a mix of interbedded and massive fine sandstone, siltstone, and mudstone sequences containing abundant features of gravity mass transport processes. Sand-rich fan lobes extend across the breadth of the ALKEWA system (Figure 5.7). Massive, sharp-based sand bodies suggest thick successions of stacked, large fan channel fills. Thinly interbedded sand and mud (serrate log patterns) include the array of fan lobe and channel margin facies. Combined seismic and core data confirm channel fill, lobe, and sheet facies (Zarra 2007; Marchand *et al.* 2015). The fan system persisted during the Middle Wilcox deposode and into the earliest Eocene Upper Wilcox. However, its thickest deposits are in the Lower Wilcox genetic supersequence (Figure 5.5).

The location and nature of the transport system that connects the shelf-margin delta and upper slope apron deposits to the fan systems is speculative. Wilcox slope deposits lie below current drill depths, and high-resolution seismic penetration of overlying thick, structurally complex strata and remnant salt canopies remains a challenge. Galloway (2007) speculated that the shelf-margin canyons may have focused, at least in part, bypass of sand downslope to the basin floor. Transport of sandy flows more than 370 km (230 miles) from the Wilcox shelf edge onto and across the abyssal plain, though initially a surprising reality, had several contributing factors. (1) The run-out lengths of the

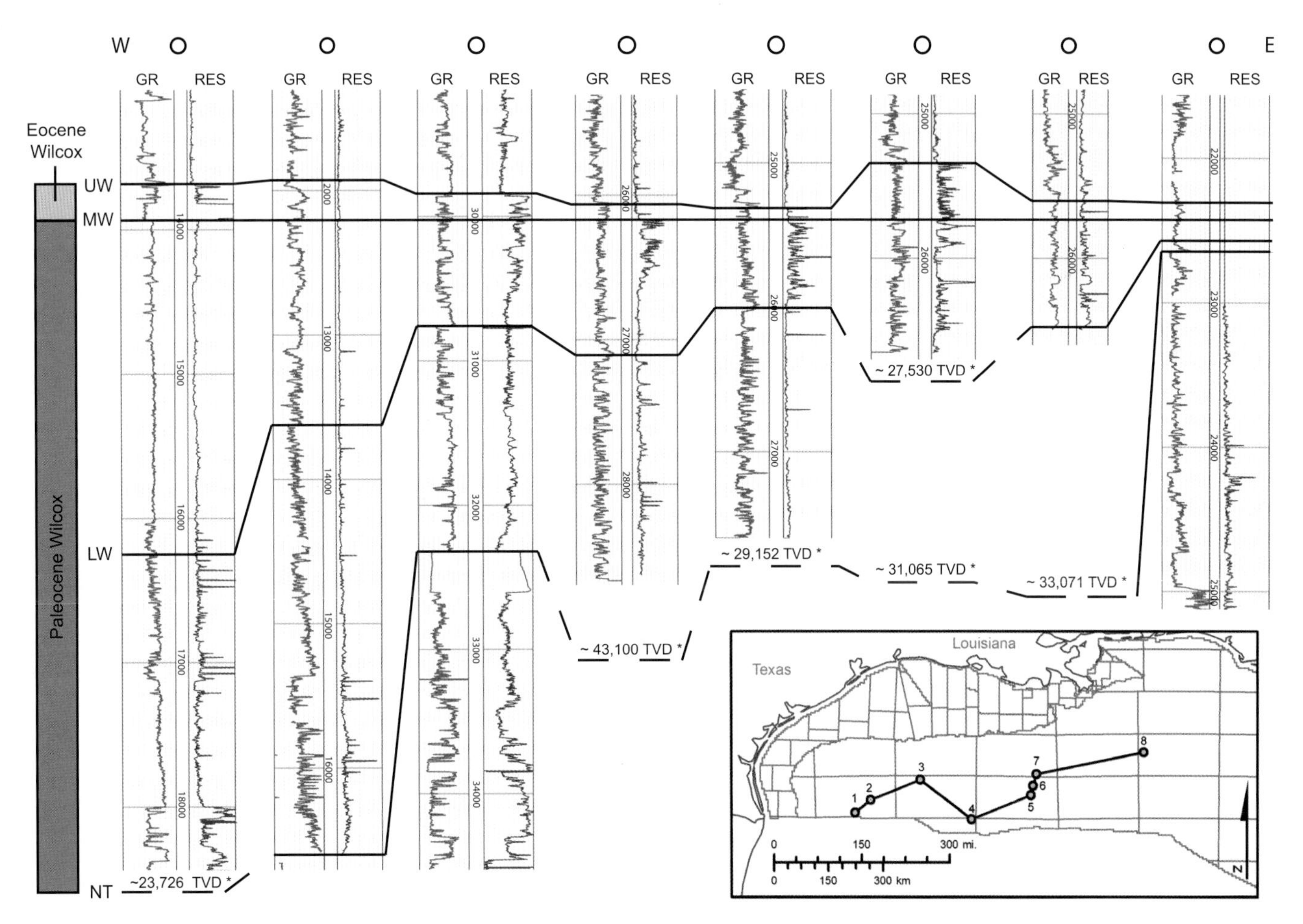

Figure 5.7 Regional strike cross-section of the Wilcox ALKEWA fan system. Datum is the top Paleocene Middle–Upper Wilcox boundary, which is typically well constrained by faunal data. The Lower–Middle Wilcox boundary is picked at a regionally correlative hemipelagic shale. Faunal tops are, however, sparse, and the top of the Lower Wilcox is commonly poorly constrained. Few wells penetrate the base of the Lower Wilcox sequence.

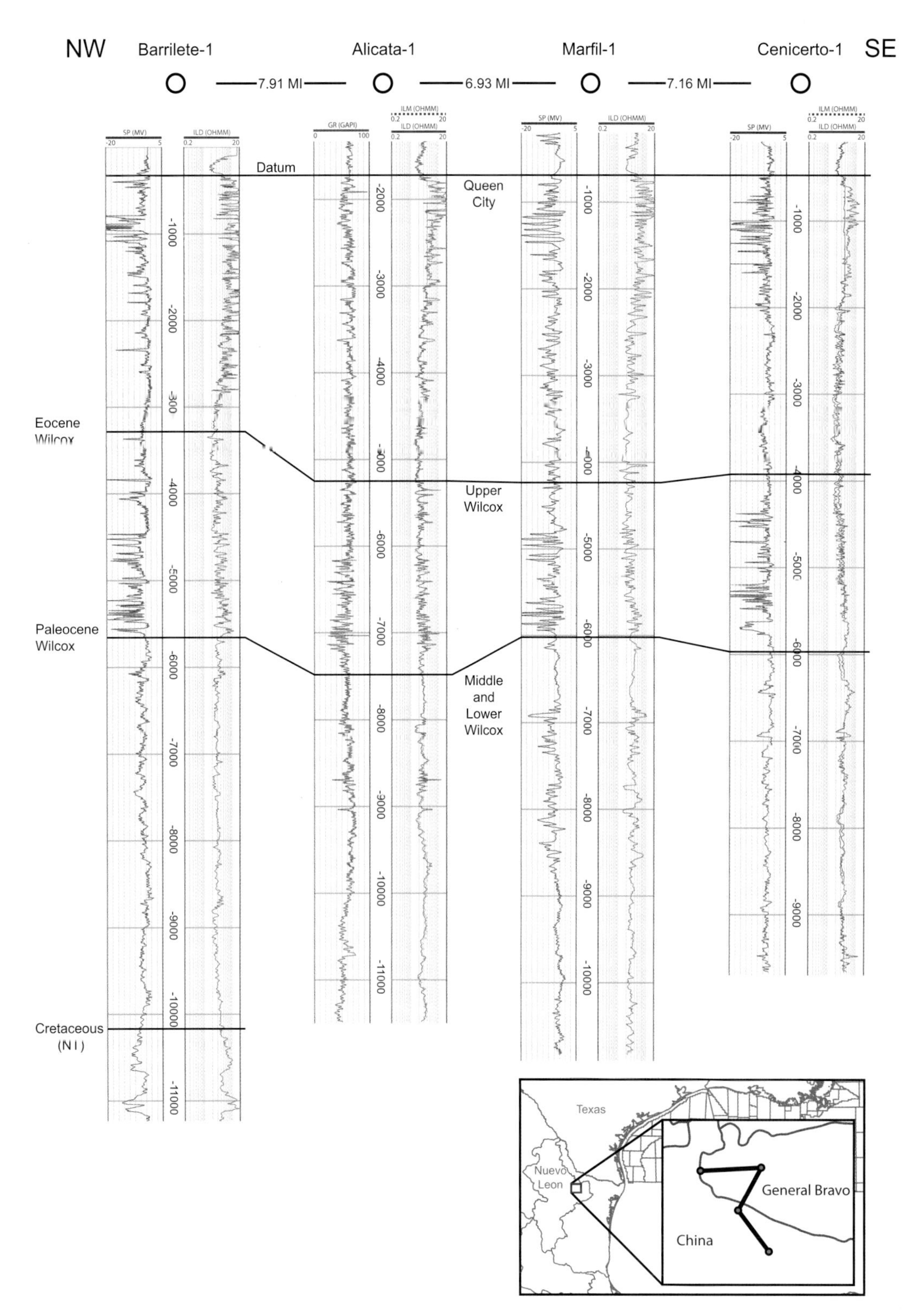

Figure 5.8 Log cross-section showing representative Lower Wilcox–Queen City supersequences in the Burgos basin. Thin sand bodies display upward-coarsening successions. Modified from De la Rocha Bascon (2016).

ALKEWA fans is commensurate with those of modern fans supplied by large continental fluvial–deltaic systems (Sweet and Blum 2012). (2) Flows remained channelized far out onto the basin floor (Lewis *et al.* 2007). Channelization enhances flow efficiency, minimizing energy loss. (3) The deltas and adjacent shore zone and shelf systems prograded onto a linear, muddy upper slope apron that passed basinward into a lower slope/continental rise. Speculatively, the broad continental rise extended the southward bathymetric gradient into the abyssal plain of the central basin. (4) Synsedimentary salt deformation produced a complex corridor bathymetry that would collect and focus subaqueous flows, as described by Steffens *et al.* (2003). (5) Silt-/clay-rich Wilcox turbidites exhibit characteristics of turbulent, stratified, and laminar flow dynamics (Kane and Ponten 2012).

The Mexican GoM paleogeography was completely different. Large river systems, coastal plains, and shore zone facies were largely absent during most Cenozoic history. Short, steep gradient streams emerged from adjacent uplands onto tectonically over-steepened basin margins; sediment bypassed coasts directly onto the submarine slope. These out-of-grade slopes were dominated by sediment bypass, local or regional erosion, and depositional onlap. Like the northern GoM margin, the depositional systems array established during the Early Paleocene persisted throughout much of the subsequent Cenozoic.

Two foreland troughs, the Chicontepec and the Veracruz basins, emerged into the GoM (Figure 5.3). The Wilcox margin of the Burgos basin was separated from the continental interior by the Parras–La Popa foreland trough. During the Paleogene, the La Popa basin was accumulating prodelta and slope mudstone (Lawton *et al.* 2009), effectively creating a depositional moat that was separated from the GoM by the peripheral Tamaulipas Arch. The Paleocene Wilcox interval is relatively thin and volumetrically minor in the Burgos basin (Figure 5.5; De la Rocha Bascon 2016; Snedden *et al.*, 2018b). Sand facies are sparse, thin, and typically cap broadly upward-coarsening facies successions (Figure 5.8), further suggesting that sediment was largely derived by longshore transport from the robust shore zone system of south Texas. The Burgos basin interval was possibly augmented by dispersion of suspended load from the foreland trough across submerged segments of the peripheral arch.

Deposition of sandy to gravelly mass transport deposits built an axial fan from northwest to southeast along the bathymetric axis of the Chicontepec Trough (Figure 5.3; Vásquez *et al.* 2014). The trough was separated from the open GoM by the high-standing Tuxpan Platform (Roure *et al.* 2009). However, the foreland opened to the Gulf both to the north and south of the platform. The landward margin of the trough was largely non-depositional to erosional, or has been removed by subsequent uplift and truncation. Erosional scour at the south end of the trough, presence of the axial fan, and paleocurrent data suggest possible bypass of mud and some bedload sediment through the saddle and onto the Gulf basin floor. Presence and morphology of the fan (Figure 5.3) is speculative, and based in part on a seismically mapped depocenter that lies to the east of the Tuxpan Platform (cross-sections 7 and 8; Figures 1.19 and 1.20).

The western flank of the Veracruz basin was uplifted by Laramide compression. Bypassing and erosional recycling across the narrow coastal systems fed a sandy to gravelly tectonic margin slope apron, the first of a nearly continuous succession of Cenozoic aprons (Gonzales and Medrano 2014). Multiple channel–lobe deposits onlapped the erosional margin. Thickness and volume of Lower Wilcox strata are very different from those found in the northern GoM depocenters. Total unit sand thicknesses typically range across 50–100+ m (i.e., <500 ft/150 m). Diversion of turbidity flows from the east-northeast continental slope gradient toward the north reflects a basinal bathymetric gradient toward the central Gulf abyssal plain.

The Yucatán Platform formed a carbonate shelf bounded by an upper slope ramp leading to a steep, erosional scarp. The Chicxulub crater (Figure 5.3), which created a deep, open-ended basin on the north platform rim, was the dominant bathymetric feature on the platform.

The Florida Platform formed the eastern margin of the Gulf. Accumulation of pure marine shelf carbonate dominated. On the south-central platform, evaporates formed within a broad salina (Figure 5.3). The platform was isolated from terrigenous clastic sediment input by the deep Suwannee channel (McKinney 1984; Popenoe *et al.* 1987; Umbarger and Snedden 2016). The channel and its influence on stratigraphic architecture are clearly imaged in regional seismic lines (Cunningham *et al.* in prep.). At the south, the Florida Strait opened to the Atlantic Ocean between the two platforms as well as the converging Cuban foreland. Together, the Suwannee channel and the Florida Straits provided two independent connections between the Gulf and the world ocean that persisted well into the Middle Eocene (Umbarger and Snedden 2016).

5.3.2 Termination and Summary

The Lower Wilcox deposode terminated with a regional, although short-lived transgressive flooding of the northern GoM margin recognized and correlated regionally as the "Big Shale" marker. The Big Shale is closely associated with last appearance datums (LADs) of *Morozovella angulata* and *Heliolithus kleinpelli*.

In summary, the deposode is important for (1) deposition of the first of a 60-million-year succession sand-rich sediment pulses fed through large rivers that drained emergent uplands within the continental interior; (2) regional depositional offlap of the continental margin tens of miles beyond its relict Cretaceous position; (3) concomitant repeated mass wasting, erosion, and filling of large slides, slumps, and submarine canyons; (4) efficient bypass of sediment, including a high percentage of fine sand and silt, onto the paleo-abyssal plain; (5) creation of tectonic basins and uplifts along the western Gulf margin that both modified the Mesozoic basin configuration and sequestered sediment derived from adjacent Mexican uplands; and (6) initiating a long history of sediment bypass onto the deep Gulf basin floor.

5.4 Late Paleocene Middle Wilcox Supersequence

The Middle Wilcox supersequence is stratigraphically bounded below by the Big Shale and above by the Yoakum Shale, which corresponds closely with the Paleocene–Eocene boundary. Late Paleocene Middle Wilcox deposits of the northern GoM record several evolutionary changes in sediment supply and paleogeography (Figure 5.1A). (1) Total volume of sediment is about 60 percent of that of the underlying Lower Wilcox. Rate of sediment supply decreased commensurately. (2) The rate of continental margin accretion also decreased dramatically. (3) The areal extent of the combined deltaic coastal plains expanded and diversified with the addition of two new fluvial/deltaic axes.

5.4.1 Paleogeography

In Texas, the Colorado fluvial–deltaic axis persisted (Figure 5.9). However, a second axis entered the far northeast Texas basin margin and flowed across a broad alluvial coastal plain that extended inland across the East Texas basin. This river produced a secondary deltaic depocenter that merged with that of the Colorado to form the composite, fluvial-dominated Calvert delta system (Xue and Galloway 1995). The deltaic depocenters contain a maximum of about 4000 ft of compacted section, reflecting the much-reduced rate of sediment supply and margin offlap. To the east, the Mississippi fluvial–deltaic axis remained active, but the deltaic deposits are relatively thin, display smaller paleo-channel dimensions, and record increased marine influence. In south Texas, the smaller, wave-dominated La Salle delta can be distinguished from the broad wave-dominated shore zones that surround it. This delta marks the growing importance of a paleo-Rio Grande drainage system derived from uplands extending across New Mexico and Arizona, and flowing eastward through the gap between the Laramide Rocky Mountain Front Range and the Sierra Madre Orientale. Preserved late Paleocene Wilcox sediments in the adjacent Burgos basin consist of narrow, wave-reworked shore zone and shelf systems plastered landward against the adjacent foreland arch. Efficient marine reworking created a broadly arcuate, prograding coastline and shelf margin.

The northern and northwestern Gulf continental slope, from central Louisiana to Tamaulipas, prograded by deposition of a broad delta- and shelf-fed slope apron (Figure 5.9). Few wells with adequate biostratigraphic control to date Paleocene slope sequences penetrate the deeply buried apron, but those that do show it to be muddy with local, sand-filled slope channels. Submarine canyons are prominent features of both the overlying Yoakum Shale and underlying Big Shale intervals. A narrow stretch of continental margin south of the Sabine Uplift retreated a few miles from its Lower Wilcox maximum; several small, mud-filled canyons cluster there. The largest of the canyons, the Yoakum, excavated the entire Middle Wilcox interval along much of its length (Figure 5.6B).

Regional and efficient bypass of sandy sediment across the shelf margin continued to nourish the large ALKEWA abyssal plain fan system (Figure 5.9). The eastern fan covers the largest area, but the thickest intervals are located beneath the western fan near the international boundary.

The mapped paleogeography of the Mexican Gulf shows only modest evolutionary changes from the Lower Wilcox. In part this reflects the fact that published interpretations rarely subdivide the Paleocene interval, here an exception being the Chicontepec Canyon (Nieto 2010). Ongoing foreland tectonism affected the landward margin of the Veracruz basin. Upland erosion and recycling nourished the tectonic margin apron. Numerous mapped sand belts define turbidite channel–lobe complexes that onlap the erosional margin and extend up to 100 km or more downslope (Gonzales and Medrano 2014). Mapped northward deflection of the longest channel fills indicates the early influence of the bathymetric Veracruz Trough (a feature that would play an important role in basinal sediment transport and deposition throughout the Cenozoic).

High-resolution mapping of the Chicontepec basin-fill differentiates a large mass transport complex along its northwest margin and deposition of an elongate, south-flowing, axial fan complex, cut by large erosional scours (Cossey *et al.* 2007; Vásquez *et al.* 2014). Further north along the chain of Laramide foreland troughs, the La Popa basin continued to shoal, filling with deltaic and coastal sediment (Figure 5.1). The mapped spill of Chicontepec fan deposits across the erosional sill at the south end of the Tuxpan Platform and into the open Gulf abyssal plain is speculative, but supported by the presence of Paleocene sand bodies penetrated in the deep Puskon 1 well lying east of the Tuxpan Platform, and seismic interpretation of a thick Wilcox interval in the western GoM (cross-sections 7 and 8; Figures 1.19 and 1.20).

5.4.2 Termination and Summary

Middle Wilcox deposition terminated with a uniquely abrupt flooding event that can be traced across the northern GoM. The transgression was recorded by deposition of a thin, widespread, marine shale marker known as the Yoakum Shale. Shorelines retreated more than 150 km (100 miles) from their prograded Middle Wilcox position, reaching the outcrop belt in many places across the modern coastal plain. Although the Paleocene Wilcox has proved to be a particularly difficult interval for dating by calcareous benthic and planktonic foraminifera, increasing chronostratigraphic resolution places the Yoakum at or very near the Paleocene–Eocene boundary and, thus, the global Paleocene–Eocene Thermal Maximum (PETM) (Dickey and Yancey 2010; Paleo-Data Inc. 2017). Detailed paleontologic, isotopic, and geochemical study has located the PETM in the context of this regional flooding event

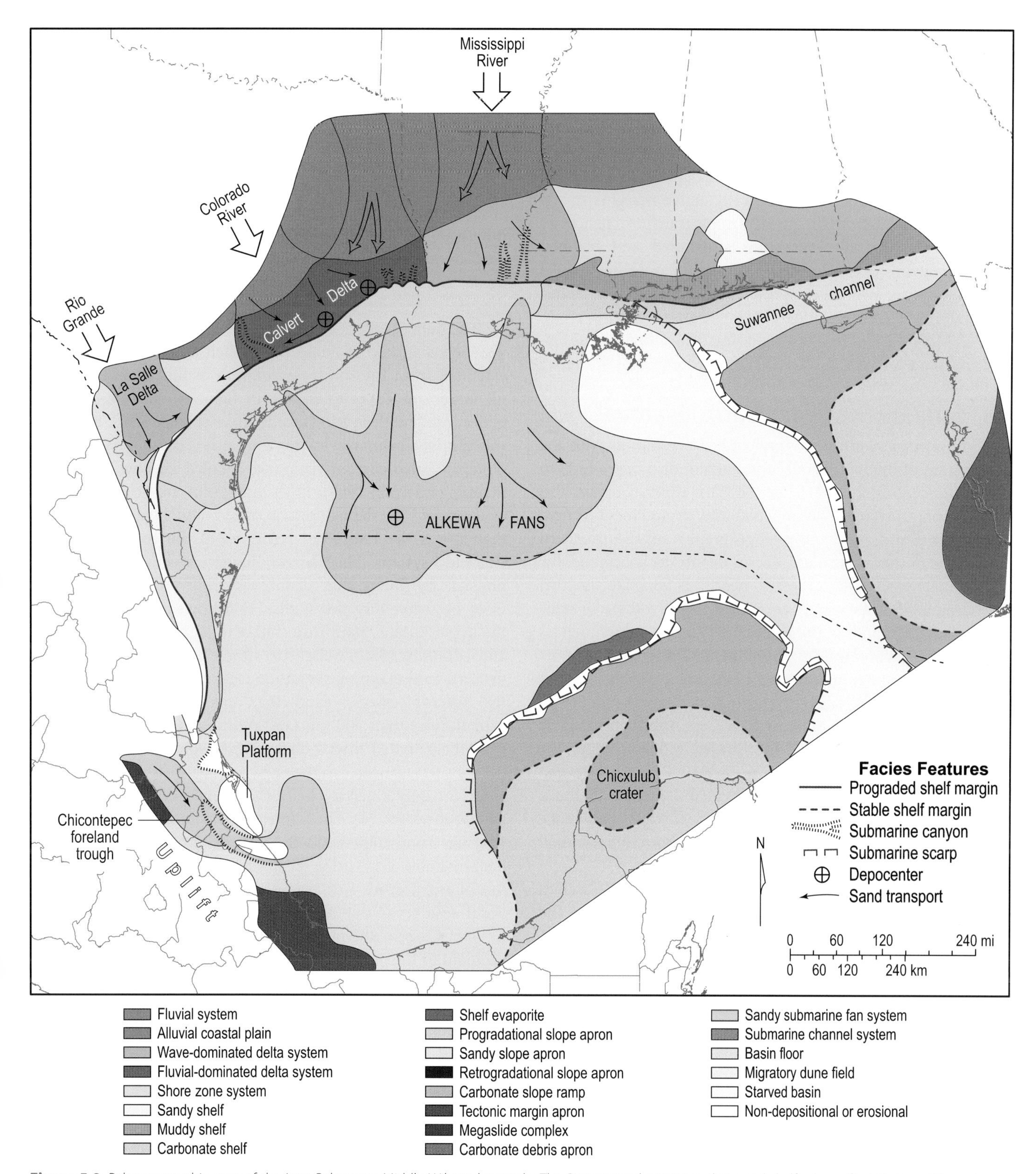

Figure 5.9 Paleogeographic map of the Late Paleocene Middle Wilcox deposode. The Depositional system outlines and shelf edge reflect their positions at maximum progradation.

at outcrop (Sluijs *et al.* 2014). In the subsurface, this boundary is closely associated with the last occurrence of *Morozovella angulata*, *Morozovella velascolensis*, and *Heliolithus kleinpelli*.

The Middle Wilcox deposode records major, but waning, sediment influx to the northern GoM and diversification of fluvial axes and delta types across the northwestern and central coastal plain. Along both the offlapping depositional slope of the northern Gulf and the tectonically active margins of the Mexican foreland troughs, gravity mass transport processes diverted volumetrically significant suspended and bedload sediment down the continental slope and far onto the abyssal plain, depositing an array of apron and submarine fan systems. Deposition was abruptly punctuated by a global oceanic event, the PETM, and its accompanying dramatic, geologically short-lived rise in sea level.

5.5 Early Eocene Upper Wilcox Deposode

In addition to its eustatically punctuated beginnings with the termination of the PETM and consequent return of sea level to its previous position, the Upper Wilcox supersequence also records major changes in the patterns of sediment influx, distribution, and resultant paleogeography of the northern GoM. The resultant Upper Wilcox deposode continued for approximately six million years, nearly as long as the two Paleocene Wilcox deposodes combined. Total volume of sediment is less than both of the Paleocene Wilcox supersequences (Figure 5.1A). Volume rate of sediment supply continued its evolutionary decline; the calculated rate was less than half that of the Lower Wilcox (Galloway *et al.* 2011; Zhang *et al.* 2018). Margin accretion rates similarly decreased. Although much of the northern Gulf margin demonstrated modest offlap, the greatest advance occurred along the northwest margin between the Burgos basin and the soon-filled and buried Yoakum Canyon. Within this elongate depoaxis, the shelf edge advanced as much as 30 km (20 miles) from its terminal Paleocene position.

Among the first depositional events was the rapid infilling of the Yoakum submarine canyon, which had followed the transgressive PETM shoreline updip to the limit of platform submergence (Galloway *et al.* 1991; White and Snedden 2016). Unpublished seismic transects of the canyon fill reveal clinoforms prograding from northeast to southwest, across the canyon axis, indicating filling of the canyon by along-shelf mud advection from the first delta lobes that lay to the north. Deposition of progradational shoreface deposits, expanded by loading compaction of the thick, underlying mud fill, completing the healing phase of the canyon's history (Figure 5.6B). Elsewhere, canyon incision and filling at a much-reduced scale continued into the Early Eocene Upper Wilcox along the northwest Gulf margin (Cornish 2013), likely supplying the western remnant of the ALKEWA fan system (Figure 5.10).

5.5.1 Paleogeography

The Early Eocene northwestern and central Gulf basin continental margins introduced dramatically differing paleogeographies and depositional architectures (Figure 5.10). The prominent Paleocene Mississippi fluvial–deltaic axis was replaced by two modest, marine-dominated deltas flanking the previous depocenter. A broad, retrogradational coastline separated the deltas and ultimately retreated northward into the mouth of the Mississippi Embayment in northeast Louisiana. Shelf and coastal plain deposits are 500–1000 ft (150–300 m) thick. The shelf margin and slope sequence averages only a modest 1000–2000 ft (300–600 m) in total thickness.

The bulk of Upper Wilcox genetic supersequence sediment was deposited along an elongate shelf-margin depoaxis that extended from the southwest margin of the Yoakum canyon fill in central Texas to the north flank of the Burgos basin, beneath the modern Rio Grande River (Figure 5.10). Sediment was supplied primarily by the sand-rich Carrizo fluvial system. Outcrop paleo-current data and detailed shallow subsurface mapping (Hamlin 1988) shows due southward flow of the river as it entered the depositional coastal plain, suggesting in turn that it was the redirected terminus of the continental-scale Colorado system. The Carrizo fluvial system exhibits several unusual attributes that distinguish it from other northern GoM Cenozoic fluvial systems. (1) It flowed diagonally across the upper coastal plain from north to south (Ayers and Lewis 1985; Hamlin 1988), oblique to simple basinward dip. (2) The detailed sand mapping reveals no clear apical point of entry. At the outcrop, coarse, sand-rich, amalgamated channel fills extend more than 300 km (180 miles) along the outcrop belt. (3) Amalgamated fluvial channel fill deposits extend far into the subsurface, nearly to the Early Eocene shelf margin, leaving room for only a narrow belt of shelf-margin wave-dominated deltaic deposits. (4) The aggregate fluvial interval is tabular, ranging from 1000–2000 ft (300–600 m) in thickness, and demonstrates very little basinward thickening. This unique fluvial architecture creates one of the largest meteoric groundwater aquifers of the Gulf coastal plain.

The Carrizo fluvial belt was flanked on the north by a remnant of the paleo-Colorado fluvial system and on the south by an increasingly prominent Rio Grande system (Figure 5.10). Together, these three rivers created a depositional coastal plain, dominated by channel fill and associated deposits that extended from the Burgos basin across most of the Texas continental margin.

This broad fluvial belt, in turn, fed an equally broad, coalesced assemblage of deltas, the Live Oak–Rosita delta system of Edwards (1981). The bulk of the coastal deltaic facies are wave-dominated and sandy. The exception is the delta of the Colorado, which displays facies of a mixed fluvial-/tidal-influenced delta (Zhang *et al.* 2016).

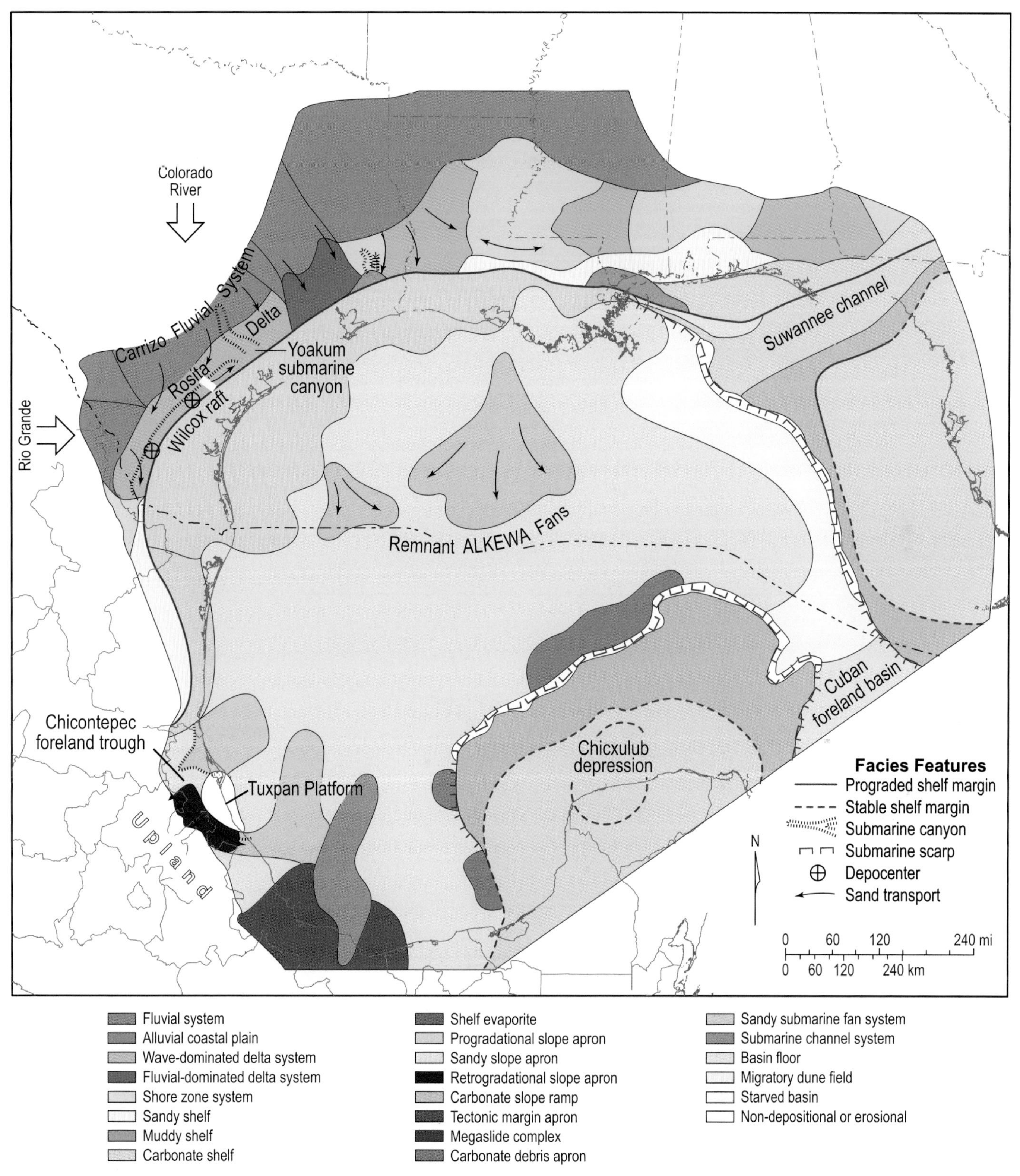

Figure 5.10 Paleogeographic map of the Early Eocene Upper Wilcox deposode. Depositional system outlines and shelf edge reflect their positions at maximum progradation. White box locates seismic line shown in Figure 5.12.

Box 5.2 Stratigraphic and Facies Architectures of a Prograding Northern Gulf Basin Continental Platform and Margin

The northern margin of the Gulf basin is first and foremost characterized by progressive offlap of a sedimentary prism composed of sub-equal volumes of sediment deposited in platform (coastal plain and shelf), continental slope, and basin (continental rise and abyssal plain) regimes. Combined loading subsidence of the underlying attenuated continental crust, compactional subsidence, subregional extension, and salt evacuation combined to produce high rates of subsidence that were commonly closely balanced with sediment supply. Consequently, supersequences typically display thick, repetitive successions of off-stepping (progradational), vertically stacking (aggradational), and back-stepping (retrogradational) facies sequences.

A highly detailed dip profile of Lower–Upper Wilcox genetic supersequences located in the structurally simple Paleogene shelf margin of central Texas illustrates these common patterns of stratal architecture and facies stacking (Zhang *et al.* 2016). Figure 5.11A shows the detailed correlation of "high-frequency" sequences bounded by maximum flooding surfaces and maximum regressive surfaces, as well as shoreline and shelf edge positions at maximum progradation. The shelf edge trajectory (here uncomplicated by growth structures) displays rapid advance as initial Lower Wilcox sequences prograded across the deep shelf landward of the subjacent, foundered Cretaceous shelf platform. As progradation approached the Late Cretaceous shelf margin, five sequences stacked vertically with little advance of the shelf edge. A thick, mud-rich prodelta/slope apron records the depositional regrading of the foundered Cretaceous platform and slope. With establishment of a stable clastic slope gradient, a series of off-stepping high-frequency sequences built the shelf edge several miles basinward. Final sequences record declining sediment input and consequent coastal retrogradation followed by Lower Wilcox deposode termination with the Big Shale transgression. With onset of the Middle Wilcox depositional episode, shorelines rapidly prograded to the shelf edge. There, sediment supply was balanced by subsidence and downslope bypass. The shelf edge aggraded, with little further margin offlap from the underlying Lower Wilcox position. Following the abrupt Yoakum (PETM) transgression, which pushed a shoreline of maximum transgression far updip, a succession of sand-rich high-frequency sequences initiated rapid advance of successive shorelines and shelf edges. The climax Upper Wilcox shelf edge lies basinward of the cross-section.

Figure 5.11B shows facies patterns within the high-frequency sequence framework. Note that maximum flooding surface boundaries were used to delimit the sequences. Locating the cross-section within Wilcox paleogeographies places the Lower

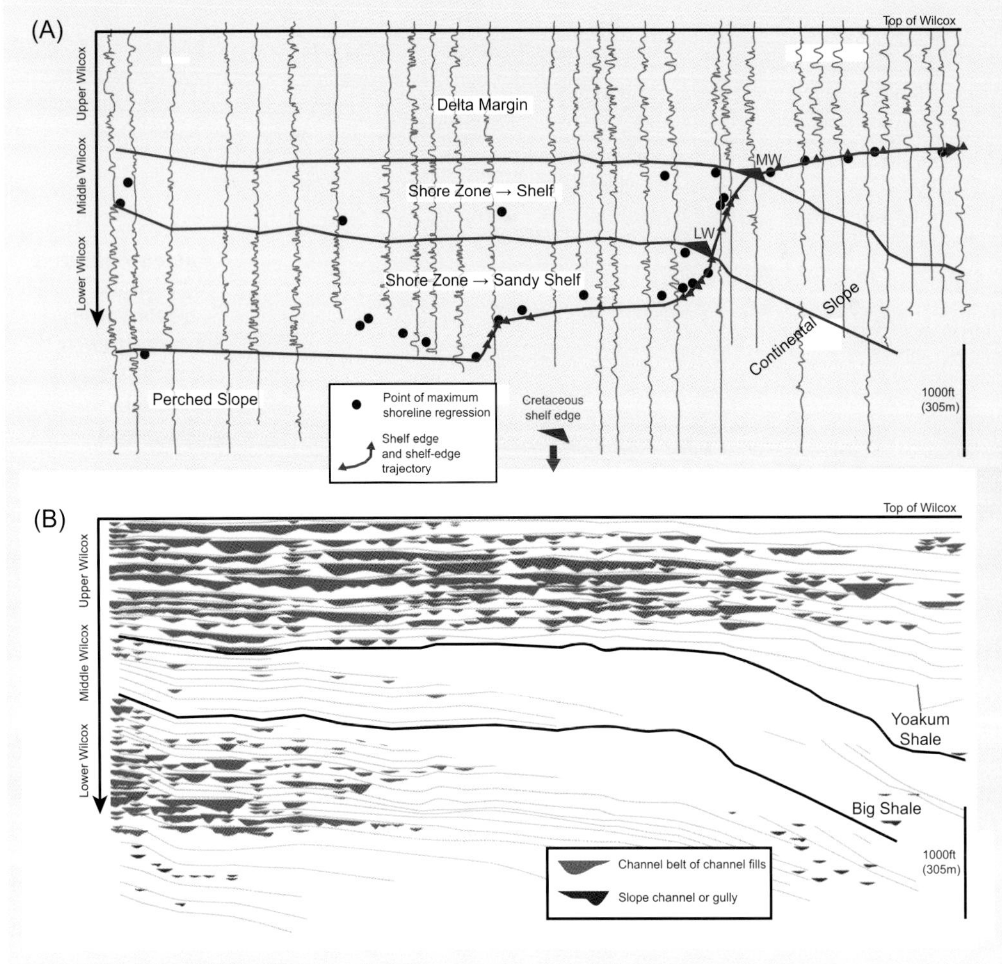

Figure 5.11 High-resolution correlation cross-section of the Lower and Middle Wilcox shelf margins, and the initial shelf edges of the Eocene Upper Wilcox. **(A)** High-frequency sequence correlations with interpreted shelf edge and shoreline trajectories. Basal Lower Wilcox supersequences prograded across the foundered Late Cretaceous–Early Paleocene shelf platform. Location of the Lower Wilcox (LW) and Middle Wilcox (MW) shelf edges at maximum offlap are shown. The terminal Upper Wilcox shelf margin (UW) lies basinward of the section termination. **(B)** Distribution of interpreted channel fill deposits within the high-resolution framework. Blue lines are flooding surface correlations. Channel fills include barrier inlet, coastal estuary, and distributary deposits, depending on the depositional system. Thin, discontinuous upper slope gully fills lie basinward of the shelf edges. Modified from Zhang *et al.* (2016).

Box 5.2 *(cont.)*

Wilcox in shore zone and sandy shelf depositional systems, the Middle Wilcox in shore zone (updip) and muddy shelf (downdip) systems, and the Upper Wilcox in the transition zone between wave-dominated delta and sandy shore zone systems. Generic channel facies shown on the cross-section are interpreted accordingly: inlet and estuary fills in shore zones and fluvial/distributary channels in the delta. Channel fill facies lie within a fabric of relatively continuous, upward-coarsening shoreface and delta front sands.

Each genetic sequence produces distinct vertical and lateral distributions of channel fills. Both Lower and Upper Wilcox genetic sequences show the common GoM pattern: initial offstepping facies, a core of vertically stacked facies, and terminal back-stepping facies tracts. The genetic supersequence records a multi-million-year episode of increasing sediment supply and consequent progradation followed by an extended period of balanced supply and accommodation. Terminal back-stepping records increasing accommodation domination of the deposodes that culminates in transgressive flooding of the coastal plain.

Finally note the visually distinct differences in abundance and distribution of sand in each of the three Wilcox deposodes. This demonstrates the significant reorganization of paleogeography and consequent facies associations across the major maximum flooding surfaces that bound northern GoM supersequences.

The northwestern Gulf margin also exhibits unique continental margin architecture. The Early Eocene extensional growth fault belt that delineates the paleo-margin is characterized by unusually dramatic displacement and expansion of delta front, prodelta, and upper slope facies sequences (Edwards 1981). Upper Wilcox supersequence deposits thicken abruptly from ~2000 ft (600 m) into the elongate depoaxis that contains >10,000 ft (3 km) of sediment. Within this depocenter, >5000 ft (1.5 km) of that interval is sand-rich. Fiduk *et al.* (2014), using regional seismic data, recognized this abrupt zone of expansion to be the infilled extensional gap created by a continental margin-scale, synsedimentary raft (Figure 5.12). The long-term accommodation volume created along the headwall of the raft formed an elongate sediment trap, limiting margin offlap and, as discussed below, sequestering most of the sediment that might otherwise have continued downslope and onto the abyssal plain. It is noteworthy that the northern margin of the raft, which constituted a shear zone, coincides with the Yoakum Canyon, itself a uniquely large and mature member of the Wilcox family of canyons. In fact, no comparable submarine canyon would be seen on the northern Gulf margin until the Pleistocene (Galloway 2005a).

On the abyssal plain, remnant ALKEWA submarine fans persisted into the Early Eocene (Figure 5.10). However, the areal extent of sandy fan channel and lobe facies decreased, total thickness of sandy interval penetrated by wells is at most a few hundred feet, and fan deposition was replaced by a regional, condensed hemipelagic drape about two million years before the Reklaw transgression terminated the Upper Wilcox deposode onshore (Figure 5.1). Only small submarine canyons cut Upper Wilcox strata, either during or subsequent to their accumulation. Seismically imaged scours of slope canyon scale

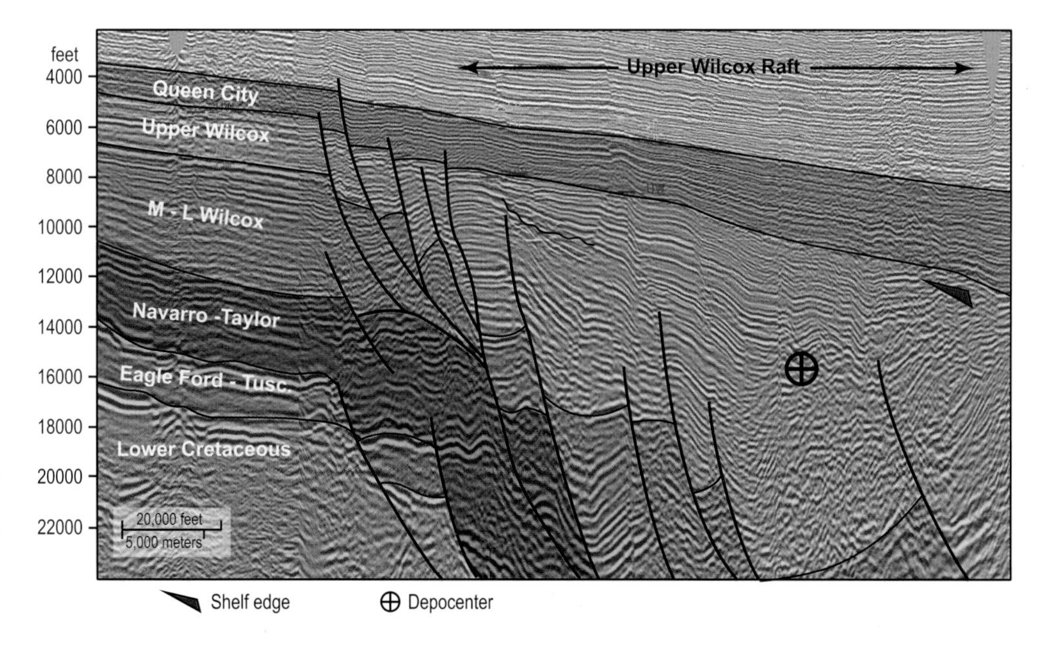

Figure 5.12 Seismic line interpretation of the headwall extensional growth faults bounding the Wilcox raft, south Texas. Expanded Upper Wilcox deposits create an Early Eocene depocenter exceeding 12,000 ft in thickness. For location, see Figure 5.10. Seismic line courtesy of ION.

at the interpreted Early Eocene slope base have been described (McDonnell *et al.* 2008).

Along the western basin margin, the Chicontepec and Veracruz forelands remained tectonically active (Vásquez *et al.* 2014). The compressional front advanced onto the western margins of the basins, resulting in erosional truncation and incision of older strata (Figure 5.10). In contrast to the sand-rich Upper Wilcox depocenter that extends along the south Texas paleo-margin, the Eocene Wilcox is relatively thin in the Burgos basin (Figure 5.8). Sand bodies are abundant, but also thin. Upward-coarsening successions are typical, suggesting longshore transport from the sand-rich deltaic headlands to the north.

Submarine scour at both the north and south ends of the Tuxpan Platform produced local erosional unconformities that opened onto the GoM. In addition to the eroded trough-floor sediment, the canyons likely bypassed bedload sediment into the Gulf. Accordingly, speculative abyssal fans are drawn on the paleogeographic map. Fan system size and run-out length were likely limited by the small drainage basins of the short, steep rivers flowing off of the Laramide foreland uplift (De la Rocha Bascon 2016; Snedden *et al.*, 2018b). A thick basin-floor Eocene seismic sequence (cross-sections 7 and 8; Figures 1.19 and 1.20) in the deep western Gulf floor suggests significant slope bypass along the western Gulf continental margin.

In contrast to the narrow, platform-constricted Chicontepec Trough, the Veracruz foreland merged into the Gulf abyssal plain. Multiple submarine channel–lobe complexes spilled eastward, forming the framework of the broad tectonic margin apron (Gonzales and Medrano 2014). Seismic mapping of sand body geometry and distribution (Figure 5.13) outlines multiple

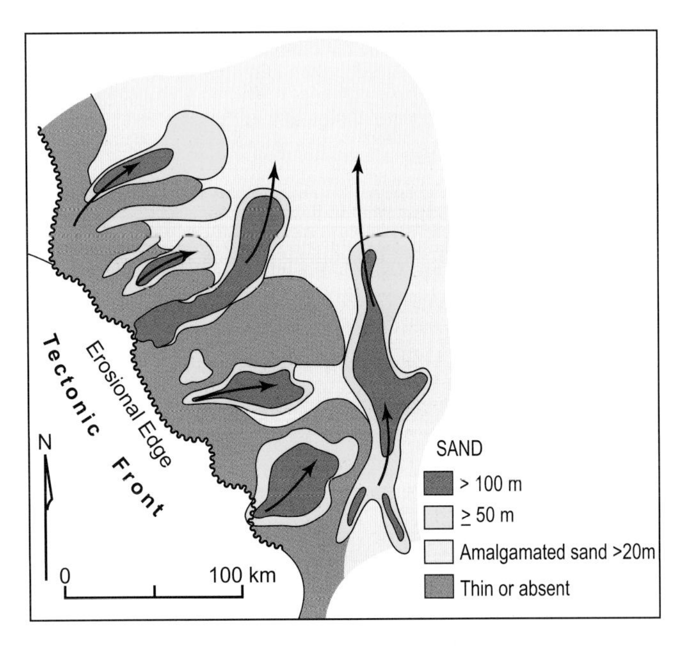

Figure 5.13 Thickness and distribution of sand facies within the Veracruz tectonic margin apron. Narrow, erosionally confined slope channels flare out into broad depositional belts and lobes. The two longest channels continue northward, becoming tributaries to the north-flowing submarine channel complex of the Veracruz Trough. Modified from Gonzales and Medrano (2014).

channel complexes onlapping the erosional upper slope. Distal ends of the two longest channel belts deflect northward, collecting into a north-flowing submarine channel-belt that extended along the axis of the deep basin Veracruz Trough.

On the Florida Platform, evaporate deposition ended. The Suwannee channel continued to separate the platform from influx of terrigenous mud from the adjacent continental landmass; pure carbonate deposition blanketed the platform and bounding ramp. At the south end of the platform, the advancing Cuban foreland basin impinged onto the southeastern-most corner of the Gulf (Escalona and Yang 2013).

5.5.2 Termination and Summary

The Upper Wilcox deposode terminated with the regional Reklaw transgressive flooding across the breadth of the northern Gulf coastal plain. The LAD of *Acarinina soldadoensis* corresponds closely to the Reklaw maximum flooding and is the most widely recognized faunal top. The extent of the resultant shallow, transgressive shelf was comparable to that seen during the Early Paleocene Midway flooding of the continent. Maximum flooding, extending to the modern outcrop belt, was relatively brief, but culminated a long-term retrogradational history. In the La Popa, Chicontepec, and Veracruz basins, Early Eocene clastic influx also waned, and the Guayabal marine mudstone ultimately dominated deposition across the two Gulf-bounding basins (Figure 5.1).

5.5.3 Wilcox Paleoceanography

The Wilcox trio of depositional episodes followed upon the cataclysmic Chicxulub impact event and bridged the globally significant PETM. The initially surprising discovery of thick, extensive, sandy submarine fan systems hundreds of kilometers basinward of the Wilcox continental margin further stimulated interest in and speculation about the paleoceanographic evolution of the early GoM. Faunal paleobathymetry confirmed that the Early Paleocene Gulf was several thousands of feet deep and provided data useful for examining oceanographic attributes of the small ocean basin.

Expedition 364 of the International Ocean Discovery Program drilled into the peak ring of the Chicxulub impact crater in the spring of 2016 (Morgan *et al.* 2016). Micropaleontologic examination of the core immediately above the impactite layer documented extremely rapid faunal recovery following the impact (Lowery *et al.* 2017). Planktonic and benthic foraminifera, calcareous nannoplankton, calcispheres, bioturbation, and geochemical proxies all indicate that organic productivity in the Chicxulub crater recovered within 30,000 years following the impact. Rate of recovery of diversity and species abundance took much longer and varied between groups. Planktonic foraminifera quickly diversified, and all common Paleocene tropical/subtropical species appeared as expected. Trace fossils rapidly reappeared. Calcareous nannoplankton were slower to recover. Diverse and abundant macro- and microbenthic organisms indicate food availability and good oxygen conditions on the sea floor. The latest Paleocene, just prior to the onset of the PETM, was characterized by a normal and diverse open marine assemblage of foraminifera and calcareous nannoplankton.

Systematic analysis of microfaunal assemblage data (Purkey Phillips, unpublished project report) and organic geochemical signatures (Cunningham *et al.*, unpublished project report) through the Wilcox and across the PETM boundary further refine paleoceanographic reconstruction. Relatively low total organic carbon values across the PETM suggest continued high rates of sediment supply, resulting in dilution, particularly in the western basin where large fans are present. Dominance of terrigenous kerogen reflects continental derivation of the organic matter. The greatest organic matter content is found in slope deposits, with local enrichment in bathymetrically isolated intraslope basins. Disoxic to anoxic conditions occurred variably in time and space within the Paleogene–Early Eocene Gulf.

Microfaunal data further refine the picture. Five regions, grouping deep basin Wilcox wells with distinct faunal signatures, were defined by Purkey Phillips (in prep.; Figure 5.14). Region 1 includes the sediment-starved eastern basin, including the Chicxulub crater on the Yucatán Platform. There, pelagic deposition dominated abyssal condensed intervals overlain by predominantly oligotrophic surface waters. Region 2, in the northeast corner of the basin, was also characterized by pelagic accumulation at abyssal depths. Surface waters were, however, eutrophic throughout deposition of the condensed Wilcox interval. Region 3 is characterized by an expanded stratal section of deepwater, hemipelagic continental slope facies. Turbid surface waters characterized the Paleocene Wilcox, when a large delta system lay to the north. Multiple disconformities reflect the depositionally active fan setting. Oligotrophic surface waters followed in the Eocene, as the Mississippi fluvial–deltaic axis was progressively abandoned. In region 4, agglutinated forams dominate the Paleocene assemblage, indicating very turbid water. The thick Lower–Middle Wilcox succession further records the very high rate of sediment influx. Following continued rapid deposition of Radiolarian-dominated sediment during the PETM (Yoakum Shale and Canyon equivalent), agglutinated forms disappear and the surface waters return to oligotrophic. Finally, region 5 wells demonstrated considerable well-to-well variability, with low total abundance of all forms in concordance with high sedimentation rates in proximal fan and slope apron settings. Faunal content in and around the PETM interval suggests a nutrient-enriched (especially SiO_2) water mass of normal salinity.

Rosenfeld and Pindell (2003) hypothesized the closure of the GoM by Paleocene convergence of Cuba and the South Florida Platform and consequent evaporative drawdown of the GoM basin. The hypothesis has retained some support as an explanation of the extensive submarine fan systems found in the Wilcox. However, we conclude that the preponderance of evidence negates the drawdown hypothesis. (1) The Cretaceous–Eocene existence of the Suwannee channel as a deep, current-swept opening to the Atlantic provided a second connection into the global ocean (Umbarger and Snedden 2016). (2) The fan systems require no special explanation. Their size and run-out lengths are consistent with those of submarine fans supplied by large, continental-scale rivers (Snedden *et al.* 2018a). (3) Although sand-rich, Wilcox fans consist dominantly of easily suspended and transported very

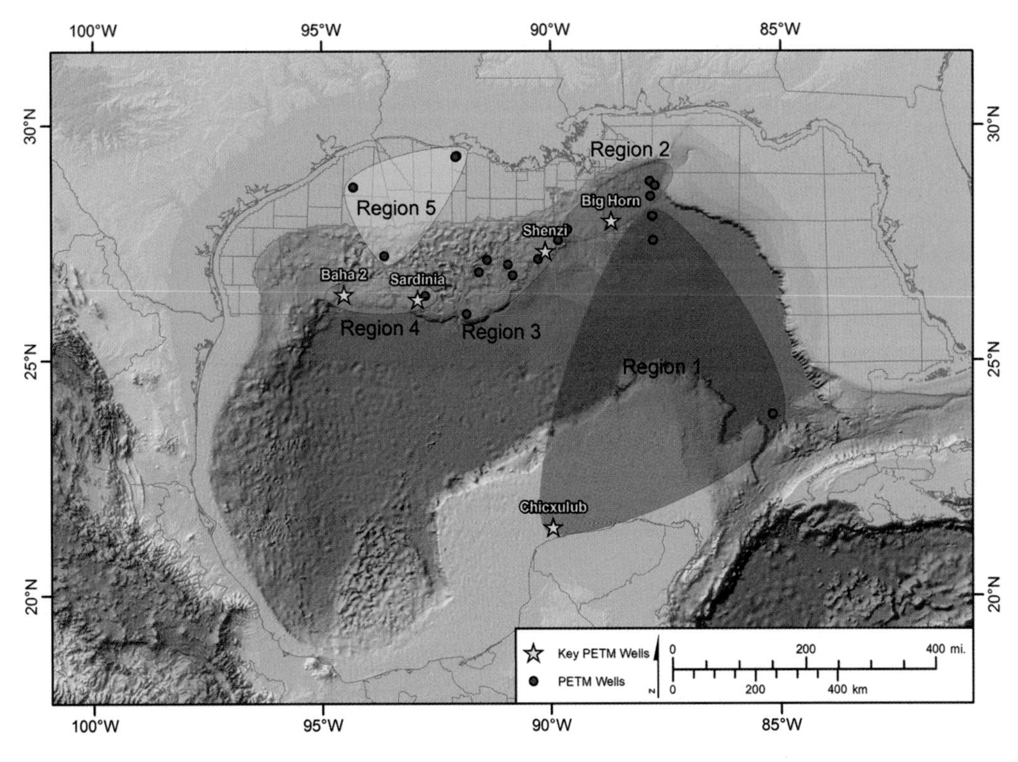

Figure 5.14 Wilcox paleoceanographic regions differentiated by analysis of faunal content in 22 deep Gulf wells. Stars show wells with most complete faunal data that typify each of the areas. From Purkey Phillips *et al.* (unpublished project report).

fine sand and silt. Wilcox rivers transported abundant coarse-to-fine sand in meandering, mixed-load channels. The observed limited grain size range of the abyssal fans requires sequestration of the coarse through fine sand load and much of the mud in contemporaneous coastal plain, coastal, shelf, and continental slope depositional systems. (4) Submarine canyons of the Wilcox evidence submarine excavation and erosion inconsistent with subaerial exposure. (5) Wilcox fan deposits were depositionally active during much of lower, middle, and early Upper Wilcox depositional episodes. Transport to the basin floor was ongoing and not constrained to a time-limited phase associated with unique paleoceanography. (6) Both faunal content and organic geochemistry record normal marine water composition and circulation, with no evidence of perturbations that would accompany isolation and partial evaporation of the saline-water mass. (7) Sea-level proxies derived from the western Gulf margin are suspect as they lie within the tectonically active foreland province where both tectonic uplift and subsidence dominate relative sea-level change.

5.6 Middle Eocene Queen City and Sparta Deposodes

By the beginning of the Middle Eocene, the Laramide uplands of the Front Range and Western Interior of the United States were deeply eroded, and their flanks buried by accumulating intermontane basin-fill. Consequently, sediment supply to the northern GoM was reduced to the lowest values since the Early Paleocene (Figure 5.1). However, Laramide tectonism, which had progressed from north to south along the North American plate, continued to impact the basins of Mexico. Supply of coarse sediment was rejuvenated in the late Middle Eocene; the Tantoyuca-Chapopote Formations reflect renewed influx of sandy to gravelly submarine flows into the forelands. Alluvial deposits of the Carroza Formation filled the remnant La Popa basin (Lawton *et al.* 2001). The Burgos basin and adjacent Rio Grande Embayment, which bridged the very different US and Mexican Gulf provinces, accumulated the thickest Middle Eocene succession, totaling more than 5000 ft (1500 m) along the continental margin depocenter. It was there that modest margin offlap occurred.

From the Burgos basin to the Mississippi Embayment, the Middle Eocene record consists of two volumetrically minor supersequences (compared to their earlier Wilcox counterparts). The Weches Formation separates the lower Queen City from the overlying Sparta genetic supersequence (Figure 5.1). The Weches is unique in northern GoM stratigraphy. It records about two million years of sediment starvation. The unit is highly fossiliferous and glauconitic, locally becoming a "green sand" exhibiting large-scale, multi-directional cross-stratification indicative of shelfal reworking and bar formation. At outcrop in east Texas, weathering of the glauconite concentrated iron sufficiently for economic iron ore extraction.

5.6.1 Paleogeography

Queen City paleogeography (Figure 5.15) is generally representative of the Middle Eocene GoM. The central and northern coastal plain remained flooded, forming a broad muddy shelf. Intra-plate stress adjustments rejuvenated the Sabine Uplift and Wiggins Arch. The Suwannee channel, though largely filled, continued to separate the pure carbonate accumulation of the Florida Platform from the muddy shelf to the north. For the first time since the Early Paleocene, a marine bay extended into the mid-continent along the length of the Mississippi Embayment. The broad shelf and funnel-shaped embayment combined to amplify tidal processes.

A mixed fluvial-/tide-dominated delta and adjacent shore zone are preserved along the northwest margin of the Mississippi Embayment and around the southern margin of the emergent Sabine Uplift (Figure 5.15). A full array of coastal and deltaic facies containing typical features of macrotidal processes characterize Queen City deposits across the east Texas outcrops and subsurface (Ramos and Galloway 1990). A broad but relatively thin delta was centered in the Houston salt basin, but sediment supply was insufficient to prograde the delta to the relict Upper Wilcox continental margin in the face on ongoing subsidence. Southwestward, a well-developed strandplain system extended to the Rio Grande axis, the largest of the Queen City fluvial–deltaic systems. In the Burgos basin, the first appearance of an integrated river system, called the Rio Bravo, is recorded by a second deltaic depocenter. This fluvial–deltaic axis confirms that the La Popa Trough and Tamaulipas Arch no longer effectively deflected large rivers draining northern Mexican uplands axially along the foreland trend or sequestered the bulk of their sediment load (Lawton *et al.* 2015). The Rio Grande delta did not prograde onto the shelf margin. However, together the two deltas provided adequate sediment to prograde the continental margin a few tens of miles. The Rio Bravo margin offlap was interrupted by subregional slope collapse, retrogradation, subsequent healing, and renewed offlap (Antunano 2009).

Following Weches flooding, the Sparta supersequence established several thin, platform deltas and marginal shore zones around the northern Gulf periphery. A renewed Mississippi delta quickly refilled the Mississippi Embayment and prograded onto the early Middle Eocene shelf. Like the Queen City, the deltas remained well shelfward of the relict continental margin except in the Burgos basin. Renewed, though still modest sediment influx along the Mississippi and Colorado fluvial axes may be explained by Middle Eocene rebound uplift in the northern Rockies (Cather *et al.* 2012). Because Middle Eocene strata are rarely differentiated in Mexico offshore literature, the Sparta-equivalent paleogeography of the peripheral Chicontepec and Veracruz foreland troughs and western GoM margin appears to be similar to that shown for the Queen City supersequence.

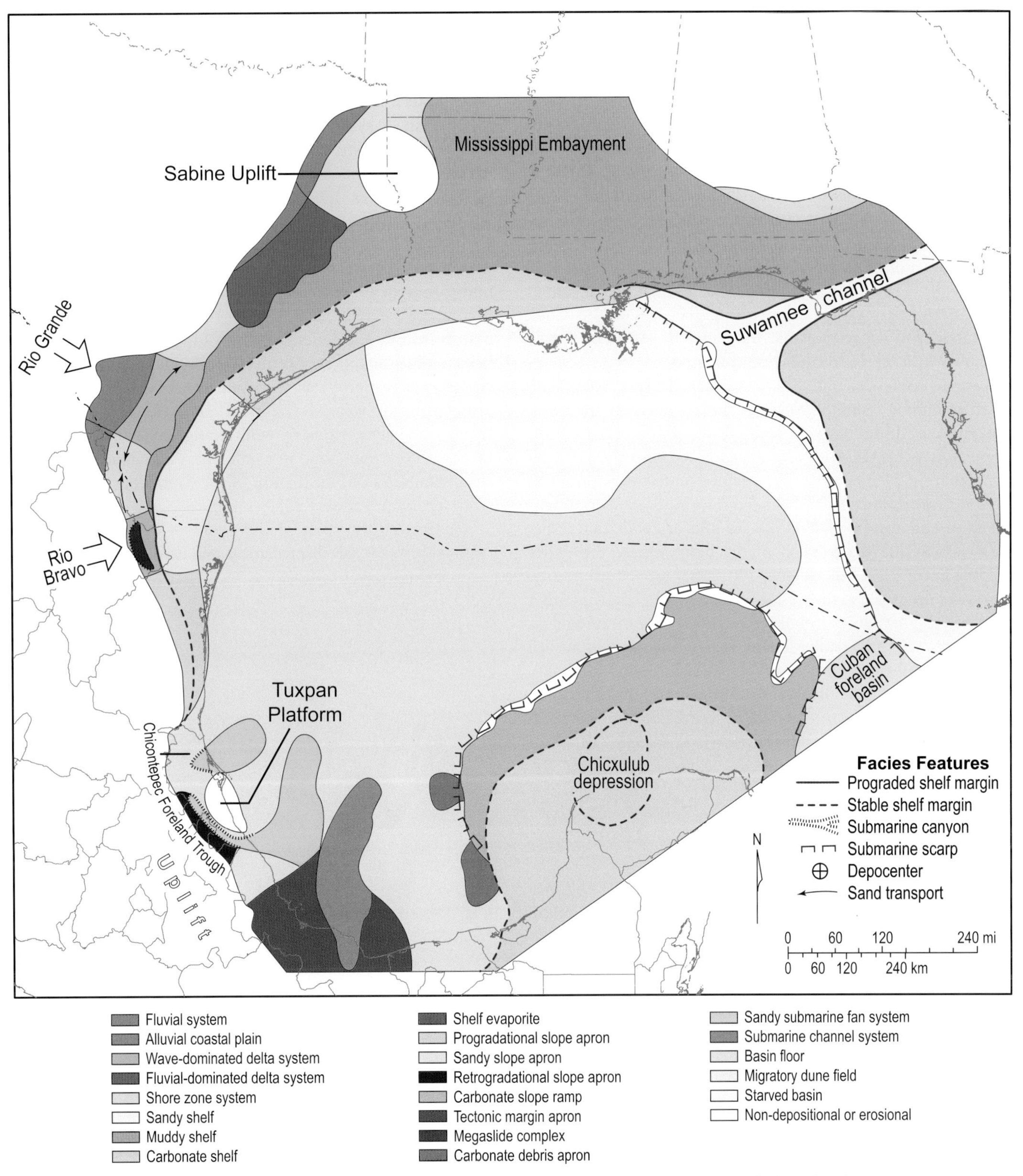

Figure 5.15 Paleogeographic map of the Middle Eocene Queen City deposode. Depositional system outlines and shelf edge reflect their positions at maximum progradation.

5.6.2 Termination and Summary

The glauconite-rich Weches Formation records an extended period of flooding and relative sediment starvation of the northern Gulf. It includes the *Chiasmolithus solitus* and *Morozovella aragonensis* LADs. Middle Eocene deposodes terminated in regional transgressive flooding of the northern basin, depositing the Cook Mountain Shale across the northern GoM. The equivalent boundary lies within the lower Tantoyuca-Chapopote interval in the Chicontepec and Veracruz basins, although exact interpretation of chronostratigraphic relationships remains somewhat fluid. Vásquez *et al.* (2014) date the mud-rich Guayabal Formation at about 38.5–40 million years, encompassing the age of the Cook Mountain and suggesting that the pause in sediment supply extended into the Mexican foreland margin.

The Queen City and Sparta supersequences, along with their bounding flooding surfaces and condensed horizons, constitute a 10-million-year interval of minimal clastic sediment influx (by GoM Cenozoic standards) into the US Gulf basin margin. Delta headlands remained on the broad, muddy, post-Reklaw transgressive shelf, and their equivalent continental slope and abyssal plain strata are thin or condensed, especially across the eastern half of the basin, where the entire Middle Eocene interval averages 100–200 ft (30–60 m) in total thickness. In contrast, the western Gulf abyssal plain contains a relatively robust succession of up to 2000 ft (600 m) of equivalent strata. This broadly lobate wedge thickens westward toward the Mexican continental slope (Galloway *et al.* 2000). Onshore Laramide deformation continued there, and tectonic uplands continued to provide active sources of both new and recycled sediment to the adjacent foreland troughs. Much of that sediment spilled beyond the troughs and into the open Gulf.

5.7 Structural Evolution

The geologically abrupt Paleocene influx of massive volumes to new sediment into the GoM basin initiated the first phase of a 65-million-year history of intra basin tectonism that rivals in diversity, complexity, rates, and magnitudes of deformation that found in any sedimentary basin, regardless of tectonic style. Recognition, delineation, and understanding of the resultant structural features have been an ongoing process spanning many decades. Regional syntheses of the northern Gulf by Diegel *et al.* (1995) and Peel *et al.* (1995) are milestones in the integration of modern concepts of gravity and salt tectonics with the abundant onshore and offshore well control and regional deep penetration reflection seismic data. As in its sedimentary history, the tectonostratigraphic evolution of the Mexican Gulf margin was dominated by crustal tectonics driven by the convergence of Pacific plates beneath the relatively narrow Mexican segment of the North American plate.

5.7.1 Northern Gulf Margin

In the northern Gulf margin clastic wedge, a family of extensional structural domains forms a broad arc extending from eastern Louisiana to the Burgos basin (Figures 5.16 and 5.17). Listric, syndepositional growth faults constitute the shallowest and most obvious structural manifestation of extension (Paleocene–Eocene detachment, cross-sections 5 and 6; Figures 1.17 and 1.18). Synsedimentary faulting was most active along the shelf margin to slope transition. Growth fault families are associated with each of the Laramide supersequences, where they broadly coincide with the updip margin of principal depocenters (i.e., Lower Wilcox, Figure 5.5). Lower and Upper Wilcox structural domain maps illustrate the general relationship between arcuate fault belts and interval depocenters. Individual faults typically coalesce and sole out at deep salt horizons, which provided a mechanically weak detachment zone. However, some fault families sole out at similarly weak stratigraphic boundaries. The Lower Wilcox fault zone in south Texas and adjacent Tamaulapas state soles out on the top of the Cretaceous (Figure 5.16; cross-section 3; Figure 1.15).

Compensatory compression was accommodated by regional primary salt displacement and deformation extending

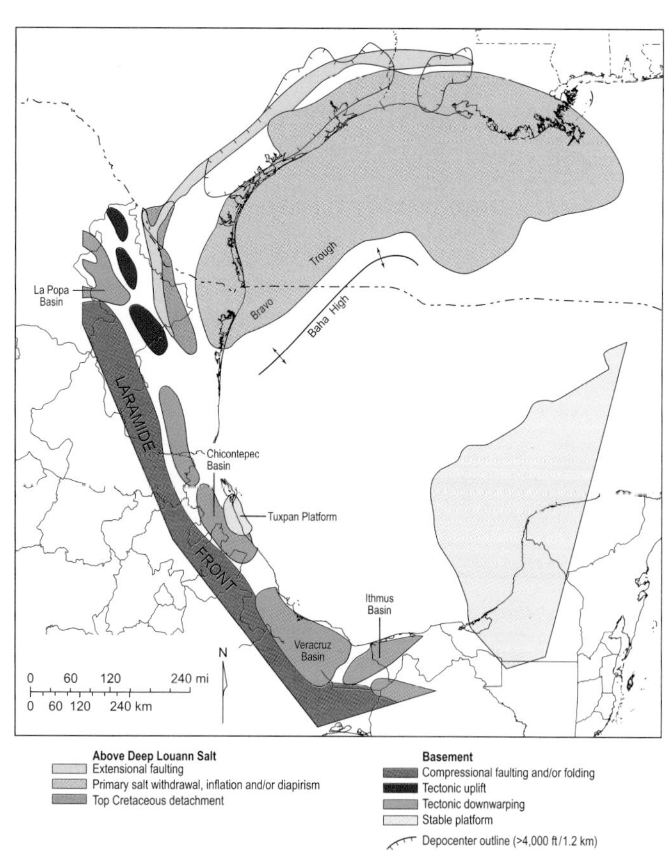

Figure 5.16 Lower Wilcox structural domains. A broadly similar array of domains persisted through the Middle to Late Paleocene. Prominent elements include the arc of extensional faulting and broad area of primary salt mobilization in the northern GoM and chain of foreland troughs along the western margin.

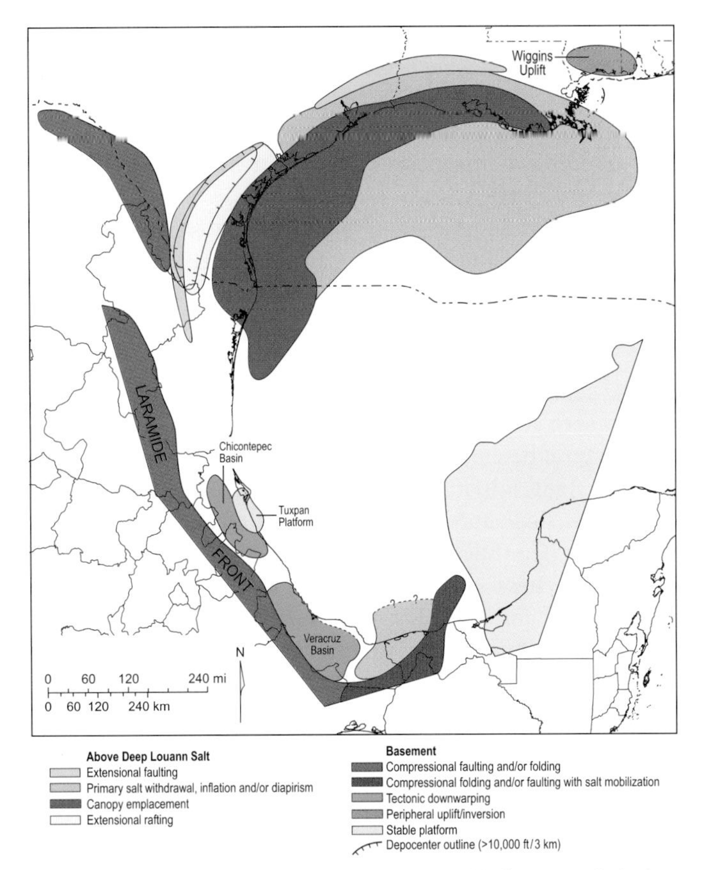

Figure 5.17 Upper Wilcox structural domains. Notable features include the arcuate domain of salt canopy emplacement, the array of peripheral tectonic elements including the Laramide front and uplifts along the Rio Grande and reactivated basement features including the Wiggins Arch and (just north of the map area) the Sabine Uplift in northeast Texas.

from the lower continental slope far out onto the deep Gulf abyssal plain (Figure 5.16). As deformation continued into the Eocene, salt was forced both out and up, initiating the first of the great Cenozoic salt canopies (Figure 5.17). Recent regional seismic surveys have revealed an inflated salt complex, up to 100 km (60 miles) or more wide and about 400 km (250 miles) long, extended along the length of the Bravo basement trough, beneath the deep Paleocene–Eocene continental rise from south Texas to the Burgos basin (Figure 5.16; Hudec *et al.* accepted). The diapiric massif likely limited accommodation space for sediment accumulation. This belt is recognized today by the seismic gap in Wilcox strata.

Upslope along the paleo-continental margin, salt evacuation upward into diapirs and basinward toward the nascent canopy created additional accommodation space for Wilcox and subsequent supersequences (Diegel *et al.* 1995). The combination of loading subsidence, extensional thinning, and salt evacuation provided the vertical space for the commanding thicknesses of sediment in the depocenters at the terminus of the principal fluvial–deltaic axes.

Direct Laramide crustal deformation played a minimal role in the northern Gulf provinces. Tilting of the proximal Gulf margin likely destabilized Paleocene and Early Eocene sediments, causing intra-formational megaslides (such as the Queen City in the Burgos basin) and regional detachment, basinward slip, and rafting in Lower and Upper Wilcox supersequences. Folding and faulting overprinted the western strata in the Burgos basin (Eguiluz de Antunano 2007). Further eastward, Early Eocene crustal stress changes reactivated the Sabine and Wiggins uplifts (Ewing 2009; Figure 5.17).

5.7.2 Western Gulf Margin

The structural fabric of the Mexican Gulf margin, from the Burgos basin to the Ithmus basin in the Bay of Campeche, was dominated by direct overprint of Laramide compression onto the Mesozoic basin margin (Alzaga-Ruiz *et al.* 2009a; Padilla y Sánchez *et al.* 2013; Padilla y Sánchez 2014). A linear chain of foreland troughs extended from the La Popa basin on the north to the Ithmus basin on the south (Figures 5.16 and 5.17). Between lay the Chicontepec and Veracruz basins. The compressional Laramide front bounded the troughs on the west. The uplands of the fold and thrust belt provided proximal sources of coarse, compositionally immature sediment. Supply was largely through short, high-gradient streams and alluvial fans. The steep tectonic slopes extended from upland directly into the subaqueous basin; gravity mass transport deposits dominate the basin-fills. Large, integrated fluvial systems with prograding coastal plains and depositional shelf margins were unlikely in such a tectonic setting. However, an integrated river with an extended drainage basin in northern Mexico flowed axially into the La Popa basin (Lawton *et al.* 2015), which lies beyond the northern terminus of the Laramide front.

A series of Late Cretaceous–Paleocene flexural arches bounded the northern troughs. The Tuxpan Platform, the southern outlier of this chain of uplifts (Figure 5.16), remained elevated above the axial trough and Gulf basin (Roure *et al.* 2009). Thus the northern foreland troughs formed a nearly continuous moat bounded on the east by the arches, effectively insulating the open Gulf from sediment input from northern Mexico uplands. This is consistent with the relatively thin (compared to the adjacent Texas segment of the Gulf margin) Paleocene and Early Eocene Wilcox successions in the Burgos basin, as well as their strike-fed coastal and marine facies architecture. As the locus of foreland compression moved into central Mexico in the Early Eocene, basin subsidence eased and the arches became inactive in the north (Figure 5.17). Fluvial input may have spilled from the La Popa basin across the remnant arch and into the Burgos/Gulf margin by the Early Eocene (De la Rocha Bascon 2016; Snedden *et al.*, 2018b); it was definitely established by the Middle Eocene Queen City deposode.

The Chicontepec, Veracruz, and Ithmus and onshore basins of the Sureste province remained active through the Middle Eocene. However, all opened at least in part to the larger GoM. Coarse gravity mass transport deposits both filled the structurally subsiding troughs and spread eastward into the GoM. The Veracruz basin was particularly active. There, sandy gravity flows constructed a broad, deepwater apron and then collected into and flowed along the north-trending Veracruz Trough, an inferred bathymetric low. In the Campeche province, Laramide compressional folding and faulting triggered initial mobilization of primary salt.

5.8 Summary: Laramide Compressional Phase

The Laramide Phase produced a 25-million-year megasequence that recorded the abrupt arrival of sediment from continental uplifts followed by a long-term decline in sediment supply and basin filling. Several regional depositional themes characterize the phase:

1. The bulk of sediment entered the western half of the basin. The eastern basin remained relatively starved, and condensed strata are widespread there.
2. Individual depositional episodes and their paleogeography were strongly controlled by evolving history and patterns of sediment supply.
3. In the northern basin, sediment input was focused through several continental-scale rivers (Galloway 2005b; Galloway *et al.* 2011). Rate and scale of continental margin offlap were greatest in the resulting largest fluvial/deltaic depocenters (Winker 1982; Galloway 2005a).
4. Widespread syndepositional gravity and salt tectonics produced a complex array of structures (Diegel *et al.* 1995; Peel *et al.* 1995).
5. The Mexican Gulf margin, in contrast, was characterized by formation and progressive filling of a chain of compressional foreland troughs.
6. Longitudinal sediment transport and filling characterized the northern La Popa and Chicontepec Troughs.
7. The southern Veracruz and Ithmus basins accumulated sediment gravity flow deposits that spilled around regional structural/bathymetric barriers onto the western GoM basin floor, initiating a long-lived assemblage of sandy tectonic slope aprons and abyssal plain channel–fan systems (Gonzales and Medrano 2014).
8. Together the foreland troughs sequestered much of the Paleocene and Early Eocene sediment derived from adjacent compressional uplifts. Nonetheless, a prominent westward-thickening sedimentary wedge lies on the western GoM abyssal plain (De la Rocha Bascon 2016; Snedden *et al.*, 2018b).

Chapter 6

Cenozoic Depositional History 2
Middle Cenozoic Geothermal Phase

6.1 Basin and Continental Framework

The Late Eocene presaged a new era in the evolution of the western North American plate. Compressional Laramide tectonics terminated. The continental landscape was remolded by regional crustal heating, volcanism, and intra-plate stress changes. New hinterland landscapes included uplifts, basins, and volcanic edifices. Important processes that impacted the GoM and its source areas included:

1. volcanism and uplift along the Sierra Madre Occidental;
2. inversion and unroofing of Mesozoic basin-fill, expanding and rejuvenating the Sierra Madre Oriental;
3. Late Eocene and Mid-Oligocene peaks in explosive volcanism in the Trans-Pecos, Mogollon, San Juan, and Great Basin volcanic/caldera complexes;
4. exhumation of basins of the Rocky Mountain front and renewed incision of Laramide uplift remnants;
5. elevation and erosion of the Edwards Plateau, adjoining the northwestern GoM basin fringe.
6. onset of tilting subsidence with peripheral elevation along the northern Gulf margin.

In addition, changes in climate impacted sediment supply to and deposition in the GoM basin. Progressive aridization across the Western Interior of continental North America (Cather *et al.* 2012) extended onto the northwest Gulf margin (e.g., Galloway 1977; Galloway *et al.* 1982c). The humid tropical climate belt was restricted to southern Mexico and Yucatán (Scotese, 2017).

The progression of Middle Cenozoic deposodes displays several regional depositional themes.

1. Sediment supply increased slowly from the late Laramide Phase minimum. Supply increased dramatically during the Oligocene and remained high into the Early Miocene as the full array of tectonic uplands matured, and continental drainage basins integrated the sources.
2. As in the earlier Laramide Phase, sediment supply was highest along the northwest Gulf margin. However, the shift of depocenters to the east-central margin was presaged by Early Miocene deposodes.
3. Several large continental rivers continued to supply the bulk of new sediment to the Gulf, building extensive, long-lived shelf-margin deltas.
4. Wave-dominated, sand-rich shore zone depositional systems attained their peak development, forming volumetrically and economically important elements of all genetic supersequences.
5. Continental margins display extensive offlap, primarily through efficient entrapment of sediment in broad, constructional continental slope aprons. Abyssal fan systems are absent or volumetrically minor components of northern GoM depositional systems tracts.
6. Continental margins prograded onto the regional salt canopy that had been mobilized by Paleocene sedimentary loading and matured during the subsequent Eocene interval of reduced slope and basin deposition. Shallow salt deformation created increasingly complex growth structures that directly modified sea floor bathymetry. This, in turn, influenced both local and regional patterns of sediment transport and accumulation.
7. The eastern Gulf margin and abyssal plain remained sediment-starved.
8. Carbonate reef and platform environments expanded from the Florida Platform to their furthest western extent on the northeast Gulf continental shelf.
9. By the end of the Eocene, the Chicontepec foreland trough was filled, and deposition along the central Mexican Gulf margin shifted to the Tampico–Misantla basin. Tilting subsidence along the continental margin initiated burial of the Tuxpan Platform beneath the prograding sedimentary wedge.
10. The Veracruz and Sureste basins continued their evolution as elements of the tectonically active margin. Tectonically over-steepened continental margins limited potential for construction or preservation of a coastal plain or continental shelf. The western Gulf margin south of the Burgos basin was characterized by near-universal submarine bypass of sediment to the continental slope and adjacent basin floor.
11. The long-lived Veracruz tectonic slope apron and derivative submarine channel depositional systems tract reached its greatest extent.

6.2 Chronostratigraphy and Depositional Episodes

The deposodes of the Middle Cenozoic Geothermal Phase extend from the latest Middle Eocene through the earliest Middle Miocene (Figure 6.1). Across the northern Gulf margin, where the genetic stratigraphic framework and chronology are well established, this interval contains five deposodes: the Yegua, Jackson, Frio, Lower Miocene 1 (LM1), and Lower Miocene 2 (LM2). Like their Laramide precursors, each deposode terminated with regional transgression and flooding of the continental platform. Functional stratigraphic boundaries for each genetic supersequence include marine shale tongues, widespread microfaunal markers, and condensed horizons. Because most marine strata do not extend to outcrop, unlike their older counterparts, they have no regionally consistent formal stratigraphic nomenclature.

The deposodes are of quite variable duration and volumetric importance (Figure 6.1A). The shortest deposode, the Yegua, lasted less than two million years. The Frio deposode bridges the entire 10+ million years of the Oligocene, and is by far volumetrically dominant.

Along the Mexican margin, with its strong tectonic overprint, regressive/transgressive cycles as currently described are more variable in timing and duration (Figure 6.1B). In the Chicontepec and Tampico–Misantla basins, the Tantoyuca regression corresponds relatively well with the combined Yegua–Jackson interval. Similarly, the Oligocene Palma Real regression, bounded by the marine Horcones and Alazán Shales, correlates with the Frio deposode. The Escolin–Coatzintla Formation encompasses the combined LM1–LM2 interval. Further south, however, basin stratigraphy clearly reflects the dominating impact of tectonically modulated sediment supply, basin deformation, and intra-basin erosion. In

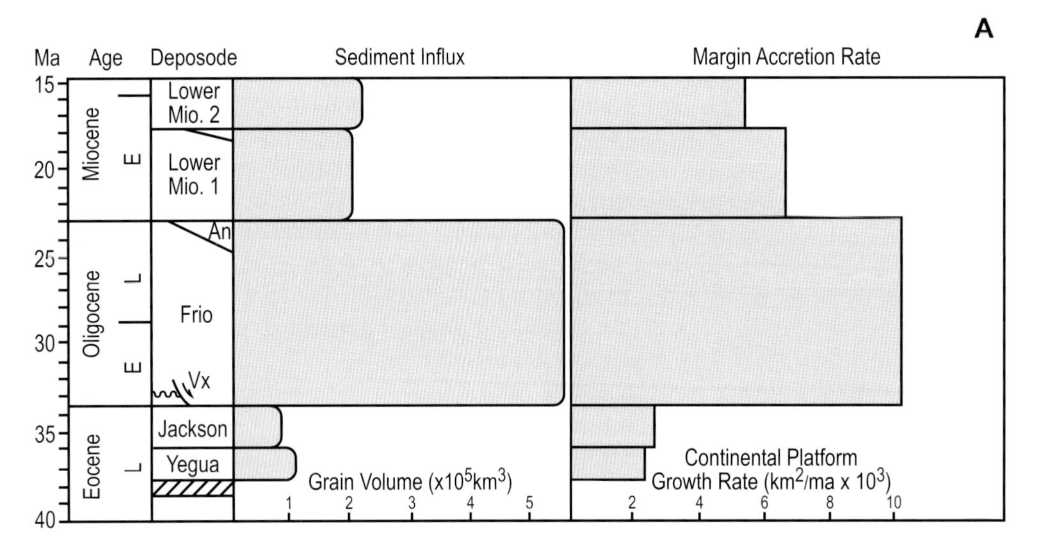

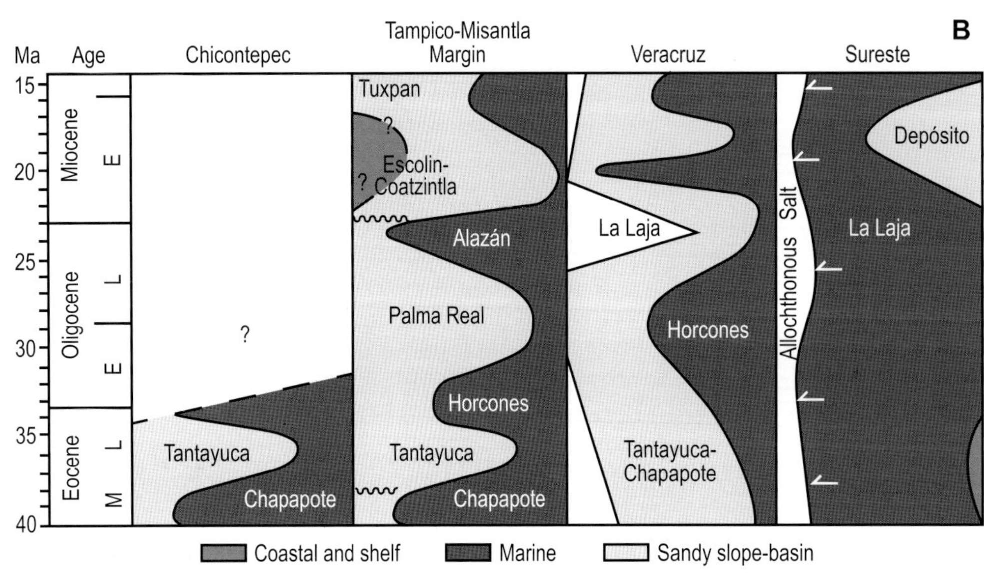

Figure 6.1 Depositional episodes and chronology of the Middle Cenozoic Geothermal Phase. **(A)** Northern Gulf (US) margin deposodes, sediment influx, calculated as total grain volume, and continental margin accretion rates. **(B)** Stratigraphic units of the Mexican Gulf margin. Patterns distinguish terrestrial, coastal and shelf, deep marine, and sandy, deep marine facies associations. Expanding continental, shallow marine, or sandy marine facies indicate regressive episodes due to increased sediment influx or to uplift to the basin margin and consequent forced regression. The Sureste area includes the Macuspana, Comalcalco, and Ithmus basins.

Campeche, the Sureste basin experienced the first incursion of sandy sediment into the southern Gulf, the Early Miocene Deposito Formation.

6.3 Previous Work

Studies of units of the Middle Cenozoic phase soon benefited from the concepts and techniques developed in the Wilcox syntheses. Regional mapping combined the extensive, public domain, subsurface database created by decades of petroleum exploration and development, description of accurately mapped outcrops, and application of modern process sedimentology and depositional facies analysis to outcrop, mine, and core exposures.

The first of many Texas Bureau of Economic Geology (BEG) depositional systems analyses to follow upon the Wilcox project focused on the Yegua and Jackson Groups (Fisher *et al.* 1970). Here, uranium was added to the mix of energy resources found in northwestern GoM units. The Jackson outcrop was the site of active uranium mines. In addition, the unit hosts petroleum and lignite resources. Outcrop to mid-depth Yegua depositional systems were interpreted and mapped as a framework for evaluating lignite resources of the Texas coastal plain (Kaiser *et al.* 1980). Yegua genetic supersequence correlation and mapping (Meckel and Galloway 1996) completed regional paleogeographic mapping and interpreted the depositional history of this, the first of the Middle Cenozoic supersequences beneath the Texas coastal plain.

A new phase of regional correlation, mapping, and depositional systems delineation of Middle–Late Cenozoic supersequences at the BEG began with the Oligocene (including both Vicksburg and Frio formations) interval. In this and subsequent studies, genetic supersequence stratigraphic correlation used widely recognized marine shales and their contained foraminiferal marker events for definition of depositional episodes. Initial work was focused on the deep Frio as a possible geopressured geothermal resource (Bebout *et al.* 1978; Loucks 1978). As the focus shifted to conventional petroleum resources, the Frio was selected as a natural laboratory for regional geologic synthesis, analysis of genetically defined oil and gas plays, and quantification of resource potential. Galloway *et al.* (1982a, 1982b) divided the Oligocene interval into three "operational units" for regional correlation and lithofacies mapping across the breadth of the Texas coastal plain and inner continental shelf. Results were combined to create an interpretive paleogeographic reconstruction and history. Galloway (1986b) updated deep Frio continental margin interpretations and maps using deep wells along the modern coast and inner continental shelf. Combes (1993) differentiated the Early Oligocene Vicksburg unit across the Texas–Louisiana coastal plain, and mapped lithofacies and depositional systems to determine their relationship to petroleum plays. Hernandez-Mendoza (2000) carried Frio–Vicksburg genetic supersequence mapping and interpretation into the Burgos basin. More recently, a series of studies at the BEG have focused on highly detailed correlation and mapping of high-frequency sequences in the Corpus Christi Bay area, south Texas (Brown *et al.* 2004). The studies have employed system tract terminology and models.

The systematic analysis then moved up-section to the Lower Miocene genetic supersequences. Rainwater (1964) had provided a systematic overview of Miocene stratigraphy and deposition across the northern GoM. Because coastal plain outcrops become increasingly sparse and non-marine, Neogene stratigraphy and chronology relies almost exclusively on subsurface correlation and micropaleontologic dating. To provide a stratigraphic framework for future work in the northwest Gulf, Morton *et al.* (1985) compiled a suite of correlation cross-sections, importantly bridging the data discontinuity commonly created by the modern shoreline. Using this framework, Galloway *et al.* (1986) completed the regional synthesis of the Texas margin paleogeography. Comparable regional syntheses for the northeastern and western Gulf margins are largely lacking. However, Early Miocene strata are included in the Veracruz basin study of Jennette *et al.* (2003).

6.4 Late Eocene Yegua and Jackson Deposodes

The Yegua and Jackson deposodes were both minor players in the context of northern GoM depositional history. Sediment volumes of each supersequence averaged 100,000 km^3 (24,000 miles^3), and rates of continental platform accretion area were comparably low (Figure 6.1). Because their depositional paleogeographies are generally similar, and Late Eocene strata are rarely subdivided in the literature of the Mexican margin, the Yegua paleogeographic map will be used for discussion of both genetic supersequences. Presence of volcanic ash beds in these supersequences has allowed refined dating using outcrop samples and traces their origin to caldera complexes in northern Mexico (Yancey *et al.* 2018). This connection demonstrates more than 800 km (500 miles) of northeastward transport of fine sand-sized grains in volcanic ash plumes.

6.4.1 Paleogeography

Three extensive fluvial–deltaic axes prograded across the broad Eocene continental platform of the northern Gulf (Figure 6.2; Meckel and Galloway 1996; Ewing and Vincent 1997). Two, the Mississippi and Houston rivers, constructed the fluvial-dominated Cockfield (named for the stratigraphic name applied to equivalent strata in Louisiana) and the Liberty deltas. The Rio Grande, flowing across the south Texas coastal plain, constructed the wave-dominated Falcon delta. The fourth, a small wave-dominated delta of the Rio Bravo river prograded into the Burgos basin. The Liberty and Falcon deltas build the principal depocenters, and modest shelf-margin offlap occurred in front of both deltas. Along-strike shoreface and shelf reworking from the Falcon deltaic headland combined with sediment supplied by the comparatively smaller Rio Bravo to extend margin offlap into the Burgos basin margin.

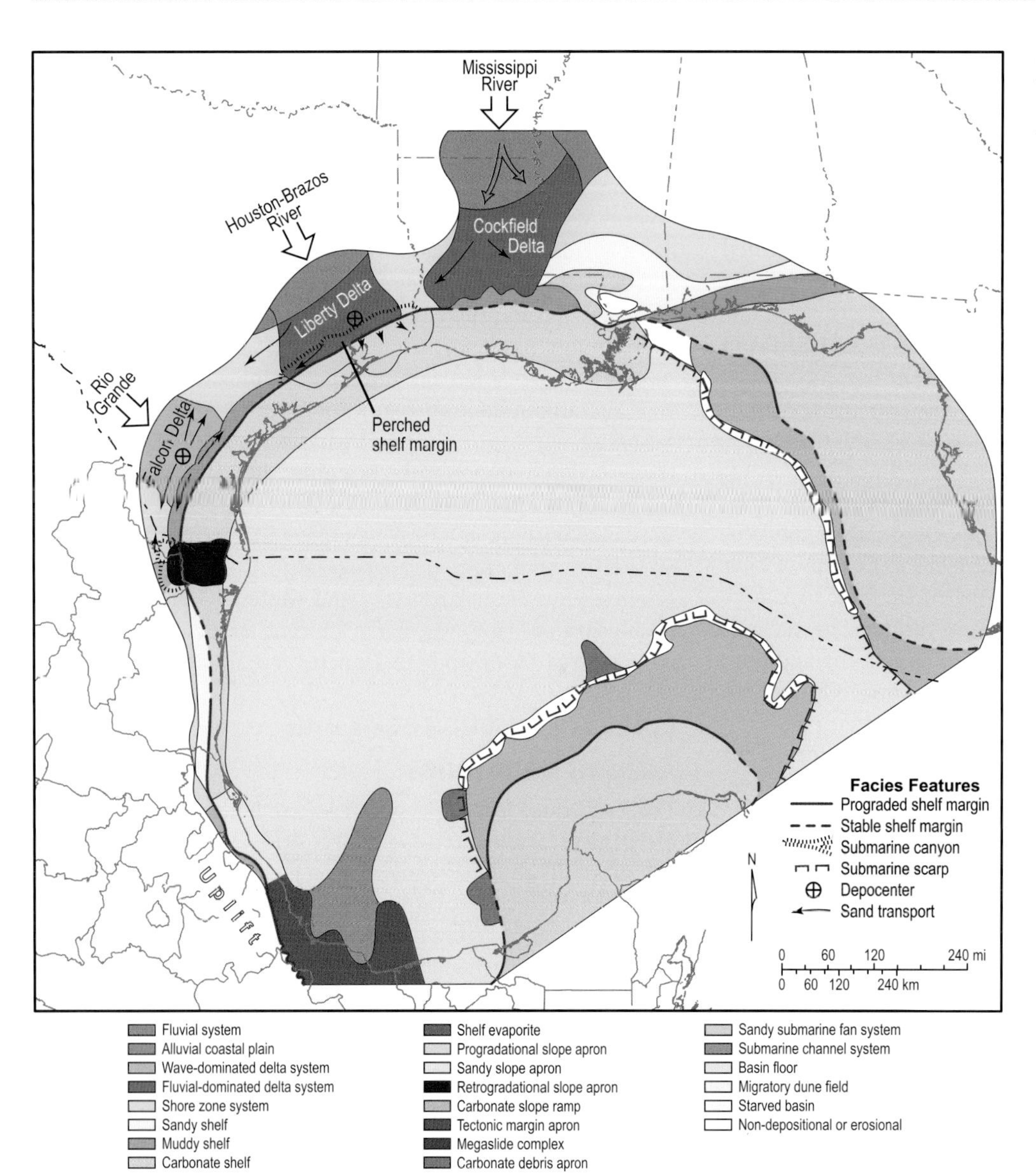

Figure 6.2 Paleogeographic map of the Late Eocene Yegua deposode. Depositional system outlines and shelf edge reflect their positions at maximum progradation.

The Cockfield delta prograded across the broad, transgressive Early Eocene shelf, which had expanded northward into the mouth of the Mississippi Embayment (and far onto stable continental crust). The delta remained on the depositional platform created by earlier deposodes. Declining supply during the subsequent Late Eocene Jackson deposode resulted in a second transgressive flooding of the embayment and deposition of thin, muddy shelf facies across the entire central Gulf continental platform. Between the Texas deltas, well-developed strandplain and barrier/lagoon shore zones deposited a sandy rim across the breadth of the northwest margin.

Although long-term shelf edge progradation characterized the northwestern Gulf margin, offlap was punctuated by regional collapse and depositional healing along both the Burgos delta front (Eguiluz de Antunano, 2007) and the Liberty delta front (Edwards 1991, 2000; Ewing and Vincent 1997). The two events had differing timing and probable causes. The Burgos margin collapse occurred within the deposode, creating an intra-Yegua supersequence unconformity. Margin failure was likely precipitated by uplift and tilting of the immediately adjacent upland that extended into the Gulf margin itself. Following the failure, renewed progradation healed the retrogradational embayment. The Liberty margin collapse occurred with initial progradation of the delta system onto the long-relict Upper Wilcox continental shelf margin. Steepening of the bathymetric slope across this foundered margin, which had remained depositionally moribund for about 10 ma, created intra-formational instability. The veneer of Middle Eocene mud combined with an evacuated salt weld provided a mechanically weak layer. More than 1000 ft (338 m) of expanded delta front and prodelta sediments prograded rapidly onto the shelf margin, triggering a coalesced series of slumps and shallow growth faults along the entire distal periphery of the delta system (Figure 6.2). Ultimately, the Yegua delta built onto and across these perched slumped deposits and onto the continental slope, constructing a depositional slope apron. The Liberty delta deposits are also notable for several incised fluvial valleys that cut tens of miles across the shelf mud that separated the highstand delta depocenter from the depositional shelf edge (Ewing and Fergeson 1991; Fang, 2000).

The latest Eocene Jackson deposode continued accumulation of the Texas deltas and produced what is probably the best-documented example of the strike-fed barrier/lagoon facies tract so prominent in the Middle Cenozoic Geothermal Phase.

Open-pit uranium mining in south Texas exposed the three-dimensional array of shoreface, barrier, inlet, and lagoonal facies characteristic of the system (e.g., Galloway *et al.* 1979). The Yegua and Jackson supersequences both contain significant commercial lignite deposits in the fluvial-dominated delta and adjacent shore zone systems (Kaiser *et al.* 1980). These are the youngest commercial lignite deposits in the basin; increasing mid-Cenozoic aridity terminated widespread marsh/swamp formation and preservation of organic matter.

The Mexican Gulf margin displays two broad depositional systems tracts (Jennette *et al.* 2003; Ambrose *et al.* 2005). Extending south from the Burgos basin, the Tampico–Misantla margin developed a narrow wave-dominated shore zone and sandy shelf constructed of reworked sediment from short, steep streams arising in adjacent uplands. Small wave-dominated deltas occurred at the south end of this system, where shelf edge offlap is suggested. Numerous slope channels and gullies extend down the steep continental slope, creating a broad sandy apron. The extent of sandy flows onto the basin floor is speculative. In the Veracruz basin, uplift of the basin margin resulted in incision of numerous submarine canyons and bypass of sediment directly from the uplands onto the lower slope and basin floor. The tectonic apron and north-flowing Veracruz Trough submarine channel system, both already in place, continued to be the principal sites of sediment transport and deposition.

6.4.2 Termination and Summary

The Yegua deposode was terminated by deposition of thin transgressive shelf muds. The Moodys Branch Marl (eastern Gulf) and Caddell Shale (northwestern Gulf) cap the Yegua supersequence. Faunal markers are few; the *Camerina moodysbranchensis* last appearance datum (LAD) is useful. The termination of the Jackson deposode and onset of Oligocene deposition produced one of the more dramatic and, at the same time, complex stratigraphic boundaries of the Gulf. Initial transgression of Jackson delta and shore zone systems was immediately followed by relative sea-level fall that included both eustatic and, in the northwestern Gulf margin, tectonic components. In the thin but relatively fossiliferous outcrop and shallow subsurface interval in Mississippi, the Eocene–Oligocene boundary is closely approximated by a maximum flooding surface, condensed interval, and superjacent disconformity dated as Early Oligocene (Echols *et al.* 2003). Clustered micropaleontologic LADs include *Hantkenina alabamensis, Uvigerina cocoaensis, Discorbis cacoensis, Globorotalia cerroazulensis, Globorotalia cocoaensis*, and *Marginulina cocoaensis*.

The Late Eocene deposodes were a transitional interval between the Laramide Phase and the ensuing eastward-progressing tidal wave of sediment supply recorded by the Oligocene Frio genetic supersequence.

6.5 Oligocene Frio Deposode

The Frio deposode constitutes one of the signature chapters in the depositional history of the GoM. The immense, supply-modulated influx of sediment is reflected in the genetic supersequence volume, distinctive mineralogies, and coarse sand and gravel content. The resultant supersequence includes the Catahoula, Vicksburg, Frio, and Anahuac stratigraphic units in the northern and northwestern Gulf and the Palma Real and bounding Hercones and Alazán strata along the Tampico–Misantla margin (Figure 6.1). Only Veracruz basin stratigraphy displays a disparate history and stratigraphic architecture. There, the La Laja Formation records a late Oligocene pulse of basin margin erosion and sediment influx.

Oligocene tectonism directly impacted the northwest Gulf basin-fill. In the Burgos basin, an unconformity separates the proximal Vicksburg from the overlying Frio (Hernandez-Mendoza, 2000). In adjacent south Texas, tilting destabilized the Early Oligocene Vicksburg succession, initiating shallow detachment at the Upper Eocene level. Consequent extensional faulting (the Vicksburg fault zone) dramatically expanded deltaic and upper slope facies successions and displaced them many miles basinward along the detachment (Langford and Combes 1994). A bit further northward, in the south Texas uranium province, subtle tilting subsidence created the low-angle discordance between Upper Eocene Jackson strata and overlapping basal Oligocene Catahoula fluvial sands that was observed in open-pit mines and closely spaced exploratory drilling (Galloway 1977). Duration of unconformity development is limited to less than one million years by zircon dating of Jackson and Frio ash beds (Yancey *et al.* 2018). Together, the updip Frio and its equivalent units record tectonic uplift and truncation in Tamaulipas state on the west, and a broad halo of more subtle effects of uplift and tilting that extended as far as the central Texas coastal plain.

Basal Frio/Vicksburg progradation was also impacted by both regional and global Eocene–Oligocene climate change and consequent global eustasy (Yancey *et al.* 2003; Miller *et al.* 2005). Impact is subtle in the northwestern Gulf, where tectonism and a high rate of initial sediment influx overwhelmed the eustatic signal. However, on the east Texas coastal plain, beyond the direct impact of tectonism and where Vicksburg sediment supply was comparatively moderate, the basal Vicksburg fluvial system excavated a valley that extends from the outcrop across tens of miles of underlying muddy shelf strata to the mid-dip delta system (Gregory 1966). Outside of the valley fill, a widespread, mature paleosol lies at the base of the Catahoula outcrop (Galloway and Kaiser 1980).

Tectonic inversion, Laramide basin exhumation, and regional explosive, caldera-forming volcanism across the southwestern USA and Mexico had multiple and diverse impacts on deposition in the GoM (Figure 6.3).

1. Although sediment clastic influx expanded as far as the Louisiana coastal plain during the Early Oligocene, its first arrival, volume, and composition varied from west to east. The Vicksburg Formation (ca. 34–32 Ma) contains the record of this progression (Combes 1993). In the Burgos and south Texas depocenter, the Vicksburg interval is thick, sand-rich, and rests directly on Late Eocene Jackson deposits. In the Upper Texas coastal plain, the Vicksburg is relatively thin, containing at most a few parasequences. In Louisiana, the Vicksburg is a shelf mudstone, reflecting continued shallow submergence of the continental platform that began with Late Eocene Jackson abandonment of the Mississippi fluvial axis.
2. Accumulation rates varied regionally during the 10-million-year deposode (Galloway and Williams 1991). The deposode ended with the long-term (approximately two million years) Anahuac Shale retrogradation, which culminated in regional maximum flooding. At this time, the *Heterostegina* limestone reefs extended westward into central Louisiana (Figure 6.3).
3. Multiple large, bedload-rich rivers (Galloway 1977) delivered sediment to the northwest and central Gulf, focused into four major fluvial–deltaic axes (Figure 6.3). These include the Mississippi, Houston, Rio Grande, and Rio Bravo. The three Oligocene depocenters are directly associated with these fluvial–deltaic axes. The Rio Grande

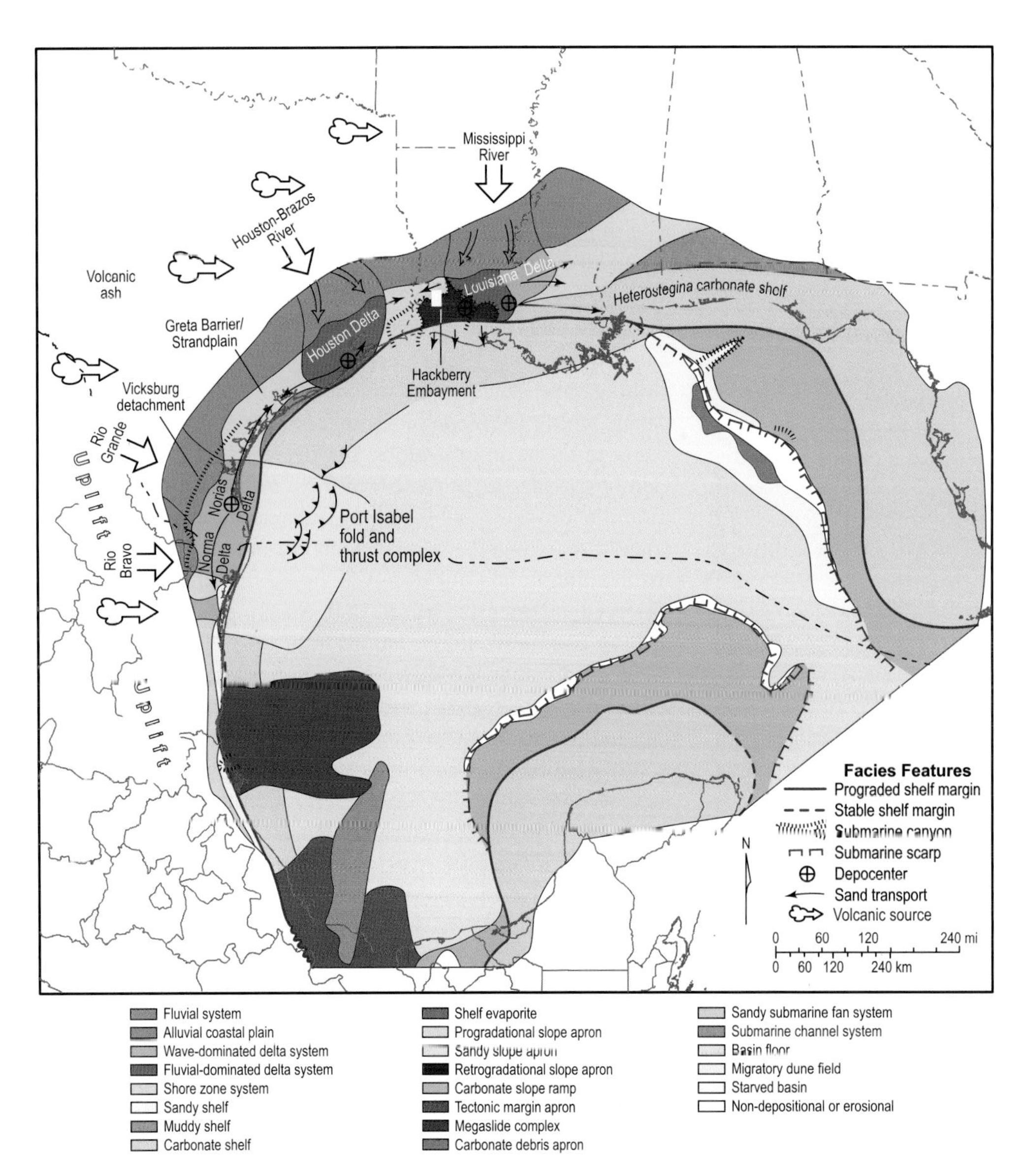

Figure 6.3 Paleogeographic map of the Oligocene Frio deposode. Depositional system outlines and shelf edge reflect their positions at maximum progradation. White box locates seismic line shown in Figure 6.4.

and Rio Bravo converged to supply a single, composite depocenter.

4. Explosive eruptions and accompanying caldera formation in the volcanic fields provided sediment in two ways (Galloway 1977). Erosion of the volcanic edifices and ejecta supplied sand and gravel directly to the drainage basins of the rivers, most notably the Rio Grande system. Volcanic ash that was swept east across continental North America by high-altitude winds blanketed the drainage basins of all the major rivers. Rapid weathering and pedogenic alteration of the airfall ash produced smectite-rich clay. Easily eroded ash and clay were, in turn, washed into rivers, adding additional suspended load. The Mississippi fluvial system, furthest removed from the western upland and volcanic sources, but collecting the airfall ash and clay that repeatedly blanketed its mid-continent drainage basin, built a particularly mud-rich depocenter.
5. Frio sands are mineralogically diverse (Galloway 1977; Loucks *et al.* 1986). A strong west-to-east decrease in abundance of volcanic grains and plagioclase feldspar (and commensurate increase in quartz) parallels the increasing distance between river drainage axes and the Oligocene volcanic uplands.

6.5.1 Paleogeography

Four principal delta systems, produced at the terminus of each of the four continental rivers, rim the northern GoM basin (Figure 6.3). Deltas of the Rio Bravo and Rio Grande fluvial systems merged to form the composite, sand-rich, wave-dominated Reynosa–Norias system, which bridges the Burgos basin in Tamaulipas and Rio Grande Embayment of south Texas. The coarse Rio Bravo fluvial channel fill complex constitutes the Norma Conglomerate, which extends from outcrop into the subsurface (Hernandez-Mendoza 2000). The south Texas outcrop and shallow subsurface Gueydan (Catahoula-equivalent) Formation consists of deposits of the Rio Grande fluvial system, which was both sandy bedload- and volcanic ash-rich (Galloway 1977). Together, these rivers constructed a robust, highly progradational continental margin that buried the basal Oligocene Vicksburg fault zone and advanced the shelf edge 95–145 km (60–90 miles) basinward of its Eocene position. Rapid accumulation triggered a succession of arcuate growth faults that advanced basinward in tandem with shelf edge progradation, greatly expanding upper slope and shelf-margin delta facies (the Frio growth fault belt). Total Frio–Vicksburg thickness exceeds 12,000 ft (3600 m) along the trajectory of the prograded continental margin from the Burgos basin to western Louisiana.

In southeast Texas, the Houston River flowed southward onto the Frio coastal plain, prograding the Houston delta. This large, fluvial-dominated delta system accomplished 45–65 km (30–40 miles) of continental margin offlap. In the depocenter, combined loading subsidence, substrate compaction, and salt evacuation accommodated more than 12,000 ft (3600 m) of section.

The Mississippi River, flowing southward from the mid-continent, constructed a large delta that spread across much of the central Louisiana coastal plain. Both the Mississippi and Houston were mixed-load rivers, but the relatively high mud: sand ratio of the Louisiana depositional systems indicate that the Mississippi transported an abundance of reworked airfall volcanic ash in its suspended sediment load. The axial delta was fluvial-dominated. However, on the eastern margin of the Mississippi delta system, sand distribution reflects significant marine reworking of the delta front as it merged into the adjacent sandy shore zone. Although its drainage basin was furthest removed from western continental uplands, the river system was large and it transported copious sediment. Continental margin offlap along the advancing delta front exceeded 50 km (30 miles), and the western side of the deltaic depocenter is more than 12,000 ft (4000 m) thick.

A broad, mixed sand–mud strandplain was constructed on the coastline between the Houston and Mississippi deltas (Figure 6.3). Progradation of this strandplain, along with the western margin of the Mississippi delta, was interrupted in Middle Oligocene (ca. 26 Ma) by collapse, slope failure, and evacuation of a large volume of the shelf margin, creating the Hackberry Embayment (Cossey and Jacobs 1992; Edwards, 2000; Galloway, 2005b). Retrogradational failure was likely triggered by rapid evacuation of salt due to sediment loading of the subjacent Eocene canopy and consequent oversteepening of the continental margin profile. The stratigraphic consequence was insertion of a wedge of muddy, deepwater sediment tens of miles updip into the offlap succession of shelf and strandplain/deltaic facies typical of the Frio offlap margin (Figure 6.4). This wedge presents a classic record of a retrogradational margin: (1) a canyon-cut basal unconformity initiated by amalgamation of slump scars; (2) an onlap wedge of slope sediment containing turbidite channel and overbank deposits; and (3) a capping succession of progradational upper slope and shelf/coastal deposits that healed the embayment and resumed regional margin offlap.

Strike reworking of sediment from the Norias and Houston deltaic headlands supplied abundant sand to the intervening shoreface. Convergence of longshore drift in the broad, interdeltaic coastal bight and the abundant supply of sand constructed the Greta barrier–strandplain system, named for the immense sand body that caps the Oligocene Frio supersequence of the middle Texas coastal plain (Galloway *et al.* 1982a). This shore zone system extends more than 240 km (150 miles) along-strike and contains at its thickest more than 5000 ft (1500 m) of sand that was deposited in the full suite of barrier island, strandplain, and sandy shelf environments (Tyler and Ambrose 1984; Galloway 1986a). Stacked, amalgamated coastal sand bodies prograded more than 80 km (50 miles), following the advance of the flanking deltaic headlands and the continental shelf edge.

Detailed sand mapping within the Frio sequence (Galloway *et al.* 1982b) quantified the dramatic facies changes that occur across the coastal plain/lagoon/barrier island/shelf

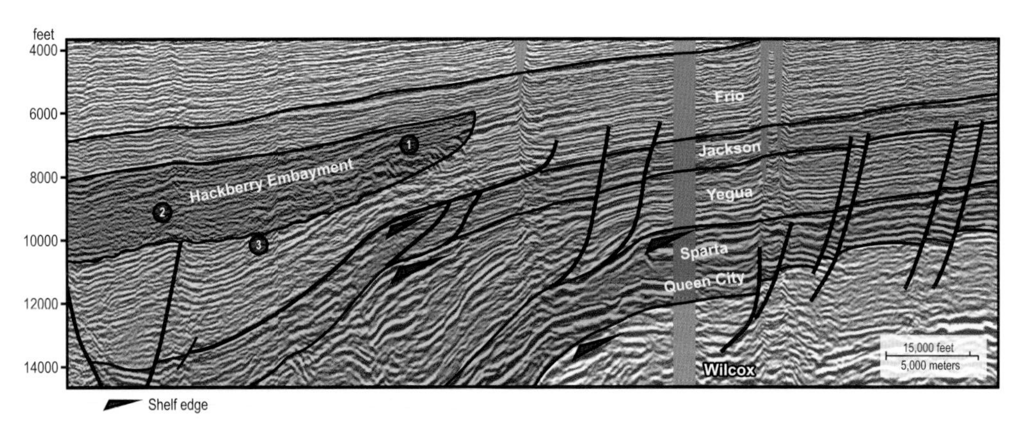

Figure 6.4 Seismic line interpretation showing expanded Frio supersequence produced by offlap over the subjacent, stacked Eocene shelf edges and onto the Paleogene salt canopy. Destabilization of the Frio slope by deep salt evacuation triggered slope failure that initiated the formation of the retrogradational Hackberry Embayment within the middle Frio succession. Imaged features typical of retrogradational margins (Figure 1.37B) include (1) rotated slump blocks of lower Frio along the headwall scarp; (2) chaotic seismic facies characteristic of the sandy lower embayment deposits; and (3) erosional truncation of underlying structure. For location, see Figure 6.3. Seismic line courtesy of ION.

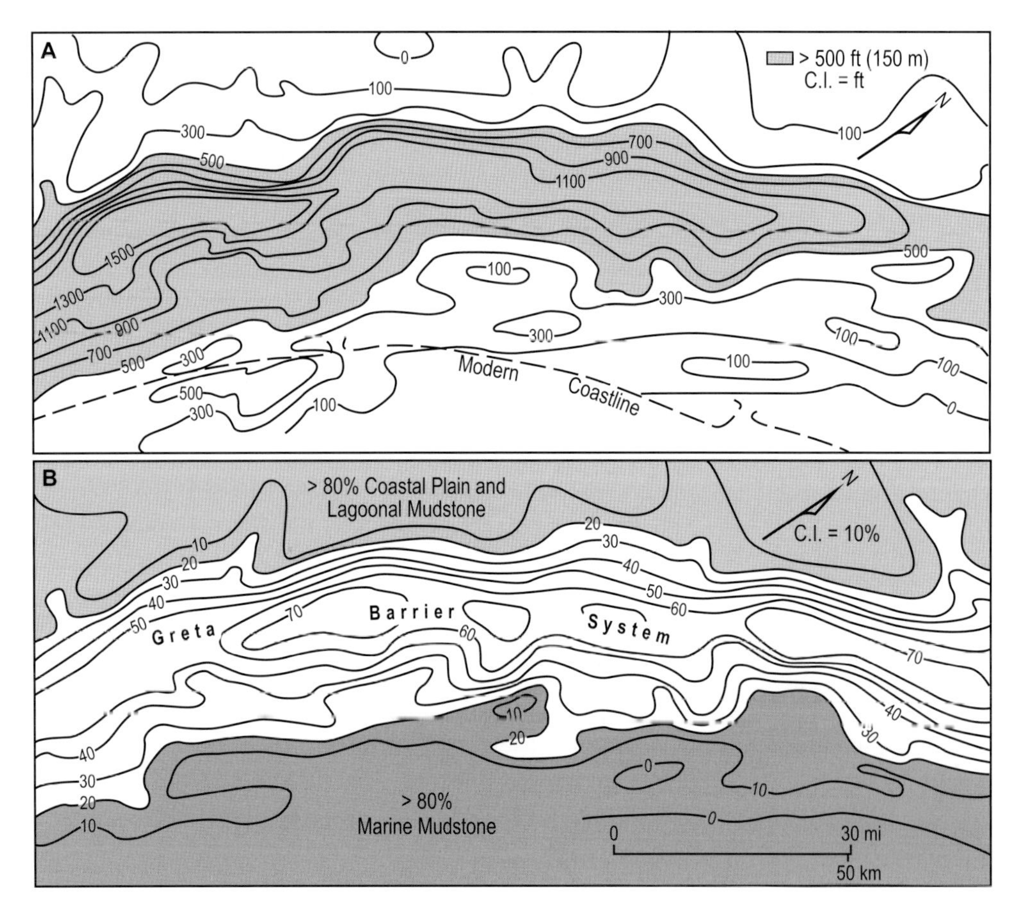

Figure 6.5 **(A)** Net sand thickness map for the middle correlative unit within the Frio genetic supersequence, axial Greta barrier/strandplain system, central Texas coastal plain. Position of the modern coastline provides a geographic reference. **(B)** Sand percentage for the middle correlative unit of the axial Greta barrier–strandplain system. Modified from Galloway *et al.* (1982b).

depositional systems tract (Figure 6.5). Vertical stacking of barrier facies created narrow, strike-elongate sand belts more than 1000–1500 ft (300–450 m) thick. Along the coastline depoaxis, sand constitutes >60 percent of the total map interval. Sand thickness and proportion decreases dramatically both landward and seaward of the shoreline axis. Basinward, shelfal mud replaces coastal sand facies; sand percentage decreases to <10 percent. Lagoon-ward, net sand percentage is commonly 20 percent or less. Digitate, dip-oriented sand fingers within the mud-dominated lagoon and coastal plain facies are channel

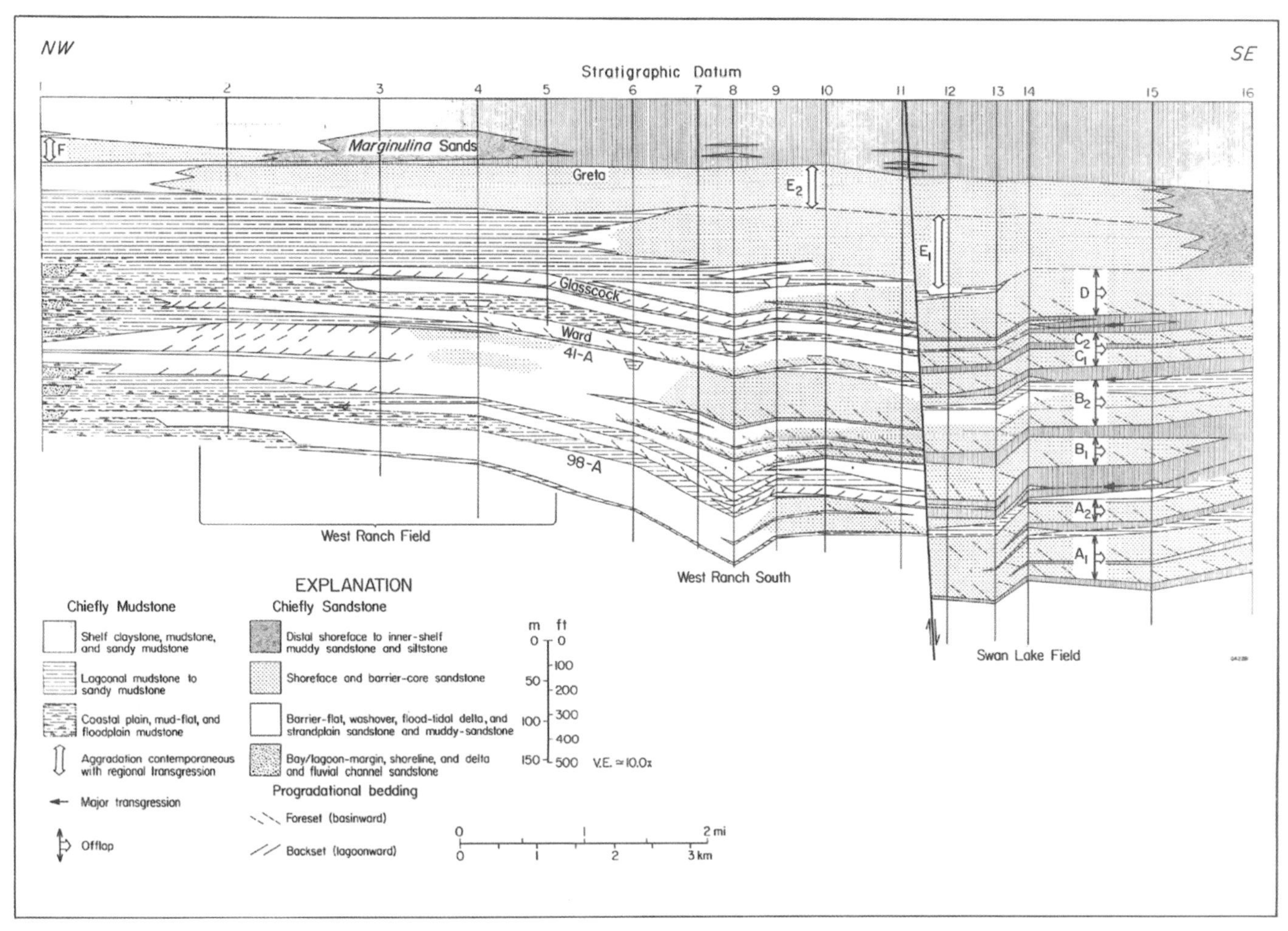

Figure 6.6 Detailed sand body correlation illustrating the facies architecture of the Greta barrier system. Successive aggradational parasequences A_1–D record repeated progradation of shoreface and barrier sands along a relatively stable long-term middle Frio coast. Over-thickened units E_1–F reflect retrogradational retreat of the coastline in the Late Oligocene. The lagoonal mudstone facies is best developed behind these barriers. The Marginulina sands are amalgamated inner-shelf storm deposits. The stratigraphic datum approximates the maximum flooding surface within the capping Anahuac Shale. 98-A, 41-A, Ward, Glasscock, and Greta sands are principal reservoirs of the giant West Ranch Field. From Galloway (1986a).

fills and lagoonal deltas of small, basin margin rivers that traversed the Catahoula coastal plain (Galloway 1977). High-resolution correlation of middle–upper Frio parasequences using the abundant well control in the giant West Ranch petroleum field (Galloway 1986a) illuminates the stratigraphic architecture of the barrier sand and bounding lagoonal and shelf mud facies (Figure 6.6). Initially, thick, aggradational back barrier, barrier core, inlet fill, and shoreface sands (units A_1 through C) record a persistent shoreline axis that prograded or transgressed only a few miles landward or basinward during accumulation of nearly 1000 ft of section. Depositional style changes with unit D. Thick, highly aggradational (100–200 ft/30–60 m thick) sand bodies reflect an extended late Frio interval of near-balance between sediment supply by longshore drift and ongoing subsidence. During deposition of unit E_1, the stable shoreline remained within a belt only about 5 km (3 miles) wide. Equally thick lagoonal and shelf muds accumulated on the landward and seaward side of the aggradational barrier complex. Unit E_2 (the widely productive Greta sand) caps the local interval, and records slow retrogradation of the barrier. Shallow shelf muds above the Greta grade landward into the shelfal *Marginulina* sands, which, in turn, thin and merge with distal shoreface sand of Frio unit F.

Box 6.1 Growth Faults and Interdeltaic Depositional Systems Tracts

Both the regional paleogeography and detailed structural, depositional facies, and stratigraphic architectures of the Oligocene Frio continental shelf and margin have been documented in the Corpus Christi area of south Texas (Galloway *et al.* 1982b; Galloway 1986a; Galloway and Morton 1989; Olariu *et al.* 2013; Figure 6.7). Here, detailed work incorporating closely spaced well log, core, and seismic data delineates multiple regressive depositional cycles across three structural subbasins (Figure 6.8). Each subbasin is bounded up and down dip by major growth faults that are elements of the regional Frio domain of extensional

Box 6.1 (*cont.*)

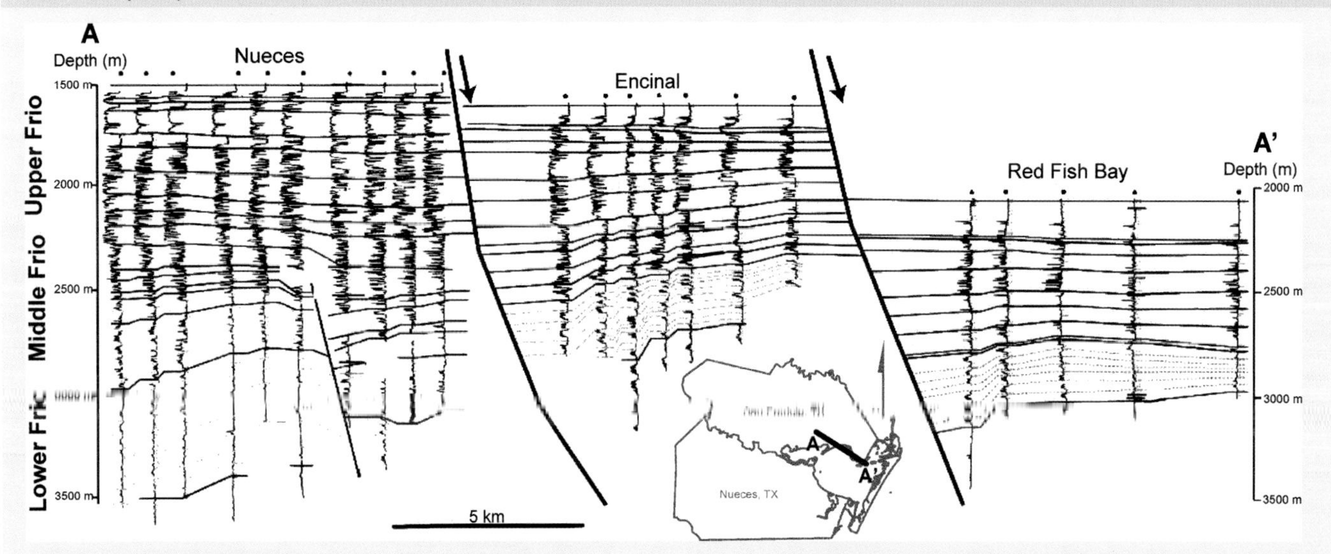

Figure 6.7 High-resolution dip cross-section of the middle–upper Frio Greta barrier–strandplain system, Nueces Bay area, south Texas. The section traverses three locally differentiated growth fault bounded subbasins: Nueces, Encinal, and Red Fish Bay. Correlation lines trace transgressive shale markers. From Olariu *et al.* (2013).

faulting and canopy loading. Individual regressive units typically expand across faults, reflecting the syndepositional fault displacement and commensurate increase in accommodation space over 100 ky timespans.

A caveat to the discussion that follows. Growth fault literature and sequence stratigraphic models commonly equate growth faulting to shelf edge delta margins. Accordingly, Olariu *et al.* (2013) interpreted Frio strata in the Corpus area to have been deposited as wave-dominated shelf edge deltas. Here, I reinterpret the cycles within their middle and upper Frio paleogeographic context to consist of shore zone and shelf facies. Several observations support this revised interpretation. (1) The Corpus Christi Bay area lies within the southern end of the regionally mapped Greta barrier–strandplain system (Figure 6.2). The wave-dominated Norias delta lies well south of Corpus Christi. Only their lower Frio interval (approximately equivalent to the middle Frio of Galloway *et al.* [1982b]) evidences mixed deltaic and shore zone facies. (2) The contemporary Frio coastal plain updip of Corpus Christi is mud-rich and contains few fluvial channel fills of minor coastal streams. Muddy coastal plain, bay, and lagoonal facies dominate. (3) Net sand and percentage sand maps for both composite intervals and individual sand bodies are dominated by strongly strike-parallel contours. Strike-oriented coastal facies tracts were produced by northward, along-strike transport of beach/shoreface sand reworked from the adjacent deltaic headland. (4) Individual sand bodies are similarly dominantly strike-oriented, linear, and narrow, and display primarily aggradational (rather than progradational) stacking. (5) Mapped shorelines consistently cluster along narrow, strike-parallel belts within each of the three fault-bounded subbasins. (6) Cores described by Galloway and Morton (1989) and Olariu *et al.* (2013) are dominated by fine-grained, extensively bioturbated sands and muds interpreted to be shoreface and shelf deposits. Shelf storm beds are common.

Basinward, in the Red Fish Bay subbasin, only the upper Frio interval is penetrated by the cross-section wells. This interval is equivalent to the sand-rich interval in the Nueces subbasin. The mud-dominated section consists of thin, interbedded, bioturbated distal shoreface and storm-dominated shelf sand bodies in a matrix of shelf mud. This association is typical of the narrow shelves that separated prograding northern GoM shore zone systems from their contemporary shelf edges.

Recurrent upward-coarsening log patterns (progradation) and ready correlation of individual progradational cycles place the sand-bearing Frio interval on the shelf platform. The Frio shelf edge trajectory lies within mud-dominated distal shelf and upper slope strata below all but the deepest wells in the Nueces subbasin and below and seaward of all wells in the Encinal and Red Fish Bay subbasins. The Frio paleo-shelf edge at maximum progradation lay nearly 30 km (20 miles) seaward of the southeast end of the cross-section (Figure 6.2). The growth faults that bound the subbasins had their origins deep in the Frio interval in and around the active paleo-margin depocenter and exhibit greatest displacements and growth there. They are a small part of the interregional extensional faulting domain that is rooted above the inboard area of the Sigsbee canopy. Regional GoM growth faults initiated at the unstable shelf edge commonly continue to be reactivated during the deposition of overlying shelf and coastal plain strata.

The Corpus Christi example demonstrates several general attributes of the northern GoM margin. (1) Growth faulting is developed along, and influences, interdeltaic segments of the continental margin, as well as the deltaic margins. (2) Shore zone systems, though commonly progradational, typically also construct a narrow, muddy shelf that separates the sandy shoreface from the depositional shelf edge. (3) A well-tuned balance between structural subsidence (accommodation) and sediment accumulation (supply) can develop and persist for geologically significant timespans, accumulating thick, aggradational facies tracts.

Box 6.1 (*cont.*)

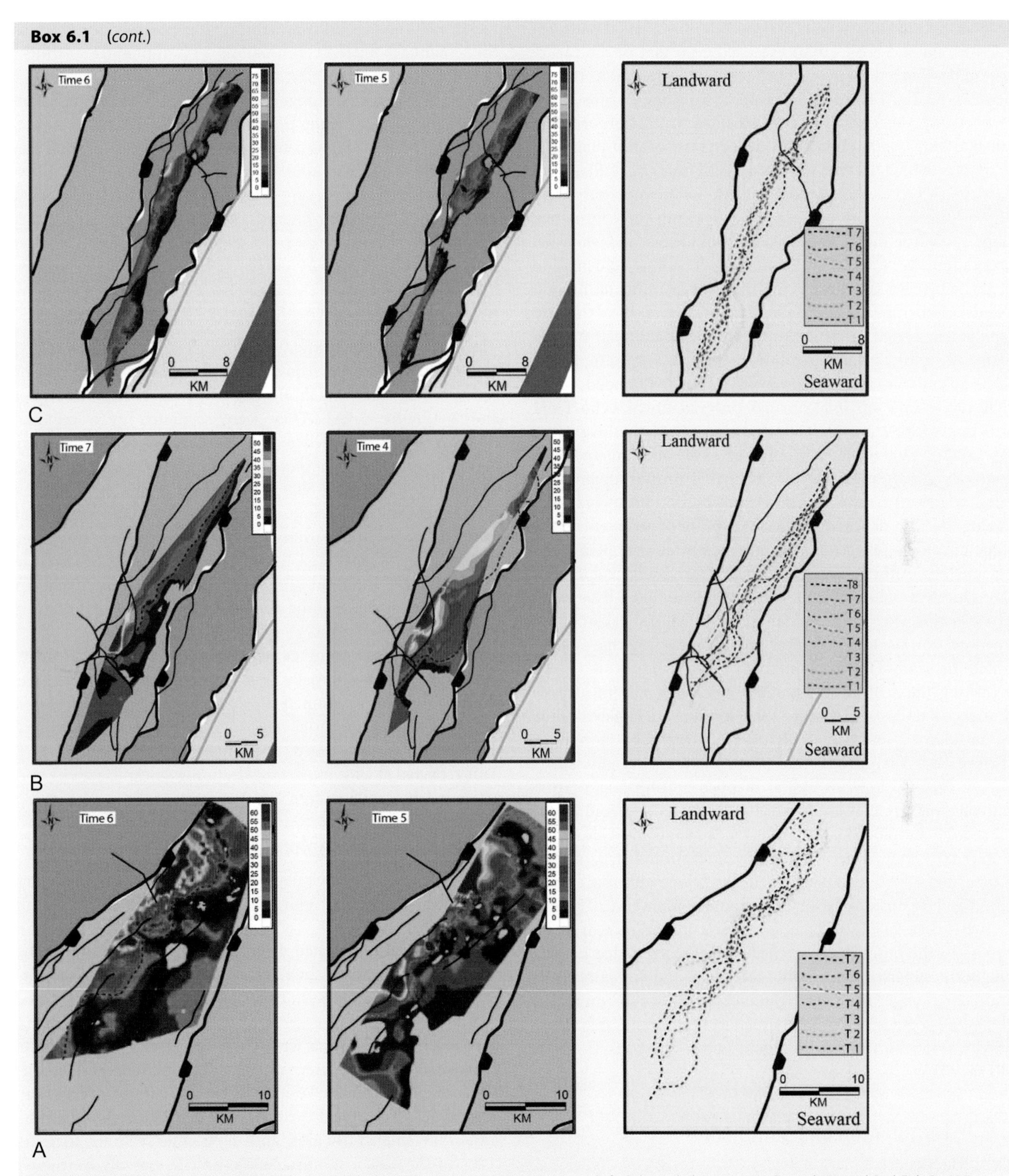

Figure 6.8 Representative sand thickness maps for individual sequences within the growth fault-bounded Nueces **(A)**, Encinal **(B)**, and Red Fish Bay **(C)** subbasins. Oldest sand units (Nueces subbasin) show greatest along-strike variability, reflecting influence of adjacent Norias delta lobes. Progressively younger units of the Encinal and Red Fish Bay subbasins display increasing strike-continuity and mapped linear shorelines typical of a barrier island or strandplain coast. The vertical stacking of the shorelines along a narrow belt within each of the broader subbasins is a dramatic reflection of the long-lived balance between sediment supply and subsidence rate. Compiled from Olariu *et al.* (2013).

During the Oligocene Frio deposode, continental platform growth rate and areal extent exceeded even that of the earlier Laramide-sourced Wilcox. Shelf edge progradation ranged between 65 and 145 km (40–90 miles) from the Burgos basin, in northern Mexico, to eastern Louisiana. In total, more than 100,000 km^2 (28,000 $miles^2$) of depositional coastal plain and shelf was added to the northern GoM margin. Offlap was particularly efficient as most of the sediment transported beyond the shelf edge was sequestered within the continental slope, constructing a coalesced belt of delta- and shelf-fed prograding slope aprons (Figure 6.3). Loading deformation of the subjacent salt canopy, and ongoing continental margin extension, manifested in the abundant growth faults of the Frio fault zone, created efficient sediment traps on the shelf-margin deltas and subjacent upper continental slope. For example, the slope apron fronting the Norias delta contains ponded, sand-rich successions of amalgamated progradational delta front and upper slope gravity mass transport beds that are more than 1000 ft (300 m) thick. Compared to the other major offlap continental margins of the GoM Cenozoic, proportionally little sandy sediment escaped the slope apron onto the abyssal plain during the Frio deposode. Efficient capture of sediment in the slope aprons is reflected in the absence of abyssal plain submarine fans. Although the basin floor supersequence is several thousand feet thick, wells to date have encountered little sand. The principal exception thus far is the basinal equivalent of the Reynosa–Norias delta-fed apron, where a few wells have penetrated the sandy Oligocene intervals.

Along the central and southern Mexican margin, depositional coastal plain and shore zone deposits are largely thin or absent due to renewed uplift of the western basin margin (Figure 6.3). A large mass transport complex, imaged seismically on regional lines, as well as mapped erosional submarine canyons indicate that uplift directly tilted, elevated, and destabilized the margin. This pattern extended southward along the Veracruz margin, where submarine erosion and bypass continued the Paleocene–Eocene history of slope apron accumulation and expansion eastward (Hernandez-Mendoza 2013). As in the earlier Paleogene, the volumetric importance of slope bypass to the basin floor is reflected in the expanded basinal Oligocene supersequence (cross-sections 7 and 8; Figures 1.19 and 1.20). Longer submarine channels of the apron converged along the Veracruz Trough, which collected and redirected gravity flows northward (Arreguín-Lopez *et al.* 2011; CNH, 2015a, 2015b).

6.5.2 Termination and Summary

The Frio deposode terminated with a progressive regional transgression that culminated at the Oligocene–Miocene boundary. This flooding was an extended phase lasting nearly two million years, reflecting the progressive late Oligocene decrease of sediment supply from upland and volcanic sources. It is recorded in dip stratigraphic cross-sections as a back-stepping, or retrogradational, succession of coastal and deltaic facies tracts and a basinward-thickening mud wedge known as the Anahuac Shale. Culmination of continental platform flooding, and minimal clastic sediment influx to the outer shelf and basin allowed late Oligocene expansion of the eastern Gulf carbonate platform along the foundering outer shelf over local structural highs as far as central Louisiana, and local patch reef development as far as southeast Texas. Stratigraphic condensation resulted in clustering of numerous paleontologic LADs, including *Globigerina ciproensis*, *Dictyococcites bisecta*, *Discorbis gravelli*, *Heterostegina* sp., *Globigerina sellii*, and *Cibicides jeffersonensis*.

6.6 Early Miocene LM1 and LM2 Deposodes

The Early Miocene stratigraphic record of the northern GoM contains genetic supersequences deposited during two deposodes, which Galloway *et al.* (2000) named Lower Miocene 1 and 2 (Figure 6.1A). They were separated by a regional flooding event, recorded in the *Marginulina* Shale, dated by benthic foram *Marginulina ascensionensis* LAD at about 18 Ma. The LM2 deposode continued about one million years into the Middle Miocene. These deposodes had somewhat more manageable durations of approximately five and three million years. Across the US Gulf Coast, LM1 and LM2 include the Oakville Formation and Fleming Group respectively. In Mexico they include all or part of the Escolin–Coatzintla, La Laja, and Depósito units (Figure 6.1B).

Although the volumes of the LM1 and LM2 supersequences are substantially less than that of the Frio, their shorter durations resulted in comparably high calculated supply rates (Galloway *et al.* 2011). The rate of continental margin accretion declined, but remained high (Figure 6.1A). This reflected at least in part the declining volcanic activity and active uplift across western North America.

6.6.1 Paleogeography

The two Lower Miocene deposodes will be discussed together as they display similar paleogeographies across the northern and western Gulf.

Two fluvial-dominated deltas prograded rapidly across the Anahuac transgressive shelf and onto the central GoM continental margin (Figure 6.9). The Mississippi fluvial–deltaic axis continued its long history as a major continental river and deltaic depocenter. To its west, a new fluvial axis, named for its similarities to the Quaternary Red River drainage basin, traversed the easternmost Texas Miocene coastal plain (Galloway *et al.* 1986). It supplied the Calcasieu delta system, which was somewhat smaller in area and depocenter volume than its sister to the east in Louisiana. Both delta systems continued as principal depositional elements of the LM2 deposode (Figure 6.10).

In south Texas, the Rio Grande fluvial–deltaic axis similarly resumed progradation of the shelf margin following the Anahuac retrogradational flooding. Amalgamated gravelly to sandy channel fill facies of the bedload-dominated river system extend from outcrops in south Texas into the shallow subsurface (Galloway *et al.* 1979). However, the proportional volumetric importance of this depocenter decreased from its Oligocene peak, and further continued its decline during the subsequent LM2 deposode. The North Padre delta system, like its Paleogene precursors, was wave-dominated and sand-rich. Although still present, the Norma fluvial–deltaic axis in the Burgos basin was small and a secondary avenue of sediment supply.

An extensive shore zone complex, named the Oakville barrier bar in older literature, bridged the interdeltaic coastal bight between the North Padre and Calcasieu deltaic depocenters (Galloway *et al.* 1986). Like the Frio, the barrier system produced thick, strike-aligned, amalgamated sand bodies separating landward lagoonal and coastal plain facies from downdip fossiliferous, muddy facies deposited on the narrow shelf. Notably, the North Padre depocenter shifted to the northeast flank of the delta system, reflecting volumetric importance of along-strike coastal reworking and longshore sediment transport from the deltaic headland to the adjacent interdeltaic bight.

Each of the three major deltaic and the interdeltaic shore zone systems actively prograded the continental margin during the LM1 deposode. By LM2, reduced sediment supply through the Rio Grande combined with the efficient longshore reworking to limit further offlap of the northwest Gulf margin. However, depositional offlap continued from the central Texas to the Louisiana continental margin. The shelf-margin deltas constructed sandy, delta-fed slope aprons. There, multiple slope channels captured and transported sand down the slope and onto the adjacent basin floor (Figures 6.9 and 6.10). The prograding shore zone/shelf/slope depositional systems tract typically separated the barrier shoreface sands from the shelf edge by the narrow but

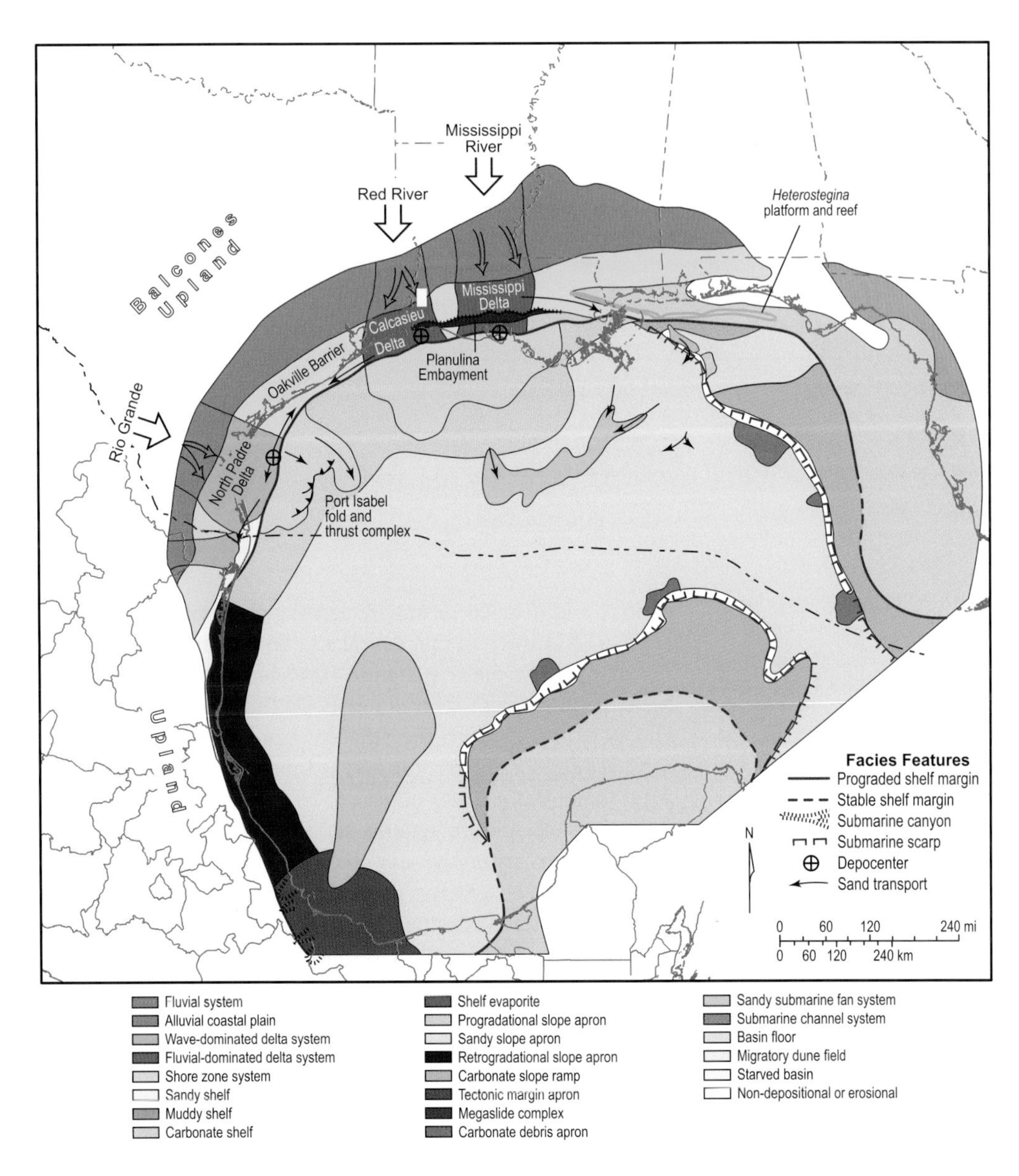

Figure 6.9 Paleogeographic map of the Early Miocene LM1 deposode. Depositional system outlines and shelf edge reflect their positions at maximum progradation. White box locates seismic line in Figure 6.11.

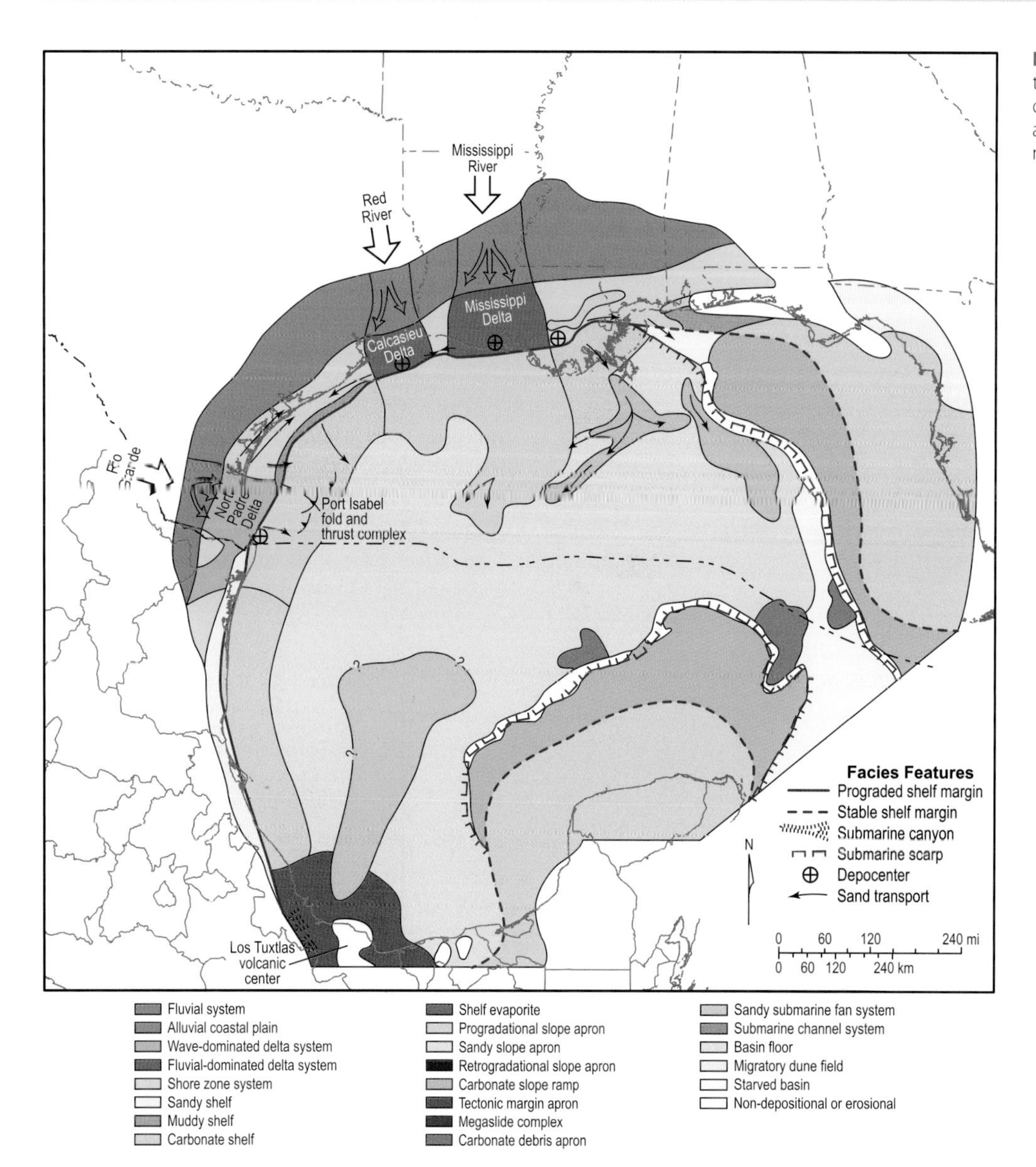

Figure 6.10 Paleogeographic map of the Early to early Middle Miocene LM2 deposode. Depositional system outlines and shelf edge reflect their positions at maximum progradation.

muddy shelf. The progradational shelf-fed slope apron is thus dominantly muddy.

For the first time since the Early Eocene Upper Wilcox supersequence, sandy, channelized sediment gravity flow facies are mapped from the lower slope onto and across the adjacent abyssal plain. In the LM1 genetic supersequence, a sandy belt containing both stacked channel fill and unconfined lobe sand bodies extends nearly 400 km (250 miles) southwestward from an interpreted apex in the northeast corner of the deep basin (Figure 6.9). In the LM2 supersequence, two channel belts are differentiated and display lobate or bifurcating patterns indicative of expanding submarine fan lobe development. One dispersal system trended south-southeast, parallel and adjacent to the base of the Florida Escarpment. The second system turns from dip-oriented to the west-southwest, along the toe of the slope apron and parallel to the trend of incipient folds of the syntectonic Mississippi Fan–Atwater fold belt. Additional well control and detailed seismic facies mapping will likely elaborate these features and may reveal additional ones. Interestingly, the mapped submarine channel–fan elements originate from the largely sediment-starved northeast corner of the Gulf basin and not directly from the adjacent sandy delta-fed apron as most sequence stratigraphic and sedimentologic models would place them. Rather, they appear to be sourced from shore zone or sandy shelf-margin systems supplied, in turn, by longshore redistribution from the east flank of the Mississippi delta. Sandy sediment gravity flows bypassed this largely relict, narrow, and steep continental slope preserved at the intersection between the carbonate-dominated Florida Platform and the now actively prograding central Gulf margin. Deposition of sandy channel fill and lobe facies occurred where flows dispersed onto the low-gradient basin floor.

Early Miocene slope offlap was interrupted in the central Gulf by formation of a broad retrogradational slope complex, the Planulina Embayment (Figure 6.11). Rapid collapse of the subjacent salt canopy with Early Miocene loading was followed by accumulation of an over-thickened depocenter containing >10,000 ft (3000 m) of slope and deltaic facies. Nearly half of the interval is sand-bearing, and individual sand bodies are up to many hundreds of feet thick.

Along the Tampico–Misantla margin, in the western Gulf, initial shelf edge retrogradation was followed by renewed offlap in late Early Miocene. A narrow, sandy shelf separated the shelf edge from the emergent uplands. In the Veracruz basin, the Lower Miocene sedimentary wedge onlapped the erosional east flank of the tectonically elevated Cordoba Platform. Marginal submarine canyons bypassed sediment gravity flows downslope onto a broad tectonic margin apron that filled the compressional trough and spilled as much as 200 km (125 miles) across the Gulf basin floor. By the LM2 deposode (Figure 6.10), uplift and accretion of the Los Tuxtlas volcanic complex formed a partial barrier between the Veracruz basin proper and the open GoM. This barrier may also have provided a local sediment source and helped focus flows from the multiple slope canyons of the apron into the integrated submarine channel system. The channelized flows produced mixed erosional and depositional channel systems that extended as much as 500 km north along the inferred Veracruz bathymetric trough (Figure 6.12). Larger channel fills are up to 2 km (1.3 miles) wide and 10–100 m (35–350 ft) thick (Winter 2018). Regional seismic facies patterns suggest development of an elongate abyssal fan at the distal end of the channel system. The fan system appears to have reached its maximum extent in the Lower Miocene supersequences. Wells drilled along the Veracruz Trough channel axis display a sand-rich lower succession. Serrate log response characterizes composite sand bodies that range from a few tens of meters to more than 100 m in thickness (Figure 6.13A). Both upward-coarsening and upward-fining sequences are common. Cores (Figure 6.13B) confirm the presence of a thin- to thick-bedded, sandy, turbidite facies association within the submarine channel fills.

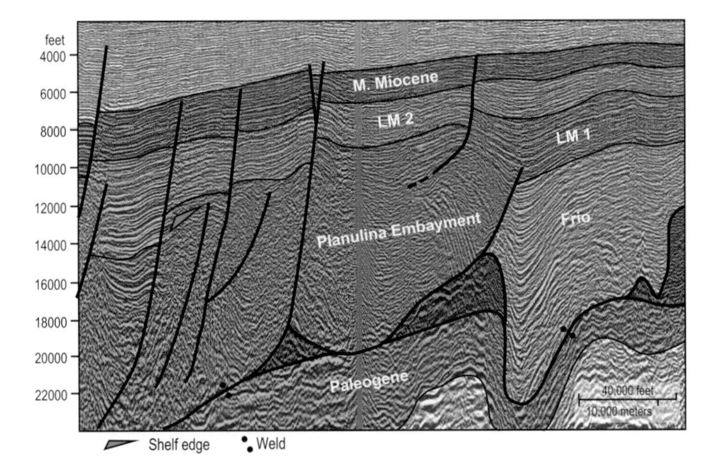

Figure 6.11 Seismic line interpretation showing the Planulina Embayment at the base of the LM1 supersequence. Large-scale salt mobilization within the subjacent salt canopy (now reflected in the remnant salt bodies and associated weld) triggered foundering of the relict Frio margin. Salt evacuation and sediment remobilization created accommodation space for fivefold expansion of the structurally complex earliest Miocene outer slope succession. For location, see Figure 6.9. Seismic line courtesy of ION.

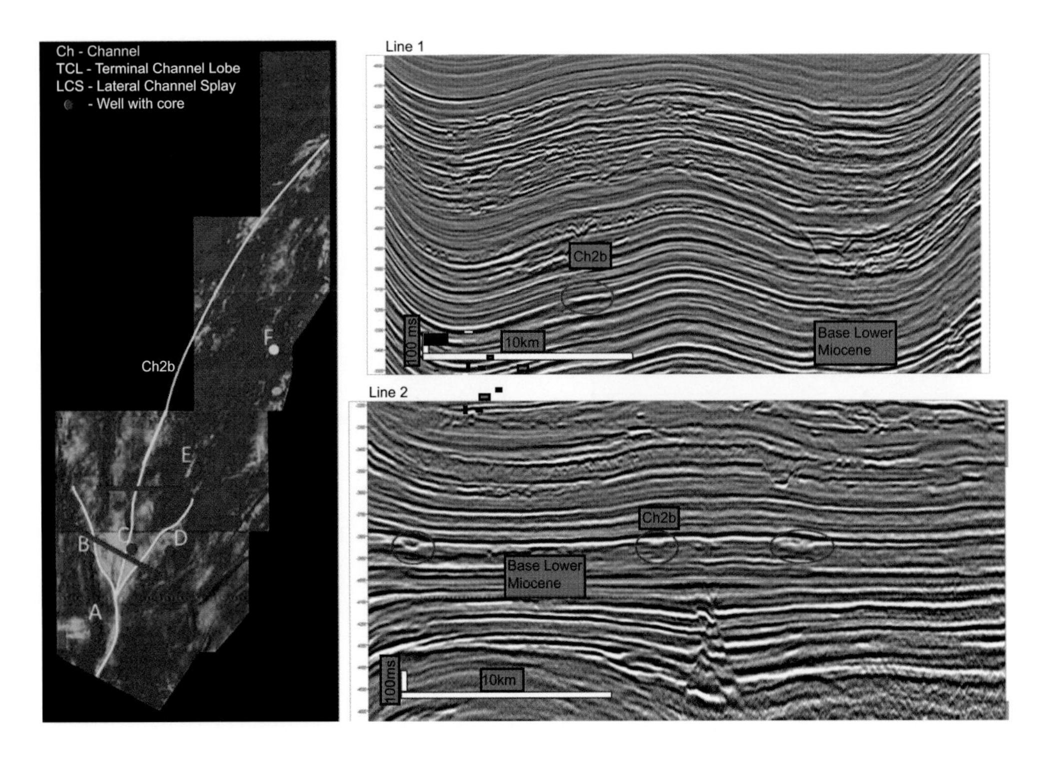

Figure 6.12 Seismic line across the Lower Miocene submarine channel system of the Veracruz Trough. Relatively small channels create modest reflection anomalies that allow attribute mapping of both channel axes and related overbank levee and splay deposits. Note increasing scale and erosional aspect of younger Miocene channels on the seismic lines. From Winter (2018) with permission from CNH.

6.6.2 Termination and Summary

The Early Miocene genetic supersequences record several evolutionary patterns that culminate the Middle Cenozoic Geothermal Phase:

1. the rise, domination, and then gradual decline of the Rio Grande River system as a major supplier of sediment to the GoM;
2. rebirth and expansion of a sandy, prograding, delta-fed slope apron fronting the Mississippi fluvial–deltaic axis;
3. appearance of organized submarine channel systems on the northeastern Gulf abyssal plain presaging the future development of large Neogene submarine fan systems;
4. continued expansion of the tectonic margin apron in the Veracruz basin and maturation of the north-flowing submarine channel system.

Across the northern Gulf basin margin, the *Amphistegina* Shale records termination of the Lower Miocene deposodes LM1 and LM2. The marker benthic foram *Amphistegina* B approximates maximum flooding and is dated slightly younger than 15 Ma.

6.7 Structural Evolution

Just as the depositional history of the Middle Cenozoic Geothermal Phase perpetuated two dramatically different

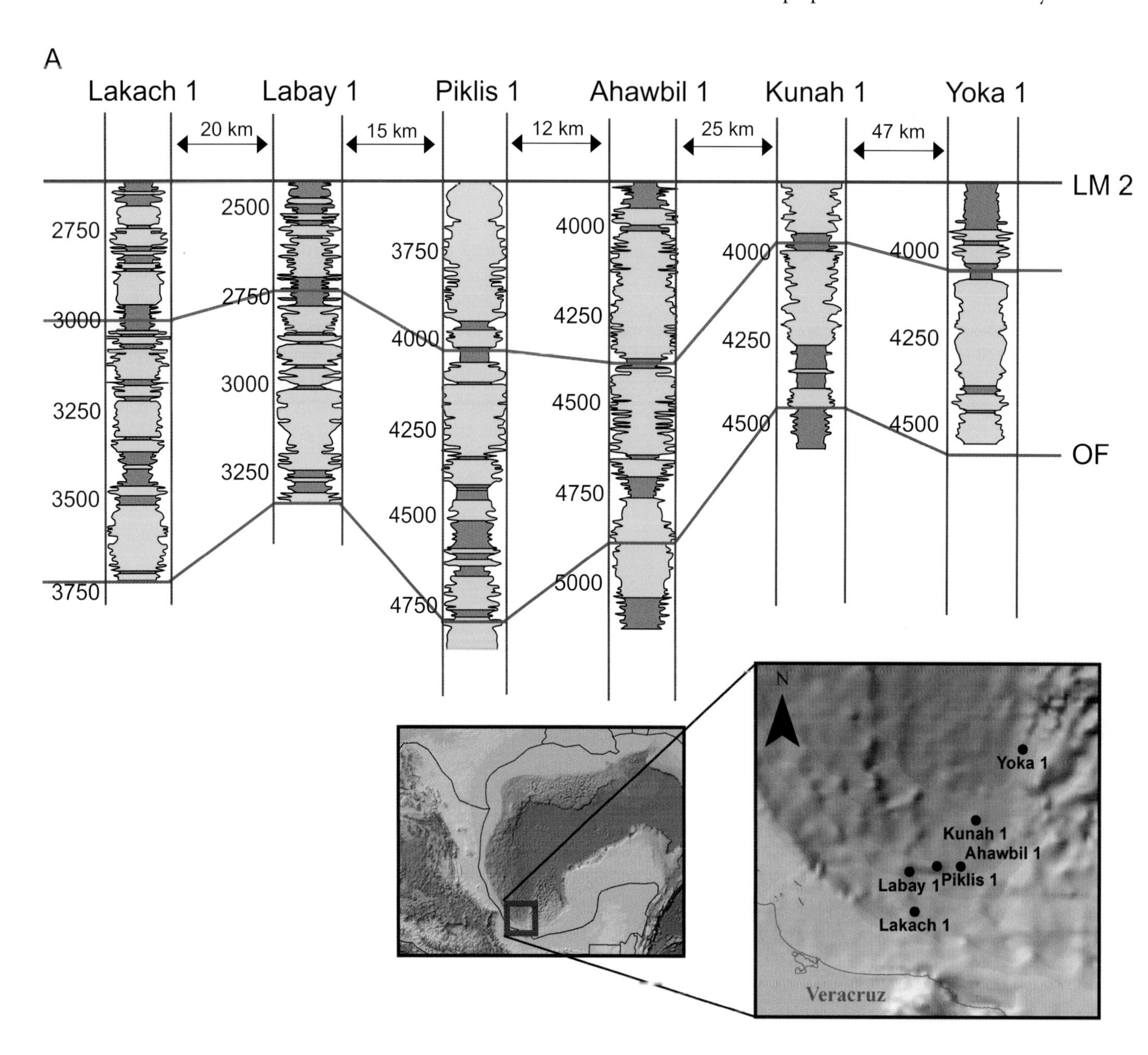

Figure 6.13 **(A)** Well log cross-section paralleling the northward transport axis of the Lower Miocene Veracruz Trough submarine channel system. Datum is the top of the LM2 supersequence. **(B)** Representative core through facies (YOKA – 1 well) of the Veracruz Trough submarine channel system. Abundant thick to thin, fine-grained turbidite beds compose the submarine channel axis deposits.

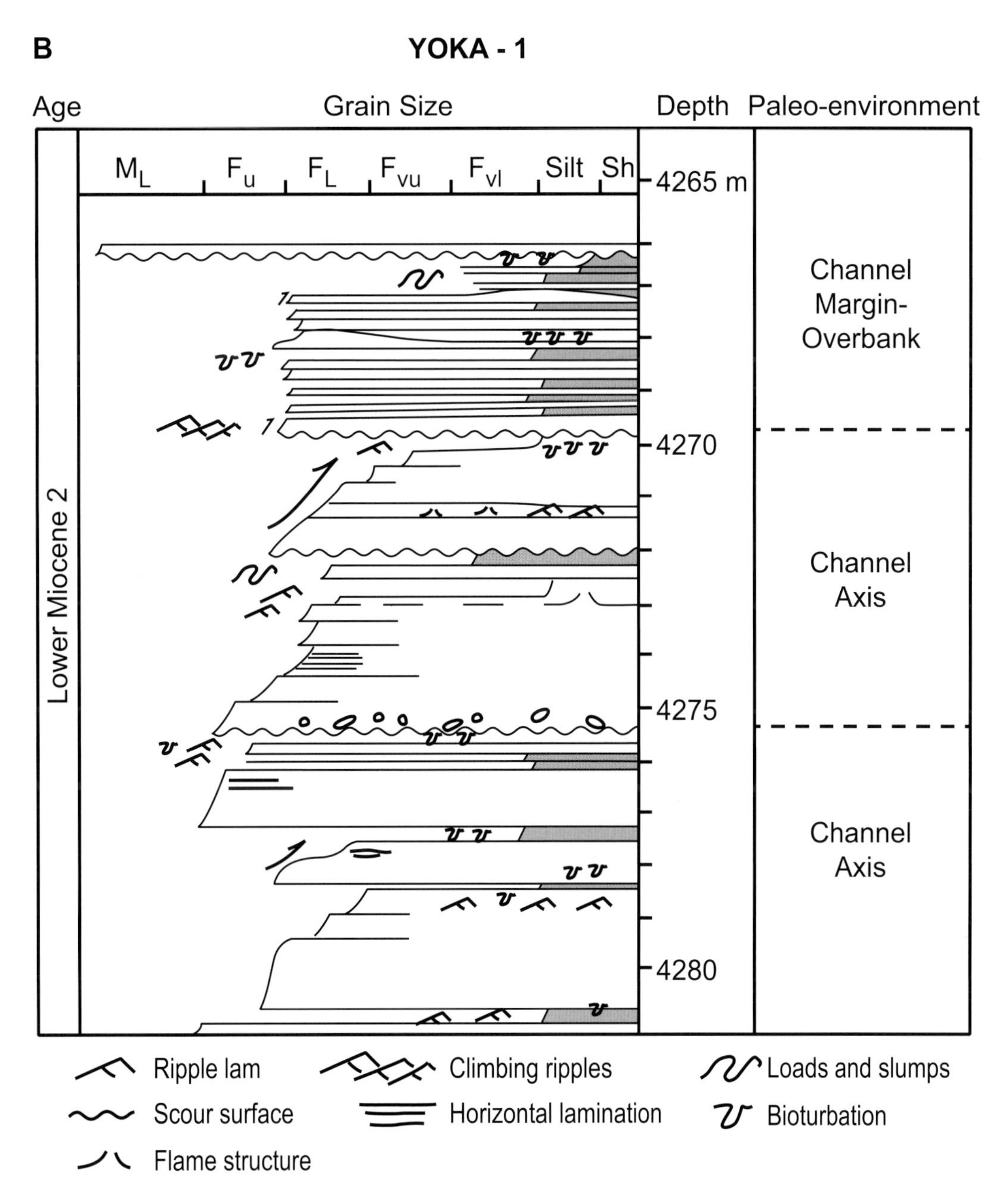

Figure 6.13 *(cont.)*

paleogeographies across the northern and western GoM, so too did the structural history create two very different assemblages of tectonic domains. (1) The northern-northwestern basin structure was dominated by intrabasinal gravity tectonics with limited impact by basement deformation. (2) The southwestern Mexican margin and Campeche Bay were directly impacted by compressional deformation driven by ongoing plate subduction along the nearby Pacific Rim. Gravity tectonics played a secondary role.

6.7.1 North-Northwestern Gulf

Three structural domain maps (Figures 6.14–6.16) capture the broad geographic patterns and evolutionary increase in diversity of tectonic regimes during the deposodes of the Middle Cenozoic Geothermal Phase.

6.7.1.1 Late Eocene

Ongoing salt deformation dominated the northern Gulf margin. More than 10 million years of slow sedimentation during the Middle Eocene had provided ample time for expulsion, advance, and amalgamation of rising salt bodies to create the broad Sigsbee salt canopy beneath the outer slope and adjacent abyssal plain of the northern GoM. Low rates of sediment supply and limited offlap allowed further inflation and expansion of the canopy. By the Late Eocene, a broad swath of shallow canopy emplacement extended beneath the

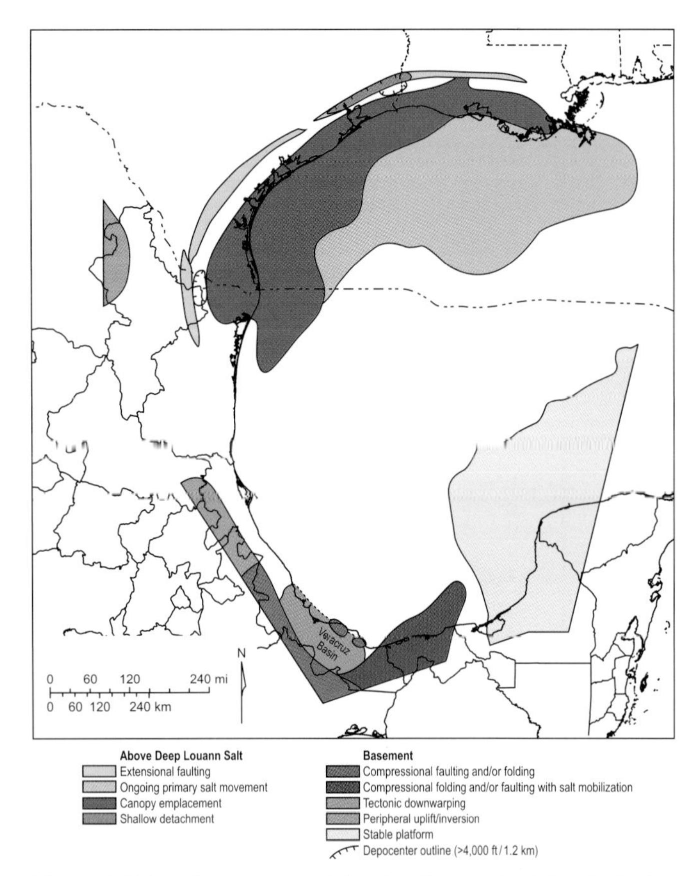

Figure 6.14 Late Eocene structural domains. Three regional domains in the northern Gulf consist of structures with roots at the Louann Salt level. The local domain of Yegua extensional growth faulting with a shallow shale detachment occurs in southeast Texas. The western and southern Gulf, in contrast, displays an array of domains rooted in tectonic compression, uplift, or inversion.

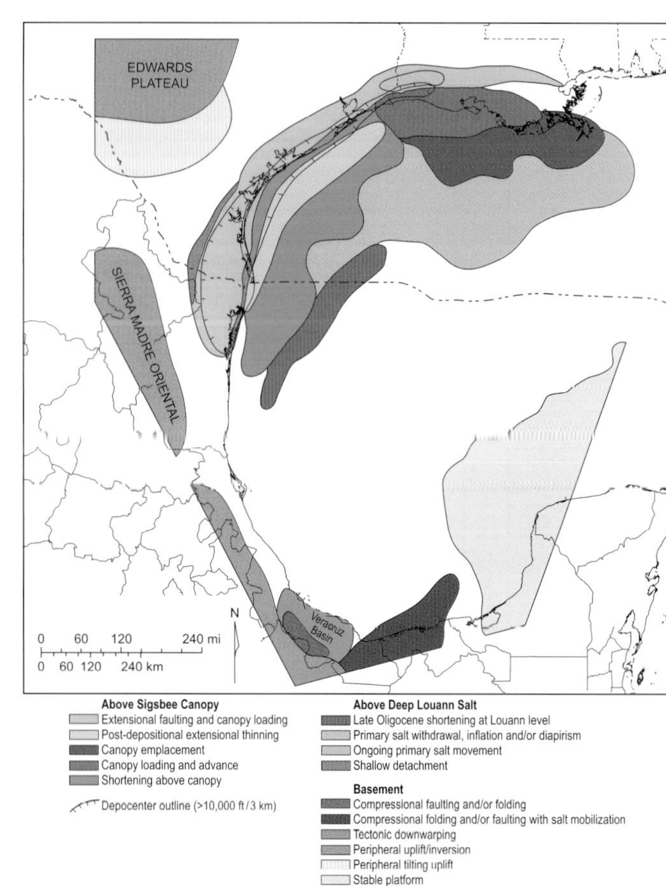

Figure 6.15 Oligocene structural domains. Deformation both above the deep Louann Salt and, basinward, above the Sigsbee salt canopy created a diverse array of domains in the northern Gulf. The broad, inboard extensional arc of growth faulting had its roots in the canopy. Compression and shortening at the Louann level initiated the Port Isabel and Perdido fold belts. A chain of basement uplifts and inversions extended along the western periphery of the Gulf basin from central Texas to Campeche. Compressional deformation continued in the Veracruz basin.

modern coast and shelf from Louisiana to Tamaulipas (Figure 6.14). Ongoing primary salt adjustments to loading continued both beneath and basinward of the canopy. Three families of extensional growth faults, which soled out at deep salt level, linked to create a nearly continuous fault zone extending from Louisiana to the Burgos basin. Shallow detachment and extensional slumping and faulting at the Yegua continental shelf edge temporarily created the retrogradational margin preserved beneath the eastern Texas coastal plain.

6.7.1.2 Oligocene

The array of tectonostratigraphic domains increased in complexity (Figure 6.15), a response to the long Oligocene interval of continental margin offlap, rapid depositional loading along nearly 1300 km (800 miles) of the continental margin, and expansion of the locus of loading onto the Sigsbee salt canopy. Two broad families of linked extension, translation, and compression characterize the genetic supersequence. Once formed, these families persisted into the Neogene. Generally strike-elongate extensional domains updip and broadly compressional fold belt domains in the basin dominate the Texas–Tamaulipas area (Oligo-Miocene canopy extensional detachment, [cross-sections 3 and 6; Figures 1.15 and 1.18]). Both deep, primary salt and the shallow salt canopy facilitated deformation and large-scale basinward mass translation, but salt structures are secondary features. The Louisiana coastal plain and shelf, in contrast, is dominated by broad domains characterized by the assemblage of salt and salt-related structures they contain (Oligo-Miocene canopy extensional detachment [cross-sections 4 and 5; Figures 1.16 and 1.17]).

In south Texas, the Early Oligocene shallow detachment created the narrow, highly extended Vicksburg growth fault zone along the Oligocene continental margin (cross-section 3; Figure 1.15). Detachment and basinward sliding within underlying Eocene shale provided a slow-motion conveyer belt that displaced shelf edge delta deposits 10 miles or more basinward from their original site of deposition (Langford and Combes 1994). Subsequent basinward off-stepping of the locus of Frio extension and canopy loading created a growth fault belt, which extends from Louisiana to the Burgos basin. Further basinward, compression above the buried canopy initiated a broad band of structural shortening by folding and reverse

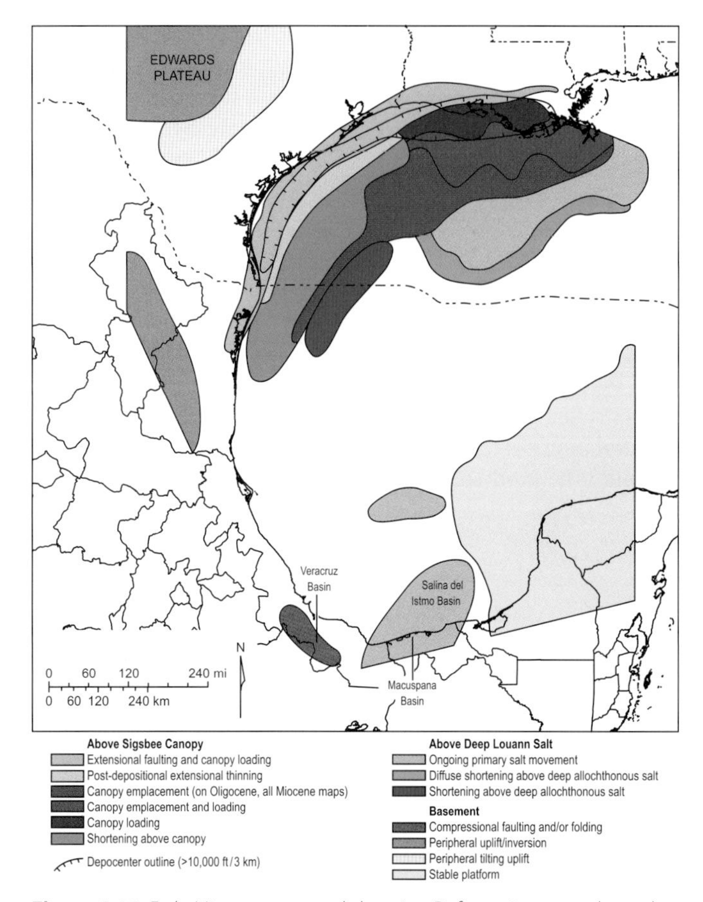

Figure 6.16 Early Miocene structural domains. Deformation at or above the Sigsby salt canopy and, basinward, above deep Louann Salt produced a suite of domains defined by their abundance and style of structures. At the basinward limit of deformation, diffuse compressional shortening above deep salt initiated the Mississippi Fan and related fold belts. Offshore Texas domains broadly continued the Late Eocene pattern of inboard extension and infilling, post-deposode extensional thinning, and basinward reverse faulting, folding, and uplift. The Mexican Gulf was structurally quiescent with the exception of compressional deformation within the Veracruz basin, and primary salt movement due to loading of deep salt in the Campeche province.

faulting of Frio strata, and initiated the Port Isabel fold belt (Oligo-Miocene canopy fold belt [cross-sections 3 and 6; Figures 1.15 and 1.18]). Inferred Oligocene evacuation of the salt body contained within the Bravo Trough provided accommodation of a highly expanded deep basin Frio succession (Hudec *et al.* in review). In addition, Oligocene compressional structural stacking produced a tectonically over-thickened succession. The resultant Frio interval is more than 18,000 ft (2200 m) thick. On the abyssal plain, Late Oligocene shortening at the Louann Salt level nucleated the first anticlinal ridges of the Perdido fold belt. Loading by continental margin deposits of the Early Miocene supersequence initiated a new zone of extension that created a strike-elongate zone of structurally attenuated Frio interval (later infilled by expanded Miocene section) that parallels the basinward margin of the Frio preserved depocenter (cross-section 6; Figure 1.18).

The Louisiana coastal plain and offshore is dominated by irregularly shaped domains defined by salt deformation at both primary and canopy levels. Here, salt structures are prominent. Updip extension was largely balanced by regional canopy deformation (Dooley *et al.* 2013). Inboard canopy loading and evacuation of salt from the Sigsbee canopy initially defined Oligocene structural evolution. By the Neogene, outboard salt canopy deformation continued the pattern. The Frio growth fault belt records the landward zone of extensional faulting and canopy loading. Progressing basinward, successive domains of canopy loading and advance, seaward-advancing canopy emplacement, and ongoing primary salt movement affected local and regional patterns of sediment accommodation.

Tectonic deformation also encroached on the structural periphery of the basin. Uplift of the Edwards Plateau in central Texas accentuated tilting uplift/subsidence across south Texas. Tectonic inversion that elevated the Sierra Madre Oriental placed uplands along the Burgos basin hinterland.

6.7.1.3 Early Miocene

The LM1 and LM2 supersequences record a similar array of tectonostratigraphic domains (Figure 6.16). The basin margin belt of extensional faulting and canopy loading forms a broad arc extending from eastern Louisiana to the Burgos basin. Combined extension, attenuation of underlying Frio strata, and canopy loading created excess accommodation space for the strike-elongate Lower Miocene depocenter. In the northwest Gulf, shortening above the canopy and, furthest basinward, above deep allochthonous Louann Salt continued in the Port Isabel and Perdido fold belts (Oligo-Miocene extensional detachment and canopy fold belt [cross-sections 3 and 6; Figures 1.15 and 1.18]). Domains dominated by canopy loading, mixed canopy emplacement and loading, and canopy emplacement occur progressively basinward in the north-central Gulf (Oligo-Miocene canopy extensional detachment and canopy fold belt [cross-sections 4 and 5; Figures 1.16 and 1.17]). Ongoing primary salt movement and initiation of a belt of diffuse shortening at the distal fringe of the Louann Salt created two additional domains on the abyssal plain, the Perdido and Atwater fold belts (cross-sections 3, 4, and 6; Figures 1.15, 1.16, and 1.18).

6.7.2 Southwestern Gulf

The Mexican Gulf, including Tampico–Misantla, Veracruz, and Sureste basins provinces, continued their history of juxtaposed compression and downwarping. During the Late Eocene and Oligocene, the western periphery was subjected to uplift both by tectonic inversion of previous downwarps and renewed compressional folding and faulting. In the Sureste province, compression extended into the Bay of Campeche, where folding and faulting mobilized deep primary salt, which added to structural complexity. The Veracruz basin continued its history of downwarping.

Uplift along the basin margin largely ended by the Early Miocene (Figure 6.16). However, compressional folding and faulting continued to modify the western flank of the Veracruz basin. In the Sureste province, the Macuspana and Salina del

Istmo basins were structurally differentiated. Primary salt movement continued in both.

6.8 Summary: Middle Cenozoic Phase

The Middle Cenozoic Geothermal Phase produced a 23-million-year megasequence recording tectonic reorganization within the North American plate and consequent resurgent sediment supply from post-Laramide western North America. New crustal uplifts and volcanic centers provided the upland sources. Sediment supply was greatest in the Veracruz basin and along the northwestern Gulf margin from the Burgos basin to the Mississippi axis in the central Gulf. The eastern Gulf remained starved of clastic sediment. The five depositional episodes that comprise the megasequence and their paleogeography were strongly modulated by changing rates of sediment supply and by basin margin tectonics. The Oligocene Frio deposode initiated the first of a succession of major continental margin offlap wedges onto the Sigsbee salt canopy. Crustal tilting and elevation of the basin fringe amplified normal basinward displacement of both the sediment prism and mobilized salt. In the northern GoM, sediment influx was largely sequestered in lower coastal plain, coastal, and slope apron depositional systems where combined extension and salt evacuation created accommodation volume commensurate with the supply. Structurally influenced complexity of continental margin bathymetry created complex bedload sediment transport pathways that favored intraslope ponded sediment accumulation. With the exception of the Veracruz tectonic margin, little sand volume escaped onto the basin floor.

Chapter 7

Cenozoic Depositional History 3

Neogene Tectono-climatic Phase

7.1 Basin and Continental Framework

The Middle Miocene ushered in a 15-million-year phase of robust sediment supply, and continental margin accretion (Figure 7.1). Ongoing high rates and regional patterns of sediment yield resulted from the interplay of two principal factors: (1) the North American landscape was modified by rifting, regional elevation, and exhumation of older basin-fills (in the USA), and creation of new tectonic and volcanic uplands adjacent to the western Gulf depositional basin margin (in Mexico); and (2) regional and global climate change, culminating in the North American ice sheet, impacted sediment yield and transport efficiency. In the marine depositional basin, increasingly high-frequency and -amplitude sea-level changes played a prominent role in stratigraphic evolution (Miller *et al.* 2005). Thus the "tectono-climatic" designation for this final phase of Gulf basin history.

Neogene extension along the Rio Grande rift both isolated the Gulf of Mexico (GoM) from southwestern uplands and elevated a chain of rift-shoulder uplifts as new sediment sources. Pliocene regional upwarping elevated the broad, domal Rocky Mountain Orogenic Plateau (McMillan *et al.* 2006). In the eastern USA, erosion of the Appalachian Mountains and adjacent Cumberland Plateau was rejuvenated in the Miocene, dramatically increasing sediment yield to eastern drainage basins (Boettcher and Milliken 1994; Liu, 2014).

In Mexico, several tectonic and volcanic uplands were active and impacted both the sediment influx and structural development of the western GoM. These include the Sierra Madre de Chiapas (Witt *et al.* 2012), the Trans-Mexican Volcanic Belt and its outliers (Paredes *et al.* 2009), the Sierra Zongolica upland (Roure *et al.* 2009), and the Cordoba Platform (Roure *et al.* 2009). Tectonism segmented the Sureste/Campeche province into several smaller basins, including the Salina del Istmo, Comalcalco, and Macuspana basins (Figure 1.11). The Veracruz basin continued its long history of subsidence and sediment accumulation, and active deposition spread into the Campeche salt basin. In the Salina del Istmo, shortening began as early as the Oligocene (23 Ma) and continues into the present day, with the most recent contraction on the far western areas near the Kunah-1 and Yoka-1 wells (Snyder and Ysaccis 2018). The total section shortening of 23 percent (27 km; 17 miles) is based on wide azimuth (WAZ) 3D seismic data, with the main contractional phase estimated to be post-Lower Miocene (Snyder and Ysaccis 2018). This, plus the interpreted Paleogene rafting of the Mesozoic interval, makes this area one of the most tectonically complex regions in the GoM basin.

An arid climate continued to dominate the southwestern USA and northwestern Gulf coastal plain (Scotese 2017). However, a humid tropical climate belt prevailed in the southern Mexican uplands and coast, providing ample runoff to large rivers there.

The Neogene Tectono-climatic Phase displays several broad depositional themes:

1. Sediment influx rates were high in the Middle Miocene, declining to still-respectable minimum values in the Late Miocene and Pliocene. Influx was resurgent in the Pleistocene, attaining values not seen since the Paleocene Wilcox maximum (Galloway *et al.* 2011).
2. Sediment influx was highest in the north-central Gulf basin, which became the increasingly dominant depocenter as continental rivers converged on the Mississippi Trough.
3. The role of shore zone depositional systems as volumetrically important sediment repositories declined dramatically (Galloway, 2002).
4. Continental margins displayed impressive rates of offlap (Figure 7.1A). Robust delta-fed aprons provided the depositional foundation for shelf edge progradation. Mass basinward transfer of salt from beneath updip depocenters further augmented construction of the continental slope by insertion of allochthonous salt from below and within. Basinward injection, inflation, and advance of large volumes of shallow salt beneath the depositional continental slope added significant volume to the slope prism.
5. Offlap was interrupted locally by subregional margin foundering and collapse at a stratigraphically significant scale (Morton 1993).
6. Continental slope bathymetry was profoundly influenced by inflation, evacuation, extrusion, and basinal flow of salt within the Sigsbee canopy and its younger, shallower offsprings. As a result, the continental slope was an obstacle course of "complex corridors" for gravity flow transport systems (Steffens *et al.* 2003). Local slope accommodation was created by rapidly subsiding minibasins (Prather *et al.* 1998; Prather, 2000; Meckel *et al.* 2002). Ponded

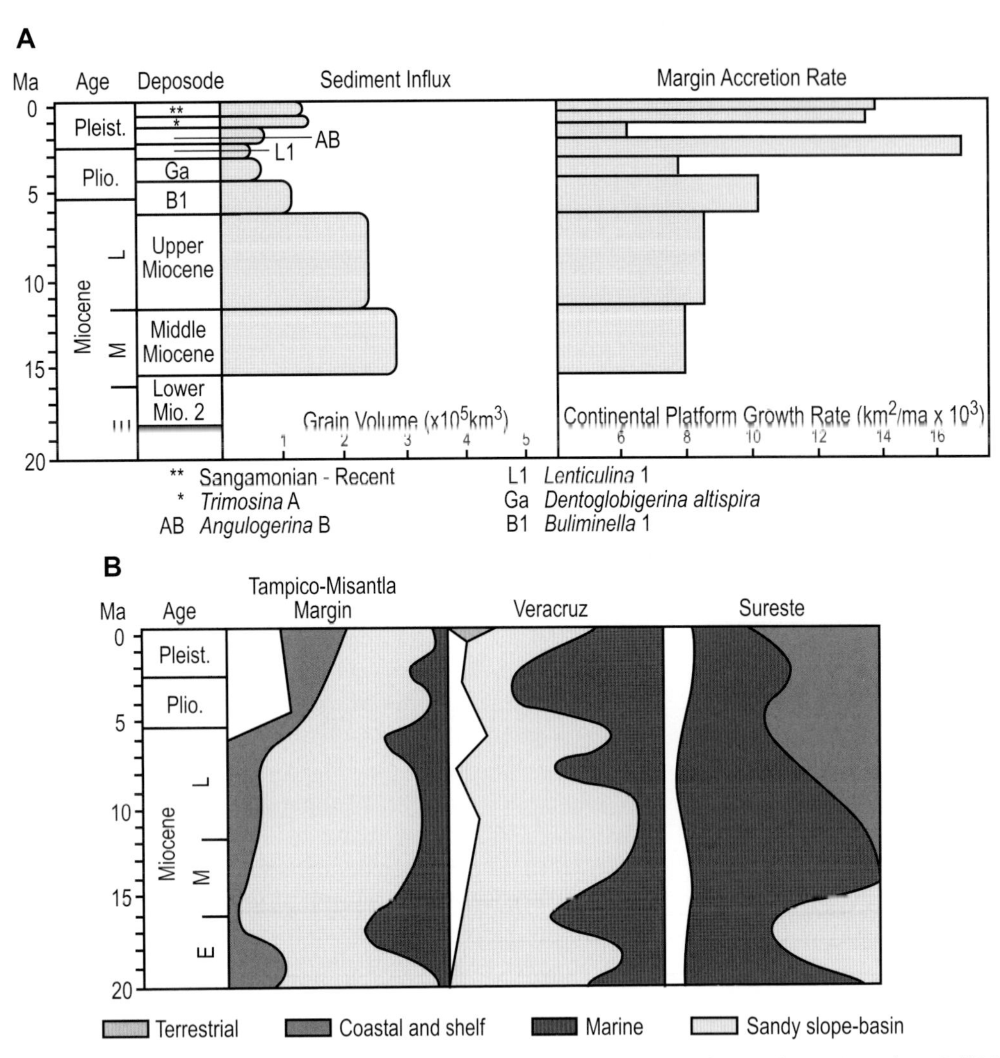

Figure 7.1 Depositional episodes and chronology of the Late Cenozoic Tectono-climatic Phase. **(A)** Northern Gulf (US) margin deposodes, sediment influx rates, calculated as total grain volume, and continental margin accretion. High-frequency Plio-Pleistocene deposodes reflect the combined impact of increasing impact of glacioeustasy and high-resolution paleontologic subdivision within the robust northern depocenter. **(B)** Stratigraphic units of the Mexican Gulf margin. Patterns distinguish terrestrial, coastal and shelf, deep marine, and sandy, deep marine facies associations. The Sureste province includes the Macuspana, Comalcalco, and Isthmus basins and the adjacent Campeche salt basin.

minibasin-fills can exceed 10,000 feet in thickness. As upslope minibasins filled, sediment spilled along active corridors into the next basin downslope.

7. Abyssal plain fan systems again assumed a role as the volumetrically important terminus of regional sediment dispersal systems. The largest submarine fans spread across the breadth of the Gulf basin. Thus, despite the complex transport pathways and locally high rates of accommodation space creation by salt evacuation and extension, a significant proportion of total sediment supply successfully bypassed to the basin floor. Mature, shelf-penetrating submarine canyons did not, however, become abundant until the Pleistocene (Galloway, 2005b).
8. The Mexican Gulf margin developed a narrow, prograding sediment prism composed of wave-dominated coastline, shelf, and progradational slope apron systems (Jennette *et al.* 2003; Ambrose *et al.* 2005; Hernandez-Mendoza, 2013). Migratory giant dune fields spread along the lower slope and adjacent basin plain (Perez 2017).
9. The Veracruz basin filled, culminating in progradation of a coastal plain and shoreline (Jennette *et al.* 2003).
10. A prograding continental margin succession of slope apron, deltaic, shore zone, and coastal plain systems advanced into and across the Camalcalco, Macuspana, and Salina basins and onto the thick evaporite interval of the southern Campeche salt basin (Garcia-Molina 1994; Ambrose *et al.* 2004).

7.2 Chronostratigraphy and Depositional Episodes

The Neogene stratigraphy of the US and Mexican Gulf margins was largely unraveled as petroleum exploration provided subsurface data, first beneath the coastal plain and then the continental shelf. The monotonous alternation of sand and mudstone initiated the use of micropaleontology as a way to identify, correlate, and date subsurface units. Thus stratigraphic nomenclature for the Neogene deposodes and supersequences is largely based on prominent faunal markers, assemblage, and interpreted age. Evolving sophistication and resolution of the micropaleontologic framework, particularly when combined with reflection seismic data, well control, and sophisticated sequence stratigraphic and depositional models, led to creation of detailed stratigraphic frameworks (e.g., Morton and Ayers 1992; Weimer *et al.* 1998, 2017; Zeng and Hentz, 2002; Martin *et al.* 2004) and chronology (e.g., Paleodata, Inc. 2017). Increasing magnitude and frequency of eustatic sea-level change and resultant environmental shifts, very high rates of accumulation in increasingly focused depocenters, and petroleum exploration needs all combined to produce uniquely high-resolution stratigraphies in some parts of the basin-fill.

Galloway *et al.* (2000), building on studies and correlations of previous workers, selected a widely recognized framework of two Miocene and six Plio-Pleistocene deposodes that can be regionally correlated across the northern GoM (Figure 7.1A). Following long-established custom, the resultant genetic supersequences are named for their age (Miocene) or the widely used paleontologic marker defining the unit top (Plio-Pleistocene). Middle and Upper Miocene deposodes correspond closely (but not exactly) to the global Middle and Late Miocene ages. The next three deposodes together approximate the Pliocene interval. Note, however, that the *Buliminella* 1 (PB1) deposode began in the latest Miocene, the *Dentoglobigerina altispira* (PGa) deposode lies within the middle of the Pliocene, and the *Lenticulina* 1 (PL1) deposode extended a few hundred thousand years into the global Early Pleistocene. The remaining Pleistocene includes three deposodes: the *Angulogerina* B (PAB), *Trimosina* A (PTA), and the Sangamonian (PS). For this basin-wide synthesis, we will group these genetic sequences into two supersequences that approximate the Pliocene and Pleistocene intervals.

In Mexico, no basin-scale stratigraphic nomenclature has been established. Correlations among basins are based largely on local faunal occurrences (e.g., Nieto 2010), interpreted age, seismic markers, and/or interpreted sequence stratigraphic ties to global sea-level curves. Broad facies patterns (Figure 7.1B) suggest a prominent transgression/onlap phase at the end of Early Miocene, followed by one or more major regressive pulses in the remaining Middle to Late Miocene. The Plio-Pleistocene record is diverse and reflects tectonic uplift of the Tampico–Misantla basin margin, much-reduced sediment supply (until Late Pleistocene) in the Veracruz basin, and a surge of supply-driven offlap in the basins of the Sureste province.

7.3 Previous Work

The advance of exploration from the Gulf coastal plain onto the continental shelf was followed by significant discoveries in progressively younger strata, culminating in Pleistocene sequences beneath the outer shelf and upper slope. With little or no outcrop and coastal plain stratigraphy consisting of non-marine facies, new stratigraphic and correlation frameworks were required. Micropaleontology, largely based on foraminifera assemblages, provided the practical basis for correlation and dating using drill cuttings. The offshore also provided the first opportunity to apply systematic regional seismic surveys to aid correlation and sort out the complex structures produced by insertion and deformation of the Sigsbee salt canopy.

Shideler (1986) and Reed *et al.* (1987) produced the first folios of regional correlation sections, incorporating both well and seismic records, for the Neogene interval beneath the continental shelf. Regional depositional systems syntheses of the Miocene followed with publications of Morton *et al.* (1988), Wu and Galloway (2002), and Combellas-Bigott and Galloway (2006). Morton *et al.* (1991) and Morton and Ayers (1992) brought Texas Bureau of Economic Geology (BEG) depositional systems analyses up through the Plio-Pleistocene sequences of the central GoM continental shelf.

Several university-based industry consortia programs produced suites of subregional reports on select areas of the continental shelf and the emerging exploration fairways beneath the continental slope. The GoM Structural and Stratigraphic Synthesis (GMS3) under the joint supervision of J. S. Watkins and W. R. Bryant (Texas A&M University) and R. T. Buffler (University of Texas at Austin) was initiated in 1987. Its main objective was to map the structure and stratigraphy of the northern GoM shelf and slope. Using 130,000 line km (80,000 line miles) of seismic data and logs and paleo-reports from over 700 wells, students mapped five Plio-Pleistocene and four Miocene horizons throughout the northern Gulf. Numerous publications resulted, culminating in GCAGS Special Publication 80 in 1996.

Following up on his 1990 publication on the seismic sequence stratigraphy of the Mississippi Fan, Paul Weimer organized a consortium at the University of Colorado, with a focus on the central GoM continental margin in the Green Canyon protraction area. Using both 2D and 3D seismic and well datasets, students and research associates applied sequence stratigraphic methodology and systems tract analysis to document Late Neogene structural and stratigraphic evolution of the continental margin. Summaries of project results were included in a special issue of the AAPG Bulletin (e.g., Weimer *et al.* 1998).

The emerging importance of minibasin-hosted hydrocarbon plays in shelf and upper slope suprasalt Neogene strata instigated several key papers that described the interplay between structural evolution and depositional history of slope turbidite systems (e.g., Armentrout *et al.* 1996; Prather *et al.* 1998; Booth *et al.* 2000, 2003; Winker and Booth 2000).

Similar regional syntheses or consortia studies of Neogene deposits in the western Gulf are few. Joint projects between Pemex and the BEG produced integrated studies of the Veracruz basin (Jennette *et al.* 2003), the Macuspana basin (Ambrose *et al.* 2004), the offshore Burgos basin (Hernandez-Mendoza *et al.* 2008b), and the Laguna Madre–Tuxpan margin (Ambrose *et al.* 2009). Chavez-Valois *et al.* (2009) provide a recent synthesis of the Sureste province.

7.4 Middle Miocene Deposode

The Middle Miocene genetic supersequence records the culmination of the shift of sediment input by continental river systems from the northwestern to the central GoM continental margin. The principal depocenter lies beneath the eastern Louisiana coast. A secondary depocenter underlies the central Texas shelf. Shelf margin accretion rate accelerated to a respectable 8000 km^2 (3100 miles^2) per million years (Figure 7.1A). Maximum shelf edge progradation of ~50 km (30 miles) occurred within the two depocenters. Three fluvial–deltaic axes dominated sediment supply (Figure 7.2). The Mississippi River continued its long history as a major continental drainage system supplying sediment to the GoM. For the first time in the Cenozoic, a major river, named the Tennessee, entered the coastal plain to the east of the Mississippi axis. This system, likely present as a regional drainage element emerging from a long-dormant and deeply eroded upland that had supplied little sediment throughout the Late Mesozoic through the Early Miocene, became the conduit for large volumes of quartz-rich sand and mud derived from resurgent Middle Miocene erosional unroofing of the Paleozoic Appalachian upland (Boettcher and Milliken 1994). To the west, a second, newly important river system occupying the approximate position of the modern Guadalupe River emerged on the central Texas coastline. It contained reworked Cretaceous grains and clasts that confirm erosion of the central Texas Balcones upland supplied part of the load of this fluvial axis, which was previously initiated as a relatively small river by the earliest Miocene (Galloway *et al.* 1982c).

Along the Mexican Gulf margin, multiple small rivers supplied sediment to the depositional coastline. The evolving

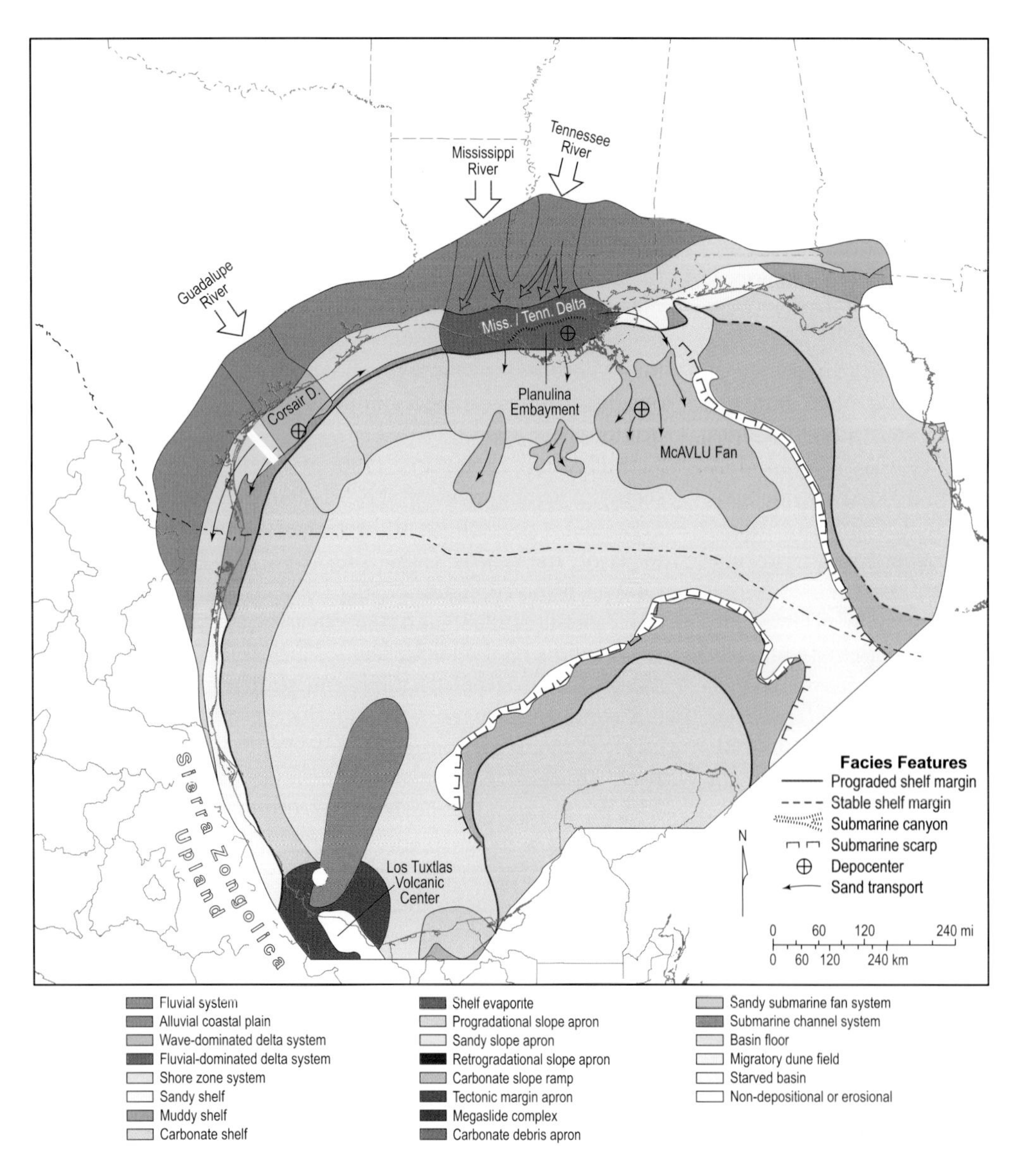

Figure 7.2 Paleogeographic map of the Middle Miocene deposode. Depositional system outlines and shelf edge reflect their positions at maximum progradation. White box locates seismic line shown in Figure 7.16.

history of multiple local drainage basins and their diverse source terranes is reflected in the temporal and geographic variation in sandstone mineralogy and in the abundance of mechanically and chemically unstable sedimentary and volcanic rock fragments (Dutton *et al.* 2002; Martinez-Medrano *et al.* 2011). Part of this sediment was redistributed along the shoreline and narrow shelf that fringed the uplands. Much sediment spilled over the shelf edge, constructing a prograding slope apron. Construction of the apron allowed the western Gulf shelf edge to advance up to a few tens of kilometers. Erosion and sediment bypass down the continental slope continued in the Veracruz basin.

7.4.1 Paleogeography

Together the three continental fluvial systems supported deposition of two large delta systems in the northern Gulf: the wave-dominated Corsair delta and the merged deltas of the adjacent Mississippi and Tennessee Rivers (Figure 7.2; Morton *et al.* 1988; Combellas-Bigott and Galloway, 2002a). Progradation of the Corsair delta onto the continental margin initiated the arcuate Corsair fault, a highly listric growth fault that soled onto the shallow, subjacent salt canopy. Expansion across the fault accommodated more than 10,000 ft (3000 m) of delta front sands and prodelta muds. The fluvial-dominated Mississippi–Tennessee delta system similarly prograded onto the upper slope. Along the front of the Tennessee, early depositional loading induced rapid salt evacuation and consequent subsidence and margin failure, creating the Harang Embayment (Combellas-Bigott and Galloway, 2002a). Like other retrogradational embayments, the Harang was healed by subsequent progradation of the delta. Margin offlap was greatest along the east flank of the Tennessee axis, into the previously undersupplied northeast corner of the Gulf.

Flanking the delta systems, wave-dominated shore zones extended eastward to south Alabama, westward along much of the Texas coast (excluding the Corsair delta), and southward along the Burgos and Tampico–Misantla coasts to a termination at the north end of the Veracruz basin (Ambrose *et al.* 2005; Hernandez-Mendoza *et al.* 2008a; Figure 7.2). Combined reworking of sediment from the multiple local streams, longshore reworking by shoreface and shelf currents from the deltaic headlands, and along-strike mud advection nourished narrow, muddy shelf systems and prograding slope aprons. Load-induced extensional faulting along the margin expanded outer shelf and upper slope facies.

Sand transported eastward from the Tennessee delta along the shoreface and shelf edge combined with sand and mud directly bypassed over the shelf edge and down the delta-fed slope apron to construct a major sand-rich submarine fan system on the northeastern Gulf abyssal plain. This fan system was named the McAVLU fan for its location within the Mississippi Canyon, Atwater Valley, and Lund OCS protraction areas (Galloway *et al.* 2000). Regional seismic lines show the fan to extend nearly 400 km (250 miles) southeastward across the basin floor. Proximal fan deposits, at the base of the continental slope, aggraded a depocenter more than 6000 ft (1800 m) thick and containing >2000 ft (600 m) of gross sand. Stacked submarine fan channel fill, levee, overbank, and splay facies are all well-developed (e.g., Reynolds, 2000; Cumming, 2002; Henry *et al.* 2017). The McAVLU fan, like its Early Miocene LM2 precursors (Figure 6.10), is displaced nearly 160 km (100 miles) eastward from the source delta's eastern margin (Snedden *et al.* 2012). This pattern, which persisted for more than 10 million years into the Late Miocene, defies conventional source–sink models that place fans directly downslope from the fluvial–deltaic axis. Like the contemporaneous shore zone systems, the offset of the fan from its source fluvial–deltaic axis dramatically demonstrates the volumetric and economic importance of regional along-strike transport pathways. Transport through coastal and shelf systems, which efficiently separate sand and mud into divergent pathways, helps to explain the sand-rich facies content of the McAVLU fan system.

Ultra-deep drilling has revealed additional, smaller fans emerging from the toe of the delta-fed slope apron beneath the Louisiana slope. These fans extend 80+ km (50+ miles) onto the Miocene abyssal plain. As drilling is sparse, and much of the Middle Miocene slope toe lies beneath thick, structurally complex salt canopies and welds, additional undrilled fans may well exist.

In the Veracruz basin, deposition of the tectonic apron continued (Figure 7.2). Channelized gravity flows scoured erosional gorges and small canyons along the basin margin; the depositional apron onlapped this bypass and scour surface (Jennette *et al.* 2003; Hernandez-Mendoza 2013). Large, erosional channel systems occur along the axis of the Veracruz Trough. The largest of these significantly exceed the scale of underlying Lower Miocene submarine channels; channel widths range across 5–10 km (3–6 miles; Winter 2018). Northward thickening of the Middle Miocene interval suggests increased sediment bypass to the west-central basin floor. The apron continued to supply sediment to the north-flowing submarine channel complex. However, both systems were smaller than their Early Miocene precursors. In the Macuspana and Comalcalco basins of the Sureste province, northward advance of a prograding deltaic platform was presaged by arrival of thick, sandy slope apron deposits, which are in turn overlain by late Middle Miocene wave-dominated delta and flanking shore zone sediments. The apron contains debrite and turbidite channel–lobe facies bounded by condensed intervals (Gutiérrez Paredes *et al.* 2017).

A notable feature of deep slope and basin Middle Miocene strata along the Mexican continental margin is the first, albeit limited, appearance of large-scale, accretionary sediment waves (which will be described in Section 7.5). These features create a distinctive seismic reflection facies suite.

7.4.2 Termination and Summary

The Middle Miocene deposode ended with regional transgression dated by several microfossil last appearance datums (LADs), including *Textularia* W, *Globorotalia soshi robusta*, *Globorotalia fohsi lobata*, and *Discoaster kuglerai* at ca. 11.8 Ma.

7.5 Late Miocene Supersequence

The Late Miocene deposode, recorded by the Upper Miocene genetic stratigraphic supersequence, lasted nearly six million years, ending about 500,000 years before the end of the Miocene (Figure 7.1A). Sediment influx remained high, and at least modest growth of the continental platform continued along the length of the US and Mexican margin. Widespread areas of erosion or non-deposition along the west Florida slope heralded establishment or a significant acceleration of the GoM loop current (Mullins *et al.* 1983).

7.5.1 Paleogeography

The principal fluvial–deltaic axes and slope systems (Figure 7.3) continued the paleogeographic framework established in the Middle Miocene deposode. The dominant depocenters lie beneath the Tennessee and Mississippi deltas, where the large, fluvial-dominated deltas prograded onto the continental slope and the shallow Sigsbee salt canopy (Wu and Galloway 2002, 2003). Secondary depocenters occur beneath the proximal McAVLU fan and the reduced, but still active wave-dominated Corsair delta system (Morton *et al.* 1988).

As in the Middle Miocene supersequence, depositional shore zone systems rimmed much of the northwestern Gulf. Relatively shallow burial beneath the Texas continental shelf allows seismic imaging of the last of the major strike-fed barrier bar systems of the northern GoM (Figure 7.4). Such sand-rich depositional systems had persisted as major elements

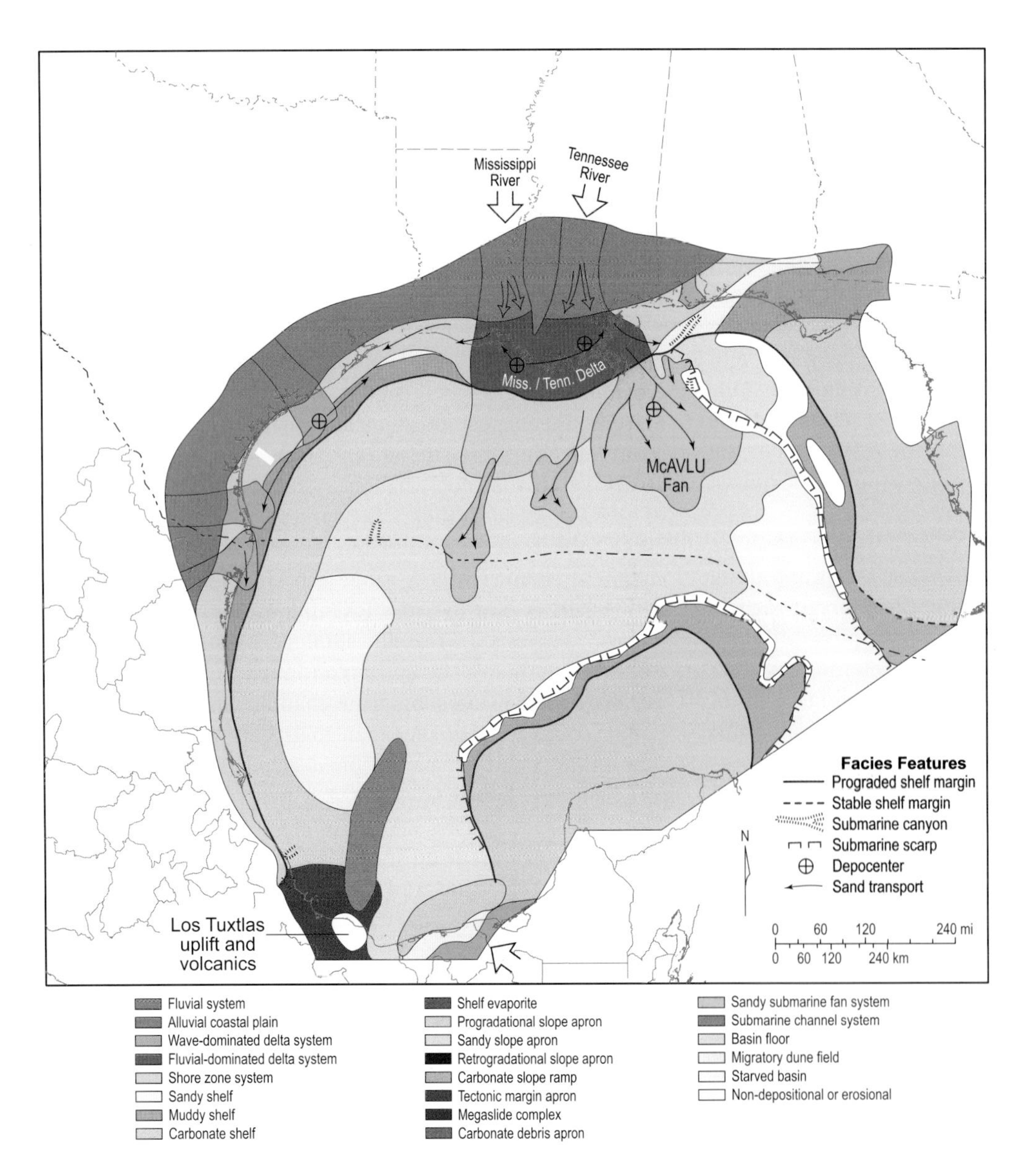

Figure 7.3 Paleogeographic map of the Upper Miocene deposode. Depositional system outlines and shelf edge reflect their positions at maximum progradation. White box locates seismic line shown in Figure 7.4.

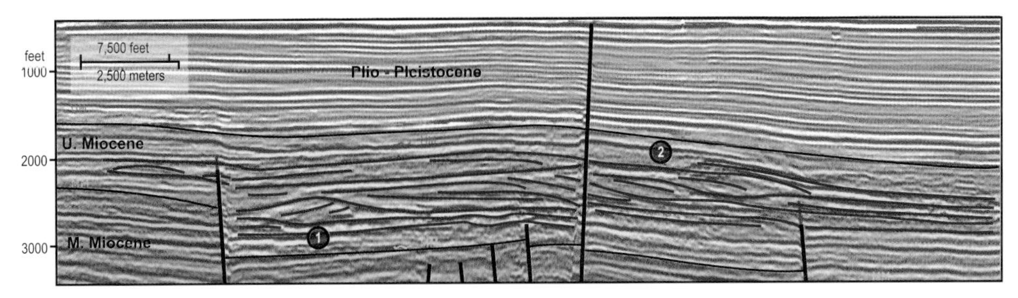

Figure 7.4 Seismic line interpretation showing internal architecture of the sand-rich shore zone system preserved within the Upper Miocene supersequence of south-central Texas shelf. Mounded reflections, sigmoidal progradational clinoforms, and lenticular elements characterize the axial barrier complex that is nearly 1000 ft thick. For location, see Figure 7.3. Seismic line courtesy of ION.

of the northwestern coastal zone throughout the Middle Cenozoic Geothermal Phase. Though shore zones were present in the later Neogene, they accumulated only thin, discontinuous facies tracts of minimal volumetric importance relative to deltaic and slope/basin depocenters (Galloway, 2002).

Abundant well penetrations and good seismic resolution provide a clear picture of the sediment transport pathways and distribution in the northeastern fluvial–deltaic depocenter and its associated submarine fan system (Figure 7.5). Combined loading, salt evacuation, and compaction subsidence accommodated more than 12,000 ft (3600 m) of sediment along the combined delta margin depocenters beneath the Louisiana shelf. The sandiest succession is displaced eastward, lying along the front of the Tennessee delta. Both interval thickness and sand distribution maps indicate the volumetric importance of the McAVLU fan system. Local depocenters in the fan exceed 8000 ft (2400 m) in thickness. Dip-elongate, bifurcating, digitate sand distribution axes containing more than 1000–2000 ft (300–600 m) gross interval of sand radiate from the fan apex along the shelf margin on the east flank of the Tennessee delta. Well control shows supersequence thickness and sand content to decrease markedly along the updip flank of the Miocene Atwater–Mississippi Fan fold belt (discussed below). However, seismic data display typical fan facies extending nearly 160 km (100 miles) further across the abyssal plain.

Emergence of much of the McAVLU fan from beneath the shallow salt canopies allows enhanced seismic resolution of its depositional architecture. Further, discovery of large, highly prolific deep basin Miocene petroleum fields, including Thunderhorse and Ram Powell, and their subsequent development has provided a uniquely rich data suite for an abyssal plain fan system. Regional well log cross-sections, such as Figure 7.6, display the complex vertical and lateral facies relationships within the fan depocenter and the adjacent slope apron. Close examination of well log patterns shows the full array of blocky, upward-fining, upward-coarsening, and serrate motifs typical of channelized turbidite flow systems. High-resolution studies, incorporating cores, 3D seismic imaging, and closely spaced production well data has characterized fan system reservoir facies in detail (Bramlett and Craig 2002; Meckel 2002; Greene and O'Neill 2005). Identified lenticular reservoir types include erosional, depositional, multilateral, and multistory channel fills, along with associated channel levee and abandonment facies. Sheet reservoirs include channel–margin splay, channelized lobe, and sheet lobe facies. Lateral facies changes, crosscutting units, and internal erosion surfaces are prominent features of the fan systems.

Despite wholesale bypass of sediment to the basin floor, the shelf edge in front of the Mississippi–Tennessee delta system advanced nearly 80 km (50 miles) over the underlying progradational delta-fed apron (Figure 7.5A). Smaller fans emerge from the slope toe in front of the delta systems (Figure 7.3).

In the Burgos basin, small wave-dominated deltas emerge from the long, but thin shore zone facies belt. There, combined sediment supply was sufficient to produce a progradational bulge in the continental margin.

Paleogeography of the Mexican Gulf margin largely continued Middle Miocene patterns. The progradational sandy shore zone, shelf, and upper slope systems display dramatic expansion across the Quetzalcoatl fault zone along the arcuate Tampico–Misantla margin. The broad, shelf-fed progradational slope apron began to develop nascent anticlinal uplifts (Salomon-Mora *et al.* 2009). Seismic attribute maps (Arreguín-Lopez *et al.* 2011; Figure 7.7) illustrate the facies architecture of the submarine channel system that flowed northward from the tectonic margin apron of the Veracruz basin. Anastomosing to highly sinuous channels lie within broad (2–15 km/1.2–9 miles) belts. Multiple local sediment sources along or near the Gulf basin margin included the Trans-Mexican Volcanic Belt and its Los Tuxtlas outlier and recycled Paleogene orogens of the Sierra Zongolica upland (Dutton *et al.* 2002; Paredes *et al.* 2009). The result was a suite of lithic-rich sandstones. Seaward, extensive migratory bedform complexes characterized much of the lower slope and abyssal plain (Arce 2017). The dunes display a range of depositional styles ranging from upslope-climbing migration to vertical accretion (Figure 7.8).

A fully developed coastal plain, deltaic, shore zone, shelf and slope apron depositional systems tract prograded northward into the basins of the Gulf of Campeche. The clastic facies graded northeastward into carbonates of the Yucatán Platform.

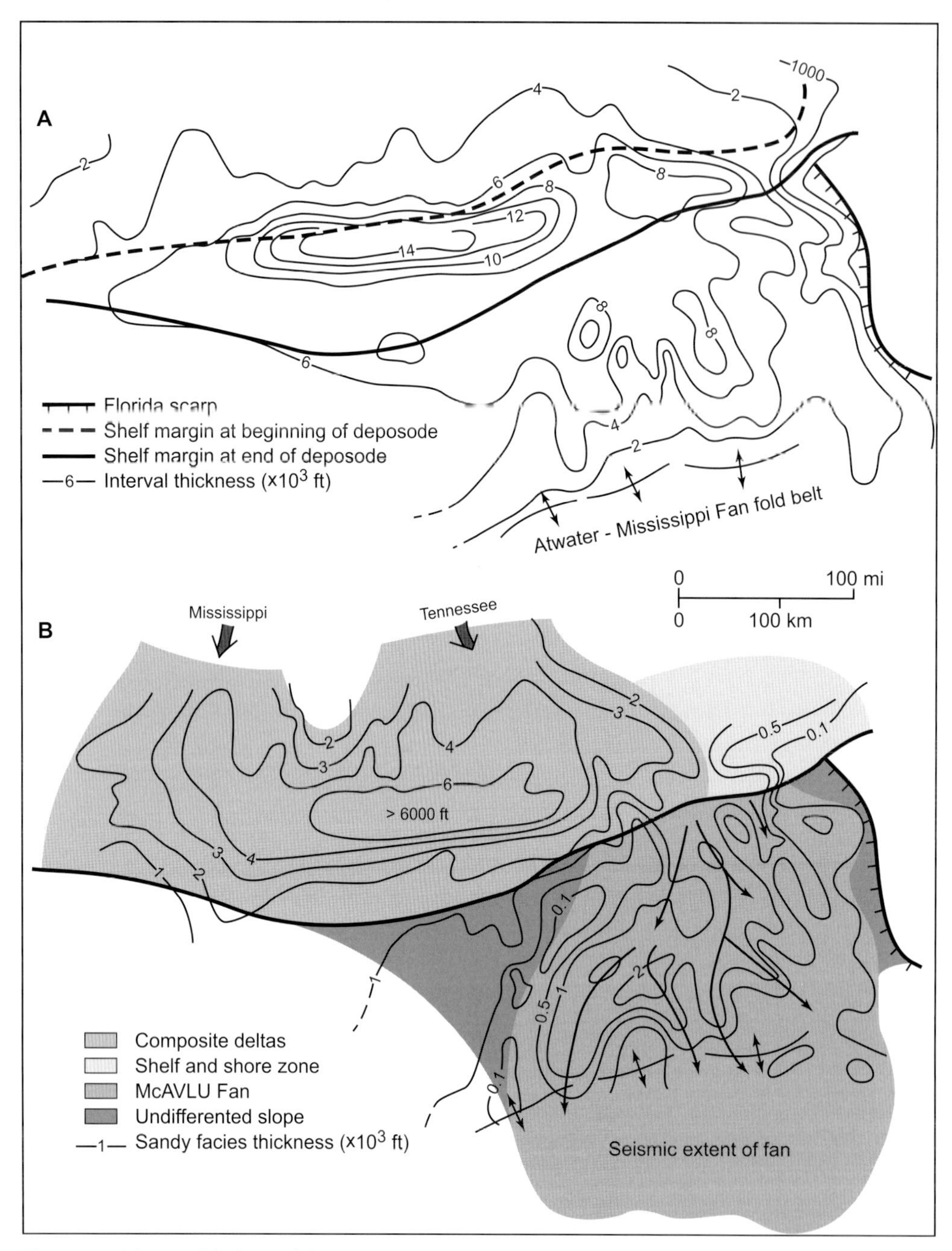

Figure 7.5 **(A)** Interval thickness of the Upper Miocene genetic supersequence in the northeastern GoM basin. **(B)** Gross thickness of sandy facies. The amalgamated Mississippi–Tennessee delta system created the depocenter outlined by the 8000 ft contour as it prograded over the inherited shelf margin. The sand depocenter is displaced eastward, along the Tennessee delta front, suggesting that the river was comparatively sandy. The lobate outline of the McAVLU fan is clearly shown by the 1000 ft sand contour.

7.5.2 Termination and Summary

The Upper Miocene deposode ended across the northern GoM with long-term regional coastal retreat and margin flooding. Benthonic foraminifera *Bigenerina* A and *Robulus* E LAD markers are widely used to date this boundary at 6 Ma, within the Messinian (Figure 7.1A). Although the *Robulus* E datum dates about 700,000 years younger than *Bigenerina* A (Paleo-Data Inc. 2017), the two tops are commonly found in close stratigraphic proximity, reflecting relatively low sediment accumulation rate and regional transgression in the early Messinian. Broadly equivalent intervals of diminished sediment influx and basin accumulation are interpreted in the Tampico–Misantla and Veracruz margins and in the Sureste province (Figure 7.1B).

Together, the Middle and Late Miocene deposodes were a relatively stable and distinctive 10-million-year phase in GoM evolution. Notable milestones include:

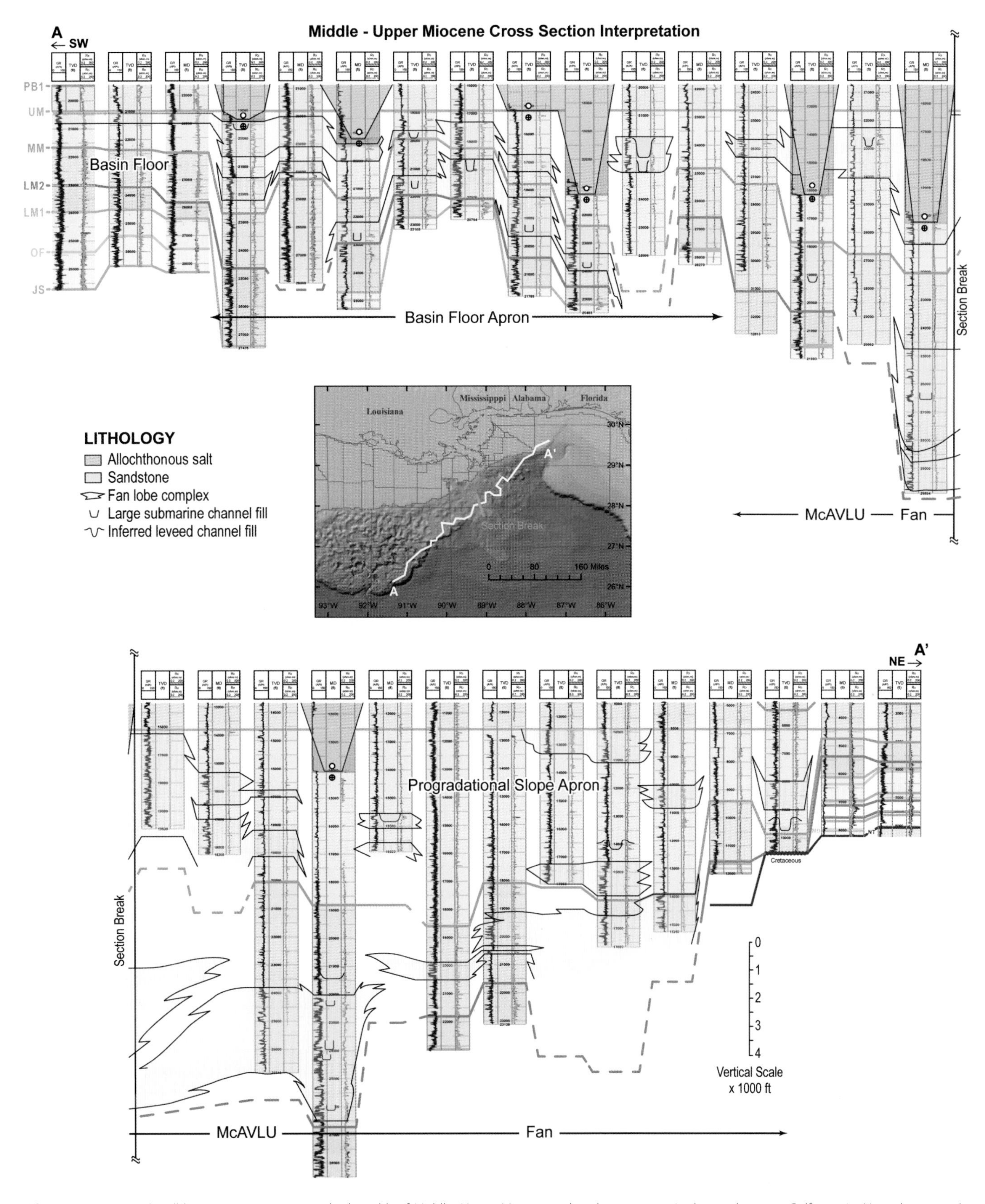

Figure 7.6 Regional well log cross-section across the breadth of Middle–Upper Miocene paleo-slope systems in the northeastern Gulf margin. Note that several wells penetrated superjacent, thick salt canopies emplaced by Plio-Pleistocene salt flow. The section traverses the transition from the Upper Miocene lower continental slope to abyssal plain and crosses the McAVLU fan on the northeast and the slope apron on the southwest.

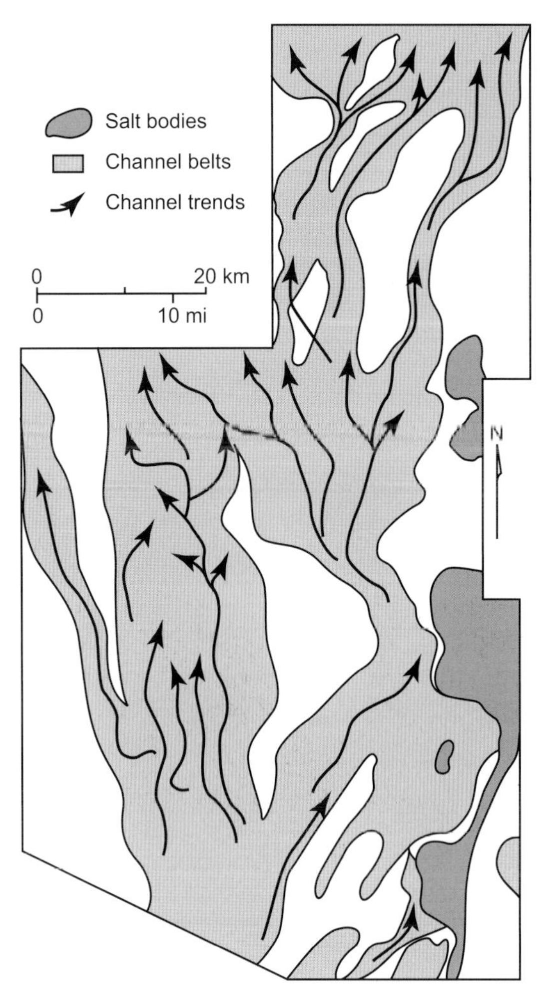

Figure 7.7 Seismically based map of interpreted sand belts and their contained axial channels in the Veracruz Trough submarine channel system. Redrawn from Arreguín-Lopez *et al.* (2011).

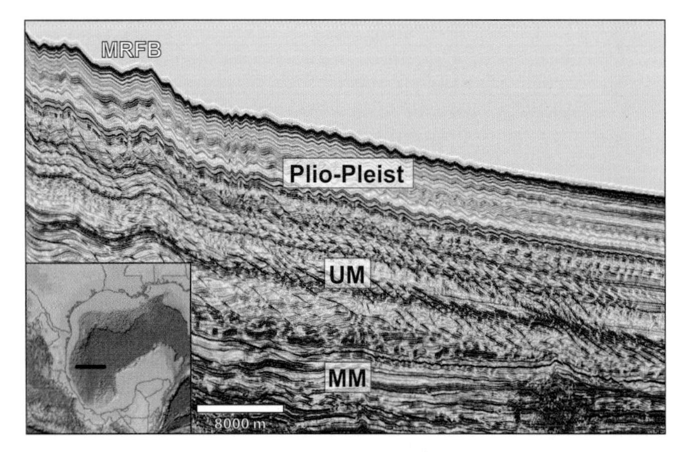

Figure 7.8 Seismic line interpretation typical of the Tampico–Misantla lower slope to rise showing the various climbing wavy bedform morphologies that dominate Late Neogene sequences. Upslope bedform climb creates apparent seaward dipping accretion sets analogous to the pseudo-stratification long recognized (at a much smaller scale) in climbing ripples. From Arce (2017).

1. The central Gulf margin emerged as the predominate depocenter. Declining supply through rivers of the northern Mexico–Texas margin further amplified the shift of depocenters from west to east.
2. With arrival of rejuvenated sediment flux from Appalachia through the Tennessee axis, the northeast corner of the basin, which had been largely sediment-starved for most of the Cenozoic, was largely infilled by combined aggradation and offlap.
3. Renewal of volumetrically significant and focused bypass of sandy sediment across the depositional continental slope established and continued to nourish a new suite of abyssal plain submarine fan systems.
4. Areas of aggrading, high-amplitude, migrating bedforms were first established and then became widespread in the deep western Gulf. Contemporaneously, accelerated deep marine current activity began to influence patterns of deep basin erosion and accumulation.

7.6 Pliocene Deposodes

Galloway *et al.* (2000) followed the regional studies of Morton *et al.* (1991) and Morton and Ayers (1992) in defining three Pliocene deposodes and their genetic sequences. From oldest to youngest there are the (1) Late Messinian–Early Pliocene *Buliminella* 1 (PB1) deposode; (2) *Dentoglobigerina altispira* (PGa) deposode; and (3) *Lenticulina* 1 (PL1) deposode, which lasted a few hundred thousand years into the Early Pleistocene, as most recently redefined (IUGS, 2011; Figure 7.1A).

PB1 punctuated the latest Miocene–Pliocene transition with the abrupt termination of deposition in the McAVLU fan system. For nearly two million years, deep marine sediment was largely sequestered in a broad, prograding slope apron that extended across the front of the combined Mississippi–Tennessee delta system from offshore Mississippi to east Texas. Abyssal plain aggradation was greatest across the west-central basin floor; limited well and seismic data suggest a coalesced sandy basin-floor apron and possible small fans.

The PGa and PL1 sequences have the least volume. By Neogene standards, rate of sediment supply was relatively low during the PGa and PL1 deposodes, culminating a long-term decline that began in the Late Miocene (Figure 7.1A). In marked contrast to all previous Cenozoic deposodes, most depocenters lie seaward of the maximum progradational shelf edge, beneath the contemporary slope apron. Depocenters were further compartmentalized by local patterns of salt canopy evacuation or inflation. Though variable through time, margin accretion rates remained high despite the moderated rate of sediment influx.

7.6.1 Paleogeography

The paleogeographic map (Figure 7.9) combines and generalizes the depositional systems and sediment transport pathways of the PGa and PL1 deposodes. Together, the two sequences mapped record ~1.6 million years of deposition, a relatively

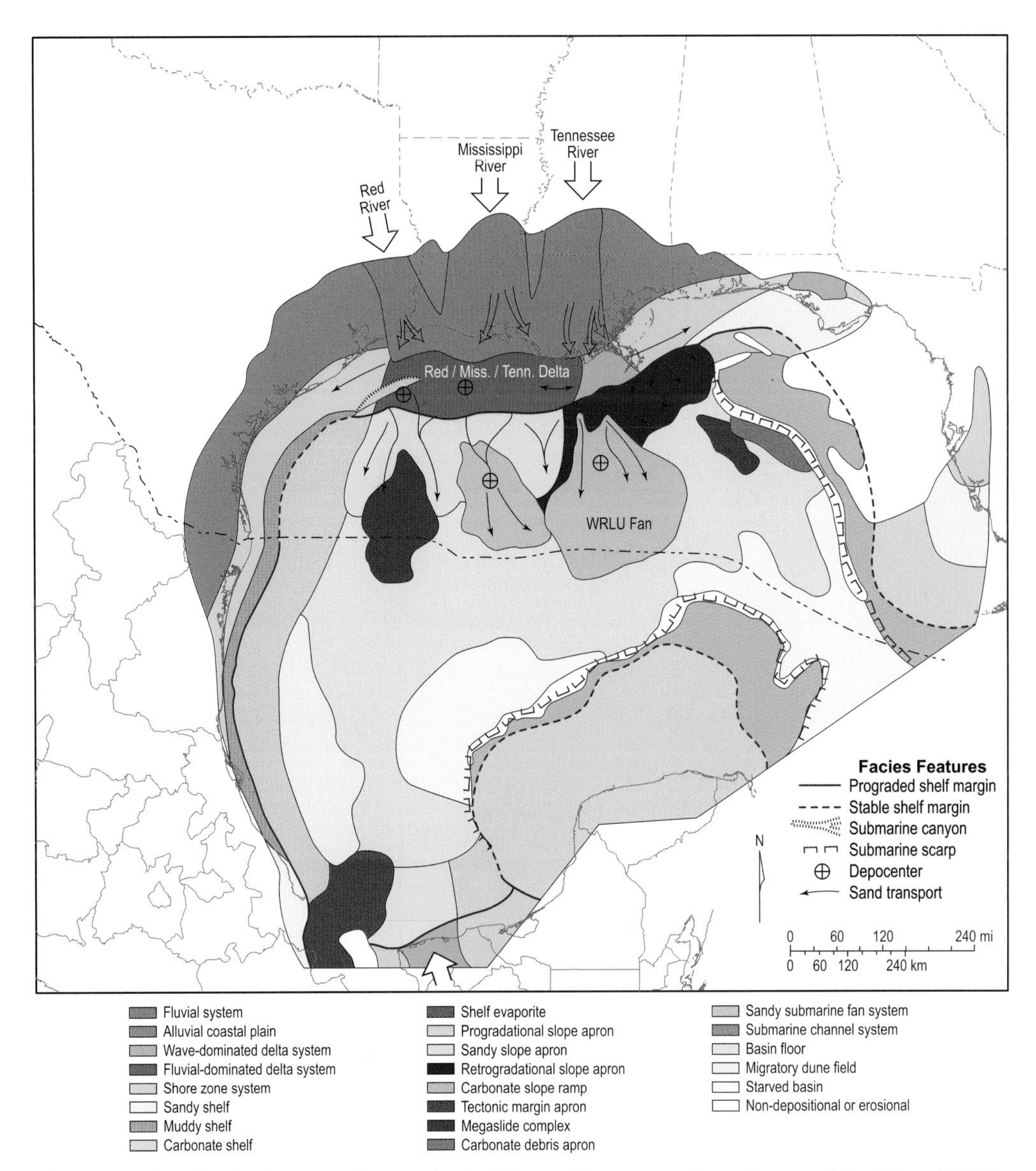

Figure 7.9 Generalized paleogeographic map for the Pliocene. The map combines elements of the PGa and PL1 deposodes. Depositional system outlines and shelf edge reflect their positions at maximum progradation.

brief interval compared to that of the earlier map intervals. The short time interval, combined with the relatively shallow burial depth of the sequences and consequent abundance of well and high-resolution seismic data, allows definition of much detail.

Sediment input to the northern GoM remained highly focused into the central margin. The Tennessee fluvial axis declined in importance, but remained active. The Mississippi influx continued unabated. A new addition, the Red River, flowed across the east Texas coastal plain. The addition of the Red River axis and continued prominence of the Mississippi axis were responses to the onset of uplift and incision of the Rocky Mountain Orogenic Plateau (Galloway *et al.* 2011). The three river systems converged onto the depositional coastal plain where they constructed a broad, composite delta system that spread 480 km (300 miles) from easternmost Louisiana to east Texas. Continental margin offlap was limited to the broad front of the shelf margin deltas. The deltas were dominantly large and fluvially dominated. However, the eastern flank of the Tennessee delta front is an exception. There, continuing decrease in sediment supply from the Tennessee River, which began in the Late Miocene, allowed marine reworking and ongoing subsidence to dominate the delta front. Local shelf margin retreat was accompanied by mass wasting. Resedimented delta front and margin sand created a narrow but sandy retrogradational slope apron (Figure 7.9).

In contrast to previous Cenozoic deposodes, the locus of sediment accumulation in Plio-Pleistocene deposodes shifted from the shelf margin basinward onto the upper–middle continental slope (Figure 7.10A). There, ongoing salt evacuation due to canopy loading combined with rapid minibasin infilling to efficiently trap sediment. Sand, however, tended to bypass the uppermost continental slope; sand depocenters commonly

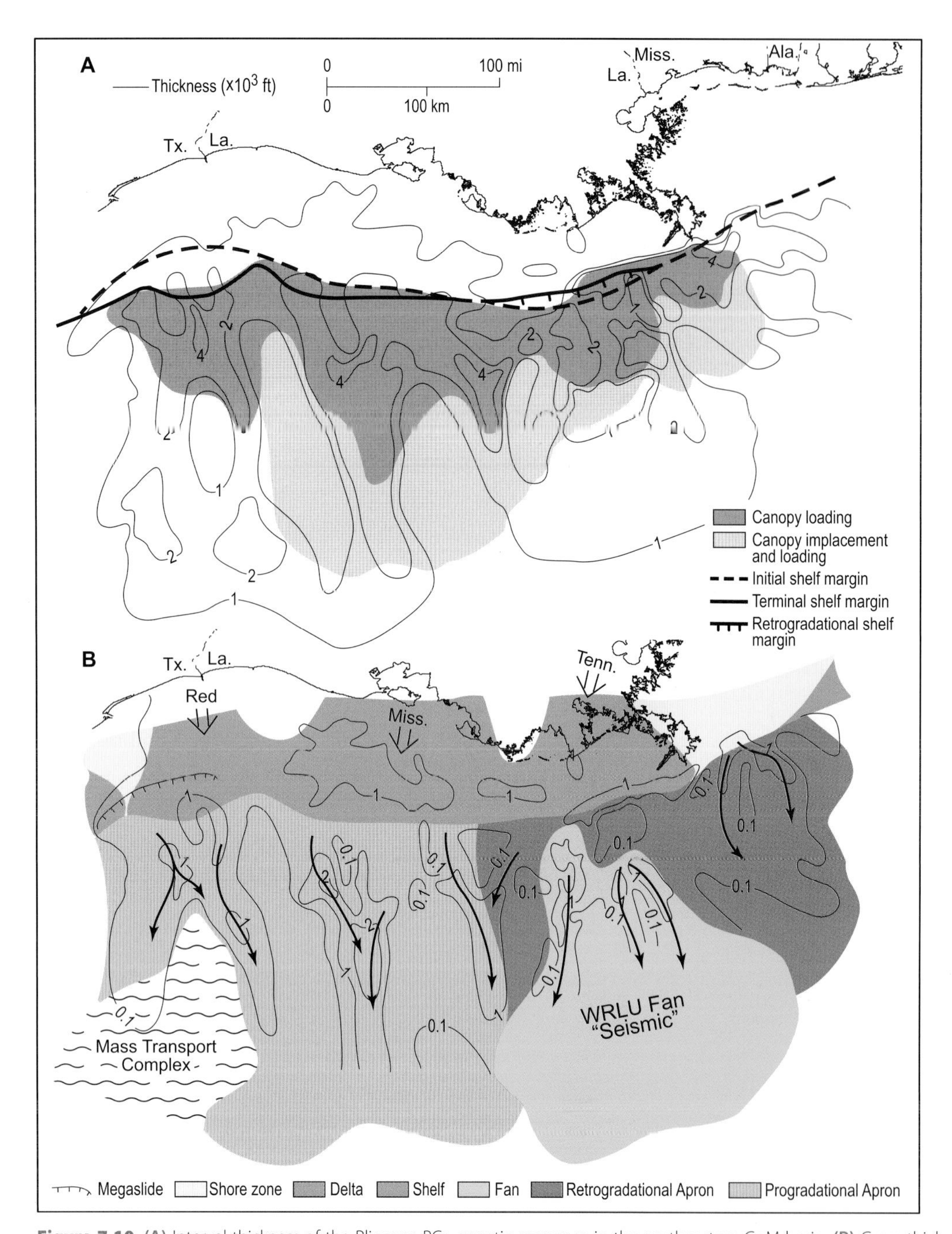

Figure 7.10 **(A)** Interval thickness of the Pliocene PGa genetic sequence in the northeastern GoM basin. **(B)** Gross thickness of sandy facies. The amalgamated Mississippi–Red River delta front supplied multiple, dip-oriented depocenters outlined by the 2000 ft contours as it prograded over the inherited shelf margin. Modest shelf edge offlap accompanied deposition of the delta-fed apron. The Tennessee delta front and shelf edge retreated several miles. The resultant retrogradational slope apron is a mix of new and recycled sediment. The lobate outline of the WRLU fan extends onto the basin floor from this apron.

traverse the delta-fed slope apron–lower continental slope (Figure 7.10B).

Along strike from the large deltaic headland, thin shore zone systems, fronted by narrow continental shelves, extend across the breadth of the northern Gulf Coast. Western rivers were small; their sediments were reworked and incorporated into shore zones. Nevertheless, shore zone and shelf sediment successions are uniformly thin, reflecting the minimal sediment storage on the continental platform that is typical of the Late Neogene Gulf (Galloway 2002).

Advance of the continental platform onto the extensive, shallow salt canopy produced complex continental slope bathymetry pocked by numerous rapidly subsiding minibasins (Diegel *et al.* 1995; Prather 2000). Minibasin "fill and spill" slope deposition characterized the prograding delta-fed apron fronting the Red River–Mississippi delta systems (Steffens *et al.*

2003). Linear "complex corridors" fronted the Tennessee delta, where lower rates of sediment supply created a comparatively stable to retrogradational margin with less loading and deformation of subjacent salt. Minibasin-fills consists of a mix of mass transport, erosional and depositional channel fills, and turbidite lobes and sheets (Armentrout and Clement 1990; Badalini *et al.* 1999; Meckel *et al.* 2002). Abundant well control for the supracanopy minibasin-fills demonstrates that a substantial proportion of sand bypassed to and was trapped in the upper to middle slope (Figure 7.10B). At least four sandy channel–lobe belts extend downslope from the Red River and Mississippi shelf-margin deltas in the progradational apron. Several smaller belts form the sandy skeleton of the Tennessee flank retrogradational apron. The depositional systems tract responsible for accumulation of the sandy fairways extending from the mud-dominated upper slope down the middle slope with locally ponded sand thicks, and onto the slope toe is graphically imaged in the seismic map shown in Figure 7.11. Leveed, depositional channel belts terminate both in intraslope basins related to active salt structures (forming the closed depocenters containing 1000–2000 ft [300–600 m] of sandy channel and lobe facies) and at the slope toe, where amalgamated lobes form a broad, sheet-like slope-toe apron. Where channel axes remain structurally focused and long-lived, they may terminate onto a defined abyssal plain submarine fan.

After nearly two million years of interrupted submarine fan system deposition on the Gulf abyssal plain during the PB1 deposode, a new fan (named the WRLU fan; Figure 7.9) system developed in the PGa deposode. This fan system, mapped using a mix of well and seismic data, emerges from the transition zone between the Mississippi progradational and Tennessee retrogradational slope apron. It forms a lobate sediment body with a broad apex containing diverging sand-rich belts on the middle slope (Figure 7.10B). Sparse subsalt and ultra-deep drilling penetrating channel fill and lobe sand bodies 100–200 ft in thickness suggest a possible second fan emerging from the broad delta-fed apron to the west of the WRLU fan system (Figure 7.12). Log patterns suggest both sand-rich lobe and channel fill facies successions in both the PGa and PL1 genetic sequences. Mounded morphology of the abyssal fans is clearly imaged on regional seismic profiles (cross-sections 3 and 5; Figures 1.15 and 1.17).

The Tampico–Misantla margin is a thick offlap succession displaying steep upper slope clinoforms (Ambrose *et al.* 2005; Fouad *et al.* 2009; cross-sections 7 and 8; Figures 1.19 and 1.20). Several small, wave-dominated deltas and an extensive shore zone supplied sediment to the narrow, sandy shelf and shelf margin. Sediment loading and ongoing displacement across the Quetzalcoatl growth fault zone accommodated a thick succession of continental margin facies. The broad progradational to aggradational slope apron contains numerous sandy channel–lobe belts trailing downslope from the shelf edge.

The Veracruz basin continued to fill with deposits of a shrinking tectonic margin slope apron overlain along its landward margin by progradational deposits of a sandy shelf and coastal plain (Jennette *et al.* 2003; Arreguín-Lopez and

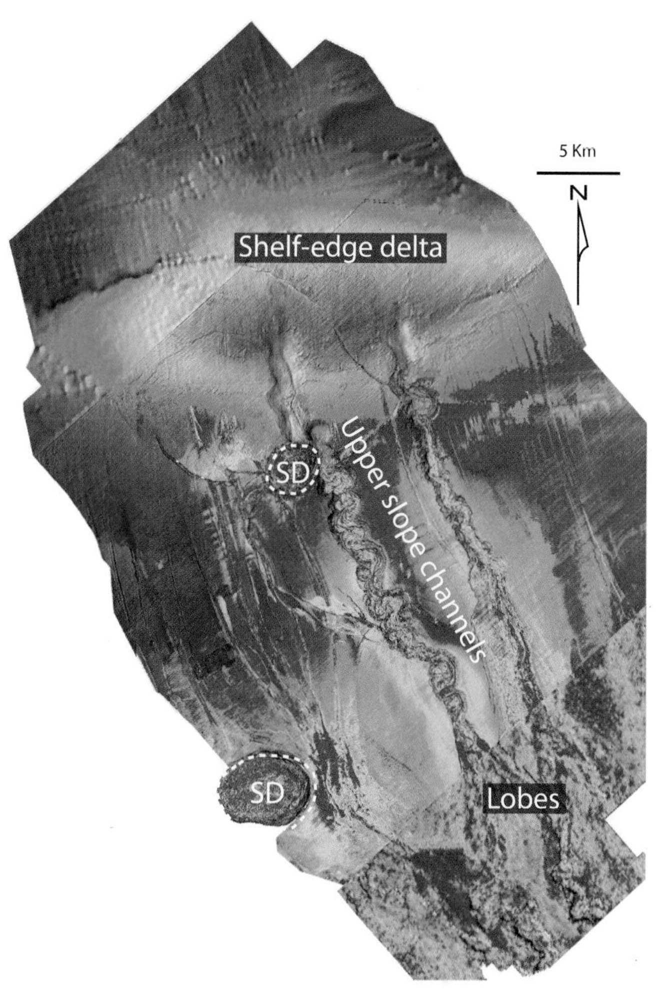

Figure 7.11 Seismic attribute map of the Einstein–Fugi slope, eastern GoM, graphically showing the depositional systems tract from a shelf-edge delta to slope apron. Erosional gorges cut into the delta front collect and redirect sandy sediment into slope channels, which evolve from erosional to depositional downslope. Muddy prodelta, mass transport, and hemipelagic sediment constructs the inter-channel upper to mid-slope. As slope gradient decreases, leveed channels open into turbidite lobes that merge to form the broad, sand-rich lower slope apron. Bathymetric irregularities caused by extensional faulting and salt stocks collect perched lobes. From Prather *et al.* (2017).

Weimer, 2004). The elevated Los Tuxtlas Platform created an uplap boundary for apron deposits along the southeastern Veracruz basin margin. Deposition of an organized submarine channel system along the Veracruz Trough ceased.

In Campeche, precursors of the Usumacinta/Grijalva fluvial systems, large rivers originating in the tropical Chiapas upland, combined to prograde a wave-dominated delta across the Salina del Istmo, Comalcalco, and Macuspana basins and into the open Bay of Campeche (Ambrose *et al.* 2003; Chavez-Valois *et al.* 2009). Delta progradation alternated with episodes of shore zone deposition and cross-shelf valley incision. Continental margin offlap by the prograding slope apron and shelf-margin deltas loaded the underlying Campeche salt. Salt stock inflation, canopy formation, and subsiding minibasins created a bathymetrically complex slope, much like that of the north-central Gulf margin.

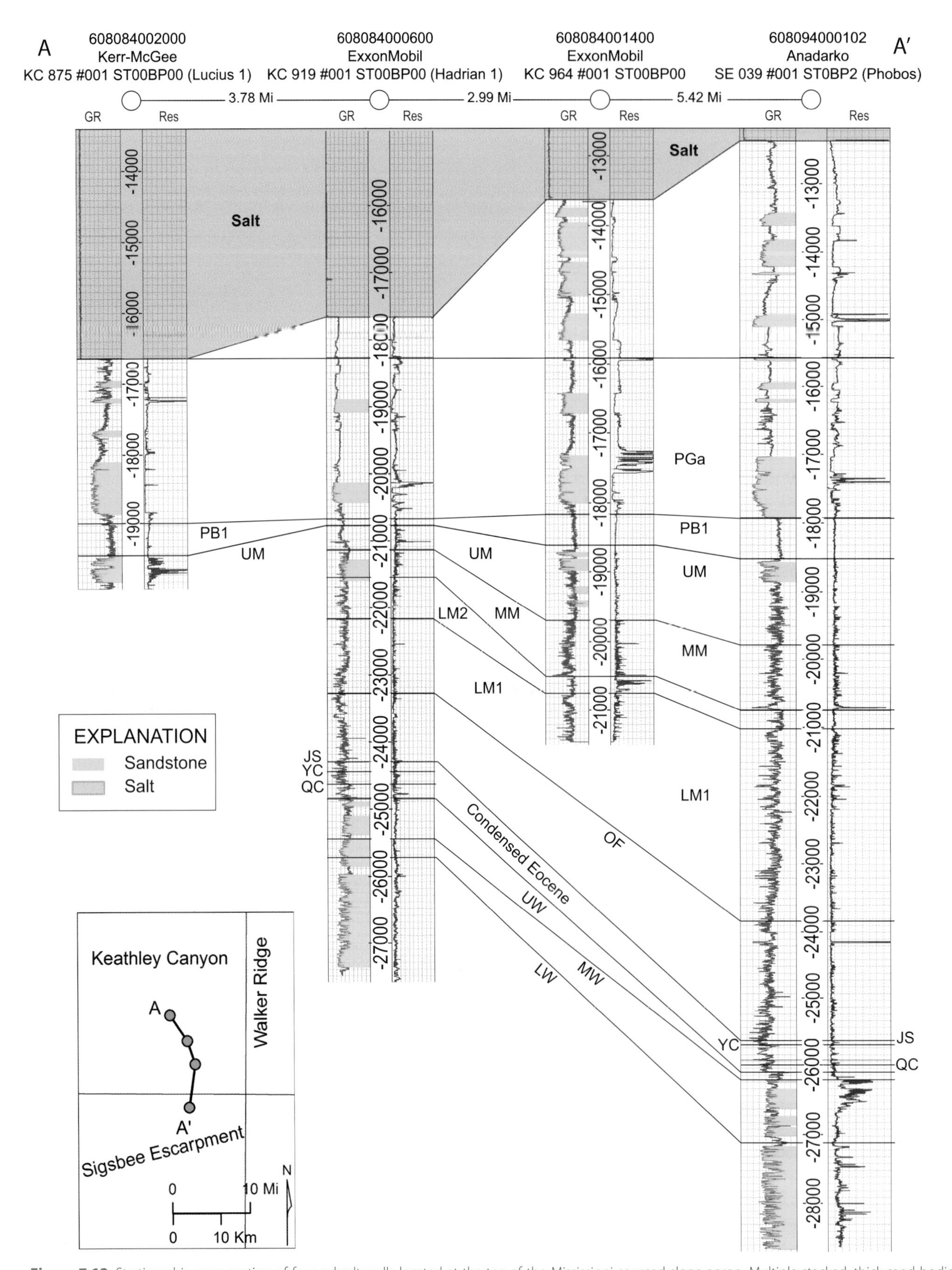

Figure 7.12 Stratigraphic cross-section of four subsalt wells located at the toe of the Mississippi-sourced slope apron. Multiple stacked, thick sand bodies in the Pliocene PGa and PL1 supersequences suggest a possible aggrading submarine fan system developed at the toe of the apron and extended onto the basin floor.

7.6.2 Termination

The combined PB1, PGa, and PL1 deposodes terminated with one of many flooding events that punctuated Late Neogene icehouse deposition. The *Lenticulina* 1 marker (ca. 2.3 Ma) is particularly useful because it approximates the Plio-Pleistocene boundary and is widely identified in wells of the northern Gulf. Several additional paleo markers cluster at this boundary (Paleo-Data Inc. 2017). Most importantly, the subsequent PAB deposode contains the first isotopic evidence of inflow of glacial meltwater into the GoM (Joyce *et al.* 1993). Meltwater incursion on a scale sufficient to alter oxygen isotope ratios of the GoM indicates the formation of the continental North American ice sheet. Continental glaciation dramatically affected runoff, fluvial geomorphology, and sediment yield of the Mississippi and possibly the Tennessee Rivers.

7.7 Pleistocene Supersequence

Galloway *et al.* (2000) differentiated three Pleistocene deposodes: *Angulogerina* B (PAB), *Trimosina* A (PTA), and Sangamonian fauna (PS). The Pleistocene deposodes experienced and record the full impact of ice sheet outwash, increasing amplitude and frequency of glacioeustasy, and interior North American drainage basin integration into a single valley-confined river system – the Mississippi. Sediment influx increased, continental margin accretion rate accelerated, and sediment bypass to the slope and basin reached its Cenozoic zenith.

7.7.1 Paleogeography

Figure 7.13 combines PTA and PS paleogeographies and stratigraphic features to generalize Pleistocene paleogeography. Alternating incision and filling of the Mississippi Valley led first to the capture of the Tennessee drainage basin (western Appalachians and Cumberland Plateau) and finally, in Late Pleistocene, to permanent capture of the Red River drainage system (eastern Rocky Mountain Orogenic Plateau; Galloway *et al.* 2011; Bentley *et al.* 2015). Delta flank shore zones and small deltas produced by a familiar family of Gulf Coast rivers, such as the Rio Grande, Brazos, and Apalachicola, accumulated comparatively minor volumes of sediment (Galloway 2002).

Rapid progradation of the fluvial-dominated Red River and Mississippi delta lobes, and repeated eustatically forced regression and cross-shelf river valley extension enhanced direct bypass of river-borne sediment onto the continental slope. Numerous structurally formed intraslope and basin-floor depocenters locally accumulated up to 10,000 ft (3000 m) of sediment. The shelf margin in front of the combined delta systems continued to actively prograde onto the foundation of the subjacent delta-fed apron. Like their Pliocene precursors, Pleistocene deposodes continued slope minibasin fill-and-spill deposition. Cold, sediment-laden outflows and inherent slope instability created numerous rapidly excavated, very large submarine canyons, particularly along the eastern margin of the delta system (Figure 7.13). Retrogradational slumping excavated many of the canyon heads tens of miles onto the shelf platform (Coleman *et al.* 1983). Giant slumps along the rapidly loaded shelf margin spread mass transport complexes far out onto the basin floor.

Focused bypass of sandy sediment down the slope and onto the basin floor through the large canyons, as well as the sediment mobilized by canyon excavation, further increased supply to the submarine fans on the Gulf basin floor. Initially, during the PAB deposode, the WRLU fan aggraded and spread across the abyssal plain. In the subsequent PL1 deposode, progressive eastward shift of the fan depocenter initiated the precursor lobes of the Mississippi Fan (which has remained the dominant fan system into the Quaternary; Weimer 1990). The proximal Mississippi Fan depocenter contains more than 6000 ft (1800 m) of sediment deposited at the slope toe. The fan spread across the basin, uplapping the base of both the Florida and Campeche scarps (cross-sections 1–3 and 9; Figures 1.13 to 1.15 and 1.21). Smaller, geologically short-lived fans record the emergence of a particularly efficient canyon or bathymetric conduit across the slope apron. The Bryant and Alaminos Fans (Figure 7.13) are examples (Morton and Weimer, 2000). The significance of channelized slope bypass and fan deposition is reflected in the observation that during the Pleistocene more than half of the Gulf abyssal plain was covered at some time by a fan system.

Along the Tampico–Misantla margin, the curvilinear shore zone/shelf/slope apron depositional systems tract maintained a dominantly aggradational continental margin. However, sediment supply and accumulation was minor (cross-sections 7 and 8; Figures 1.19 and 1.20). The shelf edge was pinned by ongoing extension along the Quetzalcoatl growth fault zone. The remnant Veracruz basin was filled by progradational offlap of a coastal plain/shore zone/shelf/slope apron systems tract to and around the Los Tuxtlas high, creating the template for modern physical geography. Continued progradation of the Usumacinta/Grijalva fluvial–deltaic axis and its flanking shore zone systems, and offlap of the delta-fed apron onto and across the salt basin (Garcia-Molina 1994) similarly presaged the modern physical geography of the Campeche province.

7.7.2 Termination and Summary

In our mapping, we attempt, where paleontologic control permits in the northern GoM depocenter, to terminate the PS genetic stratigraphic sequence with the Sangamonian transgression, equal to marine isotope stage 5 and dated at 125,000 to 75,000 BP. However, for practical purposes, the post-transgression glacioeustatic cycle strata, including the Holocene, are commonly included in the mapped unit across most of the basin. Sedimentation during this final 600,000-year phase of Gulf history essentially created the basin geography as we see it today.

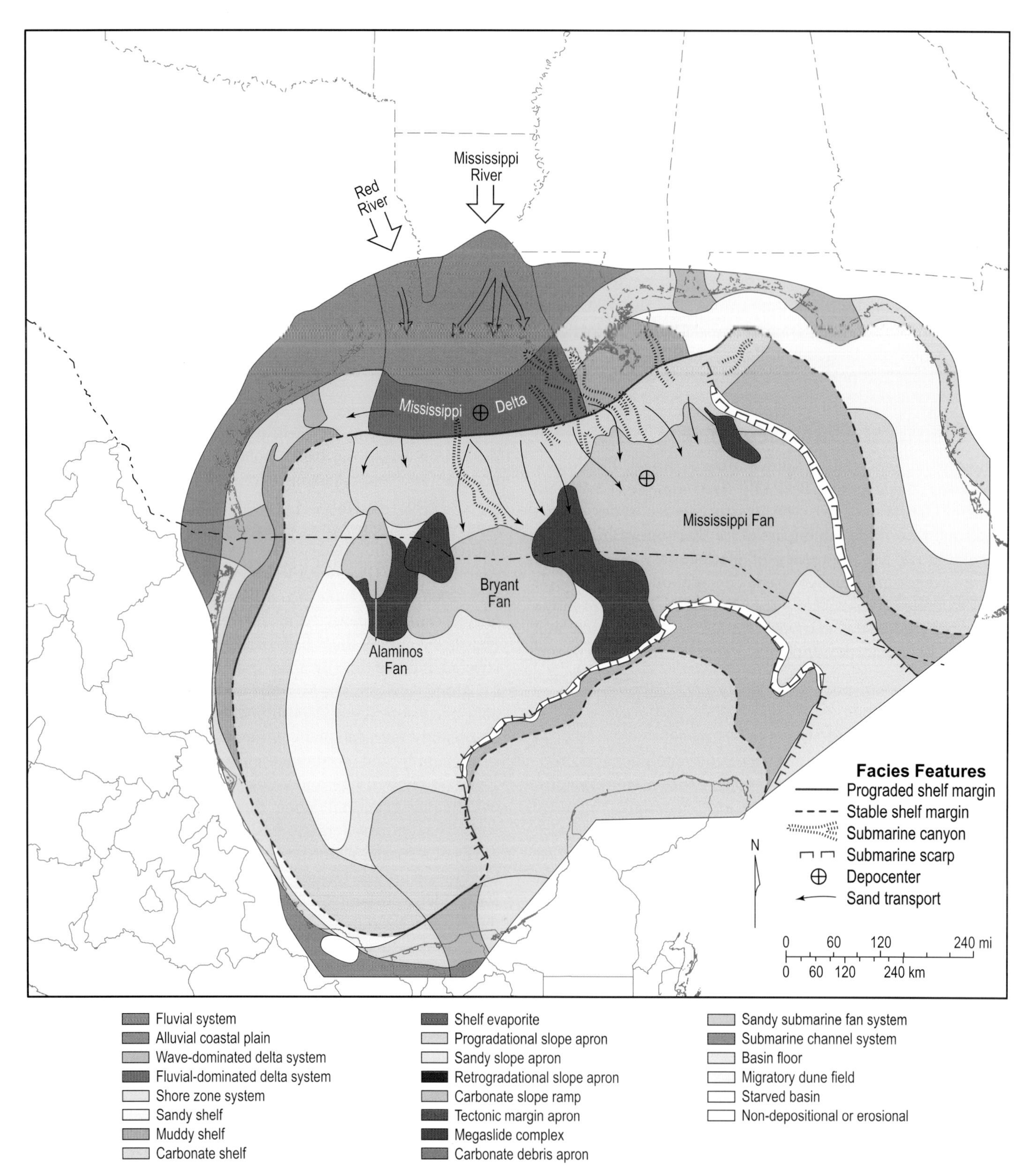

Figure 7.13 Generalized paleogeographic map for the Pleistocene. The map combines elements of the PTA and PS deposodes. Depositional system outlines and shelf edge reflect their positions at maximum progradation.

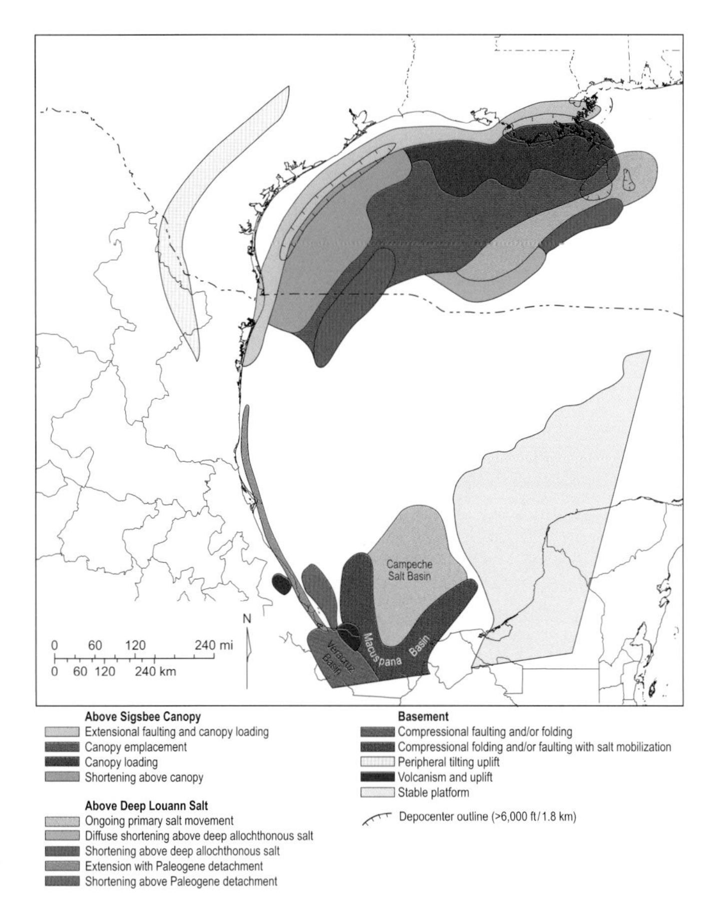

Figure 7.14 Middle Miocene structural domains. Broad patterns established during the Middle Cenozoic persisted into the Middle Miocene. Nearly continuous arcs of extensional growth faulting rim the Gulf margin from Louisiana to Veracruz. Compressional shortening extended along the distal edge of deep salt is reflected in ongoing deformation in the Mississippi Fan, Atwater, and Perdido fold belts. Canopy loading and emplacement dominated offshore Louisiana, whereas shortening above the canopy continued along the Port Isabel compressional domain. The first folds of the Mexican Ridges fold belt formed off southern Veracruz state. Basement deformation and volcanism impacted intrabasinal structures and salt deformation in the Veracruz basin and Campeche province.

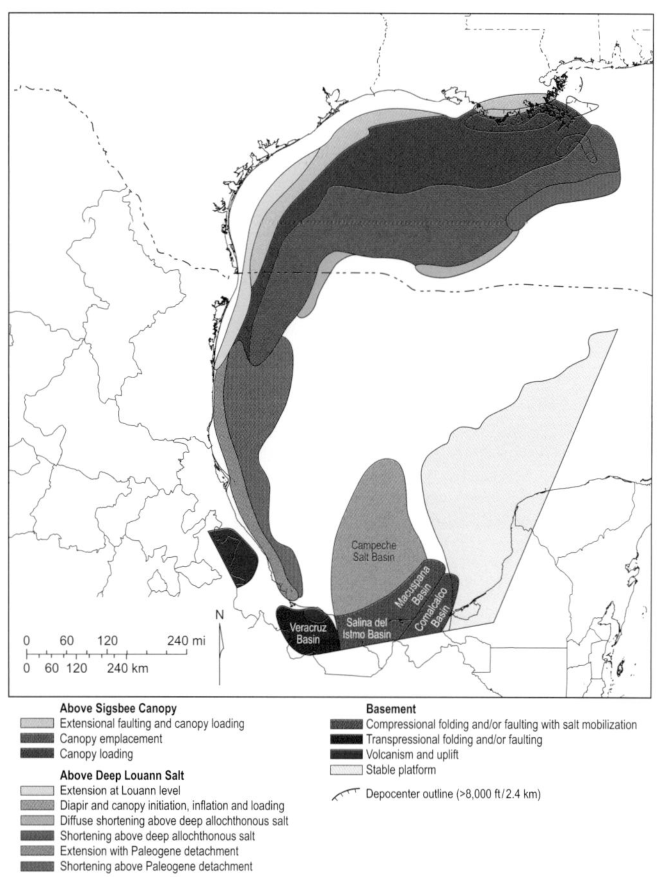

Figure 7.15 Upper Miocene structural domains. Nearly continuous arcs of extensional growth faulting rim the Gulf margin from Louisiana to Veracruz. Extensional strain was localized both at the canopy level (to the northeast) and at the Louann level (to the southwest). Extensive areas of canopy loading and basinward emplacement extended across the breadth of the northern and northwestern Gulf. Shortening continued along each of the basinal fold belts. Volcanism accompanied compressional and transpressional folding and faulting in the basins of Veracruz and Campeche. Depositional loading combined with structural deformation to drive salt mobilization at canopy and deep salt levels. Paired extensional (Quetzalcoatl fault zone) and shortening (Mexican Ridges fold belt) domains extended along the Tampico–Misantla margin.

7.8 Structural Evolution

Several broad themes are reflected in the evolving structural configuration of the Late Neogene GoM basin (e.g., Figures 7.14 and 7.15). They are discussed in the context of the three different depositional provinces: the northern Gulf – Burgos; the western Gulf – Tampico–Misantla; and the southern Gulf – Veracruz–Campeche.

7.8.1 Northern Gulf: Burgos Basin

Loading and expulsion of the proximal Paleogene Sigsbee salt canopy was accompanied by emplacement and advance of a shallower, distal canopy (Neogene canopies and basins; cross-sections 4–6; Figures 1.16–1.18). In offshore Louisiana, the area of canopy loading expanded basinward, ultimately including all but a distal fringe, where canopy emplacement continued to spread salt southward onto the abyssal plain. As deep compressional shortening above the canopy and above deep allochthonous salt ended (updip) or declined (downdip) in Late Miocene time, the broad belt of canopy emplacement expanded southward into offshore northern Mexico (Figure 7.15).

A continuous crescentric belt of extensional faulting persisted through the Miocene (Corsair–Wanda faults, cross-section 6; Figure 1.18) and played an important role in localizing depocenters (Figures 7.13 and 7.14). A progression of fault zones, beginning with the Early Miocene Clemente-Tomas, Middle Miocene Corsair, and Late Miocene Wanda, defined the structural grain across the breadth of the northwestern GoM margin (Figure 7.16). Extension was commonly accompanied by canopy loading and evacuation, but in Late Miocene time extended to the deep Louann level (Figure 7.15). The slope-toe zones of compressional shortening were well developed in Middle to Late Miocene. Both the Mississippi–Atwater and Perdido fold belts emerged as prominent

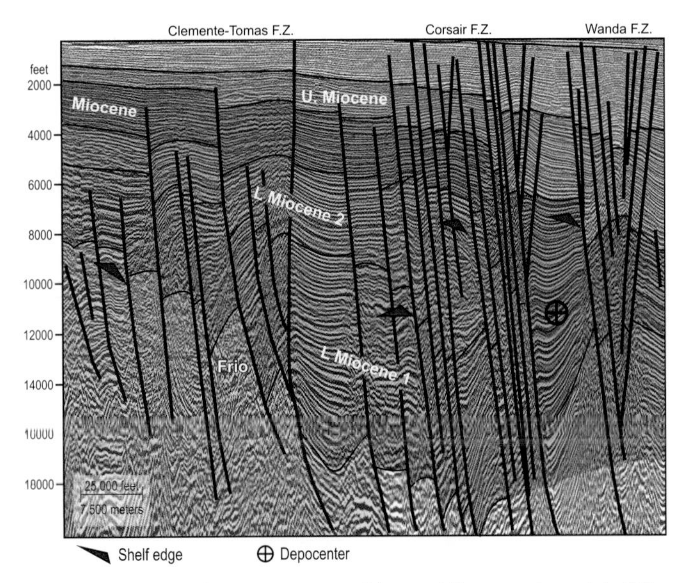

Figure 7.16 Seismic line interpretation of the south Texas continental shelf. The progressive succession of Miocene growth fault families, beginning with the Early Miocene Clemente-Tomas and ending at the shelf edge with the Late Miocene Wanda fault zone dominates the structural fabric of the northwest GoM Neogene margin. The southern end of the regionally important Middle Miocene Corsair depocenter is highlighted by the cross-hairs. For location, see Figure 7.2. Seismic line courtesy of ION.

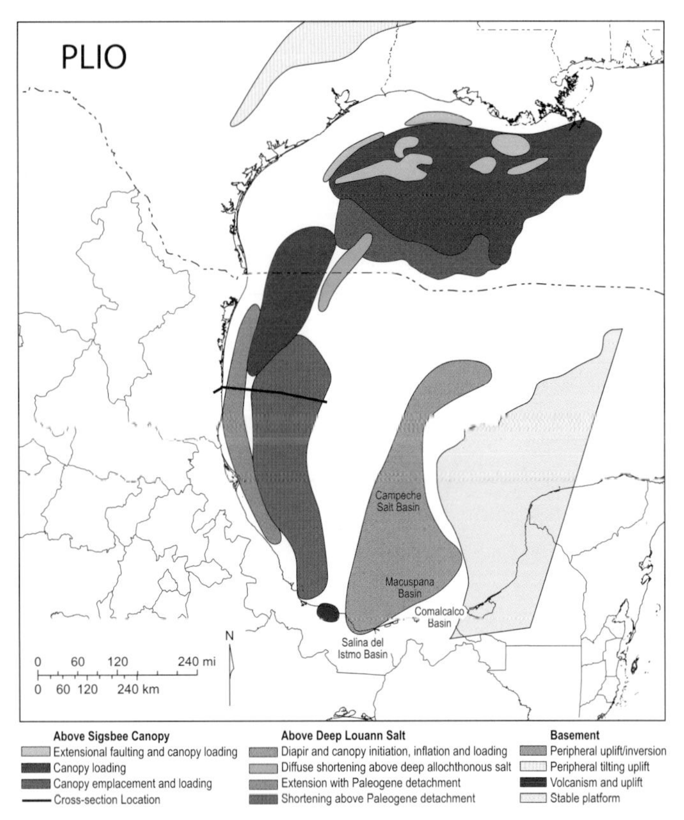

Figure 7.17 Plio-Pleistocene structural domains. Canopy loading dominated across the northern Gulf. Subregional families of extensional faults (roho faults) associated with canopy loading and salt evacuation developed within and along the updip margin of the canopy complex. Paired extensional faulting and shortening rimmed the western Gulf margin. Salt deformation spread along the length of the Campeche salt basin.

structural elements along the slope toe (cross-sections 3, 4, and 6; Figures 1.15, 1.16, and 1.18). By the Pliocene (Figure 7.17), extensional faulting clustered into localized areas at the updip periphery of and within the broad region of canopy loading in the north-central Gulf. Localized supra-salt depocenters, called roho structural systems, developed by loading-induced evacuation and basinward displacement of tabular salt bodies (Schuster 1995; cross-section 5; Figure 1.17).

7.8.2 Tampico–Misantla Margin

Two tectonostratigraphic provinces dominate the western Gulf margin (cross-sections 7 and 8; Figures 1.19 and 1.20). Updip, the Quetzalcoatl fault zone, a continuous, narrow, arcuate belt of extension with detachment within Paleogene strata, traces the Neogene depositional shelf edge (Figure 7.18). Rapid displacement across the master fault pinned the shelf margin. Up to 5 km (3.1 miles) of sediment was accommodated along the fault-bounded upper slope depocenter (Alzaga-Ruiz *et al.* 2009b; Roure *et al.* 2009). A compensatory zone of shortening, long known as the Mexican Ridges, lies downdip along the broad Mexican continental rise (Figure 7.18). Compression is manifested as faulted anticlinal folds. Middle Miocene folding was limited to a small area at the south end of the margin. Folding extended northward during the Late Miocene, and expanded into a broad belt extending from Veracruz to the Burgos. Folding continues and most folds support prominent sea floor ridges. Origin of the coupled extensional and compressional provinces remains uncertain. Le Roy *et al.* (2007) proposed deep crustal shear along the margin. Alzaga-Ruiz *et al.* (2009a) argued for intrabasinal gravity tectonics along the prograding Neogene depositional margin.

7.8.3 Veracruz–Campeche Margin

Initially the western margin of the Veracruz basin and adjacent Cardoba platform were deformed, uplifted, and further eroded. Volcanism expanded into the Veracruz basin, forming the Los Tuxtlas volcanic center (Figures 7.14 and 7.15). Volcanics are both intruded and deposited within the Neogene deposits of the Veracruz basin (Figure 7.19). Plate tectonic stress within the southern Mexican crust initiated compressional/transpressional folding and faulting across the Veracruz and Sureste Salina del Istmo, Macuspana, and Comalcalco basins (Figures 1.21 and 7.19; Catemaco fold belt). Salt mobilization triggered by compressional deformation and ongoing sediment loading led to extensive development of salt diapirs and canopies during deposition of Late Miocene and Plio-Pleistocene supersequences. By Late Neogene, depositional loading and inherent instability resulted in salt deformation along the length and breadth of the Campeche Salt basin.

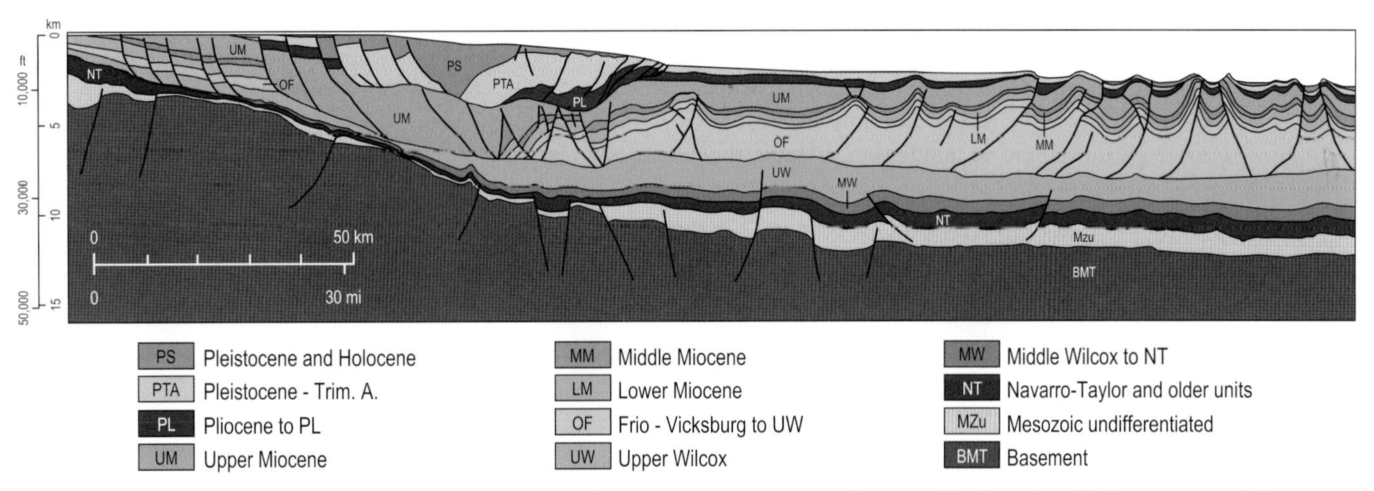

Figure 7.18 Interpreted seismically based schematic cross-section of the southern Tampico–Misantla margin. The Quetzalcoatl fault zone, centered along the shelf edge, expands the Miocene–Pleistocene supersequences of the Neogene Tectono-climatic Phase. Basinward displacement along Paleogene detachments terminates beneath the continental rise in the Mexican Ridges fold belt. Modified from CNH (2015b). For location see Figure 7.17.

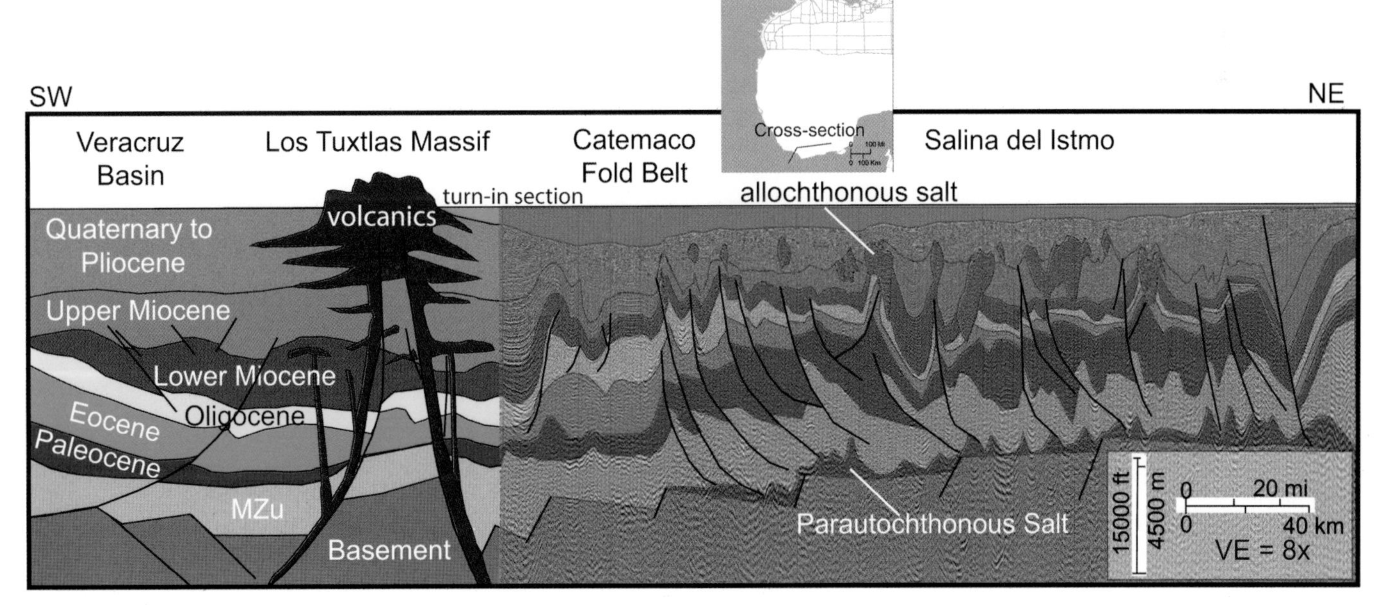

Figure 7.19 Generalized cross-section from the Veracruz basin to the Salina del Istmo (southern Campeche salt basin). The thick Neogene section includes the Veracruz basin progradational platform and associated volcanic series of the Los Tuxtlas Massif. Slope/basin systems of the Salina del Istmo lie beneath the Gulf of Campeche. Neogene compressional structures and related salt intrusion create traps. Veracruz basin and Los Tuxtlas Massif portion of section modified from Andreani *et al.* (2008). Seismic Line courtesy of ION.

7.9 Summary: Neogene Tectono-climatic Phase

The 15-million-year Neogene Tectono-climatic Phase deposited a megasequence recording moderate rates of sediment supply (with the exception of the Pleistocene) from inland continental North American sources in the northern and proximal tectonic uplands in the southern Gulf basin. During the Middle Miocene deposode, multiple large rivers were widely distributed across the northern periphery of the basin. By the Pliocene deposodes, continental North American drainage had coalesced onto three main rivers that all flowed onto the central Gulf coastal plain. By the end of the Pleistocene, the glacially fed Mississippi captured the adjacent rivers and completely dominated sediment supply to the northern GoM. For the first time, a large fluvial–deltaic complex also constructed a coastal plain and prograding margin into the Bay of Campeche. Emergence of numerous small streams draining adjacent tectonic uplands provided sufficient sediment for modest progradation of coastal and marine systems along the Veracruz–Tamaulipas margin.

Depositional loading on both the northern Gulf and Campeche continental margins mobilized primary salt and shallow canopies, creating a diverse array of syndepositional structures and a complex slope bathymetry that profoundly influenced sediment transport pathways and accumulation. Late in the phase, repeated excavation of large submarine canyons, discharge of cold, sediment-laden glacial meltwater, and high-frequency, high-amplitude sea-level cycles all contributed to wholesale bypass of sediment onto the continental slope and basin floor. Construction of large, long-lived submarine fan systems resulted. The largest fans display run-out distances of up to 600 km (370 miles) from the contemporary shelf edge.

Chapter 8 Cenozoic Depositional Synthesis and Emerging Hydrocarbon Plays

8.1 Evolving Drainage Basins and Depocenters

The wealth of data, acquired over decades of exploratory drilling and seismic acquisition, has provided a uniquely rich foundation for study of depositional and structural evolution of a small ocean basin. The Gulf of Mexico (GoM) has most recently become a laboratory for qualitative and quantitative analysis of continent-scale source-to-sink sediment transport and depositional systems.

Galloway *et al.* (2011) combined GoM synthesis with the comparably detailed literature on evolving continental North American tectonics, geomorphology, and erosional history to prepare a suite of interpretative maps connecting interior drainage basins with the individual fluvial axes of the Gulf margin. Sediment transport system interpretations were conditioned by the well-constrained paleogeographic and volumetric history of the Cenozoic deposodes. Such geologically grounded source-to-sink interpretations are further constrained by sand composition data (e.g., Loucks *et al.* 1986; Dutton and Loucks 2010; Martinez-Medrano *et al.* 2009; Paredes *et al.* 2009; Dutton *et al.* 2012; Ambrose *et al.* 2013). More recently, detrital zircon analysis has added a new and powerful tool for defining the sediment source terranes (e.g., Mackey *et al.* 2012; Craddock and Kylander-Clark 2013; Blum and Pecha 2014; Heintz *et al.* 2015; Lawton *et al.* 2015; Sharman *et al.* 2016; Xu *et al.* 2016a, 2017; Blum *et al.* 2017; Fan *et al.* 2018; Yancey *et al.* 2018).

8.1.1 Source Areas

The natural grouping of GoM Cenozoic supersequences into three tectonostratigraphic phases – the Paleogene Laramide Phase, Middle Cenozoic Geothermal Phase, and Neogene Tectono-climatic Phase – reflects the primary influences of tectonics and climate on drainage basin development and consequent sediment yield (Syvitski and Milliman 2007). Relationship of the principal drainage axes to tectonically elevated source terranes within and surrounding the interpreted GoM drainage basin are summarized in Figure 8.1A–C.

Bounding Paleocene–Early Eocene uplands include the Laramide Orogenic Belt, mountains of the Front Range Uplift, and remnant uplands of the Paleozoic Appalachian and Ouachita Uplifts (Figure 8.1A). Western tributaries to the Rio Grande and Colorado rivers extended as far as southwestern Wyoming and western Arizona. Significant changes in western upland drainage occurred at the end of the Paleocene as the area of intermontane closed basins expanded northward. The Rio Bravo drainage basin, which tapped much of the length of the Laramide Orogenic Belt, remained focused into the contemporary foreland troughs. Only with their infilling in the Early Eocene did the northern Mexico Laramide uplifts supply sediment directly into the GoM through the Rio Bravo and Rio Grande. Interpreted west–east distribution of detrital zircon populations (Blum and Pecha 2014; Blum *et al.* 2017) reflects the relative contributions of sandy sediment from the various North American sources to each of the trunk fluvial systems, and to the secondary, basin margin streams that constructed the coastal plain between them (Figure 8.2). Grenville-age zircons of eastern North America dominate the Mississippi axis and coastal plain streams to the east. Diverse zircon assemblages reflect numerous smaller streams that constructed the inter-axial coastal plain in the Ark–La–Tex. The Colorado axis contains a sub-equal mix of eastern, mid-continent, and Western Cordilleran and Laramide zircons. South Texas samples are dominated by reworked Cordilleran zircons likely recycled from basin-fringing Cretaceous deposits.

The Middle Cenozoic Thermal Phase achieved relative stability of its bounding drainage divides by the Oligocene Frio (OF) deposode (Figure 8.1B). The principal evolutionary intra-phase change was the Early Miocene migration of the continental river input in east Texas from the Houston-Brazos axis eastward to the Red River axis, with its attendant reorganization of drainage basins. Principal tectonic uplands included a series of uplifts and inversions along the eastern belt of the Mexican Laramide complex (areas 1 and 2, Figure 8.1B) and the uplifted and exhumed Chiapas Massif. Volcanic fields, accompanied by regional crustal heating and uplift, arose in Trans-Pecos Texas, along the Sierra Madre Oriental, and in the San Juan province of southwest Colorado. West–east plots of detrital zircon data for the OF (Blum *et al.* 2017; Yancey *et al.* 2018) and Early Miocene LM1 and LM2 supersequences (Xu *et al.* 2016a, 2017) document source terranes for trunk rivers and for secondary streams that deposited the inter-axial

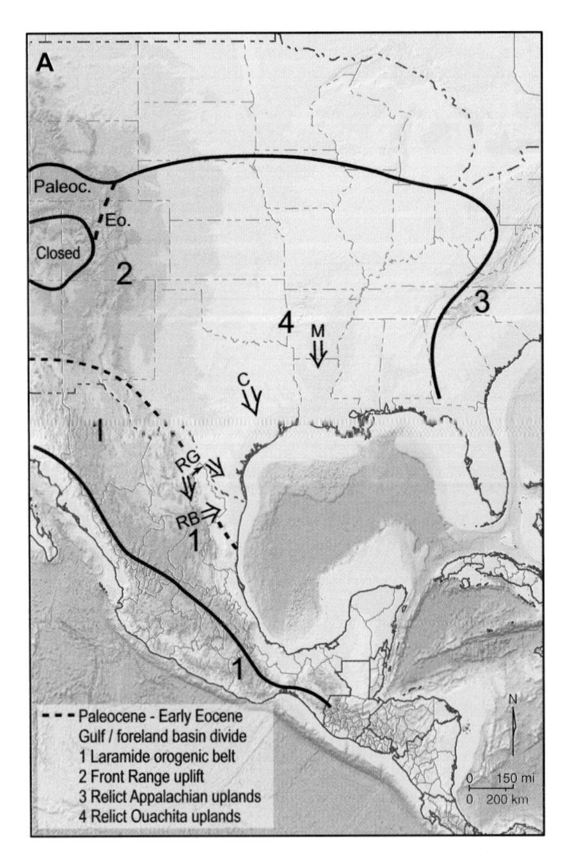

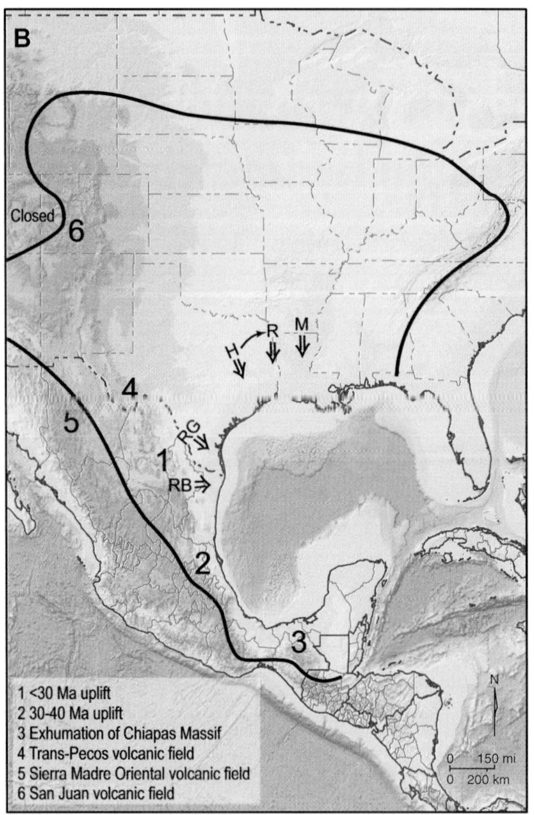

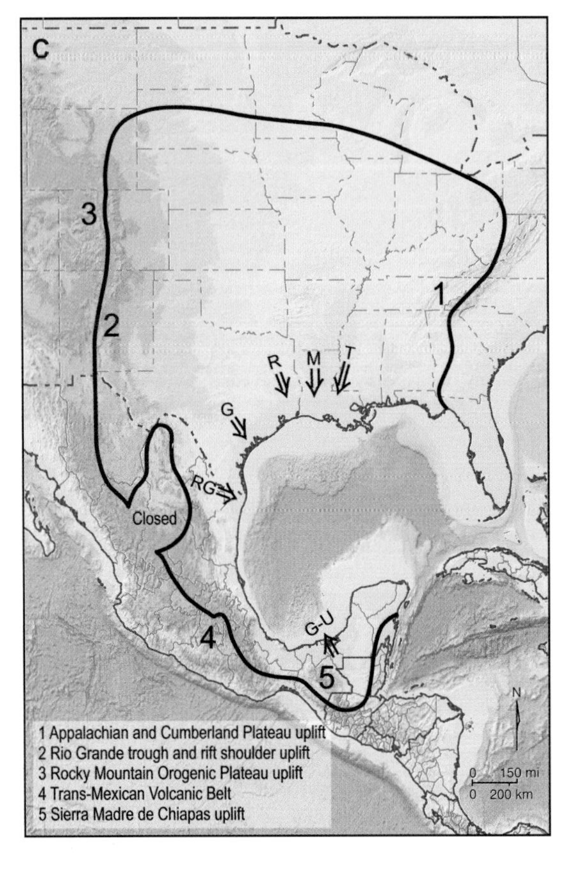

Figure 8.1 Principal upland sediment sources and entry points of the major fluvial axes for each of the three Cenozoic tectono-climatic phases. Continental drainage divides separating rivers flowing into the GoM from those flowing into the Atlantic Ocean, Hudson Bay, closed intermontane basins of the Western Interior, and the Pacific Ocean. **(A)** Paleogene Laramide Phase. Principal upland sources included (1) the Laramide Orogenic Belt, (2) the Front Range Uplift, and (3) the relict Appalachian Uplift. Note the shift of the southwestern drainage divide from its Paleocene–Early Eocene position southward into western Mexico by Mid-Eocene. **(B)** Middle Cenozoic Geothermal Phase. Upland source areas included (1) <30 Ma crustal inversion and uplift, (2) 30–40 Ma crustal uplift, (3) exhumed Chiapas Massif, (4) the Trans-Pecos volcanic field, (5) the Sierra Madre Oriental volcanic field and uplift, and (6), the San Juan volcanic field. A network of closed basins established the western drainage divide. **(C)**. Neogene Tectono-climatic Phase. Principal uplifts and source terranes included (1) elevated/rejuvenated Appalachian and Cumberland Plateau, (2) Rio Grande Trough and rift-shoulder uplift, (3) Rocky Mountain Orogenic Plateau, (4) Trans-Mexican Volcanic Belt, and (5) Sierra Madre de Chiapas Uplift.

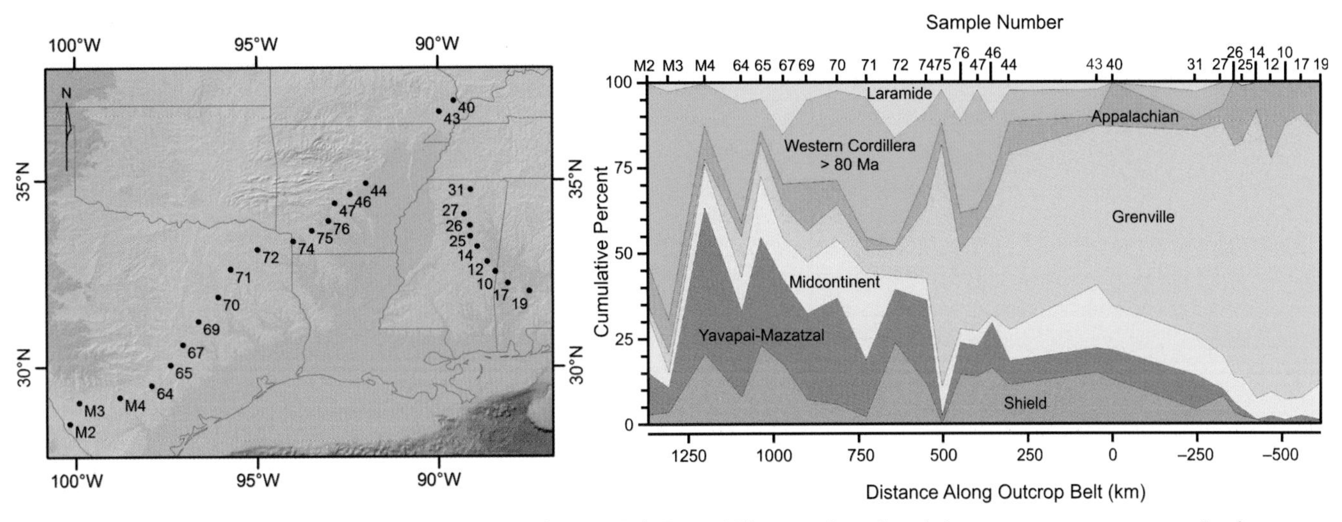

Figure 8.2 Along-strike trends in detrital zircon populations in Paleocene–Early Eocene Wilcox samples collected along an outcrop traverse extending from western Mississippi to south Texas. The plot shows spatial changes in percentage contributions of populations associated with different source terranes. Likely association to major fluvial axis is based on geographic correspondence of sample with mapped fluvial axes and inter-axial coastal plains. Modified from Blum *et al.* (2017).

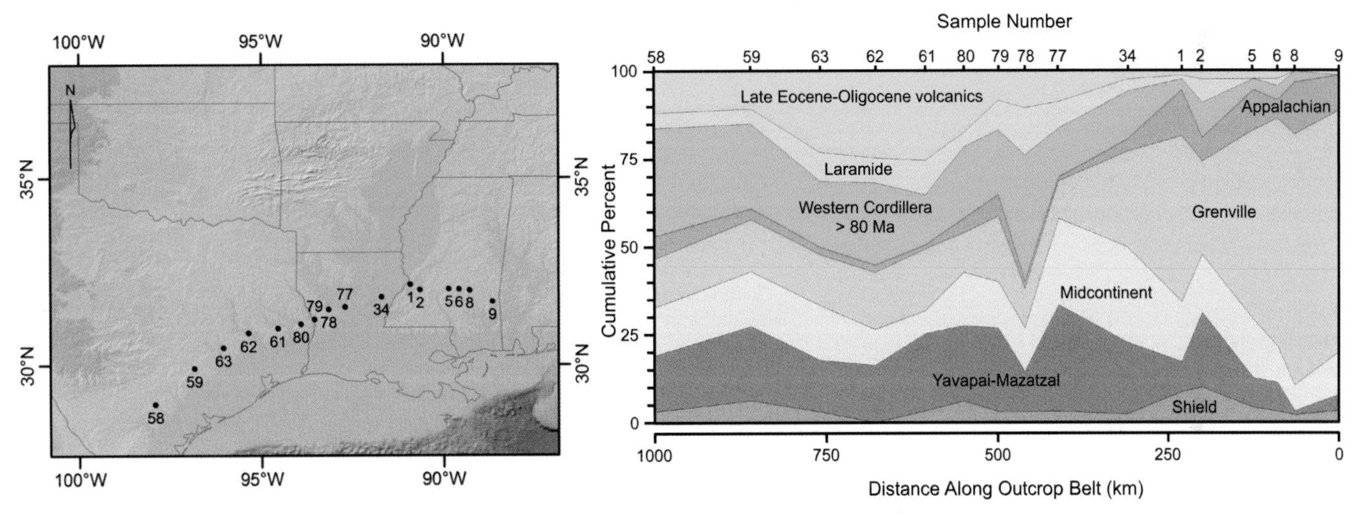

Figure 8.3 Along-strike trends in detrital zircon populations in Oligocene Frio samples collected along an outcrop traverse extending from Mississippi to south Texas. The plot shows spatial changes in percentage contributions of populations associated with different source terranes. Likely association to major fluvial axis is based on geographic correspondence of sample with mapped fluvial axes and inter-axial coastal plains. Modified from Blum *et al.* (2017).

coastal plains (Figures 8.3 and 8.4A). Penecontemporaneous Late Eocene–Oligocene volcaniclastics are prominent components in the Rio Grande and Houston-Brazos streams but are minor in the Mississippi and eastern coastal plain rivers of the Frio deposode. Zircon dates of contemporaneous reworked ash both substantiate the chronology of the Yegua–Frio deposodes and document the volumetric importance of airfall material as a component of total sediment supply during this time. Zircon populations clearly distinguish Early Miocene LM1 and LM2 Mississippi (Grenville-dominated), Red River (strong Yavapai–Mazatzal) and Rio Grande (strong Western Cordillera arc) fluvial axes. Petrographic data for the Lower Miocene sandstones further illustrate typical GoM relationships between source area and sand composition (Figure 8.4B). Recycled Mid-Continent and Appalachian sedimentary rocks are highly quartzitic; recycled and first-cycle western orogenic sources provide compositionally diverse sands. The pattern of decreasing quartz and increasing lithic and feldspar content from east (Mississippi) to west (Rio Grande) seen in the Lower Miocene sandstones is typical of all northern GoM Cenozoic supersequences. Composition, in turn, is a major determinant of regional patterns in burial diagenesis and resultant reservoir quality. A similar east–west compositional trend typifies Frio supersequence sands.

The Neogene Tectono-climatic Phase experienced a significant loss of distal Rocky Mountain landscapes as contributory sources (Figure 8.1C). Initiation of the Rio Grande rift created a chain of rapidly subsiding, closed basins that effectively

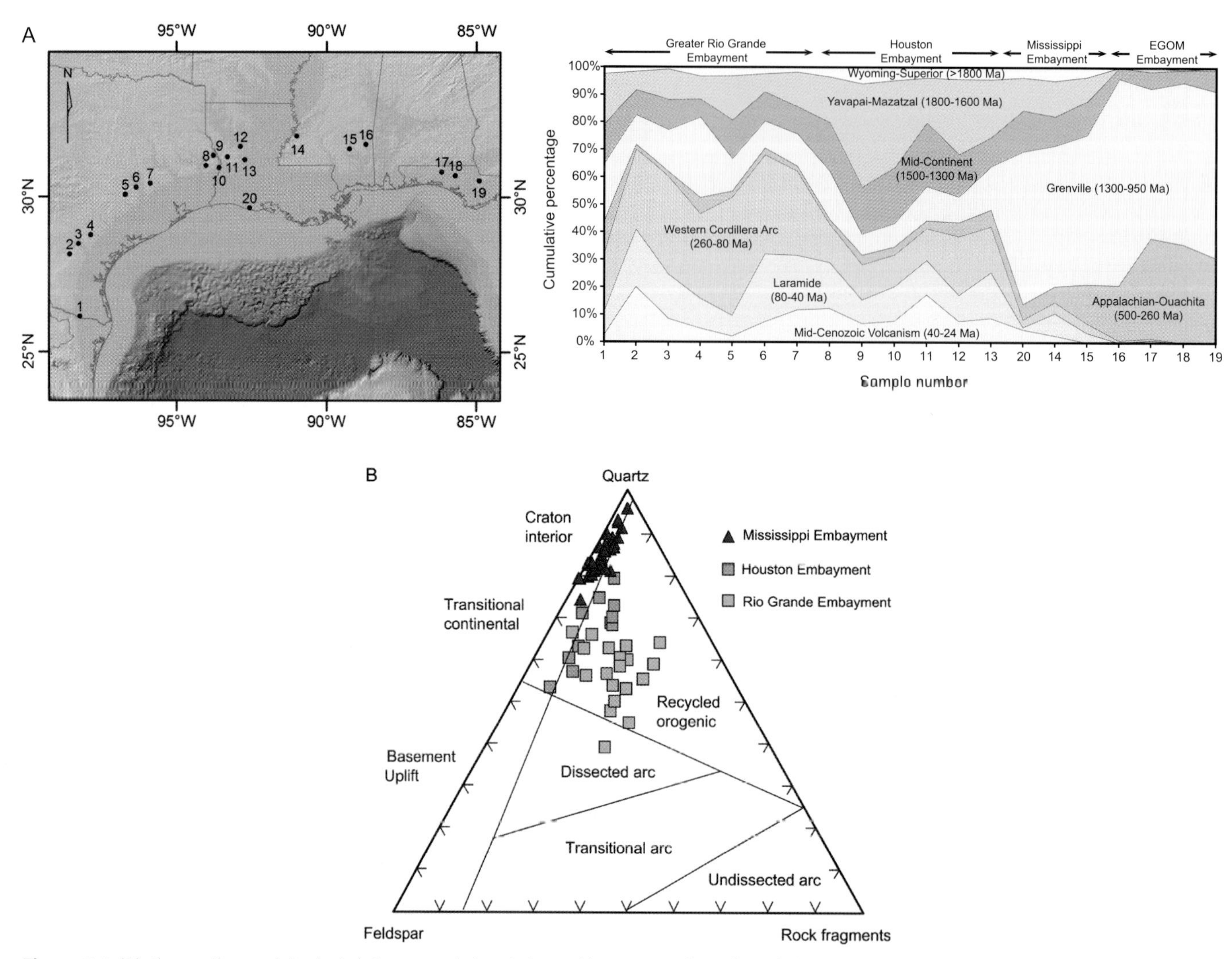

Figure 8.4 **(A)** Along-strike trends in detrital zircon populations in Lower Miocene samples collected along an outcrop traverse extending from the Florida panhandle to south Texas. The plot shows spatial changes in percentage contributions of populations associated with different source terranes. Likely association to major fluvial axis is based on geographic correspondence of the sample with mapped fluvial axes and inter-axial coastal plains. **(B)** Lower Miocene sand composition for sample suites from wells located in the Mississippi Embayment, Houston Embayment, and Rio Grande Embayment. These correspond to the Mississippi, Red River, and Rio Grande fluvial axes, respectively. Note progressive increase in feldspar (Red River) and rock fragments (Rio Grande) at the expense of quartz grains from east to west. Distinctive bulk composition and detrital zircon populations both clearly reflect the three different continental river systems and their different drainage basins.From Xu *et al.* (2017).

intercepted eastward-flowing tributaries. Only the eastern rift margin uplifts provided sediment destined for the Gulf. However, at the same time, the Appalachian and adjacent Cumberland Plateau experienced regional uplift and unroofing. In the latest Neogene, regional domal elevation of the Rocky Mountain Orogenic Plateau created the broad, east-sloping High Plains (still present today) and rejuvenated erosion along the central Front Range. In Mexico, the Trans-Mexican Volcanic Belt and the Sierra Madre de Chiapas Uplift created important new orogenic uplands that drained into the southern Gulf.

8.1.2 Drainage Basin Reconstructions

Figures 8.5 through 8.16 are our most recent reconstructions of the areal distribution of the drainage basins of the continental rivers that provided the bulk of sediment to the northern GoM (extension of maps into central Mexico remains to be done). They incorporate the accumulating database of detrital zircon analyses that has been generated since the drafting of the first-generation drainage maps by Galloway *et al.* (2011) and several recent reconstructions of continental rivers of the Western Interior. Mapped paleoflow directions and reconstructions of tributary channel segments in contemporary intracontinental basins were systematically compiled in the earlier mapping and continue to constrain the reconstruction of the tributary network. Position and scale of fluvial–deltaic depocenters in the Gulf margin establish relative size and location of the continental rivers at their entry onto the depositional coastal plain. For all maps, the position of the drainage divide separating rivers that flow into the GoM from

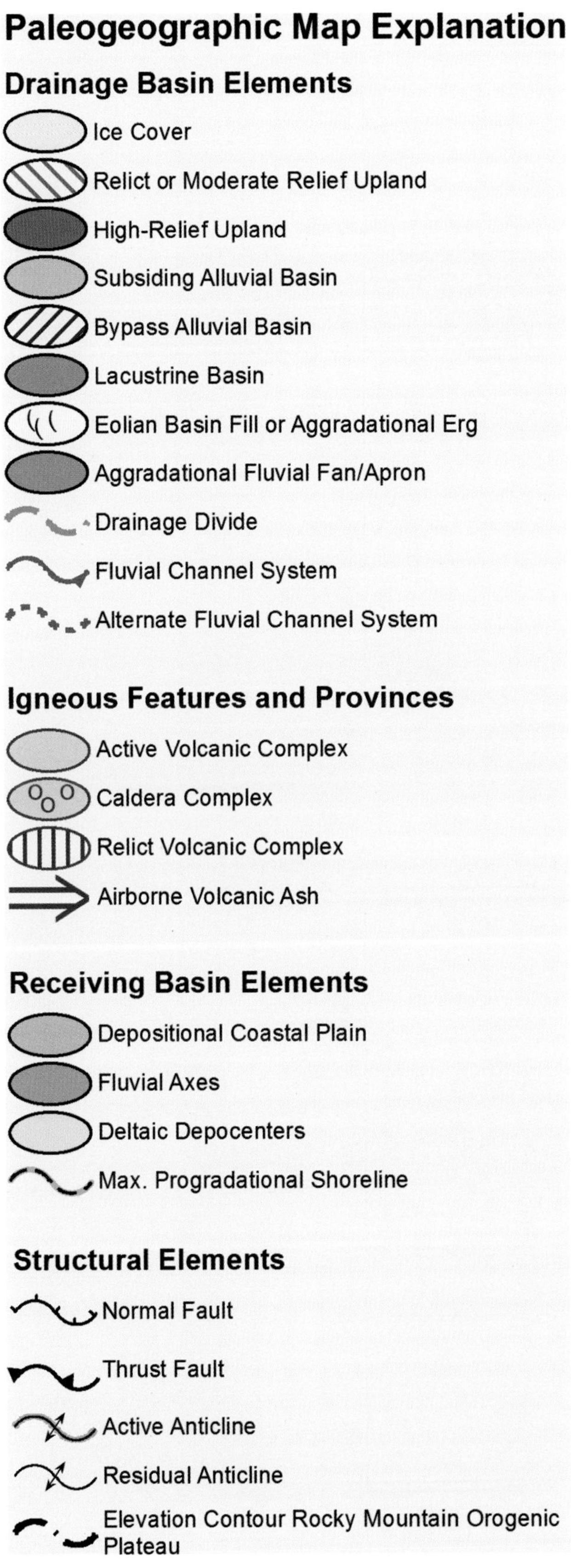

Paleogeographic Map Explanation

Drainage Basin Elements

Ice Cover
Relict or Moderate Relief Upland
High-Relief Upland
Subsiding Alluvial Basin
Bypass Alluvial Basin
Lacustrine Basin
Eolian Basin Fill or Aggradational Erg
Aggradational Fluvial Fan/Apron
Drainage Divide
Fluvial Channel System
Alternate Fluvial Channel System

Igneous Features and Provinces

Active Volcanic Complex
Caldera Complex
Relict Volcanic Complex
Airborne Volcanic Ash

Receiving Basin Elements

Depositional Coastal Plain
Fluvial Axes
Deltaic Depocenters
Max. Progradational Shoreline

Structural Elements

Normal Fault
Thrust Fault
Active Anticline
Residual Anticline
Elevation Contour Rocky Mountain Orogenic Plateau

Figure 8.5 Explanation for paleogeographic maps, Figures 8.6 to 8.16. From Galloway *et al.* (2011).

those flowing north toward Hudson Bay is poorly constrained. Figure 8.5 provides the explanation applicable to all maps.

8.1.2.1 Paleocene–Middle Eocene

Four drainage reconstructions reflect the geomorphic evolution of drainage basins through the Laramide Phase. The first map (Figure 8.6) addresses the question of why sediment supply to the Gulf was minimal during the first two million years of the Paleocene, despite the Late Cretaceous emergence of tectonically active Laramide uplands. At this time, the Gulf shoreline extended far onto the continent, well inland of the modern outcrop belt. Although global sea level was high, it was not extreme and would, in fact, remain comparatively high throughout the Laramide Phase. The map suggests several possible contributing factors. (1) The uplands of the central Front Range and Western Interior were flanked by numerous intermontane basins, all of which were actively subsiding and accumulating Early Paleocene sediment. (2) A remnant lowlands and embayment, the Cannonball Sea, extended into the northern plains states. Rivers arising from uplands in Wyoming, western Colorado, and Utah drained northeastward into this depression. (3) Tributaries arising from the Laramide front and basins of southwestern New Mexico and west Texas drained southeastward into and along the Mexican foreland trough. There they debouched into the deepwater Laramide foreland basin in northern Mexico. (4) Only the drainage arising along and west of the southern Rockies of New Mexico drained eastward toward the Gulf. The San Juan and Raton basins in northern New Mexico and southern Colorado were actively filling and likely sequestered much of the sediment load. Similarly, the Denver basin sequestered sediment derived from the Front Range that likely was a tributary to a mid-continent river system. In summary, diversion of sediment north to the Cannonball or south to the Mexican foreland trough combined with sequestration in the numerous subsiding intermontane basins to limit sediment supply to the GoM and consequent delay in coastal progradation onto and across the relict Late Cretaceous shelf.

By 62 Ma the northern Gulf margin shifted rapidly from starvation to feast. Initially two dominant continental rivers integrated tributaries arising from the northern and central Front Range (Figure 8.7). Headward expansion and piracy extended tributaries westward to the Green River basin and across New Mexico into Arizona. Perhaps of equal importance, several of the large interior basins, including the Denver, Raton, and San Juan had filled, allowing export of sediment into the east-flowing trunk streams. However, interior drainage centered on the Uinta basin of Utah, and north-flowing drainage through the Powder River basin diverted sediment into closed intermontane receiving basins. Although the alternative drainage basin mapping of Sharman *et al.* (2016) limits central Rocky Mountain input to the Mississippi system, diverting the tributaries southward to the Colorado, several observations favor the reconstruction of a large east-flowing trunk stream as shown here. (1) Detrital zircon data

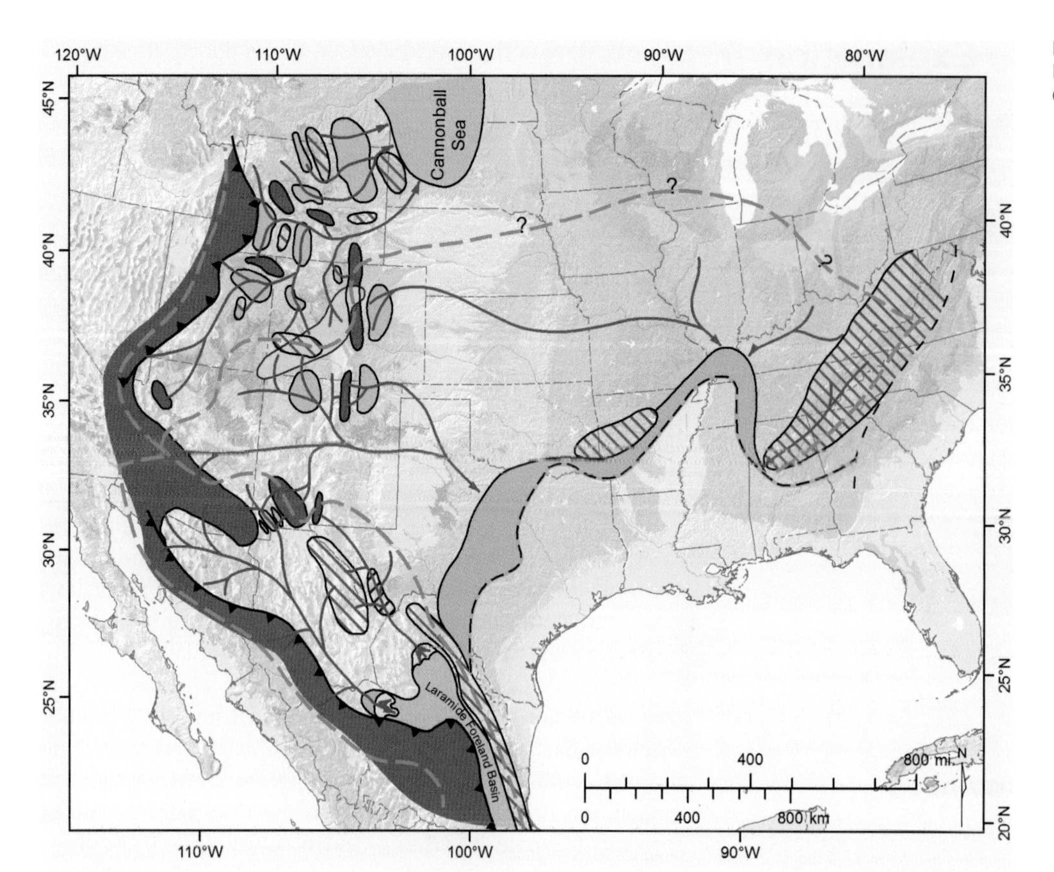

Figure 8.6 Early Paleocene drainage basin paleogeography. Modified from Galloway *et al.* (2011).

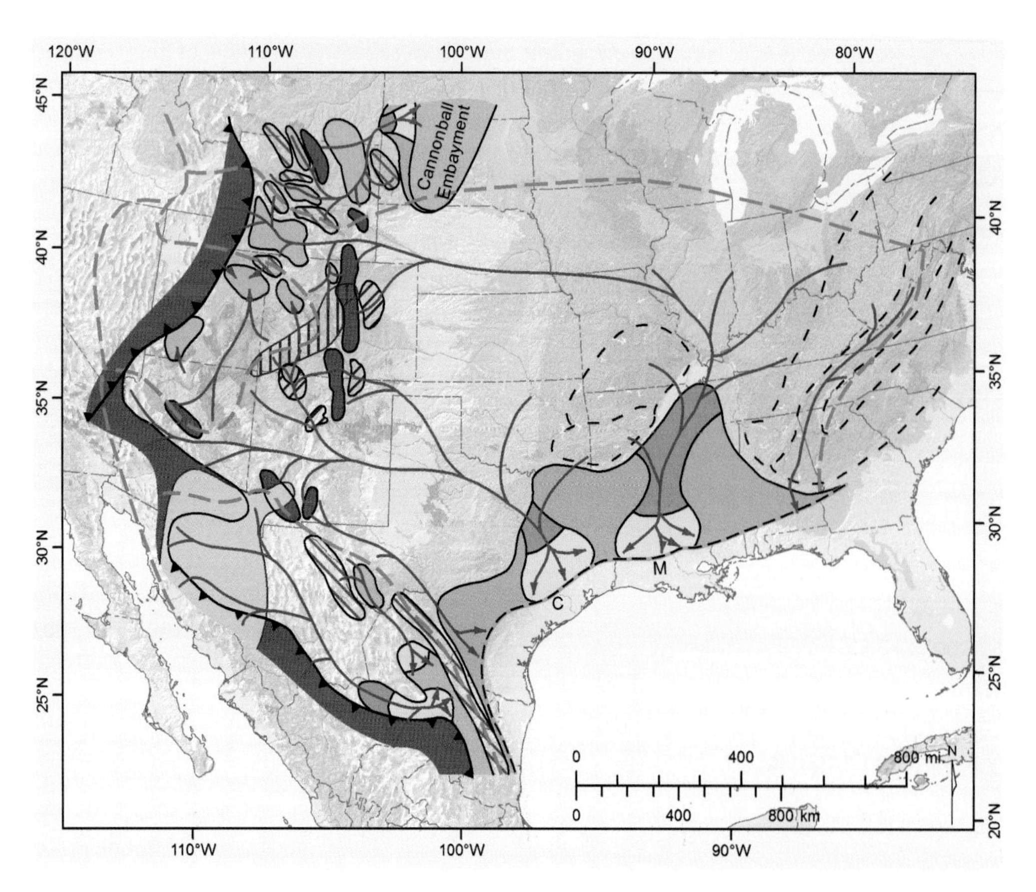

Figure 8.7 Late Paleocene drainage basin paleogeography. Modified from Galloway *et al.* (2011).

(Figure 8.2) document both Grenville (Appalachian) *and* Western Cordillera/Laramide sources for the Mississippi fluvial–deltaic axis. (2) The presence of volcanic rock fragments in Wilcox sandstones of Louisiana (Dutton and Loucks 2010) records input from the igneous Colorado Mineral Belt. (3) A large Mississippi fluvial system draining erosionally active uplands is necessitated by the paleogeographic and volumetric importance of the Holly Springs delta system and the eastern ALKEWA submarine fan complex (Figure 5.5). Although the Rio Grande was a minor fluvial system in the Early Paleocene, its impact can be differentiated by a mappable sand depocenter in the Late Paleocene Middle Wilcox supersequence (Figure 5.9). It is further distinguished as a distinct drainage system by the dominance of Western Cordillera zircon content in samples from south Texas. Small streams arising in the remnant southern Appalachian uplands are dominated by Grenville zircons (Mississippi–Missouri outcrop samples).

Early Eocene Wilcox deposode paleogeography (Figure 8.8) records significant evolution of Laramide Phase continental drainage patterns. This in part reflects the climatic changes manifested across the continental interior following the Paleocene–Eocene Thermal Maximum (PETM) (Hessler *et al.* 2017). Small, declining rivers drained into the east-central Gulf. Three large continental rivers supplied the bulk of the sediment. The central and southern Rocky Mountain Front Ranges were the principal upland source for the Carrizo and Houston-Brazos river systems. Associated basins, including the Denver, Raton, and San Juan, contain amalgamated fluvial channel fills, demonstrating both limited sediment capture and fluvial reworking and bypass into tributaries of both river systems. Headward expansion and tributary capture continued to grow the Rio Grande drainage basin westward into and across the Baca basin (New Mexico/Arizona), tapping the southwestern Laramide front. The continental drainage divide separated an area of closed drainage centered on the Uinta and Green River basins from GoM-directed tributaries. Mapped fluvial systems in the Wyoming basins continued to flow northward. Northern Mexico uplands drained into and along the foreland troughs, which continued to fill axially.

Middle Eocene fluvial systems (Figure 8.9) record the waning stage of the Laramide Phase. Sediment supply was at low ebb, culminating in regional transgressive flooding (Weches) of much of the depositional margin of the northern Gulf. Remnant Laramide uplands were eroded to their resistant crystalline cores. A broad aggradational alluvial apron surrounded the central Front Range, further limiting sediment supply through Mississippi and Houston-Brazos tributaries. The Rio Grande had the most evolved drainage network, extending into the Baca basin in east-central Arizona. However, several small basins retained some of the sediment eroded from local uplands in New Mexico. Filling and fluvial bypass of the

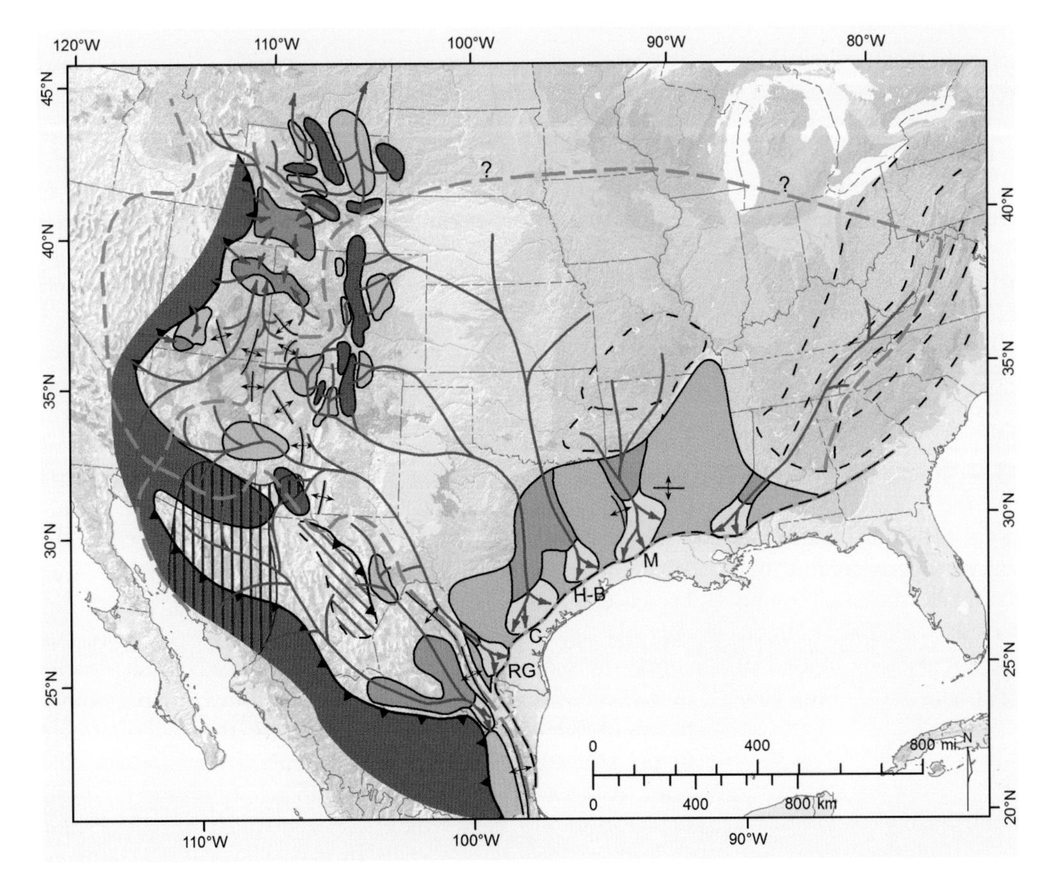

Figure 8.8 Early Eocene drainage basin paleogeography. Modified from Galloway *et al.* (2011).

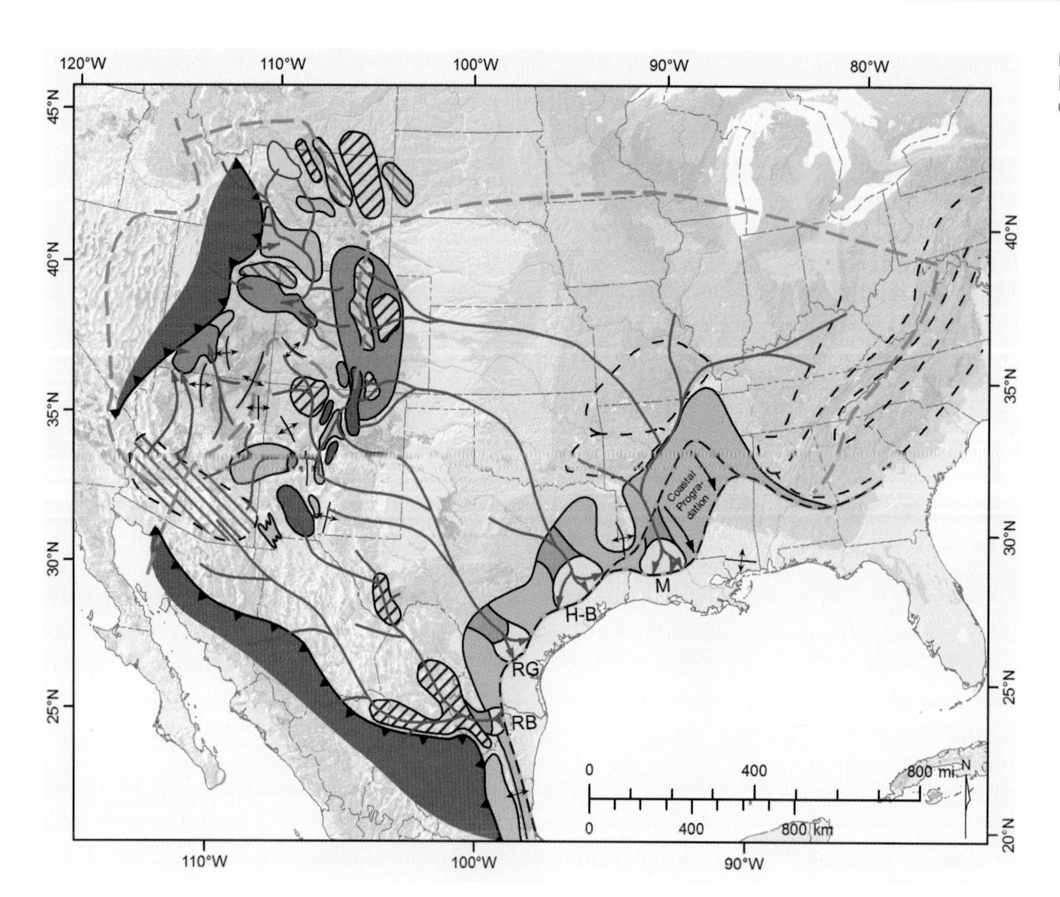

Figure 8.9 Middle Eocene drainage basin paleogeography. Modified from Galloway *et al.* (2011).

northern foreland trough in Mexico allowed the overflow of the Rio Bravo into the Gulf coastal plain of the Burgos basin.

Reconstructions of Paleocene–Early Eocene paleo-drainage basin geography combined with calculated sediment volumes in the GoM receiving basin creates a natural laboratory for quantitative source-to-sink analysis. Zhang *et al.* (2018) applied the BQART model for relationships among paleogeography, climate, and stream sediment load, combined with Monte Carlo simulation to calculate a statistical range of sediment supply to the GoM by the individual and combined Wilcox rivers. Their results demonstrated quantitatively reasonable match between calculations using two drainage basin reconstructions (Galloway *et al.* 2011; Sharman *et al.* 2016) and total sediment supply rates, and accurately predicted the observed decrease in supply across the Paleocene–Eocene boundary. Differences in calculations for individual river systems within each deposode reflects the differences in tributary reconstructions and degree of emphasis in conditioning such reconstructions based on observed volumetrics, mineralogic compositions, and detrital zircon data of individual fluvial–deltaic depocenters of the Paleocene and Eocene Gulf margin.

8.1.2.2 Late Eocene–Early Miocene

Crustal heating with consequent volcanic activity and crustal uplift rejuvenated sediment yield and outflow from the Western Interior source terranes. Initially drainage basins and trunk river axes remained similar to their lower Eocene configuration (Figure 8.10). The most distal fluvial axis, the Mississippi, supplied a significant deltaic system (Cockfield Formation), but faded during the Late Eocene Jackson as sediment from the north-central Front Range was sequestered in the Wind River alluvial apron. The Louisiana–Mississippi coastal plain was transgressed, creating a broad, muddy shelf during the Jackson deposode. The central and southern Front Range continued to supply runoff. Most importantly, episodes of explosive volcanic activity spread plumes of volcanic ash across all of the drainage basins and the Gulf margin itself, providing large volumes of suspended mud to the fluvial systems.

Peak sediment supply occurred in the Oligocene. Headwater tributaries of the Mississippi extended westward into Wyoming as the previous Eocene closed drainage was breached on the east (Figure 8.11). Mixed eolian and fluvial deposits aggraded and expanded the White River apron, but supply of sediment, particularly as reworked airfall debris, overwhelmed local storage and reestablished a prograding, but mud-rich Mississippi delta system by 31 Ma. Tributaries of the Houston-Brazos and Rio Grande Rivers drained volcanic edifices of the San Juan, Mogollon, northern Sierra Madre, and Big Bend volcanic fields, which provided sand and gravel. Reworking airfall ash that blanketed the drainage

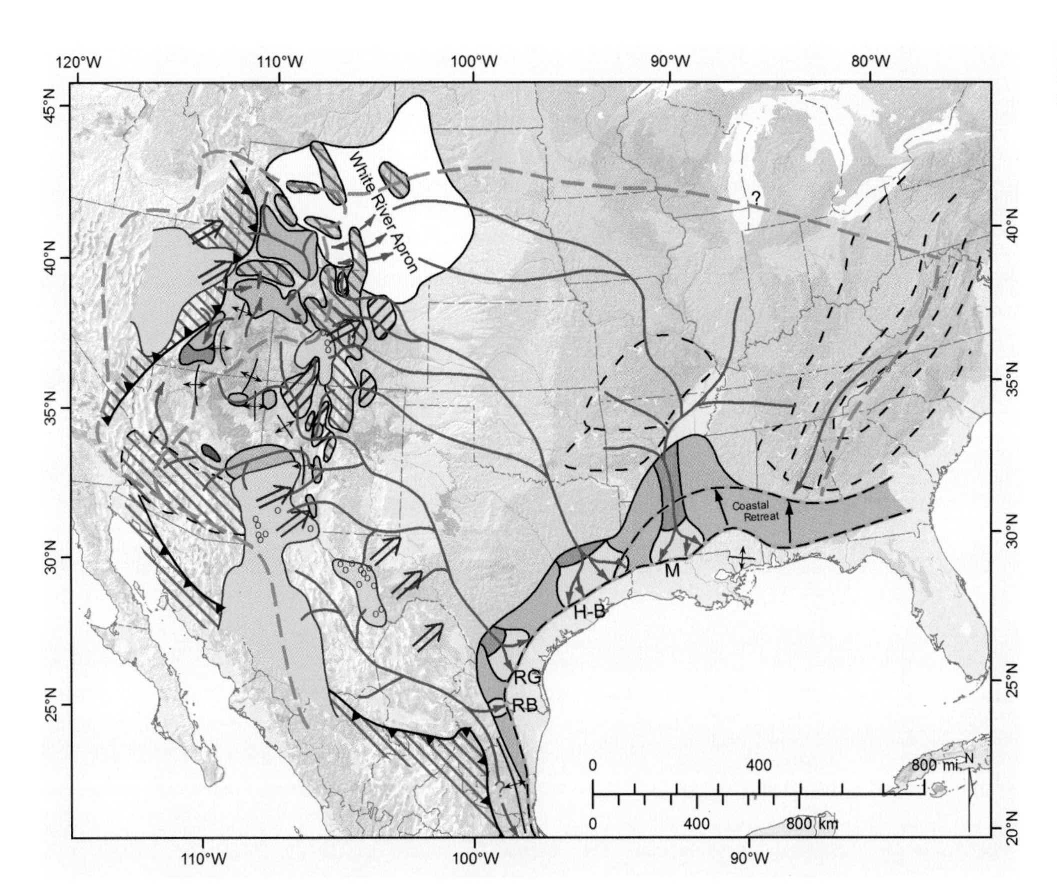

Figure 8.10 Late Eocene drainage basin paleogeography. Modified from Galloway *et al.* (2011).

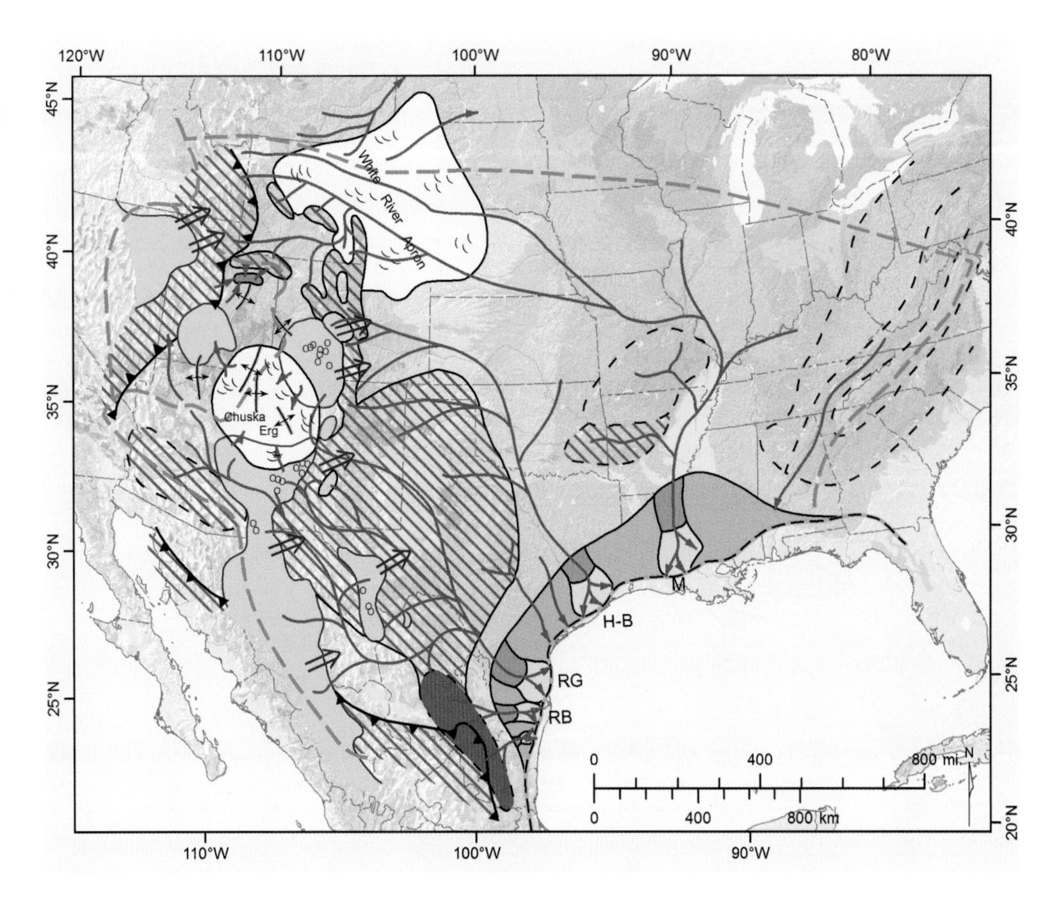

Figure 8.11 Oligocene drainage basin paleogeography. Modified from Galloway *et al.* (2011).

basins augmented sediment supply from the erosional uplands. Regional uplift precluded further accumulation in the intermontane basins, and several of them experienced ongoing erosional evacuation that was initiated in the Late Eocene. Presence of penecontemporaneous volcanic zircons is diagnostic of Frio deposode fluvial deposits (Figure 8.3). Despite the arid climate that extended from the northwestern GoM coastal plain across the southwest and Western Interior, these rivers were large and sediment-laden by the time they arrived on the Gulf margin. The relatively smaller Rio Bravo axis of the Burgos basin was likely more locally sourced by inversion and uplift of the basin rim by several thousand feet.

Early Miocene drainage patterns display some significant evolutionary changes from their Oligocene precursors (Figure 8.12). Most notable is the decrease in volcanic grains and detrital zircons (Figure 8.4). The Mississippi remained the longest river system, draining eastward across the Early Miocene Arikaree alluvial apron from headwaters in western Wyoming. The expanding area of broad uplift across inland Texas (encompassing the Edwards Plateau) may have played a role in diverting the river system draining the central Front Range (which was tectonically rejuvenated) eastward from the Houston-Brazos axis to the Red River axis. The still-significant volcanic zircon content and reconstructed drainage patterns in southern Colorado–northern New Mexico suggest capture of some former Rio Grande tributary elements by the Red River system. An extensive area of interior drainage, centered in Utah, and mapped south-directed channels in Arizona restrict Rio Grande tributaries to southern New Mexico. Speculative expansion of Rio Bravo headwaters is suggested by continued importance of this element in the Burgos basin.

8.1.2.3 Middle Miocene–Pleistocene

The Neogene Tectono-climatic Phase began with dramatic changes in continental geomorphology and consequent Middle Miocene deposode drainage patterns (Figure 8.13). On the east, uplift and unroofing of the Appalachian uplands rejuvenated a sediment source that had been largely moribund since the Mesozoic. Tributaries collected across Kentucky and Tennessee to flow southward into the northern GoM. The system paralleled but remained independent of the larger Mississippi system. Convergence of channel axes in Louisiana created the merged delta system. Headwaters of the Mississippi, now largely confined to the east-flowing tributaries arising in the central Front Range of southwestern Wyoming and eastern Colorado, arose along the toe of the aggradational Ogallala alluvial apron. Recognized precursors of the North and South Platte rivers are preserved in the veneer of alluvial deposits. The continuing arid climate produced flow-limited, energy-deficient rivers that were unable to fully flush their sediment load across the broad plains. However, sufficient sediment did bypass the apron to create a Middle Miocene Mississippi fluvial–deltaic depocenter.

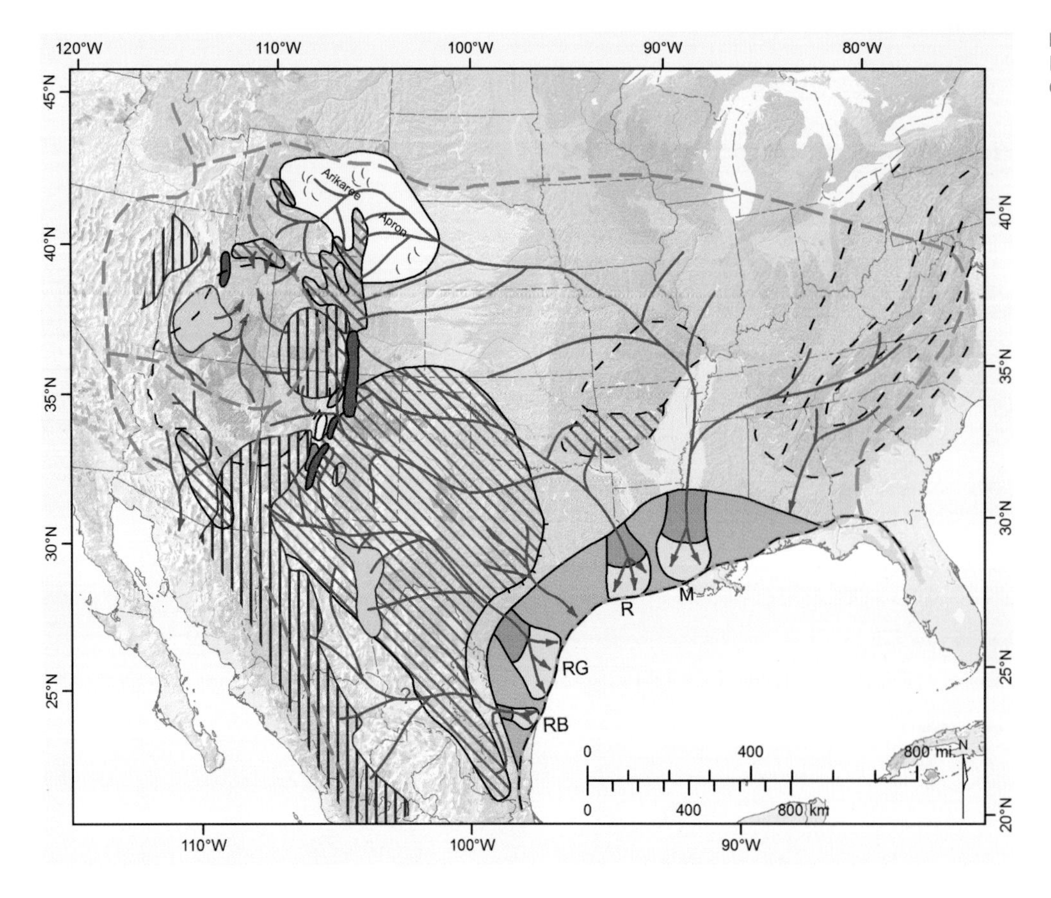

Figure 8.12 Early Miocene drainage basin paleogeography. Modified from Galloway *et al.* (2011).

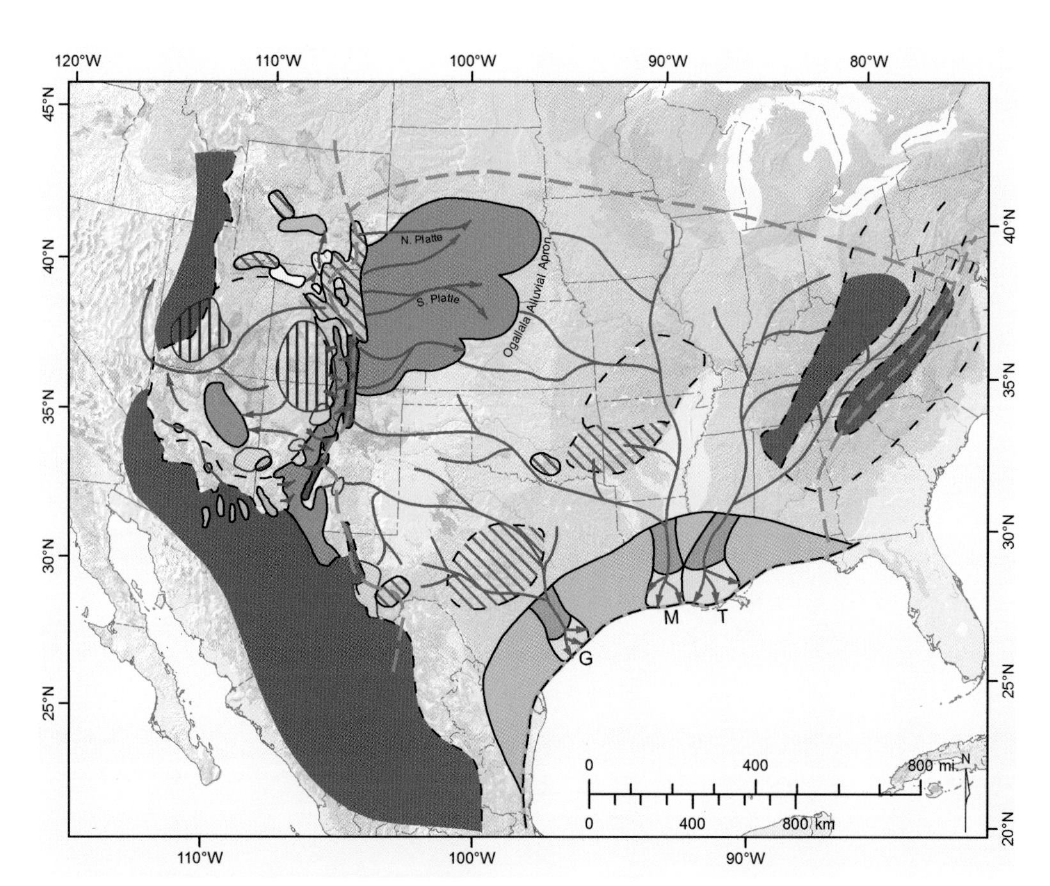

Figure 8.13 Middle Miocene drainage basin paleogeography. Modified from Galloway *et al.* (2011).

Initiation of the Rio Grande rift, extending from south-central Colorado, through central New Mexico, into Trans-Pecos Texas created a series of closed basins that trapped all sediment derived from further west. The continental drainage divide was displaced significantly east of its Paleogene locations. On the east, rift-shoulder uplifts sourced east-flowing tributaries (Figure 8.13). A reconstructed north-flowing paleo-Pecos River drained remnant volcanic uplands in the Big Bend area. In summary, tributary access to western sources was truncated and the prevailing arid paleoclimate limited outflow from the uplands that were available. The Rio Grande River effectively ceased to exist as a significant supplier of sediment to the Gulf. Remnant tributaries flowed across the Edwards Plateau of central Texas, collecting into the Guadalupe fluvial axis of the Middle–Upper Miocene supersequences. In contrast, the emergent Appalachian terrane provided an increasingly important supply of sediment to the northeastern GoM. The Cenozoic basin depocenter shifted to its easternmost position.

Upper Miocene deposode (Figure 8.14) drainage patterns remained relatively stable. Aggradation of the Ogallala apron expanded southward into the Texas panhandle, adding the paleo-Arkansas and Canadian River axes. Reemergence of a secondary depocenter in the far south Texas–Burgos basin area suggests reintegration of Rio Grande and Rio Bravo Rivers.

Pliocene continental geomorphology reflected ongoing evolution of the Neogene drainage and sediment supply patterns (Figure 8.15). Appalachian uplands contributed runoff and sediment to a shrinking Tennessee axis. To the west, epeirogenic uplift of the Western Interior province, centered in western Colorado, created the Rocky Mountain Orogenic Plateau. The broad area of uplift extended from Wyoming to southern New Mexico, with the landscape elevated 1–2 km (3200–6500 ft) above the surrounding continent. Several rivers expanded canyons along the Front Range. East-flowing rivers began to incise the Ogallala alluvial apron. However, the southern extension of the east–west drainage divide remained pinned along the east flank of the Rio Grande rift. The chain of axial rift basins filled from north to south, but the increasingly integrated south-flowing river terminated in a large lacustrine basin. Continuing integration of the north-flowing Pecos diverted outflow from most of New Mexico to the Texas panhandle. There, an east-flowing paleo-Canadian River combined with paleo-Arkansas drainage to flow into east Texas and then into the north-central Gulf as the Red River fluvial–deltaic axis.

The final evolution of North American drainage basins was one of progressive growth and, ultimately, nearly complete consolidation of the Mississippi River system. Climate played an increasing role. Formation and repeated advance of

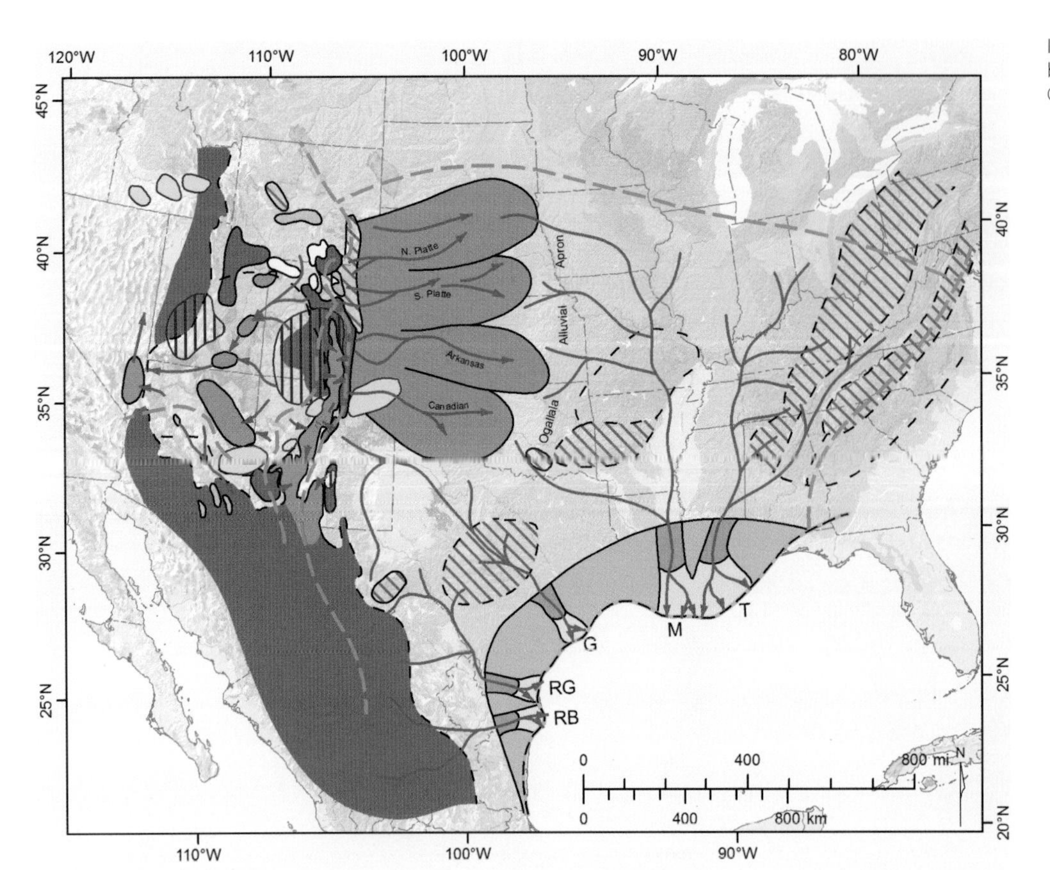

Figure 8.14 Late Miocene drainage basin paleogeography. Modified from Galloway *et al.* (2011).

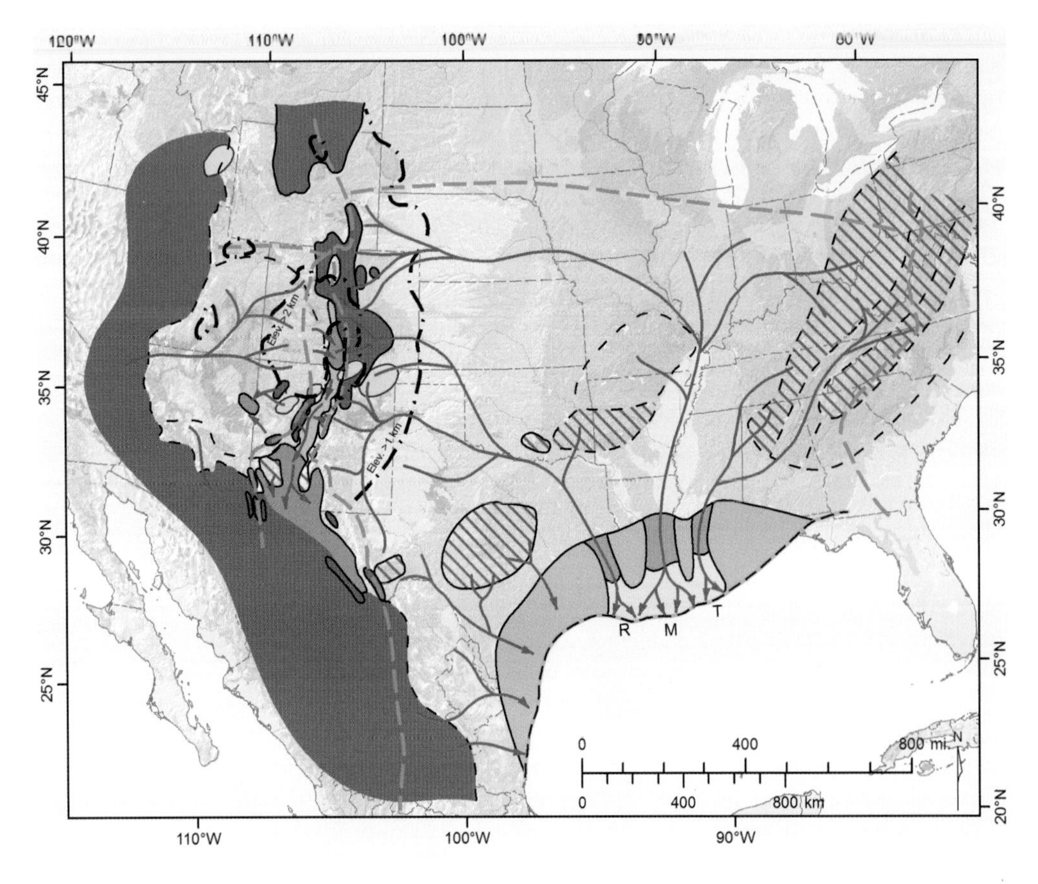

Figure 8.15 Pliocene drainage basin paleogeography. Modified from Galloway *et al.* (2011).

the North American ice sheet reorganized and expanded tributary systems across the northern continental interior (Figure 8.16). Glacial retreat sent surges of outwash-laden meltwater into and down the Mississippi. Valley incision along the Ohio and Mississippi intermittently and then permanently trapped the Tennessee and Red Rivers by the end of the Pleistocene, creating the single Mississippi axis we see today. Alpine glaciation accelerated erosional sculpting of the mountain belts across the center of the orogenic plateau, further enhancing sediment yield. Final integration of a through-flowing Rio Grande and reversal of flow in the Pecos resulted in reemergence of the Rio Grande as a significant, though relatively modest, fluvial–deltaic axis in the northwest Gulf margin.

Notably, mapped lengths of trunk streams ranged between 1000 km and 2000 km (620–1240 miles) throughout the Cenozoic. Changes in trunk-stream lengths through time were proportionally modest relative to total length values. Drainage basin areas, though more subjective in interpretation, are generally largest for rivers flowing into the mapped Gulf margin depocenters. One of the most dramatic quantitative changes in drainage basin area occurred with the Early Eocene capture of the central Rocky Mountain front drainage, originally directed across the mid-continent to the Mississippi axis by tributaries to the southeast-flowing Houston-Brazos and Colorado axes (Figures 8.7 and 8.8). Multiple, sub-equal drainage basins supported the rivers that flowed into the Gulf from Mississippi to northern Tamaulipas state during the Eocene through Lower Miocene deposodes. Beginning in the Middle Miocene, trunk rivers converged toward the Louisiana coastal plain, ultimately creating the single, amalgamated drainage basin of the Quaternary Mississippi.

The dramatic changes in sediment influx volume from deposode to deposode, especially in the Paleogene, despite modest changes in trunk stream length and drainage basin area, argue for the importance of tectonically generated uplift and climate as equally important controls on sediment yield. Large submarine fan systems were produced, as expected, where continental rivers with large sediment loads entered the northern Gulf. However, presence of large rivers, while necessary, was not sufficient for abyssal fan system development. Oligocene Frio rivers, though among the largest in area and sediment load, produced only a prograding slope apron, as did many individual fluvial axes of other deposodes. Further, the largest of the Neogene fan systems lie at the flanks of the contemporary deltaic depocenter, documenting how significant along-strike sediment transport separated fluvial–deltaic and slope components of the source-to-sink system.

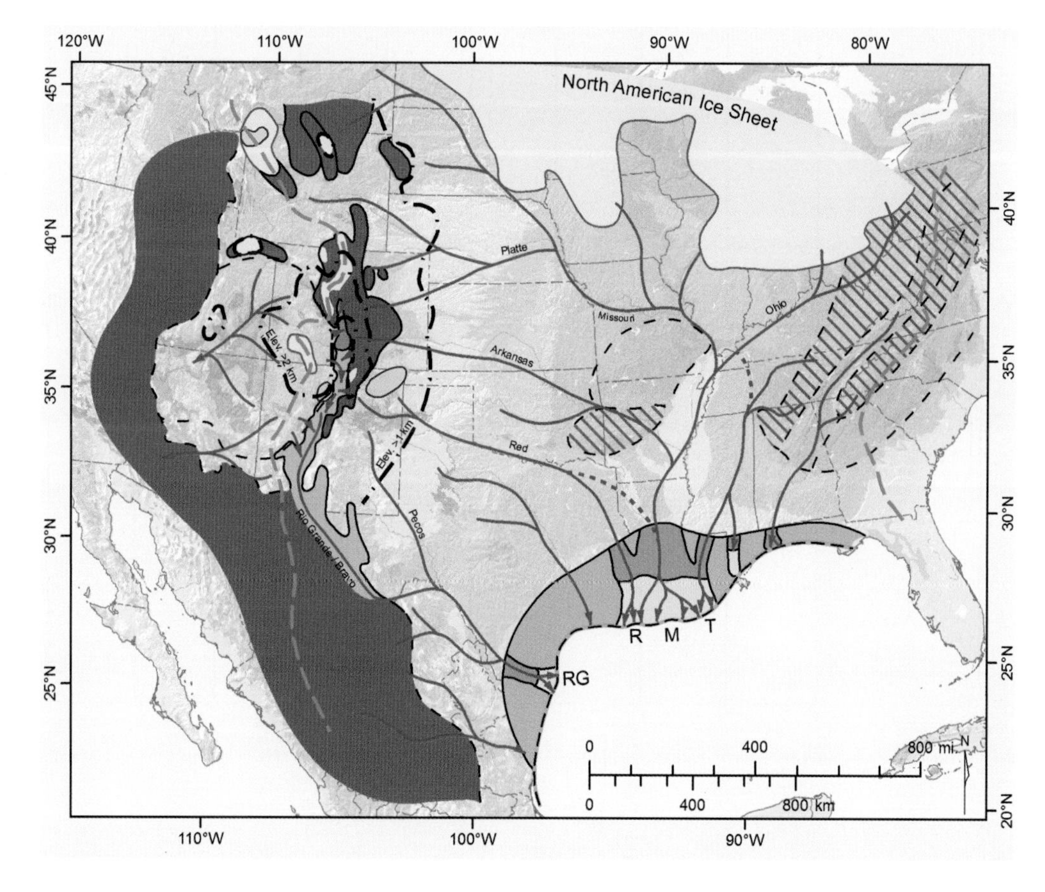

Figure 8.16 Pleistocene drainage basin paleogeography. Modified from Galloway *et al.* (2011).

8.1.3 Fluvial–Deltaic Axes

Eight continental fluvial–deltaic axes dominated sediment influx to the northern divergent margin of the Gulf (Figure 8.17). From west to east, they are named Rio Bravo, Rio Grande, Guadalupe, Colorado, Houston-Brazos, Red, Mississippi, and Tennessee after modern rivers having similar locations on the modern coastal plain and/or broadly similar drainage basins (Galloway *et al.* 2011). Only the Rio Bravo lies on the northern Mexican basin margin. A ninth fluvial axis, the Grijalva–Usumacinta, likely emerged on the Campeche margin in the Late Neogene, following the Chiapanecan orogeny (Witt *et al.* 2012). The remainder of the Cenozoic Mexican Gulf margin was dominated by proximal uplands with local, small drainage basins. Active crustal deformation produced over-steepened gradients that favored direct bypass of sediment into deepwater depositional systems; preserved depositional coastal plains were absent or small until the Late Neogene.

The broad, northern Gulf sedimentary prism was largely constructed by the progradation of a succession of large delta systems to the shelf margin (Figure 8.18). These successive delta systems created off-stepping clusters that together prograded the continental platform 320 km (200 miles) or more from the relict Cretaceous shelf margin to its Holocene position. The most persistent deltaic progression fronts the Mississippi axis. In contrast, the large deltas of the Texas–Tamaulipas margin show broad lateral dispersion, reflecting the evolutionary succession of multiple, shifting fluvial axes. Nonetheless, each fluvial axis supplied a major delta system for multiple deposodes.

8.2 Growth of the Continental Margins

Compilation of the mapped locations of shelf edges at maximum progradation of each of the deposodes summarizes the history of continental margin offlap (Figure 8.19; Galloway 2005a). Several generalizations emerge:

1. Lower Wilcox, Oligocene Frio, Lower Miocene 1, Middle Miocene, Upper Miocene, and the composite Plio-Pleistocene deposodes accomplished the bulk of margin

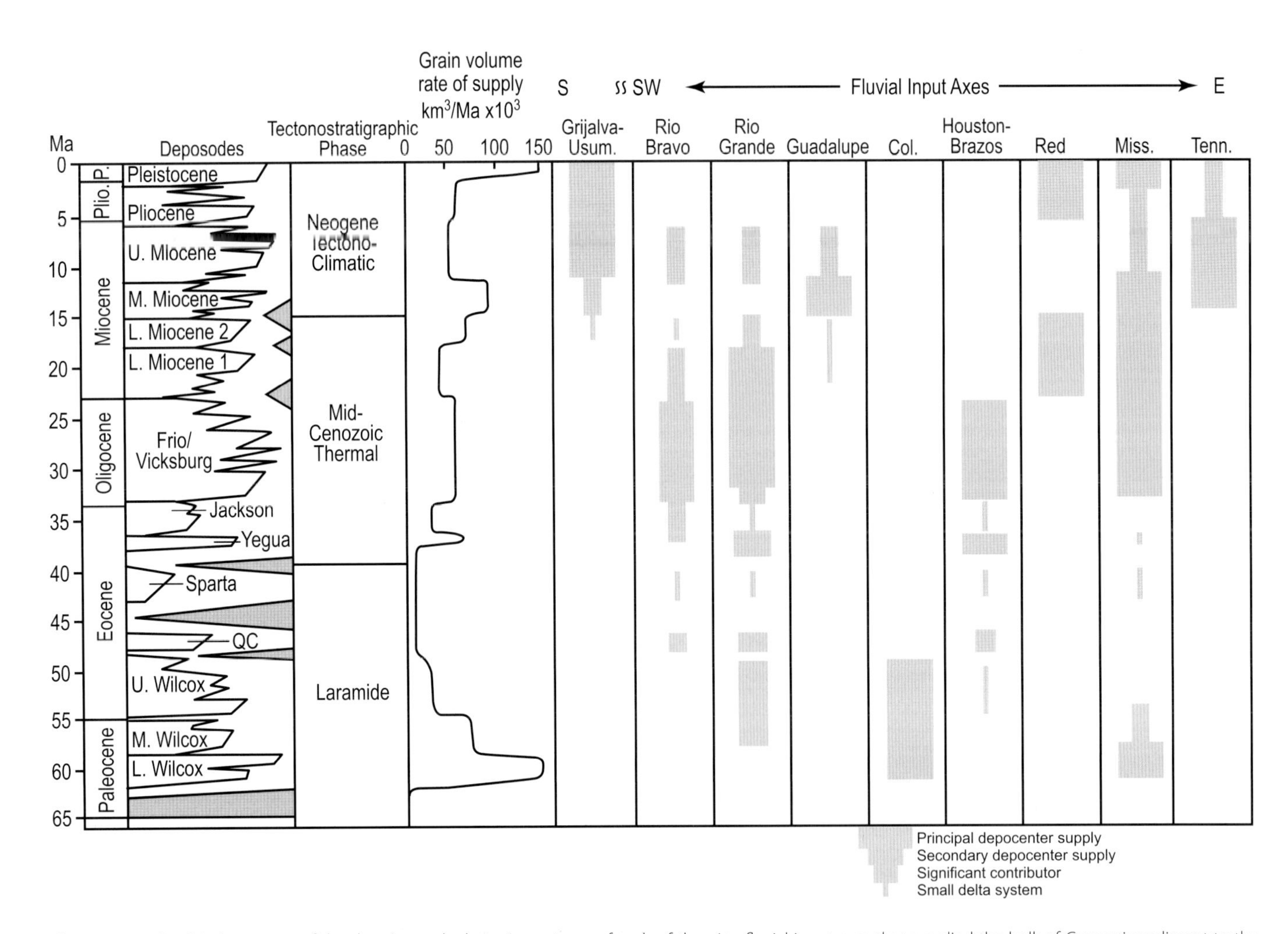

Figure 8.17 Graphical summary of the duration and relative importance of each of the nine fluvial input axes that supplied the bulk of Cenozoic sediment to the GoM basin. Modified from Galloway *et al.* (2011).

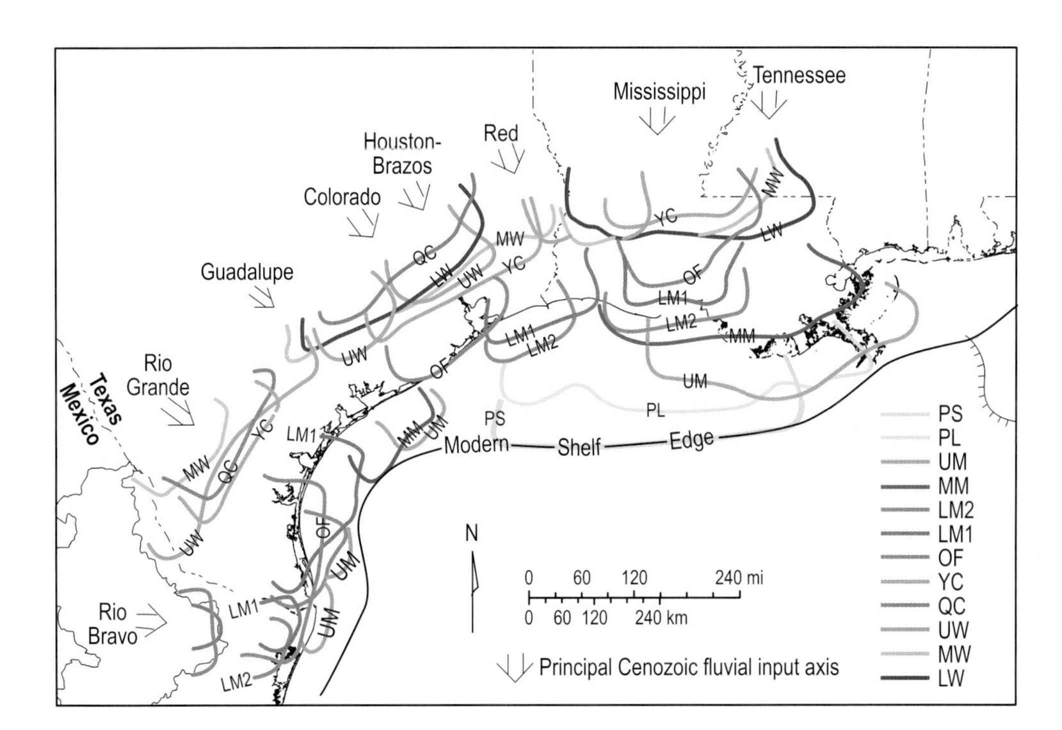

Figure 8.18 Generalized outline of principal Cenozoic delta systems. Deltas advanced basinward through time. Clusters of deltas reflect the entry locations of the eight fluvial axes, from the Tennessee to the Rio Bravo, that supplied them.

offlap. This is broadly consistent with the history of sediment influx.

2. Peak rates of shelf edge advance occurred in or close to principal deltaic depocenters.
3. Peak shelf edge advance rates were very high by global basin standards. The Lower Wilcox shelf edge advanced 20–30 km (12–19 miles) per million years; Oligocene Frio 16–20 km (10–12 miles) per million years; Middle and Upper Miocene 16–20 km (10–12 miles) per million years; and the Plio-Pleistocene 30–40 km (18–25 miles) per million years (Galloway 2005a).

8.3 Continental Slope and Basin Evolution

The continental slope and basinal depositional systems are the principal theaters of ongoing and potential future hydrocarbon exploration and development in the GoM. In this analysis, we have differentiated and mapped a suite of depositional systems that comprise potential reservoir-bearing, deepwater sediment dispersal systems (Figure 1.35). These include (1) progradational slope aprons; (2) retrogradational margin aprons; (3) tectonic margin aprons; (4) submarine fan systems; (5) megaslide complexes; (6) submarine channel systems; and (7) migratory dune fields. The first five are documented hosts of reservoir sand bodies that support major hydrocarbon plays. Submarine channel systems in the northern Gulf are an emerging play type. Composition of migratory deepwater dune fields is unknown, though they are likely mud-dominated or alternating mud and sand.

Figure 8.20 compiles the mapped (and for a few elements, speculative) distribution of collapsed continental margins, submarine channel complexes, and abyssal plain fan systems. Several themes are apparent:

1. Paleogene margin collapses are clustered in the northwest Gulf. They correspond in time and location with contemporary tectonic uplift and/or tilting.
2. Neogene collapse due to rapid salt evacuation and consequent foundering was centered on the central Gulf margin.
3. Submarine fans of all ages cover a large portion of the Gulf abyssal plain. Wholesale shelf edge bypass, with focused sand transport onto and across the abyssal plain, was a major feature of both greenhouse (Paleocene–Early Eocene) and icehouse (Miocene–Pleistocene) worlds. Nearly two-thirds of the modern abyssal plain is underlain by one or more fan systems.
4. Paleogene submarine fans accumulated within the Chicontepec foreland trough, and speculatively spilled around the north and south margins of the Tuxpan Platform and onto the western Gulf abyssal plain.
5. Recurrent sediment gravity flows that formed the Veracruz basin tectonic apron coalesced into the north-sloping Veracruz Trough, where they were redirected northward. Transport along the trough extended 300–500 km (180–310 miles), with possible terminal fan formation at its terminus in the west-central Gulf. This sediment transport system was continually present from the Eocene through the Miocene, a timespan approaching 50 million years.

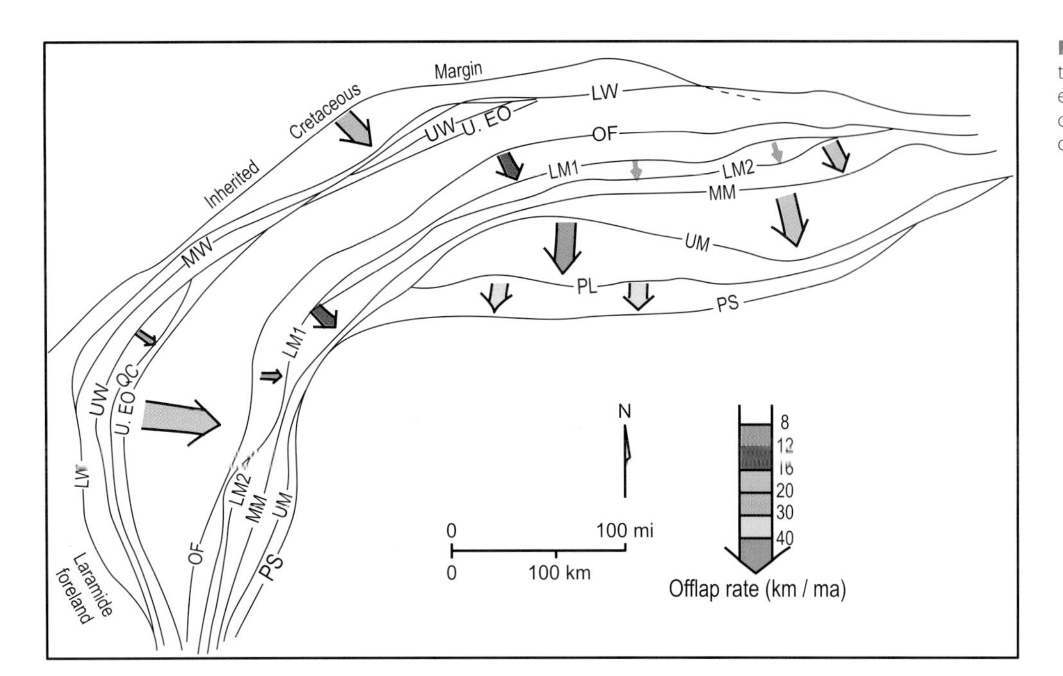

Figure 8.19 Shelf edge locations at the end of each of the depositional episodes. Arrows show, by length and color, the maximum continental margin offlap rate (km./Ma) and its location.

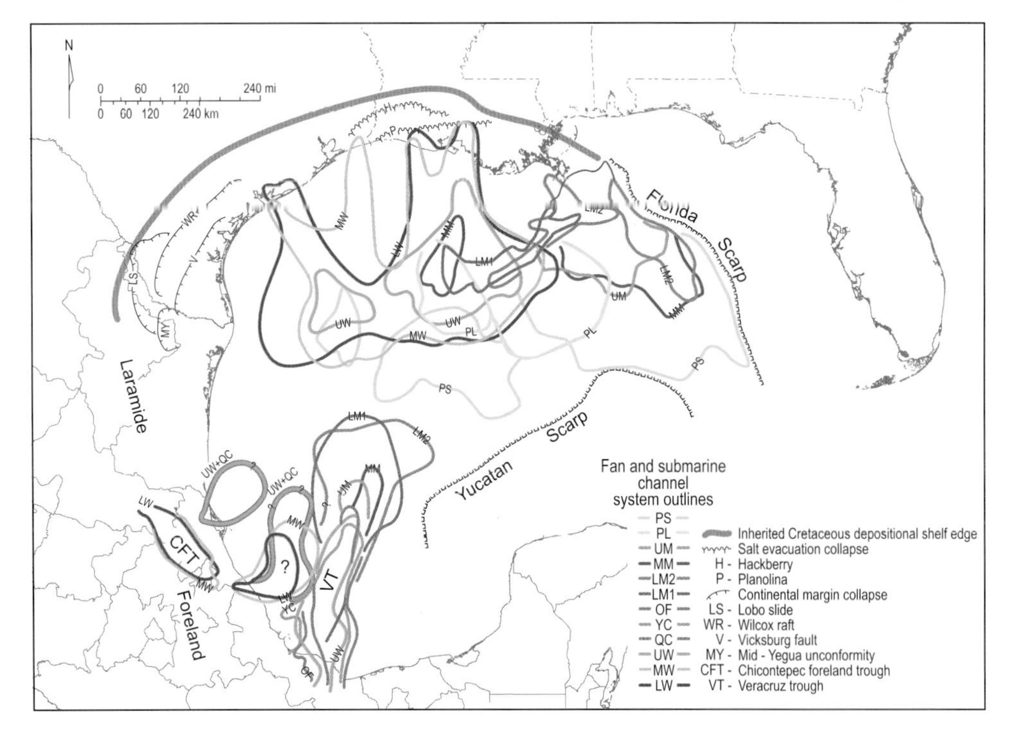

Figure 8.20 Compiled distribution of (1) collapsed continental margins, (2) submarine channel complexes, and (3) abyssal plain fan systems. The features are collected within boundaries of the Cenozoic deepwater Gulf, including the inherited Cretaceous depositional shelf edge, Florida scarp, Campeche scarp and Laramide foreland. Large submarine fans are concentrated in the northern Gulf. Small-to-medium fans and stacked submarine channel complexes occupy the southwest Gulf.

Slope aprons, which are not shown in Figure 8.20, underlie essentially all of the modern continental slope and shelf around the Gulf rim. Many of the aprons contain multiple channel–lobe complexes spread along-strike across the width of the apron. Within aprons, channel–lobe complexes commonly stack vertically. Large and/or long-lived channel–lobe elements extend beyond the slope toe onto the adjacent abyssal plain, creating smaller versions of the abyssal plain fan systems. Note that our differentiation of fan systems versus channel–lobe complexes by scale is arbitrary. Our objective was to differentiate and map the *deposode- and basin-scale* fans that transported sand far onto the abyssal plain, deposited thick, aggradational turbidite successions, and, consequently, created *play-scale* reservoir systems.

Part

IV

Petroleum Habitat

This part contains a standalone chapter that transitions from foundational science to discussion of the petroleum habitat associated with each depositional unit or unit aggregate (supersequence) described in Parts II and III. This includes the USA, Mexico, and Cuba. Beginning with an overview of the current GoM resource size and spatial distribution, ensuing sections cover Mesozoic conventional exploration and unconventional resource plays. Key petroleum habitats are considered for Paleogene and Neogene intervals as well, concluding with comments on the unique adaptation of seismic technology to meet the subsalt challenge in the GoM.

Chapter 9

GoM Petroleum Habitat

9.1 Background

The term *petroleum habitat* has often been used to incorporate all of the elements that are crucial to a prospect, discovery, or field. We employ the concept in the same fashion as Lewis Weeks did in 1958, when he introduced one of the first comprehensive volumes on the critical geologic factors inherent to 50 global basins (Weeks 1958). Our understanding of the Gulf of Mexico (GoM) petroleum habitat has greatly benefited from major advances in technology and understanding that follows from exploration. In addition, exploration risk and uncertainty have been greatly reduced by placing plays, prospects, discoveries, appraisal wells, and field extensions in a geologic context (Galloway 2009).

As our focus is the depositional evolution of the basin, we focus on reservoir-related factors in this chapter, but also consider the other elements that are pertinent to success or failure in upstream phases. We consider established, emerging, and potential future exploration plays in our basin-scale depositional context. It is also useful to describe discoveries and producing fields from a similar perspective. Paralleling the geochronological structure of the book, we begin with pre-salt frontier plays and progress to the most mature Pleistocene reservoir intervals of the suprasalt realm. This is preceded by a look at the current state of plays in terms of upstream maturity and then a discussion of the evolving seismic technology that is so important to exploration, development, and field production.

In this chapter, we also consider both conventional and unconventional reservoirs. In fact, GoM reservoirs span a spectrum from shale (source rock) plays to hybrid plays, to classic carbonate and sandstone conventional reservoirs. While geologic factors often overlap across this reservoir spectrum, we discuss these in separate sections for the purpose of clarity.

Since source rocks are the primary reservoirs in unconventional plays, discussion will focus on basin-scale to regional factors that enhance organic enrichment and preservation, relevant therefore to both reservoir types. The paleogeographic maps described in preceding chapters are key considerations in understanding both conventional and unconventional plays. As mentioned, it is also useful to assess established production trends as these attest to the quality of the overall petroleum system, including reservoir, source, seal, and trap elements.

9.2 Gulf of Mexico Undiscovered Resources

The most recent assessment of risked undiscovered technically recoverable oil and gas resources (UTRR) for the US GoM federal waters shows a mean of 73.60 BBOE, with a range of 61.55–86.93 BBOE (BOEM 2017; Figure 9.1). The mean is about 50 percent of the UTRR for the entire country. Total mean conventional endowment (produced + undiscovered) is assessed as 152 BBOE, including the Florida Straits (BOEM 2016). This does not include unconventional onshore oil and gas plays, potential tight gas plays, or gas hydrates. Overall, 65 percent of the UTRR is oil for the US GoM offshore. In spite of the long exploration history and substantial proved reserves, the US GoM continues to attract exploration investment designated for evaluating new plays, new models, and concepts. Unconventional resource estimates for onshore plays are assessed by the USGS and are discussed in unit-specific sections of this chapter.

Since opening the country to international exploration in 2015, Mexico has held numerous competitive bidding rounds,

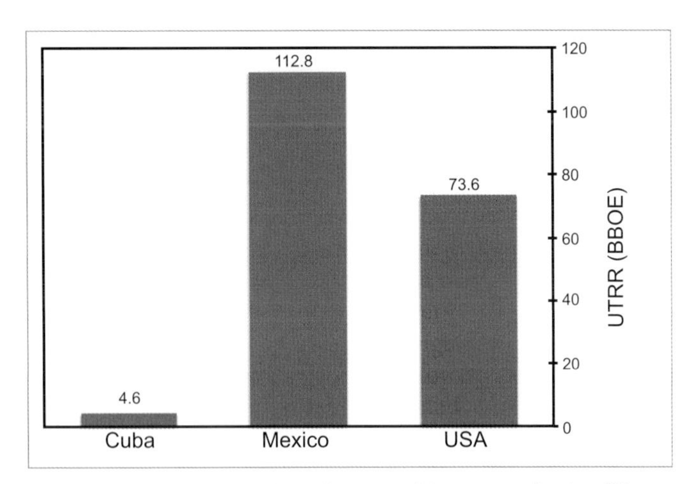

Figure 9.1 Undiscovered technically recoverable resources for the offshore USA, Mexico, and Cuba. Note that Mexico includes onshore "prospective reserves" though the bulk of future potential is clearly offshore. The 2014 US GoM federal resources are from BOEM (2017); Mexico resources from CNH (2018); and Cuba resources from USGS (2005).

mainly in the offshore area. The National Hydrocarbon Commission of Mexico (CNH) estimates prospective resources of 112.8 BBOE, including both conventional (50.6 BBOE) and unconventional resources (60.2 BBOE) (Guzmán 2018; Figure 9.1). About half of the prospective resources are in oil, with substantial estimated resources of undiscovered unconventional gas in the onshore Burgos and Sabinas–Burro–Picahos basins (CNH 2018).

In Cuba, the USGS assessed mean UTRR of 4.6 BBO, 9.8 TCGF, and 900 million barrels of natural gas liquids (USGS 2005; Figure 9.1). Three Mesozoic plays are considered the petroleum habitat for these resources: (1) carbonate platform; (2) foreland basin; and (3) fold and thrust belt. The north Cuba thrust belt is thought to encompass the bulk of these resources. Cuba has large proven resources of heavy oil (e.g., Varadero Field), though ultimate recoveries are likely to be low and gas estimates are somewhat speculative.

9.3 Spatial Distribution of Current GoM Discoveries

In the greater GoM basin, Mesozoic reservoirs are located in a roughly concentric band around the more basinward Cenozoic discoveries (Figure 9.2). This is logical, given the long-term control of Mesozoic shelf margins by the subsidence associated with the transition from thick to thin continental crust. With the exception of the Poza Rica trend of Mexico, few carbonate slope and basinal systems have economically viable reservoir properties. By contrast, slope and basinal siliciclastics are prolific hydrocarbon reservoirs in areas basinward of the Mesozoic shelf margins and dominate outer shelf and deepwater discoveries and reserves.

There are some exceptions, of course. The Mesozoic Norphlet trend of discoveries, including Shell's Appomattox (MC 392 #1), Rydberg (MC 525 #2) prospects, and most recently Chevron's Ballymore (MC 607 #1) prospects represent

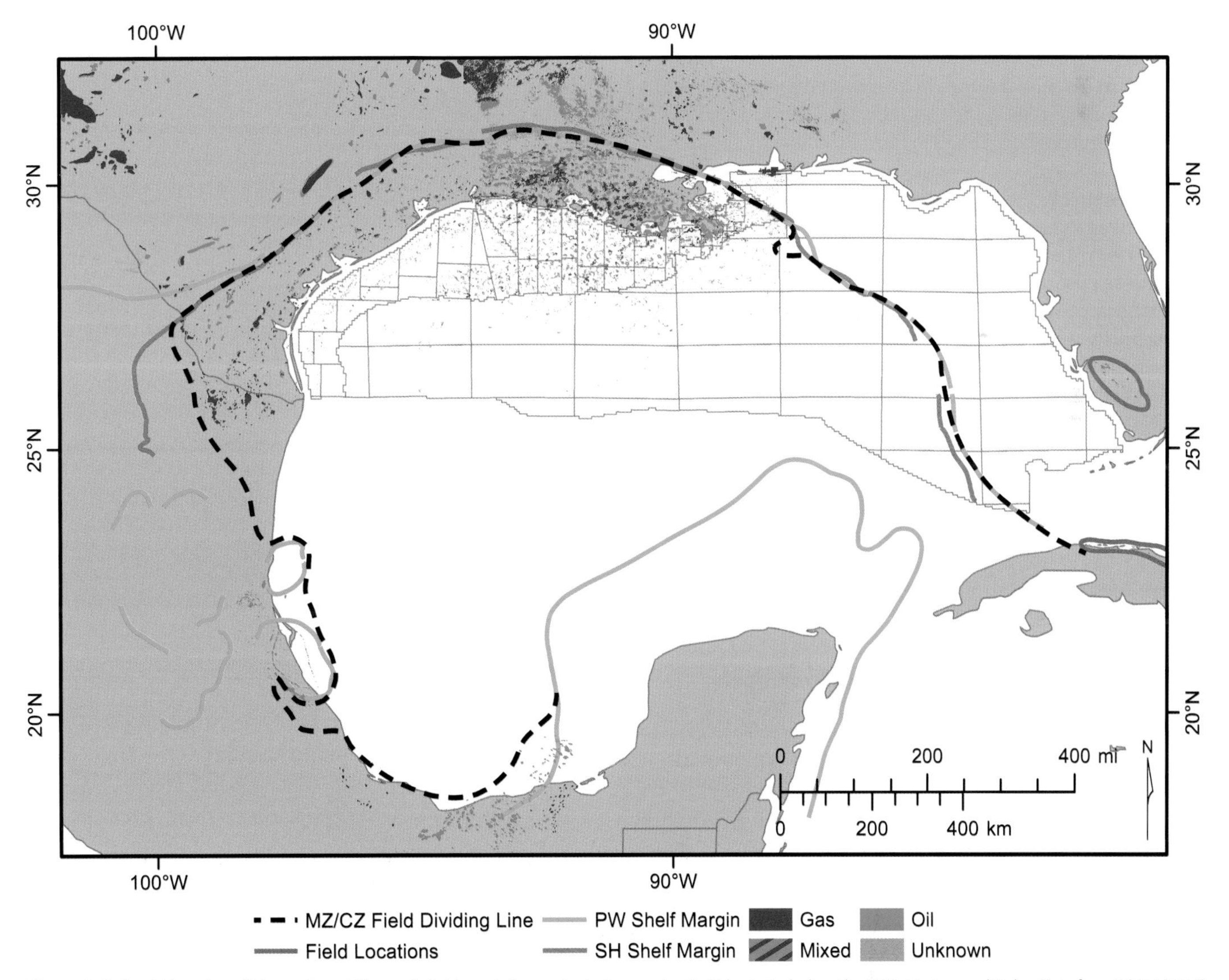

Figure 9.2 Spatial location of Mesozoic and Cenozoic fields and discoveries in the greater GoM basin including the USA, Mexico, and Cuba. Data from BOEM (2016), CNH (2018), and USGS (2005).

an eolian sandstone play in the present-day US eastern GoM deepwater (Figure 9.2). The westward salient of Mesozoic discoveries here is accentuated by post-depositional rafting of the Norphlet, as explained in Section 3.3.4.

In Mexico, there are several other deviations from the concentric band around the basin (Figure 9.2). The Macuspana basin is located landward of the Mesozoic/Cenozoic field dividing line, where shale evacuation has set up a series of shallow extensional structures that largely trap natural gas in the Miocene and Pliocene (Ambrose *et al.* 2003). Further west, Miocene deepwater sandstones are oil-bearing in the onshore Veracruz basin (Jennette *et al.* 2003). A Cenozoic shift eastward lies south and west of the giant Golden Lane Mesozoic carbonate platform fields, where the Chicontepec field complex produces from Paleogene reservoirs (Cossey *et al.* 2007; Figure 9.2).

The Mesozoic/Cenozoic Field dividing line can also be extended to Cuba, where a series of heavy oil fields reservoired in Mesozoic carbonates contain about 97 percent of that country's proved reserves (Schenk 2008). Future potential discoveries would likely also be contained with Mesozoic reservoirs in the fold and thrust belt, foreland basin, and carbonate platform (USGS 2005). Small Mesozoic discoveries and fields of the Sunniland trend of Florida are also understandably inboard of the Mesozoic/Cenozoic dividing line (purple outline in Figure 9.2).

9.4 Synopsis of Current GoM Exploration Plays

A 2014 breakdown of proved and probable reserves shows nine proven exploration plays dominate the northern GoM (USA) offshore sector (Figure 9.3). Only one Mesozoic reservoir (Jurassic Norphlet) is the habitat for significant resources (>500 MMBOE) in offshore areas, including the giant Appomattox discovery (Godo 2017). However, Mesozoic reservoirs are important producers, source rocks, and unconventional plays in onshore areas, not shown here. In addition, it is possible that future Mesozoic plays may develop, including combination and hybrid reservoirs.

For the US Cenozoic, the Paleogene Wilcox subsalt exploration play (as of 2014), is the largest habitat, with an estimated UTRR of 21.6 BBOE from 26 discovered pools. A number of large Paleogene discoveries are not yet developed, but production of the Wilcox at Chinook-Cascade, Great White, and Jack-St. Malo by operators Petrobras, Shell, and Chevron, respectively, has been ongoing for a number of years.

Several newer plays have significant expected growth, including the Lower Miocene of the Green Canyon protraction block, where exploration is among the highest in the northern GoM. Tahiti Field, discovered in 2002, went on production in May 2009 (Carreras *et al.* 2006; Moore 2010). New production also went onstream from the Lower Miocene of Holstein Deep (GC643) in 2016 and the Shell-operated Vito Field (GC 940, 941, 984, 985) has been sanctioned for development.

The Pleistocene (449 pools), Pliocene (627 pools), and Upper Miocene (556 pools) exploration plays are quite mature, having been the primary play in the suprasalt exploration phase in the 1980s and 1990s (see Prather *et al.* 1998). Little or no new drilling targeting these suprasalt reservoirs has occurred in recent years. The deep shelf, sub-weld Miocene play has largely been abandoned due to high pressures and temperatures and related drilling costs.

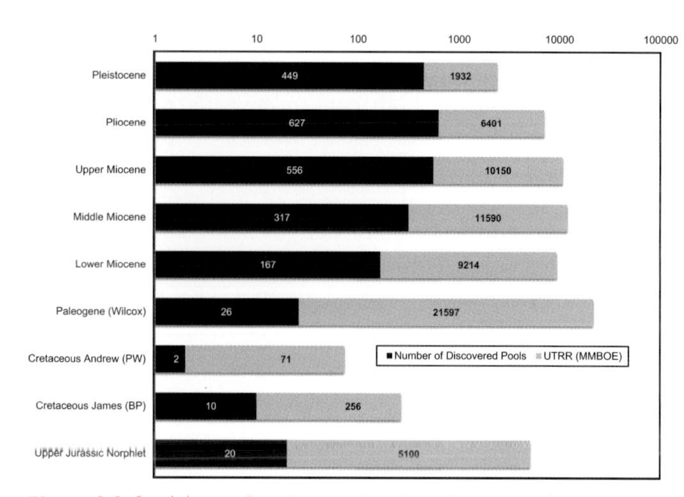

Figure 9.3 Breakdown of undiscovered technically recoverable resources (UTRR) and discovered pools by play in US GoM federal waters. The 2014 assessment data are from BOEM (2017).

In Mexico, Mesozoic carbonate fields contain the majority of the discovered hydrocarbons, with just five fields (Ku, Maloob, Zaap, Xanab, and Xux) providing over half of the current production (as of June 2017, https://portal.cnih.cnh.gob.mx/dashboard-reservas.php). The K–Pg breccia is the largest reservoir by production volumes followed by the Kimmeridgian (Haynesville–Buckner, HVB) interval of the Pilar de Akal contractional belt (Sandrea *et al.* 2018). Cenozoic deepwater sandstone discoveries have been made in the Perdido fold belt of Mexico and offshore Veracruz Trough, but are not yet on production (Colmenares and Hustedt 2015). However, future exploration for Cenozoic siliciclastic reservoirs is being built on a solid program of seismic acquisition, geochemical seabed sampling, and tender rounds through the end of 2018.

In subsequent sections, we comment on the diverse universe of GoM petroleum habitats by reservoir intervals, from pre-salt to Pleistocene, from conventional to unconventional. This discussion also spans the wide spectrum from frontier concepts (e.g., pre-salt, Lower Tuscaloosa) to emerging plays (Norphlet rafts), to reservoirs with limited future potential (Pleistocene suprasalt).

9.5 Pre-salt Petroleum Habitat

There are a limited number of wells targeting pre-salt reservoirs in the greater GoM basin, primarily onshore USA. In Cass County of northeast Texas, initial reports indicated a small discovery in the Eagle Mills interval, although uncertainty exists regarding the age of the oil reservoir and the zone in question did not produce oil for longer than a few weeks (Aubrey 1984). Nearby wells such as Shell #B-3 McDonnell and Primary Fuels #1 Ellingston Trust that were sampled for detrital zircon provenance analyses (see Section 2.3.3) had poor reservoir properties due to extensive burial diagenesis.

Offshore, the Gainesville 707 (GV707 #1) well was drilled through a pre-salt interval to test Ordovician sandstones in a large structure. We estimate from seismic and well data that about 1900 ft (580 m) of Eagle Mill interval was present above the Paleozoic basement interval, consistent with unpublished reports by the Florida Geological Survey as well as seismic cross-sections shown by Mohn and Bowen (2012). Mud logs from GV707 #1 record reddish to buff-colored sandstones with no visible porosity or oil shows, interbedded with reddish siltstones and variable amounts of anhydrite.

However, previous reports of potential pre-salt-sourced oils (Schumacher and Parker 1990) and Triassic lacustrine source rocks and oil family in the northern rim of the GoM basin (James *et al.* 1993; Hood *et al.* 2002) may indicate that broader syn-rift lacustrine source potential exists. Similar non-marine geochemical indicators were among the first clues to the presence of what later turned out to be important pre-salt source rocks in both Brazil and Angola (Dickson and Schiefelbein 2012).

In Mexico, pre-salt reservoirs and source rocks are of interest due to a distinctive seismic signature suggestive of marine and marginal marine character below the Campeche and Yucatán salt (Miranda Peralta *et al.* 2014). Surface oil seeps and chemosynthetic communities are present near the Campeche Knolls, indicating a mature Jurassic oil source (Naehr *et al.* 2007). CNH has reported Pemex plans to drill the Yaaxtaab-1 well for a pre-salt objective in the 2018–2019 period (CNH 2017a).

9.6 Smackover–Norphlet Supersequence

One of the oldest reservoirs with established production in the northern GoM is the Norphlet Jurassic play. Onshore, the Norphlet produces oil mainly in Mississippi, Alabama, and Florida (Godo 2017), but reserves are much smaller in comparison to the more prolific Smackover. In 1979, Mobil discovered the Mary Ann deep gas field in Mobile Bay of Alabama state waters (Marzano *et al.* 1988; Frost 2010). Here, thick, linear sand dunes depress the underlying Louann Salt (Figure 9.4). This creates unusual relief on top and base reservoir that causes local separation of gas and water between eolian dunes (Mankiewicz *et al.* 2009). In spite of the deep depths (>20,000 ft) and reservoir age, the Norphlet reservoir quality is excellent, with chlorite cementation mitigating the degrading effects of quartz cementation and secondary porosity is well-developed (Ajdukiewicz *et al.* 2010).

While production in the shallow water of Mobile Bay began in the 1980s, offshore exploration failed to find economically viable discoveries in spite of numerous wells drilled in the Destin Dome or Pensacola protraction blocks. In 2003, Shell moved about 175 km (108 miles) seaward into the deepwater of DeSoto Canyon and made a significant find in the Shiloh prospect (DC 269 #1; Table 9.1). Interest in the area was generated by a US lease round featuring the first new eastern Gulf blocks up for bid since 1998 (Godo 2017). Presumably basin modeling results indicated that cold deepwater paleo-environments and lower heat flow in transitional crust would be favorable for an oil play downdip of a high-pressure, high-temperature gas play at Mobile Bay.

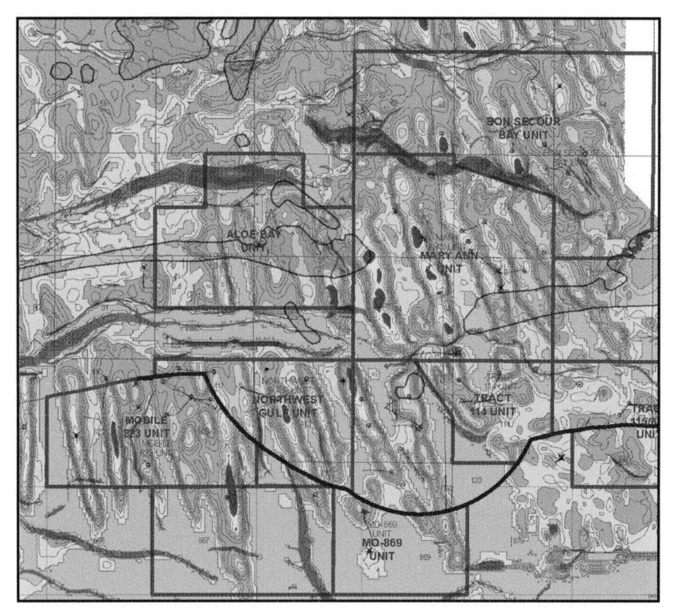

Figure 9.4 Norphlet linear dunes, Mobile Bay area of offshore Alabama. Modified from Frost (2010).

Reservoir objectives included the Cotton Valley–Bossier (CVB), HVB, and Norphlet (Godo 2017).

Like Mobile Bay, the reservoirs in the deepwater Norphlet play are eolian sandstones with dune-scale cross-beds and chlorite cements which apparently mitigate diagenetic reduction of porosity and permeability. Oil is sourced from the Oxfordian Shale of the Lower Smackover and sealed by the tight Smackover Limestone and Haynesville Shales (Godo 2017). There are multiple discoveries and Appomattox (MC 392 #1) is the largest, approved for development by Shell and its partners (Table 9.1).

This and other discoveries are centered upon large salt structures that have rafted apart, creating synkinematic growth in the overlying Haynesville and Cotton Valley interval (Table 3.1). The Norphlet play area is based on depositional and structural trends and current well control. The lateral limits are based on mapping the transition from eolian erg to carbonate shoreline (western limit) and from erg to fluvial/wadi (eastern limit). The Cheyenne well (LL399 #1; Figure 3.10) is a key well, as the Norphlet is apparently absent, suggesting its location in the coeval marine seaway. The northern limit is defined by the transition from early salt raft structures in the deepwater fairway to late salt structures typical of the failed Destin Dome shelf exploration area. The southern limit is inferred to occur where the Mesozoic interval generally thins southeastward. The Madagascar (DC 757 #1) and Swordfish wells are dry holes, but this may or may not define the western or southeastern limits of the play. Sake (DC 726 #1) was a dry hole, defining the Norphlet on the Middle Ground Arch as having a poor reservoir (Figure 3.10).

In recent years, some interest has surrounded the Anadarko reports of Smackover play in the Raptor well (DC 535 #1), but porosity in the Smackover was not laterally continuous, as demonstrated in the sidetrack, and the lease was relinquished. The lease for the Gettysburg West (DC 398) discovery

Table 9.1 Comparison of Oxfordian plays, northern GoM

	Norphlet deepwater	Smackover onshore
Predecessor	Mobile Bay Mary Ann Field (Mobil)	Smackover fields of Alabama, Texas, and other states
Play opener/ year	Shiloh (DC 269 #1)	Numerous discoveries in 1960s–1970s
Major operator(s)	Shell, Chevron	Numerous independents
Largest discoveries (year, block)	ExxonMobil-operated Mary Ann Gas Field (1979); Shell-operated oil discoveries at Appomattox (2010, MC 392, 391, 348), Vicksburg (MC 393), and Ryberg (MC 525); Chevron-operated oil discovery at Ballymore (MC 607). Discovery under evaluation.	Jay Field (1970), Santa Rosa Co., Florida and other small onshore fields. No offshore production
Reservoir/EOD	Norphlet eolian sandstones	Thrombolite (microbialite) build-ups and grainstone banks
Trap types	Four-way and three-way raft structures	Four-way closures on basement highs
Source	Oxfordian	Smackover
Seal	Smackover Tight Limestones and Haynesville Shale	Haynesville Shale or Buckner Anhydrite
Hydrocarbon type	Oil	Oil, gas
Challenges	Oil quality (solid bitumen), trap failure, reservoir limits undetermined	Porosity development, deep burial diagenesis, pervasive cementation in marine environments
Positives	Chlorite content of sandstones and low heat flow mitigates reservoir diagenesis	Dolomitization related to meteoric diagenesis or exposure
Status	Established play: Appomattox funded for development, Ballymore discovery under evaluation	Mature Play. Little Cedar Creek (1994) and nearby Brooklyn Field (2011)
Comment	Extent of play still being delineated but numerous dry holes to southeast and west	Unable to extend onshore play into offshore areas due to unfavorable porosity development

was terminated by BOEM due to a lack of development activity by the operator Shell. The Chevron-operated Ballymore well (MC 607 #1), reportedly a 2017 hub-class (>500 MMBOE) discovery, extends the deepwater Norphlet play 22+ km (14+ miles) to the southwest (Figure 3.9).

In Mexico, Oxfordian sandstone reservoirs are present in the Ek-Balam field just east of the large Cantarell complex of the Campeche basin (Angeles-Aquino and Cantú-Chapa 2001; Ricoy-Paramo 2005; Mitra *et al.* 2007; Figure 3.13). Coastal dunes landward of a restricted sabkha are the primary reservoir paleo-environment (Roca-Ramisa and Arnabar 1994). Reported porosities are in the 10–22 percent range (Murillo-Muneton *et al.* 2002). Examination of cores from the Balam 101 by the senior author indicates preserved intergranular porosity in Jurassic-age and deeply buried (>4500 m) sandstones (Figure 3.13E). Juarez (2001) reported results of internal Pemex studies that identified secondary porosity as a result of feldspar dissolution. Anhydrite and dolomite cements are common. Oils of around 26 degree API are thought to be derived from oil-prone type I–II Oxfordian sources (Guzmán-Vega and Mello 1999). A gross-comparison to the Norphlet of the northeastern GoM can be made, though our reconstructions suggest a more limited eolian development than large erg system of the US sector. Polished grains and high-angle cross-beds (>30 degrees (Figure 3.13D, E) suggest eolian processes, with conglomerates in the Nix-1 well indicating fluvial wadi systems accessing nearby source terranes.

Production from Smackover Limestone (and local dolostones) are restricted to onshore areas of the northern GoM (Figure 9.4; Table 9.1). On the western flank of the East Texas salt basin in Texas, a broad shallow marine carbonate ramp (inner and outer ramp of Figure 3.15) hosts three subplays of producing Smackover fields: (1) fault line; (2) salt structures; and (3) basement highs (Kosters *et al.* 1989). The downdip limit of these Smackover discoveries is defined by where grainstones/packstone shoals are no longer present in the deeper portion of the ramp. The fault line trend relates to the breakaway zone at the Louann pinch-out (see cross-section 5; Figure 1.17). The largest fields are grainstone banks or microbialite build-ups developed over basement highs, dolomitized and exposed to meteoric fluids and dissolution (Harwood and Fontana 1984).

This distribution is paralleled by the set of three Smackover subplays in the onshore areas of Mississippi and Alabama (Figure 9.5): (1) peripheral fault line; (2) interior salt basin; and (3) Jackson Dome igneous complex (Mancini 2010). However, Jackson Dome Smackover reservoirs are dominantly sandstones, fed by the paleo-Mississippi River system (Figure 3.15). New work suggests that thrombolytic boundstone reservoirs are more important than the grainstone banks of the northwestern side of the basin (Mancini *et al.* 2004). Diagenetically enhanced porosity and permeability values of 20 percent and nearly 8 darcies are documented in the Little Cedar Creek Field of Southwest Alabama (Haddad and Mancini 2013).

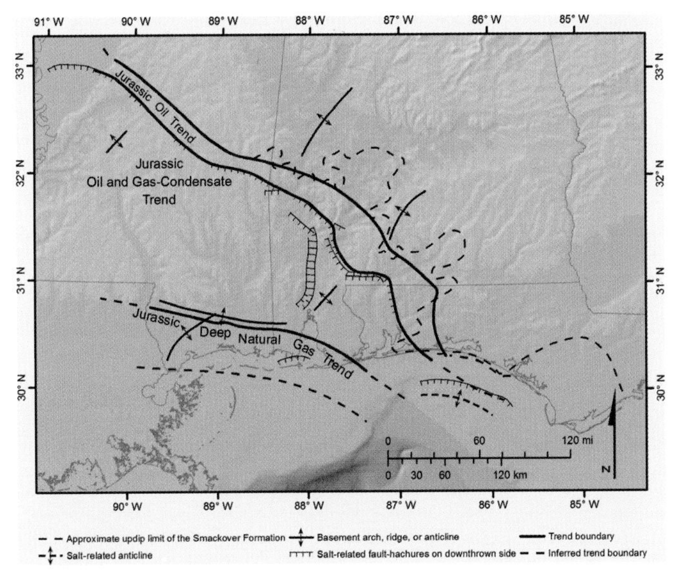

Figure 9.5 Eastern GoM Smackover plays. Modified from Mink *et al.* (1985).

Current Smackover production in the northeastern GoM can be divided into the deep gas trend, the oil and gas-condensate trend, and the relatively narrow oil trend across Florida, Alabama, and Mississippi (Figure 9.5). These were first discovered in the 1960s and 1970s. It was inevitable that this trend would be taken offshore, toward Mobile Bay. Unlike the Norphlet, the Smackover currently does not produce in offshore areas, in spite of the fact that it contains the primary Oxfordian source rock (Godo 2017; Table 9.1). Onshore traps are set up by northeast–southwest trending arches and basement highs that provide closure complemented by reservoir pinch-outs and grainstone build-up limits (Figure 9.6). The major reservoir types are grainstone banks and thrombolite (microbialite) build-ups (Mink *et al.* 1985; Mancini *et al.* 2006; Haddad and Mancini 2013).

The failure of industry exploration to carry the Smackover play offshore into Destin Dome protraction block and adjacent areas is based on two primary factors. First, the large salt structures of Destin Dome developed quite late, as observed in salt evacuation that affected the Miocene and younger strata (Bowman 2012). Thus, trap formation likely post-dated source rock maturation and migration (Pashin *et al.* 2016). Second,

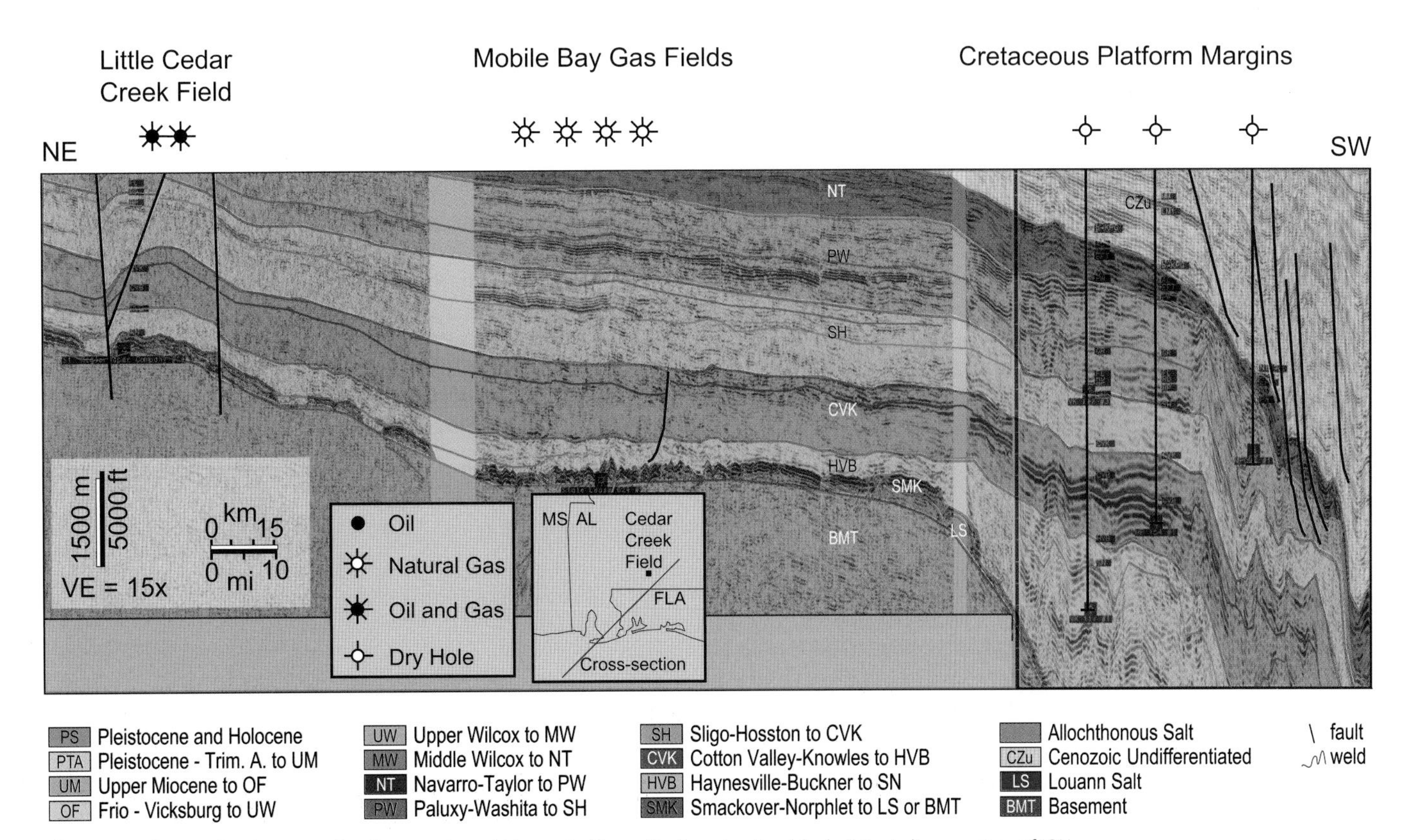

Figure 9.6 Seismic line interpretation from onshore Alabama to Viosca Knoll protraction block. Seismic line courtesy of ION.

Smackover porosity development is linked to dolomitization, either through exposure or meteoric diagenesis (Mancini *et al.* 2004). Some of the best fields, such as Cedar Creek in onshore Alabama, are developed on small, local basement highs where the exposure and meteoric processes are focused (Haddad and Mancini 2013; Figure 9.6). Offshore, particularly on transitional crust, small local basement structures are rare.

In Mexico, Oxfordian-age limestones and dolostones roughly coeval to the Smackover, called the Zuloaga (Oivanki 1974; Figure 3.2), have limited offshore production, in comparison to the prolific overlying Kimmeridgian interval. Oxfordian dolostones are secondary reservoirs in the Sitio Grande Field of the Macuspana basin (CNH 2018).

9.6.1 Oxfordian Source Rocks

The Mesozoic of the GoM is also the habitat for several world-class source rock intervals, including the Oxfordian (Cunningham *et al.* 2016). Early work on the geochemistry of northern GoM oil seeps and reservoir oils indicated active source rocks in the Oxfordian, Tithonian, Ceno-Turonian stages and within Cenozoic intervals, as well as minor sources in other intervals (Aptian and Albian; Comet 1992). Wenger *et al.* (1994) and Hood *et al.* (2002) compiled this data and proposed a map distribution of semi-concentric bands of source rocks around the basin, younging toward the basin center (Figure 9.7). Work since then has better defined the source rock distributions (Cunningham *et al.* 2016) and we now recognize major organic enrichment within the Mesozoic Ceno-Turonian (Eagle Ford), Tithonian-centered (upper Haynesville and Lower Cotton Valley), and Oxfordian (Smackover-equivalent) sources (note use of northern GoM stratigraphic names here).

Understanding source richness and organic facies patterns over the offshore and Gulf rim are key factors in assessing exploration risk and uncertainty in the entire basin. Maps of Mesozoic source rocks are important to play element mapping and basin modeling. Here in the GoM, as elsewhere globally, paleo-environmental trends can set up source deposition.

Exploration in the onshore salt basins pointed to the Oxfordian Smackover Formation as a source for carbonate-sourced oils in the northern GoM (Oehler 1984; Sassen *et al.* 1987; Sassen 1988, 1990; Claypool and Mancini 1989; Wenger *et al.* 1994; Cunningham *et al.* 2016). Discoveries in the northern GoM shelf and Norphlet deepwater play of the eastern GoM have also been linked to the Oxfordian source, locally known as the "Brown Dense" (Wenger *et al.* 1994; Hood *et al.* 2002; Godo 2006; Mankiewicz *et al.* 2009; Godo *et al.* 2011). The distinctive geochemical signature of oils from this source identifies a separate GoM petroleum system (Wenger *et al.* 1994; Hood *et al.* 2002; Ferworn *et al.* 2003). It may also contribute as a secondary source for mixed oils in reservoirs and surface seeps (Cole *et al.* 2001; Hood *et al.* 2002).

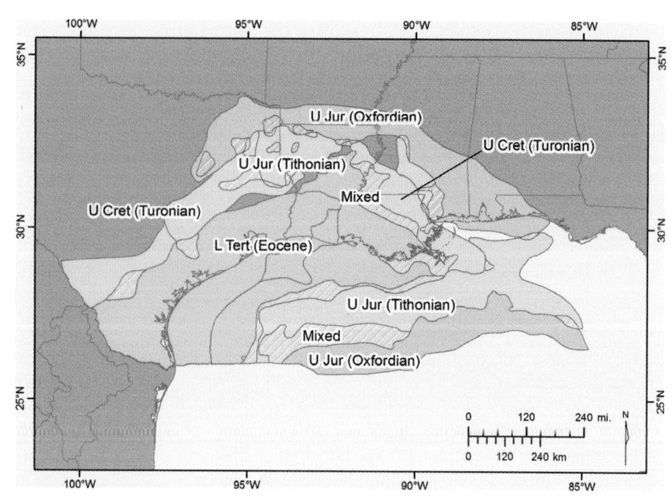

Figure 9.7 Main petroleum systems and source rock ages based on the map of Hood *et al.* (2002). In this interpretation, the source for most of the reservoired oil on the GoM slope is considered to be centered on the Tithonian (Wenger *et al.* 1994; Hood *et al.* 2002). Mixed systems are shown in hachures.

One way to evaluate basin-scale trends in the Oxfordian (Smackover-equivalent) source rock is through source rock mapping including geochemical parameter maps (Cunningham *et al.* 2016; see Box 9.1). Weight-percentage total sulfur for the Oxfordian-sourced oil family is an especially illuminating parameter as it is a well-documented proxy for carbonate versus more marly or siliciclastic influenced source facies (Figure 9.8; Cunningham *et al.* 2016).

Box 9.1 Source Rock Mapping

Source rock mapping is a standard technique used in both conventional and unconventional reservoir exploration and evaluation. An approach that has proven successful is the use of geochemical parameter maps. One important parameter to map regionally is total sulfur (Figure 9.8), as discussed above. In addition to total sulfur, other geochemical parameter maps can be employed, with up to 16 biomarker ratios that have specificity for and susceptibility to a variety of source bed depositional environment, thermal maturity, and geologic age controls. Mapping these parameters using proprietary or public domain interpretive ranges can help distinguish carbonate-rich from clay-rich mudrocks and the redox conditions in the water column affecting the preservation of organic matter. The geochemical parameter maps were constructed by gridding (ArcMap natural neighbor) over the data extent for each of the Mesozoic and Cenozoic oil family and source rock age grouping interpretations included in the GeoMark RFDbase for the produced oils and oil seep datasets. It is important to note that large gaps in data coverage, especially between the northern and southern GoM, interpolated lithofacies, or other paleo-environmental trends over gap areas should be viewed cautiously and only accepted if supported by other datasets.

Locations of wells and piston core seeps used in the geochemical parameter mapping components in the study of Cunningham *et al.* (2016) were rotated back to paleo-positions based on the most recent PLATES model for the GoM. This is particularly important for geochemical parameter maps produced for

Box 9.1 *(cont.)*

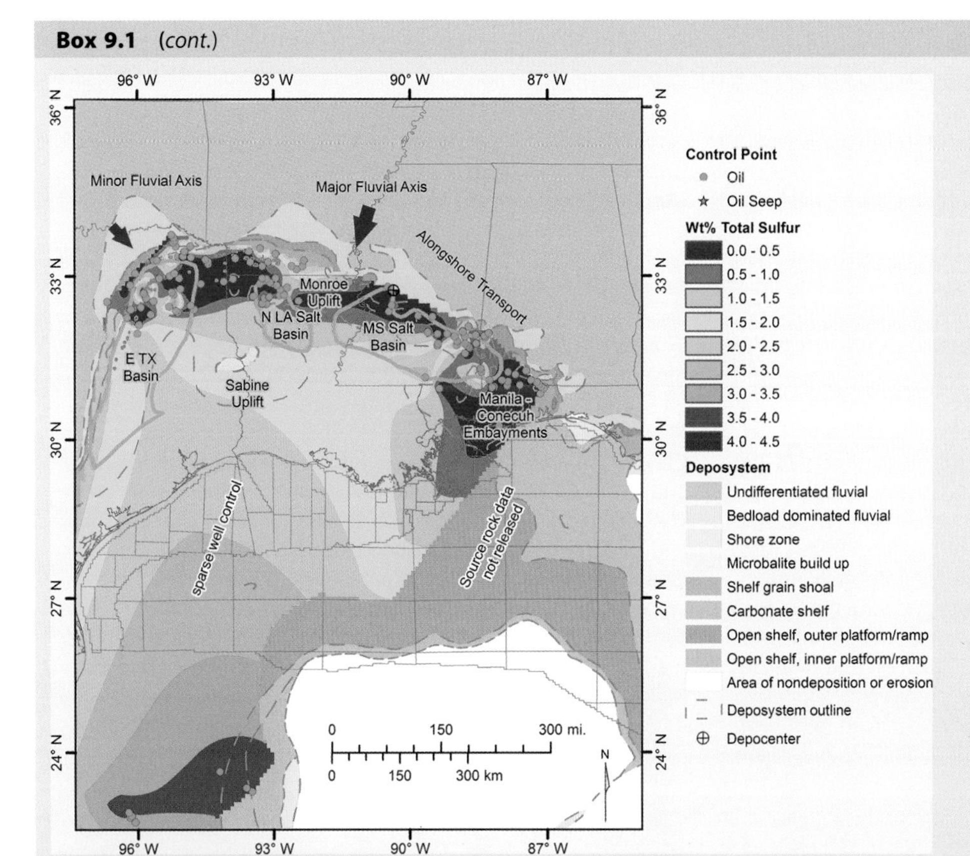

Figure 9.8 Oil sulfur content map posted on the Smackover paleogeography.

the Jurassic Smackover (Oxfordian) and Cotton Valley–Bossier (CVB; Tithonian), and Early Cretaceous Cotton Valley–Knowles (CVK; Berriasian–Valanginian) units that were deposited during sea floor spreading. Additional source maps were produced for the GBDS units deposited subsequent to sea floor spreading, including a broad Lower Cretaceous interval centered on the Bexar/Pine Island–Rodessa (BP-RD) and the Mid–Upper Cretaceous Eagle Ford–Tuscaloosa (EFT); Cenomanian–Coniacian.

The bulk of the sulfur in crude oil occurs bonded in organic compounds; however, small amounts can occur as elemental sulfur in solution and as hydrogen sulfide gas. As a general rule, sulfur contents of 0.5 wt.% or less are indicative of "sweet crudes" sourced by marine, paralic, or lacustrine siliciclastic source rocks with algal or terrigenous higher plant kerogen. Sulfur contents greater than 0.5 wt.% are more indicative of "sour crudes" sourced by marine carbonate source rocks with algal kerogen (Moldowan *et al.* 1985) or hypersaline environment source rocks at sulfur levels above 1–1.5 wt.% (Blanc and Connan 1993). The primary reason for this difference is the lack of iron in carbonate environments to scavenge the reduced sulfur produced during sulfate reduction either in the sediments or overlying waters to then create pyrite. Therefore, the reactive sulfur is incorporated into the kerogen and eventually the generated oil.

The oil sulfur content map[1] for the Smackover-sourced oil family posted on the Smackover paleogeography shows areas of lower and higher value throughout the trend of the northern Gulf rim interior salt basins (Figure 9.8). Although they are spatially broad, the lower oil sulfur value areas are centered near the Sabine Uplift, Monroe Uplift, and the Manila and Conecuh Embayments. These coincide with fluvial and broad shore zone areas of siliciclastic sediment entry onto a carbonate shelf. The narrower intervening areas of higher sulfur content occur in the eastern Mississippi Interior, the North Louisiana, and East Texas salt basins in carbonate shelf and grain shoal environments. Differences in Smackover-sourced oil sulfur content, pristane/phytane ratio, and carbon isotopes of saturated and aromatic hydrocarbons have been attributed to higher salinity, lower oxygen content, and less input of clay minerals in the marine environment of the Mississippi Interior Salt Basin compared with the Manila–Conecuh Embayments (Claypool and Mancini 1989; Sassen 1990). The main effective source facies in the Smackover has been identified as transgressive mudstones and marls of the lower member, the Brown Dense (Oehler 1984; Sassen *et al.* 1987; Sassen 1988; Claypool and Mancini 1989). Evidence for clastic influx in the Conecuh Embayment occurs as black laminated shale interbeds cumulatively over 60 ft (18 m) thick and enriched in terrigenous

[1] The data used to produce this map was donated to the project by GeoMark and are included in their worldwide Rock and Fluid Database (RFDbase). It is GeoMark's interpretation of oil families for the Gulf of Mexico that has been used to produce this map. For more in-depth explanation and discussion of the original geochemical interpretations of samples, the reader is directed to contact GeoMark or access licensed reports for the central/eastern and western Gulf of Mexico Surface Geochemical Exploration (SGE) program reports of GeoMark released in 2005. See Box 9.1 for further comments on map construction.

Box 9.1 *(cont.)*

plant material which occur in the Lower Smackover (Baria *et al.* 2008). The sulfur content map suggests similar siliciclastic influx points may exist in the Lower Smackover to the west, as well.

Paleogeography exerts the initial control on the sulfur content of Smackover crudes, but secondary effects such as biodegradation, water washing, thermal cracking, and thermochemical sulfate reduction (TSR) are important modifiers. Biodegradation and water washing generally increase sulfur content while thermal cracking decreases sulfur content and TSR may or may not increase total sulfur in oil or condensate, depending on the extent of the TSR reaction and organic reactants (Claypool and Mancini 1989; Manzano *et al.* 1997; Machel 2001). Thermochemical sulfate reduction and thermal cracking of kerogen and petroleum are located in reservoirs nearer the depocenters of the Smackover in the Gulf rim trend, whereas biodegradation and water washing are more localized along the margins of the depocenters and along the updip limit of the Smackover. Also, facies variation in the Smackover, the overlying Buckner Anhydrite, and the underlying Pine Hill Anhydrite and structuring may also influence the extent of TSR through mineralogic compositions and juxtaposition of reactants (Claypool and Mancini 1989). At this point, details of all the competing processes governing sulfur distribution in Smackover-sourced oils have not been unraveled. It appears, however, that along the northern Gulf rim, the primary control Smackover paleogeography exerts on organic facies, and thus sulfur content, is still evident.

Some oil sulfur data exists for the southern GoM, allowing the total sulfur map to be extended into Mexico. The Smackover–Norphlet play continues to the southeast from the northern Gulf rim basins into the offshore at the Mobile Bay gas-condensate field and south to deepwater discoveries in the Mississippi Canyon and DeSoto Canyon protraction areas. Alteration processes such as TSR (Mobile Bay) and solid bitumen impregnation in reservoir porosity (deepwater) appear to continue into the offshore (Mankiewicz *et al.* 2009; Godo *et al.* 2011). Oxfordian Smackover-equivalent-sourced oils are also noted offshore of Mexico in the Campeche shelf basin and onshore in the Chiapas-Tabasco and Tampico–Mislanta basins (Guzmán-Vega and Mello 1999; Guzmán-Vega *et al.* 2001; Valdés et al. 2009). The Campeche and Chiapas-Tabasco oils have >2 percent sulfur and through molecular data on oils and correlative rock molecular and lithofacies information are interpreted as sourced from an anoxic shelf carbonate–marl unit (Guzmán-Vega *et al.* 2001). The Tampico–Mislanta oils have been correlated to deeper-water marine marls of the Oxfordian Santiago Formation and have molecular characteristics of sourcing from a more clay-rich marl deposited under somewhat more oxygenated conditions (Guzmán-Vega *et al.* 2001).

9.7 Haynesville–Buckner Supersequence

The HVB supersequence includes both conventional and unconventional reservoirs. Currently, conventional reservoirs of the HVB are of lesser economic importance than the underlying Smackover–Norphlet or overlying CVB (Kosters *et al.* 1989; Mancini 2010). The USGS assessed the undiscovered technically recoverable resources of the Haynesville as 1.1 BBO and 196 TCFG in onshore lands and state waters of the Gulf Coast (Paxton 2017a).

Like the underlying Smackover, grainstone bank reservoirs are developed on salt structures of the west flank of the East Texas salt basin. A second trend is present around the west side of the Sabine Uplift (Presley and Reed 1984; Kosters *et al.* 1989). Oolitic grainstones with favorable diagenetic dissolution are among the best reservoirs (Ahr 1981).

In the Ark–La–Tex region (Arkansas/Louisiana/Texas), Haynesville grainstone banks, called the Buckner “B,” are developed as secondary reservoirs in both stratigraphic traps and in banks centered on structural features (Moore *et al.* 1992). The primary Smackover reservoirs have greater ultimate recoveries. West of the Sabine Uplift, a number of Gilmer carbonate shoals exist but the best porosity in the leached ooid grainstone facies can be quite vertically restricted, often to 10–15 ft (3.0–4.6 m) in a well (Ahr 1981).

The Gilmer Platform margin reef facies Haynesville is porous but appears to be largely devoid of hydrocarbons, in spite of its prominent development across the eastern GoM (Figures 3.19 and 3.23). The failure of the top seal above the platform margin reef at Sake #2 (DC 726 #2) may be both a local and a regional concern.

An exception to the rather poor HVB conventional carbonate production is an areally limited deep gas play, the poorly named “Cotton Valley lime pinnacle reef play” of the 1990s (Montgomery 1996). Isolated reefal build-ups, located on salt structures in the East Texas salt basin, were active targets in a drilling campaign that yielded at least four new fields in Texas (Figure 9.9). Reef build-ups were relatively small (200–800 acres) and scattered over an 80 km (50 miles) trend (Montgomery 1996), suggesting that a major platform margin reef was not present west of the mapped platform western termination (cf. Figure 3.23). The stratigraphic position below the Bossier shale (basal part of the CVB supersequence) and Kimmeridgian age suggested by Montgomery (1996) indicate these reefal carbonates are actually part of the HVB supersequence and not the “Cotton Valley.”

The Kimmeridgian interval of Mexico, coeval with the HVB supersequence, is one of the major reservoir intervals of Mexico, producing in the Comalcalco, Pilar de Akal contractional belt (Akal–Reforma trend), Macuspana and Tampico–Misantla basins (Angeles-Aquino and Cantú-Chapa 2001; Mitra *et al.* 2006). Offshore, it is formally designated as the Akimpech Formation (Cantú-Chapa 2009). Like the HVB of the northern GoM, it is carbonate-rich. Oolitic limestones formed in grain shoals and banks are among the best reservoirs, as moldic and vuggy porosity are developed in the bank crest while cemented primary pores and limited microporosity are generally found in bank flanks and interbank limestones (Treviño García 2012). Dolomitization also plays a key role in improving reservoir quality of the bank crest facies (Angeles-Aquino and Cantú-Chapa 2001). Miocene shortening also elevated extensional rollovers, creating structural traps (Chernikoff *et al.* 2006).

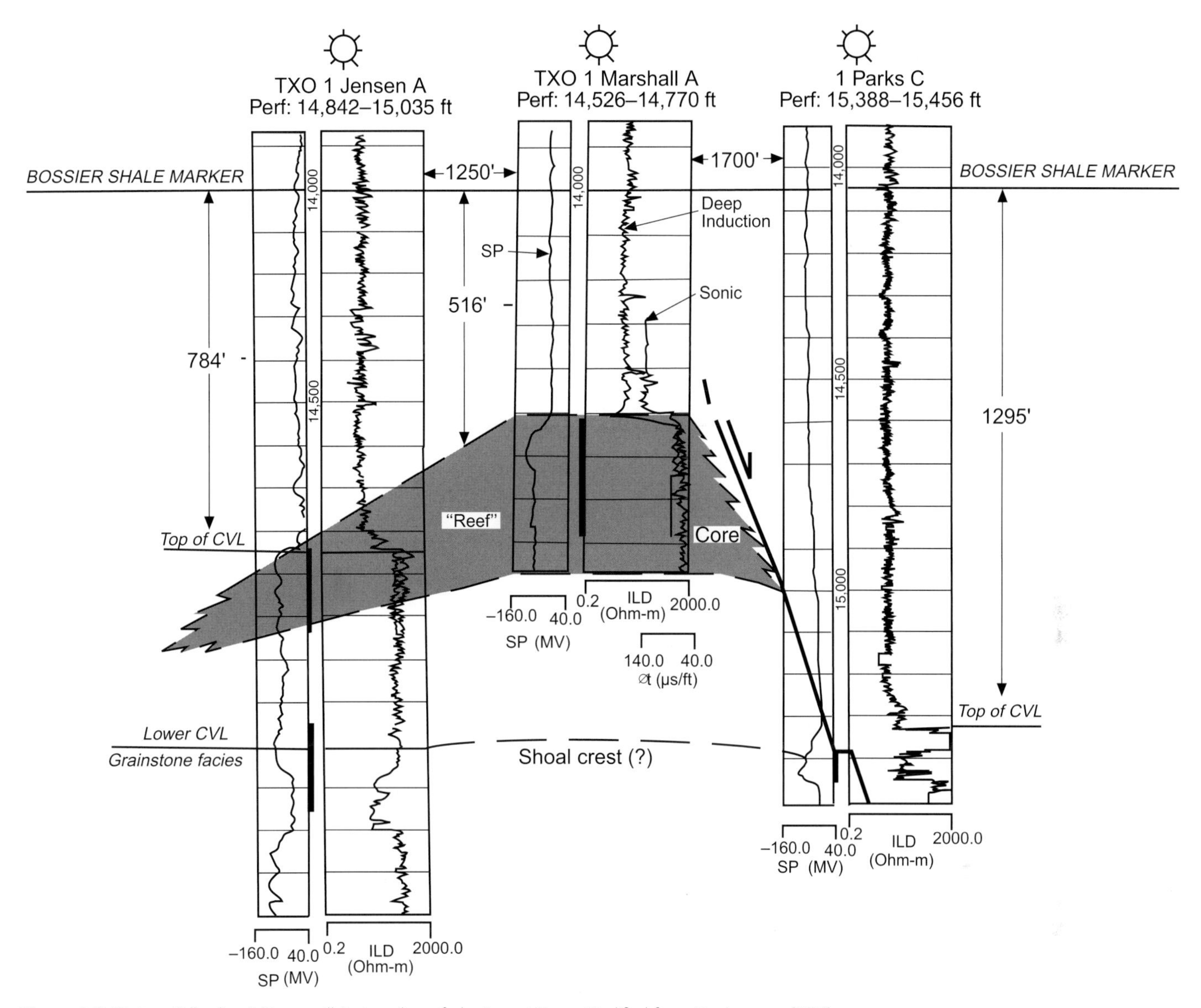

Figure 9.9 "Cotton Valley lime" (Haynesville) pinnacle reef play in east Texas. Modified from Montgomery (1996).

Tectonic factors also play a key role in hydrocarbon trap formation and reservoir porosity development. Three-dimensional seismic and structural analysis in offshore Campeche of Mexico has revealed evidence of Mesozoic extension that was overprinted by the Cenozoic compressional events (Chernikoff *et al.* 2006; Figure 9.10). Listric growth faults detaching on Callovian salt were later reactivated during the Middle Miocene compressional event to form structurally high traps, which are prolific producers in the Pilar de Akal contractional belt.

Systematic changes in carbonate lithofacies (grainstone shoals) around these Mesozoic extensional faults suggest an association that departs from the Wilson (1975) model of Jurassic ramps. Shoals develop on the elevated footwalls, and are time-transgressive as each footwall subsides below the wave base (Figure 9.10). The lithofacies changes around faults parallel trends in the porosity and permeability, which are consistently diminished to the east and west of the faults (Chernikoff *et al.* 2006). Many Mesozoic extensional structures were reactivated during Miocene shortening, for example, forming the traps of the Cantarell complex (Mitra *et al.* 2005).

Further north and west, Horbury *et al.* (2003) noted local tectonic uplift resulting in angular unconformities and karst surface formation at the base and within the Kimmeridgian interval of the Lamprea-2 well. This is in close proximity to the Nayada-1 well, where a prominent Neocomian (Valanginian stage) unconformity is observed.

Kimmeridgian (HVB-equivalent unit) represents the second-most important oil-producing reservoir in Mexico, behind only the K–Pg breccia, the main reservoir at Supergiant Cantarell Field (Magoon *et al.* 2001). However, reservoir porosities are sometimes quite low (5–13 percent) in deeper fields (>3 km) and production through fracture systems associated with folding and thrusting is an important element (e.g., Mitra *et al.* 2006). Key fields in the Campeche shelf with Kimmeridgian reservoirs include Och–Uech–Kax complex, May, Ixtal, Ku–Maloob–Zapp complex, Sinian,

A

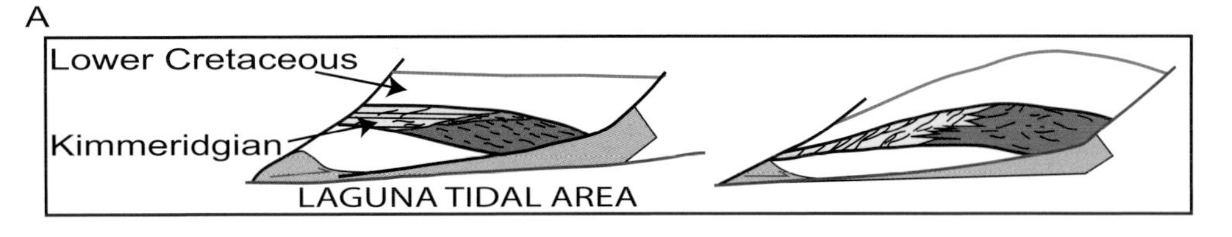

B

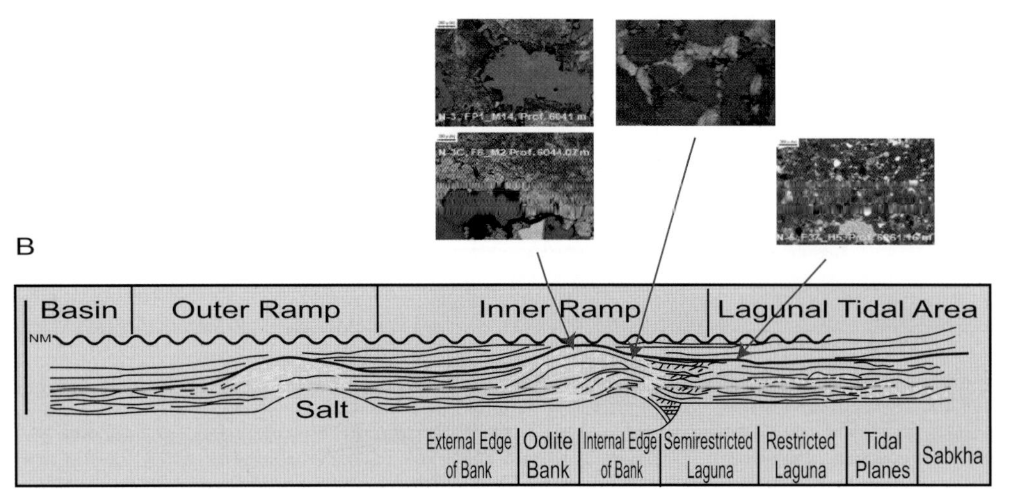

Figure 9.10 (**A**) Akimpech (HVB) grainstone bank reservoirs formed traps as Miocene shortening elevated extensional rollovers. Modified from Chernikoff *et al.* (2006). (**B**) Variation in porosity types and quality from grainstone bank crest (high porosity moldic/vuggy pores in photomicrographs) to bank flank and interbank (Treviño García 2012).

Taratunich, and Jujo–Tecominoacan (Magoon *et al.* 2001; Clark *et al.* 2003; Guzmán 2013). Agava, A.J. Bermudez, Bellota, the Cactus–Nispero–Rio Nueva complex, Caparrosos–Pijije–Escuintle complex, Cardenas, Chinchorro, Eden–Jolote, Jacinto, Luna–Palapa, Mora, Paredon, Sitio Grande of the Comalcalco basin also have Kimmeridgian production (Acevedo and Dautt 1980; Luneau *et al.* 2003; Ysaccis *et al.* 2006; CNH 2018). The HVB supersequence is productive at considerable depths (4400–5300 m) in comparison to the shallower Top Cretaceous (K–Pg breccia; Acevedo and Dautt 1980; Magoon *et al.* 2001).

In the Tampico–Misantla basin, Kimmeridgian reservoirs are present in Tamaulipas-Constituciones, Arenque, San Andres, and Lobina Fields (Guzmán 2001; Magoon *et al.* 2001; Salter *et al.* 2005; CNH 2018). Arenque Field is an example of a Kimmeridgian oolite/bioclastic bank, which onlaps the flank of a prominent basement high. The best reservoir quality is found in the oolitic bioclastic facies with median values of 20 percent porosity and 5 md permeability (Horbury *et al.* 1996). Abundant microporosity (pores <2 nm) is also documented in the Kimmeridgian reservoirs at Arenque (Horbury *et al.* 2003).

9.8 Cotton Valley–Bossier Supersequence

The CVB supersequence contains both conventional and unconventional reservoirs. Tight gas sandstones are common objectives in the East Texas salt basin (Klein and Chaivre 2002). Source rock (shale) unconventionals are also concentrated in a similar area (Cicero *et al.* 2010). Risked mean undiscovered conventional resources in two shelf plays are assessed as 2.9 BBO and 57 TCFG. Unconventional resources undiscovered as of 2017 are listed as a risked mean of 52 TCFG (Paxton 2017b).

In Mexico, Tithonian strata generally include the primary source rock, called the Pimienta, the source part of the Pimienta–Tamabra(!) petroleum system (Magoon *et al.* 2001). Conventional carbonate and siliciclastics reservoirs are present in a number of Mexico onshore and offshore fields within the equivalent CVB supersequence. The La Casita is the most important reservoir in the Sabinas basin (Eguiluz de Antunano 2001). The middle member of the La Casita Formation produces gas in the Sabinas basin from tight (3–10 percent porosity) fractured quartz arenites (Dyer and Bartolini 2004). The Tithonian La Casita Formation of the Sabinas basin of Mexico has notable hydrocarbon production. The Lampazos Field produces gas from coeval shore zone siliciclastics (Guzmán 2001). Similar fracture-aided permeability facilitates gas production at the Merced Field (Veltman *et al.* 2012). Fracture density is greater in sandstones and dolostones than in limestones and is oriented northwest–southeast along the basin axis.

Carbonate reservoirs, largely dolomitized limestones and dolostones, produce hydrocarbons in Jacinto, Mora, Paredon, Sitio Grande, and Sen Fields of the Comalcalco basin (Acevedo and Dautt 1980). Most traps here are compressional horsts or footwalls and fracturing is essential, given the low *in-situ* porosities (often <5 percent). Extensive fracturing is associated with strike–slip motion in the Jacinto and Paredon field area (Gonzalez-Posades *et al.* 2005). Here, Tithonian strata are mainly finely crystalline dolomudstones and limestone wackestones, likely formed in open to deep marine paleo-environments. Intensely sheared and brecciated rocks are common (Gonzalez-Posades *et al.* 2005). Though the fields are in close proximity, they have different fluid contacts and hydrocarbon types.

9.8.1 Jurassic Petroleum Systems and Source Rocks

Tithonian-centered source rocks, referring to zones of organic enrichment at the base of the CVB and top of the HVB supersequences, were recognized early on from seep and field oil analyses in both the northern GoM and Mexico (Comet *et al.* 1993; Guzmán-Vega *et al.* 2001; Clara Valdés *et al.* 2009). Biomarker and other geochemical analyses demonstrated that Tithonian-centered sources were clearly different, more marly or siliciclastic-rich source beds, and thus a separate petroleum system than the Oxfordian (Cole *et al.* 1999, 2001). Tithonian-equivalent source rocks are also thought to have charged the Varadero heavy oil field in Cuba (Schenk 2008).

In spite of the voluminous geochemical evidence of a Tithonian-centered source in deep offshore areas, there was remarkably little calibration until quite recently. Cunningham *et al.* (2016) employed **Δ** Log R, a petrophysical approach using logs calibrated against total organic carbon (TOC) measurements from sidewall cores and cuttings (see Box 9.2), to evaluate Tithonian shales in deepwater areas of the northern GoM. These were compared against oil family interpretations based on reservoir oil and sea floor seeps, a database constructed by Geomark Ltd. and TDI-Brooks International Inc. (2005a, 2005b). Data locations were restored to the pre-sea floor spreading and pre-rafted locations, following the same approach used here for the CVB paleogeographic map (Figure 3.25).

Box 9.2 Δ (Delta) Log R Technique

The Δ (Delta) Log R technique for identifying zones of organic enrichment in sedimentary rocks and quantifying TOC content was originally conceived using a qualitative log overlay approach. In the late 1970s and early 1980s, Esso geoscientists exploring in the North Sea recognized that the covariance of porosity curves (normally sonic transit time) and deep resistivity curves could be used to define organic-rich trends in the Kimmeridge Clay Formation. With proper scaling, the transit time and resistivity curves in shales were observed to overlie or track in non-source intervals and separate in organic-rich intervals. This approach was applied in play and prospect mapping in different basins of the world and found to be predictive for source rocks deposited in a range of paleo-environments and with differing lithologies.

Passey *et al.* (1990) formalized and quantified the technique by developing empirical equations between the degree of log separation, called Δ Log R, derived from combinations of resistivity versus sonic, density, and neutron porosity logs and TOC as a function of thermal maturity (Figure 9.11). Quantitative Δ Log R analysis requires that intervals of tracking at lower maturity be defined, so that a baseline condition for the logs can be created to allow calculation of the Δ Log R. The thermal maturity range for accurate calculation of TOC from Δ Log R occurs over vitrinite reflectance (Ro) 0.42–0.9 or level of organic metamorphism (LOM) 6–10.5; however, calibrations at the limits of this range can be applied in lower and higher thermal maturity settings (Passey *et al.* 2010). At low levels of thermal maturity, below the generative windows for various kerogen types, the Δ Log R separation is related primarily to porosity curve response to low-density and -velocity kerogen

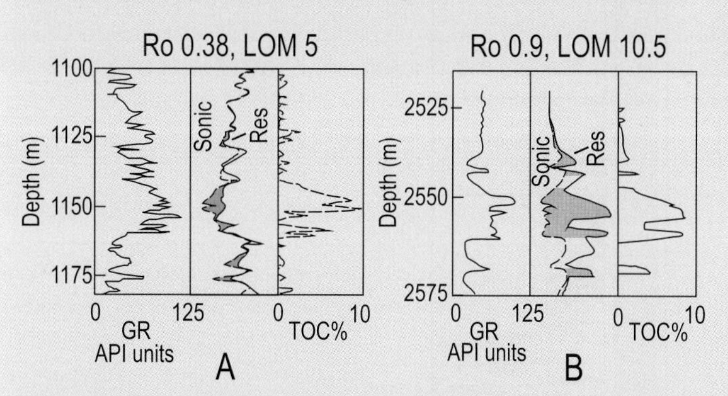

Figure 9.11 Sonic and resistivity overlay and calculated total organic carbon profiles for example Canadian carbonate source rocks. (**A**) Immature source rocks (Ro = 0.38; LOM = 5). (**B**) Peak maturity source rocks (Ro = 0.9; LOM = 10.5). Modified from Passey *et al.* (1990). Abbreviations: GR, gamma-ray; Sonic, sonic transit time; TOC%, total organic carbon weight percent; Ro, vitrinite reflectance; LOM, level of organic metamorphism; Res, resistivity.

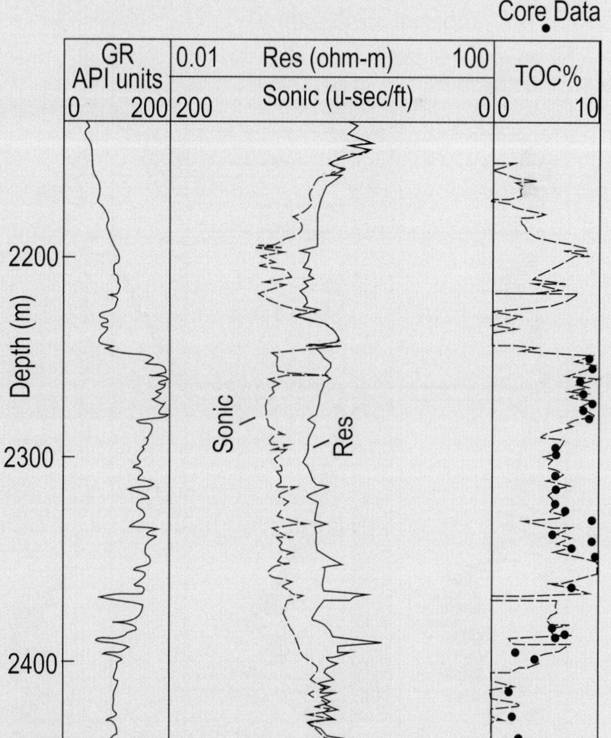

Figure 9.12 Sonic and resistivity overlay and calculated total organic carbon (TOC) compared against measured TOC (solid circles) from well samples. Modified from Passey *et al.* (1990). Abbreviations: GR, gamma-ray; Sonic, sonic transit time; TOC%, total organic carbon weight percent; Res, resistivity.

Box 9.2 *(cont.)*

(Figure 9.11A). With increasing thermal maturity, generated hydrocarbons lead to increases in the resistivity response, enhancing the Δ Log R separation (Figure 9.11B). Predictability within low-TOC shales (<1 wt.%) fails due to low kerogen and generated hydrocarbon volumes; however, these remain excellent intervals for base-lining curves. Conversely, log-calculated TOC values of coals are also not accurate using the shale-based calibration data, but coal beds do display a significant Δ Log R separation.

Although empirical relationships have been observed between TOC and the differing porosity log responses and also gamma-ray or spectral gamma, these tend to be region-specific and require local calibration datasets for prediction. A benefit of the Δ Log R technique is that it can be used broadly in a qualitative fashion to identify source rock intervals and quantitatively even in undrilled settings, assuming thermal maturity can be estimated using basin models or regional geothermal or maturation gradient information. Total organic carbon and thermal maturity calibration with core or plug data (Figure 9.12) are necessary to increase accuracy and confidence in predictions, as are the ruling out of anomalous Δ Log R separation due to borehole washouts, saline-water or hydrocarbon-bearing reservoir intervals, tight or uncompacted intervals, and certain lithologies such as igneous rocks or evaporates. Complementary gamma-ray and caliper logs should be used to block out these zones of anomalous influence.

In recent years, the Δ Log R technique has also been widely applied in evaluation of unconventional shale reservoir plays. With sufficient well control and known thermal maturity trends, sweet spots of enhanced TOC can be mapped out and exploited for horizontal drilling and artificial fracture stimulation (Passey *et al.* 2010).

A key calibration well is the Norton prospect (GBB 754 #1; Figure 9.13). Though Norton and the correlative well Wrigley (EW922 #1) are carapace penetrations (see Box 1.1), biostratigraphy suggests stratigraphic condensation did not occur until later in the salt structure's history and the Upper Jurassic is relatively complete. The Wrigley interval was overturned by salt later in its history (Cunningham *et al.* 2016). Both wells show the log-derived TOC profile increasing upward from the top HVB surface to the CVB midpoint, with values exceeding 10 percent TOCc and then declining through the upper CVB and through the overlying CVK interval (Figure 9.13). Both wells show net interval thicknesses >300 ft (91 m), with organic enrichment of more than 5 percent TOC (log-derived), indicating excellent source quality and stronger oxygen deficiency in the CVB depositional environment. Further, organic extracts on the Norton Upper Jurassic samples show high levels of the biomarker bisnorhopane (D. Jarvie, pers. comm., 2012), which is enriched in source rocks deposited under anoxic conditions

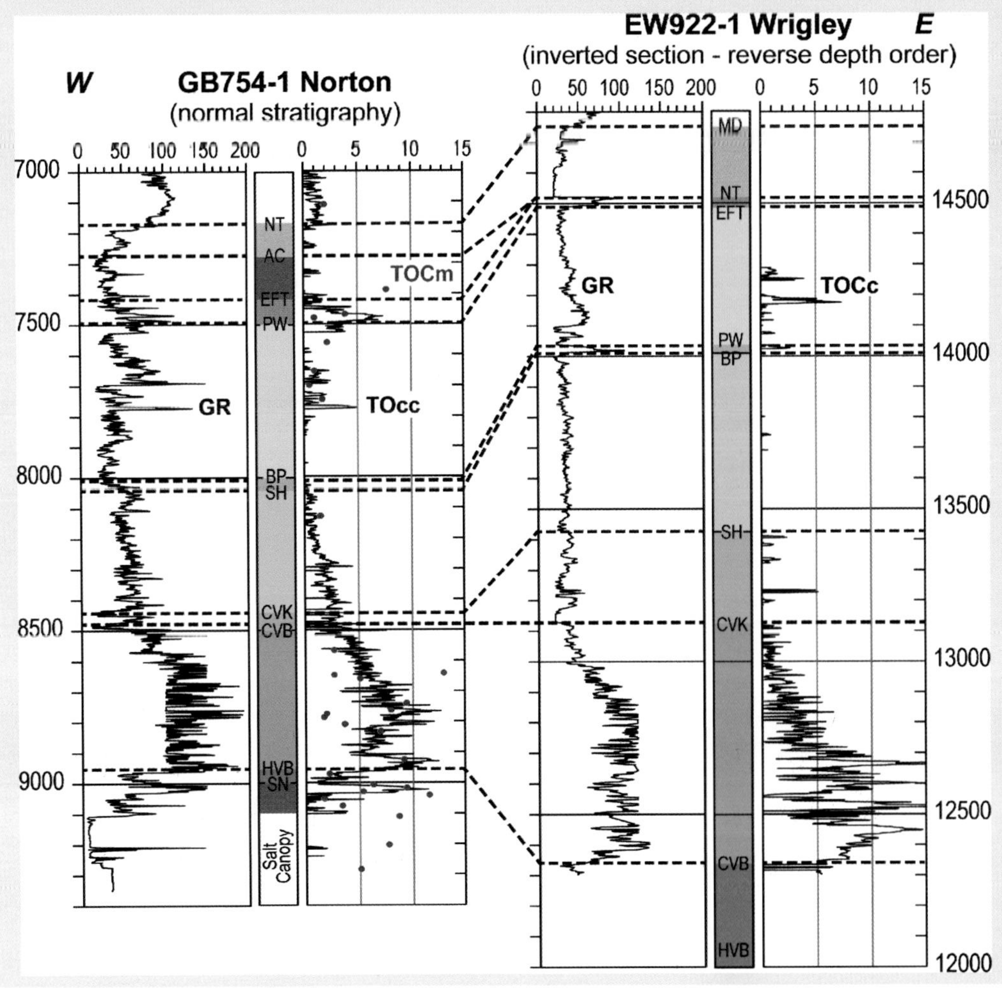

Figure 9.13 Δ Log R analyses and calculated (c) total organic carbon percentage with TOCm (TOC measured) from core plug and cuttings from the Norton (GB 754 #1) and Wrigley (EW 922 #1) wells. Modified from Cunningham *et al.* (2016).

Box 9.2 (*cont.*)

(Peters *et al.* 2005). Given the similar log patterns and levels of organic enrichment at Norton and Wrigley, anoxia is suggested to have existed broadly over this starved basin area.

Plotting of log-derived TOC averages against the CVB reconstruction reveals several important trends in the northern GoM (Figure 9.14). Results suggest that progressive ventilation of the GoM may have occurred over the CVB interval, with eastern areas showing increased oxygen levels and reduced preservation of organic matter first and higher organic preservation lingering later in the western GoM. This progressive ventilation of the Gulf basin may have been paced by sea floor spreading and resulting changes in paleobathymetry at the eastern side of the basin. Most reconstructions limit the connection of the western GoM to the global ocean with the positioning of the Chortis, Chiapas,

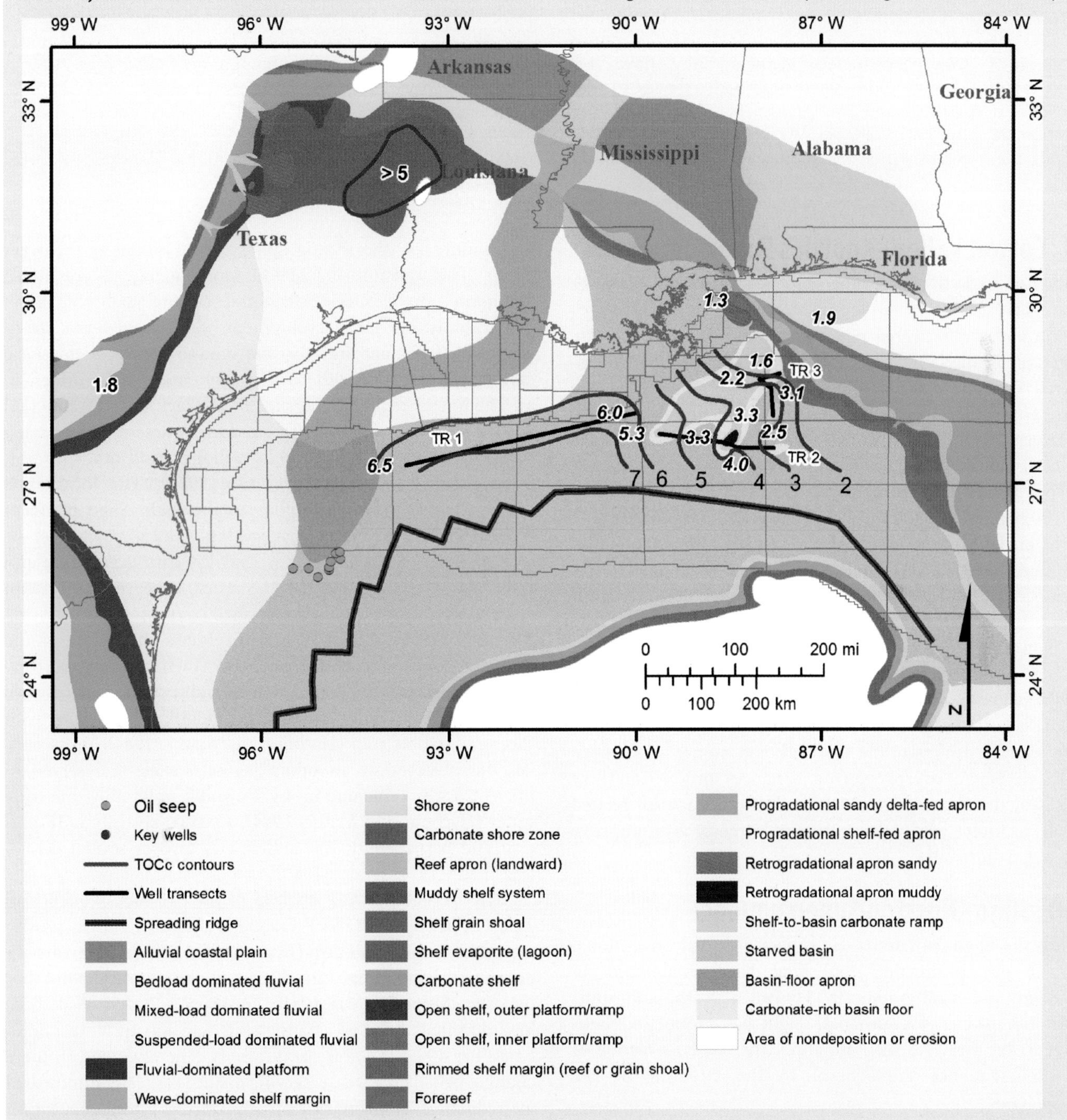

Figure 9.14 Total organic carbon (wt% TOC) for the Tithonian-centered source in the northern and southern GoM displayed on CVB paleogeography. Contours in the northern GoM are based on the average calculated (Δ Log R) total organic carbon (>1 percent TOCc) for the CVB unit at well locations. Two additional offshore wells to the north of the transect wells are displayed with TOCc values of 1.3 percent (VK 117 #1) and 1.9 percent (DD 422 #1). The distribution of the organic-rich upper Lower Bossier facies is after Hammes and Frébourg (2012). Total organic carbon polygons in the southern GoM are based on the measured data from (A) the Tampico–Mislanta basin (after USGS Mexico Assessment Team 2014) and (B) the southeastern basins of Mexico (after Clara Valdés *et al.* 2009).

Box 9.2 (*cont.*)

and other structural blocks (Cunningham *et al.* 2016). Some-dilution by carbonate debris coming from the platform margin is also evident in the trend of values over the talus apron (shelf-to-basin carbonate ramp) extending across the Destin Dome protraction block (Figure 9.14).

A zone of organic enrichment onshore along the East Texas basin border with Louisiana is based on known high TOC intervals in the CVB and HVB supersequences of the famous Haynesville Shale gas play area (Cicero *et al.* 2010; Hammes *et al.* 2011). Our reconstructed paleogeography shows this area to be part of a muddy silled shelf basin, similar to that envisioned for the Eagle Ford in the Maverick basin (see Section 4.8). The basinward sill was provided by the Sabine Island, a basement high at the south of the small shelf basin (Hammes *et al.* 2011). Cores show the Lower Bossier to be intensely bioturbated, suggesting deposition under oxic to slightly dysoxic waters. However, a more organic-rich facies attributed to relatively high marine productivity was deposited during the upper Lower Bossier within this silled restricted embayment (Hammes and Frébourg 2012). Any anoxic bottom water mass or oxygen minimum layer existing beyond the sill in the more open GoM during Tithonian CVB deposition apparently was not able to breach the sill and further enhance organic enrichment (Cunningham *et al.* 2016).

Of course, access to well-oxygenated surface and bottom waters may have also controlled shelf-margin reef development. The lack of reef development in western parts of the GoM may relate to longer periods of anoxia than further eastward.

9.9 Cotton Valley–Knowles Supersequence

Production from the CVK supersequence is from the Knowles carbonates and the Terryville Sandstones, though current production and undiscovered potential oil and gas resources are far greater in the latter unit. The USGS estimates that undiscovered conventional (including tight gas) resources range from 127 to 1157 BCFG, with a mean assessment of 547 BCFG, the largest resources within the interior salt basins of Texas and Louisiana (Dyman and Condon 2006a).

The Terryville Sandstone of east Texas is quartz-rich ($Q > 80$ percent), but has locally low porosity and permeability due to burial diagenesis including extensive quartz cementation that filled both primary pores and secondary pores developed during carbonate dissolution (Bailey 1983). Excluding fracture-related values, matrix permeabilities are generally less than 1 md and porosities range from 2 to 14 percent, which is typical of the tight gas sandstones of the Harrison and Panola counties study area of Bailey (1983).

The development of massive hydraulic fracturing in the 1980s spurred drilling of the Terryville Sandstones in northern Louisiana where **overpressuring** enhances production flow rates (Coleman and Coleman 1981). Natural fractures related to salt tectonics and other structural trends also improve natural gas flow rates.

9.10 Sligo–Hosston Supersequence

Within the Sligo–Hosston (SH) supersequence, conventional oil and gas production has long been established in the Sligo and stratigraphically equivalent carbonate units (Pettit Limestone of Louisiana). Tight gas production continues in the Travis Peak Formation, the basal sandstone of the supersequence and coeval to the Hosston of the eastern Gulf.

One of most significant conventional discoveries in the Sligo is the Black Lake Field of Natchitoches Parish, Louisiana. The 1964 discovery was important due to the size (estimated original oil in place of 156 MMBO and 0.9 TCFG) and nature of the stratigraphic trap (White and Sawyer 1966). The operator Placid Oil found an areally restricted "bioherm" on a subtle structural closure with an updip permeability barrier. Development wells proved the stratigraphic entrapment and also the unique nature of this local carbonate build-up, with 34 dry exploration wells drilled in an attempt to replicate the success at Black Lake (White and Sawyer 1966). The carbonate build-up is described as having rudistid pelecypods, gastropods, miliolids, and other forams, though oolites were noted (Bailey 1978). The location is about 72 km (45 miles) updip of the main Sligo platform margin reef (Figure 4.9), suggesting a local patch reef or shelf grain shoal development. Detailed core description and interpretation confirms that the Black Lake reservoir is an isolated build-up in an inner-shelf setting (Harbour and Mathis 1984). In contrast to early views that most of the porosity is secondary (Bailey 1978), high primary porosity between carbonate skeletal fragments and ooid grains is well preserved by the unique diagenetic conditions associated with this stratigraphic trap (Harbour and Mathis 1984). In particular, the presence of an updip lagoonal shale seal served to prevent early cementation by downdip migrating meteoric water (Harbour and Mathis 1984). Average porosity is 16 percent and permeability 100 millidarcies to over several darcies (Hermann 1971; Krafve 1980).

The basal sandstone interval of the SH supersequence in Texas is called the Travis Peak and is a major tight gas producer in the East Texas basin (Li and Ayers 2008). Reservoirs are dominantly found in mixed-load, high-sinuosity point bar deposits (Upper Travis Peak) and Lower Travis Peak bed-load-dominated fluvial channel fills and amalgamated braid bars (Tye 1992). Lateral continuity is generally good in these channel belts. However, reservoir quality is greatly diminished by reduction of primary porosity during burial diagenesis. Over a relatively short depth range of 6000–10,000 ft (1830–3050 m), porosity declines from 17 to 5 percent and permeability decreases by four orders of magnitude, from 10 md to 0.001 md (Dutton and Diggs 1992). While sandstones are present further basinward (Li and Ayers 2008), the decline in permeability is a major limiting factor to this tight gas play.

In their overall evaluation of the undiscovered oil and gas resources, the USGS combined the Travis Peak of east Texas with the coeval units of Louisiana, Arkansas, Mississippi, and Alabama (Dyman and Condon 2006b). The total conventional undiscovered resources were assessed as a mean of 29 MMBO, 1.1 TCFG, and 21 MMBNGL, with the bulk of oil resources in updip areas and gas resources in downdip areas of greater source rock maturity and reduced reservoir quality.

In Mexico, Sabinas basin discoveries were made in Cupido Limestone in Anahuac and Totonaca Fields on the shelf margin of the SH. However, the Sligo carbonate fields are relatively small in comparison to younger Albian platform and slope fields.

9.11 Bexar–Pine Island Supersequence

Conventional production from the Bexar–Pine Island (BP) supersequence is mainly from onshore patch reefs and grainstone banks in the interior shelf, particularly in the areas near the Sabine Uplift. An example is the Fairway Field of east Texas that produces high-gravity oil from the James Limestone (Webster *et al.* 2008). Skeletal (rudist) grainstones represent the primary reservoir pay zones. Reef core stromatoporiod boundstones can be cemented and form local barriers and baffles (Figure 9.15). The field is located on a turtle structure and has produced since its discovery in 1960. It has benefited from infill drilling, new 3D seismic data surveys, horizontal drilling, and enhanced oil recovery techniques to extend field life over 50 years (Webster *et al.* 2008). About 52 percent of the

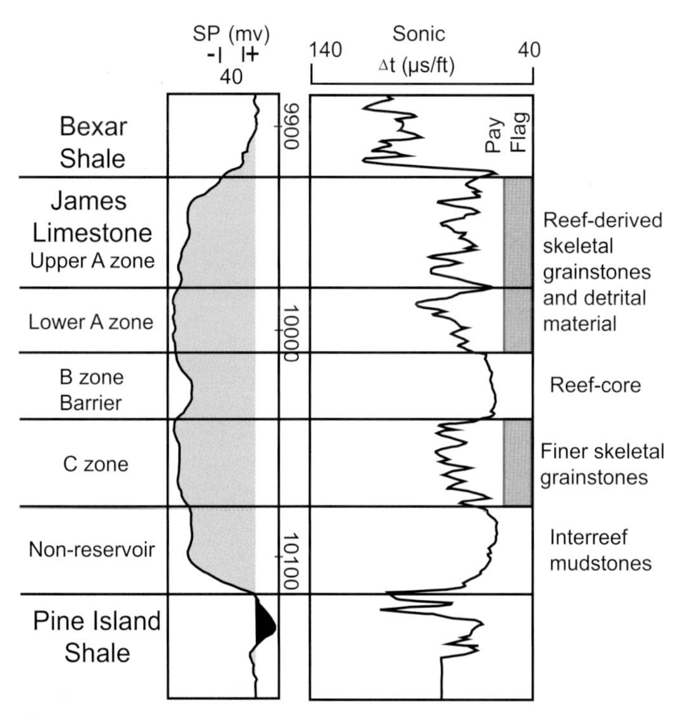

Figure 9.15 Reservoir zonation in the James Limestone (BP supersequence) and interpreted carbonate deposystems. Modified from Webster *et al.* (2008).

410 MMBO originally in place has been recovered. Other similar James Limestone are producers that are present in the East Texas basin (Kosters *et al.* 1989). In the north Louisiana salt basin, oil in the producing James Limestone of the Chatham Field is trapped by closure setup by underlying salt structures (Bebout *et al.* 1992).

In federal waters, the BP supersequence back-reef reservoirs are better reservoirs and better traps than the coeval platform margin reef and forereef itself. Shelf grain shoals, which includes grainstone banks and patch reefs, are James Limestone gas producers in Viosca Knoll 69, 252, and 256, and Mobile 991 blocks (Petty 1999). Porosity is well developed in shelf rim apron (back-reef) grainstones landward of the platform margin in DeSoto Canyon wells DC 512 #1 and FM 456 #1, but were unfortunately water-bearing. Improved porosity in back-reef aprons and shoals over the coeval platform margin is probably due to exposure to meteoric water or a lack of marine cements common to the submerged platform margin (Petty 1999). BOEM has assessed the UTRR of the James Limestone as around 256 MMBOE, with 10 discovered pools to date (Figure 9.3).

In Florida, the BP interval includes the "Brown Dolomite" zone (Applegate 1984), associated with the large benthic foram *Choffatella decipiens* that is an important biohorizon in the BP supersequence (Olson *et al.* 2015). The Brown Dolomite unit is within the Lehigh Acres Formation, below the main anhydrite-bearing Punta Gorda, as is the BP supersequence further north and west. The porous dolostone zone, up to 100 ft (33 m) in thickness, is present in several wells onshore and offshore and is known to extend as far as Marquesas Key in south Florida.

In spite of deposition during OAE1a and OAE1b, the BP interval is considered to be a minor source rock for conventional reservoirs. In south Texas, the TOC parameter reported for the BP source intervals is relatively low, an average 0.86 percent (weight percent). But values are clearly lowered by maturation, with Ro values in the range of 1.2–2.2 percent, so original values could have been higher (Hackley 2012). Organic facies from rock-eval is also affected by maturation and appears Type III, terrigenous, but again probably higher in the Type II range if corrected for maturation. Further eastward, the organic content does not improve much, again in the range of 0.17–1.08 percent for even immature samples, and generative capacity is weak (Enomoto *et al.* 2012). Offshore, the estimated organic content improves to peak values of 2 percent, but measurements are in the thin condensed intervals of several salt carapace wells (AT 182/183 Sturgis and AT 026 Big Horn).

The potential of the BP as an unconventional reservoir has been considered in several studies (Hull and Loucks 2010; Hull 2011; Enomoto *et al.* 2012; Hackley 2012). Measured parameters for BP shales are generally poor relative to better-known source rock plays (Figure 9.16A). The TOC content is about half of the TOC content of shales in the core areas of the Eagle Ford and Tithonian-centered interval (CVB, HVB) of the GoM. Percentage of clay, a detriment to so-called "fracability" or favorable brittle response to artificial fracture stimulation, is

A. BP

Shale Reservoir Comparisons

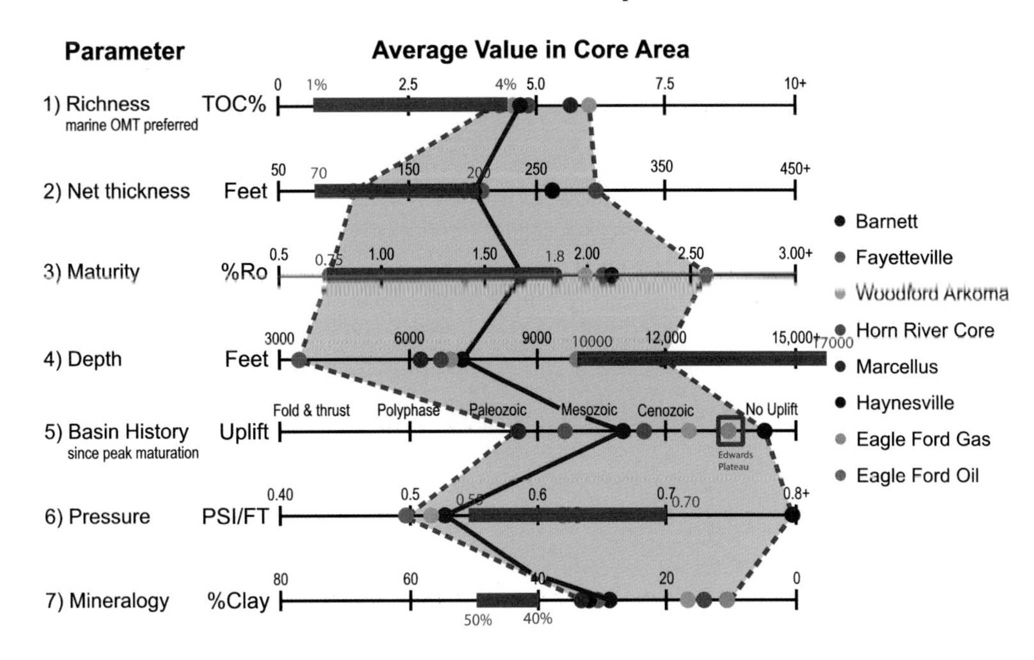

B. TMS

Shale Reservoir Comparisons

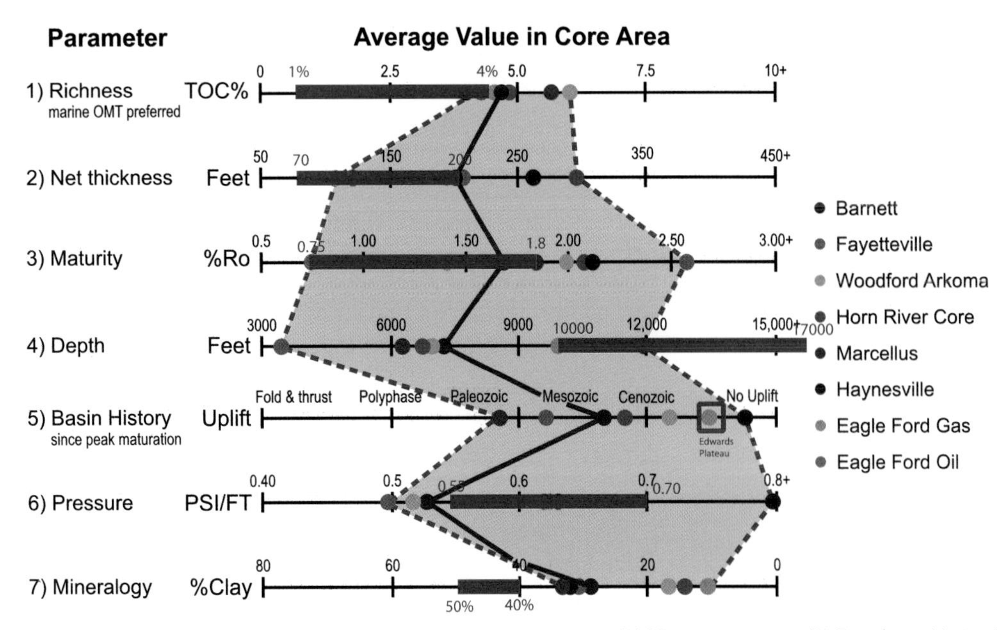

Figure 9.16 Shale reservoir comparisons on key screening parameters: (**A**) BP supersequence: (**B**) Tuscaloosa Marine Shale.

high. The depth to BP shales for south Texas is outside the range when the Miocene inversion and uplift is considered. Other measured values fall in the range (overpressuring, net thickness, etc.) but low organic richness and high clay percentages have to be major concerns.

In several respects, the BP shales are comparable to the Tuscaloosa Marine Shale (TMS), as discussed in Section 9.15.1 and Figure 9.16B, and by Lowery *et al.* (2017). Like the TMS, the Pearsall Shale (BP supersequence) has had a few productive wells (see Hackley 2012), but drilling results did not generate much exploration interest prior to the recent downturn in oil and gas prices (Hull 2011).

9.12 Rodessa Supersequence

As mentioned, the Rodessa is an areally restricted but productive unit, limited to the northern GoM between the East Texas salt basin and Florida, landward of the shelf margin. However, it can be locally well-developed and produces from both carbonate and siliciclastic reservoirs. In east Texas, the Rodessa is a skeletal to ooid grainstone bank initiated over underlying salt pillows west of the Sabine Uplift (Bebout *et al.* 1992). The Rodessa Limestone producers are generally less prolific than the underlying Pettit (Sligo) Limestone (Bebout *et al.* 1992). The Rodessa of Trawick Field of Texas is an anticlinal entrapment of gas in oolitic grainstones, but also secondary in economic terms to the underlying Sligo. Net pays are thin (10–30 ft; 3–9 m) and show considerable local variation (Kosters *et al.* 1989).

Sandstone reservoirs formed in fluvial–deltaic systems of Mississippi and Alabama are nicknamed the Hazlehurst, Collin, and Richton lobes (Reese 1976). However, one caveat is that the maps upon which these lobes are named includes underlying sandstones of the BP supersequence. The biggest Rodessa sandstone producing fields (>90 BCFG as of 1992) are all centered on salt structures of the Mississippi salt basin (Duckworth *et al.* 1992).

9.13 Glen Rose Supersequence

The Sunniland Formation is notable due to the well-documented productive trend established across the south Florida counties of Collier, Lee, Hendry, and Dade (Mitchell-Tapping 1986, 2002). As discussed earlier, these prominent features are discrete caprinid rudistid reef build-ups formed in the regressive period of Glen Rose (GR) deposition and these transition to carbonate beach/shoals at the top of the Sunniland. Top seal is provided by the Lake Trafford anhydrite beds formed in a sabkha (hypersaline shelf paleoenvironment). Dolomitization of the caprinid bafflestone and ooid grainstones provides excellent porosity and permeability (Loucks and Crump 1985; Liu 2015). Field sizes range from 160 to 7500 acres (0.65–30 km^2; Figure 4.18; Liu 2015). Net porous oil-bearing intervals can be relatively thin (<20 ft; 6 m). Many of the fields are in steep decline or have since been abandoned due to high water cut. Total oil production from 57 wells was 110 MMBO and 9.7 TCFG at the end of 2000 (Mitchell-Tapping 2002).

The shale units that were deposited on top of the drowned SH carbonate system record the global positive carbon isotope excursions associated with the Aptian OAE1a and the Aptian–Albian OAE1b (Phelps *et al.* 2015). OAE1a occurs in the first shales deposited on top of the drowned SH carbonate platform. Northern Tethyan carbonate platforms also drowned during OAE1A (Föllmi *et al.* 1994; Weissert *et al.* 1998), suggesting this was a global response to either the OAE or the eustatic highstand with which it coincided, both of which were ultimately driven by the emplacement of the Ontang–Java Plateau in the western equatorial Pacific (e.g., Leckie *et al.* 2002). Neither OAE1A nor OAE1B are associated with substantial organic carbon burial on the Gulf shelf, unlike the younger Cenomanian–Turonian OAE2, which occurs just above the high TOC rocks of the Lower Eagle Ford Shale in Texas (e.g., Lowery *et al.* 2014). However, Leg 77 of the Deep Sea Drilling Program (DSDP) recovered organic-enriched sediments in the southeastern Gulf, with up to 5 percent TOC, roughly equivalent to the SH and GR supersequences (Buffler *et al.* 1984). The potential therefore exists for GR-aged source rocks in the deepwater, most likely in proximity to Atlantic Ocean-connected areas of the basin.

9.14 Paluxy–Washita Supersequence

Hydrocarbon production from the Paluxy–Washita (PW) supersequence was initially established in the 1960s with dry gas discoveries in the shelf margin portion of the Edwards Formation, locally known as the Stuart City "trend" (Bebout *et al.* 1977). However, low porosities did dampen activity for a period of time until horizontal drilling revitalized the play in the late 1990s (Waite 2009). A trend of eight or more fields today extends across south Texas. One of the larger fields, Pawnee (Bee and Live Oak counties of Texas) produces from back-reef, reef core, and forereef carbonates, with porosities typically less than 8 percent and permeabilities commonly less than a millidarcy (Loucks *et al.* 2013). Macropores were largely filled by carbonate cements during burial, but micropores were more resistant to cementation and today form the bulk of the *in-situ* pore system. Current improvements in horizontal drilling and hydraulic fracturing continue to benefit exploitation efforts (Waite 2009).

In the eastern GoM (Viosca Knoll and Main Pass protraction blocks), hydrocarbon production has proved quite limited, fewer than 10,000 barrels total, and other wells found were wet or with limited pay (Petty 1999). Interest eventually shifted to the underlying James Limestone (part of the BP supersequence) that had several subsequent discoveries in shelf and back-reef facies. Currently, BOEM estimates a UTRR of just 71 MMBOE with two known discoveries (Figure 9.3).

Though only a formal division of the Cretaceous into an Upper and Lower exists, informally the Albian to Ceno-Turonian strata in Mexico are often called "Middle Cretaceous" (Lehmann *et al.* 1999). Middle Cretaceous reservoirs are noted in a number of Mexican onshore and offshore fields of the Veracruz basin (Mata Pionche, Mecayucan fields),

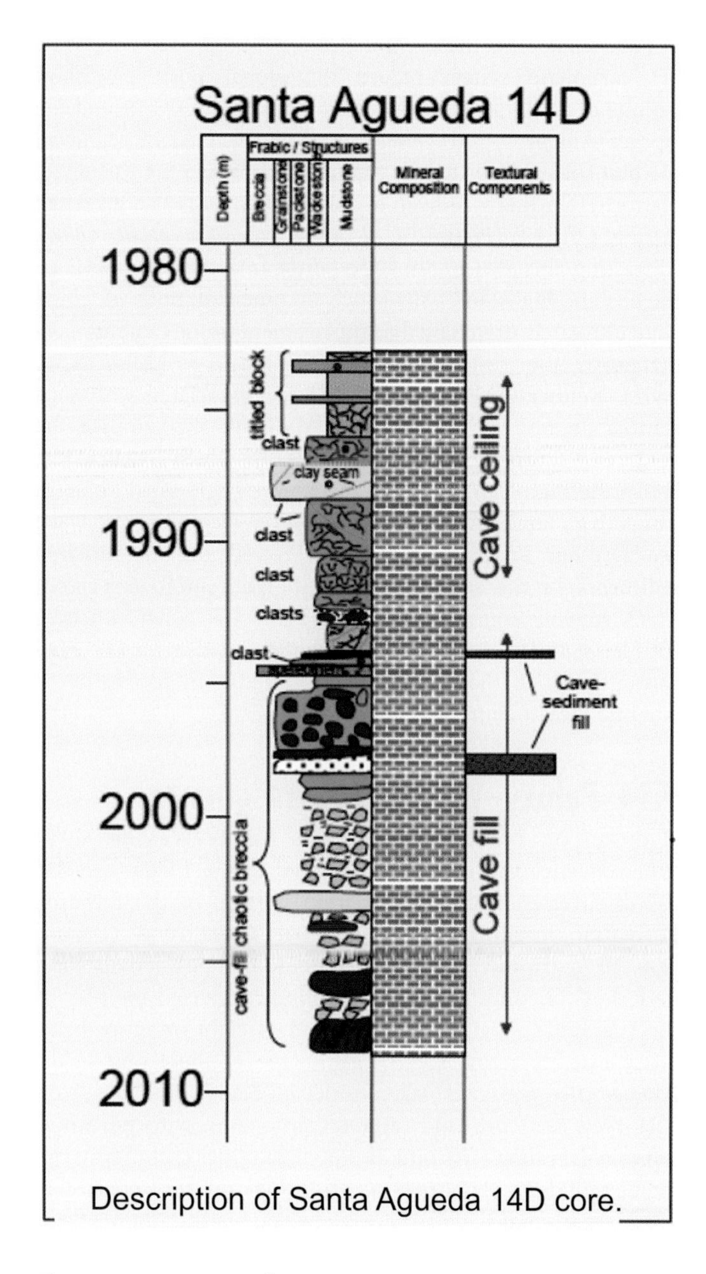

Figure 9.17 Description of Santa Agueda 14D core from the Golden Lane platform showing cave-fill chaotic breccia. Modified from Janson *et al.* (2011).

Tampico–Misantla (Poza Rica), Campeche shelf (Sinan), Comalcalco basin (Catedral, Giraldas, Jacinto, Yagual Fields), and Canru Field of onshore Sureste–Chiapas (Acevedo and Dautt 1980). These are largely fractured and/or breccia reservoirs, though dolomitization enhances both porosity and fracture intensity (Williams-Rojas and Hurley 2001). The Middle Miocene Chiapanecan compressional stage is key to generation of fractures and breccia but also local salt diapirism (Gutteridge *et al.* 2019).

The Poza Rica Field is stratigraphically and genetically linked to the Golden Lane Platform of Central Mexico (Chen *et al.* 2001; Janson *et al.* 2011). The El Abra Formation of the platform is time-equivalent to, and the source of, the carbonate sediment gravity flows found in the Tamabra Formation of Poza Rica (Janson *et al.* 2011). Carbonate breccias are common but do not have a similar genesis to the K–Pg breccias formed by the Chicxulub impact. Platform margin canyons and high relief at Golden Lane are an indication of the high paleo-relief and considerable paleo-slope adjacent to Poza Rica (Janson *et al.* 2011). Karst towers and sink holes are commonly observed on 3D seismic data (Janson *et al.* 2004). Karst features in core include dissolution breccias, crackled clasts, speleothems (cave features). Cave-fill chaotic breccias are common in cores of wells drilled in the Santa Agueda field, located in the platform interior trend of the Golden Lane Platform (Janson *et al.* 2004; Figure 9.17). The enhanced porosity and permeability from platform interior exposure, meteoric diagenesis, and syndepositional fracturing associated with the cave-fill contribute to high well and field productivity at fields like Santa Agueda (Chen *et al.* 2001). Here and at Poza Rica, slope reservoirs show a considerable range of porosity (1–25 percent) and permeability (0.1–700 md), reflecting complex depositional processes and post-depositional diagenesis and tectonics (Magoon *et al.* 2001).

9.15 Eagle Ford–Tuscaloosa Supersequence

The EFT supersequence is the host for hydrocarbons found in both conventional and unconventional reservoirs. Conventional production from the Lower Woodbine sandstone commenced in the 1920s with discovery of the Mexia field and the supergiant east Texas field in 1930 (Ambrose *et al.* 2009). Exploration migrated south to the "downdip Woodbine" of southeast Texas in the late 1960s and 1970s (Adams and Carr 2010). Today, secondary and tertiary recovery efforts continue in older fields (Ambrose *et al.* 2009) and new deep traps are occasionally being pursued (Adams and Carr 2010).

East of the Sabine Uplift, the coeval Tuscaloosa Formation is also a prolific producer. The mid-1970s surge in drilling activity and deep gas discoveries focused in the "deep Tuscaloosa" of south Louisiana (Barrell 1997). Trap styles (and field discoveries) range from salt-cored anticlines (Port Hudson), expanded three-way fault closures (e.g., Morganza), and expanded and faulted rollover anticlines (e.g., Judge Digby; Figure 9.18). A key success element in the deep Tuscaloosa play is the preservation of porosity at great depth due to the presence of chlorite clay rim cement, which acts to limit formation of deep burial quartz cementation (Ryan and Reynolds 1997; Woolf 2012). A significant example is the Port Hudson Field, where thick reservoirs range from bedload-dominated fluvial to slope channel deposits (Barrell 2000).

Unlike the downdip Woodbine play, which is limited to onshore areas, Tuscaloosa conventional discoveries extend into offshore areas. The ultra-deep, high-temperature, high-pressure Tuscaloosa play was opened with the Highlander prospect discovery by Freeport McMoRan in St. Martin Parish of southern Louisiana (Rynott 2015). Further downdip is the deep shelf Tuscaloosa gas discovery in Davy Jones #2 of South Marsh Island (Moffett 2015), and the deepwater Tiber (KC 109 #1) discovery (Horn 2012; Snedden *et al.* 2016b). The

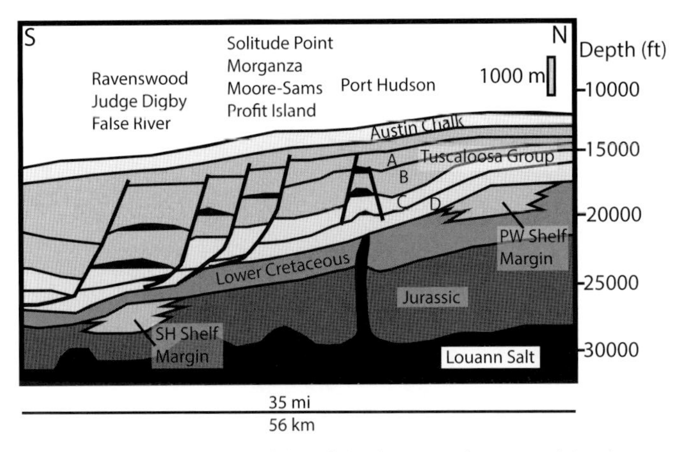

Figure 9.18 Trap styles and fields of the deep Tuscaloosa trend, South Louisiana. Black polygons depict gas accumulations. Modified from Barrell (1997).

Tuscaloosa also has a greater potential upside than the Woodbine, as large salt-related structures remain untested in deepwater areas (Harding *et al.* 2016; Snedden *et al.* 2016b). However, one key uncertainty for deepwater Tuscaloosa plays is the presence of **chlorite** minerals and development of secondary porosity in these deeply buried, submarine fan-type reservoirs.

9.15.1 Eagle Ford and Tuscaloosa Marine Shale Source Rocks

Source rocks for both conventional and unconventional reservoirs are well documented in the Ceno-Turonian of the northern GoM as well as the adjacent Western Interior Seaway (Arthur *et al.* 1987; Hood *et al.* 2002; Dubiel *et al.* 2003). The known Mesozoic source rock play, documented through extensive drilling in Texas, is the Eagle Ford Shale (Denne *et al.* 2014). Sedimentology from core interpretation, regional reconstructions, and seismic interpretation indicates that the most organically enriched part of the Eagle Ford is the Lower Eagle Ford (Hammes *et al.* 2016; Figure 9.19). Deposition occurred in a shelfal paleo-environment behind the older Edwards (PW) shelf margin, which restricted circulation and oxygenation, and enhanced organic matter preservation (Hentz and Ruppel 2011; Alnahwi *et al.* 2018).

The Eagle Ford Shale, now a mature unconventional hydrocarbon play, occurs in an approximately 50-mile wide belt extending from the Mexican border northeastward to the Texas–Louisiana border (Hammes *et al.* 2016; Figure 9.19). Exploration and production are focused in south Texas, where Eagle Ford wells produce oil, condensate gas, or dry gas (Tian *et al.* 2012). Variable production is thought to be a function of organic content, lithology (as relates to “fracability”), and depth that controls hydrocarbon phase (Hentz and Ruppel 2011).

The Eagle Ford is often divided into two major units, in descending order, the Upper Eagle Ford (UEF) and Lower Eagle Ford (LEF). Detailed petrophysical analysis indicates the TOC is higher in the LEF than in the UEF (Hammes *et al.* 2016). This is interesting, as deposition of the LEF in south Texas is known to have preceded the global OAE2 that was established from work in the Western Interior Seaway (Arthur *et al.* 1987; Lowery *et al.* 2017). The OAE2 actually occurs near the base of the UEF (Phelps *et al.* 2015; Alnahwi *et al.* 2018).

Local structural and paleogeographic elements that are relevant to Eagle Ford deposition were discussed by Hammes *et al.* (2016; Figure 9.19). The Maverick basin, located north of the Edwards (PW supersequence) platform margin contains higher average TOC in comparison to areas to the south and east (Figure 9.19). Thinning of the Eagle Ford south of the Edwards Platform margin is observed on 2D seismic lines (Figure 4.31B), particularly in the area of the Cretaceous detachment (see also cross-section 3; Figure 1.15). The latter areas, where one might expect upwelling of deep basinal waters, is actually quite lean in both the UEF and LEF from an organic enrichment standpoint (Alnahwi *et al.* 2018; Figure 9.19). This points to another mechanism necessary for organic enrichment/ preservation in the Eagle Ford, as discussed below.

The USGS recently updated its assessment of the total undiscovered resource for the Eagle Ford, with mean resources of 8.5 BBO, 17.2 TCFG, and 349 MMBNGL (Whidden *et al.* 2018). This places the Eagle Ford as one of the top five largest continuous oil and gas resources assessed by the USGS in the USA. About 5 BBO of the 8.5 BBO is located in one assessment unit, the Eagle Ford Marl Continuous Oil AU, which is delineated by the USA–Mexico border, the 25-percent-clay line, and the thermal maturity window for oil (0.6–1.3 percent modeled vitrinite reflectance; Whidden *et al.* 2018). The potential deep gas resources of the Eagle Ford or Tuscaloosa seaward of the coeval shelf margin south to the limit state waters were not quantitatively assessed due to a lack of data.

A lesser-known source rock, confined to eastern onshore Louisiana and southern Mississippi, is the TMS. The TMS is the middle unit of the Tuscaloosa Group, lying above the basal Tuscaloosa Sandstone and below the upper Tuscaloosa Sandstone (Woolf 2012). The TMS has not been quantitatively assessed by the USGS (Whidden *et al.* 2018).

Lowery *et al.* (2017) conducted a comprehensive study of the TMS, using a combination of core sedimentology, micropaleontology, and carbon isotopes from the Sun #1 Spinks well located in Pike County, Mississippi (Figure 9.20). The results and conclusions have broader implications for the Ceno-Turonian of the northern GoM. Beds of enhanced organic enrichment, ranging from 1 to 4.6 percent, are most common above than the zone showing positive carbon isotope excursion typically associated with OAE2. In this way, it is similar to the Lower Eagle Ford organically enriched unit that also does not align with OAE2. However, these enriched units also are not stratigraphically equivalent between the TMS and LEF. The TMS itself is largely Turonian and thus younger than the OAE2 at the Ceno-Turonian stage boundary. Like the LEF, the TMS was formed in a shelfal paleo-environment, as foram assemblages are mainly neritic, deepening upward to the shelf (Lowery *et al.* 2017).

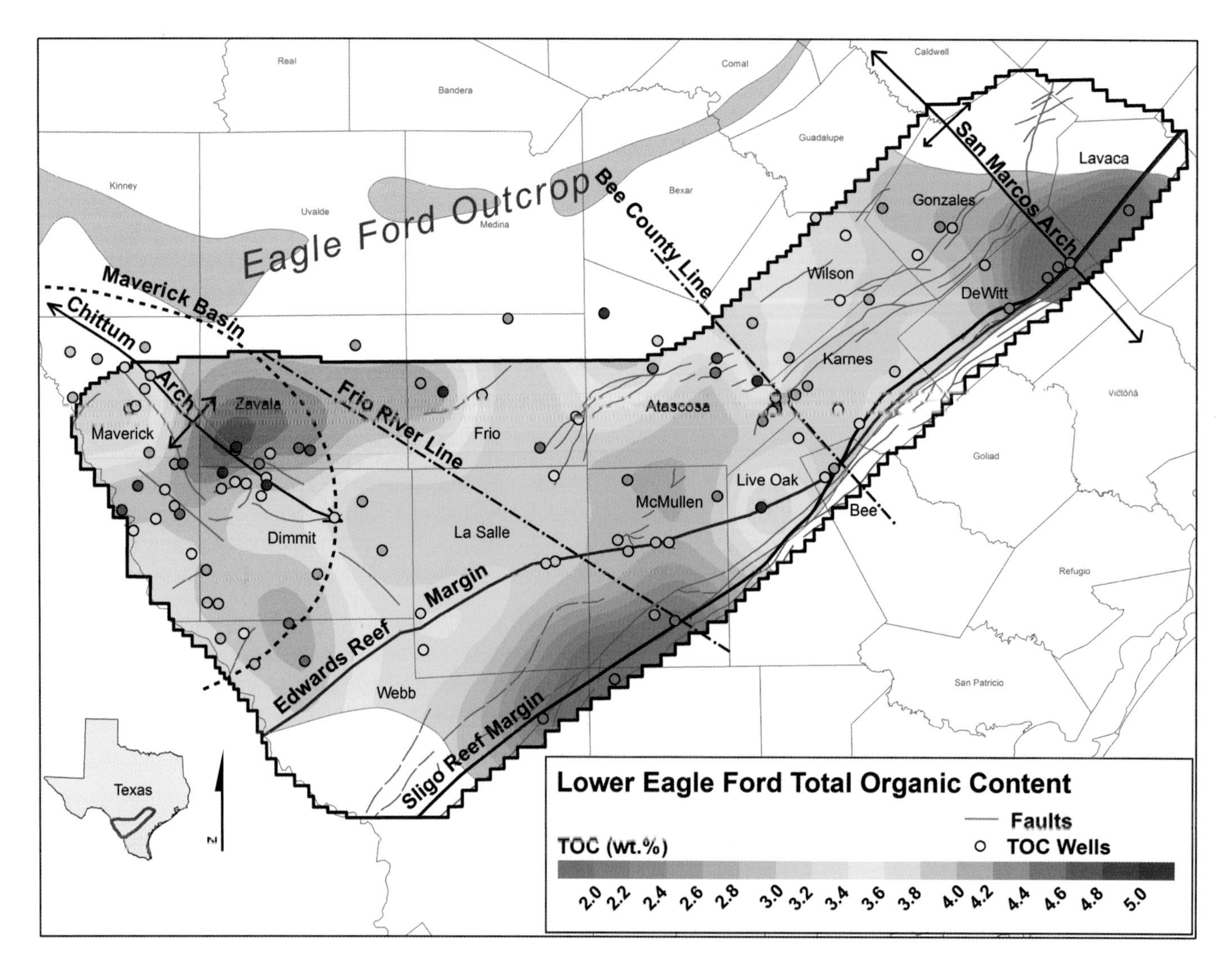

Figure 9.19 Lower Eagle Ford average TOC maps calculated from logs. Modified from Hammes *et al.* (2016).

Comparison of the LEF and TMS in shelfal areas with Δ Log R analyses (see Box 9.2) of the coeval interval in key GoM wells facilitates a basin-scale model for source rock development, timing of organic enrichment, and influence of the Western Interior Seaway (Figure 9.21; Lowery *et al.* 2017). Following Tuscaloosa Sandstone lowstand deposition in Louisiana and roughly similar Woodbine Formation deposition in Texas (Figure 9.21A), eustatic sea-level rise began, continuing through the global OAE2 event (Figure 9.21B) at the Ceno-Turonian boundary. Some shelf areas (Texas) and northern GoM deepwater areas experienced deposition of organically enriched shales, but this is not the case in the TMS play area. Lower Eagle Ford organic-rich shales were deposited just prior to the OAE2 event in the silled Maverick basin. Western Interior Seaway outflow apparently ventilated the Maverick basin, reducing organic matter preservation (Figure 9.21B). This increase in oxygenation is also supported by interpretation of trends in X-ray fluorescence (XRF) major elements of the Eagle Ford presented by Alnahwi *et al.* (2018). Tuscaloosa Marine Shale organic enrichment does not take place until the maximum flooding event, when nutrient-rich, low-oxygen waters encroached upon the shelf of Mississippi and Louisiana, initiating a timeframe of maximum organic matter preservation (Figure 9.21C). The continued influence of Western Interior Seaway outflow waters reduced preservation of organic matter and the UEF was deposited in Texas (Lowery *et al.* 2017; Alnahwi *et al.* 2018). The final phase was marked by sandstone deposition in the TMS area and accumulation of more organically deficient shales (UEF) and chalks (Austin Formation) in Texas (Figure 9.21D).

This model and the observations presented above underline the importance of placing unconventional reservoirs like the Eagle Ford and TMS in a paleogeographic context. Appeals to standard upwelling models for source rock enhancement may not be appropriate, as alternatives may be revealed by considering careful reconstruction of the paleogeography and paleoceanography.

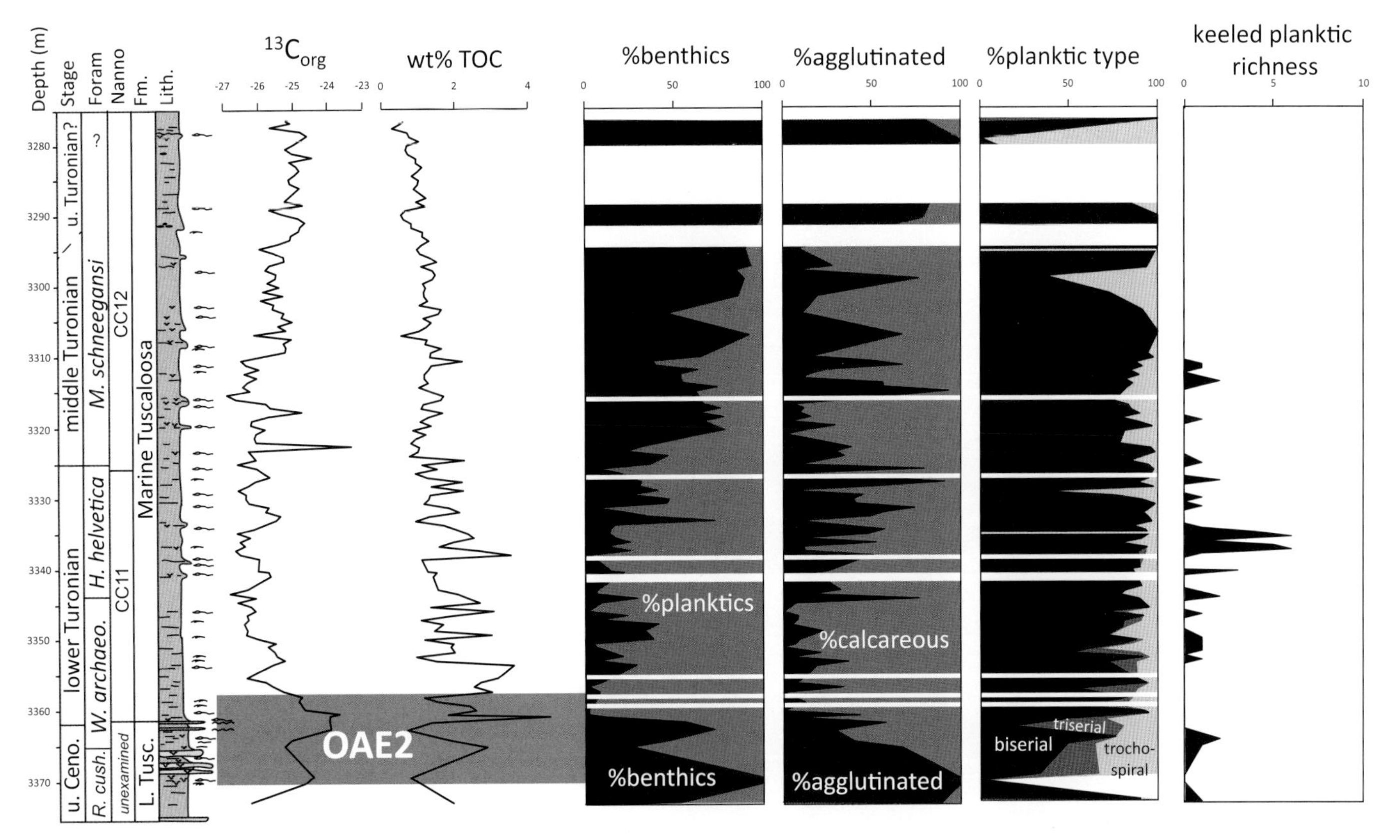

Figure 9.20 Carbon isotopes, TOC, and major foraminiferal population data through the Spinks #1 well core. From Lowery *et al.* (2017).

By comparison to Texas, exploitation of the Eagle Ford equivalent (Agua Nueva) source rock in Mexico is still in its infancy. The first six horizontal wells targeting source rock reservoirs were drilled in 2012, versus over 5000 wells at the same point in the USA (Parra *et al.* 2013).

Interest in unconventional reservoirs is focused on Agua Nueva in the Burgos region and Pimienta (Tithonian or CVB equivalent) in Tampico–Misantla. In the Burgos basin, the same organic-rich LEF interval (Hawkville facies of Stoneburner 2015) is known to extend across the border (Cruz Luque *et al.* 2018). The richest zone is below the OAE2, as it is in the Maverick basin (Cruz and Aguilera 2018). Similar to the trend in Texas, the thickest interval of high TOC (>1–2 percent) is likely to be located updip of the relict Albian (PW) platform margin reefs (Figures 4.31 and 9.19, respectively).

However, a number of non-geologic factors will determine the success of the play in Mexico, including water and road access, infrastructure (pipelines and refineries), and impact on local farms (Meneses-Scherrer *et al.* 2017). CNH has laid out a general plan for development, and lease rounds are being formulated at the time of this writing (CNH 2017b).

9.16 Austin Chalk Supersequence

As a reservoir, the Austin Chalk (AC) enjoyed brief but robust periods of exploration activity in the late 1970s and early 1980s in central Texas as a fractured chalk oil play. Past success in the AC play of Texas largely depended on the presence of oil in chalk micropores and natural fracture systems, aided by short horizontal wells and limited stimulation (Haymond 1991). However, extended reach horizontal drilling and hydraulic fracturing represents a new opportunity to exploit the AC in Texas and its equivalents to the east (e.g., Tokio Chalk of Louisiana).

As discussed in Section 4.9, Dravis (1981) studied Texas cores and outcrops to recognize two depositional facies: shallow-water chalks and deeper-water, possibly basinal chalks. The two paleobathymetrically distinct chalk facies exhibit pronounced differences in chemical and isotopic composition, diagenesis, and reservoir properties. Shallow-water chalks show evidence of meteoric water dissolution or original aragonite and/or early cementation (Dravis 1981). Low-bulk iron (average 370 ppm) and strontium (average 620 ppm) concentrations confirm freshwater diagenesis. Porosities are relatively high (average 20 percent) due to favorable diagenesis and shallow burial. The deeper-water chalk facies has lower porosity (often less than 16 percent) and matrix permeability (excluding fractures) due to aragonite preservation, burial compaction, and pervasive cementation (Dravis 1981). Preserved porosity is largely microporosity. Chemical data, including high iron and strontium content, suggest burial diagenesis dominated (Dravis 1981). Porosity declines more steeply with depth of burial than for equivalent

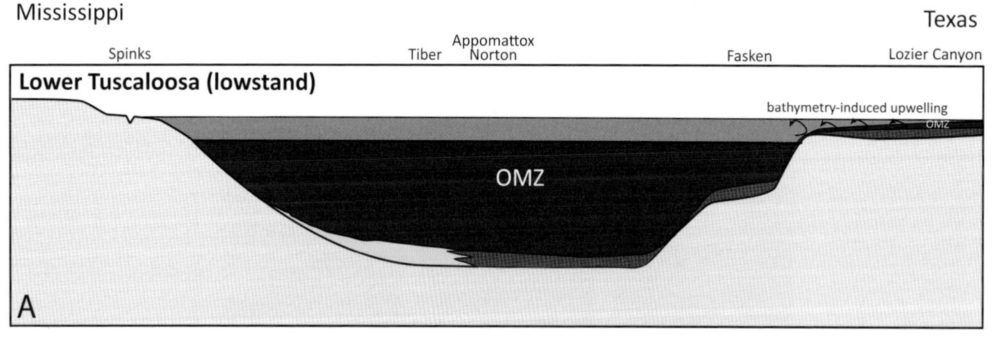

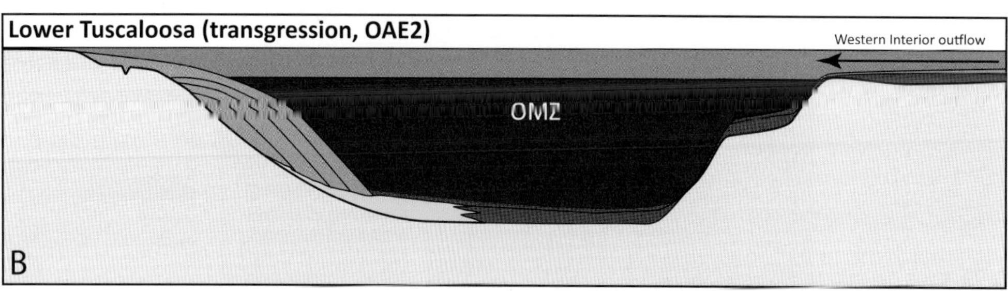

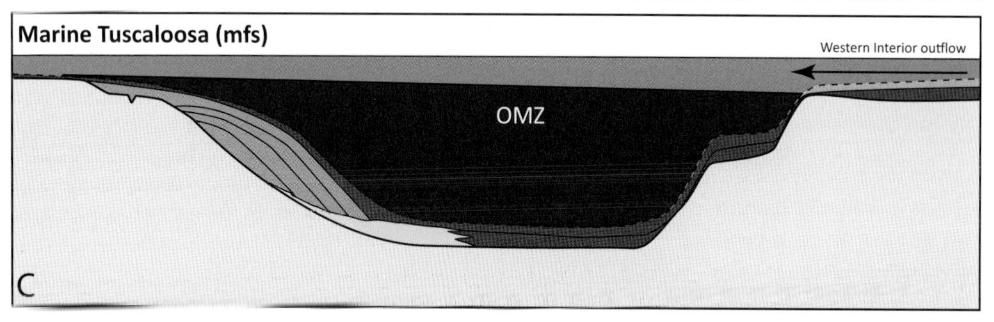

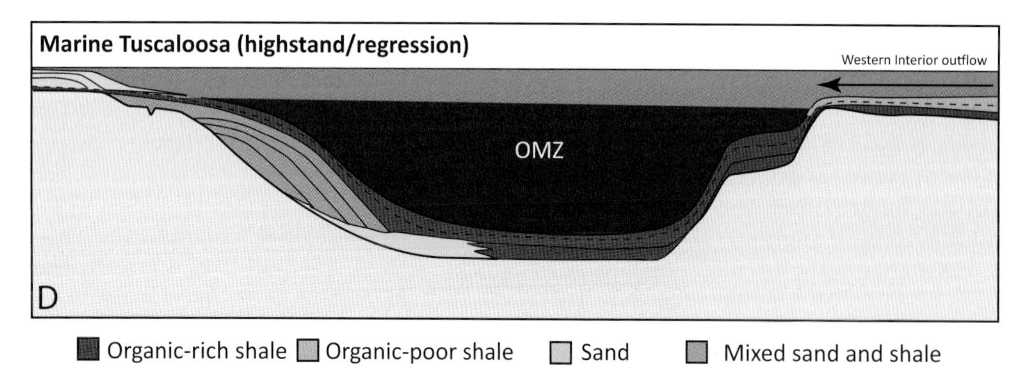

Figure 9.21 Summary of deposition across the northern GoM before, during, and after the OAE2, comparing data from Mississippi shelf (Spinks core), the deepwater GoM (Tiber, Appomattox, and Norton wells; this study), the shallow paleo-shelf of Texas (Lozier Canyon; see Lowery *et al.* 2014), and the deep shelf of Texas. OMZ, oxygen minimum zone. Modified from Lowery *et al.* (2017).

chalks in the North Sea or Ekofisk field of the Norway (Figure 9.22).

The location and distribution of natural fracture systems thus is a critical factor for exploring and exploiting the AC and its equivalents. The AC is effectively a "hybrid" reservoir, with characteristics of both conventional and unconventional play targets. Hydrocarbons exist in chalk micropores of coccoliths and micritic matrices (Dravis 1981). Natural fractures, either enhanced by stimulation or not, help deliver oil to the borehole.

Natural fractures in the AC have been studied for many years, driven by this economic motive. Natural fracture density and distribution in Texas, at least, are controlled by the mechanical stratigraphy of the formation and the regional structural position. Three mechanical-stratigraphic units were identified from experimental deformation tests by Corbett *et al.* (1987) on Texas outcrop samples: (1) an upper brittle fracture massive chalk (Big House Chalk member); (2) a middle ductile-chalky marl (Dessau and Burditt members); and (3) a lower brittle fractured chalk

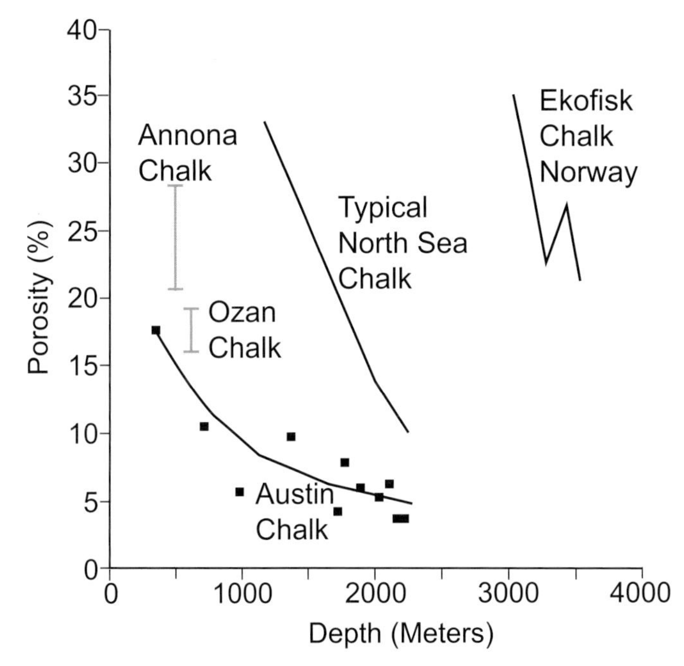

Figure 9.22 Porosity trend with depth for Austin Chalk, North Sea, and Ekofisk Field of Norway. Modified from Dravis (1981).

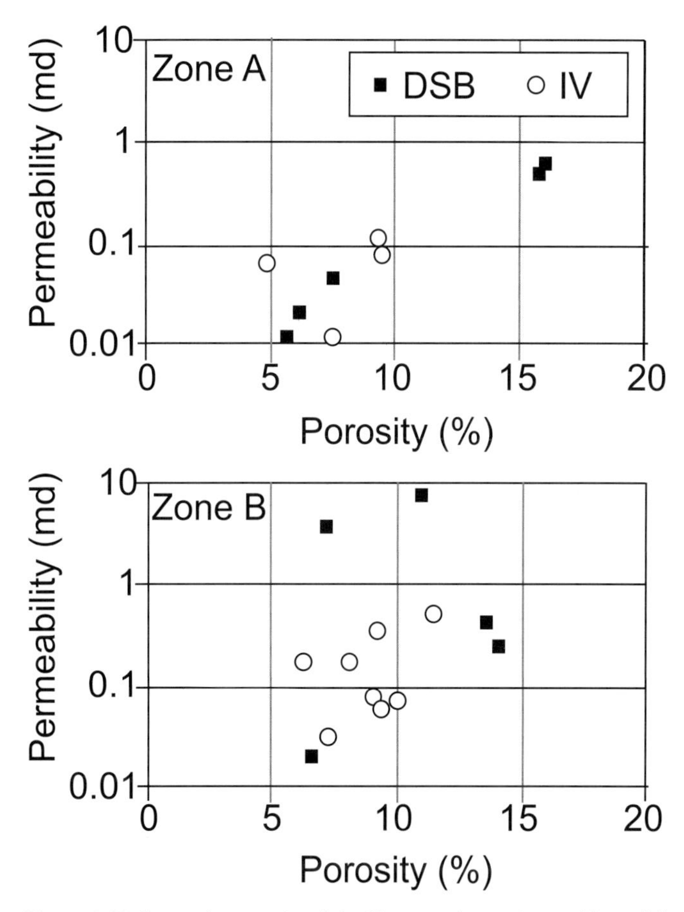

Figure 9.23 Reservoir properties of the Olmos sandstones in two "downdip" wells; DSB = discrete storm beds; IV = intervening siltstone lithofacies. Modified from Snedden and Jumper (1990).

(Atco member). Fracture strength correlated with porosity and clay (smectite) content, with more ductile response from the middle Dessau and Burditt members versus the more brittle over- and underlying chalks. This fit outcrop observations of greater fracture intensity in the brittle chalks (Corbett *et al.* 1987).

The AC fracture systems were mapped from outcrops near San Antonio and showed a clear relationship to regional extensional fault systems (Ferrill *et al.* 2017). Natural fractures are clearly parallel with or orthogonal to mapped fault systems. These fault systems in turn reflect local stress evolution, development of fault relays, etc. Thus, understanding the local stress field is a key to fracture prediction (Ferrill *et al.* 2017).

The lower portion of the AC in southeast Texas shows 0.5–3.5 percent TOC, with localized zones exceeding 20 percent (Grabowski 1981). The higher TOC zones are confined to the deeper-water chalks, which fits with the presence of trace fossils known to prefer low-oxygen conditions, as observed by Dravis (1981).

Later work (Tian *et al.* 2012) suggested that much of the AC oil was derived from the underlying Eagle Ford source rock, which itself became a prominent shale oil play after 2010. It is likely that both self-sourced and migrated oil are present in the unit, with maturity and biodegradation playing a role in oil gravity trends (Grabowski 1981).

A recent USGS assessment for the AC and its equivalents in four assessment units supports a conventional mean UTRR of 78 MMBOE, 2.3 TCFG, and 257 MMBNGL. Mean UTRR for unconventional or continuous resources is far higher at 879 MMBO, 1.3 TCFG, and 106 MMBNGL (Pearson *et al.* 2011). No assessment is currently available for the AC equivalent (San Felipe) in Mexico. AC age-equivalent reservoirs produce oil in the Copite Field of Mexico, mainly through fractured limestones.

9.17 Navarro–Taylor Supersequence

From a petroleum geology standpoint, the Navarro–Taylor (NT) supersequence is one of the lesser reservoir intervals of the GoM. Escondido, Olmos, and San Miguel Formations produce variable amounts of oil and gas in south Texas, dating back to the 1940s (Weise 1980). The Olmos (Maastrichtian) Formation is currently the largest producer here, as described below.

Olmos hydrocarbon discoveries are segmented into what is called the "updip" and "downdip" Olmos (Snedden and Kersey 1982). Updip Olmos sandstones have higher porosity and permeability (20–28 percent, 2 –400 md) than downdip sandstones (Tyler and Ambrose 1986). Sandstones of the downdip Olmos, Escondido, and San Miguel Formations produce tight oil (low permeability) and gas in south Texas (Tyler and Ambrose 1986). Permeabilities are often less than 1 md, with porosities ranging from 5 to 15 percent on average (Figure 9.23). The high clay content of these lower shoreface to shelfal storm-generated sandstone beds explains the low

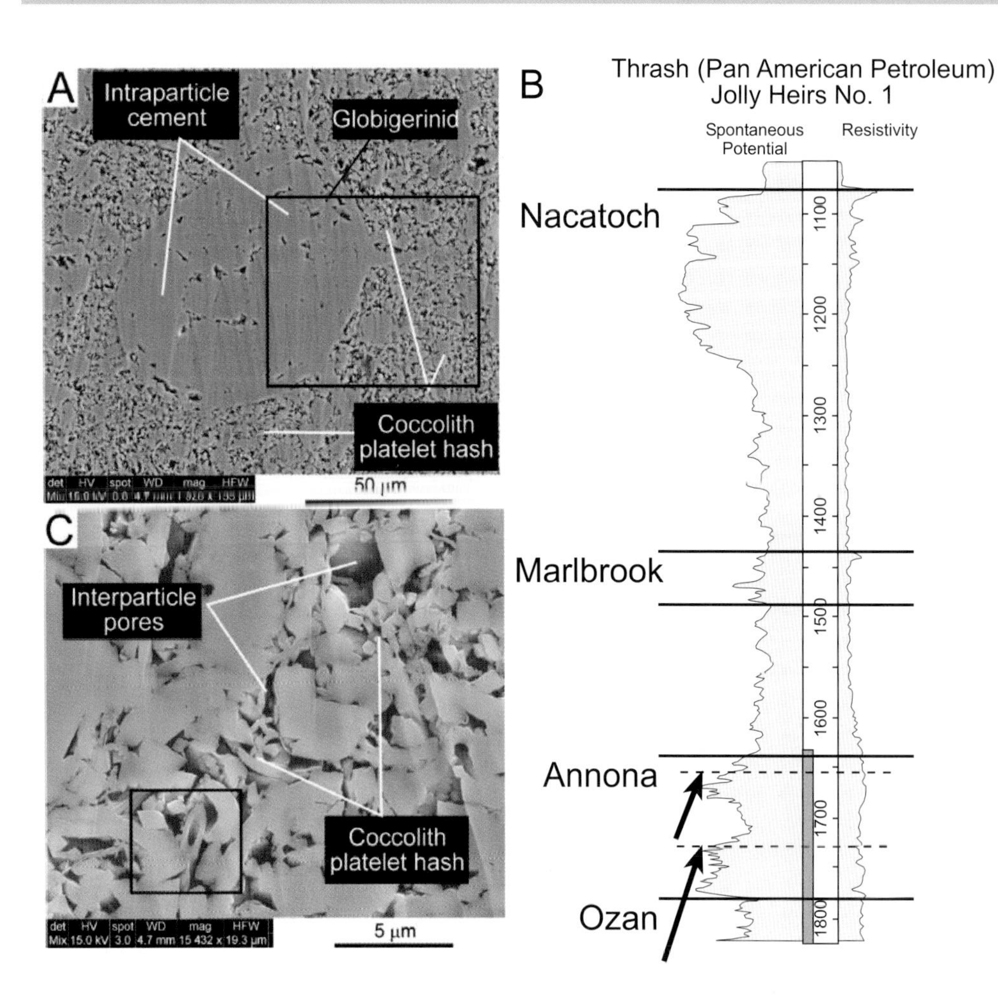

Figure 9.24 Annona–Ozan reservoirs. (**A**) Microporosity; (**B**) key well log showing units; (**C**) interparticle pores and coccolith hash. Modified from Loucks *et al.* (2017a).

permeability and low primary production flow rates of these downdip fields (Snedden and Jumper 1990).

Most downdip Olmos fields are small, with few producing more than 250 KBOE on primary recovery (Tyler and Ambrose 1986). An exception is the AWP Field of McMullen County, Texas that has produced over 50 MMBOE after artificial stimulation of horizontal wells (Swift and Mladenka 1997). In fact, the Olmos Sandstone play of the entire downdip areas has benefited from techniques used on unconventional or tight reservoirs like the Eagle Ford, including long multilateral wells and large hydraulic fractures (Terrace Energy at www.terraceenergy.net).

In north Texas, deltaic and shelf sandstones of the Nacatoch Formation are reservoirs flanking East Texas basin salt domes and in traps along the Mexia–Talco fault zone (McGowen and Lopez 1983). In areas immediately west of the Nacatoch, NT-equivalent chalks are called the Annona and Ozan Formations. Although production from the Annona and Ozan began as early as the 1900s, the overall recovery is still less than 5 percent of original oil in place. In shallow-buried chalks, porosity can be quite high (20–28 percent) but permeabilities are usually low (often less than 1 md; Figure 9.22) due to a lack of extensive natural fracture systems (Loucks *et al.* 2017a; Figure 9.24). However, new technology involving artificial stimulation and horizontal drilling could allow access to the remaining 94 percent of oil in place.

9.18 K–Pg Boundary Deposits

Cretaceous–Paleogene (K–Pg) boundary unit breccias and other Chicxulub-impact-related sedimentary units are one of the most important conventional oil reservoirs in offshore Mexico. K–Pg reservoirs produce or have previously produced hydrocarbons in Akal, Ek-Balam, Ixtoc, Abkatum-Pol-Chuc, Caan, May, Cantarell-Sihil, Taratunich, and Ku-Maloob-Zaap of the Campeche basin (Acevedo 1980; Mitra *et al.* 2005, 2006, 2007). There is no known production elsewhere in Mexico. Because of the size and importance of the K–Pg breccia unit at Cantarell-Sihil, a number of models have been proposed to explain its excellent reservoir quality. Clearly tectonic deformation associated with the Middle Miocene Chiapanecan event has enhanced fractures in the area. Samples and logs from the Reforma-Akal trend, where the K–Pg boundary deposit is an important, if not primary, reservoir, emphasize the favorable post-depositional diagenetic and burial history. Many cores show carbonate breccia facies cross-cut by fracture sets generated during the Middle Miocene

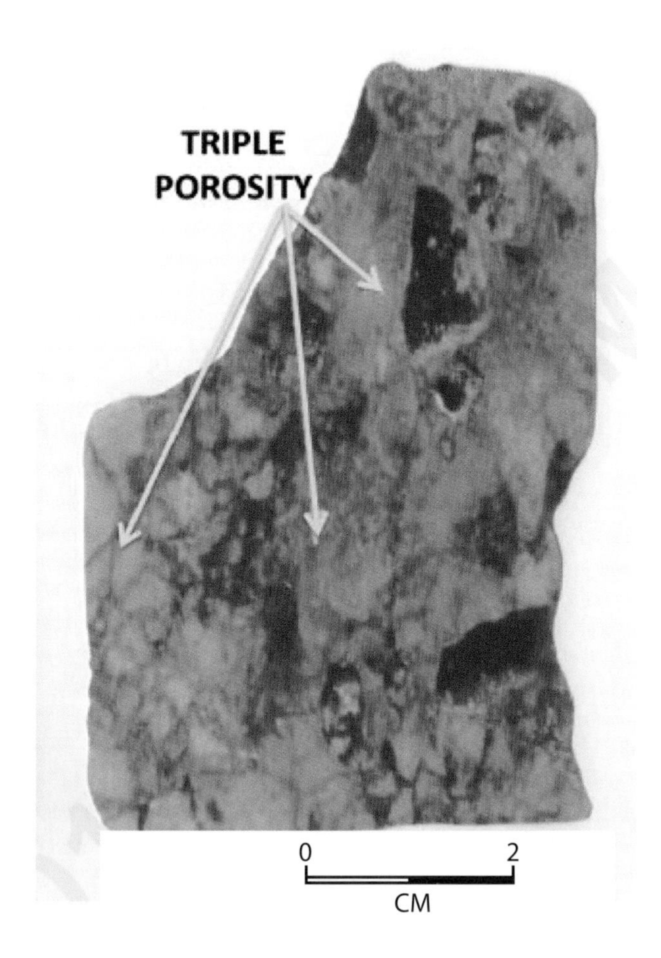

Figure 9.25 Triple-porosity (matrix porosity, vugs, and fracture-induced pores) in a K–Pg carbonate breccia sample. From Stabler (2016).

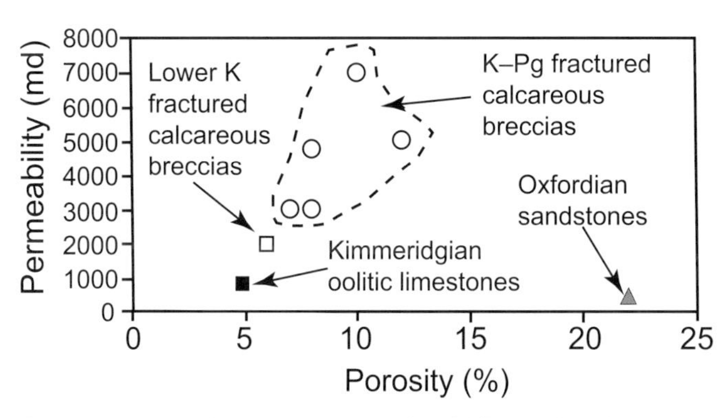

Figure 9.26 Permeability versus porosity plot of different oil-producing zones in the Cantarell, Ek-Balam, Ixtoc, and Ku–Maloob–Zaap Fields. Modified from Murillo-Muneton *et al.* (2002).

Chiapanecan orogeny (Murillo-Muneton *et al.* 2002). The breccia is dolomitized, which creates a triple-porosity system of matrix porosity, vugs, and fracture-induced pores (Figure 9.25). The reservoir permeability is heterogeneous on a small to medium scale and thus it has been difficult to predict reservoir performance (Stabler 2016). Porosity varies from 8 to

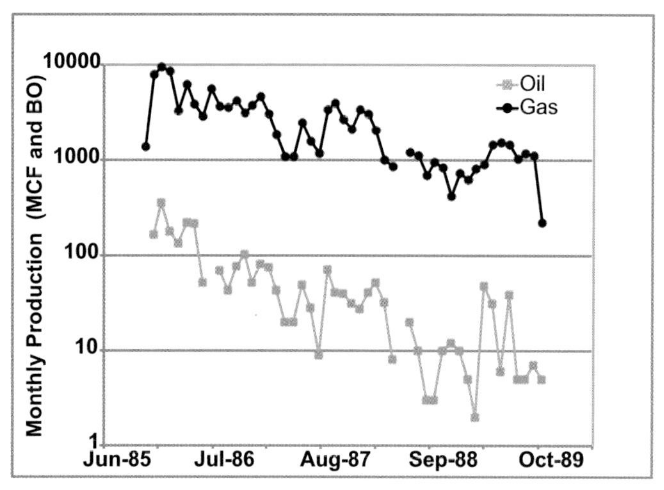

Figure 9.27 K–Pg reservoir production history at Tom Walsh Field, Webb County, Texas.

12 percent, but permeabilities are in the 3–5 darcy range (Grajales-Nishimura *et al.* 2000; Murillo-Muneton *et al.* 2002; Figure 9.26).

An unusual model for karst-enhanced fractures and porosity was suggested in the unpublished dissertation of Ricoy-Paramo (2005). The concept was Neogene thrusting resulting in a 1 km (0.6 mile) uplift of the central Akal Block, putting K–Pg breccias above the sea surface and allowing exposure to meteoric water. Unfortunately, this work was only supported by an unpublished internal Pemex report (Horbury 2000) and suggested isotopic geochemistry was not carried out (Ricoy-Paramo 2005).

In the northern GoM, there is limited production from contemporaneous K–Pg. Sandstones derived from the impact-related tsunami reworking of the coeval siliciclastic shoreline form thin sheet reservoirs in Webb County, Texas (Figure 4.50). At the Tom Walsh Field, oil and natural gas flowed for a short period, less than five years, prior to abandonment (Figure 9.27). The steep decline here may be due to the limited reservoir thickness, though tight oil and gas reservoirs generally exhibit short production plateaus. In north Louisiana, chalks displaced and transported following impact produce limited volumes of hydrocarbons at the IPHN central Field (G. Kinsland, pers. comm.) though no production data is available.

9.19 Implications for Mesozoic Exploration

Our correlations, mapping, paleogeographic reconstructions, and interpretation of the Mesozoic units lead us to some interesting trends that may be considered for future industry exploration. Some plays are emerging at the time of this writing.

A shelf margin delta of the Nacatoch–Ozan slope and basinal sandstone trend is present in east Texas, as indicated by the NT paleogeographic map (Figure 4.35) and cross-section (Figure 4.37). It may have slope or even basinal equivalents. At least six wells have penetrated the NT with thickness exceeding 200 ft (61 m). We would expect the deepwater fairway to be centered on southeast Texas, including Newton,

Orange, and Jefferson Counties of Texas, and Beauregard, Calcasieu, and Cameron Parishes of Louisiana. However, drilling depths would probably exceed 24,000–30,000 ft (7317–9146 m), based on regional mapping. There may be areas where the Nacatoch Sandstone could be combined with Annona Chalk reservoirs as a geographically focused "stack" play. The NT-equivalent Annona Chalk has historically had low oil recovery, which may be mitigated by new drilling and hydrofracturing techniques (Loucks *et al.* 2017a).

The AC supersequence includes sandstones formed as shelf sand ridges under accelerated current flow in the relatively narrow passageway between the Western US (Cretaceous) Interior Seaway and the GoM (Figure 4.34). These ridges are discrete, stratigraphically isolated sand bodies that could form hydrocarbon traps within shelf muds. These also could be combined with shallow AC or NT chalk reservoirs to form a geographically focused stack play.

Identification of two pathways for Ceno-Turonian sands to enter the paleo-deepwater GoM (Figure 4.25) is important to future exploration in Mississippi Canyon and other areas where large undrilled salt structures are present. The post-2016 rebound in oil prices may renew interest in this play, particularly if a new well is safely drilled and a large oil discovery is established.

With Ballymore discovery by Chevron in 2017, the Norphlet play potential appears to be enhanced, even as the most prospective areas of the eolian erg center near Appomattox are leased or already drilled. The less explored "fore-erg" (downwind erg margin) may have more limited reservoir quality as grain size and net sand are expected to decrease, and increased non-erg facies interbedding may be typical.

The HVB supersequence, important as an onshore, unconventional (source rock) play, may be less prospective in the northern GoM due to continued low natural gas prices. Well penetration of the coeval Gilmer rimmed shelf reef have not yielded large discoveries to date and numerous dry holes have been drilled (e.g., recent Sake prospect, DC 726 #2). However, there is renewed interest in the Kimmeridgian of Mexico, where several lease blocks near the Lamprea-2 well, a Jurassic test, are currently being evaluated.

9.20 Synopsis of Cenozoic Petroleum Habitat

Cenozoic petroleum exploration in the GoM has played a major role in the global history of oil and gas exploration and development. Besides introducing explorationists to new trap families, including salt diapirs and growth faults, expansion of exploration plays and consequent field development onto the continental shelf and ultimately down the continental slope to the abyssal plain established several milestones. Notable firsts include (BOEM 2016):

1. 1947: first well drilled out of sight of land, about 12 miles off the coast.
2. 1975: first deepwater well, MC 194, drilled in 1022 ft of water, resulting in the Cognac discovery.
3. 1988: first subsea completion.
4. 1989: first tension leg platform (TLP) in 1760 ft of water, Joliett Field.
5. 1990: first subsalt petroleum discovery, Mica Field.
6. 1990: first successful ultra-deepwater Wilcox well, the BAHA II prospect, in 7790 ft of water. The well reached its Mesozoic objectives but was dry.

Recent US Bureau of Ocean Energy Management assessment of undiscovered resources (UTRR) for the continental shelf and slope provides insight into the relative prospectivity of Cenzoic supersequences (Figure 9.3). The Paleocene Laramide Phase supersequences are assigned a potential of nearly 22 billion BOE, all in undifferentiated Wilcox supersequence plays and most in deepwater. Middle Cenozoic Geothermal Phase potential, about 9 billion BOE, is placed largely in the Lower Miocene supersequences. About 85 percent is assigned to deepwater plays. Assessed potential totaling nearly 22 billion BOE lies in the Middle and Upper Miocene supersequences of the Late Cenozoic Tectono-climatic Phase. More than 75 percent of the total is placed in deepwater plays. Pliocene deepwater potential adds a further 6 billion-plus BOE, dominantly deepwater as well.

9.20.1 Common Geologic Attributes

Cenozoic petroleum habitats display several general attributes that characterize many of the principal producing plays (Figure 9.28). First, thermally mature source rocks predominantly lie below the producing reservoirs. Vertical migration, commonly along fault or salt structure-related conduits, charged overlying reservoirs. Consequently, multiple stacked reservoirs commonly occur within a single field. Stacked reservoirs may extend through several thousand feet of interval and bridge two or more genetic supersequences. Growth faults in particular create expanded stratigraphic intervals and vertically persistent traps hosting multistory reservoirs (Figure 9.29). The full array of sand bodies deposited within lower coastal plain, deltaic, shore zone, and shelf systems provide reservoirs in a variety of onshore and continental shelf plays (Figure 9.30). As wells reached deeper, both vertically and in terms of water depth, sandy facies of slope aprons, submarine canyon fills, and abyssal plain submarine fan systems provided abundant, commercially viable reservoirs.

In contrast to the Mesozoic petroleum systems, reservoirs are almost entirely sandstone, and there are no unconventional shale plays being pursued. Consequently, burial diagenesis plays a major role in determining reservoir quality (Loucks *et al.* 1986; Ambrose *et al.* 2013). Shallow to moderately deep Neogene reservoirs lose porosity primarily by compaction. Moderately to deeply buried reservoirs, particularly those of the Laramide and Geothermal Phase supersequences, undergo significant chemical diagenesis that includes precipitation of quartz and/or calcite cement, grain alteration and dissolution, and creation of secondary porosity. Dutton *et al.* (2012) described the basin-wide variation in sand composition for Lower Miocene sands; the regional patterns they documented are reasonably typical of other Cenozoic units. In the

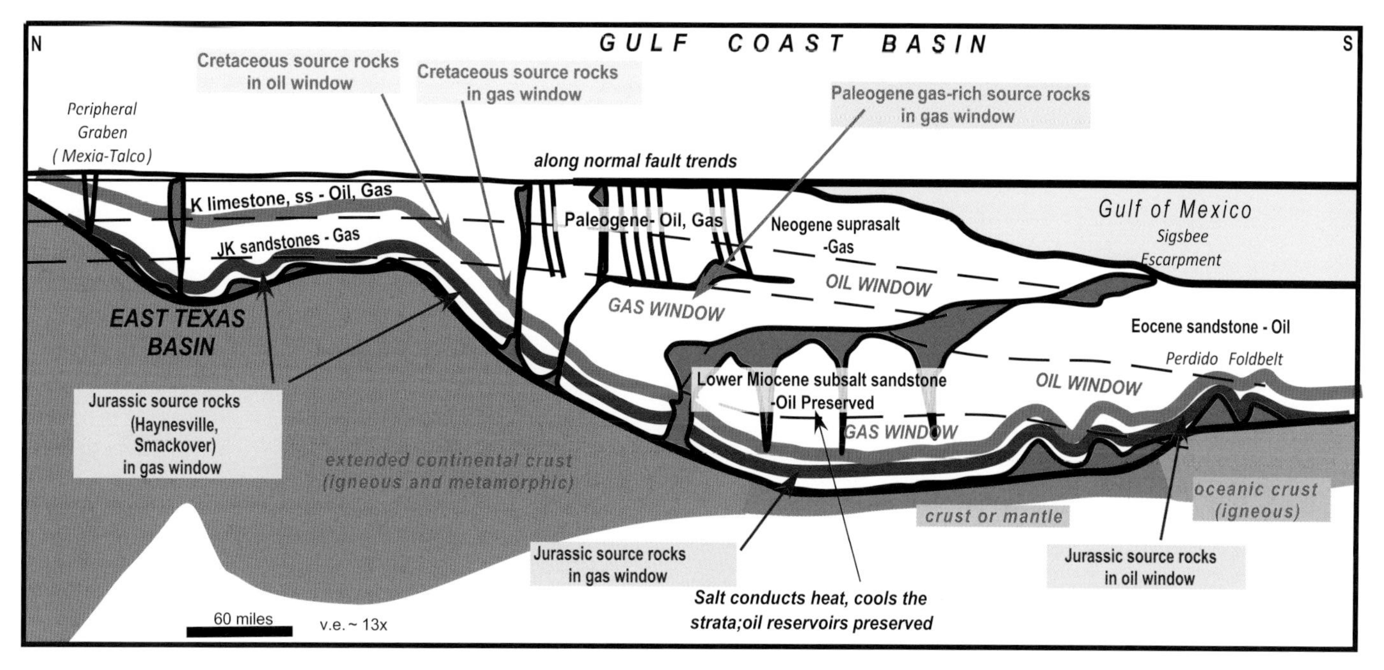

Figure 9.28 Generalized petroleum system framework of the northern GoM margin. The section generally extends from the East Texas basin southward across the coastal plain, continental shelf, slope, and onto the abyssal plain. Modified from Ewing (2016).

northeastern Gulf, sands are relatively quartz-rich, with subordinate feldspar, and minor rock fragment grains. In the north-central Gulf (southeast Texas), quartz still dominates, but feldspar content becomes significant. Moving to the northwest Gulf, including the Burgos basin, the combined feldspar and rock fragment percentage approximates that of quartz. In the Veracruz basin, quartz percentage increases, and feldspar grains dominate over rock fragments. In short, sandstones of the northwest and western Gulf basin are diverse and relatively mineralogically immature, making them susceptible to diagenetic alteration.

Rapid deposition of relatively young sediments along the Gulf basin margins has created pervasive overpressure in much of the basin-fill. In the northern GoM, most onshore wells drilled 8000–10,000 ft (2500–3000 m) encounter substantially overpressured reservoirs. Off shore, the top of overpressure is typically even shallower. Overpressuring has retarded physical compaction in some areas, preserving porosity in unconsolidated sands. Onshore, the highly pressured Frio and Wilcox intervals have been objects of study for development of geopressured/geothermal energy (Loucks 1978; Bebout *et al.* 1982; Winker *et al.* 1983).

In contrast to many petroleum basins, where reserves are dominated by a few giant fields, the Cenozoic interval of the GoM is characterized by many hundreds of medium to small fields dispersed in both 3D space and geological age of the host reservoirs (Figure 9.2). Systematic analyses of petroleum plays have utilized several approaches to organizing this plethora of oil and gas fields. The simplest, but least geologically significant, is based on position relative to the coastline, further differentiated offshore by position relative to the modern shelf edge. Another common approach has been to define plays by primary tectonostratigraphic trends. Stable paleo-coastal plain and shelf, expanded fault zone, and continental slope and basin floor fairways are differentiated (e.g., Hackley and Ewing 2010). Galloway *et al.* (1982b) applied the depositional system/trap style play concept. A subsequent series of unit-specific studies and petroleum atlas projects at the Texas Bureau of Economic Geology (Galloway *et al.* 1983; Kosters *et al.* 1989; Morton and Ayers 1992) followed this approach. More recently, analysis of offshore, and especially deepwater, plays has stressed the important relationship of reservoirs and traps to the extensive salt canopy and compressional fold belts (Figure 9.31; Weimer *et al.* 2017). Deepwater fields of the northern GoM are placed in one of four provinces. (1) *Basins province* fields lie within salt- or weld-bounded basins formed on the salt canopy. (2) *Subsalt province* fields lie below the salt canopy or its weld. The subsalt position creates technological challenges both to seismic imaging and drilling. (3) *Fold belt province* fields lie along the Mississippi Fan, Keathley–Walker, and Perdido fold belts. (4) *Abyssal plain* fields lie beneath relatively flat basin floor basinward of the salt and fold belts.

Drilling into abyssal plain, fold belt, and subsalt provinces introduced an array of trap configurations rarely seen previously (Figure 9.32). Salt canopies commonly play a major role, providing both seals and structural discontinuities. Welds may also create traps. Deformed deep autochthonous salt forms anticlines and drape structures. Residual salt bodies also create superjacent drape structures. Localized salt evacuation forms temporary depocenters that later become elevated as remaining salt is expelled, forming turtle structures. Complex salt migration histories can encapsulate large bodies of sediment within the salt canopies or composite canopies and welds.

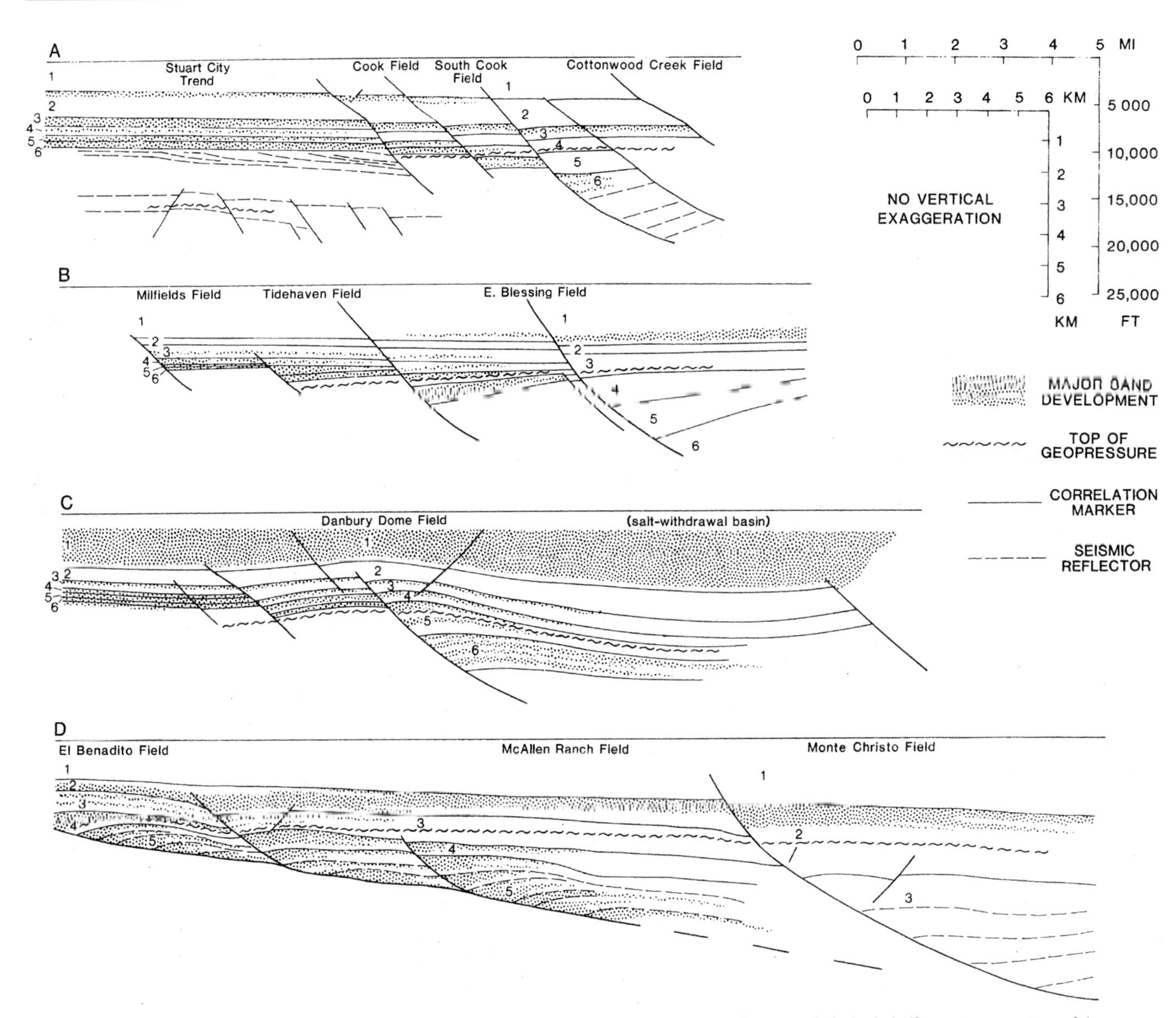

Figure 9.29 Representative dip cross-sections showing the structural styles and sand distribution in four growth-faulted, shelf-margin successions of the northern Gulf Coast. (**A**) Faulted Lower Wilcox, south Texas. Faults sole out onto deep salt. (**B**) Listric faults with consequent landward dip reversal. Frio, central Texas. (**C**) Listric master fault with adjustment faults and rollover anticline over deep-seated salt. Frio, southeast Texas. (**D**) Highly listric faults with rollover anticlines. Vicksburg (Early Oligocene) play, south Texas. From Galloway and Hobday (1996).

9.21 Petroleum Habitat of the Laramide Phase Supersequences

All of the major depositional episodes of the Cenozoic created supersequences containing major petroleum plays. Reservoirs range in age from Early Paleocene through Pleistocene. The bulk of the hydrocarbon reserves have been found in the thick, offlapping sedimentary prism of the northern Gulf basin. As a general rule, the total volume of discovered hydrocarbons is directly proportional to the total volume of sand within the sequence, reflecting the primary role of the Cenozoic siliciclastics as reservoirs.

9.21.1 Wilcox Supersequences

Sandstones of all three Wilcox supersequences have been prolific producers of gas and oil along a broad, strike-parallel belt extending from eastern Louisiana to the Burgos basin (Figure 9.33). The combined mature Wilcox plays have an estimated cumulative production of more than 14 BBOE and extend from the Burgos basin to Mississippi (Figures 1.15–1.18). Reservoirs include the full suite of sandstone facies of the deltaic and shore zone systems. Most traps are produced by growth fault-related structures and include rollover anticlines and various fault traps along the Wilcox fault zone

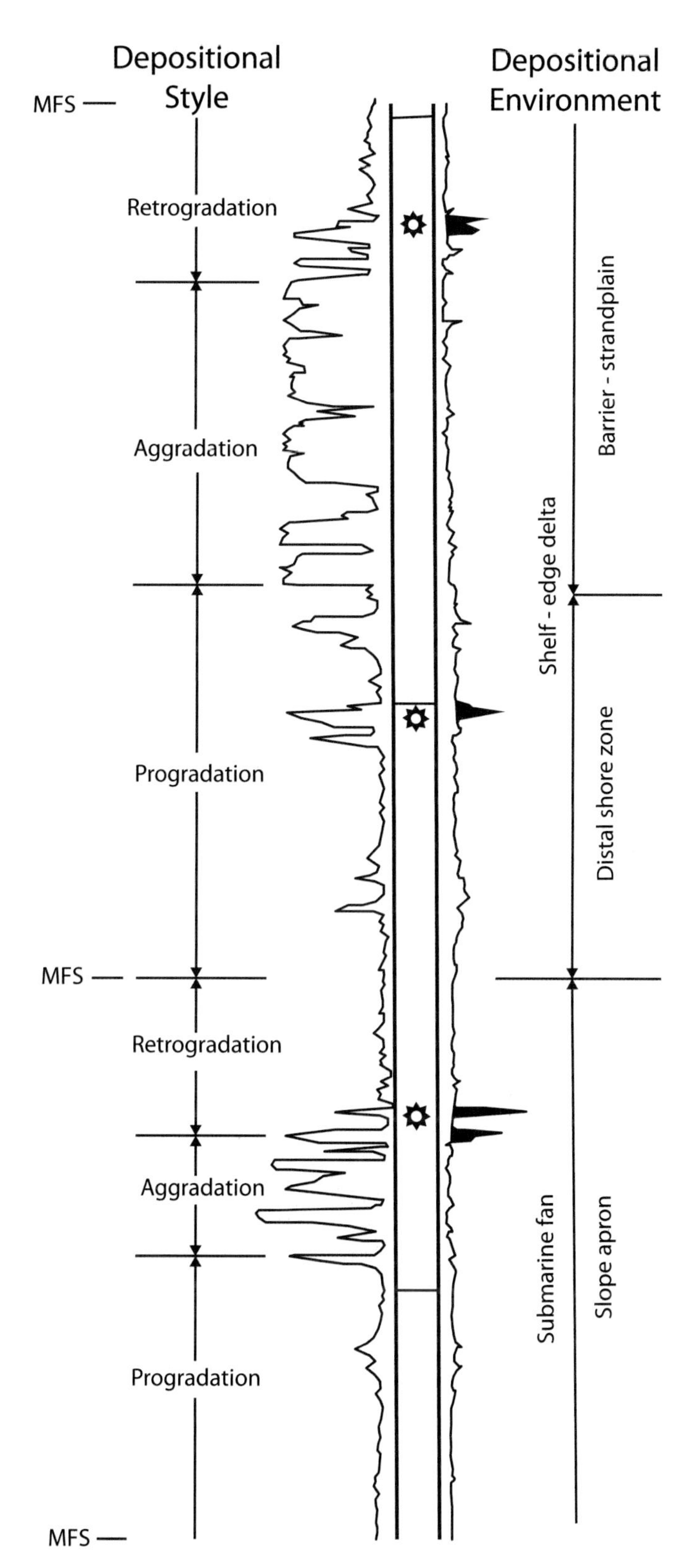

Figure 9.30 Generalized vertical succession of reservoir facies associations typical of the northern GoM coastal plain and shelf plays. Left log profile generalizes the vertical succession typical of deltaic depocenters. Thin transgressive veneers cap expanded retrogradational, aggradational, and progradational delta front and delta plain parasequences. Prodelta muds merge downward into muddy upper slope deposits of the delta-fed slope apron containing discontinuous turbidite channel fill, overbank, and lobe facies. On the right, the corresponding facies succession found in a shore zone barrier bar setting contains thick, aggradational barrier sand bodies, as well as mud-rich lagoon fill and well-developed shoreface facies. Basal shelf muds grade into the underlying, muddy slope apron. Erosionally inset submarine canyon fill consists largely of mud with onlapping basal turbidite sands. All sand facies create important reservoirs in multiple Cenozoic plays of the northern Gulf. From Galloway and Hobday (1996).

(Figure 9.29A). In the Houston and South Louisiana salt provinces, deep salt stocks create domal and faulted anticlinal structural traps (including large rollover anticlines and related faults).

Three plays defined by their unique tectonostratigraphic settings are worthy of note (Figure 9.33). (1) The Lobo megaslide created a play dominated by complexly faulted basal Wilcox sand bodies (Long 1985). Several unconformities further complicate the field architectures. Reservoirs are structurally segmented shelf and shoreface sandstones. The play has yielded several TCF of gas. (2) In eastern Louisiana and western Mississippi, an outlier of mid-dip fields straddles the Mississippi River. Reservoirs are delta front and delta plain sand bodies of the Middle Wilcox Holly Springs delta. Structures are subtle; most traps are stratigraphic or have a stratigraphic component. (3) The Early Eocene Wilcox raft, which is manifested by the high-displacement Upper Wilcox fault zone extending from the Rio Grande to the middle Texas coastal plain, created the third fairway. Reservoirs are highly expanded, stacked delta front sands of the Rosita delta system (Edwards 1981).

The downdip limit of the Wilcox plays is created by their deep burial beneath younger Paleogene units, poor reservoir quality due to diagenetic alteration, high temperatures, and geopressure. Dutton and Loucks (2010) determined that secondary porosity dominates pore space in Wilcox sands lying below the 300°C isotherm.

Discovery of oil in thick Paleocene to earliest Eocene submarine fan deposits in the Perdido fold belt dramatically expanded the area of Wilcox prospectivity (Box 9.3). Subsequent wells rapidly demonstrated the widespread extent of basinal sandstone and yielded multiple hydrocarbon discoveries (Zarra 2007). Currently 28 hydrocarbon discoveries in Wilcox strata are documented (Weimer *et al.* 2017). Reserve estimates exceed 4 BBOE. The fields and discoveries group into two broad plays (Figure 9.33). In the northwestern Gulf, several discoveries extend along the Perdido fold belt into Mexican waters (Colmenares and Hustedt 2015). A larger area, which includes numerous subsalt discoveries, occupies the west-central Gulf continental slope. Productive structures lie along the Keathley–Walker and Atwater fold belts and extend northward beneath the extensive salt canopy. Subsalt traps are complex (Figure 9.32), and include three-way dip closed against salt stocks or welds, four-way closures formed by structural contraction, and four-way closure created by differential subsidence onto an underlying autochthonous

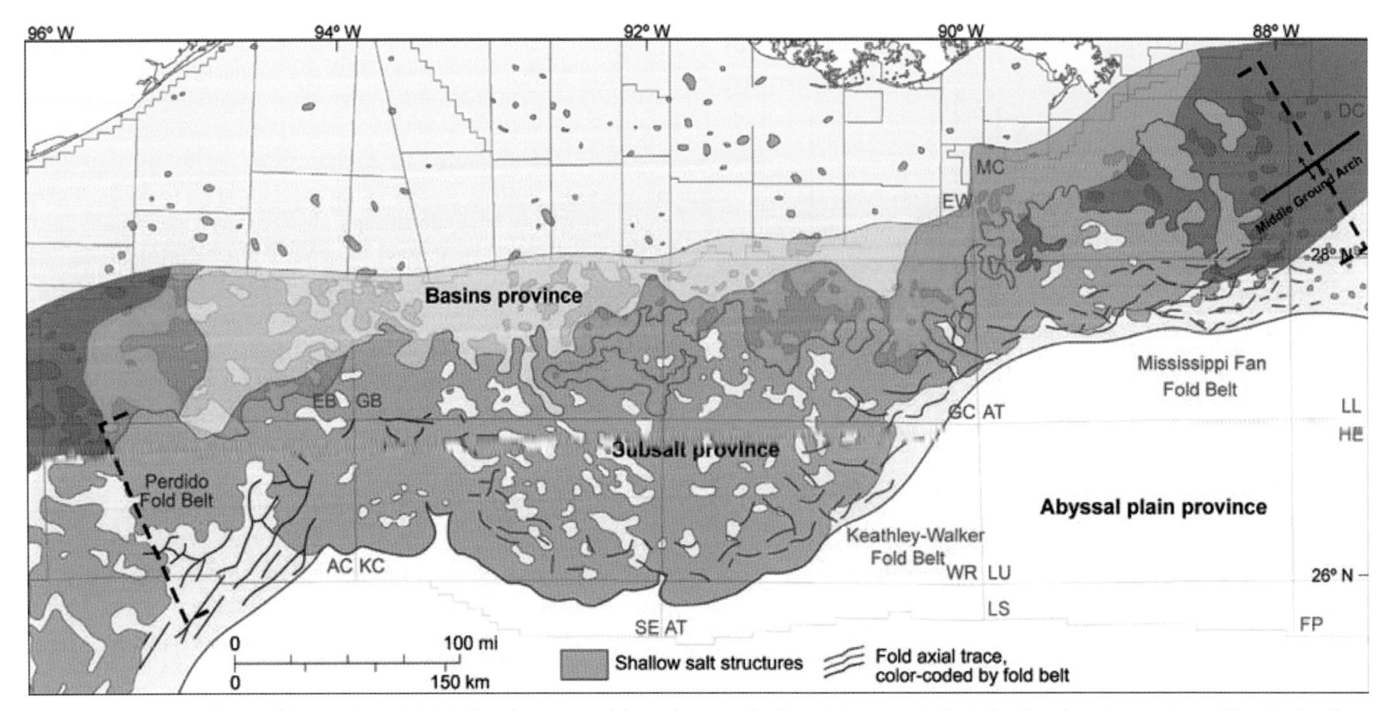

Figure 9.31 Regional map of the northern GoM shelf and continental slope showing the four major tectonically defined exploration provinces. Blue shades show age of fill in the basins province: dark blue = Miocene; medium blue = Pliocene; light blue = Pleistocene. From Weimer *et al.* (2016).

salt surface (Pilcher *et al.* 2014). Reservoir facies consist of lithic arkoses and feldspathic litharenites; grain size ranges from fine sand to silt, reflecting their deposition at the distal end of a highly evolved continental sediment dispersal system. Reservoir facies include channel fills and lobes (Marchand *et al.* 2015). Grain size and reservoir quality decrease from channel to proximal lobe and then to distal sheets. Upper Jurassic source rocks are interpreted to provide hydrocarbon charge to these deepwater plays (Figure 9.7).

Box 9.3 Deepwater Exploration in the GoM and Significance of the BAHA Wells

The history of exploration in the northern GoM basin reflects a mixture of conservative geologic thinking punctuated by bold initiatives. Exploration has both benefited from advances in technology and driven many of those advances as frontiers expanded into deep water and below allochthonous salt. The BAHA wells are a recent example of how new data overturned conventional wisdom and triggered major exploration plays.

Through most of the twentieth century, principal Cenozoic reservoir facies were found in continental fluvial, deltaic, and coastal origin. Beginning in the 1950s with the API Research Project 51 study of nearshore sediments in the northern GoM, several industry-funded research projects set out to describe modern reservoir analogs such as the Mississippi and Brazos rivers and their deltas. However, downdip and the deepest wells generally penetrated thick, commonly overpressured mudrock successions, suggesting that sand was sequestered in those familiar shallow marine and coastal plain systems. Shallow marine coring confirmed the conventional wisdom concerning the failure of sand transport onto or across the Holocene continental shelf to the shelf edge.

However, the presence of the Mississippi Fan provided tantalizing evidence that sand transport pathways could break the bounds of the shelf mud prism and spread potential reservoir facies onto and across the slope and abyssal plain. Deepwater sediment wedges, such as the Hackberry and Planulina Embayments, yielded the initial examples of documented "turbidite" reservoirs.

As drilling spread offshore, most reservoir targets remained sands of the familiar shallow-water and coastal origin. However, the downdip limit of potential reservoir sand facies expanded significantly as wells on the continental shelf began to penetrate intraslope basin-fills containing ponded Plio-Pleistocene submarine channel (lenticular) and lobe (sheet) sands. As continued success pushed drilling over the modern shelf margin and down the continental slope, deep wells in the northeast Gulf opened up Miocene plays in lower slope and basinal reservoirs fans.

However, potential of the slope and basinal extensions of Paleogene supersequences, so productive across the northern GoM coastal plain, remained problematic. They were deeply buried and obscured beneath structurally complex Neogene depocenters and allochthonous salt bodies. Only at the modern lowermost continental slope and abyssal plain did they reemerge from the Sigsbee scarp to potentially drillable burial depths and seismic visibility. That placed them 200 miles or more from their contemporaneous shelf edges and the nearest well penetrations. The potential and nature of reservoirs was highly speculative at best.

Box 9.3 (*cont.*)

There the story remained into the middle 1990s. Beginning in 1994, the **Energy and Minerals Applied Research Center (EMARC)** of the Department of Geological Sciences at the University of Colorado at Boulder began a reevaluation of the Perdido fold belt in the northwestern deep GoM. The project was supported by four companies (Shell, Texaco, Amoco, Mobil) that had leased the mineral rights to the fold belt during the mid-1980s. Although reservoir potential retained high risk, the seismic data clearly revealed large, anticlinal structures. Paleogeographic maps produced by the GBDS project at the Institute for Geophysics, University of Texas at Austin, had mapped, based on interpolation between onshore paleogeography and deep GoM seismic data collected by UTIG, widespread deposition of sandy slope and basin floor aprons across the north-central and northwest Gulf basin (Galloway 2002). *Offshore Magazine*, on April 1, 1996, announced: "A group led by Shell Oil is drilling a wildcat well in a world record water depth of 7,625 ft. Designated BAHA, the prospect is located about 200 miles southeast of Corpus Christi, Texas ... Located in the Alaminos Canyon area, the BAHA prospect lies beneath water depths of 6,500 to nearly 9,000 ft." Drilling of the BAHA #1 well in Alaminos Canyon 600 took place from April to June 1996. Spudded in 7612 ft of water, the well soon experienced drilling problems and reached total depth at 3596 ft. However, it had penetrated sand within the Frio interval containing 14 ft of producible petroleum (per MMS classification). Both the potential for Paleogene basinal sand reservoirs and reality of a functioning petroleum system had been demonstrated.

A second well, BAHA #2, on the BAHA structure (AC 557) was delayed until late 2000/early 2001. BAHA #2 was also a dry hole, but in reaching its objective of fractured Mesozoic carbonates, the well penetrated nearly 2000 ft of sand-bearing Wilcox strata. Not only was the presence of sand hundreds of miles from its contemporary shelf edge, far onto the Paleocene abyssal plain, documented, but high net-to-gross and sand body thickness soon led to the informal name "Whopper Sand." Between 2001 and 2007, 12 announced Wilcox discoveries extended the Wilcox fairway from Trident at the southern margin of the Alaminos Canyon area Cascade in the northeast Walker Ridge area (Meyer *et al.* 2007).

The regional distribution of thick successions, dominantly Paleocene Wilcox sandstone, dramatically changed the paleogeographic reconstruction of Wilcox megasequences. The scale of sand distribution so surprised many Gulf basin geologists that soon some interpreters suggested that a unique basin history was called for. The principal catastrophic model that emerged proposed partial desiccation of the Gulf at some time or times during Wilcox deposition (Rosenfeld and Pindell 2003). However, as documented by Sweet and Blum (2012), the scale of the canyons and fans as currently mapped is commensurate with the Wilcox continental drainage basins and fluvial systems that sourced them. No special history is needed; indeed, quantitative scaling indicates that they could have been expected. Once again, the GoM had surprised and benefited explorationists with its efficient sand dispersal and reservoir creation.

The Wilcox of deepwater areas presents multiple recovery challenges, a function of both depositional and post-depositional histories. The basinal Wilcox reservoir facies consist of fine- to very fine-grained sand and silt, which constrains permeability even without the complications of diagenetic reduction. A particular issue for the "outboard" Wilcox of Alaminos Canyon, Keathley Canyon, and southern Walker Ridge protraction blocks is the high silt content in submarine fan fringe and avulsion splays (terminology of Power *et al.* 2013). These small, rigid grains of quartz and feldspar partially or completely block pore throats and increase flow tortuosity (Marchand *et al.* 2015). A threshold silt content of >30 percent in a sandstone results in very low permeability, typically below 1–5 md (Figure 9.34). Although abundant silt also impacts porosity, its effect is not as profound as on fluid flow. It is not uncommon to see interbedding of silt-rich, non-fluorescent sandstones and silt-poor, oil-bearing sandstones in pay zones (Power *et al.* 2013).

The "inboard" Wilcox of Green Canyon, Garden Banks, and northern Walker Ridge protraction blocks has shown a considerable range in reservoir quality, mainly related to the burial history of these earlier-formed primary basins (terminology of Pilcher *et al.* 2011). Reservoir properties can be excellent in basins where early salt emplacement has mitigated heat flow, or quite poor where late salt emplacement did not. Reservoir heterogeneity and compartmentalization present additional challenges to efficient recovery.

The Wilcox "deep shelf" play, pioneered by Freeport McMoRan in the early 2000s, ultimately resulted in dry holes and small, non-commercial gas discoveries. The Davy Jones I (SMI 230) ultra-deep Wilcox test encountered very high pressures and temperatures near total depth (TD), conditions so extreme that testing and completion were unsuccessful. The BP Will K well (HI-119A #1), which targeted a deep (28,000 ft) sub-weld structure on the shelf, found the Wilcox to have low porosity and permeability (Blankenship *et al.* 2010). Will K remains at the time of this writing the highest combination pressure–temperature well in the GoM. A further limitation of sub-weld potential is the absence of salt to mitigate heat flow and thus reduce thermal gradient.

In addition to their petroleum endowment, the Laramide Phase supersequences provide a further contribution of great importance to many of the coastal plain and shelf petroleum systems: source rocks. Regional synthesis of Wilcox–Sparta sediments shows large areas, and consequently large volumes, of mudstones with 1–2 percent measured TOC (Figure 9.35; Cunningham, in prep.). The buried organic matter was dominated by land-derived plant material, creating Types II and III kerogen. High TOC values group around major shelf-margin deltaic depocenters, where distributaries flushed abundant macerated plant debris from coastal marshes and riverine swamps into the Gulf. Slope facies are also rich in TOC, and submarine fans contain up to 2 percent TOC. Types II and III, as well as mixed

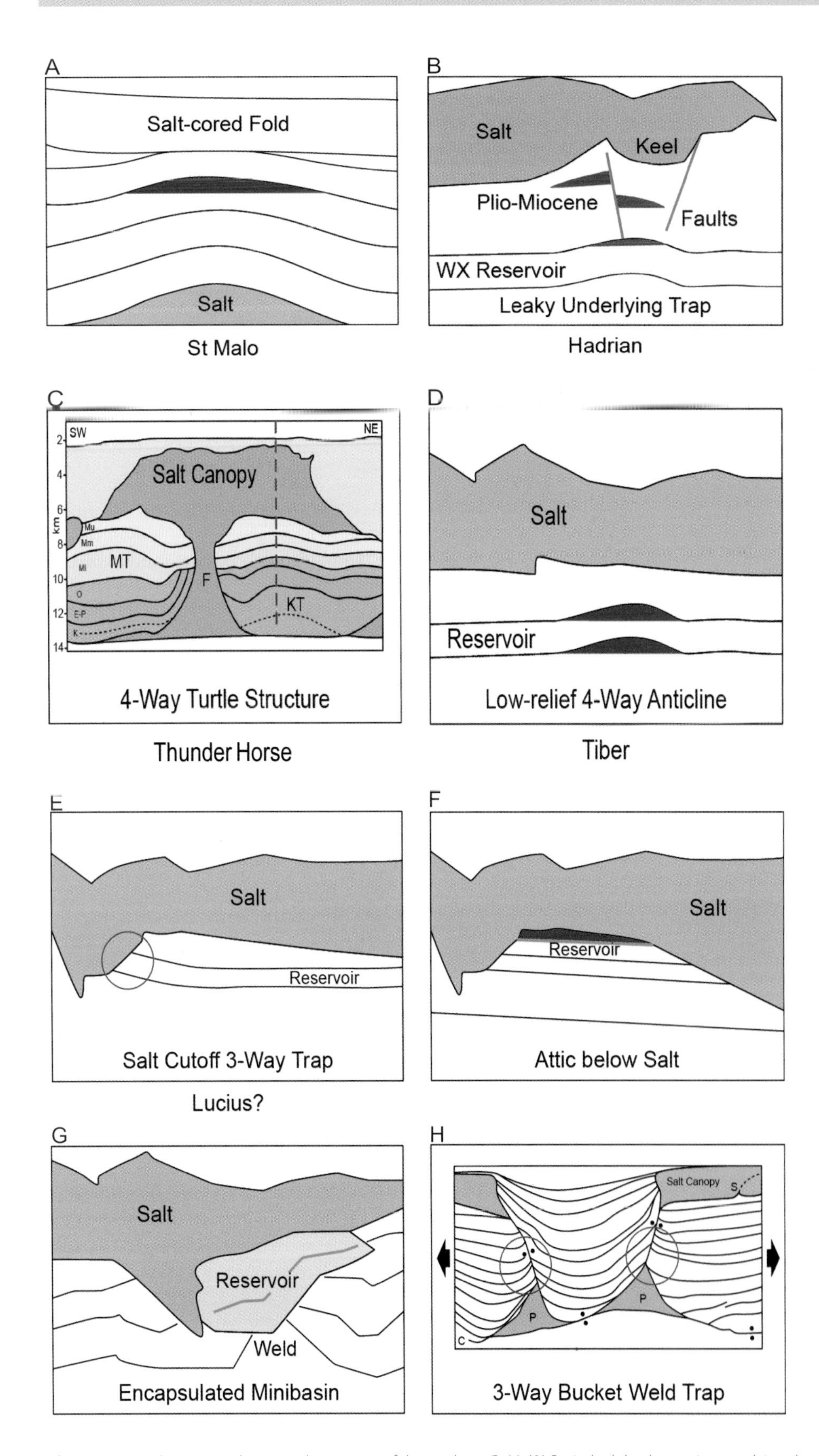

Figure 9.32 Salt-associated structural trap types of the northern GoM. (**A**) Buried salt body creating overlying dome. (**B**) Salt canopy concealing underlying four-way closure with associated faulting. (**C**) Turtle structure created by salt evacuation around a former withdrawal syncline. Overlying canopy may or may not be present. (**D**) Salt canopy concealing underlying simple four-way closure. (**E, F**) Salt canopy forming a seal in one or more directions to create closure. (**G**) Complex salt deformation history resulting in an earlier-formed minibasin-fill becoming encapsulated within the salt body and its related welds. (**H**) Welds within or on flanks of a minibasin creating updip seal. (**B**) and (**H**) are adapted from Pilcher *et al.* (2011).

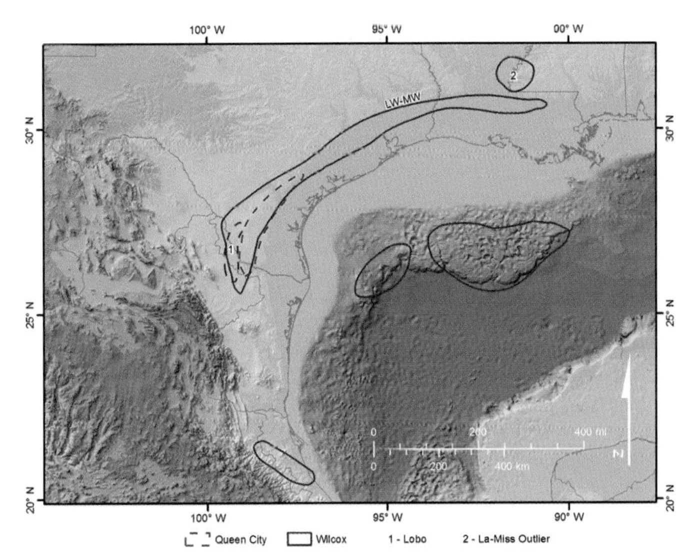

Figure 9.33 Principal hydrocarbon fairways within Paleogene Laramide Phase supersequences.

kerogens, are also present in the Sureste province in the southern GoM. Because the Paleocene–Eocene continental margin developed irregular bathymetry in response to ongoing salt movement, dysoxic–anoxic conditions occurred variably along the slope in isolated depressions. This depositional history potentially created pockets of organically enriched mudstone with up to several percent TOC. The resultant immense volume of moderately rich source rocks is interpreted to be an important contributor to the oil and gas fields of the northern GoM coastal plain and continental shelf (Figure 9.7).

In the deepwater areas where the Wilcox is not self-sourced, there is a regional variation in oil quality related to the underlying source rock trends. Oil viscosity, a key factor in flow rate, increases north and westward from 23 degrees API at Jack Field to 28 degrees API at the Tiber discovery (KC 109 #1) and then to 37 degrees API at Great White Field (Eikrem *et al.* 2010).

The Queen City and Sparta supersequences are relatively minor hydrocarbon producers. The Queen City hosts a modest gas-dominated play that extends from the Burgos basin into the south Texas coastal plain.

In the western GoM, Laramide Phase, slope and basinal sandstones of the Chicontepec host an extensive play that extends the length of the Chicontepec basin (Figure 9.33). This play has been described as a "giant field" measuring 123 × 25 km (76 × 16 miles). Low-permeability reservoirs are very coarse to silty litharenite pervasively cemented by calcite and quartz (Bitter 1993). Late dissolution of calcite created secondary porosity. Carbonate rock fragments dominate lithic grains.

9.21.2 Potential Wilcox Play Expansion

The greatest focus for the past decade has been on deepwater ALKEWA fan system reservoirs. The Trion well has confirmed the extension of the Perdido fold belt play into the Mexican outer continental shelf (OCS). The emergent play is expected to display similar trap and reservoir attributes as its US precursor and lies in the fold belt and subsalt fairways. The presence of thick salt canopies and stocks reduces the local thermal gradient, helping both to preserve reservoir quality and to depress the oil window (Dutton and Loucks 2010; Davison and Cunha 2017). Improving subsalt seismic resolution (see Section 9.25) and advances in drilling and completion technologies are expected to lead to additional discoveries in the deep subsalt fairway beneath the GoM continental slope. Reservoir quality and reservoir continuity are limiting factors in economic development of some of the discoveries.

A few ultra-deep wells drilled on the lower coastal plain and adjacent inner shelf have encountered sandstone bodies within the thick delta-fed aprons of the Wilcox continental margin. The Davy Jones well encountered gas in highly overpressured Wilcox strata. The highly expanded distal delta front and upper slope apron intervals continue as a potential exploration fairway for natural gas (Warwick 2017). Limitations for development of an expanded continental margin play include the high temperature, geopressure, and low porosity and permeability of sandstones due to diagenesis (Dutton and Loucks 2010; Ambrose *et al.* 2013).

9.22 Petroleum Plays of the Middle Cenozoic Geothermal Phase Supersequences

With the exception of the volumetrically minor Jackson supersequence, all Geothermal Phase supersequences host regionally extensive petroleum fairways across the northern GoM coastal plain and shallow shelf from Louisiana to the Burgos basin (Figure 9.36). The subjacent Paleogene strata provide the primary source rock for the largely gas-prone suite of plays. Onshore plays and their field and individual reservoir attributes are compiled in major petroleum atlases (Galloway *et al.* 1983; Kosters *et al.* 1989; Bebout *et al.* 1992).

9.22.1 Yegua and Jackson Supersequences

Although modest in volume relative to the overlying Frio and Lower Miocene supersequences, the Yegua, and to a lesser degree Jackson, supersequences have a long history of oil and gas production. More than 400 fields have been developed along the coast-parallel fairway (Figure 9.36). They produce in a series of stratigraphic and reactivated growth fault-related traps in the Burgos basin and along the south Texas coastal plain (Hackley and Ewing 2010). Reservoirs include shoreface sand bodies of both Falcon delta and adjacent shore zones. In the Houston salt basin and southern Louisiana coastal plain, delta front and distributary facies of the Liberty and Cockfield delta systems combine with traps associated with deep salt stocks to form a significant play. The narrow zone of extensional growth faults created by margin slides hosts the most recent of the gas plays, primarily along the shelf–slope transition at the distal front of the Liberty and Cockfield deltas. With the aid of amplitude versus offset (AVO) and seismic attribute

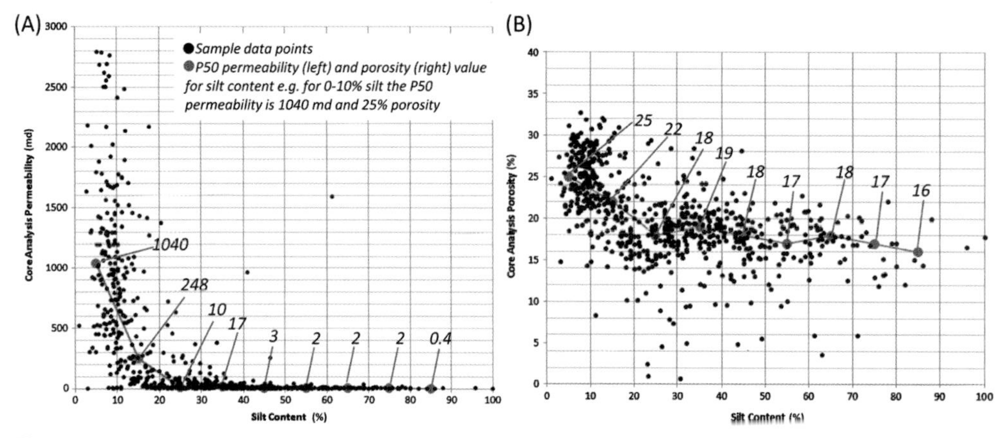

Figure 9.34 Effect of increasing silt content on measured reservoir permeability (A) and porosity (B). From Marchand *et al.* (2015).

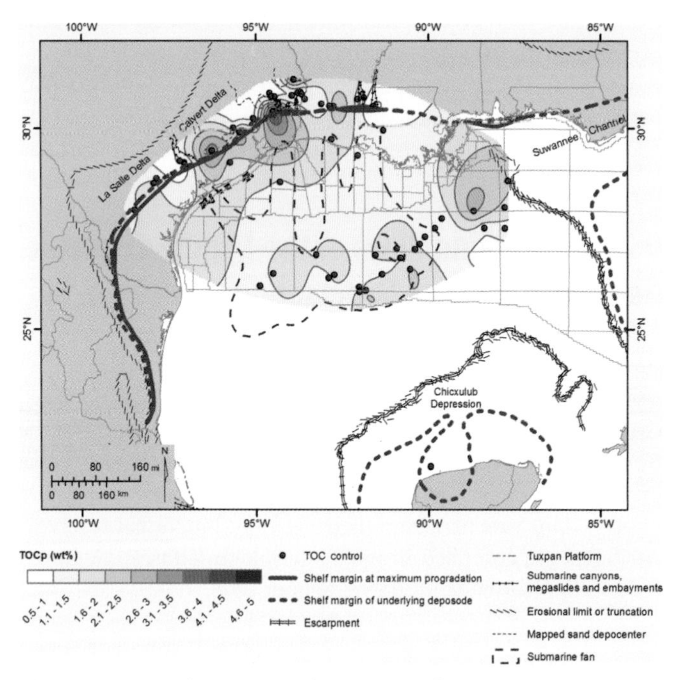

Figure 9.35 Map showing interval average TOC for Middle Wilcox mudrocks of the northern Gulf. Highest values are associated with deltaic depocenters along the Texas margin. From Cunningham *et al.* (unpublished project report).

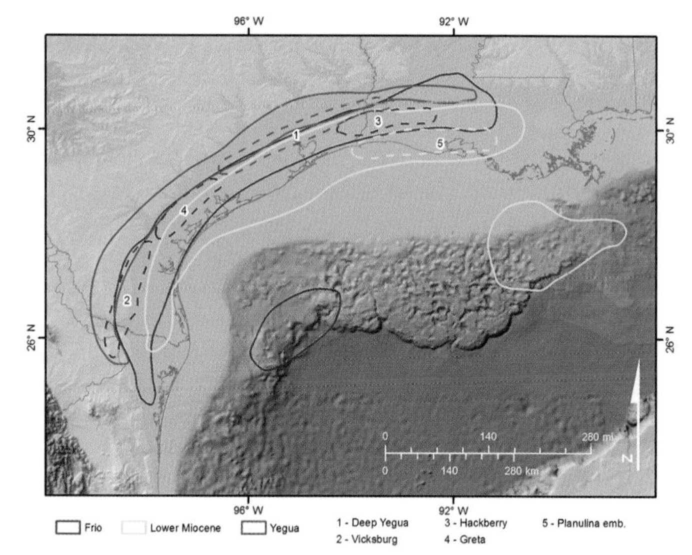

Figure 9.36 Principal hydrocarbon fairways within Middle Cenozoic Geothermal Phase supersequences.

technologies, nearly 3.5 TCF of gas has been found in the expanded Yegua interval of this play (Ewing 2007).

9.22.2 Frio Supersequence

The mature plays of the northwestern GoM Frio continental shelf and extensional continental margin have been the subject of detailed analyses, both because of the volume of hydrocarbons produced and the massive subsurface database created by decades of exploratory drilling and field development (Galloway *et al.* 1982a, 1986; Hernandez-Mendoza 2000). The fairway forms a broad, strike-parallel belt extending from the Burgos basin, across the Texas coastal plain, and into eastern Louisiana (Figure 9.36). Together, Frio plays of this fairway have produced more than 10 TCF of gas and one billion barrels of natural gas liquids. The deep Frio (>10,000 ft) has also been a major focus of research and testing of geopressured/geothermal and entrained methane potential (John *et al.* 1998).

Like earlier supersequences, the Frio can be broadly divided into stable shelf, expanded extensional margin, and continental slope fairways (Swanson *et al.* 2013). Deltaic distributary channel fill, delta front, mouth bar, and coastal barrier facies provide reservoirs in the Norma, Norias, Houston, and Louisiana delta systems. The Greta barrier/strandplain has been especially prolific and contains two of the giant GoM Cenozoic fields: Tom O'Conner and West Ranch. Like the Wilcox and Yegua before it, traps are generally associated with reactivated older faults and folds, extensional faults with associated dip reversal and rollover (Figure 9.29B), and, in the east Texas–Louisiana coastal plain, deep salt structures (Figure 9.29C). Stratigraphic traps created by updip facies change of sandy barrier and beach ridge facies into lagoonal and coastal plain mudstone are an important element in updip Greta barrier/strandplain plays.

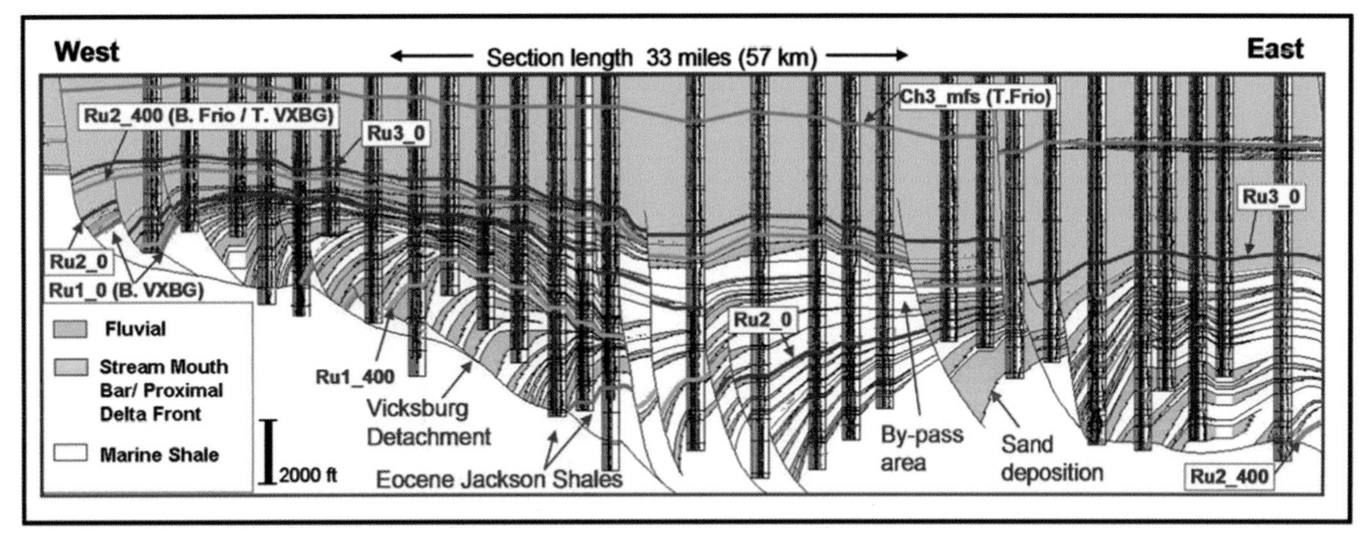

Figure 9.37 Stratigraphic cross-section of the Vicksburg (basal Frio supersequence) and lower Frio section of south Texas. The listric, low-angle Vicksburg detachment creates dramatic basinward displacement, thickening, repetition, and rollover anticline structure of lower Oligocene delta front sands. Potential for distal delta front bypass across the elevated toe of the fault compartment may supply sand to fault-bounded interslope basins. Similar interval and facies expansion of the superjacent Frio delta front section occurs on the east (downdip) end of the section at the first of the progression of deeply rooted Frio faults. Section datum approximates the maximum flooding surface within the Anahuac Shale capping the Frio supersequence. From Feragen *et al.* (2007).

Three plays within the Frio fairway are created by unique subregional structural or stratigraphic elements: the Vicksburg fault zone, Hackberry Embayment, and Greta sand (Figure 9.36). (1) Vicksburg fields are created by rollover and fault seals aligned along the highly listric Vicksburg fault (Figures 9.29D and 9.37). Reservoirs are dominantly expanded, stacked delta front and upper slope frontal splays (Combes 1993; Langford and Combes 1994). (2) The middle Frio Hackberry retrogradational apron contains multiple fields in turbidite channel and lobe sands (Ewing and Reed 1984; Cossey and Jacobs 1992). Traps include faulted anticlines and stratigraphic pinch-outs. The Texas portion of the play has produced more than 2 TCF of gas. Downdip limit to production is largely controlled by economic rather than geological constraints. (3) The Greta sand, an informal name given to thick, amalgamated, back-stepping barrier bar facies in the upper Frio shore zone, is a very different play (Figure 6.6). Traps are gentle structures created by reactivation of older Paleogene faults and regional pinch-out. Reservoir facies are dominantly barrier and beach ridge sands, and their related shoreface, inlet fill, and back-barrier sand-sheets (Tyler and Ambrose 1984; Galloway and Cheng 1985). The mature play has yielded more than four billion barrels of oil.

Like the Wilcox, deepwater drilling along the Perdido fold belt revealed the presence of Frio basinal slope apron sands containing hydrocarbons (Eikrem *et al.* 2010). Reservoir facies at Great White and Silvertip Fields include unconsolidated, incised submarine channel fill, depositional channel fill, and sheet lobe sand bodies. High porosity and permeability characterizes these shallowly buried reservoirs. The presence of volcanic ash and zeolite cement typifies the volcanogenic Frio sands. However, unlike the basinal Wilcox, the basinal Frio interval is comparatively thin and grades eastward into a calcareous, muddy, sediment-starved interval. Thus the deepwater play appears to be reservoir constrained, although the Oligocene-age reservoir discovery at Supremus-1 has expanded the play southward into Mexican waters (Vallejo *et al.* 2012; Colmenares and Hustedt 2015).

9.22.3 Lower Miocene Supersequences

The Lower Miocene fairway includes reservoirs in both the Lower Miocene 1 (LM1) and Lower Miocene 2 (LM2) genetic supersequences; LM2 fields generally lay basinward of LM1 fields, reflecting the long-lived Early Miocene history of continental margin progradation. Like the Paleogene fairways, Lower Miocene fields form an arcuate fairway extending along the lower coastal plain from the Rio Grande River to eastern Louisiana (Figure 9.36). Numerous fields also lie beneath the inner continental shelf in state and federal waters. Also like their Paleogene precursors, Lower Miocene fields form updip, relatively shallow, stable shelf and deeper, extensional margin plays. Fault-related structures dominate the south Texas coastal plain. Structures related to salt stocks as well as growth faults form traps in the Houston Embayment and south Louisiana. Houston Embayment fields produce from deltaic and coastal sands of the Calcasieu delta system and adjacent shore zones (Galloway 1989b). Reservoirs of the North Padre delta and broad Oakville shore zone systems are gas-prone, commonly producing in structures created by reactivation of underlying Frio faults. Together the Texas plays have produced more than 4 BBOE.

In south Louisiana, the LM1 Planulina Embayment interval contains both expanded deltaic and slope channel and lobe reservoir facies (Figure 9.36). It is a gas-prone play, discovered in 1945.

Beneath the continental shelf, Lower Miocene slope apron and submarine fan depositional systems have hosted numerous recent discoveries (Weimer *et al.* 2016). Fourteen fields, including Mad Dog, Neptune, Shenzi, and Tahiti, produce from sheet lobes and slope channel–lobe sand bodies. Typical traps include anticlines and three-way closures against salt. Development of the subsalt Tahiti and Mad Dog Fields posed a significant challenge, given illumination issues and unexpected reservoir compartmentalization (Rivas *et al.* 2009; Thacher *et al.* 2013).

9.22.4 Potential Fairway Expansions

The well-documented presence of abundant potential reservoir sand bodies in slope, base-of-slope, and basin plain depositional systems of Upper Eocene–Lower Miocene supersequences, combined with the complex and diverse array of trap configurations has inspired several exploration concepts, and some frontier drilling.

1. The Chevron-operated Lineham Creek well in southwest Louisiana has confirmed the presence of deep, geopressured, gas sands within the Yegua/Cockfield slope apron.
2. Observed bypass of distal delta front and upper slope sand into syndepositional fault-generated intraslope basins of the Vicksburg–Frio Norias delta system suggests further potential of the south Texas Oligocene continental slope apron (Galloway 1986b; Feragen *et al.* 2007; Ambrose *et al.* 2013). However, reservoir quality, extreme depth, and geopressure pose ongoing challenges for the play.
3. The BAHA II and Supremus-1 discoveries demonstrate the presence of both reservoir and charge elements of a potential larger petroleum system in the slope base and abyssal plain facies of the northwestern Gulf Frio. Both fold belt and subsalt plays are present.
4. Large volumes of subsalt Lower Miocene Calcasieu and Mississippi delta-fed slope apron and nascent abyssal plain fan systems remain untested in the central and northeastern GoM.
5. In Mexico, channel and lobe facies of the Veracruz tectonic margin apron are suggested targets (Jennette *et al.* 2003). Similarly, submarine channel fills and adjacent splay and lobe facies of the Veracruz Trough submarine channel system are largely untested targets (CNH 2015a). Winter (2018) used 3D seismic data to map relatively a channelized deepwater fairway from the Veracruz margin northward (Figure 6.12). Six wells drilled by Pemex targeting the Lower Miocene in this fairway have found mixed results, with non-commercial gas discoveries and likely gas development at Lakach. Terminal lobes at the ends of these deepwater channels are relatively small (10–12 km long by 6–22 km wide; Winter 2018) in comparison to coeval fans of the northern GoM (90–225 km long by 130–380 km wide; Snedden *et al.* 2018a).

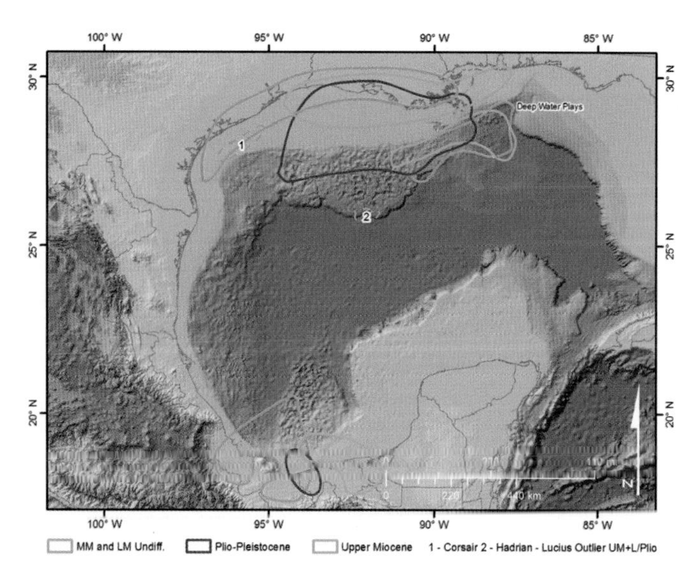

Figure 9.38 Principal hydrocarbon fairways within Neogene Tectono-climatic Phase supersequences. Abbreviation: MM and LM Undiff., Middle Miocene and Lower Miocene Undifferentiated.

9.23 Petroleum Plays of the Neogene Tectono-climatic Phase Supersequences

The areal pattern of hydrocarbon fairways changed dramatically in the Neogene. Just as the locus of fluvial–deltaic sediment supply was increasingly focused onto the central and northeast Gulf margin, so too does production shift to the eastern Gulf continental shelf and continental slope (Figure 9.38). Middle Miocene–Early Pliocene strata of the Veracruz basin and Sureste province have proved to be prolific hosts for numerous fields.

In the northern Gulf, all Neogene supersequences are productive in one or more fairways. As offshore drilling confirmed that slope aprons and fans contain abundant reservoir targets, commonly with excellent reservoir quality, exploration expanded over the shelf edge and into the deepwater of the modern continental slope and basin. The geology of shelf plays was described by Morton and Ayers (1992), Morton and Jirik (1989), and Morton *et al.* (1988, 1991). Weimer *et al.* (2016, 2017) carried the geologic synthesis into the deepwater Gulf.

Fields of the inner-shelf plays display differing geographic patterns. On the Texas shelf, only the Middle Miocene supersequence is productive (Figure 9.38). The broad Louisiana shelf (extending westward into the offshore Texas High Island and Galveston Island protraction areas) hosts plays in Middle Miocene through Pleistocene sequences. In both areas, shallow production extends northward onto the lower coastal plain.

Production of all Neogene supersequences also extends basinward onto the continental slope. Fields of the slope fairways also display two differing geographic patterns (Figure 9.38). The Plio-Pleistocene fairway extends along the breadth of the middle–upper continental slope of the northern Gulf. In contrast, the Middle and Upper Miocene supersequences are dominantly productive beneath the continental slope of the northeast Gulf margin, centered in and around the Green Canyon and Mississippi Canyon protraction areas. Here, the producing fields are grouped by Weimer *et al.* (2017) into the basins, subsalt, and abyssal plain provinces, depending on their relation to the extensive salt canopy (Figure 9.31). Each province produces from multiple genetic supersequences and is further characterized by structural styles and depositional system associations:

1. The basins province contains Middle Miocene–Pleistocene reservoirs. Thick Neogene intervals fill the deep basins inserted into and through allochthonous salt. Multistory lobes (sheet sands) and channels (lenticular sands) deposited in slope aprons provide reservoirs. Traps include three-way closures against salt flanks and faults, and pinch-out and onlap terminations against bounding salt or welds. Augur field, discovered in 1994, pioneered development in the fairway (Dean *et al.* 2002).
2. The subsalt province is dominated by Neogene reservoirs deposited both in lower slope apron and submarine fan depositional systems. Reservoir sand bodies can be quite thick due to vertical amalgamation and size of fan channels and lobes. The rugose morphology of the allochthonous salt top creates a variety of trap configurations as well as providing a highly efficient lateral seal (Figure 9.32). The Mars–Ursa field complex is an early, well-described representative of the play (Meckel *et al.* 2002). Mars field, discovered in 1989, had initial reserves of 800 MBOE.
3. The fold belt and abyssal plain provinces are dominated by Miocene submarine fan reservoirs. Traps are relatively simple compressional anticlines and compactional drapes over deep salt. The BP-operated Atlantis Field (GC 743) is an excellent example (Mander *et al.* 2012). High porosity (26–32 percent) and permeability (500–1500 md) Middle Miocene reservoirs contain hydrocarbons in a salt-cored structural trap below the allochthonous salt canopy (Figure 9.39). The reservoirs exhibit very high net to gross ratio and have little clay or cement. However, the high quartz content and complex multiphase tectonic history have resulted in localized development of deformation bands and sub-seismic faults that have induced reservoir compartmentalization.

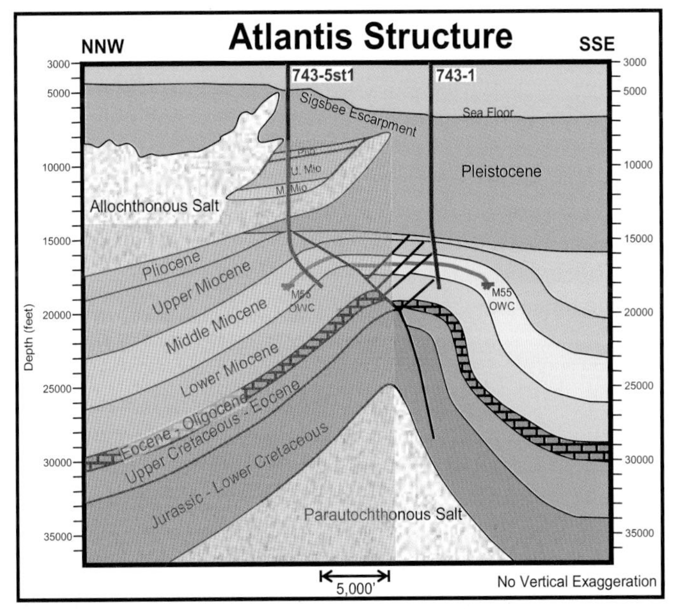

Figure 9.39 Structural cross-section of the Atlantis Field, which lies partially below the allochthonous salt canopy. Middle Miocene reservoirs are lobe deposits of a submarine fan system fronting the slope apron of the Tennessee delta system. Transparent grey area shows the low seismic illumination area below salt that obscures interpretation. From Mander *et al.* (2012).

In all the fairways, reservoirs are typically very fine sand, reflecting their deposition at the distal end of highly evolved sediment dispersal systems. Overpressure is encountered at shallow depths, and has enhanced reservoir porosity and permeability by retarding burial compaction (Taylor *et al.* 2010).

New syntheses, incorporating geochemical wells from numerous wells that sampled the Mesozoic interval, reveal a diverse but orderly array of dominant source rock ages (Figure 9.7). The Paleocene Laramide Phase source for the coastal plain and inner continental shelf petroleum systems give way to Cretaceous, mixed, and, beneath the deep continental slope, Upper Jurassic source rocks. Mixing of sources is also common.

9.23.1 Middle and Upper Miocene Supersequences

The Middle Miocene genetic supersequence is the youngest unit with significant onshore petroleum production in the northwestern Gulf. A shallow gas play on the Texas coastal plain lies in fault and anticlinal structures above petroleum fields in older Miocene and Frio supersequences (Morton *et al.* 1988). The middle Miocene Corsair fault system created a structural fairway along the inner shelf. Of note is the Brazos Ridge play on the central Texas shelf. There, the large, highly listric extensional Corsair fault created an expanded succession of shelf-edge delta front sands of the Corsair delta (Figure 7.2). Rollover and antithetic faults created traps for the gas play.

In the Upper Miocene, the areas of the onshore and shelf plays contract to the central Gulf, reflecting the increasing dominance of sediment supply through the Mississippi and Tennessee fluvial-deltaic axes. Reservoir facies are dominantly deltaic, coastal, and sandy shelf sand bodies. Extensional fault and salt stock-related structures dominate.

As discussed above, Middle–Upper Miocene reservoirs dominate discoveries and production in the deepwater provinces. More than 50 fields have been found in each of the supersequences. Thunder Horse field, discovered in 1999 and put on stream in 2008 remains one of the most prolific (Arnold *et al.* 2010). Reserves exceed one billion barrels of oil. Peak production approached 250,000 barrels per day. Reservoirs include channel and lobe deposits of the McAVLU fan system as well as younger Plio-Pliocene sand

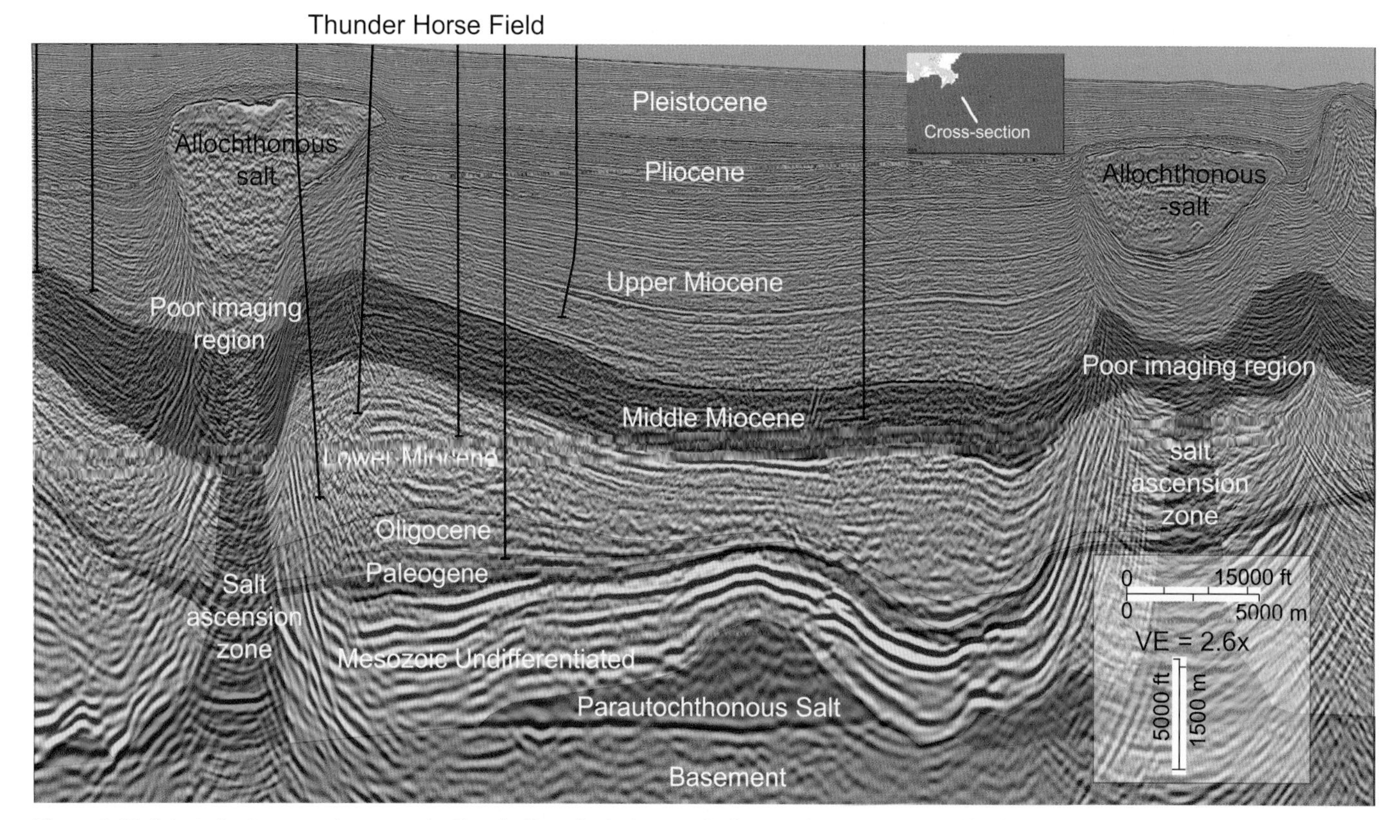

Figure 9.40 Seismic line interpretation across the Thunder Horse basin showing the large turtle structure created by deep salt evacuation (prominent thickening of 66 Ma to 15 Ma supersequences) followed by salt body evacuation into rising salt stocks and structural inversion (beginning at about 11 Ma). The large four-way turtle structure creates the trap for contained Middle Miocene reservoirs of Thunder Horse Field. Seismic line courtesy of ION.

bodies (Cepeda *et al.* 2010). The traps at Thunder Horse include a three-way trap closure against salt/weld and a four-way structural closure created by post-Lower Miocene salt evacuation to form a faulted turtle structure at the Upper and Middle Miocene levels (Figure 9.40). The field is offset from an underlying Mesozoic salt pillow structure, which is yet untested.

Sediments of Middle and Upper Miocene deposodes are important gas reservoirs in the Veracruz, Salina del Istmo, and Macuspana basins of Mexico (Chavez-Valois *et al.* 2009; Sosa Patron *et al.* 2009; Martinez-Medrano *et al.* 2011). The Salina del Istmo alone had produced more than 1.5 billion barrels of oil and 2 TCF of gas from 40 fields by 2009. In the Veracruz basin, turbidite channel and lobe sand bodies were deposited in the western margin slope apron. Sands are litharenites with varying percentages of carbonate, volcanic, and metamorphic rock fragments, depending on their specific upland source (Dutton *et al.* 2002; Martinez-Medrano *et al.* 2009). Source rocks include Tithonian and Cenomanian–Turonian shale and limestone. Biogenic gas was generated in Oligocene and Miocene basinal mudrocks (Prost and Aranda-Garcia 2001). Traps are dominated by compressional structures.

In the Sureste province, rapid Neogene burial of Tithonian source rocks beneath the prograding Grijalva–Usumacinta deltaic axis initiated hydrocarbon generation in the late Oligocene. Lower Miocene deltaic and marine muds provided a shallow source rich in Types II and III kerogen. Traps were formed by Late Neogene structural deformation and salt migration in time to intercept hydrocarbons, which continued to be generated. Migration along normal faults occurred in the Late Neogene.

Miocene petroleum system reservoirs dominate the Macuspana and Comalcalco basins. Reservoir facies are found in the deltaic, shore zone, and slope apron depositional systems. Sands range from subarkose to quartzose arkose. Salt deformation created a variety of structures, much like those seen in the northern Gulf, including minibasins and local canopies. Traps include compressional folds, salt structures, and stratigraphic pinch-outs.

9.23.2 Plio-Pleistocene Supersequences

The Plio-Pleistocene fairway encompasses coastal Louisiana, the entire width and breadth of the central Gulf continental shelf, and the adjacent upper continental slope (Figure 9.38). Weimer *et al.* (2016) identified more than 100 fields and discoveries in Pliocene and Early Pleistocene reservoirs (the majority being Pliocene). Only five reservoirs are of Late Pleistocene age. The fairway can be further separated into two end members. Coastal plain and inner-shelf reservoirs are sandy facies of the combined Red, Mississippi, or Tennessee delta systems (depending on specific age and location).

Traps include an array of anticlines, domes, and fault-related structures typical of the salt canopy province.

On the outer shelf and continental slope, slope apron channel, lobe, channel levee, and slump deposits dominate reservoir sand bodies (Pulham 1993; Box 9.4). Here, shelf-edge delta and slope apron deposition encountered the rugose bathymetry and dynamic deformation that define the minibasin province. Most of the deepwater Plio-Pleistocene fields are associated with minibasin-fills or roho fault systems. In the mid-1980s exploration expansion onto the upper slope established the "Flex" trend, which dominated deepwater exploration for the next decade. Many fields were discovered using the newly developed "bright spot" seismic technology (increased amplitudes linked to hydrocarbons). Most reservoirs were small, gas-charged, and discontinuous, but easily delineated using the bright spot technique. Success at Auger, Mars, and other Pliocene–Pleistocene fields was founded on the use of well-log calibrated seismic facies classifications (e.g., Prather *et al.* 1998; Dean *et al.* 2002).

Box 9.4 Impacts of Large Mass Transport Complexes on Petroleum Systems

The abundance and scale of large mass transport complexes (MTCs) were clearly demonstrated by the earliest seismic traverses of late Cenozoic continental slope sequences of the northern Gulf margin. Shallow seismic data allows regional mapping of youngest slide and debris flow deposits, further demonstrating their importance as an element of Neogene slope and basinal depositional systems (Damuth and Olson 2015). Their dynamic origin, scale, abundance, and variable compositional and petrophysical attributes make them important as potential hydrocarbon seals, migration pathways, and reservoirs, and as shallow geohazards for drilling and completion activities.

MTC deposits range from sand-prone to mud-prone. The idealized mass flow unit includes three domains (Figure 9.41). (1) The headwall domain lies at the upslope end of the MTC. (2) The toe domain comprises the down-flow terminus of MTC deposit. (3) Between is the translational domain, characterized by downslope displacement of sediment (Bull *et al.* 2009). Fluidized sandy debris flows may evolve into turbulent flows with consequent run-out beyond the compressional toe of the MTC. The headwall is characteristically extensional; the toe compressional. Internal and boundary shear dominates the translational domain. Mass transport deposits (MTDs) commonly erode underlying sediments, creating an unconformity with significant relief (Diaz *et al.* 2011). Dimensions are variable and can be quite large, with lengths of muddy MTDs ranging from a few kilometers to hundreds of kilometers (Meckel 2010). Thickness values of 10–400 m are common. Allochthonous salt deformation and consequent sea floor bathymetry played a major role in both initiation of slope failure and morphology of the resultant MTC complex. Anatomy of a typical, large Pleistocene example, the Ursa MTD, has been described using well and seismic data by Gutiérrez (2018). Despite the relatively lithologic uniformity of this MTC, seismic response is quite variable (Figure 9.42), reflecting the complexity of internal deformation, petrophysical properties, and grain fabric created by the dynamics of MTC processes.

Sand-prone Neogene MTC deposits have been interpreted to form significant reservoir facies in several northern GoM fields, including Joliett, Neptune, K2, Shenzi, Mad Dog, Gunnison, and Thunder Horse (Meckel 2010). However, these reservoirs have a checkered production history. While a few have performed as expected, the internal discontinuity, textural mixing, and resultant poor sorting and heterogeneity typical of MTDs have resulted in low recovery in others. Cardona *et al.* (2016) observed

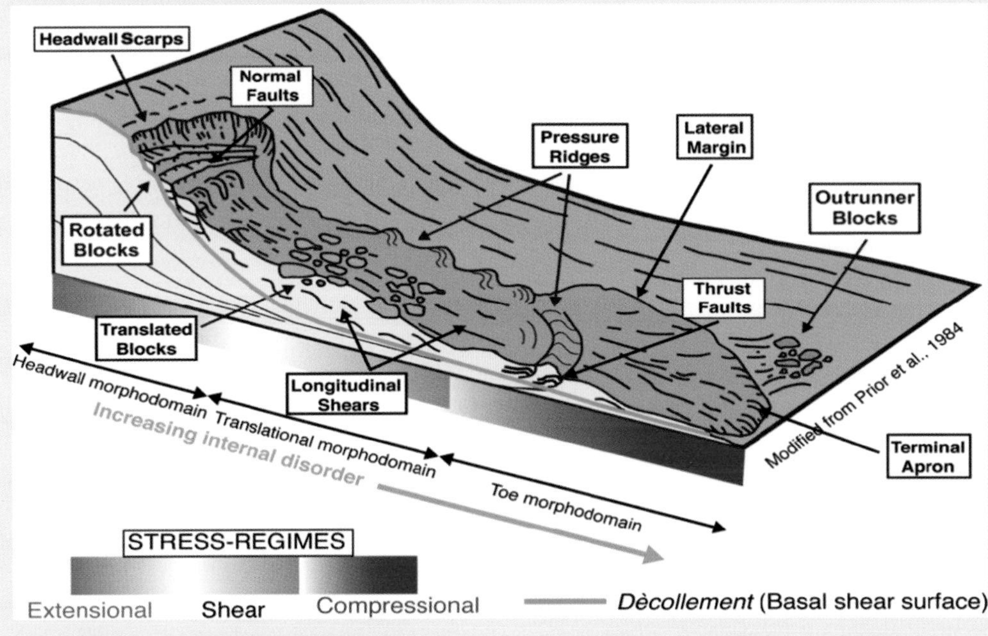

Figure 9.41 Generalized depositional model of a large mass transport deposit. From Cardona *et al.* (2016).

Box 9.4 *(cont.)*

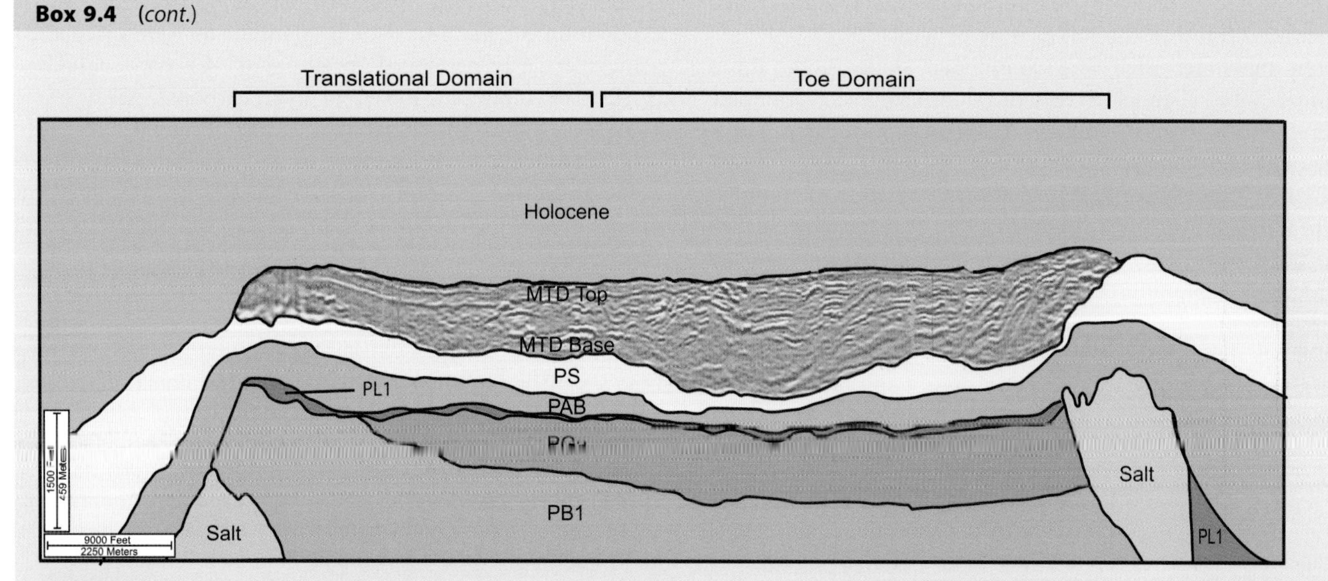

Figure 9.42 Longitudinal seismic profile of the Ursa mass transport deposit. The extensional headwall domain grades basinward into the translational domain, dominated by basinward transport, which grades beneath the modern slope into the compressional domain. Salt deformation triggered Late Pliocene shelf margin failure. From Gutiérrez (2018). Seismic line courtesy of TGS.

mass wasting surfaces and MTC deposits in the sand-rich Neogene shelf margin and slope succession in the Mexican Gulf margin. They propose that resultant base-of-slope MTC deposits are a potential reservoir target and that erosional remnants of sandy shelf margin units may create traps.

The sealing potential of MTDs is high, given their thickness and areal extent. However, extensional and compressional faults and shears in both the headwall and toe domains can compromise seal integrity (Cardona *et al.* 2016). Moscardelli and Wood (2016) suggest that shear deformation during MTD flow can align clay minerals and can enhance seal capacity.

Late Pliocene and Early Pleistocene reservoirs extend the fairway into the Garden Banks protraction area, seaward of the southeast Texas coastal plain. The westward expansion of the fairway, compared to the Middle–Upper Miocene limits, reflects the arrival of the Late Neogene Red River fluvial–deltaic axis (and resultant sandy shelf-edge delta and slope apron depocenter), and the concomitant decline of sediment influx through the Tennessee axis.

Pliocene fields also characterize southern Mexico hydrocarbon plays (Chavez-Valois *et al.* 2009; Sosa Patron *et al.* 2009; Figure 9.38). In the Macuspana basin, deltaic and coastal Pliocene sands are the principal reservoirs. Reservoirs in the Comalcalco basin include Pliocene through Pleistocene fluvial, deltaic, shore zone, and slope apron sand facies. Vertical migration along Late Neogene normal faults charged the shallow, young reservoirs.

9.24 Implications for Cenozoic Exploration

Although the Cenozoic succession in the GoM basin has been an exploration target for more than a century, and the number of exploratory wells drilled is immense, the sheer volume of proven and potentially productive reservoir sandstone in multiple depocenters several kilometers thick results in the reality that large sedimentary volumes remain sparsely drilled or entirely untested. Furthermore, much of that sediment volume lies below deep waters of the continental slope and basin floor and has experienced widespread salt, gravity, or convergent plate margin tectonic deformation. Together, the tectonic history created diverse, superimposed, and commonly complex arrays of structures. The recent flurry of large, initially unexpected discoveries of oil fields in ultra-deepwater Paleogene reservoirs demonstrated yet again that the multiple petroleum systems of the Gulf remain viable twenty-first century exploration targets.

9.24.1 Northern Gulf

Principal frontiers in the northern GoM lie either at extreme drilling depths beneath the coastal plain and continental shelf, or in the ultra-deepwater beneath the modern continental slope and basin floor.

Deep drilling in deltaic headlands of the major progradational deposodes (Wilcox, Yegua, Frio–Vicksburg, Lower Miocene) has documented the presence of sandy slope channel and ponded channel–lobe sands within delta-fed aprons. Potential gas plays in these continental slope and basin floor sand bodies have been suggested by several analyses of undiscovered resources (e.g., Hackley and Ewing 2010; Warwick 2017). The Davy Jones I well encountered gas in expanded delta front and upper slope apron sands beneath the southwest Louisiana coast. The Lineham Creek well in southwest Louisiana similarly confirmed the presence of gas reservoirs in the Yegua/Cockfield slope apron. Downdip limits of production in the Hackberry (Frio) and Planulina (Lower Miocene) retrogradational aprons are constrained by economics rather than by loss of reservoir or trap potential.

Development of commercial deep slope apron gas plays faces challenges in their inherent high temperatures, geopressures, and diagenetic constraints on reservoir quality. In addition, the proximal slope is characterized by sand bypass, creating a mud-dominated succession in offlap continental margin successions in the upper slope apron sediment prism (Figure 9.30). Local structural/bathymetric ponding of upper slope apron channel–lobe sands within intraslope basins could create exceptions to this pattern (Prather 2000).

Ultra-deep drilling, both beyond and beneath the regional salt canopy, has established several emergent plays with large discoveries. These include subsalt fairways in Paleocene–Pliocene supersequences and subtle structural and stratigraphic closures beyond the distal canopy margin. Although some of the plays have numerous well penetrations, much subsalt sediment volume, including lower slope apron and abyssal fan systems, remains untested because of depth and technological challenges to seismic imaging of the complex structure that lies beneath the salt canopy. Discoveries of both Lower Miocene and Pliocene submarine channel and fan systems reservoirs in the Hadrian and Lucius Fields (2009, Keathley Canyon area) created a producing outlier far to the west of the Miocene fairways and seaward of the Plio-Pleistocene fairway. This suggests that Neogene reservoir potential extends across the Keathley Canyon and Walker Ridge areas. Reservoir targets include the western extension of the WRLU fan system and smaller fans and sandy slope apron lobes fronting the fluvial–deltaic depocenter of the Miocene supersequences.

The Paleogene Wilcox and Frio fairways extend southward beneath the Mexican continental slope (Figures 9.33 and 9.36). The Supremus and Trion wells established potentially commercial fields in both Wilcox and Frio reservoirs. Regional mapping places Wilcox reservoirs within the ALKEWA fan system (Figure 5.9). Frio reservoirs are interpreted to be stacked turbidite channel–lobe sands of the Norias–Norma delta-fed slope apron (Figure 6.3). Expansion of these reservoir systems into the Mexican Ridges fold belt would significantly expand the fairway.

9.24.2 Western and Southwestern Gulf

Potential plays of the Mexican Gulf are much less constrained by offshore exploratory drilling than those of the USA. Like the better-known onshore basins, their geologic history and framework are also very different. As reviewed in the previous chapters, the basins of the western margin are tectonically defined by crustal compression and strike–slip. Sediment source areas were local and adjacent to the basin margin. Coastal plains were narrow or nonexistent; sediment crossed the shoreline directly onto and down the continental slope. With the exception of the Neogene Sureste and Campeche salt basins, northern Gulf geologic models have limited applicability.

South of the offshore Burgos basin, the narrow, strike-aligned Neogene sedimentary prism stretches the length of the modern Laguna Madre–Tuxpan shelf. Ambrose *et al.* (2005) analyzed the exploration potential of the trend. CNH (2015b) extended the analysis into the deepwater Mexican Ridges fold belt. Primary prospectivity lies in sandstones of the Middle Miocene–Pliocene slope aprons. Potential traps include the array of structures produced by listric extensional faults and their compensatory compressional structures at the slope toe, the Mexican Ridges fold belt. Underlying Tithonian source rocks provide both oil and gas. However, most oil generation likely pre-dated formation of the geologically young structural traps.

Frontiers in the Veracruz continental slope and adjacent basin floor include the succession of older tectonic margin aprons and Neogene progradational slope aprons. In the deep basin, the north-trending submarine channel system that followed a bathymetric trough between the western tectonic continental margin and the eastern Campeche salt structures provides a potentially broad, sand-rich reservoir fairway that extends hundreds of kilometers along the Veracruz Trough.

In the offshore Salina del Istmo and Campeche salt basins, a variety of potential plays, expanding upon the shelf and onshore production, have been suggested (CNH 2015b). The Miocene delta-fed slope aprons and eastern flank of the Veracruz Trough submarine channel complex provide reservoir facies. Complex salt structures, including canopy complexes and minibasins, provide abundant potential traps. The Zama discovery in 2017 supports expansion of this Neogene fairway northward.

9.25 Seismic Technology Evolution in the GoM

As GoM exploration transitioned from identification of prospects above the allochthonous salt canopy to those targets below the salt, the seismic industry evolved its technology to meet the new challenges. Salt has special properties that make imaging more difficult than for sedimentary intervals. It is more generally ductile than siliciclastics or carbonates, which can result in unusual geometries at the salt–sediment interface (steep dips, closely spaced fault networks, variable stresses and pore pressures in adjacent rocks; Jackson and Hudec 2017). Major challenges faced the seismic companies tasked with imaging the subsalt domain: (1) out-of-plane energy around complex salt bodies; (2) areas of poor illumination; (3) steep dips, especially on salt flanks and megaflaps; and (4) variable velocities within salt bodies due to impurities (Herron 2011, 2014).

The first step was the shift from 2D to 3D seismic data as a primary exploration tool. Beyond noise reduction and improved resolution, 3D seismic data, properly migrated, allows complex 3D geometries to be correctly positioned and out-of-plane energy (sideswipe) distortions to be greatly reduced (Brown 2011).

Strata below the salt canopy also were poorly imaged due to poor illumination: Ray paths on older, short streamer lines were largely near-vertical, resulting in shadow zones below salt (Figure 9.43A). Subsalt areas well away from the Sigsbee margin of salt were particularly troublesome to image. In a typical 3D marine seismic survey the vessel traverses the surface in a predetermined direction above the subsurface target. Because most of the recorded seismic signals travel nearly parallel to the sail line, at a small azimuth, the survey is called a narrow azimuth or NAZ survey. Azimuth is the angle at the source location between the sail line and the direction to a given receiver. The target essentially is illuminated from one direction in NAZ surveys. Reprocessing of the original seismic data provides some improvement in reflection continuity but shadow zones often persist.

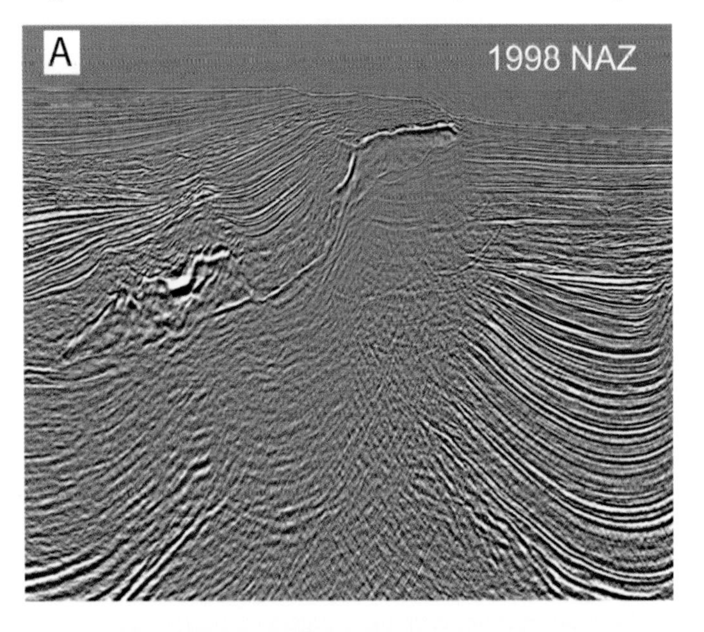

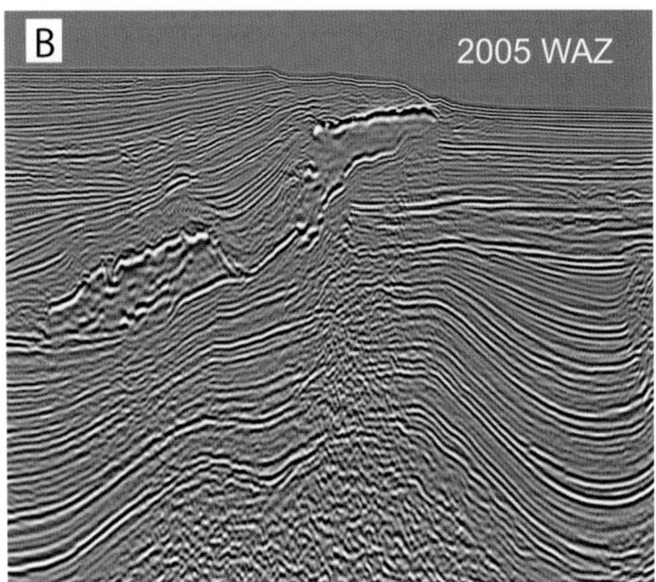

Figure 9.43 Imaging Improvement at Mad Dog Field, GoM. (**A**) Narrow azimuth; (**B**) wide azimuth. From Smith (2013). Courtesy of BP and Mad Dog partners and GEO ExPro.

This was addressed in a major change in seismic acquisition, by shooting wider azimuth surveys using multiple source and receiver boat configurations. Wide azimuth (WAZ) surveys are designed to widen the azimuth distribution over the target in one preferred single direction (Figure 9.43B). Different designs are available, but they require at least two source vessels in addition to the streamer vessel. Each source line is shot multiple times with increasing lateral offset. To improve acquisition efficiency, some WAZ surveys are acquired with multiple streamer vessels. The WAZ data provides a significant uplift in illumination, but also has benefits in suppression of multiples, especially from salt suture diffractions (Figure 9.43B). In the most difficult areas, such as the Alaminos Canyon protraction blocks, full-azimuth seismic data is acquired using a distinctive pattern of overlapping circles, thereby improving fold but also reducing the downtime associated with vessel turns in a rectangular grid (Amundsen and Landro 2008).

The problem of steep dips adjacent to and variable velocities within salt are best addressed with pre-stack depth migration. Older time-migrated seismic data is often marred by notable time pull-ups or pull-downs due to lateral velocity variations. The problem is particularly acute around salt bodies due to impurities such as sedimentary rock incorporated into salt during upward and lateral salt movement. The so-called "dirty salt" plagued early depth-migrated sections, particularly using older and cheaper Kirchoff migration algorithms. Steeply dipping salt flank strata were also poorly imaged. This resulted in poor well positioning and exploration failures. The seismic industry responded with new migration algorithms including reverse time migration and Gaussian beam migration (Herron 2014). These newer, but also more expensive, processing flows better handled multipathing (intersection of seismic wavefields) due to complex salt geometries (see detailed discussion by Chaikin in Jackson and Hudec 2017).

Processing workflows have become more efficient and faster as computing power has increased. As Jackson and Hudec (2017) superbly illustrate, however, pre-stack depth migration around salt is still a highly interpretative procedure of picking proper velocity models for salt and sediment and iterating these velocity models until the appropriate top and base of salt surfaces are well defined and stratal interfaces and faults are enhanced. In more complex areas, this process can

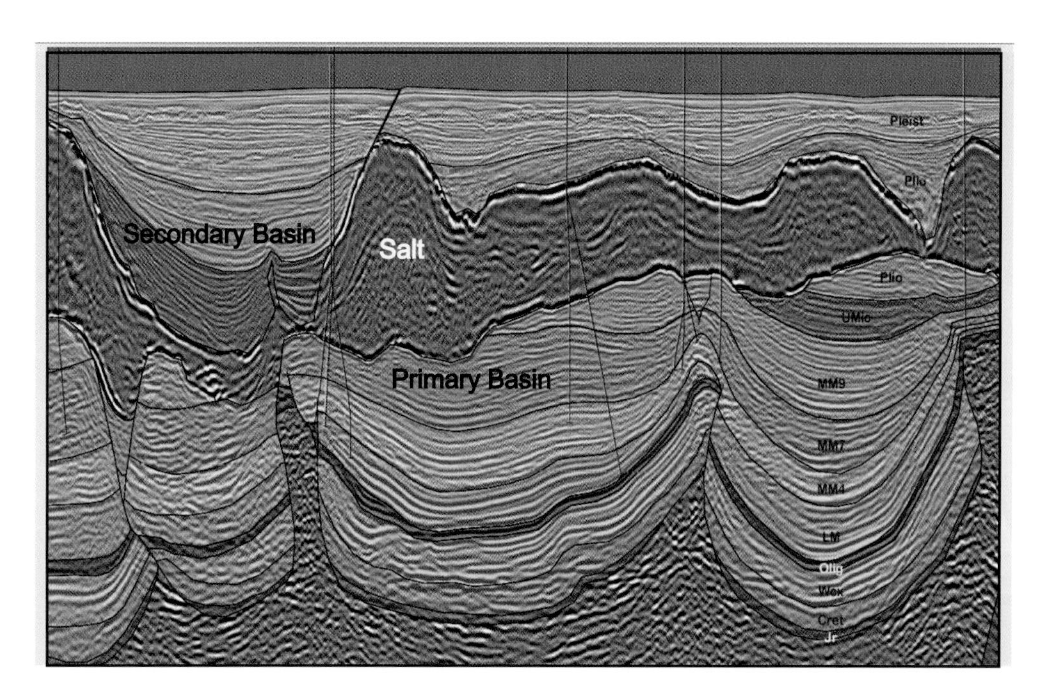

Figure 9.44 Example of imaging improvement with wide azimuth seismic data below the salt canopy in the northern GoM (from Leyendecker *et al.* 2014).

take up to two or more years from acquisition to final product delivery. Reprocessing is common, particularly after wells are drilled and new velocity information is obtained. For more detailed information on processing workflows for better imaging of salt structures, see Jackson and Hudec (2017).

The success of these techniques greatly improved exploration success ratios in many subsalt plays, including the Paleogene Wilcox (Meyer *et al.* 2007), as the risk of poorly imaged traps has declined. An example of the superior trap imaging is evident in a WAZ seismic line described in Leyendecker (2014; Figure 9.44). Subsalt primary basins are well-imaged, with excellent reflection amplitude and continuity even below more complex salt bodies. Base of salt, the most difficult to resolve, is nicely imaged, with the subsalt cutoffs being an important trap style below the salt canopy in the northern GoM (Figure 9.32E).

Glossary

Italic type indicates a cross reference to another glossary entry. Defined terms are highlighted in bold text in the chapter where the term is introduced. Terms defined in the text or figures are not included here.

3D seismic. Seismic data collected with the goal of providing three-dimensional images of an area, trap, or reservoir by acquiring a grid of closely spaced 2D seismic lines with sufficiently small bin sizes to properly migrate reflections to the correct position in 3D space. Wide azimuth 3D seismic acquisition, often used for imaging subsalt structures, is carried out by deploying multiple seismic vessels to record reflections out to the side of the recording spread. This improves signal-to-noise ratio of subsalt imaging, reservoir illumination, and velocity models needed for depth-imaging.

Abyssal plain fan. Point-sourced, basinal depositional systems that extend from the lower slope onto and across the deep basin floor. They are characterized by their lobate geometry and well-defined depocenter.

Allochthonous salt. Salt that, as a result of loading and deformation, has migrated upward (and often seaward) to reside at a higher stratigraphic level. Salt canopies, even if attached to the autochthonous salt source, are allochthonous salt. The sedimentary interval below allochthonous salt is termed "subsalt" strata. Traps below the salt canopy can be formed where the base of the salt truncates reservoir-bearing strata, at the subsalt cutoff.

Anastomosing channel. A morphologic type of subaerial (river, delta distributary) or submarine channel system consisting of multiple, coexisting, interweaving channels separated by overbank deposits.

Autochthonous salt. Salt that presently resides near its original depositional position, stratigraphically above "pre-salt" sedimentary rocks or basement. Autochthonous salt that has been structurally modified or mildly deformed, often with base of salt near-horizontal, has been referred to as *parautochthonous* salt (Jackson and Hudec 2017). Autochthonous salt can also overlap oceanic or continental crust.

Basement. A general, often informal, term that refers to crust below sedimentary rocks of exploration interest. In the GoM, basement can be crystalline rocks (igneous or metamorphic rocks) or in some areas metasedimentary rocks of Paleozoic age. While there is no current production from basement rocks in the basin, fractured basement rocks are reservoirs for oil and gas in other parts of the world. Basement can be continental, transitional, or oceanic crust and is separated from overlying strata by an unconformity.

Bedload river. A river characterized by substantial sand and gravel sediment load. Bedload rivers are characterized by low to moderately sinuosity channels, braiding, and tabular, sandy channel fill deposits.

Biostratigraphy. Stratigraphic analysis based on study of the biologic components of sedimentary successions. In industry practice, biostratigraphy is largely based on microfossils, either microfauna (e.g., foraminifers) or microflora (e.g., ancient spores and pollen) as these can be recovered from well cuttings during drilling. Calcareous nannofossils have become particularly important in Neogene deepwater exploration because of their occurrence in marine sediments and often-short age ranges that are of great utility in dating sequences and supersequences.

Bouma sequence. The ideal vertical succession of sedimentary structures that are thought to represent the product of deposition of low-density turbidity flows, a type of sediment gravity flow common in deepwater environments, but also found in lakes and other settings. Flume experiments indicate decelerating flows produce (in ascending order), a structureless to fining-upward bed (Bouma a), planar horizontal laminated bed (b), ripple-laminated bed (c), planar horizontal siltstone (d), and ungraded mudstone (e). Arnold Bouma first described the succession, though often incomplete, from European outcrops, but these have been recognized in exposures and cores around the world. High-density turbidity flow deposits, particularly those with a high clast content, are better described by the Lowe classification scheme.

Breccia. A type of sedimentary deposit dominated by rock clasts exceeding 2 mm in size. Angular to subangular, sometimes randomly oriented, these are not transported long distances as traction load by currents but more likely as debris flows or mass flows where turbulence is dampened out and rounding of these large clasts is hampered. Breccia deposits associated with the Cretaceous–Paleogene boundary impact event are dominated by limestone rock clasts due to the predominance of GoM carbonate paleo-environments at the Chicxulub impact site and around the entire Gulf basin, where margin failure is known to have occurred with the large seismic wave originating at the point of impact (Sanford *et al.* 2016; *see* Section 4.11.3).

Bucket weld. A special type of salt weld (see *salt weld*) where primary and secondary minibasins are joined together following suprasalt minibasin loading. The sediment loading within the secondary basin forces salt evacuation and juxtaposition of basins of different ages along a near-vertical weld.

Caprinid rudistids. Rudists are a type of ancient invertebrate pelecypod (bivalve) that lived from the Late Jurassic to the end of the Mesozoic and were a major component of the Cretaceous platform margin reefs that extended from the northern to southern GoM, as well as large portions of the Atlantic margin and Tethys Ocean. Caprinid rudists, a special family of rudist bivalves, were widespread, adapting well to the higher seawater temperatures of the Cretaceous period. High porosity associated with rudist reefs, particularly in back-reef and patch reef paleo-environments, makes these prolific hydrocarbon reservoirs.

Chlorite. Iron-rich alumno-silicate clay mineral that can form a distinctive rim of blades that can coat quartz and other detrital mineral surfaces in a sandstone reservoir. Development of chlorite rim cements may extract silica from pore waters that would have otherwise ended up in pore-occluding quartz cement. Known to be important in preserving porosity in deeply buried sandstones in the northern GoM.

Chronostratigraphy. Time-rock units. Age information, typically from biostratigraphy or absolute dating (or both) is integrated with identification of sequences from seismic data, well logs, outcrops, or cores to define the stratigraphy of an area or well. Seismic reflections usually follow chronostratigraphic surfaces.

Clinoform. Large-scale sedimentary strata formed in a sigmoidal geometry, usually reflecting progradation of a depositional system or an entire continental margin. Originally defined to include the fondoform (top strata), clinoform (inclined strata), and undaform (basal strata), many use the term clinoform to refer to the entire stratal package.

Continental crust. Sedimentary and crystalline strata. In the northern GoM, continental crust has an average thickness of 35 km and lies at depths in the range of 10–16 km (6–10 miles).

Continental margin. Shallow to deepwater areas adjacent to continental landmasses. Collectively includes the continental shelf, slope, and rise, but excludes the abyssal plain. The shelf margin or platform margin separates the continental slope and shelf. In the unstable and commonly structurally active Cenozoic sequences it is more commonly a narrow transition zone that includes the outermost shelf, shelf edge proper, and uppermost slope.

Conventional reservoir. Siliciclastic or carbonate reservoirs where hydrocarbons have migrated into the pore spaces from an external source rock. Entrapment of hydrocarbons requires adequate seal capacity, a trap with preserved structural integrity, a clear pathway from source rock to reservoir, and trap development that precedes hydrocarbon migration. Reservoir permeability must be sufficient to allow some flow to the borehole without substantial stimulation (exception: tight gas sandstones) and a threshold level of porosity to provide an economic volume of oil or gas.

Delta system. A delta system forms where a river transporting significant sediment load flows into a marine or lacustrine receiving basin. Deltas are characterized by distributary channel patterns, the unique environment of the channel mouth bar, and peripheral reworking by basin processes, including waves and tides. Depending on the relative balance between sediment deposition on the delta plain and distributary mouth and marine reworking, deltas may be fluvial- (river), wave-, or tide-dominated. As focused locations of sediment supply, deltas typically create long-term depocenters on the depositional basin margin.

Deltaic headland. A seaward-convex coastline created by the progradation of a delta system. Like all headlands, a deltaic headland refracts and focuses wave energy flux, accentuating wave reworking and longshore sediment transport toward the delta flanks.

Depocenter. The locus of thickest sediment accumulation and preservation within a genetic stratigraphic unit. It is recognized and delineated by an isopach or isochore map.

Depositional architecture. The arrangement, both geographically and vertically, of facies and stratal units that compose a genetic unit. Progradational (offlapping), aggradational (vertically stacking), and landward retreating (back-stepping or retrogradational) architectures are commonly recognized at the sequence and supersequence scales. At the facies scale, erosional, abrupt, gradational, and interbedded boundaries record sedimentary processes and geomorphic patterns. Multilateral and multistory facies associations reflect dynamics of the depositional system.

Depositional episode. A depositional episode (sometimes abbreviated as deposode) is the time-stratigraphic name of a geologic interval bounded by maximum regional transgressive flooding of the basin margin. The term was introduced by Frazier (1974) to describe the depositional history of late Cenozoic strata of the northern GoM. Strata of a depositional episode were recognized by bounding transgressive facies and, basinward, sediment starvation surfaces created as the coastline and locus of sediment accumulation retreated landward. An idealized depositional episode produces a vertical succession of progradational, aggradational, and retrogradational/transgressive strata.

Depositional system. A depositional system (sometimes abbreviated as deposystem) is a three-dimensional body of sediment deposited by a contiguous suite of process-related sediment environments. An array of depositional systems forms the physical geography of a basin margin at any point in time.

Detachment surface. A surface where listric faults curve into a near-horizontal orientation, usually along a contact with ductile shale or salt. Also known as a *decollement.*

Diagenesis. Changes in a reservoir that occur after deposition, including burial-related compaction, cementation, development of secondary porosity, and other processes.

Diapir. An intrusive body, usually mobile salt or shale, that is forced into overlying strata due to tectonics, sedimentary loading, or pronounced density differences that leads to buoyant upward movement into weak overlying intervals. Diapirs have a variety of shapes and sizes. Traps above and on the flanks of salt diapirs were among the first successful oil fields in the GoM, an example being the 1901 Spindletop oil discovery near Beaumont, Texas.

Erg. Eolian (aeolian) arid sand sea. Large dryland systems that develop in areas with access to an abundant supply of loose, transportable silt and sand and sufficient wind to build large dunes.

Eustatic sea-level changes. Global changes in sea level due to changes in the volume of water (e.g., higher or lower sea level due to climatic variations, ice volume increases related to global warming) or volume of the ocean basin container (related to oceanic ridge creation, for example). This differs from local variations in sea level (relative sea-level changes) due to sediment input and other autocyclic factors.

Evaporites. Sedimentary minerals including halite, anhydrite, gypsum, and other deposits formed by concentration and precipitation by evaporation of seawater or other saline fluids. Evaporites can form in oceans, lagoons, and in standing bodies of water such as lakes.

Expansion zone. An area where strata thicken substantially, often associated with a major growth fault and higher sedimentation rates in and around deltaic point sources.

Expulsion rollover. A large salt tectonic structure formed through expulsion of salt by a large prograding stratal succession. Stratal surfaces that intersect salt diapir can collapse downward on a weld (see *salt weld*), following salt evacuation. Geometrically similar in some orientations to a depositional clinoform, these are invariably orders of magnitude larger than even a continental margin clinoform.

Fault family. An assemblage of temporally related and spatially contiguous extensional faults formed along the depositionally active continental margin during a depositional episode.

Foreland trough. A tectonic depression created by tectonic loading of the crust along a compressional structural boundary. More simply, these are sedimentary basins lying at the front of a mountain chain and the adjacent craton.

Foundered margin; foundered shelf. Foundering is the process and history of progressive submergence, often over a geologically significant interval of time. Both continental margins and shelves may be subject of foundering during periods of declining sediment supply, increased subsidence rate, or eustatic sea-level rise. Accompanying transgressive retreat of the shoreline commonly creates positive feedback as offshore sediment supply is further reduced.

Genetic stratigraphic sequence. The stratigraphic unit created by a depositional episode. Genetic sequences are bounded by surfaces of maximum transgression (the maximum flooding surface) and sediment starvation (condensed intervals). The generalized genetic sequence contains a lower interval of progradational facies sequences, a middle interval of aggradational facies sequences, and a cap of retrogradational and transgressive facies sequences. The paradigm recognizes the coequal importance of tectonism, sediment supply, and eustasy, as controls of sequence development.

Geopressure. Extreme overpressure has been called geopressure, especially in the Gulf Coast. The potential of geopressured formation water production as an alternative energy source was the objective of considerable research there. See also *overpressuring*.

Geothermal or crustal heating phase. The geothermal gradient is the combination of convective and advective heat flow in the crust. Advective heat flow results from mass movement of hot fluids (water or magma) or rock from depth. Deep crustal heating and consequent upwelling of magma heats the shallow crust, and is manifested in both uplift of warm crust and extrusive volcanism.

Grain volume. The total volume of sediment particulate solids. In sedimentary strata it is calculated by compacting the sediment to zero porosity.

Gravity mass transport. The array of processes that remobilize and transport sediment downslope. In submarine settings, it includes gravity sliding, slumping, debris/mud/sand flow, and turbidity current flow.

Gravity tectonics. The array of deformational processes driven by release of gravitational potential rather than by lateral stress or crustal deformation. The term includes both halokinetic salt deformation and gravity spreading. Gravity tectonism is particularly prominent along rapidly prograding clastic continental margins.

Greenhouse world. The periods of Earth history lacking extensive continental ice sheets. Greenhouse conditions encompassed GoM history from its origin until the end of the Eocene.

Growth faults. A type of normal fault, usually listric (see *listric fault*), where the hanging wall strata show significant thickening relative to the stratigraphically equivalent footwall interval. Common in high-accommodation siliciclastic continental margins.

Icehouse world. The periods of Earth history during which extensive continental ice sheets were well developed. The latest icehouse interval began with the onset of the Oligocene, and initiation of the Frio–Vicksburg deposode.

Interdeltaic bight. A seaward-concave coastline created lying between headlands. Bights typically lie between major deltas, and are dominated by deposition in shore zone and shelf depositional systems. Much sediment may be imported by longshore transport. Tidal range is typically amplified within bights.

Last appearance datum (LAD). The extinction point of a microfossil that indicates an important marker horizon or stratigraphic "top." It is the first downhole occurrence in well cuttings and thus less prone to downhole caving than the first appearance datums (FADs). LADs provide reliable age information that supports exploration.

Listric fault. Normal faults where the fault plane curves such that the shallow portion is steeper than the base of the fault. The lower portion of a listric fault approaches a near-horizontal plane at the point of detachment, often on a ductile horizon such as shale or salt. This causes hanging wall strata to rotate and usually expand in thickness relative to the footwall. Common in zones of expansion related to deltaic or continental margin deposition.

Lithostratigraphy. Stratigraphy defined strictly by lithologic rock units such as lithofacies. Because lithofacies often are time-independent and can repeat in a stratigraphic column, these are not useful for correlation, particularly over long distances.

Log motif. The vertical pattern observed on logs, typically the gamma-ray or spontaneous potential log but sometimes on resistivity logs, that gives an indication of the trend in grain size or shale content. This trend often reflects depositional environments – for example, a bell-shaped log motif may indicate a fining-upward trend associated with fluvial channel fills. Used in conjunction with paleogeographic information, log motifs can be useful in mapping depositional systems over an area.

Mantle. The layer within the deep earth between the crust and outer core. In the northern GoM, the depth to the mantle ranges from 34 km (21 miles) to as shallow as 15 km (9 miles). The top of the mantle, also known as the Moho or Mohorovičić discontinuity, is marked by a significant increase in seismic velocity (e.g., from 6 to over 8 km/s) that is notable on seismic refraction data.

Megaslides. Extremely large submarine slides. In the GoM, megaslides create structural-scale faults, and displace stratigraphically significant sediment sheets.

Microbialite. Ancient forms of microbial communities that produced organosedimentary deposits in shallow water by the trapping, binding, and cementation of sedimentary grains by biofilms of benthic microorganisms. Thrombolites are one type of microbialite, typified by irregular clotted structures in comparison to more laminated stromatolites, which can form large carbonate bodies or build-ups. Smackover thrombolite build-ups produce oil and gas in a number of eastern GoM coastal plain fields (Mancini *et al.* 2004, 2006).

Microfossil. Microscopic fossil tests or cells of microfauna and microflora that are utilized by paleontologists to define the age, ecology, and in some cases the specific paleo-environment of the sedimentary interval within which these are found. Because microfossils can be obtained from drill well cuttings, these have become the primary tool for biostratigraphers (see *biostratigraphy*).

Minibasin. Small intrasalt basin largely surrounded by and subsiding into allochthonous (secondary minibasin) or autochthonous (primary minibasin) salt. Upwelling salt surrounds the minibasin as a network of salt walls, massifs, and welds.

Oceanic crust. The crust underlying ocean basins which is distinctive from continental crust in its composition (mafic rocks enriched in iron and magnesium) and density/velocity as revealed by seismic refraction studies. It is generally thinner than continental crust, usually less than 10 km thick, and denser (3.0 versus 2.7 g/cm^3). Oceanic crust formed in the GoM in the Jurassic as the Yucatán (Mayan) block rotated counterclockwise to open the basin. Oceanic crust forms at a spreading center (see *spreading center*). Oceanic crust can be overlapped by autochthonous salt with subsequent gravity gliding. As oceanic crust is denser than continental crust, it subsides more than continental crust, creating, in some areas, a significant paleo-slope. This slope may have contributed to salt rafting in the eastern GoM Norphlet play area.

Out-of-grade slope. A continental slope that is out of depositional grade due to gravity tectonic deformation, or change in sediment supply rate or caliber. Slope declivity may be too high or too low. Regrading is accomplished by erosion, deposition, or both.

Overpressuring. Subsurface reservoir pressures that exceed hydrostatic pressures (0.465 psi/ft). Overpressuring occurs when the weight of the sedimentary column causes the pore fluids to compress and there are insufficient pathways for fluids to escape a reservoir and reduce the subsurface pressures.

Paleocene–Eocene Thermal Maximum (PETM). A geologic peak in global warming and the largest in the Cenozoic. The PETM is thought to have occurred around 55 Ma, in the GoM during the Middle to Upper Wilcox unit transition. The relatively short duration PETM (~200 kyr) is attributed to a massive influx of isotopically light greenhouse carbon into the ocean–atmosphere system. This resulted in a global negative carbon isotope excursion (CIE) in organic and inorganic carbon recorded in marine and terrestrial settings (Cunningham *et al.* in prep.).

Paleogeographic map. An interpretive map displaying the physical geography of the area surveyed. In the context here, the maps display the array of depositional systems that dominated the history of a deposode. In some cases, it is necessary to correct for post-depositional tectonics (e.g., salt rafting) and plate tectonic movement.

Parautochthonous salt. Autochthonous salt that has been somewhat deformed but has not moved significantly from its original depositional location or stratigraphic position.

Petroleum system. All of the elements that are part of a hydrocarbon-bearing system. Reservoir, seal, source rock, and migration and trap elements can individually fail, causing a prospect to not trap oil and gas. A more narrow definition refers to the sedimentary characteristics that permit the generation of hydrocarbons from shale source rocks, leading to migration and charging of traps.

Platform margin. Refers to both a modern-day physiographic feature (e.g., Yucatán Platform margin) or an ancient depositional or structural entity that is the boundary between deep and shallow water. Long-lived reefs often are part of or form an ancient rimmed shelf or platform margin. Some are related to episodic failure of an over-steepened margin (e.g., Florida Platform margin or escarpment).

Ponded minibasin-fill. Sediment fill trapped within the salt-structure-bounded confines of a minibasin. Strata show common uplap onto bounding structures that formed bathymetric highs.

Pre-salt. The sedimentary interval that lies below autochthonous salt. In the case of the Louann Salt, estimated to have formed about 170 Ma, pre-salt strata are thought to range in age from Middle Jurassic to Triassic, though ages are not well-constrained due to poor fossil content or a lack of well penetrations. Pre-salt is different than subsalt, which refers to strata that lies below the allochthonous salt canopy. See *subsalt.*

Prospect. In the oil and gas exploration business, a prospect is a candidate trap for hydrocarbons that has been identified by detailed study of seismic and/or well data. Depending on the quality of the data and interpretations, a prospect carries some degree of risk and uncertainty, which can result in a total lack of oil and gas, entrapment of subeconomic quantities of hydrocarbons, or a successful well that spurs development and eventually a producing field. Prospects in offshore areas are given prospect names (e.g., Norton, Cheyenne) prior to drilling, which may or may not become the field name. Sometimes the prospect name is used as a well name (e.g., in Mexico, Supremus-1).

Raft. A slab of sediment or fault block that has extensionally separated from its original footwall and which lies on a surface of subregional *decollement* (see Section 1.3.8).

Retrogradational slope. Continental slope characterized by collapse, mass wasting, erosion, and/or foundering. Commonly accompanied by shelf edge retreat, deposition of an apron of remobilized sediment on the lower slope, or initiation of submarine canyon cutting.

Rift system. A roughly linear zone, 10–100+ km in scale, where the lithosphere is pulled apart by extensional tectonics usually reflecting basin-scale crustal processes. Rift systems are dominated by grabens, largely half-grabens that can change polarity (dip direction) along structural strike, separating segments of the rift. Adjacent uplifted footwalls can be eroded subaerially or subaqueously to provide transverse sediment input in the form of alluvial or submarine fans. Axial fluvial drainage in subaerial rift systems can also occur. Large-scale rifting can influence depositional patterns over an entire basin, particularly in zones of high accommodation. In the GoM, rifting preceded sea floor spreading and was thought to control sedimentation in the pre-salt interval (e.g., South Georgia rift system), though an alternative hypothesis for the areas to the west in Texas is discussed in Section 2.3.2.

Roho system. A group of basinward-dipping growth faults that sole out onto an allochthonous salt sheet, canopy, or weld (see Section 1.3.4).

Rollover anticline. Anticlinal fold created by bending of strata toward the bounding, updip, listric growth fault. A major trap style in northern GoM petroleum fields.

Sabkha. Salt flats developed in coastal or continental areas dominated by evaporation under arid conditions. Coastal sabkhas in Namibia, for example, separate the large eolian sand sea (erg) from the ocean. The proximity to a seawater source and the arid climate promote deposition of evaporite minerals including gypsum (hydrous form of anhydrite) and carbonate minerals such as aragonite. Ancient sabkhas have been identified in the Jurassic Norphlet and Smackover of the GoM, the Permian of west Texas, and Ordovician of the Williston Basin.

Salina. A landlocked body of water where evaporative concentration leads to precipitation of sodium chloride, anhydrite, and other salts.

Salt evacuation. Highly mobile salt such as the Louann of the GoM is known to have moved or been evacuated from the original position in places by depositional loading. Deflation (thinning) of salt during withdrawal may not entirely evacuate the salt body (Jackson and Hudec 2017).

Salt roller. Listric fault-bounded parautochthonous salt bodies usually occurring in a series of triangular-shaped features on a basinward-dipping surface. These can set up three-way closures such as the Norphlet deepwater play.

Salt weld. A discontinuity surface where strata, originally separated by allochthonous or autochthonous salt, are joined after salt evacuation or other processes removing salt. Joined strata are usually stratigraphically and lithologically different. See numerous examples in Jackson and Hudec (2017). Welds can also serve as a detachment surface for later listric faults. Subsurface pressures sometimes occur across a weld.

Sediment gravity flows. Fluidized subaqueous flows. The subset of gravity mass transport processes that includes debris flows, turbidity flows, and hybrid flows. See *Bouma sequence*; *gravity mass transport.*

Sediment waves. Bedforms found on the modern sea floor in many shelf, slope, and deepwater environments, a product of near-bottom traction processes. Sediment waves in deepwater settings are a particular conundrum due to the limited influences of tides and waves that form shelf ridges and similar moderate water depth bedforms. Sediment waves are clearly bedforms, symmetrical to asymmetrical upslope-facing crests or downslope dipping crests. The latest work suggests that deepwater sediment waves are formed by sediment-laden supercritical sediment gravity flows that form either long-wavelength antidunes or cyclic steps (Fedele *et al.* 2016).

Seismic reflection data. Data collected to measure the time it takes for a seismic wave to travel from a source (e.g., vibroseis, airgun, etc.), reflect off an interface between contrasting acoustic impedances, and be detected by a receiver (e.g., geophone, hydrophone). When properly corrected for offset (normal moveout correction), traces are stacked to form an acoustic representation of the subsurface. Seismic reflections show significant boundaries, either lithologic or fluid properties (or both), that can be structurally and stratigraphically interpreted and mapped. Seismic reflection data is the primary tool of exploration, with a large industry of seismic companies focused on collection, processing, and even interpretation of this type of data.

Seismic refraction data. Refracted seismic waves will move along a surface of differing acoustic impedance before returning to the surface for collection by geophones or on-bottom sensors (OBS). Compressional seismic velocities (a proxy for density) revealed by seismic refraction data provide greater insights into the deep earth structure as well-designed surveys allow refracted waves to travel much deeper than normal seismic reflection waves. Modern OBS surveys acquire data down to 30 km (18 miles) and often indicate the top of the mantle (see *mantle*), location of oceanic/continental crustal boundaries, base of sedimentary crust, and other important deep structural features.

Shelf edge. The inflection point where the gently sloping shelf gradient increases to the steeper gradient of the slope.

Shelf-margin delta. A delta system that has prograded across the breadth of the continental shelf to the continental margin. Shelf-margin deltas are characterized by increased abundance and scale of gravity mass transport deposits, growth faults, thick delta front and prodelta facies successions, and increased clinoform gradient.

Shore zone system. Depositional systems encompassing the high-energy transitional environment that extends from wave base landward to the limit of marine influence. The shore zone includes strandplain (prograding beach ridge and mudflat), barrier island/lagoon, and macrotidal systems.

Slope apron. The family of depositional, line-fed slope systems. Depending on the source system that lies updip, slope aprons may be shelf-fed or delta-fed. Along tectonically active margins, slope segments may be over-steepened, bypassing sediment to the lower slope or directly onto the basin floor. Margins undergoing erosion, collapse, and retreat create retrogradational aprons consisting of remobilized debris from the retreating slope and shelf. See Box 1.3.

Source rock. Oil and gas normally originate in mudstones that contain sufficient organic material of marine or terrestrial origin. Kerogen in these mudstones is thermally converted during burial to hydrocarbons. In conventional reservoirs, these hydrocarbons are expelled during migration into porous sandstones and carbonate reservoirs. In unconventional reservoirs, hydrocarbons may be retained in ultra-low-permeability shales until liberated by artificial stimulation and horizontal drilling. The most effective and important source rocks typically have more than 1–3 percent total organic carbon, depending on maturity. Unconventional source rocks may be overmature but still retain gas in organic or shale pore structures.

Source terrane. The area from which sediment is originally derived, often an exposed highland or mountain belt where the headwaters of significant rivers can erode, and transport sediment down-system to a basinal sink. Source-to-sink analysis attempts to identify the source terrane and the pathway to the depositional basin. Detrital zircons geochronology attempts to discern the source terrane from the distinctive age spectra of different basement blocks whose temporal accretion to global continents is well established.

Spreading center. The point where oceanic crust is created at mid-ocean ridges, though the morphology of the spreading center varies from a distinct axial high or ridge feature at high rates of spreading (6.5–16 cm/year) to an axial valley at low rates of spreading (1–4 cm/year). In the eastern GoM, seismic reflection observations (Snedden *et al.* 2014; Lin *et al.* 2019) show the spreading center to have axial valley morphology indicating a slow spreading rate, which fits estimates of 2.2 cm/year from seismic refraction studies (Christeson *et al.* 2014). See also *oceanic crust.*

Submarine channel system. Basin-floor depositional system characterized by an extended channel network collecting sediment along the slope base and transporting it far out onto the abyssal plain. It is the submarine analog of a trunk river. The channel may or may not support a fan at its terminus.

Subsalt. The sedimentary interval that lies stratigraphically below allochthonous salt, though the contact can be near-horizontal or show truncation. Subsalt strata in the northern GoM can be Mesozoic or Cenozoic in age, with the Cenozoic subsalt interval younging toward the Sigsbee Escarpment, where allochthonous salt meets the sea floor. Initial exploration in the subsalt GoM plays (Wilcox and younger interval) was hampered by poor imaging and illumination under thick salt, particularly complex salt bodies with sutures and inclusions. This problem was largely solved by development of new seismic technologies for acquisition (e.g., wide azimuth data, long streamers, multi-vessel source and receiver deployments) and processing (and reprocessing). See *pre-salt.*

Successor basin. A zone of significant deposition developed following but largely independent of a major tectonic orogeny. A successor basin is not directly linked to the orogeny, but sedimentary patterns may reflect the configuration of the underlying, earlier-formed structure. Reinterpretation of the Texas–Louisiana Eagle Mills Formation as part of a successor basin-fill above the Quachita–Marathon orogenic belt differs with prior views of the Eagle Mills being deposited in a Triassic rift system like the South Georgia rift.

Supersequence. A stratigraphic aggregation of depositional sequence sets. The hierarchical arrangement consists of sequences stacking to form sequence sets that in turn make up supersequences. Like depositional sequences, sequence sets are bounded by unconformities and can be split out into highstand, lowstand, and transgressive components reflecting relative sea-level changes (a function of subsidence, sediment supply, and eustasy). Supersequences are generally long duration cycles (±10 million years) clearly reflecting long-term tectonic trends in the basin (e.g., subsidence) and source terrane hinterland (impacting sediment supply). In the GoM basin Mesozoic interval, long-duration, low-frequency supersequences are the norm, with each supersequence containing embedded sequence sets and sequences. In the Cenozoic interval, supersequences are somewhat shorter in duration (possibly due to higher sediment flux and salt tectonics), but equally complex assemblages of reservoir, seal, and source rocks. Tectonostratigraphic units as described in this book are considered at the same hierarchical level as supersequences in most cases.

Supracanopy. Strata above the allochthonous salt canopy. In the GoM, supracanopy deposition often is localized to minibasins, known as secondary basins versus primary basins formed above allochthonous salt.

Systems tract. Linked, contemporaneous depositional systems. First defined by Brown and Fisher (1977), who recognized, in the GoM and other basins, the coeval progression of depositional systems from source terranes to deepwater basins (e.g., alluvial rivers to coastal plains to deltas to shelf to slope channels to basinal fan). The abundant well control of the onshore GoM demonstrated this pattern was a key component of the depositional sequence as earlier defined by Vail and co-workers, who worked primarily on seismic reflection datasets.

Tectonostratigraphy. Tectonostratigraphy refers to the stratigraphic framework and depositional trends that can be best understood in terms of long-term tectonic processes, including basinal subsidence, salt evacuation, and processes in the sediment source terrane that provide sediment to the basin.

Terrigenous. Sediment derived from weathering of rocks exposed above sea level. It is dominated by – but not exclusively – siliciclastic sediment.

Thrombolite. See *microbialite.*

Tilting subsidence. Subsidence across a hinge line or fulcrum with increasing subsidence basinward and potential uplift landward to the hinge. Typical post-Eocene pattern of subsidence along the northern GoM margin.

Transitional crust. Continental crust attenuated by rifting. In the northern GoM, often with a thickness of 10–17 km (6–11 miles) on the landward side.

Unconventional reservoir. Shale-dominated reservoirs where the pore space within an ultra-low-permeability mudrock contains hydrocarbons that are typically liberated by means of artificial fracture stimulation and horizontal wells. Hybrid reservoirs are a mix of conventional and unconventional reservoirs, a prime example being the Bakken Shale of North Dakota, where a thin calcareous zone between two mudstones provides a pathway for oil to reach the borehole.

Zircon. See Box 1.2.

References

Acevedo, J. S., 1980, Giant fields of the southern zone, Mexico, in M. T. Halbouty, ed., *Giant Oil and Gas Fields of the Decade 1968–1978*: American Association of Petroleum Geologists, Memoir 30, 339–385.

Acevedo, J. S., and O. M. Dautt, 1980, Giant fields in the southeast of Mexico: *Gulf Coast Association of Geological Societies Transactions*, 30, 1–33.

Achauer, C. A., 1977, Contrasts in cementation, dissolution, and porosity development between two Lower Cretaceous reefs of Texas, in D. G. Bebout and R. G. Loucks, eds., *Cretaceous Carbonates of Texas and Mexico: Applications to Subsurface Exploration*: The University of Texas at Austin, Bureau of Economic Geology, Report of Investigations 89, 127–137.

Adams, R. L., and J. P. Carr, 2010, Regional depositional systems of the Woodbine, Eagle Ford, and Tuscaloosa of the U.S. Gulf Coast: *Gulf Coast Association of Geological Societies Transactions*, 60, 3–27.

Adams, S., 1985, Lithofacies of the middle Glen Rose reef buildup, Lower Cretaceous shelf margin, east Texas and Louisiana, in D. Bebout and D. Ratcliff, eds., *Lower Cretaceous Depositional Environments from Shoreline to Slope*: Society of Economic Paleontologists and Mineralogists, Lower Cretaceous Core Workshop, 13–22.

Ahr, W. M., 1973, The carbonate ramp: an alternative to the shelf model: *Gulf Coast Association of Geological Societies Transactions*, 23, 221–225.

Ahr, W. M., 1981, The Gilmer Limestone: oolite tidal bars on the Sabine uplift: *Gulf Coast Association of Geological Societies Transactions*, 31, 1–6.

Ajdukiewicz, J. M., P. H. Nicholson, and W. L. Esch, 2010, Prediction of deep reservoir quality from early diagenetic process models in the Jurassic eolian Norphlet formation, Gulf of Mexico: *American Association of Petroleum Geologists Bulletin*, 94, 1189–1227.

Alcocer, J. A. E., 2012, Avances y Resultados de la Exploracion en la Porcion Mexicana del Golfo de Mexico Profundo: Pemex Subdirection of Exploration, 30 p.

Alegret, L., I. Arnillas, J. Arz, *et al.*, 2005, Cretaceous–Paleogene boundary deposits at Loma Capiro, central Cuba: evidence for the Chicxulub impact: *Geology*, 33, 721–724, doi:10.1130/G21573.1.

Alexander, C. I., 1951, History of discovery and development of Woodbine oil fields in East Texas, in *The Woodbine and Adjacent Strata: History of Discovery and Development of Woodbine Oil Fields in East Texas*: Dallas Petroleum Geologists, Fondren Science Series, 12–20.

Alnahwi, A., R. G. Loucks, S. C. Ruppel, R. W. Scott, and N. Tribovillard, 2018, Dip-related changes in stratigraphic architecture and associated sedimentological and geochemical variability in the Upper Cretaceous Eagle Ford Group in south Texas: *American Association of Petroleum Geologists Bulletin*, 102, 2537–2568.

Alvarez, L. W., Alvarez, W., Asaro, F., and H. V. Michel, 1980, Extraterrestrial cause for the Cretaceous–Tertiary extinction, *Science*, 208, 1095–1108.

Alzaga-Ruiz, H., L. Michel, F. Roure, and M. Seranne, 2009a, Interactions between the Laramide Foreland and the passive margin of the Gulf of Mexico: tectonics and sedimentation in the Golden Lane area, Veracruz State, Mexico: *Marine and Petroleum Geology* 26, 951–973.

Alzaga-Ruiz, H., D. Granjeon, M. Lopez, M. Seranne, and F. Roure, 2009b, Gravitational collapse and Neogene sediment transfer across the western margin of the Gulf of Mexico: insights from numerical models: *Tectonophysics*, 470, 21–41.

Ambrose, W. A., T. F. Wawrzyniec, K. Fouad, *et al.*, 2003, Geologic framework of Upper Miocene and Pliocene gas plays of the Macuspana Basin, southeastern Mexico: *American Association of Petroleum Geologists Bulletin*, 87, 1411–1435.

Ambrose, W. A., R. H. Jones, K. Fouad, *et al.*, 2004, *Sandstone Architecture of Upper Miocene and Pliocene Shoreface, Deltaic, and Valley-Fill Complexes, Macuspana Basin, Southeastern Mexico*: Bureau of Economic Geology, Report of Investigations 270, 37 p.

Ambrose, W. A., T. F. Wawrzyniec, K. Fouad, *et al.*, 2005, Neogene tectonic, stratigraphic, and play framework of the southern Laguna Madre-Tuxpan continental shelf, Gulf of Mexico: *American Association of Petroleum Geologists Bulletin* 89, 725–751.

Ambrose, W. A., T. F. Hentz, F. Bonnaffee, *et al.*, 2009, Sequence-stratigraphic controls on complex reservoir architecture of highstand fluvial-dominated deltaic and lowstand valley-fill deposits in the Upper Cretaceous (Cenomanian) Woodbine Group, East Texas field: regional and local perspectives: *American Association of Petroleum Geologists Bulletin*, 93, 231–269, doi:10.1306/09180808053.

Ambrose, W. A., R. G. Loucks, and S. P. Dutton, 2013, *Depositional Systems and Controls on Reservoir Quality (Determined from Core Data) in Deeply Buried Tertiary Strata in the Texas–Louisiana Gulf of Mexico*: The University of Texas at Austin, Bureau of Economic Geology, Report of Investigations 278, 80 p.

Ambrose, W. A., R. G. Loucks, and S. P. Dutton, 2015, Sequence-stratigraphic and depositional controls on reservoir quality in lowstand incised-valley fill and highstand shallow-marine systems in the Upper Cretaceous (Cenomanian) Tuscaloosa Formation, Louisiana, U.S.A.: *Gulf Coast Association of Geological Societies Journal*, 4, 43–66.

Amundsen, L., and M. Landro, 2008, Seismic imaging technology, Part II: lessons from wide azimuth subsalt imaging in deepwater Gulf of Mexico: *GEO ExPro*, May, 60–62.

Anderson, E. G., 1979, *Basic Mesozoic Study in Louisiana: the Northern Coastal Region and the Gulf Basin Province*: Louisiana Geological Survey, Folio Series 3, 1–58.

Anderson, T. H., and V. A. Schmidt, 1983, The evolution of Middle America and the Gulf of Mexico–Caribbean Sea region during Mesozoic time: *Geological Society of America Bulletin*, 94, 941–966.

Angeles-Aquino, F. J., and A. Cantú-Chapa, 2001, Subsurface Upper Jurassic stratigraphy in the Campeche shelf, Gulf of Mexico, in C. Bartolini, R. T. Buffler, and A. Cantú-Chapa, eds., *The Western Gulf of Mexico Basin: Tectonics, Sedimentary Basins, and Petroleum Systems*: American Association of Petroleum Geologists, Memoir 75, 343–352.

Antunano, S. E., 2009, The Yegua Formation: gas play in the Burgos Basin, Mexico, in C. Bartolini and J.R. Roman Ramos, eds., *Petroleum Systems in the Southern Gulf of Mexico*: American Association of Petroleum Geologists, Memoir 90, 49–78.

Applegate, A. V., 1984, The Brown Dolomite zone of the Lehigh Acres Formation (Aptian) in the South Florida Basin: a potentially prolific producing horizon offshore: *Gulf Coast Association of Geological Societies Transactions*, 34, 1–5.

Applegate, A. V., 1987, Part II: The Brow Dolomite Zone of the Lehigh Acres Formation (Aptian) in the South Florida Basin – a potentially prolific producing horizon offshore: *State of Florida Department of Natural Resources Information Circular*, 104, 43–72.

Aquino Lopez, J. A., and G. O. Gonzalez, 2001, The Sihil Reservoir: the fifth block of Cantarell Field: *Offshore Technology Conference*, 1–15.

Arbouille, D., V. Andrus, T. Piperi, and T. Xu, 2013, Sub-salt and pre-salt plays: how much are left to be discovered?: *American Association of Petroleum Geologists, Search and Discovery* 10545, 1–7.

Arce, L. E. P., 2017, *Neogene Current-Modified Submarine Fans and Associated Bed Forms in Mexican Deep-water Areas*: MS thesis, The University of Texas at Austin, 105 p.

Armentrout, J. M. and J. F. Clement, 1990, Biostratigraphic calibration of depositional cycles: a case study in High Island–Galveston–East Breaks areas, offshore Texas: Society of Economic Paleontologists and Mineralogists Gulf Coast Section, Stratigraphic Analysis Utilizing Advanced Geophysical, Wireline and Borehole Technology for Petroleum Exploration and Production, Seventeenth Annual Research Conference, Program and Abstracts, 21–51.

Armentrout, J. M., S. J. Malecek, P. Braithwaite, and C. R. Beeman, 1996, Intraslope basin reservoirs deposited by gravity-driven processes: south Ship Shoal and Ewing Bank areas, offshore Louisiana: *Gulf Coast Association of Geological Societies Transactions*, 46, 443–448.

Arnold, G. A., S. R. Cavalero, P. J. Clifford, *et al.*, 2010, Thunder Horse takes reservoir management to the next level: Offshore Technology Conference, Paper 20396, 10 p.

Arreguín-Lopez, M. A., and P. Weimer, 2004, Regional sequence stratigraphic setting on Miocene–Pliocene sediments, Veracruz Basin, Mexico: *Gulf Coast Association of Geological Societies Transactions*, 54, 25–40.

Arreguín-Lopez, M. A., G. Reyna-Martínez, H. Sánchez-Hernández, A. Escamilla-Herrera, and A. Gutiérrez-Araiza, 2011, Tertiary turbidite systems in the southwestern Gulf of Mexico: *Gulf Coast Association of Geological Societies Transactions*, 61, 45–53.

Artemieva, N., and J. Morgan, 2009, Modeling the formation of the K–Pg boundary layer, *Icarus*, 201(2), 768–780, doi:10.1016/j.icarus.2009.01.021.

Artemieva, N., J. Morgan, and Expedition 364 Science Party, 2017, Quantifying the release of climate-active gases by large meteorite impacts with a case study of Chicxulub: *Geophysical Research Letters*, 44, 180–188, doi:10.1002/2017GL074879.

Arthur, M. A., Schlanger, S. T., and H. C. Jenkyns, 1987, The Cenomanian–Turonian Oceanic Anoxic Event, II: palaeoceanographic controls on organic-matter production and preservation, in J. Brooks and A. J. Fleet, eds., *Marine Petroleum Source Rocks*, Geological Society, London, Special Publications 26, 401–420.

Aubrey, J., 1984, Recent Jurassic discoveries in Southeastern Cass County, Texas: *Gulf Coast Association of Geological Societies Transactions*, 34, 445–451.

Ayers, W. B., Jr., and A. H. Lewis, 1985, *The Wilcox Group and Carrizo Sand (Paleogene) in East-Central Texas: Depositional Systems and Deep-Basin Lignite*: The University of Texas at Austin Bureau of Economic Geology, 19 p.

Badalini, G., B. Kneller, and C. D. Winker, 1999, Late Pleistocene Trinity-Brazos turbidite system: depositional processes and architectures in a ponded mini-basin system, Gulf of Mexico continental slope: *American Association of Petroleum Geologists Bulletin*, 83, 1296–1346.

Bailey, J., 1978, Black Lake Field Natchitoches Parish, Louisiana: A Review: *Gulf Coast Association of Geological Societies Transactions*, 28, 11–24.

Bailey, J. W., 1983, Stratigraphy, environments of deposition, and petrography of the Cotton Valley Terryville Formation in Eastern Texas: unpublished MS thesis, The University of Texas at Austin, 229 p.

Barboza-Gudiño, J. R., A. Zavala-Monsiváis, G. Venegas-Rodríguez, and L. D. Barajas-Nigoche, 2010, Late Triassic stratigraphy and facies from northeastern Mexico: Tectonic setting and provenance: *Geosphere*, 6, 621–640, doi: 10.1130/GES00545.1.

Baria, L. R., E. Heydari, and B. G. Winton, 2008, Shale layers in the Alabama Smackover Formation and their implications for sea-level change and regional correlation: *Gulf Coast Association of Geological Societies Transactions*, 58, 67–75.

Barrell, K. A., 1997, Sequence stratigraphy and structural trap styles of the Tuscaloosa Trend: *Gulf Coast Association of Geological Societies Transactions*, 47, 27–34.

Barrell, K. A., 2000, Conducting a field study with GIS: Port Hudson field, Tuscaloosa trend, East Baton Rouge Parish, Louisiana, in T. C. Coburn and J. M. Yarus, eds., *Geographic Information Systems in Petroleum Exploration and Development*: American Association of Petroleum Geologists, Computer Applications in Geology 4, 187–194.

Barros, J. A., 1987, *Stratigraphy, Structure, and Paleogeography of the Jurassic–Cretaceous Passive Margin in the Western and Central Cuba*: MS thesis, University of Miami.

Bay, A. R., 1985, Carbonate shoaling cycles in the Lower Glen Rose Formation (Lower Cretaceous), South Texas, in D. G. Bebout, ed., *Annual Meeting of the Gulf Coast Association of Geological Societies and Society of Economic Paleontologists and Mineralogists Gulf Coast*, Section Core Workshop 4: Lower Cretaceous Depositional Environments from Shoreline to Slope, 37–46.

Bebout, D. G., 1977, Sligo and Hosston depositional patterns, subsurface of South Texas, in D. G. Bebout and R. G. Loucks, eds., *Cretaceous Carbonates of Texas and Mexico: Applications to Subsurface Exploration*: The University of Texas at Austin, Bureau of Economic Geology, Report of Investigations 89, 79–96.

Bebout, D. G. and D. R. Gutiérrez, 1982, *Regional Cross Sections, Louisiana Gulf Coast (Western Part)*: Louisiana Geological Survey, Folio Series 5.

Bebout, D. G. and D. R. Gutiérrez, 1983, *Regional Cross Sections, Louisiana Gulf Coast (Eastern Part)*: Louisiana Geological Survey, Folio Series 6.

Bebout, D. G., R. A. Schatzinger, and R. G., Loucks, 1977, Porosity distribution in the Stuart City Trend, Lower Cretaceous, South Texas in Bebout, D. G. and R. G. Loucks, eds., *Cretaceous Carbonates of Texas and Mexico: Applications to Subsurface Exploration*: The University of Texas at Austin, Bureau of Economic Geology, Report of Investigations 89, 234–256.

Bebout, D. G., R. G. Loucks, and A. R. Gregory, 1978, *Frio Sandstone Reservoirs in the Deep Subsurface along the Texas Gulf Coast*: The University of Texas at Austin, Bureau of Economic Geology, Report of Investigations 91, 92 p.

Bebout, D. G., B. R. Wise, A. R. Gregory, and M. B. Edwards, 1982, *Wilcox Sandstone Reservoirs in the Deep Subsurface along the Texas Gulf Coast: Their Potential for Production of Geopressured Geothermal Energy*: The University of Texas at Austin, Bureau of Economic Geology, Report of Investigations 117, 125 p.

Bebout, D. G., W. A. White, C. M. Garrett, and T. F. Hentz, 1992, *Atlas of Major Central and Eastern Gulf Coast Gas Reservoirs*: Gas Research Institute, Chicago, 88 p.

Bentley, S. J., M. D. Blum, J. Maloney, L. Pond, and R. Paulsell, 2015, The Mississippi River source-to-sink system: perspectives on tectonic, climatic, and anthropogenic influences, Miocene to Anthropocene: *Earth-Science Reviews*, 153, 1–35.

Bhattacharya, J. P., Copeland, P., Lawton, T. F., and Holbrook, J., 2016, Estimation of source area, river paleo-discharge, paleoslope, and sediment budgets of linked deep-time depositional systems and implications for hydrocarbon potential: *Earth-Science Reviews*, 153, 77–110, doi:10.1016/j.earscirev.2015.10.013.

Bitter, M. R., 1993, Sedimentation and provenance of Chicontepec sandstones with implications for uplift of the Sierra Madre Oriental and Teziutlan Massif, east-central Mexico: *Society of Economic Paleontologists and Mineralogists Gulf Coast Section, 13th Annual Research Conference*, 155–172.

Blanc, P., and J. Connan, 1993, Crude oils in reservoirs: the factors influencing their composition, in M. L. Bordenave, ed., *Applied Petroleum Geochemistry*: Editions Technip, 149–174.

Blanco, A., F. Maurrasse, F. Duque-Botero, and A. Delgado, 2011, Anoxic–dysoxic–oxic conditions in the Cenomanian Agua Nueva Formation (Upper Cretaceous) in central Mexico and their relation to Oceanic Anoxic Event 2 (OAE 2): *Geologic Society of America Annual Meeting, Minneapolis*, 172–174.

Blankenship, C. L., D. H. Knight, D. A. Kercho, *et al.*, 2010, Will K: another step in the evolution of the U.S. Gulf of Mexico deep gas play: *American Association of Petroleum Geologists, Search and Discovery* 90104 (abs.)

Blum, M. D., and Hattier-Womack, J., 2009, Climate change, sea-level change, and fluvial sediment supply to deepwater depositional systems: a review, in B. Kneller, O.J. Martinsen, and B. McCaffrey, eds., *External Controls on Deep-Water Depositional Systems*: SEPM (Society for Sedimentary Geology), Special Publication, 92, 15–39, doi:10.2110/sepmsp.092.015.

Blum, M., and M. Pecha, 2014, Mid-Cretaceous to Paleocene North American drainage reorganization from detrital zircons: *Geology*, 42(7), 607–610, doi:10.1130/G35513.1.

Blum, M. D., K. T. Milliken, M. A. Pecha, *et al.*, 2017, Detrital-zircon records of Cenomanian, Paleocene, and Oligocene Gulf of Mexico drainage integration and sediment routing: implications for scales of basin-floor fans: *Geosphere*, 13, 1–37, doi:10.1130/GES01410.1.

BOEM, 2016, Resource Evaluation Program: table of historical assessments, 1 p. www.boem.gov/national-assessment-history.

BOEM, 2017, *Assessment of Technically and Economically Recoverable Hydrocarbon Resources of the Gulf of Mexico Outer Continental Shelf as of January 1, 2014: OCS*, Report BOEM 2017-005, 50 p. www.boem.gov/BOEM-2017-005.

Boettcher, S. S., and K. L. Milliken, 1994, Mesozoic–Cenozoic unroofing of the southern Appalachian Basin: apatite fission track evidence from Middle Pennsylvanian sandstones: *The Journal of Geology*, 102, 655–668.

Bohor, B. F., E. E. Foor, P. J. Modreski, and D. M. Triplehorn, 1984, Mineralogic evidence for an impact event at the Cretaceous–Tertiary boundary: *Science*, 224, 867–869.

Bolli, H. M., J. B. Saunders, and K. Perch-Nielsen, 1989, *Plankton Stratigraphy: Volume 1, Planktic Foraminifera, Calcareous Nannofossils and Calpionellids*: CUP Archive, 607 p.

Booth, J. R., A. E. Duvernay III, D. S. Pfeiffer, and M. J. Styzen, 2000, Sequence stratigraphic framework, depositional models, and stacking patterns of ponded and slope fan systems in the Auger Basin: central Gulf of Mexico slope, in P. Weimer, R. M. Slatt, J. Coleman, *et al.*, eds., *Deep-Water Reservoirs of the World*: Society of Economic Paleontologists and Mineralogists Gulf Coast Section, 20th Annual Bob F. Perkins Research Conference, 82–102.

Booth, J. R., M. C. Dean, A. E. DuVernay III, and M. J. Styzen, 2003, Paleo-bathymetric controls on the stratigraphic architecture and reservoir development of confined fans in the Auger basin: central Gulf of Mexico slope: *Marine and Petroleum Geology*, 20, 563–586.

Bovay, A., 2015, *New Models of Early Cretaceous Source-To-Sink Pathways in the Eastern Gulf of Mexico*: MS thesis, The University of Texas at Austin, 105 p.

Bowman, S. A., 2012, Exploration targets of offshore Western Florida: *Gulf Coast Association of Geological Societies Transactions*, 62, 13–26.

Boyd, D. R., and B. F. Dyer, 1964, Frio barrier bar system of South Texas: *Gulf Coast Association of Geological Societies Transactions*, 14, 309–322.

Bralower, T. J., M. A. Arthur, R. M. Leckie, *et al.*, 1994, Timing and paleoceanography of oceanic dysoxia/anoxia in the Late Barremian to Early Aptian (Early Cretaceous): *Palaios*, 9, 335–369.

Bralower, T. J., C. K. Paul, and R. M. Leckie, 1998, The Cretaceous–Tertiary boundary cocktail: Chicxulub impact triggers margin collapse and extensive sediment gravity flows: *Geology*, 26, 331–334, doi:10.1130/0091-7613(1998)026<0331:TCTBCC>2.3.CO;2.

Bralower, T. J., E. Cobabe, B. Clement, *et al.*, 1999, The record of global change in mid-Cretaceous (Barremian–Albian) sections from the Sierra Madre, northeastern México: *Journal of Foraminiferal Research*, 29, 418–437.

Bramlett, K. W., and P. A. Craig, 2002, Core characterization of slope-channel and channel–levee reservoirs in Ram Powell field, Gulf of Mexico: *Society of Economic Paleontologists and Mineralogists Gulf Coast Section, Foundation Deep-Water Core Workshop, Northern Gulf of Mexico, Houston, TX*, 1–18.

Brand, J. H., 2016, Stratigraphy and mineralogy of the Oxfordian Lower Smackover Formation in the Eastern Gulf of Mexico, in C. M. Lowery, J.W. Snedden, and M. Blum, eds., *Mesozoic of the Gulf Rim and Beyond: New Progress in Science and Exploration of the Gulf of Mexico Basin*: 35th Annual Gulf Coast Section SEPM Foundation Perkins Rosen Research Conference, 14–35.

Braunstein, J., G. W. Field, J. Garst, *et al.*, 1949, Mesozoic cross-section from concordia parish, Louisiana, to Walton County, Florida: *Mississippi Geological Society and Mesozoic Committee*, 1.

Brown, A., 2011, *Interpretation of Three-Dimensional Seismic Data*: American Association of Petroleum Geologists, Memoir 42, 646 p, doi:10.1190/1.9781560802884.

Brown, Jr., L. B., R. G. Loucks, R. H. Treviño, and U. Hammes, 2004, Understanding growth-faulted, intraslope sub basins by applying sequence-stratigraphic principles: examples from the south Texas Oligocene Frio Formation: *American Association of Petroleum Geologists Bulletin*, 88, 1501–1522.

Brown, L.F., and W.L. Fisher, 1977, Seismic stratigraphic interpretation of depositional systems: examples from Brazilian rift and pull apart basins, in C. Payton, ed., *Seismic Stratigraphy: Applications to Hydrocarbon Exploration*: American Association of Petroleum Geologists, Memoir 26, 213–248.

Bryant, W. R., J. Lugo, C. Cordova, and A. Salvador, 1991, Physiography and bathymetry, in A. Salvador, ed., *The Gulf of Mexico Basin: The Geology of North America*: Geological Society of America, 13–30, doi:10.1130/DNAG-GNA-J.13.

Buffler, R. T., F. J. Shaub, R. Huerta, A. B. K. Ibrahim, and J. S. Watkins, 1981, A model for the early evolution of the Gulf of Mexico basin: *Oceanologica Acta*, 3, 129–136.

Buffler, R. T., W. Schlager, and K. A. Pisciotto, 1984, *Introduction and Explanatory Notes: Deep Sea Drilling Project, Volume LXXVII, Leg 77 Glomar Challenger*: National Science Foundation, 740 p, doi:10.2973/dsdp.proc.77.101.1984.

Bull, S., J. Cartwright, and M. Huuse, 2009, A review of kinematic indicators from mass-transport complexes using 3D seismic data: *Marine and Petroleum Geology*, 26, 1132–1151.

Byerly, G. R., 1991, Igneous activity, in A. Salvador, ed., *The Gulf of Mexico Basin:* Geological Society of America, 91–108.

Camerlo, R. H., and E. F. Benson, 2006, Geometric and seismic interpretation of the Perdido fold belt: Northwestern deep-water Gulf of Mexico: *American Association of Petroleum Geologists Bulletin*, 90, 363–386, doi:10.1306/10120505003.

Cantú-Chapa, A., 1998, Las transgresiones jurásicas en México: *Revista Mexicana de Ciencias Geológicas,* 15, 25–37.

Cantú-Chapa, A., 2009, Upper Jurassic stratigraphy (Oxfordian and Kimmeridgian) in petroleum wells of the Campeche shelf, in C. Bartolini and J. R. Roman Ramos, eds., *Petroleum Systems in the Southern Gulf of Mexico*: American Association of Petroleum, Memoir 90, 79–91, doi:10.1306/13191079M903801.

Cantú-Chapa, A., and R. Landeros-Flores, 2001, The Cretaceous–Paleocene boundary in the subsurface Campeche shelf, southern Gulf of Mexico, in C. Bartolini, R. T. Buffler, and A. Cantú-Chapa, eds., *The Western Gulf of Mexico Basin: Tectonics, Sedimentary Basins, and Petroleum Systems*: American Association of Petroleum Geologist, Memoir 75, 389–395.

Cantú-Chapa, A., and E. Ortuño-Maldonado, 2003, The Tithonian (Upper Jurassic) Edzna Formation, an important hydrocarbon reservoir of the Campeche shelf, Gulf of Mexico, in C. Bartolini, R. T. Buffler, and J. Blickwede, eds., *The Circum-Gulf of Mexico and the Caribbean: Hydrocarbon Habitats, Basin Formation, and Plate Tectonics*: American Association of Petroleum Geologists, Memoir 79, 305–311.

Cardona, S., L. J. Wood, R. J. Day-Stirrat, and L. Moscardelli, 2016, Fabric development and pore-throat reduction in a mass-transport deposit in the Jubilee gas field, eastern Gulf of Mexico: consequences for the sealing capacity of MTDs, in, G. Lamarche, J. Mountjoy, S. Bull, T. Hubble, I. Pecher, and S. Woelz, eds., *Submarine Mass Movements and Their Consequences*, Springer, 27–37, doi:10.1007/978-3-319-20979-1.

Carreras, P. E., S. G. Johnson, S. E., Turner, 2006, *Tahiti: Assessment of Uncertainty in a Deepwater Reservoir Using Design of Experiments*: Society of Petroleum Engineers, SPE 102988, 16 p.

Castillon, M., and J.P. Larrios, 1963, *Salt Deposits of the lsthmas of Tehuantepec*: Northern Ohio Geological Society, 263–280.

Cather, M. C., C. E Chapin, and S. A. Kelley, 2012, Diachronous episodes of Cenozoic erosion in southwestern North America and their relationship to surface uplift, paleoclimate, paleodrainage, and paleoaltimetry: *Geosphere* 8, 1177–1206.

Cepeda, R., P. Weimer, and G. Dorn, 2010, 3D seismic stratigraphic interpretation of the Upper Miocene to Lower Pleistocene deepwater sediments of the Thunder Horse-Mensa area, Southern Mississippi Canyon, northern deep Gulf of Mexico: *Gulf Coast Association of Geological Societies Transactions*, 60, 119–132.

Chasteen, H. R., 1983, Re-evaluation of the lower Tuscaloosa and Dantzler formations (Mid-Cretaceous) with emphasis on depositional environments and time-stratigraphic relationships: *Gulf Coast Association of Geological Societies Transactions*, 33, 31–40.

Chavez-Valois, V. M. C., L. C. Valdes, J. I. J. Placencia, *et al.*, 2009, A new multidisciplinary focus in the study of tertiary plays in the Sureste Basin, Mexico, in C. Bartolini and J. R. Roman Ramos, eds., *Petroleum Systems in the Southern Gulf of Mexico*: American Association of Petroleum Geologists, Memoir 90, 155–190.

Chen, J., C. Acosta Aduna, F. Sanchez Lu, J. Patino, and M. Olivella, 2001, Petrophysical characteristics, depositional systems, and model of geological evolution in the Golden Lane carbonate sequences: *Gulf Coast Association of Geological Societies Transactions*, 51, 31–44.

Chernikoff, A., J. G. Hernandez, and R. Schatzinger, 2006, Mesozoic extensional tectonics: its impact on oil accumulations in Campeche Sound, Gulf of Mexico: *The Leading Edge*, 25(10), 1224–1234, doi:10.1190/1.2360609.

Christeson, G. L., Y. Nakamura, R. T. Buffler, J. Morgan, and, M. Warner, 2001, Deep crustal structure of the Chicxulub impact crater: *Journal of Geophysical Research*, 106, 21751–21769.

Christeson, G. L., H. J. A. Van Avendonk, I. O. Norton, J. W. Snedden, and D. R. Eddy, 2014, Deep crustal structure in the eastern Gulf of Mexico: *Journal of Geophysical Research: Solid Earth*, 119(9), 6782–6801, doi:10.1002/2014JB011045.

Christeson, G. L., S. P. S. Gulick, J. V. Morgan, *et al.*, 2018, Extraordinary rocks from the peak ring of the Chicxulub impact crater: P-wave velocity, density, and porosity measurements from IODP/ICDP Expedition 364, *Earth and Planetary Science Letters*, 495, 1–11.

Cicero, A. D., I. Steinhoff, T. McClain, K. A. Koepke, and J. D. Dezelle, 2010, Sequence stratigraphy of the Upper Jurassic mixed carbonate/siliciclastic Haynesville and Bossier shale depositional systems in east Texas and northern Louisiana: *Gulf Coast Association of Geological Societies Transactions*, 60, 133–148.

Clara Valdés, M. D. L., L. V. Rodriguez, and E. C. Garcia, 2009, Geochemical integration and interpretation of source rocks, oils, and natural gases in southeastern Mexico, in C. Bartolini, and J. R. Roman Ramos, eds., *Petroleum Systems in the Southern Gulf of Mexico*: American Association of Petroleum Geologists, Memoir 90, 337–368, doi:10.1306/13191091M903337.

Clark, W. J., F. O. Iwere, O. Apaydin, *et al.*, 2003, Integrated modeling of the Taratunich Field, Bay of Campeche, Southern Mexico: *American Association of Petroleum Geologists, Search and Discovery* 20014, 1–6.

Claypool, G. E., and E. A. Mancini, 1989, Geochemical relationships of petroleum in Mesozoic reservoirs to carbonate source rocks of Jurassic Smackover Formation, Southwestern Alabama: *American Association of Petroleum Geologists Bulletin*, 73, 904–924.

Cleaves, A. W., 1977, Middle Glen Rose (Cretaceous) facies mosaic, Blanco and Hays Counties, Texas: *American Association of Petroleum Geologists, Search and Discovery* 90967, 1 p.

CNH, 2014, Southeast offshore basins shallow water area: petroleum geological synthesis: CNH, 1–64.

CNH, 2015a, Saline Basin: petroleum geological synthesis: CNH, 1–47. Available at http://portal.cnih.cnh.gob.mx. June 2017.

CNH, 2015b, Perdido Fold belt, subsalt belt, Mexican Ridges: petroleum geological synthesis, CNH, 1–53. Available at http://portal.cnih.cnh.gob.mx., June 2017.

CNH, 2017a, Pemex to explore Mexico's pre-salt, CNH. Available at www.upstreamonline.com/hardcopy/1367799/pemex-preparing-pre-salt-test, October 2017.

CNH, 2017b, Geological Atlas Sabina-Burgos Basin, CNH, 1-53. Available at https://portal.cnih.cnh.gob.mx/downloads/en_US/info/Geological_Atlas_Sabinas-Burgos_Basins_V3.pdf, February, 2018.

CNH, 2018, Prospective resources, CNH, 1–2. Available at https://portal.cnih.cnh.gob.mx/downloads/en_US/estadisticas/Prospective%20Resources.pdf, September 2017.

Cobiella-Reguera, J. L., E. M. Cruz-Gámez, S. Blanco-Bustamante, *et al.*, 2015, Cretaceous–Paleogene boundary deposits and paleogeography in western and central Cuba: *Revista mexicana de ciencias geológicas*, 32(1), 156–176.

Coccioni, R., V. Luciani, A. Marsili, 2006, Cretaceous oceanic anoxic events and radially elongated chambered planktonic foraminifera: paleoecological and paleoceanographic implications: *Palaeogeography, Palaeoclimatology, Palaeoecology*, 235, 66–92.

Cole, G. A., A. G. Requejo, A. Yu, *et al.*, 1999, The geochemical and basin modeling aspects of the Jurassic to Lower Cretaceous sourced petroleum system, deepwater to ultra-deepwater Gulf of Mexico, offshore Louisiana: *3rd AMGP/American Association of Petroleum Geologists International Conference Proceedings*, 1–8.

Cole, G. A., A. Yu, F. Peel, *et al.*, 2001, Constraining source and charge risk in deepwater areas: *World Oil*, 222, 69–77.

Coleman, J. L., Jr., and C. J. Coleman, 1981, Stratigraphic, sedimentologic and diagenetic framework for the Jurassic Cotton Valley Terryville massive sandstone complex, northern Louisiana: *Gulf Coast Association of Geological Societies Transactions*, 31, 71–80.

Coleman, J. M., D. B. Prior, and J. F. Lindsay, 1983, Deltaic influences on shelfedge instability processes, in D. J. Stanley and G. T. Moore, eds., *The Shelfbreak: Critical Interface on Continental Margins*: Society of Economic Paleontologists and Mineralogists, Special Publication 33, 121–137.

Colmenares, M., and J. Hustedt, 2015, Overview of the deepwater geology of the Mexican Gulf of Mexico: round one of bidding in the energy reform: *American Association of Petroleum Geologists, Search and Discovery* 10738, 1–22.

Combellas-Bigott, R. I., and W. E. Galloway, 2002a, Depositional history and genetic sequence stratigraphic framework of the middle Miocene depositional episode, south Louisiana: *Gulf Coast Association of Geological Societies Transactions*, 52, 139–150.

Combellas-Bigott, R. I., and W. E. Galloway, 2002b, Origin and evolution of the middle Miocene submarine-fan system, east-central Gulf of Mexico: *Gulf Coast Association of Geological Societies Transactions*, 52, 151–163.

Combellas-Bigott, R. I., and W. E. Galloway, 2006, Depositional and structural evolution of the middle Miocene depositional episode, east-central Gulf of Mexico: *American Association of Petroleum Geologists Bulletin*, 90, 335–362, doi:10.1306/10040504132.

Combes, J. M., 1993, The Vicksburg formation of Texas: depositional systems distribution, sequence stratigraphy, and petroleum geology: *American Association of Petroleum Geologists Bulletin*, 77, 1942–1970, doi:10.1306/BDFF8F88-1718-11D7-8645000102C1865D.

Comet, P. A., 1992, Maturity mapping of northern Gulf of Mexico oils using biomarkers: *Gulf Coast Association of Geological Societies Transactions*, 42, 433–448.

Comet, P. A., J. K. Rafalska, and J. M. Brooks, 1993, Sterane and triterpene patterns as diagnostic tools in the mapping of oils, condensates, and source rocks of the Gulf of Mexico region: *Organic Geochemistry*, 20, 1265–1296.

Conrad, C. P., 2013, The solid Earth's influence on sea level: *Geological Society of America Bulletin*, 125, 1027–1052, doi:10.1130/B30764.1

Coogan, A. H., D. G. Bebout, and C. Maggio, 1972, Depositional environments and geologic history of Golden Lane and Poza Rica trend, Mexico: an alternative view: *American Association of Petroleum Geologists Bulletin*, 56, 1419–1447.

Corbett, K., M. Friedman, and J. Spang, 1987, Fracture development and mechanical stratigraphy of Austin Chalk, Texas: *American Association of Petroleum Geologists Bulletin*, 71, 17–28.

Cordona, S., L. Wood, and L. Moscardelli, 2015, Sealing capacity of mass transport deposits: depositional model for a deepwater reservoir in Jubilee gas field, eastern Gulf of Mexico: *American Association of Petroleum Geologists, Search and Discovery* 90216.

Cornelius, S. and J. P. Castagna, 2018, Variation in salt-body interval velocities in the deepwater Gulf of Mexico: Keathley Canyon and Walker Ridge areas: *Interpretation*, 6, T15–T17, doi:10.1190/INT-2017-0069.1.

Cornish, F., 2013, Do Upper Wilcox canyons support Paleogene isolation of the Gulf of Mexico?: *Gulf Coast Association of Geological Societies Transactions*, 63, 183–204.

Cossey, S. P. J., and M. Bitter, 2016, The KPg Impact Deposits in the Tampico–Misantla Basin, Eastern Mexico: *Society of Economic Paleontologists and Mineralogists Gulf Coast Section, 35th Annual Research Conference*, 479–486.

Cossey, S., and R. Jacobs, 1992, Oligocene Hackberry Formation of southwest Louisiana: sequence stratigraphy, sedimentology, and hydrocarbon potential: *American Association of Petroleum Geologists Bulletin* 76, 589–606.

Cossey, S. P. J., J. Pindell, and J. Rosenfeld, 2007, Recent geological understanding of the Chicontepec Erosional "paleocanyon," Tampico–Misantla Basin, Mexico: The Paleogene of the Gulf of Mexico and Caribbean Basins: *Processes, Events, and Petroleum Systems, Society of Economic Paleontologists and Mineralogists Gulf Coast Section, 27th Annual Research Conference Program and Extended Abstracts*, 273–283, doi:10.5724/gcs.07.27.0273.

Covault, J. A., and S. A. Graham, 2010, Submarine fans at all sea-level stands: tectono-morphologic and climatic controls on terrigenous sediment delivery to the deep sea: *Geology*, 38, 939–942, doi:10.1130/G31081.1.

Cox, R. T., and R. B. Van Arsdale, 2002, The Mississippi Embayment, North America: a first order continental structure generated by the Cretaceous superplume mantle event: *Journal of Geodynamics*, 34(2), 163–176, doi:10.1016/S0264-3707(02)00019-4.

Craddock W. H., and A. R. C. Kylander-Clark, 2013, U–Pb ages of detrital zircons from the Tertiary Mississippi River Delta in central Louisiana: insights into sediment provenance: *Geosphere*, 9, 1–20.

Crane, M. J., 1965, Upper Cretaceous ostracodes of the Gulf Coast area: *Micropaleontology*, 11, 191–254, doi:10.2307/1484517.

Cregg, A. K., and W. M. Ahr, 1983, Depositional framework and reservoir potential of an upper Cotton Valley (Knowles Limestone) patch reef, Milam County, Texas: *Gulf Coast Association of Geological Societies Transactions*, 33, 55–68.

Crevello, P.D., and P.M. Harris, 1984, Depositional models for jurassic reefal buildups, in P. Ventress, ed., *The Jurassic of the Gulf Rim*: Society of Economic Paleontologists and Mineralogists Gulf Coast Section Foundation 3rd Annual Research Conference, 57–102.

Cruz, M., and Aguilera, R., 2018, Eagle Ford and Pimienta shales in Mexico: a case study: Society of Petroleum Engineers: SPE Canada Unconventional Resources Convention, 1–21, doi:0.2118/189797-MS.

Cruz Luque, M. M., E. Urban-Rascon, R. F. Aguilera, and R. Aguilera, 2018, Mexican unconventional plays: geoscience, endowment, and economic considerations: *Society of Petroleum Engineers*, 21, 533–549, doi:10.2118/189438-PA.

Cruz-Mercado, M. A., J. C. Flores-Zamora, R. León-Ramírez, *et al.*, 2011, Salt provinces in the Mexican portion of the Gulf of Mexico: structural characterization and evolutionary model: *Gulf Coast Association of Geological Societies Transactions* 61, 93–103.

Cumming, E. W., 2002, Core, log, and seismic characteristics of a high-performance turbidite reservoir in a salt-withdrawal minibasin: the Upper Miocene yellow sand, Mars Field, Gulf of Mexico: *Deep-Water Core Workshop, Northern Gulf of Mexico Gulf Coast Section, Society of Economic Palentologists and Mineralogists*, 19–30.

Cunningham, R. C., J. W. Snedden, I. O. Norton, *et al.*, 2016, Upper Jurassic Tithonian-centered source mapping in the deepwater northern Gulf of Mexico: *Interpretation*, 4(1), SC97–SC123, doi:10.1190/INT-2015-0093.1.

Cunningham, R. C., M. P. Phillips, J. W. Snedden, *et al.*, in prep., The Paleocene–Eocene Thermal Maximum (PETM) in Deepwater Gulf of Mexico: a new Paleogene source rock and basin-scale paleoceanographic model: *Paleoceanography*.

Curry, M. A. E., F. J. Peel, M. R. Hudec, and I. O. Norton, 2018, Extensional models for the development of passive-margin salt basins, with application to the Gulf of Mexico: *Basin Research*, 30, 1–20, doi:10.1111/bre.12299.

Damuth, J., and H. C. Olson, 2015, Latest Quaternary sedimentation in the northern Gulf of Mexico Intraslope Basin Province: I. Sediment facies and depositional processes: *Geosphere*, 11, 1689–1718. doi:10.1130/GES01090.1.

Davison, I., and T. A. Cunha, 2017, Allochthonous salt sheet growth: thermal implications for source rock maturation in the deepwater Burgos basin and Perdido fold belt, Mexico: *Interpretation*, 5, T11–T21, doi:10.1190/INT-2016 0035.1.

Davison, I., O'Bierne, E., T. Faull, and I. Steel, 2015, Vast potential, Mexico's Round Uno: *GeoExpro*, 12, 68–71.

Day, S., and M. Maslin (2005), Linking large impacts, gas hydrates, and carbon isotope excursions through widespread sediment liquefaction and continental slope failure: the example of the KT boundary event: Geological Society of America, Special Paper 384, 239–258.

Dean, M. C., J. R. Booth, and B. T. Mitchell, 2002, Multiple fields within the sequence stratigraphic framework of the Greater Auger Basin, Gulf of Mexico: *Gulf Coast Section SEPM, 22nd Annual Research Conference*, 661–680.

De la Rocha Bascon, L., 2016, *Southern Gulf of Mexico Wilcox Source-to-Sink: Investigating Siliciclastic Sedimentation in Mexico Deep-Water*: MS thesis, The University of Texas at Austin, 88 p.

Denne, R. A., and R. H. Blanchard, 2013, Regional controls on the formation of the ancestral DeSoto Canyon by the Chicxulub Impact: *Gulf Coast Association of Geological Societies Journal*, 2, 17–28.

Denne, R. A., E. D. Scott, D. P. Eickhoff, *et al.*, 2013, Massive Cretaceous–Paleogene boundary deposit, deep-water Gulf of Mexico: new evidence for widespread Chicxulub-induced slope failure: *Geology*, 41, 983–986, doi:10.1130/G34503.1.

Denne, R. A., R. E. Hinote, J. A. Breyer, *et al.*, 2014, The Cenomanian–Turonian Eagle Ford Group of South Texas: insights on timing and micropaleontological analyses: *Paleogeography, Palaeoclimatology, Palaeoecology*, 413, 2–28, doi:10.1016/j.palaeo.2014.05.029.

Dessellcs, R., M. Wilson, and J. Barminski, 2016a, *Assessment of Undiscovered Oil and Gas Resources of the Nation's Outer Continental Shelf*: Bureau of Ocean Energy Management, National Assessment of Oil and Gas Resources, 1–8.

Diaz, J., P. Weimer, R. Bouroullec, and G. Dorn, 2011, 3-D seismic stratigraphic interpretation of Quaternary mass-transport deposits in the Mensa and Thunder Horse intraslope basins, Mississippi Canyon, northern deep Gulf of Mexico, U.S.A., in C.R. Shipp, P. Weimer, and H. Posamentier, eds., *Mass Transport Deposits in Deep-Water Settings*: Society for Sedimentary Geology, Special Publication, 96, 127–149.

Dickey, R. L., and T. E. Yancey, 2010, Palynological age control of sediments bracketing the Paleocene–Eocene Boundary, Bastrop, Texas: *Gulf Coast Association of Geological Societies Transactions* 60, 717–724.

Dickinson, W. R., and Gehrels, G. E., 2008, U–Pb ages of detrital zircons in relation to paleogeography: Triassic paleodrainage networks and sediment dispersal across southwest Laurentia: *Journal of Sedimentary Research*, 78, 745–764, doi:10.2110/jsr.2008.088.

Dickinson, W. R., and Gehrels, G. E., 2009, Use of U–Pb ages of detrital zircons to infer maximum depositional ages of strata: a test against a Colorado Plateau Mesozoic database: *Earth and Planetary Science Letters*, 288, 115–125, doi:10.1016/j.epsl.2009.09.013.

Dickinson, W. R., G. E. Gehrels, and R. J. Stern, 2010, Late Triassic Texas uplift preceding Jurassic opening of the Gulf of Mexico: evidence from U–Pb ages of detrital zircons: *Geosphere*, 6, 641–662, doi:10.1130/GES00532.1.

Dickson, W., and C. Schiefelbein, 2012, Girassol-Angola's first deepwater pre-salt discovery?: *American Association of Petroleum Geologists, Search and Discovery* 20142, 1–4.

Diegel, F. A., J. F. Karlo, D. C. Schuster, R. C. Shoup, and P. R. Tauvers, 1995, Cenozoic structural evolution and tectono-stratigraphic framework of the northern Gulf Coast continental margin, in M. P. A. Jackson, D. G. Roberts, and S. Snelson, eds., *Salt Tectonics: A Global Perspective*: American Association of Petroleum Geologists, Memoir 65, 109–151.

Dixon, J. F., R. J. Steel, and C. Olariu, 2012, Shelf-edge delta regime as a predictor of deep water deposition: *Journal of Sedimentary Research*, 82, 681–687, doi:10.2110/jsr.2012.59.

Dobson, L. M., 1990, *Seismic Stratigraphy and Geologic History of Jurassic Rocks, Northeastern Gulf of Mexico*, MA thesis; The University of Texas at Austin, 170 p.

Dodge, M. M., and J. S. Posey, 1981, *Structural Cross Sections, Tertiary Formations, Texas Gulf Coast*: The University of Texas at Austin, Bureau of Economic Geology, 42 p.

Dohmen, T. E., 2002, Age dating of expected MC3B seismic event suggests that it is the K/T Boundary: *Gulf Coast Association of Geological Societies Transactions*, 52, 177–180.

Donovan, A. D., T. Scott Staerker, A. Pramudito, *et al.*, 2012, The Eagle Ford outcrops of West Texas: a laboratory for understanding heterogeneities within unconventional mudstone reservoirs: *Gulf Coast Association of Geological Societies Journal*, 1, 162–185.

Dooley, T. P., M. P. A. Jackson, and M. R. Hudec, 2013, Coeval extension and shortening above and below salt canopies on an uplifted, continental margin: application to the northern Gulf of Mexico: *American Association of Petroleum Geologists Bulletin*, 97, 1737–1764, doi:10.1306/03271312072.

Doughty-Jones, G., M. Mayall, and L. Lonergan, 2017, Stratigraphy, facies, and evolution of deep-water lobe complexes within a salt controlled intraslope minibasin: *American Association of Petroleum Geologists Bulletin*, 101, 1879–1904.

Douglas, S. W., 2010, *The Jurassic Norphlet Formation of the Deep-Water Eastern Gulf of Mexico: A Sedimentologic Investigation of Aeolian Facies, their Reservoir Characteristics, and their Depositional History*: MS thesis, Baylor University, 68 pp.

Dravis, J., 1981, Depositional setting and porosity evolution of the Upper Cretaceous Austin Chalk Formation, South-Central Texas: *South Texas Geological Society Bulletin*, 24, 4–14.

Dubiel, R. F., J. K. Pitman, and D. Steinshouer, 2003, Seismic-sequence stratigraphy and petroleum system modeling of the downdip Tuscaloosa-Woodbine, LA and TX: *Gulf Coast Association of Geological Societies Transactions*, 52, 193–203.

Dubiel, R. F., J. K. Pitman, O. N. Pearson, *et al.*, 2012, Assessment of undiscovered oil and gas resources in conventional and continuous petroleum systems in the Upper Cretaceous Eagle Ford Group, U.S. Gulf Coast Region, 2011: US Geological Survey, Fact Sheet 2012-3003, 2 p.

Duckworth, C. J., S. G. Dowty, and D. G. Bebout, 1992, KS-8: Trinity Group Sandstone—Mississippi interior salt basin, in Bebout, D. G., W. A. White, C. M. Garrett and T. F. Hentz, eds. *Atlas of Major Central and Eastern Gulf Coast Gas Reservoirs*: Gas Research Institute, Chicago, 88 p.

Durham, C. O., Jr., and S. B. Hall, 1991, The Austin Chalk: bed by bed through Central Texas: in Austin Chalk exploration symposium: geology geophysics and formation evaluation. San Antonio TX, South Texas Geological Society, 25–40.

Dutton, S., and T. N. Diggs, 1992, Evolution of porosity and permeability in the Lower Cretaceous Travis Peak Formation, East Texas: *American Association of Petroleum Geologists Bulletin*, 76, 252–269.

Dutton, S. P., and R. G. Loucks, 2010, Diagenetic controls on evolution of porosity and permeability in lower Tertiary Wilcox sandstones from shallow to ultradeep (200–6700 m) burial, Gulf of Mexico Basin, U.S.A.: *Marine and Petroleum Geology*, 27, 69–81.

Dutton, S. P., D. C. Jennette, W. A. Ambrose, and M. Martin, 2002, Petrography and reservoir quality of Tertiary deepwater sandstones in the Veracruz Basin, Mexico: *Gulf Coast Association of Geological Societies Transactions*, 52, 229–240.

Dutton, S. P., R. G. Loucks, and R. J. Day-Stirrat, 2012, Impact of regional variation in detrital mineral composition on reservoir quality in deep to ultradeep lower Miocene sandstones, western Gulf of Mexico: *Marine and Petroleum Geology*, 35, 139–153.

Dyer, M. J., and C. Bartolini, 2004, Sabinas Basin Lower Cretaceous to Jurassic production: comparison to South Texas equivalents: *Gulf Coast Association of Geological Societies Transactions*, 54, 169–184.

Dyman, T. S., and Condon, S. M., 2006a, Assessment of undiscovered conventional oil and gas resources: Upper Jurassic–Lower Cretaceous Cotton Valley Group, Jurassic Smackover interior salt basins total petroleum system, in the East Texas Basin and Louisiana–Mississippi Salt Basins Provinces: U.S. Geological Survey, Digital Data Series DDS-69-E, chapter 2, 52 p.

Dyman, T. S., and Condon, S. M., 2006b, Assessment of undiscovered conventional oil and gas resources: Lower Cretaceous Travis Peak and Hosston Formations, Jurassic Smackover interior salt basins total petroleum system, in the East Texas Basin and Louisiana–Mississippi Salt Basins Provinces: U.S. Geological Survey, Digital Data Series DDS-69-E, chapter 5, 142 p.

Echols, D. J., and D. Malkin, 1948, Wilcox (Eocene) stratigraphy, a key to production: *American Association of Petroleum Geologists Bulletin*, 32, 11–33.

Echols, R. J., J. M. Armentrout, S. A. Root, *et al.*, 2003, Sequence stratigraphy of the Eocene/Oligocene boundary interval: southeastern Mississippi: in D. R. Prothero, L. C. Ivany, and E. A. Nesbitt eds., *From Greenhouse to Icehouse: The Marine Eocene–Oligocene Transition*: Columbia University Press, 189–222.

Eddy, D. R., J. A. Van Avendonk, G. L. Christeson, *et al.*, 2014, Deep crustal structure of the northeastern Gulf of Mexico: implications for rift evolution and seafloor spreading: *Journal of Geophysical Research: Solid Earth*, 119, 6802–6822, doi:10.1002/2014JB011311.

Eddy D. R., H. J. A. V. Avendonk, G. L. Christeson, and I. O. Norton, 2018, Structure and origin of the rifted margin of the northern Gulf of Mexico: *Geosphere*, 14, 1–14.

Edwards, M. B., 1981, The Live Oak delta complex, an unstable shelf/edge delta in the deep Wilcox of South Texas: *American Association of Petroleum Geologists Bulletin*, 77, 1942–1970.

Edwards, M. B., 1991, Control of depositional environments, eustasy, gravity, and salt tectonics in sandstone distribution in an unstable shelf edge delta, Eocene Yegua Formation, Texas and Louisiana: *Gulf Coast Association of Geological Societies Transactions*, 41, 237–252.

Edwards, M. B., 2000, Origin and significance of retrograde failed shelf margins: tertiary northern Gulf Coast basin: *Gulf Coast Association of Geological Societies Transactions*, 50, 81–93.

Egedahl, K., G. L. Kinsland, and D. Han, 2012, Seismic facies study of 3D seismic data, northern Louisiana, Wilcox Formation: *Gulf Coast Association of Geological Societies Transactions*, 62, 73–91.

Eguiluz de Antunano, S., 2001, Geologic evolution and gas resources of the Sabinas in Northeastern Mexico: in C. Bartolini, R. T. Buffler, and A. Cantú-Chapa, eds., *The Western Gulf of Mexico Basin: Tectonics, Sedimentary Basins and Petroleum Systems*, American Association of Petroleum Geologists, Memoir 75, 241–270.

Eguiluz de Antunano, S., 2007, Laramide deformation in the Burgos Basin, Northeastern Mexico: the Paleogene of the Gulf of Mexico and Caribbean Basins: *Processes, Events and Petroleum Systems, Society of Economic Paleontologists and Mineralogists Gulf Coast Section, 27th Annual Bob F. Perkins Research Conference*, 688–702.

Eikrem, V., R. Li, M. Medeiros, *et al.*, 2010, Perdido development project: great white WM12 Reservoir and Silvertip M. Frio field development plans and comparison of recent well results with pre-drill models. *Offshore Technology Conference*, 1–10, doi:10.4043/20879-MS.

Elderbak, K., R.M. Leckie, and N.E. Tibert, 2014, Paleoenvironmental and paleoceanographic changes across the Cenomanian–Turonian Boundary Event (Oceanic Anoxic Event 2) as indicated by foraminiferal assemblages from the eastern margin of the Cretaceous Western Interior Seaway: *Palaeogeography, Palaeoclimatology, Palaeoecology*, 413, 29–48.

Eldrett, J. S., C. Ma, S. C. Bergman, *et al.*, 2015, Origin of limestone–marlstone cycles: astronomic forcing of organic-rich sedimentary rocks from the Cenomanian to early Coniacian of the Cretaceous Western Interior Seaway, USA: *Earth and Planetary Science Letters*, 423, 98–113, doi:10.1016/j.epsl.2015.04.026.

Ellis, B., K. R. Johnson, and R. E. Dunn, 2003, Evidence for an in situ early Paleocene rainforest from Castle Rock, Colorado: *Rocky Mountain Geology*, 38, 73–100, doi:10.2113/gsrocky.38.1.173.

Enomoto, C. B., K. R. Scott, B. Valentine, *et al.*, 2012, Preliminary evaluation of the shale gas prospectivity of the Lower Cretaceous Pearsall Formation in the onshore Gulf Coast region, United States: *Gulf Coast Association of Geological Societies Transactions*, 62, 93–115.

Epstein, S. A., and D. Clark, 2009, Hydrocarbon potential of the Mesozoic carbonates of the Bahamas: *Carbonates and Evaporites*, 24, 97–138.

Escalona, A., and W. Yang, 2013, Subsidence controls on foreland basin development of northwestern offshore Cuba, southeastern Gulf of Mexico: *American Association of Petroleum Geologists Bulletin*, 97, 1–25, doi:10.1306/06111212002.

Essex, C. W., D. M. Robinson, and A. L. Weislogel, 2016, Regional correlation of lithofacies within the Haynesville Formation from onshore Alabama: analysis and implications for provenance and paleostructure, in C. M. Lowery, J. W. Snedden, and N. C. Rosen, eds., *Gulf Rim and Beyond: New Progress in Science and Exploration of the Gulf of Mexico Basin*: Society of Economic Paleontologists and Mineralogists Gulf Coast Section, 35th Annual Research Conference Program and Extended Abstracts, 309–343.

Ewing, T. E., 1991, Structure framework, in A. Salvador, ed., *The Gulf of Mexico Basin: The Geology of North America*: Geological Society of America, 31–52, doi:10.1130/DNAG-GNA-J.31.

Ewing, T. E., 2007, Fairways in the Downdip and Middip Yegua Trend: a review of 25 years of exploration: *Gulf Coast Association of Geological Societies Transactions*, 57, 22 p.

Ewing, T. E., 2009, The ups and downs of the Sabine Uplift and the northern Gulf of Mexico Basin: Jurassic basement blocks, Cretaceous thermal uplifts, and Cenozoic flexure: *Gulf Coast Association of Geological Societies Transactions*, 59, 253–269.

Ewing, T. E., 2010, Pre-Pearsall geology and exploration plays in South Texas: *Gulf Coast Association of Geological Societies Transactions*, 60, 241–260.

Ewing, T. E., 2016 *Texas Through Time: Lone Star Geology, Landscapes, and Resources*, Texas Bureau of Economic Geology, 431 p.

Ewing, T. E., and G. Fergeson, 1991, Stratigraphic framework, structural styles and seismic signatures of Downdip Yegua Gas-condensate fields, Central Wharton County, Texas Gulf Coast: *Gulf Coast Association of Geological Societies Transactions*, 41, 255–275.

Ewing, T. E., and R. F. Lopez, 1991, Principal structural features, Gulf of Mexico Basin, in A. Salvador, ed., *The Gulf of Mexico Basin, Plate 2; The Geology of North America, Plate 2*: Geological Society of America.

Ewing, T. E., and R. S. Reed, 1984, *Depositional Systems and Structural Controls of Hackberry Sandstone Reservoirs in Southeast Texas*: The

University of Texas at Austin, Bureau of Economic Geology, Geological Circular 84-7, 48 p.

Ewing, T. E., and F. S. Vincent, 1997, Foundered shelf edges: examples from the Yegua and Frio, Texas and Louisiana: *Gulf Coast Association of Geological Societies Transactions*, 47, 149–158.

Fan, M., E. Brown, and L. Li, 2018, Cenozoic drainage evolution of the Rio Grande paleoriver recorded in detrital zircons in South Texas: *International Geology Review*, doi:10.1080/00206814.2018.1446368.

Fang, G., 2000, *Biostratigraphic and Sequence Stratigraphic Analysis of the Yegua Formation, Houston Salt Embayment, Northern Gulf of Mexico*: Ph. D. dissertation; The University of Texas at Austin, 284 p.

Faust, M., 1984, *Seismic Stratigraphy of the Middle Cretaceous Unconformity (MCU) in the Central Gulf of Mexico Basin*: Master's thesis, The University of Texas at Austin, 171 p.

Fedele, J. J., D. Hoyal, Z. Barnaal, J. Tulenko, and S. Awalt, 2016, Bedforms created by gravity flows, in D. A. Budd, E. A. Hajek, S. J. Purkis, eds., *Autogenic Dynamics and Self-Organization in Sedimentary Systems*: SEPM Special Publication 106, 95–121, doi:10.2110/sepmsp.106.12.

Feragen, E., D. Millman, H. Feldman, R. Bierley, and S. Perkins, 2007, South Texas sub-regional evaluation: area-wide integrated structural and stratigraphic framework of the Frio and Vicksburg Yields new plays and leads: *Gulf Coast Association of Geological Societies Transactions*, 57, 249–253.

Ferrill, D. A., K. J. Smart, R. N. McGinnis, A. P. Morris, and K. D. H. Gulliver, 2017, Influence of structural position on fracturing in the Austin Chalk: *Gulf Coast Association of Geological Societies Journal*, 6, 189–200.

Ferworn, K., J. Zumberge, and S. Brown, 2003, Integration of geochemistry and reservoir fluid properties: *Bureau of Economic Geology PTTC Workshop, First Annual Fluids Symposium Reservoir Fluids 2003-PVT and Beyond*, 51 p.

Fiduk, J. C., P. Weimer, B. D. Trudgill, *et al.*, 1999, The Perdido Fold Belt, northwestern deep Gulf of Mexico, part II: seismic stratigraphy and petroleum systems: *American Association of Petroleum Geologists Bulletin*, 83, 578–612.

Fiduk, J. C., L. E. Anderson, and M. G. Rowan, 2004: The Wilcox Raft: an example of extensional raft tectonics in south Texas, northwestern onshore Gulf of Mexico, in P. J. Post, D. L. Olson, K. T. Lyons, *et al.*, eds., *Salt–Sediment Interactions and Hydrocarbon Prospectivity: Concepts, Applications, and Case Studies for the 21st Century*: Society of Economic Paleontologists and Mineralogists Gulf Coast Section, 24th Annual Bob Perkins Research Conference, 293–314.

Fiduk, J. C., M. Clippard, S. Power, *et al.*, 2014, Origin, transportation, and deformation of Mesozoic carbonate rafts in the northern Gulf of Mexico: *Gulf Coast Association of Geological Societies Journal*, 3, 20–23.

Finneran, J. B., 1984, Zuloaga Formation (Upper Jurassic) Shoal Complex, Sierra de Enfrente, Coahuila, Northeast Mexico: *American Association of Petroleum Geologists Bulletin*, 68, 476 (abst).

Finneran, J. M., R. W. Scott, G. A. Taylor, and G. H. Anderson, 1984, Lowermost Cretaceous ramp reefs: Knowles Limestone, Southwest Flank of the East Texas Basin, in W. P. S. Ventress, D. G. Bebout, B. F. Perkins, and C. H. Moore, eds., *The Jurassic of the Gulf Rim*: Society of Economic Paleontologists and Mineralogists Gulf Coast Section Foundation, 3rd Annual Research Conference, 3, 125–133.

Fisher, W. L., and J. H. McGowen, 1967, Depositional systems in the Wilcox Group of Texas and their relationship to occurrence of oil and gas: *Gulf Coast Association of Geological Societies Transactions*, 17, 105–125.

Fisher, W. L., C. V. Proctor Jr., W. E. Galloway, and J. S. Nagle, 1970, *Depositional Systems in the Jackson Group of Texas: Their Relationship to Oil, Gas and Uranium*: The University of Texas at Austin, Bureau of Economic Geology, Geological Circular 70-4, 27 p.

Fitz-Díaz, E., T. F. Lawton, E. Juárez-Arriaga, and G. Chávez-Cabello, 2018, The Cretaceous–Paleogene Mexican orogen: structure, basin development, magmatism and tectonics: *Earth Science Reviews*, 183, 56–84.

Flawn, P., A. Goldstein Jr., P. King, and C. E. Weaver, 1961, *The Ouachita System*, The University of Texas at Austin, Bureau of Economic Geology, 401 p.

Fletcher, R. C., M. R. Hudec, and I. A. Watson, 1995, Salt glacier and composite salt–sediment models for the emplacement and early burial of allochthonous salt sheets, in M. P. A. Jackson, D. G. Roberts, and S. Snelson, eds., *Salt Tectonics: A Global Perspective*: American Association of Petroleum Geologists, Memoir 65, 77–108.

Föllmi, K. B., H. Weissert, M. Bisping, and H. Funk, 1994, Phosphogenesis, carbon-isotope stratigraphy, an carbonate platform evolution along the Lower Cretaceous northern Tethyan margin: *American Association of Petroleum Geologists Bulletin*, 106, 729–746.

Forgotson, J. M., and J. M. Forgotson Jr., 1976, Definition of Gilmer Limestone, Upper Jurassic formation, northeast Texas: *American Association of Petroleum Geologists Bulletin*, 60, 1119–1123.

Foss, D. C., 1979, Depositional environment of Woodbine sandstones, Polk County, Texas: *Gulf Coast Association of Geological Societies Transactions*, 29, 83–94.

Fouad, K., L. F. Brown Jr., W. A. Ambrose, *et al.*, 2009, Genetic facies analysis using seismic geomorphology and seismic attributes in the continental shelf of Eastern Mexico: petroleum systems in the southern Gulf of Mexico, in C. Bartolini and J. R. Roman Ramos, eds., *Petroleum Systems in the Southern Gulf of Mexico*: American Association of Petroleum Geologists, Memoir 90, 471–476.

Frazier, D. E., 1974, *Depositional Episodes: Their Relationship to the Quaternary Stratigraphic Framework in the Northwestern Portion of the Gulf Basin*: The University of Texas at Austin, Bureau of Economic Geology, Geological Circular 74-1, 28 p.

Frébourg, G., S. C. Ruppel, R. G. Loucks, and J. Lambert, 2016, Depositional controls on sediment body architecture in the Eagle Ford/Boquillas system: insights from outcrops in west Texas, United States: *American Association of Petroleum Geologists Bulletin*, 100, 657–682.

Frederick, B. C., M. D. Blum and J. W. Snedden, in review, Pre-Salt Eagle Mills Formation and coeval siliciclastic sources and sinks, Northern Gulf of Mexico basin: *Geological Society of America Bulletin.*

Fredrich, J. T., A. F. Fossum, and R. J. Fossum, 2007, Mineralogy of deepwater Gulf of Mexico salt formations and implications for constitutive behavior: *Journal of Petroleum Science and Engineering*, 57, 354–374.

Freeman-Lynde, R. P., 1983, Cretaceous and Tertiary samples dredged from the Florida Escarpment: *Gulf Coast Association of Geological Societies Transactions*, 29, 91–100.

Fritz, D. A., T. W. Belsher, J. M. Medlin, *et al.*, 2000, New exploration concepts for the Edwards and Sligo Margins, Cretaceous of Onshore Texas: *American Association of Petroleum Geologists Bulletin*, 84, 905–922.

Frost, W. G., 2010, The somewhat accidental discovery of the Mobile Bay gas field: a story of perseverance and good fortune: *American Association of Petroleum Geologists, Search and Discovery* 110133, 27 p.

Frush, M. P., and D. L. Eicher, 1975, Cenomanian and Turonian foraminifera and paleoenvironments in the Big Bend region of Texas and Mexico, in W. G. E. Caldwell, ed., *The Cretaceous System in the Western Interior of North America*, The Geological Society of Canada Special Paper 13, 277–301.

Fryberger, S. G., A. M. Al-Sari, and T. J. Clisham, 1983, Eolian dune, interdune, sand sheet, and silica clastic sabkha sediments of an offshore prograding sand sea, Dhahran area, Saudi Arabia: *American Association of Petroleum Geologists Bulletin*, 67, 280–312.

Galloway, W. E., 1968, Depositional systems of the Lower Wilcox Group, north-central Gulf Coast basin: *Gulf Coast Association of Geological Societies Transactions*, 18, 275–289.

Galloway, W. E., 1977, *Catahoula Formation of the Texas Coastal Plain: Depositional Systems, Mineralogy, Structural Development, Ground-Water Flow History, and Uranium Distribution*: The University of Texas at Austin, Bureau of Economic Geology, Report of Investigations 87, 59 p.

Galloway, W. E., 1986a, Reservoir facies architecture of microtidal barrier systems: *American Association of Petroleum Geologists Bulletin*, 70, 787–808.

Galloway, W. E., 1986b, *Depositional and Structural Framework of the Distal Frio Formation, Texas Coastal Zone and Shelf*: The University of Texas at Austin, Bureau of Economic Geology, Geological Circular 86-8, 16 p.

Galloway, W. E., 1989a, Genetic stratigraphic sequences in basin analysis I: architecture and genesis of flooding-surface bounded depositional units: *American Association of Petroleum Geologists Bulletin*, 73, 125–142, doi:10.1306/703C9AF5-1707-11D7-8645000102C1865D.

Galloway, W. E., 1989b, Genetic stratigraphic sequences in basin analysis II: application to Northwest Gulf of Mexico Cenozoic Basin: *American Association of Petroleum Geologists Bulletin*, 73, 143–154, doi:10.1306/703C9AFA-1707-11D7-8645000102C1865D.

Galloway, W. E., 1998, Siliciclastic slope and base-of-slope depositional systems: component facies, stratigraphic architecture, and classification: *American Association of Petroleum Geologists Bulletin*, 82, 569–595.

Galloway, W. E., 2002, Cenozoic evolution of sedimentation accumulation in deltaic and shore-zone depositional systems, northern Gulf of Mexico basin: *Marine and Petroleum Geology*, 18, 1031–1040.

Galloway, W. E., 2005a, Cenozoic evolution of the northern Gulf of Mexico continental margin, in P. J. Post, N. C. Rosen, D. L. Olson, *et al.*, eds., *Petroleum Systems of Divergent Continental Margins*: Society of Economic Paleontologists and Mineralogists Gulf Coast Section, 25th Annual Bob F. Perkins Research Conference, 613–633.

Galloway, W. E., 2005b, Gulf of Mexico Basin depositional record of Cenozoic North American drainage basin evolution, in M. D. Blum, S. B. Marriott, and S. F. Leclair, eds., *Fluvial Sedimentology VII: International Association of Sedimentologists Special Publication 35*, Wiley-Blackwell, 409–423.

Galloway, W. E., 2007, Wilcox canyons: distribution, attributes, origins and relationship to basinal sands, in L. Kennan, J. Pindell, and N.C. Rosen, eds., *The Paleogene of the Gulf of Mexico and Caribbean Basins: Processes, Events and Petroleum Systems*: Society of Economic Paleontologists and Mineralogists Gulf Coast Section, 27th Annual Bob F. Perkins Research Conference, 101–102.

Galloway, W. E., 2008, Depositional evolution of the Gulf of Mexico sedimentary basin, in K. J. Hsu, ed., *Sedimentary Basins of the World*: Elsevier, 5, 505–549, doi:10.1016/S1874-5997(08)00015-4.

Galloway, W. E, 2009, The Gulf of Mexico: geologic foundations of a petroleum mega-province: *Geo Expro*, 6, 22–26.

Galloway, W. E., and E. S. Cheng, 1985, *Reservoir Facies Architecture in a Microtidal Barrier System: Frio Formation, Texas Gulf Coast*, The University of Texas at Austin, Bureau of Economic Geology, Report of Investigations 144, 36 p.

Galloway, W. E., and D. K. Hobday, 1996, Terrigenous clastic depositional systems: *Applications to Fossil Fuel and Groundwater Resources*, 489, doi:10.1007/978-3-642-61018-9.

Galloway, W. E., and W. R. Kaiser, 1980, *Catahoula Formation of the Texas Coastal Plain: Origin, Geochemical Evolution, and Characteristics of Uranium Deposits*: The University of Texas at Austin, Bureau of Economic Geology, Report of Investigations 100, 81 p.

Galloway, W. E., and T. A. McGilvery, 1995, *Facies of a Submarine Canyon Fill Reservoir Complex, Lower Wilcox Group (Paleocene), Central Texas Coastal Plain*: Society of Economic Paleontologists and Mineralogists, Special Publication 20, 1–23.

Galloway, W. E., and R. A. Morton, 1989, Geometry, genesis, and reservoir characteristics of shelf sandstone facies, Frio Formation (Oligocene), Texas Coastal Plain, in R.A. Morton and D. Nummedal, eds., *Shelf Sedimentation, Shelf Sequences, and Related Hydrocarbon Accumulation*: Society of Economic Paleontologists and Mineralogists Foundation Gulf Coast Section 7th Annual Research Conference, 89–115.

Galloway, W. E., and T. A. Williams, 1991, Sediment accumulation rates in time and space: Paleogene genetic stratigraphic sequences of the northwestern Gulf of Mexico basin: *Geology*, 19, 986–989.

Galloway, W. E., R. J. Finley and C. D. Henry, 1979, *South Texas Uranium Province: Geologic Perspective*: The University of Texas at Austin, Bureau of Economic Geology, Guidebook 18, 81 p.

Galloway, W. E., D. K. Hobday, and M. Kinji, 1982a, Frio Formation of Texas Gulf Coastal Plain: depositional systems, structural framework, and hydrocarbon distribution: *American Association of Petroleum Geologists Bulletin*, 66, 649–688.

Galloway, W. E., D. K. Hobday, and M. Kinji, 1982b, *Frio Formation of Texas Gulf Coastal Basin: Depositional Systems, Structural Framework, and Hydrocarbon Origin, Migration, Distribution, and Exploration Potential*: The University of Texas at Austin, Bureau of Economic Geology, Report of Investigations 122, 78 p.

Galloway, W. E., C. D. Henry, and G. E. Smith, 1982c, *Depositional Framework, Hydrostratigraphy, and Uranium Mineralization of the Oakville Sandstone (Miocene), Texas Coastal Plain*: The University of Texas at Austin, Bureau of Economic Geology, Report of Investigations 113, 51 p.

Galloway, W. E., T. E. Ewing C. M. Garrett, N. Tyler, and D. G. Bebout, 1983, *Atlas of Major Texas Oil Reservoirs*: The University of Texas at Austin, Bureau of Economic Geology, 139 p.

Galloway, W. E., L. A. Jirik, R. A. Morton, and J. R. DuBar, 1986, *Lower Miocene (Fleming) Depositional Episode of the Texas Coastal Plain and Continental Shelf: Structural Framework, Facies, and Hydrocarbon Resources*: The University of Texas at Austin, Bureau of Economic Geology, Report of Investigations 150, 50 p.

Galloway, W. E., D. G. Bebout, W. L. Fisher, *et al.*, 1991a, Cenozoic, in A. Salvador, ed., *The Gulf of Mexico Basin, The Geology of North America*, Geological Society of America, 245–324.

Galloway, W. E., W. F. Dingus, and R. E. Paige, 1991b, Seismic and depositional facies of Paleocene–Eocene Wilcox Group submarine canyon fills, northwest Gulf Coast, U.S.A., in P. Weimer and M. H. Link, eds., *Seismic Facies and Sedimentary Processes of Submarine Fans and Turbidite Systems*: Springer-Verlag, 247–271.

Galloway, W. E., X. Liu, D. T. Neuberger, and L. Xue, 1994, *Reference High-Resolution Correlation Cross Sections, Paleogene Section, Texas Coastal Plain*: The University of Texas at Austin, Bureau of Economic Geology.

Galloway, W. E., P. E. Ganey-Curry, X. Li, and R. T. Buffler, 2000, Cenozoic depositional history of the Gulf of Mexico basin: *American Association of Petroleum Geologists Bulletin*, 84, 1743–1774.

Galloway, W. E., T. L. Whiteaker, and P. E. Ganey-Curry, 2011, History of Cenozoic North American drainage basin evolution, sediment yield, and accumulation in the Gulf of Mexico basin: *Geosphere*, 7, 938–973.

Garcia-Molina, G., 1994, *Structural Evolution of Southeastern Mexico (Chiapas-Tabasco-Campeche) Offshore and Onshore*: Ph.D. dissertation; Rice University, 161 p.

Gawloski, T., 1983, Stratigraphy and environmental significance of the continental Triassic rocks of Texas: *Baylor Geological Studies Bulletin*, 41, 1–48.

Ge, H., M. P. A. Jackson, and B. C. Vendeville, 1997, Kinematics and dynamics of salt tectonics driven by progradation: *American Association of Petroleum Geologists Bulletin*, 81, 398–423.

Gehrels, G., and M. Pecha, 2014, Detrital zircon U–Pb geochronology and Hf isotope geochemistry of Paleozoic and Triassic passive margin strata of western North America, *Geosphere*, 10, 49–65, doi:10.1130/GES00889.1.

GeoMark Research Ltd. and TDI-Brooks International Inc., 2005a, 1997, 1999, 2005 Western Gulf of Mexico surface geochemical exploration programs: geochemical correlation of seep oils with offshore petroleum systems: Proprietary study.

GeoMark Research Ltd. and TDI-Brooks International Inc., 2005b, 1997–2001 Central/Eastern Gulf of Mexico surface geochemical exploration programs: geochemical correlation of seep oils with offshore petroleum systems: Proprietary study.

Gleason, J. D., G. E. Gehrels, W. R. Dickinson, P. J. Patchett, and D. A. Kring, 2007, Laurentian sources for detrital zircon grains in turbidite and deltaic sandstones of the Pennsylvanian Haymond Formation, Marathon Assemblage, West Texas, U.S.A: *Journal of Sedimentary Research*, 77, 888–900, doi:10.2110/jsr.2007.084.

Glennie, K. W., 1972, Permian Rotliegendes of Northwest Europe interpreted in light of modern desert sedimentation studies: *American Association of Petroleum Geologists Bulletin*, 56, 1048–1071.

Godo, T. J., 2006, Identification of stratigraphic traps with subtle seismic amplitude effects in Miocene channel/levee sand systems, NE Gulf of Mexico, in M. R. Allen, G. P. Goffey, R. K. Morgan, and I. M. Walker, eds., *The Deliberate Search for the Stratigraphic Trap*: Geological Society, Special Publication 254, 127–151, doi:10.1144/GSL.SP.2006.254.01.07.

Godo, T., 2017, The Appomattox field: Norphlet aeolian sand dune reservoirs in the deep-water Gulf of Mexico, in R. K. Merrill and C. A. Sternbach, eds., *Giant Fields of the Decade 2000–2010*: American Association of Petroleum Geologists, Memoir 113, 29–54.

Godo, T. J., E. Chuparova, and D. E. McKinney, 2011, Norphlet aeolian sand fairway established in the deep water Gulf of Mexico: *American Association of Petroleum Geologists, Search and Discovery* 90124, 2 p.

Goggin, K., and J. Rine, 2014, *Geological Characterization of a Portion of the South Georgia Rift Basin for Source Proximal CO_2 Storage Based on Analysis of Samples from Rizer #1 Test Borehole, Colleton County, SC and USGS Clubhouse Crossroads Test Hole no. 3, Dorchester County, SC*: Weatherford Laboratories, reports HH-57745 and HH-59364.

Gohrbandt, K. H., 2002, Potential gas resources under the West Florida shelf and slope and their development: *Gulf Coast Association of Geological Societies Transactions*, 52, 351–360.

Goldhammer, R. K., and C. A. Johnson, 2001, Middle Jurassic–Upper Cretaceous paleogeographic evolution and sequence-stratigraphy framework of the Northwest Gulf of Mexico Rim, in C. Bartolini, R. T. Buffler, and A. Cantú-Chapa, eds., *The Western Gulf of Mexico Basin: Tectonics, Sedimentary Basins, and Petroleum Systems*: American Association of Petroleum Geologists, Memoir 75, 45–81.

Gomez, E., M. Doe, S. Villarroel, *et al.*, 2018, Recent Yucatan seismic survey revealing a new frontier exploration in the Gulf of Mexico: *American Association Petroleum Geologists, Search and Discovery* 30582, 1–6.

Gomez-Cabrera, P. T., and M. P. A. Jackson, 2009a, Regional Neogene salt tectonics in the Offshore Salina del Istmo Basin, Southeastern Mexico, in C. Bartolini and J.R. Roman Ramos, eds., *Petroleum Systems in the Southern Gulf of Mexico*: American Association of Petroleum Geologists, Memoir 90, 1–28.

Gomez-Cabrera, P. T., and M. P. A. Jackson, 2009b, Neogene stratigraphy and salt tectonics of the Santa Ana area, Offshore Salina del Istmo Basin, Southeastern Mexico, in C. Bartolini and J. R. Roman Ramos, eds., *Petroleum Systems in the Southern Gulf of Mexico*: American Association of Petroleum Geologists, Memoir 90, 237–255, doi:10.1306/13191086M9037.

Gong, C., R. J. Steel, Y. Wang, C. Lin, and C. Olariu, 2016, Grain size and transport regime at shelf edge as fundamental controls on delivery of shelf-edge sands to deepwater: *Earth-Science Reviews*, 157, 32–60.

Gonzales, E., and M. Medrano, 2014, Structural slope fans resulting from Paleogene compression in the Veracruz Basin, Mexico: *American Association of Petroleum Geologists, Search and Discovery* 50963, 17 p.

Gonzalez-Posadas, J. F., S. Avendano-Lopez, J. Molina, 2005, Strike–slip model for the Jacinto and Paredon Fields of the Chiapas-Tabasco Region, South East Basin, Mexico: *American Association of Petroleum Geologists, Search and Discovery* 20028, 1–7.

Gordon, M. B., P. Mann, D. Caceres, and R. Flores, 1997, Cenozoic tectonic history of the North America–Caribbean plate boundary zone in western Cuba: *Journal of Geophysical Research*, 102, 10055–10082, doi:10.1029/96JB03177.

Grabowski, G. J., 1981, Source-rock potential of the Austin Chalk, Upper Cretaceous, Southeast Texas: *Gulf Coast Association of Geological Societies Transactions*, 31, 105–113.

Gradmann, S., C. Beaumont, and M. Albertz, 2009, Factors controlling the evolution of the Perdido Fold Belt, northwestern Gulf of Mexico, determined from numerical models: *Tectonics*, 28, doi:10.1029/2008TC002326.

Gradstein, F. M., J. G. Ogg, M. Schmitz, and G. Ogg, 2012, *The Geologic Time Scale 2012*: Elsevier, 1176 p.

Grajales-Nishimura, J. M., E. Cedillo-Pardo, C. Rosales-Domínguez, *et al.*, 2000, Chicxulub impact: the origin or reservoir and seal facies in the southeastern Mexico oil fields: *Geology*, 28, 307–310, doi:10.1130/0091-7613(2000)28<307:CITOOR>2.0.CO;2.

Gray, G. G., R. J. Pottorf, D. A. Yurewicz, *et al.*, 2001, Thermal and chronological record of syn- to post- Laramide burial and exhumation, Sierra Madre Oriental, Mexico, in C. Bartolini, R. T. Buffler, and A. Cantú-Chapa, eds., *The Western Gulf of Mexico Basin: Tectonics, Sedimentary Basins, and Petroleum Systems*: American Association of Petroleum Geologists, Memoir 75, 159–181.

Greene, T. J., and B. E. O'Neill, 2005, Building a facies-based permeability model for deep-water Miocene reservoirs, eastern Gulf of Mexico: *Gulf Coast Association of Geological Societies Transactions*, 55, 268–277.

Gregory, J. L., 1966, A lower Oligocene delta in the subsurface of southeastern Texas: *Gulf Coast Association of Geological Societies Transactions*, 16, 214–227.

Grice, K., C. Cao, G. D. Love, *et al.* 2005, Photic zone euxinia during the Permia–Triassic superanoxic event: *Science*, 307, 706–709.

Gulick, S. P. S., P. J. Barton, G. L. Christeson, *et al.*, 2008, Importance of pre-impact crustal structure for the asymmetry of the Chicxulub impact crater: *Nature Geoscience*, 1, 131–135, doi:10.1038/ngeo103.

Gulick, S. P. S., G. L. Christeson, P. J. Barton, *et al.*, 2013, Geophysical characterization of the Chicxulub impact crater: *Review of Geophysics*, 51, 31–52, doi:10.1002/rog.20007.

Gutiérrez, M. A., 2018, *Systematic Lithologic Characterization of Pleistocene Mass-Transport Deposit, Mississippi Canyon of the Northern Gulf of Mexico, USA*: MS thesis, The University of Texas at Austin, 102 p.

Gutiérrez Paredes, H. C., O. Cantuneanu, and U. Hernandez Romano, 2017, Sequence stratigraphy of the Miocene section, southern Gulf of Mexico: *Marine and Petroleum Geology*, 86, 711–732, doi:10.1016/j.marpetgeo.2017.06.022.

Gutiérrez-Puente, N. A., 2006, *Estudio Micropaleontologico y Bioestratigrafico de la Columna del Pozo Cupelado – 10 (Cuenca Tampico Misantla)*: Master's thesis, Universidad Nacional Autonoma de Mexico, 102 p.

Gutteridge, P., Y. Poprawski, and A. Horbury, 2019, Salt–carbonate interactions in the Sureste Basin, SE Mexico: depositional models and analogs for Cretaceous carbonate breccias: The Geological Society, London, Petroleum Geology of Mexico and the Northern Caribbean Conference (abstract).

Guzmán, 2001, Exploration and production in México: challenges and opportunities: *American Association of Petroleum Geologists, Search and Discovery*, 10222, 47 p.

Guzmán, A. E., 2013, Petroleum history of Mexico: how it got to where it is today: *American Association of Petroleum Geologists, Search and Discovery* 10530, 27 p.

Guzmán, A. E., 2018, The upstream in México under the new energy reform: *American Association of Petroleum Geologists, Search and Discovery* 703, 19 p.

Guzmán-Vega, M. A., and M. R. Mello, 1999, Origin of oil in the Sureste Basin, Mexico: *American Association of Petroleum Geologists Bulletin*, 83, 1068–1095.

Guzmán-Vega, M. A., L. Castro Ortíz, J. R. Román-Ramos, *et al.*, 2001, Classification and origin of petroleum in the Mexican Gulf Coast Basin: an overview, in C. Bartolini, R. T. Buffler, and A. Cantú-Chapa, eds., *The Western Gulf of Mexico Basin: Tectonics, Sedimentary Basins, and Petroleum Systems*, American Association of Petroleum Geologists, Memoir 75, 127–142.

Hackley, P. C., 2012, Geological and geochemical characterization of the Lower Cretaceous Pearsall Formation, Maverick Basin, south Texas: a future shale gas resource?: *American Association of Petroleum Geologists Bulletin*, 96, 1449- 1482, doi:10.1306/11221111071.

Hackley, P. C., and T. E. Ewing, 2010, Assessment of undiscovered conventional oil and gas resources, onshore Claiborne Group, United States part of the northern Gulf of Mexico Basin: *American Association of Petroleum Geologists Bulletin*, 94,1607–1636, doi:10.1306/04061009139.

Haczewski, G., 1976, Sedimentological reconnaissance of the San Cayetano Formation: an accumulative continental margin in the Jurassic of western Cuba: *Acta Geologica Polonica*, 26.2, 331–353.

Haddad, S. A., and E. A. Mancini, 2013, Reservoir characterization modeling, and evaluation of Upper Jurassic Smackover microbial carbonate and associated facies in Little Cedar Creek Field, Southwest Alabama, Eastern Gulf Coastal Plain of the United States: *American Association of Petroleum Geologists Bulletin*, 97, 2059–2083.

Halbouty, M. T., and G. C. Hardin, 1956, Genesis of salt domes of Gulf Coastal Plain: *American Association of Petroleum Geologists Bulletin*, 40, 737–746.

Hall, D. J., B. E. Bowen, R. N. Rosen, S. Wu, and A. W. Bally, 1993, Mesozoic and Early Cenozoic development of the Texas Margin: a new integrated cross-section from the Cretaceous Shelf Edge to the Perdido Fold Belt: *Society of Economic Paleontologists and Mineralogists Gulf Coast Section 13th Annual Research Conference*, 21–31.

Hamlin, H. S., 1988, *Depositional and Ground-Water Flow Systems of the Carrizo-Upper Wilcox, South Texas*: The University of Texas at Austin, Bureau of Economic Geology, Report of Investigations 175, 61 p.

Hammes U., and G. Frébourg, 2012, Haynesville and Bossier mudrocks: a facies and sequence stratigraphic investigation, East Texas and Louisiana, USA: *Marine and Petroleum Geology*, 31, 8–26, doi:10.1016/j.marpetgeo.2011.10.001.

Hammes, U., H. S. Hamlin, and T. E. Ewing, 2011, Geologic analysis of the Upper Jurassic Haynesville Shale in east Texas and west Louisiana: *American Association of Petroleum Geologists Bulletin*, 95, 1643–1666, doi:10.1306/02141110128.

Hammes, U., R. Eastwood, G. McDaid, *et al.*, 2016, Regional assessment of the Eagle Ford Group of South Texas, USA: insights from lithology, pore volume, water saturation, organic richness, and productivity correlations: *Interpretation*, 4, 125–150, doi:10.1190/int-2015-0099.1.

Handford, C. R., and Loucks, R. G., 1993, Carbonate depositional sequences and systems tracts: responses of carbonate platforms to relative sea-level changes, in R. G. Loucks, and J. F. Sarg, eds., *Carbonate Sequence Stratigraphy: Recent Developments and Applications*: American Association of Petroleum Geologists, Memoir 57, 3–42.

Haq, B. U., 2014, Cretaceous eustasy revisited: *Global and Planetary Change*, 113, 44–58, doi:10.1016/j.gloplacha.2013.12.007.

Haq, B. U., 2017, Jurassic sea-level variations: a reappraisal: *GSA Today*, 28, 7, doi: 10.1130/GSATG359A.1.

Haq, B. U., J. Hardenbol, and P. R. Vail, 1987, Chronology of fluctuating sea levels since the Triassic: *Science*, 235, 1156–1167.

Harbour, J. L., and R. L. Mathis, 1984, Sedimentation, diagenesis and porosity evolution of carbonate sands in the Black Lake Field of Central Louisiana, in P. M. Harris, ed., *Society of Economic Paleontologists and Mineralogists Core Workshop 5: Carbonate Sands*, 306–333.

Hardenbol, J., J. Thierry, M. B. Farley, *et al.*, 1998, Mesozoic and Cenozoic sequence chronostratigraphic framework of European basins, in P. C. de Graciansky, J. Hardenbol, T. Jacquin, and P. Vail, eds., *Mesozoic and Cenozoic Sequence Stratigraphy of European Basins*: Society of Economic Paleontologists and Mineralogists, Special Publication 60, 3–14.

Harding, A., L. Walker, S. Ehlinger, and T. Chapman, 2016, The siliciclastic Upper Cretaceous play of Eastern Mississippi Canyon, in C. M. Lowery and J. W. Snedden, eds. *Mesozoic of the Gulf Rim and Beyond: New Progress in Science and Exploration of the Gulf of Mexico*: Society of Economic Paleontologists and Mineralogists Gulf Coast Section, 35th Annual Research Conference Program and Extended Abstracts.

Harris, P. M., 2008, Stratigraphic framework and new exploration concepts for the Lower Cretaceous Shelf Margin carbonates of Texas: *American Association of Petroleum Geologists, Search and Discovery* 40303, 41 p.

Harris, P. M., and W. S. Kowalik, 2005, *Satellite Images of Carbonate Depositional Settings*, American Association of Petroleum Geologists, Methods in Exploration 11, digital reprint.

Hart, W., J. Jacek, and A. Martin, 2004, Recognition and exploration significance of supra-salt stratal carapaces, in P. J. Post, D. L. Olson, K. T. Lyons, *et al.*, eds., *Salt-Sediment Interactions and Hydrocarbon Prospectivity: Concepts, Applications, and Case Studies for the 21st Century*: Society of Economic Paleontologists and Mineralogists Gulf Coast Section, 21st Annual Research Conference Program and Extended Abstracts, 166–199.

Hartley, A. J., G. Weissman, and L. Scuderi, 2015, Controls on the apex location of large deltas: *Journal of the Geological Society*, 174, 10–13.

Harwood, G., and C. Fontana, 1984, Smackover deposition and diagenesis and structural history of the Bryan's Mill Area, Cass and Bowie Counties, Texas: *Society of Economic Paleontologists and Mineralogists Gulf Coast Section, 3rd Annual Research Conference Proceedings*, 135–147.

Haymond, D., 1991, The Austin Chalk: an overview: *Houston Geological Society Bulletin*, 24, 21–27.

Heatherington, A. L., and P. A. Mueller, 2003, Mesozoic igneous activity in the Suwannee Terrane, Southeastern USA: petrogenesis and Gondwanan Affinities: *Gondwana Research*, 6, 296–311, doi:10.1016/S1342-937X(05)70979-5.

Heffner, D. M., 2013, *Tectonics of the South Georgia Rift*: Ph.D. dissertation, The University of South Carolina, 178 p.

Heintz, M. L., T. E. Yancey, B. V. Miller, and M. T. Heizler, 2015, Tephrochronology and geochemistry of Eocene and Oligocene volcanic ashes of east and central Texas: *Geological Society of America Bulletin*, 127, 770–780.

Helland-Hansen, W., Sømme, T. O., Martinsen, O. J., Lunt, I., and Thurmond, J., 2016, Deciphering Earth's natural hourglasses: perspectives on source-to-sink analysis: *Journal of Sedimentary Research*, 86, 1008–1033, doi:10.2110/jsr.2016.56.

Henry, L. C., J. A. Wadsworth, B. Hansen, 2017, Visualizing a sub-salt field with image logs: image facies, mass transport complexes, and reservoir implications from Thunder Horse, Mississippi Canyon, Gulf of Mexico: *American Association of Petroleum Geologists, Search and Discovery* 10938, 21 p.

Hentz, T. F., and S. C. Ruppel, 2011, Regional stratigraphic and rock characteristics of Eagle Ford Shale in its play area: Maverick Basin to East Texas Basin: *American Association of Petroleum Geologists, Search and Discovery* 10325, 20 p.

Hermann, L. A., 1971, Lower Cretaceous Sligo reef trends in central Louisiana: *Gulf Coast Association of Geological Societies Transactions*, 21, 187–198.

Hernandez Calvento, L., D. W. T. Jackson, A. Cooper, and E. Perez-Chacon, 2017, Island-encapsulating eolian sedimentary systems of the Canary and Cape Verde archipelagos: *Journal of Sedimentary Research*, 87, 117–125, doi:10.2110/jsr.2017.6

Hernandez-Mendoza, H. S., 2013, *Stratigraphic Characterization and Evolution of a Mid-Tertiary Age Deep Water System, Holok Area, SW Gulf of Mexico*: Ph.D. dissertation: University of Aberdeen, 358 p.

Hernandez-Mendoza, J. J., 2000, *Interpretation and Distribution of Depositional Systems: Oligocene Frio Depisode in the Subsurface of Burgos Basin, Northeastern Mexico*: MS thesis, The University of Texas at Austin, 102 p.

Hernandez-Mendoza, J. J., M. V. DeAngelo, T. F. Wawrzyniec, and T. F. Hentz, 2008a, Miocene chronostratigraphy, paleogeography, and play framework of the Burgos Basin, southern Gulf of Mexico: *American Association of Petroleum Geologists Bulletin*, 92, 1501–1535.

Hernandez-Mendoza, J. J., M. V. DeAngelo, T. F. Wawrzyniec, and T. F. Hentz, 2008b, Major structural elements of the Miocene section, Burgos Basin, northeastern Mexico: *American Association of Petroleum Geologists Bulletin*, 92, 1479–1499.

Herries, R. D., 1993, Contrasting styles of fluvial-aeolian interaction at a downwind erg margin: Jurassic Kayenta–Navajo transition, northeastern Arizona US, in C. P. North and D. J. Prosser, eds., *Characterization of Fluvial and Aeolian Reservoirs*, London Geological Society, Special Publication 73, 199–218.

Herron, D. A., 2011, *First Steps in Seismic Interpretation*: Society of Exploration Geophysicists, Geophysical Monograph 16, 217 p., doi:10.1190/1.9781560802938.

Herron, D. A., 2014, Thoughts and observation on interpreting depth-imaged data in the Jurassic Norphlet Play, Deepwater Eastern Gulf of Mexico: *American Association of Petroleum Geologists, Search and Discovery* 41342, 16 p.

Hessler, A. M., J. Zhang, J. Covault, and W. Ambrose, 2017, Continental weathering coupled to Paleogene climate changes in North America: *Geology*, 45, 911–914.

Hessler, A. M., J. A. Covault, D. Stockli, and A. Fildani, 2018, Late Cenozoic cooling favored glacial over tectonic controls on sediment supply to the western Gulf of Mexico: *Geology*, 46, 955–998, doi:10.1130/G45528.1.

Hildebrand, A. R., G. T. Penfield, D. A. Kring, *et al.*, 1991, Chicxulub crater: a possible Cretaceous/Tertiary boundary impact crater in the Yucatán Peninsula, Mexico. *Geology*, 19, 867–871, doi:10.1130/0091-7613(1991)019<0867:CCAPCT>2.3.CO;2.

Hood, K. C., L. M. Wenger, O. P. Gross, and S. C. Harrison, 2002, Hydrocarbon systems analysis of the northern Gulf of Mexico: delineation of hydrocarbon migration pathways using seeps and seismic imaging, in D. Schumacher and L. A. LeSchack, eds., *Surface Exploration Case Histories: Applications of Geochemistry, Magnetics, and Remote Sensing*: American Association of Petroleum Geologists, 25–48.

Horbury, A. D., 2000, Report detailing the carbonate sequence stratigraphy and reservoir geology of the Cantarell Field, Cuidad Carmen, Campeche: Pemex internal report.

Horbury, A. D., J. Celestino, N. Oxtoby, A. Soto, and S. Johnson, 1996, Diagénesis y evolución de la porosidad en el campo Petrolifero Arenque, costa afuera de Tajypjco, Tamaulipas, Mexico: *Boletín de la Asociación Mexicana de Geólogos Petroleros*, 45, 58–80.

Horbury, A. D, S. I. Fall, F. Gonzalez-L., *et al.*, 2003, Tectonic sequence stratigraphy of the western margin of the Gulf of Mexico in the late Mesozoic and Cenozoic: less passive than previously imagined, in C. Bartolini, R. T. Buffler, and J. Blickwede, eds., *The Circum-Gulf of Mexico and the Caribbean: Hydrocarbon Habitats, Basin Formation, and Plate Tectonics*: American Association of Petroleum Geologists, Memoir 79, 184–245.

Horn, B. W., 2012, Identifying new exploration fairways in the Gulf of Mexico Deepwater Tuscaloosa/Woodbine play: *Gulf Coast Association of Geological Societies Transactions*, 61, 245–256.

Hovorka, S. D., and H. S. Nance, 1994, Dynamic depositional and early diagenetic processes in a deep-water shelf setting, Upper Cretaceous Austin Chalk, North Texas: *Gulf Coast Association of Geological Societies Transactions*, 44, 269–276.

Hudec, M. R., and M. P. A. Jackson, 2011, *The Salt Mine: a Digital Atlas of Salt Tectonics*: American Association of Petroleum Geologists, Memoir 99, 305 p.

Hudec, M. R., and I. O. Norton, 2018, Upper Jurassic structure and evolution of the Yucatán and Campeche subbasins, southern Gulf of Mexico: *American Association of Petroleum Geologists Bulletin*, doi: 10.1306/11151817405.

Hudec, M. R., I. O. Norton, M. P. A. Jackson, and F. J. Peel, 2013a, Jurassic evolution of the Gulf of Mexico salt basin: *American Association of Petroleum Geologists Bulletin*, 97, 1683–1710, doi:10.1306/04011312073.

Hudec, M. R., M. P. A. Jackson, and F. J. Peel, 2013b, Influence of deep Louann structure on the evolution of the northern Gulf of Mexico: *American Association of Petroleum Geologists Bulletin*, 97, 1711– 1735, doi:10.1306/04011312074.

Hudec, M.R., T.P. Dooley, F.J. Peel, and J.I. Soto, accepted, Structure and evolution of the Salina del Bravo region, northeastern Mexico continental slope: *Geological Society of America Bulletin*.

Hull, D. C., 2011, *Stratigraphic Architecture, Depositional Systems, and Reservoir Characteristics of the Pearsall Shale-Gas System, Lower Cretaceous, South Texas*: Master's thesis, The University of Texas at Austin, 192 p.

Hull, D., and R. Loucks, 2010, Depositional systems and stratal architecture of the Lower Cretaceous (Aptian) Pearsall Formation in South Texas: *Gulf Coast Association of Geological Societies Transactions*, 60, 901–906.

Humphrey, W. E., and T. Diaz,, 2003, *Jurassic and Lower Cretaceous Stratigraphy and Tectonics of Northeast Mexico*, The University of Texas at Austin, Bureau of Economic Geology, 167 p.

Hunt, B. W., 2013, *Regional Norphlet Facies Correlation, Analysis and Implications for Paleostructure and Provenance, Eastern Gulf of Mexico*: Master's thesis, The University of Alabama, 112 p.

Hunt, B., D. M. Robinson, A. L. Weislogel, and R. C. Ewing, 2017, Sediment source regions and paleotransport of the Upper Jurassic Norphlet Formation, eastern Gulf of Mexico: *American Association of Petroleum Geologists Bulletin*, 101, 1519–1542.

Iannello, C., 2001, *Regional Characteristics, Timing, and Significance of Dissolution and Collapse Features in Lower Cretaceous Carbonate Platform Strata, DeSoto Canyon Area, Offshore Alabama–Florida*: Master's thesis, Texas A&M University, 73 p.

Ice, R. G., and C. L. McNulty, 1980, Foraminifers and calcispheres from the Cuesta del Cura and Lower Agua Nueva(?) Formations (Cretaceous) in east-central Mexico: *Gulf Coast Association of Geological Societies Transactions*, 30, 403–425.

Imbert, P., and Y. Phillippe, 2005, The Mesozoic opening of the Gulf of Mexico: Part 2, integrating seismic and magnetic data into a general opening model: *Society of Economic Paleontologists and Mineralogists Gulf Coast Section, 25th Annual Research Conference Program and Extended Abstracts*, 1151–1189, doi:10.5724/gcs.05.25.1151.

IUGS International Commission on Stratigraphy, 2011, Global chronostratigraphical correlation table for the last 2.7 million years: stratigraphic charts for the Quaternary, 2011.

Jackson, M. P. A., and C. Cramez, 1989, Seismic recognition of salt welds in salt tectonics regimes in *Gulf Coast Section of the Society of Economic Paleontologists and Mineralogists Tenth Annual Research Conference Program and Abstracts*, 66–71.

Jackson, M. P. A., and M. R. Hudec, 2017, *Salt Tectonics: Principles and Practice*: Cambridge University Press, 498 p.

Jackson, M. P. A., and S. J. Seni, 1984, *Atlas of Salt Domes in the East Texas Basin*: The University of Texas at Austin, Bureau of Economic Geology, Report of Investigations 140, 102 p.

Jackson, M. P. A., B. C. Vendeville, and D. D. Schultz-Ela, 1994, Structural dynamics of salt systems: *Annual Review of Earth and Planetary Science*, 22, 93–117, doi:10.1146/annurev.ea.22.050194.000521.

Jackson, M. P. A., T. Dooley, M. Hudec, and A. McDonnell, 2011, The pillow fold belt: a key subsalt structural province in the Northern Gulf of Mexico: *American Association of Petroleum Geologists, Search and Discovery* 10329, 23 p.

Jacobo Albarabn, J., M. Garduño, F. Innocenti, M. G. Pasquare, and S. Tonarini, 1992. Datos sobre el vulcanismo neogénico-reciente del complejo volcánico de Los Tuxtlas, Edo. de Veracruz, México: *Evolución petrológica y geovulcanológica*, 11.o Convención Geológica Nacional, Veracruz, Libro de Resúmenes, 97–98.

Jacques, J. M., and H. Clegg, 2002, Late Jurassic source rock distribution and quality in the Gulf of Mexico: inferences from plate tectonic modelling: *Gulf Coast Association of Geological Societies Transactions*, 52, 429–440.

James, A. T., L. M. Wenger, M. B. Melia, A. H. Ross, and C. P. Kuminez, 1993, Recognition of a new hydrocarbon play in a mature exploration area through integration of geochemical, palynologic, geologic, and seismic interpretations (onshore northern Gulf of Mexico) (abs.): *American Association of Petroleum Geologists, Search and Discovery* 90987, 123 p.

Janson, X., 2004, Golden Lane Platform Northeastern Margin: large scale margin erosion and karst features: American Association of Petroleum Geologists Annual Convention and Exposition, Dallas, Texas (Abstract).

Janson, X., R. Loucks, C. Kerans, A. Marhx, and C. Reyes, 2004, Karstification of the Lower Cretaceous Tuxpan Detached Platform (abs.): *American Association of Petroleum Geologists, Search and Discovery* 90026, 1 p.

Janson, X., C. Kerans, R. Loucks, *et al.*, 2011, Seismic architecture of a Lower Cretaceous platform-to-slope system, Santa Agueda and Poza Rica fields, Mexico: *American Association of Petroleum Geologists Bulletin*, 95, 105–146, doi:10.1306/06301009107.

Jenkyns, H.C., 2010, Geochemistry of oceanic anoxic events: *Geochemistry, Geophysics, Geosystems*, 11, Q03004.

Jennette, D., T. Wawrzyniec, K. Fouad, *et al.*, 2003, Traps and turbidite reservoir characteristics from a complex and evolving tectonic setting, Veracruz Basin, southeastern Mexico: *American Association of Petroleum Geologists Bulletin*, 87, 1599–1622.

John, C. J., G. Maciasz, and B. J. Harder, 1998, *Gulf Coast Geopressured-Geothermal Program Summary Report Compilation: Volumes I, II A, II B, III, and IV*: Basin Research Institute, Louisiana State University, DOE contract DE-FG07-95ID13366.

Jordan, L., E. Applin, E. Caldwell, *et al.*, 1949, Mesozoic cross sections: Southeastern Geological Society, cross section, 4 p.

Joyce, J. E., L. R. C. Tjalsma, and J. M. Prutzman, 1993, North American glacial meltwater history for the past 2.3 m.y.: Oxygen isotope evidence from the Gulf of Mexico: *Geology*, 21, 483–486.

Juarez, M.A.G., 2001, *Evaluacion del Sistema Petrolifero en la Parte Oriental de los Campos Ek-Balam, Chac y Takin*, Thesis, Universidad Nacional Autonoma de Mexico, 377 p.

Judice, P. C., and S. J. Mazzullo, 1982, The gray sandstones (Jurassic) in Terryville Field, Louisiana: basinal deposition and exploration model: *Gulf Coast Association of Geological Societies Transactions*, 32, 24–43.

Kaiser, W. R., W. B. Ayers Jr., and L. W. LaBrie, 1980, *Lignite Resources in Texas:* The University of Texas at Austin, Bureau of Economic Geology, Report of Investigations 104, 52 p.

Kane, I. A., and A. S. M. Ponten, 2012, Submarine transitional flow deposits in the Paleogene Gulf of Mexico: *Geology*, 40, 1119–1122.

Kennard, J. M., and N. P. James, 1986, Thrombolites and stromatolites: two distinct types of microbial structure: *Palaios*, 1, 492–503.

Kidwell, A. L., 1951, Mesozoic igneous activity in the northern Gulf Coastal Plain: *GCAGS Transactions*, 1, 182–199.

Kiessling, W., E. Flufel, and J. Golonka, 1999, Paleoreef maps: evaluation of a comprehensive database on Phanerozoic reefs: *American Association of Petroleum Geologists Bulletin*, 83, 1552–1587.

Kinsland, G. L., and J. W. Snedden, 2016, Comparison of a portion of the K/Pg boundary deposits in two locations: Webb County, Texas, and LaSalle Parish, Louisiana: *Gulf Coast Association of Geological Societies Transactions*, 66, 789–797.

Kinsland, G. L., K. Shellhouse, E. Muchiri, J. W. Snedden, and J. W. Virdell, 2017, Midway Shale: Post-Cretaceous/Paleogene boundary deposition: *Gulf Coast Association of Geological Societies* 67, 177–185.

Kiyokawa, S., R. Tada, M. Iturralde-Vinent, *et al.*, 2002, Cretaceous–Tertiary boundary sequence in the Cacarajicara formation, western Cuba, in C. Koeberl and K. G. MacLeod, eds., *Catastrophic Events and Mass Extinctions: Impacts and Beyond*: Geological Society of America, Special Paper 356, 125–144.

Klein, G. D., and K. R. Chaivre, 2002, Sequence and seismic stratigraphy of the Bossier Formation (Tithonian), Western East Texas Basin: *Gulf Coast Association of Geological Societies Transactions*, 52, 551–561.

Klemme H. D., and G. F. Ulmishek, 1991, Effective petroleum source rocks of the world: stratigraphic distribution and controlling depositional factors: *American Association of Petroleum Geologists Bulletin*, 75, 1809–1851.

Kneller, E. A., and C. A. Johnson, 2011, Plate kinematics of the Gulf of Mexico based on integrated observations from the Central and South Atlantic: *Gulf Coast Association of Geological Societies Transactions*, 61, 283–299.

Kocurek, G., and K. G. Havholm, 1993, Eolian sequence stratigraphy: a conceptual framework, in P. Weimer and H. Posamentier, eds., *Siliciclastic Sequence Stratigraphy: Recent Developments and Applications*: American Association of Petroleum Geologists, Memoir 58, 393–409.

Kosters, E., D. Bebout, S. Seni, *et al.*, 1989, *Atlas of Major Texas Gas Reservoirs*: The University of Texas at Austin, Bureau of Economic Geology, 175 p.

Krafve, A., 1980, Field Study: Black Lake Field – Pettet, Natchitoches Parish, Louisiana, Report on Selected Oil and Gas Fields – North Louisiana and South Arkansas, Shreveport Geological Society, 7 p.

Krapf, C. B. E., H. Stollhofen, I. G. Stanistreet, 2003, Contrasting styles of ephemeral river systems and their interaction with dunes of the Skeleton Coast erg (Namibia): *Quaternary International*, 104, 41–52.

Ladd, J. W., and R. E. Sheridan, 1987, Seismic stratigraphy of the Bahamas: *American Association of Petroleum Bulletin*, 71, 719–736.

Land, L. S., R. A. Eustice, L. E. Mack, and J. Horita, 1995, Reactivity of evaporites during burial: an example from the Jurassic of Alabama: *Geochimica et Cosmochimica Acta*, 59, 3765–3778.

Langford, R. P., and J. M. Combes, 1994, *Depositional Environments of Unstable Shelf-Margin Deltas of the Oligocene Vicksburg Formation, McAllen Ranch Field, South Texas*: The University of Texas at Austin, Bureau of Economic Geology, Report of Investigations 219, 60 p.

Laubach, S. E., and M. L. W. Jackson, 1990, Origin of arches in northwestern Gulf of Mexico basin: *Geology*, 18, 595–598, doi:10.1130/0091-7613(1990)018<0595:OOAITN>2.3.CO;2.

Lawless, P. N., R. H. Fillon, and R. G. Lytton III, 1997, Gulf of Mexico Cenozoic biostratigraphic, lithostratigraphic, and sequence stratigraphic event chronology: *Gulf Coast Association of Geological Societies Transactions*, 47, 275–282.

Lawton, T. F., and R. S. Molina-Garza, 2014, U–Pb geochronology of the type Nazas Formation and superjacent strata, northeastern Durango, Mexico: implications of a Jurassic age for continental-arc magmatism in north-central Mexico, Bull: *Geological Society of American Bulletin*, 126, 1181–1199.

Lawton, T. F., and J. Pindell, 2017, Upper Triassic–Middle Jurassic strata of Plomosas Uplift and Sierra Samalayuca, Chihuahua, Mexico: onshore record of Syn-rift Gulf of Mexico fault history: *American Association of Petroleum Geologists, Search and Discovery* 90291, 1–6.

Lawton, T. F., F. J. Vega, K. A. Giles, and C. Rosales-Domingues, 2001, Stratigraphy and the origin of the La Popa Basin, Nuevo León and Coahuila, Mexico, in C. Bartolini, R. T. Buffler, and U. Cantú-Chapa, eds., *The Western Gulf of Mexico Basin: Tectonics, Sedimentary Basins, and Petroleum Systems*: American Association of Petroleum Geologists, Memoir 75, 219–240.

Lawton, T. F., J. A. Bradford, F. J. Vega, G. E. Gehrels, and J. M. Amato, 2009, Provenance of Upper Cretaceous–Paleogene sandstones in the foreland basin system of the Sierra Madre Oriental, northeastern Mexico, and its bearing on fluvial dispersal systems of the Mexican Laramide province: *Geological Society of America Bulletin*, 121, 820–836, doi:10.1130/B26450.1.

Lawton, T. F., J. Pindell, A. Beltran-Triviño, *et al.*, 2015, Late Cretaceous–Paleogene foreland sediment-dispersal systems in Northern and Eastern Mexico: interpretations from preliminary detrital-zircon analysis: *American Association of Petroleum Geologists, Search and Discovery* 30423, 1–37.

Lawton, T. F., J. E. Ruiz Urueña, L. A. Solari, *et al.*, 2018, Provenance of Upper Triassic–Middle Jurassic strata of the Plomosas uplift, east-central Chihuahua, Mexico, and possible sedimentologic connections with Colorado Plateau depositional systems, in R. V. Ingersoll, T. F. Lawton, and S. A. Graham, eds., *Tectonics, Sedimentary Basins, and Provenance: A Celebration of William R. Dickinson's Career*: Geological Society of America, Special Paper 540, 481–507.

Le Roy, C., and C. Rangin, 2008, Cenozoic crustal deformation of the offshore Burgos basin region (NE Gulf of Mexico): a new interpretation of deep penetration multichannel seismic reflection lines: *Bulletin de la Société Géologique de France*, 178, 161–174.

Le Roy, C., C. Rangin, X. Le Pichon, *et al.*, 2007, Neogene crustal shear zone along the western Gulf of Mexico margin and its implications for gravity sliding processes: evidences from 2D and 3D multichannel seismic data: *Bulletin de la Société Géologique de France*, 178(2), 175–185.

Leckie, R. M., T. J. Bralower, and R. Cashman, 2002, Oceanic anoxic events and plankton evolution: biotic response to tectonic forcing during the mid-Cretaceous: *Paleoceanography*, 17, 1–29, doi:10.1029/2001PA000623.

Lehmann, C., D. A. Olseger, I. P. Montanez, *et al.*, 1999, Evolution of Cupido and Coahuila carbonate platforms, Early Cretaceous, northeastern Mexico: *Geological Society of America Bulletin*, 111, 7.

Lehmann, C., D. A. Osleger, and I. Montanez, 2000, Sequence stratigraphy of Lower Cretaceous (Barremian–Albian) carbonate platforms of northeastern Mexico: regional and global correlations: *Journal of Sedimentary Research*, 70, 373–391, doi:10.1306/2DC40917-0E47-11D7-8643000102C1865.

Leinfelder, R. R., D. U. Schmid, M. Nose, and W. Werner, 2002, Jurassic reef patterns: expression of a changing globe, in W. Kiessling, E. Flugel, and J. Golonka, eds., *Phanerozoic Reef Patterns*: Society of Economic Paleontologists and Mineralogists, Special Publication 72, 465–520.

Lewis, J., S. Clinch, D. Meyer, *et al.*, 2007, *Exploration and Appraisal Challenges in the Gulf of Mexico Deep-Water Wilcox: Part 1 – Exploration Overview, Reservoir Quality, and Seismic Imaging*: Society of Economic Paleontologists and Mineralogists Gulf Coast Section Foundation, 398–414.

Leyendecker, E. A., 2014, The Gulf of Mexico advantage: *American Association of Petroleum Geologists, Search and Discovery* 110175, 21 p.

Li, Y., and W. B. Ayers, 2008, Hydrocarbon potential of the deep Travis Peak Formation and underlying strata, western margin of the East Texas Basin: *Gulf Coast Association of Geological Societies Transactions*, 58, 607–621.

Lin, P., D.E. Bird, and P. Mann, 2019, Crustal structure of an extinct, late Jurassic-to-earliest Cretaceous spreading center and its adjacent oceanic crust in the eastern Gulf of Mexico: *Marine Geophysical Research*, 1–24. DOI: 10.1007/s11001-019-09379-5.

Lisi, A. F., 2013, *Provenance of the Upper Jurassic Norphlet and Surrounding Formations from U–Pb Detrital Zircon Geochronology*: Master's thesis, West Virginia University, 148 p.

Liu, C., 2015, *Stratigraphy, Depositional History, and Pore Network of the Lower Cretaceous Sunniland Carbonates in the South Florida Basin*: Master's thesis, The University of Texas at Austin, 69 p.

Liu, L., 2014, Rejuvenation of Appalachian topography caused by subsidence-induced differential erosion: *Nature Geoscience*, 7(7), 518–523.

Liu, L., 2015, The ups and downs of North America: evaluating the role of mantle dynamic topography since the Mesozoic, *Reviews of Geophysics*, 53, 1022–1049, doi:10.1002/2015RG000489.

Locklin, J. A., 1985, A Rodessa stratigraphic trap (Ingram-Trinity Field), Lower Cretaceous, East Texas Basin: *Annual Meeting of the Gulf Coast Association of Geological Societies and Society of Economic Paleontologists and Mineralogists Gulf Coast Section: Lower Cretaceous Depositional Environments From Shoreline to Slope – A Core Workshop*, 85–91.

Long, J., 1985, The Eocene Lobo gravity slide, Webb and Zapata counties, Texas, in *Contributions to the Geology of South Texas*: South Texas Geological Society, 270–293.

Longman, M. W., B. A. Luneau, and S. M. Landon, 1998, Nature and distribution of Niobrara lithologies in the Cretaceous Western interior seaway of the Rocky Mountain region: *Rocky Mountain Association of Geologists: The Mountain Geologist*, 35, 137–170.

Longoria, J. F., and M. A. Gamper, 1977, Albian planktonic foraminifera from the Sabinas Basin of northern Mexico: *Journal of Foraminiferal Research*, 7, 196–215, doi:10.2113/gsjfr.7.3.196.

Lopez Ramos, E., 1982, *Geología de México*: Consejo Nacional de Ciencia y Technología, México, D. F., 2, 454 p.

Lore, G. L., D. A. Marin, E. C. Batchelder, *et al.*, 2001, *2000 Assessment of Conventionally Recoverable Hydrocarbon Resources of the Gulf of Mexico and Atlantic Outer Continental Shelf as of January 1, 1999, Outer Continental Shelf Report*: U.S. Department of the Interior Minerals Management Service, 15–25.

Loucks, R. G., 1977. *Porosity Development and Distribution in Shoal-Water Carbonate Complexes: Subsurface Pearsall Formation (Lower Cretaceous) South Texas*. Texas Bureau of Economic Geology.

Loucks, R. G., 1978, Sandstone distribution and potential for geopressured geothermal energy production in the Vicksburg Formation along the Texas Gulf Coast: *Gulf Coast Association of Geological Societies Transactions*, 28, 239–271.

Loucks, R. G., and J. O. Crump, 1985, Vertical facies sequences of the Sunniland and Punta Gorda Formations in the Lower Cretaceous South Florida embayment: Natural Resource Management Corporation No. 31-2 Alico Core, in D. Bebout, and D. Ratcliff, eds., Annual Meeting of the Gulf Coast Association of Geological Societies and Society of Economic Paleontologists and Mineralogists Gulf Coast Section: Lower Cretaceous Depositional Environments From Shoreline to Slope – A Core Workshop, 111–117.

Loucks, R. G., and M. W. Longman, 1987. Lower Cretaceous Ferry Lake Anhydrite, Fairway Field, east Texas: product of shallow-subtidal deposition: *Society of Economic Paleontologists and Mineralogists Core Workshop 3: Depositional and Diagenetic Spectra of Evaporites*, 130–173.

Loucks, R. G., M. M. Dodge, and W. E. Galloway, 1986, *Controls on Porosity and Permeability of Hydrocarbon Reservoirs in Lower Tertiary Sandstones along the Texas Gulf Coast*: The University of Texas at Austin, Bureau of Economic Geology, Report of Investigations 149, 78 p.

Loucks, R. G., J. F., Lucia, and L. E. Waite, 2013, Origin and description of the micropore network within the Lower Cretaceous Stuart City trend tight-gas limestone reservoir in Pawnee Field in South Texas: *Gulf Coast Association of Geological Societies Journal*, 2, 29–41.

Loucks, R. G., G., Frébourg, and H. D. Rowe, 2017a, Upper Cretaceous (Campanian) Ozan and Annona Chalks in Caddo-Pine Island Field, Northwestern Louisiana: depositional setting, lithofacies, and nanopore/micropore network: *Gulf Coast Association of Geological Societies Journal*, 6, 73–91.

Loucks, R. G., C. Kerans, H. Zeng, and P. A. Sullivan, 2017b, Documentation and characterization of the Lower Cretaceous (Valanginian) Calvin and Winn carbonate shelves and shelf margins, onshore north-central Gulf of Mexico: *American Association of Petroleum Geologists Bulletin*, 101, 119–142, doi:10.1306/06281615248.

Lovell, T. R., 2013, *Detrital Zircon U–Pb Age Constraints on the Provenance of the Late Jurassic Norphlet Formation, Eastern Gulf of Mexico: Implications for Paleogeography*: MS thesis, The University of Alabama, 179 p.

Lovell, T., and A. L. Weislogel, 2010, Detrital zircon U–Pb age constraints on the provenance of the Upper Jurassic Norphlet Formation, eastern Gulf of Mexico: implications for paleogeography: *Gulf Coast Association of Geological Societies Transactions*, 60, 443–460.

Lowery, C. M., M. J. Corbett, R. M Leckie, *et al.*, 2014, Foraminiferal and nannofossil paleoecology and paleoceanography of the Cenomanian – Turonian Eagle Ford Shale of southern Texas: *Paleogeography, Paleoclimatology, Palaeoecology*, 413, 49–65, doi:10.1016/j.palaeo.2014.07.025.

Lowery, C. M., R. Cunningham, C. D. Barrie, T. J. Bralower, and J. W. Snedden, 2017, The northern Gulf of Mexico during OAE2 and the relationship between water depth and black shale development: *Paleoceanography*, 32, 1316–1335, doi:10.1002/2017PA003180.

Lowery. C. M., T. J. Bralower, J. D. Owens, *et al.* 2018, Rapid recovery of life at ground zero of the end- Cretaceous mass extinction: *Nature*, 558, 288–291, doi:10.1038/s41586-018-0163-6.

Lundquist, J. J., 2000, *Foraminiferal Biostratigraphic and Paleoceanographic Analysis of the Eagle Ford, Austin, and Lower Taylor Groups (Middle Cenomanian Through Lower Campanian) of Central Texas*: Ph.D. dissertation, The University of Texas at Austin, 545 p.

Lundquist, J. J., 2015, Austin Chalk (UK): stratigraphic, geophysical, hydrogeological, and hydrocarbon exploration/production characteristics: *Austin Geological Society, 50th Anniversary Field Symposium*, 140 p.

Luneau, B., M. Doe, J. Leif Colson, *et al.*, 2003, Constructing a static model of a fractured reservoir with disparate data sets: Antonio J. Bermudez Complex, Reforma District, Southern Mexico: *American Association of Petroleum Geologists, Search and Discovery* 90013.

Machel, H. G., 2001, Bacterial and thermochemical sulfate reduction in diagenetic settings, old and new insights: *Sedimentary Geology*, 140, 143–175.

Mackey, G. N., B. K. Horton, and K. L. Milliken, 2012, Provenance of the Paleocene–Eocene Wilcox Group, western Gulf of Mexico basin: evidence for integrated drainage of the southern Laramide Rocky Mountains and Cordilleran arc: *Geological Society of America Bulletin*, 124, 1007–1024.

MacRae, G., 1994, *Mesozoic Development of the DeSoto Canyon Salt Basin in the Framework of the Early Evolution of the Gulf of Mexico*: Ph.D. dissertation, Texas A&M University, 152 p.

Magoon, L. B., T. L. Hudson, and H. E. Cook, 2001, Pimienta-Tamabra(!): a giant supercharged petroleum system in the southern Gulf of Mexico, onshore and offshore Mexico, in C. Bartolini, R. T. Buffler, and A. Cantú-Chapa, eds., *The Western Gulf of Mexico Basin: Tectonics, Sedimentary Basins, and Petroleum Systems*: American Association of Petroleum Geologists, Memoir 75, 83–125.

Mancini, E. A., 2010, Jurassic depositional systems, facies, and reservoirs of the northern Gulf of Mexico: *Gulf Coast Association of Geological Societies Transactions*, 60, 481–486.

Mancini, E. A., and T. M. Puckett, 1995, Upper Cretaceous sequence stratigraphy of the Mississippi–Alabama area: *Gulf Coast Association of Geological Societies Transactions*, 45, 377–384.

Mancini, E. A., R. M. Mink, B. L. Bearden, and R. P. Wilkerson, 1985, Norphlet Formation (Upper Jurassic) of Southwestern and Offshore Alabama: environments of deposition and petroleum geology: *American Association of Petroleum Geologists Bulletin*, 69, 881–898.

Mancini, E. A., M. Aurell, J. C. Llinas, *et al.*, 2004, Upper Jurassic thrombolite reservoir play, northeastern Gulf of Mexico: *American Association of Petroleum Geologists Bulletin*, 88, 1573–1602.

Mancini, E. A., W. C. Parcell, and W. M. Ahr, 2006, Upper Jurassic Smackover thrombolite buildups and associated nearshore facies, southwest Alabama: *Gulf Coast Association of Geological Transactions*, 56, 551–563.

Mander, J., J. d'Ablaing, K. Wells, *et al.*, 2012, *21st Century Atlantis: Incremental Knowledge from a Staged Approach to Development, Illustrated by a Complex, Deepwater Field*: Gulf Coast Section of the Society for Sedimentary Geology, 1–18.

Mandujano-Velazquez, J. J., and J. D. Keppie, 2009. Middle Miocene Chiapas fold and thrust belt of Mexico: a result of collision of the Tehuantepec Transform/Ridge with the Middle America, in J. B. Murphy, J. D. Keppie, and A. J. Hynes eds., *Ancient Orogens and Modern Analogues*: London Geological Society, Special Publication 327, 1, 55–69.

Mankiewicz, P. J., R. J. Pottorf, M. G. Kozar, and P. Vrolijk, 2009, Gas geochemistry of the Mobile Bay Jurassic Norphlet Formation: thermal controls and implications for reservoir connectivity: *American Association of Petroleum Geologists Bulletin*, 93, 1319–1346, doi:10.1306/05220908171.

Manzano, B. K., M. G. Fowler, and H. G. Machel, 1997, The influence of thermochemical sulfate reduction on hydrocarbon composition in Nisku Reservoirs, Brazeau River Area, Alberta, Canada: *Organic Geochemistry*, 27, 507–521.

Marchand, M. E., G. Apps, W. Li, and J. R. Rotzien, 2015, Depositional processes and impact on reservoir quality in deepwater Paleogene reservoirs, US Gulf of Mexico: *American Association of Petroleum Geologists Bulletin*, 99, 1635–1648.

Martin, J., P. Weimer, and R. Bouroullec, 2004, Sequence stratigraphy of upper Miocene to upper Pliocene sediments of west-central Mississippi Canyon and northern Atwater Valley, northern Gulf of Mexico: *Gulf Coast Association of Geological Societies Transactions*, 54, 425–441.

Martin, R. G., and J. E. Case, 1975, Geophysical studies in the Gulf of Mexico, in A. Nairn and F. Stehli, eds., *The Gulf of Mexico and Caribbean, Ocean Basins and Margins, 3*, Plenum Press, 65–106.

Martinez-Medrano, M., R. Vega-Escobar, F. Flores-Cruz, D. Angeles-Marin, and C. Lopez-Martinez, 2009, Integrated seismic and petrographic analysis of the sandstone reservoirs of the Tertiary Veracruz Basin, Mexico, in C. Bartolini and J. R. Roman Ramos, eds., *Petroleum Systems in the Southern Gulf of Mexico*: American Association of Petroleum Geologists, Memoir 90, 217–235.

Martinez-Medrano, M., E. Gonzalez Mercado, and E. Fernandez-Avendano, 2011, Stratigraphy, petrology, and provenance of Neogene sandstones of the Veracruz basin, Mexico: *Gulf Coast Association of Geological Societies Transactions*, 61, 607–620.

Martini, M., and F. Ortega-Gutiérrez, 2016, Tectono-stratigraphic evolution of eastern Mexico during the break-up of Pangea: *Earth-Science Reviews*, 183, 38–55, doi:10.1016/j.earscirev.2016.06.013.

Martini, M., L. Solari, M. López Martínez, 2014, Correlation of the Arperos Basin from Guanajuato, central Mexico, to Santo Tomás, southern Mexico: Implications for the paleogeography and origin of the Guerrero terrane: *Geosphere*, 10, 1385–1401, doi:10.1130/GES01055.1.

Marton, G. L., and R. T. Buffler, 1999, Jurassic–Early Cretaceous tectono-paleogeographic evolution of the southeastern Gulf of Mexico Basin, in P. Mann, ed., *Caribbean Basins: Sedimentary Basins of the World*: Elsevier Science, 63–91.

Marzano, M. S., G. M. Pense, and P. Andronaco, 1988, A comparison of the Jurassic Norphlet Formation in Mary Ann Field, Mobile Bay, Alabama to onshore regional Norphlet trends: *Gulf Coast Association of Geological Societies Transactions*, 38, 85–100.

Maxwell, W. G. H., and J. P. Swinchatt, 1970, Great Barrier Reef: regional variation in a terrigenous-carbonate province: *Geological Society of America Bulletin*, 81, 691–724, doi:10.1130/0016-7606(1970)81[691:GBRRVI]2.0.CO;2.

McArthur, J. M., R. J. Howarth, and T. R. Bailey, 2001, Strontium isotope stratigraphy: LOWESS Version 3 – best fit to the marine Sr-isotope curve for 0–509 Ma and accompanying look-up table for deriving numerical age: *The Journal of Geology*, 109, 155–170.

McBride, B. C., M. G. Rowan, and P. Weimer, 1998, The evolution of allochthonous salt systems, northern Green Canyon and Ewing Bank (offshore Louisiana), northern Gulf of Mexico: *American Association of Petroleum Geologists Bulletin*, 82, 1013–1036.

McDonald, K. C., 1982, Mid-ocean ridges: fine-scale tectonic, volcanic and hydrothermal processes within the plate tectonic boundary zone: *Annual Reviews of the Earth and Planetary Sciences*, 10, 155–190.

McDonnell, A., R. G. Loucks, and W. E. Galloway, 2008, Paleocene to Eocene deep-water slope canyons, western Gulf of Mexico: further insights for the provenance of deep-water offshore Wilcox Group plays: *American Association of Petroleum Geologists Bulletin*, 92, 1169–1189, doi:10.1306/05150808014.

McDonnell, A., M. Jackson, and M. Hudec, 2010, *Salt–sediment Interactions during the Jurassic to Miocene of Mississippi Canyon Area, Northern Gulf of Mexico*: The University of Texas at Austin, Bureau of Economic Geology, Applied Geodynamics Laboratory, 34 p.

McFarlan, E., 1977, Lower Cretaceous sedimentary facies and sea level changes, U.S. Gulf Coast, in D. G. Bebout, and R. G. Loucks, eds., *Cretaceous Carbonates of Texas and Mexico: Applications to Subsurface Exploration*: The University of Texas at Austin, Bureau of Economic Geology, Report of Investigations 89, 5–12.

McFarlan, E., Jr., and L. S. Menes, 1991, Lower Cretaceous, in A. Salvador, ed., *The Gulf of Mexico Basin: The Geology of North America*: Geological Society of America, 181–204, doi:10.1130/DNAG-GNA-J.181.

McGowen, M. K., and C. M. Lopez, 1983, *Depositional Systems in the Nacatoch Formation (Upper Cretaceous), Northeast Texas and Southwest Arkansas*: The University of Texas at Austin, Bureau of Economic Geology, Report of Investigations 137, 63 p.

McKinney, M. L., 1984, Suwannee Channel of the Paleogene coastal plain: support for the "carbonate suppression" model of basin formation, *Geology*, 12, 343–345.

McMillan, M. E., P. L. Heller, and S. L. Wing, 2006, History and causes of post-Laramide relief in the Rocky Mountain orogenic plateau: *Geological Society of America Bulletin*, 118(3–4), 393–405.

Meckel III, L. D., 2002, Core, log, and seismic characteristics of a high rate amalgamated channel reservoir in a salt-withdrawal minibasin: the upper Miocene "above Magenta" sand, Ursa field, northern Gulf of Mexico, in P. Weimer, M. Sweet, M. Sullivan, *et al.*, eds., *Deep-Water Core Workshop, Northern Gulf of Mexico*: Society of Economic Paleontologists and Mineralogists Gulf Coast Section, 61–74.

Meckel III, L. D., and W. E. Galloway, 1996, Formation of high-frequency sequences and their bounding surfaces: case study of the Eocene Yegua Formation, Texas Gulf Coast, USA: *Sedimentary Geology*, 102, 155–186.

Meckel III, L. D., G. A. Ugueto, D. H. Lynch, *et al.*, 2002, Genetic stratigraphy, stratigraphic architecture, and reservoir stacking patterns of the Upper Miocene–Lower Pliocene Greater Mars–Ursa intraslope basin, Mississippi Canyon, Gulf of Mexico, in J.M. Armentrout and N.C. Rosen, eds., *Sequence Stratigraphic Models for Exploration and Production: Evolving Methodologies, Emerging Models, and Application Histories*: Society of Economic Paleontologists and Mineralogists Gulf Coast Section 22nd Annual Bob F. Perkins Research Conference, 113–147.

Meckel, T., 2010, Classifying and characterizing sand-prone mass-transport deposits: *American Association of Petroleum Geologists, Search and Discovery* 50270.

Melbana Energy, 2017, Petroleum potential of Block 9 PSC: Havana, Cuba, *Geosciencias Conference*, 1–18.

Meneses-Rocha, J. J., 2001, Tectonic evolution of the Ixtapa graben, an example of a strike–slip basin in southeastern Mexico: implications for regional petroleum systems, in C. Bartolini, R. T. Buffler, and A. Cantú-Chapa, eds., *The Western Gulf of Mexico Basin: Tectonics, Sedimentary Basins, and Petroleum Systems*: American Association of Petroleum Geologists, Memoir 75, 183–216.

Meneses-Scherrer, E. J., G. Gülen, and S. W. Tinker, 2017, Non-geologic factors necessary to develop a shale industry in Mexico: *American Association of Petroleum Geologists, Search and Discovery* 70000, 1–3.

Meyer, D., L. Zarra, and J. Yun, 2007, From BAHA to Jack, evolution of the Lower Tertiary Wilcox Trend in the deepwater Gulf of Mexico: *The Sedimentary Record*, 5, 4–9.

Meyerhoff, A. A., and C. W. Hatten, 1968, *Diapiric Structures in Central Cuba*, American Association of Petroleum Geologists, Memoir A153, 315–357.

Mickus, K., R. J. Stern, G. R. Keller, and E. Y. Anthony, 2009, Potential field evidence for a volcanic rifted margin along the Texas gulf coast: *Geology*, 37, 387–390, doi:10.1130/G25465A.1.

Miller, K. G., Kominz, M. A., Browning, J. V., *et al.*, 2005, The Phanerozoic record of global sea-level change: *Science*, 310, 1293–1298, doi:10.1126/science.1116412.

Milliken, J. V., 1988, *Late Paleozoic and Early Mesozoic Geologic Evolution of the Arklatex Area*: Master's thesis, Rice University, 286 p.

Milliken, K. T., Blum, M. D., Snedden, J. W., and Galloway, W. E., 2018, Application of fluvial scaling relationships to reconstruct drainage-basin evolution and sediment routing for the Cretaceous and Paleocene of the Gulf of Mexico: *Geosphere*, 14,1–19, doi:10.1130/GES01374.1.

Mink, R. M., B. L. Bearden, and E. A. Mancini, 1985, *Regional Jurassic Geological Framework of Alabama Coastal Waters Area and Adjacent Federal Waters Area*: Geological Survey of Alabama and State Oil and Gas Board, Final Interim Report, 83 p., doi:10.1016/0025-3227(89)90112-6.

Miranda Peralta, I. L. R., A. C. Alvarado, R. M. Villalón, *et al.*, 2014, Play hipotético pre-sal en aguas profundas del Golfo de México: *Ingeniería Petrolera*, 5, 256–266.

Mitchell-Tapping, H. J., 1986, Exploration petrology of the Sunoco Felda trend of South Florida: *Gulf Coast Association of Geological Societies Transactions*, 36, 241–256.

Mitchell-Tapping, H. J., 2002, Exploration analysis of basin maturity in the South Florida Sub-Basin: *Gulf Coast Association of Geological Societies Transactions*, 52, 753–764.

Mitchum, R. M., Jr., and J. C. Van Wagoner, 1991, High-frequency sequences and their stacking patterns: sequence-stratigraphic evidence of high-frequency eustatic cycles: *Sedimentary Geology*, 70, 131–160, doi:10.1016/0037-0738(91)90139-5.

Mitra, S., G. C. Figueroa, J. H. Garcia, and A. M. Alvarado, 2005, Three-dimensional structural model of the Cantarell and Sihil structures, Campeche Bay province, Mexico: *American Association of Petroleum Geologists Bulletin*, 89, 1–26.

Mitra, S., J. D. Gonzalez, J. H. Garcia, S. Hernandez, and S. Banerjee, 2006, Structural geometry and evolution of the Ku, Zaap, and Maloob structures, Campeche Bay, Mexico: *American Association of Petroleum Geologists Bulletin*, 90, 1565–1584.

Mitra, S., J. Gonzalez, J. Garcia, and K. Ghosh, 2007, Ek-Balam field: a structure related to multiple stages of salt tectonics and extension: *American Association of Petroleum Geologists Bulletin*, 91, 1619–1636, doi:10.1306/06260706112.

Mixon, R. B., 1963, *Geology of the Huizachal Redbeds, Sierra Madre Oriental, Mexico*: Ph.D. dissertation, Louisiana State University, 128 p., https://digitalcommons.lsu.edu/gradschool_disstheses/819

Moffett, J. R., 2015, Discovering the missing piece of the Gulf of Mexico geologic puzzle: *American Association of Petroleum Geologists, Search and Discovery* 110198, 22 p.

Mohn, K., and B. Bowen, 2012, Florida: the next US frontier – revisiting an old exploration region of the Gulf of Mexico with Modern 3D Data: *GeoExpro*, 9, 74–78.

Moldowan J. M., W. K. Seifert, and E. J. Gallegos, 1985, Relationship between petroleum composition and depositional environment of petroleum source rocks: *American Association of Petroleum Geologists Bulletin*, 69, 1255–1268.

Moldowan, J. M., Dahl, J., Huizinga, B. J., *et al.*, 1994, The molecular fossil record of oleanane and its relation to angiosperms, *Science*, 265, 768–771.

Molina, E., L. Alegret, I. Arenillas, *et al.*, 2006, The global boundary stratotype section and point for the base of the Danian Stage (Paleocene, Paleogene, "Tertiary", Cenozoic) at El Kef, Tunisia – original definition and revision: *Episodes*, 29, 263–272.

Montgomery, S., 1996, Cotton Valley Lime Pinnacle Reef Play: Branton Field: *American Association of Petroleum Geologists Bulletin*, 80, 617–629.

Moore, C. H., 1984, The Upper Smackover of the Gulf Rim: depositional systems, diagenesis, porosity evolution and hydrocarbon production, the Jurassic of the Gulf Rim: *Society of Economic Paleontologists and Mineralogists Gulf Coast Section Foundation 3rd Annual Research Conference*, 283–307.

Moore, G. T., D. N. Hayashida, C. A. Ross, and S. R. Jacobson, 1992, Paleoclimate of the Kimmeridgian/Tithonian (Late Jurassic) world: I. Results using a general circulation model: *Palaeogeography, Palaeoclimatology, Palaeoecology*, 93, 113–150, doi:10.1016/0031-0182(92)90186-9.

Moore, M. G., 2010, Exploration, appraisal, and development of turbidite reservoirs in the Western Atwater Foldbelt, Deep Water Gulf of Mexico: *American Association of Petroleum Geologists, Search and Discovery* 90104.

Moore, V., and D. Hinton, 2013, Secondary basins and sediment pathways in Green Canyon, Deepwater Gulf of Mexico: *American Association of Petroleum Geologists, Search and Discovery* 10499, 4 p.

Moretti, I., R. Tenreyro, E. Linares, *et al.*, 2003, Petroleum system of the Cuban northwest offshore zone, in C. Bartolini, R. T. Buffler, and J. Blickwede, eds., *The Circum-Gulf of Mexico and the Caribbean: Hydrocarbon Habitats, Basin Formation, and Plate Tectonics*: American Association of Petroleum Geologists, Memoir 79, 675–696.

Morgan, J. V., S. Gulick, T. Bralower, *et al.*, 2016, The formation of peak rings in large impact craters: *Science*, 354, 878–882, doi:10.1126/science.aah6561.

Morton, C. H., and P. Weimer, 2000, Sequence stratigraphy of the Alaminos Fan (Upper Miocene–Pleistocene), northwestern deep Gulf of Mexico, in P. Weimer, R.M. Slatt, J. Coleman, *et al.*, eds., *Deep-Water Reservoirs of the World*: Society of Economic Paleontologists and Mineralogists Gulf Coast Section 20th Annual Bob F. Perkins Research Conference, 667–685.

Morton, R. A., 1993, Attributes and origins of ancient submarine slides and filled embayments: examples from the Gulf Coast basin: *American Association of Petroleum Geologists Bulletin*, 77, 1064–1081.

Morton, R. A., and W. B. Ayers, Jr., 1992, *Plio-Pleistocene Genetic Sequences of the Southwestern Louisiana Continental Shelf and Slope: Geologic Framework, Sedimentary Facies, and Hydrocarbon Distribution*: The University of Texas at Austin, Bureau of Economic Geology, Report of Investigations, 210, 77 p.

Morton, R. A., and L. A. Jirik, 1989, *Structural Cross Sections, Plio-Pleistocene Series, Southeastern Texas Continental Shelf*: The University of Texas at Austin, Bureau of Economic Geology.

Morton, R. A., L. A. Jirik, and R. Q. Foote, 1985, *Structural Cross Sections, Miocene Series, Texas Continental Shelf*: The University of Texas at Austin, Bureau of Economic Geology, 26 p.

Morton, R. A., L. A. Jirik, and W. E. Galloway, 1988, *Middle–Upper Miocene Depositional Sequences of the Texas Coastal Plain and Continental Shelf: Geologic Framework, Sedimentary Facies, and Hydrocarbon Plays*: The University of Texas at Austin, Bureau of Economic Geology, Report of Investigations 174, 40 p.

Morton, R. A., R. H. Sams, and L. A. Jirik, 1991, *Plio-Pleistocene Depositional Sequences of the Southeastern Texas Continental Shelf and Slope: Geologic Framework, Sedimentary Facies, and Hydrocarbon Distribution*: The University of Texas at Austin, Bureau of Economic Geology, Report of Investigations 200, 80 p.

Moscardelli, L., and L. J. Wood. 2016. Morphometry of mass-transport deposits as a predictive tool: *Geological Society of America Bulletin*, 128, 47–80.

Moscardelli, L., L. Wood, R. Torres-Vargas, J. Bermudez, and G. Lopez-Leyva, 2008, Processes of Late Tertiary-Age mass transport and associated deposits along the eastern Mexico margin, southern Gulf of Mexico: *American Association of Petroleum Geologists, Search and Discovery* 90078.

Mullins, H. T., A. F. Gardulski, S. W. Wise, Jr., and J. Applegate, 1983, Middle Miocene oceanographic event in the eastern Gulf of Mexico: implications for seismic stratigraphic succession and loop current/Gulf stream circulation: *Geological Society of America Bulletin*, 98, 702–713.

Mullins, H. T., A. F. Gardulski, and A. C. Hine, 1986, Catastrophic collapse of the west Florida carbonate platform margin: *Geology*, 14, 167–170, doi:10.1130/0091-7613(1986)14<167:CCOTWF>2.0.CO;2.

Murillo-Muneton, G., J. Grajales-Nishimura, E. Cedillo-Pardon, *et al.*, 2002, Stratigraphic architecture and sedimentology of the main oil-producing interval at the Cantarell Oil Field: the K-T boundary sedimentary succession: *Society of Petroleum Engineers International Petroleum Conference and Exhibition, Villahermosa, Mexico*, 7 p., doi:10.2118/74431-MS.

Murray, G., 1961, *Geology of the Atlantic and Gulf Coastal Province of North America*: Harper and Brothers, 692 p.

Myczynski, R., O. Federico, and A. B. Villaseñor, 1998, Revised biostratigraphy and correlations of the Middle–Upper Oxfordian in the Americas (southern USA, Mexico, Cuba, and northern Chile): *Neues Jahrbuch für Geologie und Paläontologie*, 207, 185–206.

Naehr, T. H., I. R. MacDonald, G. Bohrmann, and E. E. Briones, 2007, Biogeochemistry of hydrocarbon seeps on the Campeche Escarpment, southern Gulf of Mexico: *Gulf Coast Association of Geological Societies Transactions*, 57, 599–604.

Nelson, T. H., 1991, Salt tectonics and listric-normal faulting, in A. Salvador, ed., *The Gulf of Mexico Basin: The Geology of North America*: Geological Society of America, 73–89, doi:10.1130/DNAG-GNA-J.73.

Nguyen, L. C., and P. Mann,, 2016, Gravity and magnetic constraints on the Jurassic opening of the oceanic Gulf of Mexico and the location and tectonic history of the Western Main transform fault along the eastern continental margin of Mexico: *Interpretation*, 4, SC23–SC33, doi:10.1190/INT-2015-0110.1.

Nicholas, R. L., and D. E. Waddell, 1989, The Ouachita system in the subsurface of Texas, Arkansas, and Louisiana: *The Geological Society of America* 1989, 1–12.

Nieto, J. O., 2010, Analisis estratigrafico de la secuencia sedimentaria en el grupo Chicontepec: unpublished dissertation, Universidad Nacional Autonoma de Mexico, 200 p.

Norton, I. O., D. T. Carruthers, and M. R. Hudec, 2015, Rift to drift transition in the south Atlantic salt basins: a new flavor of oceanic crust: *Geology*, 44, 55–58, doi:10.1130/G37265.1.

Norton, I. O., L. A. Lawver, and J. W. Snedden, 2016, Gulf of Mexico tectonic evolution from Mexico deformation to oceanic crust, in C. M. Lowery and J. W. Snedden, eds., *Mesozoic of the Gulf Rim and Beyond: New Progress in Science and Exploration of the Gulf of Mexico Basin*: Society of Economic Paleontologists and Mineralogists Gulf Coast Section, 35th Annual Research Conference, 1–12.

Norton, I. O., L. Lawver, J. Snedden, 2018, Rift to drift transition in the Gulf of Mexico: AGU Fall Meeting, https://agu.confex.com/agu/fm18/meetingapp.cgi/Paper/358043.

Núñez-Useche, F., R. Barragán, J. A. Moreno-Bedmar, and C. Canet, 2014, Mexican archives for the major Cretaceous oceanic anoxic events: *Boletín de la Sociedad Geológica Mexicana*, 66, 491–505.

Núñez-Useche, F., C. Canet, R. Barragan, and P. Alfonso, 2016, Bioevents and redox conditions around the Cenomanian–Turonian anoxic event in Central Mexico: *Palaeogeography, Palaeoclimatology, Palaeoecology*, 449, 205–226.

Nyberg, B., W. Helland-Hansen, R. L. Gawthorpe, *et al.*, 2018, Revisiting morphological relationships of modern source-to-sink segments as a first-order approach to scale ancient sedimentary systems: *Sedimentary Geology*, 373, 111–133.

Nyberg, J., and J. A. Howell, 2016, Global distribution of modern shallow marine shorelines: implications for exploration and reservoir analogue studies: *Marine and Petroleum Geology*, 71, 83–104.

Oehler, J. H., 1984, Carbonate source rocks in the Jurassic Smackover Trend of Mississippi, Alabama and Florida, in J. G. Palacas, ed., *Petroleum Geochemistry and Source Rock Potential of Carbonate Rocks*: American Association of Petroleum Geologists, Studies in Geology 18, 63–69.

Officer, C. B., and C. L. Drake, 1985, Terminal Cretaceous environmental events. *Science*, 227, 1161–1167.

Ogg, J. G., G. M. Ogg, and F. M. Gradstein, 2016, *A Concise Geologic Time Scale*: Elsevier, 230 p.

Oglesoba, O. C., W. A. Ambrose, and R. G. Loucks, 2018, Application of instantaneous-frequency attribute and gamma-ray wireline logs in the delineation of lithology in Serbin field, Southeast Texas: a case study: *Interpretation*, 6, T1023–T1043, doi:10.1190/int-2018-0041.1.

Oivanki, S. M., 1974, Paleodepositional environments in the Upper Jurassic Zuloaga Formation (Smackover), northeastern Mexico: *Gulf Coast Association of Geological Societies Transactions*, 34, 258–278.

Olariu, M. I., U. Hammes, W. A. Ambrose, 2013, Depositional architecture of growth-fault related wave-dominated shelf edge deltas of the Oligocene Frio Formation in Corpus Christi Bay: *Marine and Petroleum Geology*, 48, 428–440.

Oloriz, F., A. B. Villasenor, C. Gonzalez-Arreola, 2003, Major lithostratigraphic units in land-outcrops of north-central Mexico and the subsurface along the northern rim of Gulf of Mexico Basin (Upper Jurassic–lowermost Cretaceous): a proposal for correlation of tectono-eustatic sequences: *Journal of South American Earth Sciences*, 16, 119–142, doi:10.1016/S0895-9811(03)00049-X.

Olson, H. C., J. W. Snedden, and R. Cunningham, 2015, Development and application of a robust chronostratigraphic framework in Gulf of Mexico Mesozoic exploration: *Interpretation*, 3, SN39–SN58, doi:10.1190/INT-2014-0179.1.

Olsen, P. E., 1997, Stratigraphic record of the Early Mesozoic breakup of Pangea in the Laurasia Gondwana Rift System: *Annual Review of Earth and Planetary Sciences*, 25, 337–401, doi:10.1146/annurev.earth.25.1.337.

Ortega-Flores, B., L. Solari, T. F. Lawton, and C. Ortega-Obregón, 2014, Detrital-zircon record of major Middle Triassic–Early Cretaceous provenance shift, central Mexico: demise of Gondwanan continental fluvial systems and onset of back-arc volcanism and sedimentation: *International Geology Review*, 56, 237–261, doi:10.1080/00206814.2013.844313.

Ortuno-Arzate, S., H. Ferket, M.-C. Cacas, R. Swennen, and F. Roure, 2003, Late Cretaceous carbonate reservoirs in the Cordoba Platform and Veracruz Basin, eastern Mexico, in C. Bartolini, R. T. Buffler, and J. Blickwede, eds., *The Circum-Gulf of Mexico and the Caribbean: Hydrocarbon Habitats, Basin Formation, and Plate Tectonics*: American Association of Petroleum Geologists, Memoir 79, 476–514.

Oxley, M. L., and E. D. Minihan, 1969, Alabama exploration underway. Pt. 1: *Oil and Gas Journal*, 67, 207–212.

Padilla y Sánchez, R. J., 2007, Evolucion Geologica del Sureste Mexicano desde el Mesozoico al presente en el contexto regional del Golfo de Mexico: *Boletín de la Sociedad Geologica Mexicana*, 59, 19–42.

Padilla y Sánchez, R. J., 2014, Tectonics of Eastern Mexico: Gulf of Mexico and its hydrocarbon potential, *American Association of Petroleum Geologists, Search and Discovery* 10622, 54 p.

Padilla y Sánchez, R. J., and R. Jose, 2016, Late Triassic–Late Cretaceous paleography of Mexico and the Gulf of Mexico, in C. M. Lowery, and J. W. Snedden, eds., *Mesozoic of the Gulf Rim and Beyond: New Progress in Science and Exploration of the Gulf of Mexico Basin*: Society of Economic Paleontologists and Mineralogists, Gulf Coast Section, 35th Annual Research Conference, 1–30.

Padilla y Sánchez, R. J., I. Dominguez Trejo, A. G. Lopez Azcarraga, *et al.*, 2013, Tectonic map of Mexico: Division de Ingenieria en Ciencias de la

Tierra Facultad de Ingenieria Universidad Nacional Autonoma de Mexico [map].

Paine, W. R., 1971, Petrology and sedimentation of the Hackberry sequence of southwest Louisiana: *Gulf Coast Association of Geological Societies Transactions*, 21, 37–55.

Paleo-Data Inc., 2017, PDI Neogene biostrat chart of the Gulf Basin US.

Pardo, G., 1975, Geology of Cuba, in A. E. M. Nairn and F. G. Stehli, eds., *The Ocean Basins and Margins: The Gulf of Mexico and the Caribbean*: Plenum, 3, 553–615.

Pardo G., 2009, *The Geology of Cuba*: American Association of Petroleum Geologists, Studies in Geology 58, 311–341.

Paredes, H. C. G., M. M. Medrano, and H. L. Sessarego, 2009, Provenance for the Middle and Upper Miocene sandstones of the Veracruz Basin, Mexico, in C. Bartolini and J.R. Roman Ramos, eds., *Petroleum Systems in the Southern Gulf of Mexico*: American Association of Petroleum Geologists, Memoir 90, 397–407.

Parra, P. A., N. Rubio, C. Ramirez, *et al.* 2013, Unconventional reservoir development in Mexico: lessons learned from the first exploratory wells: *Society of Petroleum Engineers Unconventional Resources Conference*, 1–14, doi:10.2118/164545-MS.

Parrish, J. T., and F. Peterson, 1988, Wind directions predicted from global circulation models and wind directions determined from eolian sandstones of the western United States – a comparison: *Sedimentary Geology*, 56, 261–282.

Pashin, J. C., J. Guohai, and D. J. Hills, 2016, Mesozoic structure and petroleum systems in the DeSoto Canyon salt basin in the Mobile, Pensacola, Destine Dome, and Viosca Knoll Areas of the MAFLA Shelf: *Society of Economic Paleontologists and Mineralogists Gulf Coast Section, 35th Annual Research Conference*, 416–449.

Passey, Q. R., S. Creany, J. B. Kulla, F. J. Moretti, and J. D. Stroud, 1990, A practical model for organic richness from porosity and resistivity logs: *American Association of Petroleum Geologists Bulletin*, 74, 1777–1794.

Passey, Q. R., K. Bohacs, W. L. Esc, R. Klimentidis, and S. Sinha, 2010, From oil-prone source rock to gas-producing shale reservoir-geologic and petrophysical characterization of unconventional shale gas reservoirs: *International Oil and Gas Conference and Exhibition in China: Society of Petroleum Engineers*, 1–29.

Paxton, S. T., 2017a, Assessment of undiscovered oil and gas resources in the Haynesville Formation, U.S. Gulf Coast, 2016: United States Geological Survey: Fact Sheet 2017-3016, 1–2, doi:10.3133/fs20173015.

Paxton, S. T., 2017b, Assessment of undiscovered oil and gas resources in the Bossier Formation, U.S. Gulf Coast, 2016: United States Geological Survey, Fact Sheet 2017-3015, 1–2, doi:10.3133/fs20173015.

Pearson, K., R. F. Dubiel, O. N. Person, and J. K. Pitman, 2011, Assessment of undiscovered oil and gas resources of the Upper Cretaceous Austin Chalk and Tokio and Eutaw Formations, Gulf Coast, 2010: U.S. Geological Survey National Assessment of Oil and Gas Fact Sheet 3046, 1–2.

Peel, F. J., J. R. Hossack, and C. J. Travis, 1995, Genetic structural provinces and salt tectonics of the Cenozoic offshore U.S. Gulf of Mexico: a preliminary analysis, in M. P. A. Jackson, D. G. Roberts, and S. Snelson, eds., *Salt Tectonics: A Global Perspective*: American Association of Petroleum Geologists, Memoir 65, 153–175.

Pemex Exploración y Producción, 2013, Provincia Petrolera Golfo de Mexico Profundo: Pemex, 1–26.

Penfield, G. T., and A. Camargo-Zanoguera, 1981, Definition of a major igneous zone in the central Yucatan platform with aeromagnetics and gravity, in *Technical Program, Abstracts and Bibliographies: Society of Exploration Geophysicists: 51st Annual Meeting* (Abs.), 37.

Pepper, F., 1982, Depositional environments of the Norphlet Formation (Jurassic) for Southwestern Alabama: *Gulf Coast Association of Geological Societies Transactions*, 32, 17–22.

Perez, L. E., 2017, *Neogene Current-Modified Submarine Fans and Associated Bed Forms in Mexican Deep-Water Areas*: MS thesis, University of Texas at Austin, 91 p.

Perfit, M. R., and W. W. Chadwick, Jr., 1998, Magmatism at mid ocean ridges: constraints from volcanological and geochemical investigations, in W. R. Buck, P. T. Delaney, J. A. Karson, and Y. Lagabrielle, eds., *Faulting and Magmatism at Mid-Ocean Ridges:* American Geophysical Union, 59–115, doi:10.1029/GM106p0059.

Perkins, B. F., 1985, Caprinid reefs and related facies in the Comanche Cretaceous Glen Rose Limestone of Central Texas: Annual Meeting of the Gulf Coast Association of Geological Societies and Society of Economic Paleontologists and Mineralogists Gulf Coast Section: Lower Cretaceous Depositional Environments From Shoreline to Slope – A Core Workshop, 129–140.

Peters, K. E., C. C. Walters, and J. M. Moldowan, 2005, *The Biomarker Guide, Volume 2: Biomarkers and Isotopes in Petroleum Exploration and Earth History*, Cambridge University Press, 300 p.

Pettigrew, R., C. Priddy, A. Elson, *et al.*, 2017, Fluvial–aeolian-evaporitic interactions in arid continental basins: implications for basin-scale migration and reservoir characterisation [poster]: *American Association of Petroleum Geologists, Search and Discovery* 70291.

Petty, A. J., 1995, Ferry Lake, Rodessa, and Punta Gorda anhydrite bed correlation, Lower Cretaceous, offshore eastern Gulf of Mexico: *Gulf Coast Association of Geological Societies Transactions*, 45, 487–493.

Petty, A. J., 1999, Petroleum exploration and stratigraphy of the Lower Cretaceous James Limestone (Aptian) and Andrew Formation (Albian): Main Pass, Viosca Knoll, and Mobile area, northeastern Gulf of Mexico: *Gulf Coast Association of Geological Societies*, 49, 440–450, doi:10.1306/E4FD3F45-1732-11D7-8645000102C1865D.

Petty, A. J., 2008, Stratigraphy and petroleum exploration history of the Cotton Valley Group (Lower Cretaceous to Upper Jurassic) and Haynesville Group (Upper Jurassic), offshore northeastern Gulf of Mexico: *Gulf Coast Association of Geological Societies Transactions*, 58, 713–728.

Phelps, R. M., 2011, *Middle-Hauterivian to Lower-Campanian Sequence Stratigraphy and Stable Isotope Geochemistry of the Comanche Platform, South Texas*: Ph.D. dissertation, The University of Texas at Austin, 227 p.

Phelps, R. M., C. Kerans, R. G. Loucks, *et al.*, 2014, Oceanographic and eustatic control of carbonate platform evolution and sequence stratigraphy on the Cretaceous (Valanginian–Campanian) passive margin, northern Gulf of Mexico: *Sedimentology*, 61, 461–496, doi:10.1111/sed.12062.

Phelps, R. M., C. Kerans, R. Da-Gama, *et al.*, 2015, Response and recovery of the Comanche carbonate platform surrounding multiple Cretaceous oceanic anoxic events, northern Gulf of Mexico: *Cretaceous Research*, 54, 117–144, doi:10.1016/j.cretres.2014.09.002.

Phillips, S., 1987, *Shelf Sedimentation and Depositional Sequence Stratigraphy of the Upper Cretaceous Woodbine-Eagle Ford Groups, East Texas*: Ph.D. dissertation, Cornell University, 507 p.

Pierce, J. D., 2014, *U-Pb Geochronology of the Late Cretaceous Eagle Ford Shale: Defining Chronostratigraphic Boundaries and Volcanic Ash Source*: MS thesis, The University of Texas at Austin, 144 p.

Pilcher, R. S., B. Kilsdonk, and J. Trude, 2011, Primary basins and their boundaries in the deep-water northern Gulf of Mexico: origin, trap types, and petroleum system implication: *American Association of Petroleum Geologists Bulletin*, 95, 219–240, doi:10.1306/06301010004.

Pilcher, R. S., R. T. Murphy, and J. M. Ciosek, 2014, Jurassic raft tectonics in the northeastern Gulf of Mexico: *Interpretation*, 2, 39–55, doi:10.1190/INT-2014-0058.1.

Pindell, J., and L. Kennan, 2001, Kinematic evolution of the Gulf of Mexico and Caribbean, in R. H. Fillon, N. C. Rosen, P. Weimer, *et al.*, eds., *Petroleum Systems of Deep-Water Basins*: Society of Economic Paleontologists and Mineralogists Gulf Coast Section, 21st Annual Research Conference, 193–220.

Pindell, J., E. C. Miranda, A. Ceron, and L. Hernandez, 2016, Aeromagnetic map constrains Jurassic–Early Cretaceous syn-rift, break up, and rotational seafloor spreading history in the Gulf of Mexico, in C. D. Lowry and J. W. Snedden, eds., *Mesozoic of the Gulf Rim and Beyond: New Progress in Science and Exploration of the Gulf of Mexico Basin*: Society of Economic Paleontologists and Mineralogists, Gulf Coast Section, 35th Annual Research Conference, 1–24.

Pitman, J. K., 2014, *Reservoirs and Petroleum Systems of the Gulf Coast*: U. S. Geological Survey, 4 p.

Pittman, J. G., 1989, Stratigraphy of the Glen Rose Formation, western Gulf of Mexico: *Gulf Coast Association of Geological Societies Transactions*, 39, 247–264.

Pogge von Strandmann, P. A. E., H. C. Jenykyns, and R. Woodfine, 2013, Lithium isotope evidence for 13C enhanced weathering during Oceanic Anoxic Event 2: *Nature Geoscience*, 6, 668–672.

Pollastro, R. M., 2001, *1995 USGS National Oil and Gas Play-Based Assessment of the South Florida Basin, Florida Peninsula Province: National Assessment of Oil and Gas Project: Petroleum Systems and Assessment of South Florida Basin*, U.S. Geological Survey, Digital Data Series 69-A, 1–17.

Pollastro, R. M., C. J. Schenk, and R. R. Charpentier, 2001, *Assessment of Undiscovered Oil and Gas in the Onshore and State Waters Portion of the South Florida Basin, Florida – USGS Province 50: National Assessment of Oil and Gas Project: Petroleum Systems and Assessment of South Florida Basin*, U.S. Geological Survey, Digital Data Series 69-A, 1–17.

Popenoe, P., V. J. Henry, and F. M. Idris 1987, Gulf trough: the Atlantic connection: *Geology*, 15, 327–332.

Porres Luna, A. A., 2018, Plenaria 5 "Desarrollo de Yacimientos No Convencionales y en Aguas Profundas" Situacion Actual de los Yacimientos No Convencionales y en Aguas Profundas en Mexico y Tecnologias necesarias para desarrollarlos, *Comision Nacional de Hidrocarburos*, 31.

Potter-McIntyre, S. L., J. R. Breeden, and D. H. Malone, 2018, A Maastrichtian birth of the ancestral Mississippi River system: evidence from the U–Pb detrital zircon geochronology of the McNairy Sandstone, Illinois, USA: *Cretaceous Research*: 91, 71–79, doi:10.1016/j.cretres.2018.05.010.

Power, B., J. Covault, A. Fildani, *et al.*, 2013, Facies analysis and interpretation of argillaceous sandstone beds in the Paleogene Wilcox Formation, deepwater Gulf of Mexico: *Gulf Coast Association of Geological Societies Transactions*, 63, 575–578.

Prather, B. E., 1992, Evolution of a late Jurassic carbonate/evaporate platform, Conecuh Embayment, Northeastern Gulf Coast, USA: *American Association of Petroleum Geologists Bulletin*, 76, 164–190.

Prather, B. E., 2000, Calibration and visualization of depositional process models for above-ground slopes: a case study from the Gulf of Mexico: *Marine and Petroleum Geology*, 17, 619–638.

Prather, B. E., J. R. Booth, G. S. Steffens, and P. A. Craig, 1998, Classification, lithologic calibration, and stratigraphic succession of the seismic facies of intraslope basins, deep-water Gulf of Mexico: *American Association of Petroleum Geologists Bulletin*, 82, 701–728.

Prather, B. E., C. O'Byrne, C. Pirmez, and Z. Sylvester, 2017, Sediment partitioning, continental slopes and base-of-slope systems: *Basin Research*, 29: 394–416. doi: 10.1111/bre.12190.

Presley, M. W., and C. H. Reed, 1984, Jurassic exploration trends of East Texas, in M. W. Presley, ed., *The Jurassic of East Texas*, East Texas Geological Society, 11–22.

Prost, G., and M. Aranda-Garcia, 2001, Tectonics and hydrocarbon systems of the Veracruz Basin, Mexico, in C. Bartolini, R. T. Buffler, and A. Cantú-Chapa, eds., *The Western Gulf of Mexico Basin: Tectonics, Sedimentary Basins, and Petroleum Systems*: American Association of Petroleum Geologists, Memoir 75, 271–291.

Puga-Bernabéu, A., J. M. Webster, R. J. Beaman, and V. Guilbaud, 2013, Variation in canyon morphology on the Great Barrier Reef margin, northeastern Australia: The influence of slope and barrier reefs: *Geomorphology*, 191, 35–50, doi:10.1016/j.geomorph.2013.03.001.

Pulham, A. J. 1993, Variations in slope deposition, Pliocene–Pleistocene, offshore Louisiana, northeast Gulf of Mexico, in P. Weimer and H. W. Posamentier, eds., *Siliciclastic Sequence Stratigraphy: Recent Developments and Applications*: American Association of Petroleum Geologists, Memoir 58, 199–234.

Purkey Phillips, M., in prep., The first regional biostratigraphic characterization of the Paleocene/Eocene Thermal Maximum (PETM) in the Gulf of Mexico.

Radovich, B. J., J. Moon, C. D. Connors, and D. Bird, 2007, Insights into structure and stratigraphy of the northern Gulf of Mexico from 2D pre-stack depth migration imaging of mega-regional onshore to deep water, long-offset seismic data: *Gulf Coast Association of Geological Societies Transactions*, 57, 633–637.

Radovich B., B. Horn, P. Nuttall, and A. McGrail, 2011, The only complete regional perspective: *GEO ExPro*, 8(2), 36–38.

Rahl, J. M., P. W. Reiners, I. H. Campbell, S. Nicolescu, and C. M. Allen, 2003. Combined single-grain (U–Th)/He and U/Pb dating of detrital zircons from the Navajo Sandstone, Utah: *Geology*, 31, 761–764, doi:10.1130/G19653.1.

Rainwater, E. H., 1964, Regional stratigraphy of the Gulf Coast Miocene: *Gulf Coast Association of Geological Societies Transactions*, 14, 81–124.

Ramos, A., and W. E. Galloway, 1990, Facies and sand-body geometry of the Queen City (Eocene) tide-dominated delta-margin embayment, NW Gulf of Mexico Basin: *Sedimentology*, 37, 1079–1098.

Randazzo, A. F., 1997, The sedimentary platform of Florida: Mesozoic to Cenozoic, in A. F. Randazzo and D. S. Jones, eds., *The Geology of Florida*, University Press of Florida, 39–56.

Raymond, D. E., W. E. Osborne, C. W. Copeland, and T. L. Neathery, 1988, *Alabama Stratigraphy*. Geological Survey of Alabama, Circular, 140, 1–102.

Reading, H. G., and M. Richards, 1994, Turbidite systems in deep-water basin margins classified by grain size and feeder system: *American Association of Petroleum Geologists Bulletin*, 78, 792–822.

Reed, J. C., C. L. Leyendecker, A. S. Khan, *et al.*, 1987, *Correlation of Cenozoic Sediments, Gulf of Mexico Outer Continental Shelf. Part I: Galveston Area, Offshore Texas Through Vermilion Area, Offshore Louisiana*: Gulf of Mexico OCS Regional Office, Report MMS 87-0026.

Reese, D., 1976, Pre-ferry lake lower Cretaceous deltas of south Mississippi and producing trends: *Gulf Coast Association of Geological Societies Transactions*, 26, 59–63.

Reiners, P. W., I. H. Campbell, S. Nicolescu, *et al.*, 2005, (U–Th)/(He–Pb) double dating of detrital zircons: *American Journal of Science*, 305, 259–311.

Renne, P. R., A. L. Deino, F. J. Hilgen, *et al.*, 2013, Time scales of critical events around the Cretaceous–Paleogene boundary: *Science*, 339, 684–687.

Reynolds, T., 2000, Reservoir architecture in the Mars field, deepwater Gulf of Mexico, USA: the implications of production, seismic, core and well-log data: *Society of Economic Paleontologists and Mineralogists Gulf Coast Section 20th Annual Research Conference*, 877–892.

Ricoy-Paramo, V., 2005, *3D Seismic Characterisation of the Cantarell Field, Campeche Basin, Mexico*: Master's thesis, Cardiff University, 418 p.

Riggs, N. R., T. M. Lehman, G. E., Gehrels, and W. R. Dickinson, 1996, Detrital zircon link between headwaters and terminus of the Upper Triassic Chinle–Dockum paleoriver system: *Science*, 273, 97–100.

Riller, U., M. H. Poelchau, A. S. P. Rae, *et al.*, 2018, Rock fluidization during peak-ring formation of large impact structures: *Nature*, 11, 511–518, doi:10.1038/s41586-018-0607-z.

Rine, J. M., 2014, Reconstruction of diagenetic and burial history of South Georgia Rift Basin: analysis of sandstones from the Rizer #1 South Carolina: *Houston Geological Society Bulletin*, 57, 11–14.

Rine, J. M., B. E. Hollon, R. Fu, N. Houghton, and M. Waddell, 2014, Diagenetic and burial history of a portion of the Late Triassic South Georgia Rift Basin based on petrologic and isotopic (d18O) analyses of sandstones from test borehole rizer #1, Colleton County, SC: *American Association of Petroleum Geologists, Search and Discovery* 51016, 31 p.

Rios Lopez, J. J., 1996, Disposicion de Los Cuerpos de Areniscas y su Relacion en el Mantemiento de La Produccion del Campo Ek-Balam, Campeche: *Asociacion Mexicana de Geologos Petroleros*, 65, 1–38.

Rios Lopez, J. J., and A. Cantú-Chapa, 2009, Stratigraphy and sedimentology of Middle Eocene Kumaza calcarenites "member" in the Ku, Maloob, and Zaap oil fields, offshore Campeche, Mexico, in C. Bartolini and J.R. Roman Ramos, eds., *Petroleum Systems in the Southern Gulf of Mexico*: American Association of Petroleum Geologists, Memoir 90, 257–277.

Rivas, L. F., J. Sanclemente, and W. K. Rickets, 2009, Tahiti subsurface: drilling and completion technology challenges and accomplishments: Offshore Technology Conference, 1–15. doi:10.4043/19861-MS.

Roca-Ramisa, L., and D. Arnabar, 1994, Geological and geomechanical reservoir characterization: The Ek-Balam Field Study in Mexico: *Congresso Brasileiro de Petroleo*, 1–28.

Rodriguez Hernandez, M. L. A., P. L. Oritz Gomez, C. B. Sanchez, *et al.*, 2003, Distribution of depositional facies and reservoir properties from middle Cretaceous carbonates of the Cordoba Platform: *American Association of Petroleum Geologists, Search and Discovery* 90020, 1 p.

Roesink, J. G., P. Weimer, and R. Bouroullec, 2004, Sequence stratigraphy of Miocene to Pleistocene sediments of east-central Mississippi canyon, northern Gulf of Mexico: *Gulf Coast Association of Geological Societies Transactions*, 54, 587–601.

Rogers, R., 1987, A palynological age determination for the Dorcheat and Hosston Formations: the Jurassic–Cretaceous boundary in northern Louisiana: *Gulf Coast Association of Geological Societies Transactions*, 37, 447–456.

Rosenfeld, J. and J. Pindell, 2003, Early Paleogene isolation of the Gulf of Mexico from the world's oceans? Implications for hydrocarbon exploration and eustasy, in C. Bartolini, R. T. Buffler, and J.F. Blickwed, eds., *The Circum-Gulf of Mexico and Caribbean: Hydrocarbon Habitats, Basin Formation, and Plate Tectonics*: American Association of Petroleum Geologists, Memoir 79, 89–103.

Roure, F., H. Alzaga-Ruiz, J.-P. Callot, *et al.*, 2009, Long lasting interactions between tectonic loading, unroofing, post-rift thermal subsidence and sedimentary transfers along the western margin of the Gulf of Mexico: some insights from integrated quantitative studies: *Tectonophysics*, 475, 119–189.

Rowan, M. G., 1995, Structural styles and evolution of allochthonous salt, Central Louisiana outer shelf and upper slope, in M. P. A. Jackson, D. G. Roberts, and S. Snelson, eds., *Salt Tectonics: a Global Perspective*: American Association of Petroleum Geologists, Memoir 65, 199–228.

Rowan, M. G., 2018, The South Atlantic and Gulf of Mexico salt basins: crustal thinning, subsidence and accommodation for salt and presalt strata, in K. R. McClay and J. A. Hammerstein, eds., *Passive Margins: Tectonics, Sedimentation and Magmatism*, London, Geological Society, Special Publication 476, 1–44. doi:10.1144/SP476.6.

Rowan, M. G., and K. F. Inman, 2011, Salt-related deformation recorded by allochthonous salt rather than growth strata: *Gulf Coast Association of Geological Societies Transactions*, 61, 379–390.

Rowan, M. G. and P. Weimer, 1998, Salt–sediment interaction, Northern Green Canyon and Ewing Bank (Offshore Louisiana), Northern Gulf of Mexico: *American Association of Petroleum Geologists Bulletin*, 82, 1055–1082.

Rowan, M. G., B. D. Trudgill, and J. C. Fiduk, 2000, Deepwater, salt-cored fold belts: lessons from the Mississippi Fan and Perdido fold belts, northern Gulf of Mexico, in W. Mohriak and M. Talwani, eds., *Atlantic Rifts and Continental Margins*: American Geophysical Union, Geophysical Monograph, 115, 173–191, doi:10.1029/GM115p0173.

Rowan, M. G., F. J. Peel, and B. C. Venderville, 2004, Gravity-driven fold belts on passive margins, in K. R. McClay, ed., *Thrust Tectonics and Hydrocarbon Systems*: American Association of Petroleum Geologists, Memoir 82, 152–182.

Rowan, M. G., K. A. Giles, T. E. Hearon IV, and J. C. Fiduk, 2016, Megaflaps adjacent to salt diapirs: *American Association of Petroleum Geologists Bulletin*, 100, 1723– 1747, doi:10.1306/05241616009.

Rozendal, R. A., and W. S. Erskine, 1971, Deep test in Ouachita Structural Belt of Central Texas, *American Association of Petroleum Geologists Bulletin*, 55, 2008–2017.

Ryan, P. C. and R. C. Reynolds Jr., 1997, The chemical composition of serpentine/chlorite in the Tuscaloosa Formation, United States Gulf Coast: EDX vs. XRD determinations, implications for mineralogic reactions and the origin of anatase: *Clays and Clay Minerals* 45(3), 339–352.

Rynott, T., 2015, Gulf of Mexico Inboard Lower Tertiary and Cretaceous: plays and potential (abs.): *American Association of Petroleum Geologists, Search and Discovery* 90219.

Salomon-Mora, L. E., M. Aranda-Garcia, and J. R. Roman Ramos, 2009, Contractional growth faulting in the Mexican Ridges, Gulf of Mexico, in C. Bartolini and J.R. Roman Ramos, eds., *Petroleum Systems in the Southern Gulf of Mexico*: American Association of Petroleum Geologists, Memoir 90, 93–116.

Salter, R., M. Beller, D. Shelander, *et al.*, 2005, Increasing accuracy and resolution in porosity prediction for carbonate reservoirs using high-resolution seismic technology in the Arenqu-Lobina Area, Offshore Mexico: *American Association of Petroleum Geologists, Search and Discovery* 90039.

Salvador, A., 1987, Late Triassic–Jurassic paleogeography and origin of the Gulf of Mexico Basin: *American Association of Petroleum Geologists Bulletin*, 71, 419–451.

Salvador, A., ed., 1991a, The Gulf of Mexico basin, in *The Geology of North America, V.* Geological Society of America, 568 p.

Salvador, A., 1991b, Origin and development of the Gulf of Mexico basin, in A. Salvador, ed., *The Geology of North America V: The Gulf of Mexico Basin*: Geological Society of America, 389–344.

Salvador, A. and J. M. Quezada-Muneton, 1989, Stratigraphic correlation chart, the Gulf of Mexico Basin, in A. Salvador, ed., *The Geology of North America, V: The Gulf of Mexico Basin*, Geological Society of America, plate 5.

Sánchez-Hernández, H., 2013, *Stratigraphic Characterization and Evolution of a Mid-Tertiary Age Deep Water System, Holok area, SW Gulf of Mexico*: Ph.D. dissertation, The University of Aberdeen, 358 p.

Sanchez-Montes de Oca, R., 1980, Geología petrolera de la Sierra de Chiapas: *Boletin de la Asociacion Mexicana de Geologos Petroleros*, 31, 67–97.

Sandrea, I., R. Sandrea, M. Limon, K. Vazquez, A. Horbury, and M. Shann, 2018, *Mexico History of Oil Exploration, its Amazing Carbonates, and Untapped Oil Potential*: Pennwell, 144p.

Sandwell, D. T., R. D. Müller, W. H. F. Smith, E. Garcia, and R. Francis, 2014, New global marine gravity model from CryoSat-2 and Jason-1 reveals buried tectonic structure: *Science*, 346, 65–67, doi:10.1126/science.1258213.

Sanford, J. C., J. W. Snedden, and S. P. S. Gulick, 2016, The Cretaceous–Paleogene boundary deposit in the Gulf of Mexico: large-scale oceanic basin response to the Chicxulub impact: *Journal of Geophysical Research: Solid Earth*, 121, 22 p., doi:10.1002/2015JB012615.

Santamaria Orozco, D. M., 2000, *Organic Geochemistry of Tithonian Source Rocks and Associated Oils from the Sonda De Campeche, Mexico*: Ph.D. dissertation, Rhine-Westphalia Institute of Technology, 201 p.

Sassen, R., 1988, Geochemical and carbon isotopic studies of crude oil destruction, bitumen precipitation, and sulfate reduction in the Deep Smackover Formation: *Organic Geochemistry*, 12, 351–361.

Sassen, R., 1990, Geochemistry of carbonate source rocks and crude oils in Jurassic salt basins of the Gulf Coast, in J. Brooks, ed., *Classic Petroleum Provinces*, London Geological Society, Special Publications 50, 265–277.

Sassen, R., C. H. Moore, and F. C. Meendsen, 1987, Distribution of hydrocarbon source potential in the Jurassic Smackover Formation: *Organic Geochemistry*, 11.5, 379–383.

Saucier, A. E., 1985, Geologic framework of the Travis Peak (Hosston) Formation of east Texas and northern Louisiana, in R. J. Finley, S. P. Dutton, Z. S. Lin, and A. E. Saucier, eds., *The Travis Peak (Hosston) Formation: Geologic Framework, Core Studies, and Engineering Field Analysis*: The University of Texas at Austin, Bureau of Economic Geology, report prepared for the Gas Research Institute under contract no. 5082-211-0708, 233 p.

Saunders, M., L. Gieger, K. Rodriguez, and P. Hargreaves, 2016, The delineation of pre-salt license blocks in the Deep Offshore Campeche-Yucatan Basin: *American Association of Petroleum Geologists, Search and Discovery* 10667, 8 p.

Sawyer, D. S., R. T. Buffler, and R. H. Pilger Jr., 1991, The crust under the Gulf of Mexico basin, in A. Salvador, ed., *The Gulf of Mexico Basin: The Geology of North America*: Geological Society of America, 53–72, doi:10.1130/DNAG-GNA-J.53.

Schenk, C. J., 2008, Jurassic–Cretaceous composite total petroleum system and geologic models for oil and gas assessment of the North Cuba Basin, Cuba, in U.S. Geological Survey North Cuba Basin Assessment Team, *Jurassic–Cretaceous Composite Total Petroleum System and Geologic Assessment of Oil and Gas Resources of the North Cuba Basin, Cuba*: U.S. Geological Survey, Digital Data Series DDS-69-M, 94 p.

Schlager, W., R. T. Buffler, D. Angstadt, and R. L. Phair, 1984, Geologic history of the southeastern Gulf of Mexico, in R. T. Buffler and W. Schlager, eds., *Deep Sea Drilling Project Initial Reports 77*, US GPO, 715–738.

Schlanger, S. O., and H. C. Jenkyns, 1976, Cretaceous oceanic anoxic events: causes and consequences: *Geologie en Mijnbouw*, 55, 179–184.

Schlische, R. W., 2003, Progress in understanding the structural geology, basin evolution, and tectonic history of the Eastern North American Rift System, in P. M. LeTourneau and P. E. Olsen, eds., *The Great Rift Valleys of Pangea in Eastern North America*: Columbia University Press, 21–64.

Schulte, P. J., L. Alegret, I. Arenillas, *et al.*, 2010, The Chicxulub asteroid impact and mass extinction at the Cretaceous–Paleogene boundary: *Science*, 327, 1214–1218, doi:10.1126/science.1177265.

Schulte, P., J. Smit, A. Deutsch, *et al.*, 2012, Tsunami backwash deposits with Chicxulub impact ejecta and dinosaur remains from the Cretaceous–Palaeogene boundary in the La Popa Basin, *Mexico: Sedimentology*, 59, 737–765, doi:10.1111/j.1365-3091.2011.01274.x.

Schumacher, D., and R. M. Parker, 1990, Possible pre-Jurassic origin for some Jurassic-reservoired oil, Cass Co., Texas, in *Gulf Coast Oils and Gases, Their Characteristics, Origin, Distribution, and Exploration and Production Significance*: Society of Economic Paleontologists and Mineralogists Gulf Coast Section, 9th Annual Research Conference, 59–68.

Schuster, D. C., 1995, Deformation of allochthonous salt and evolution of related salt-structural systems, eastern Louisiana Gulf Coast, in M. P. A. Jackson, D. G. Roberts, and S. Snelson, eds., *Salt Tectonics: A Global Perspective*: American Association of Petroleum Geologists, Memoir 65, 177–198.

Scotese, C. R., 2017: Paleomap project, www.scotese.com/Default.htm.

Scott, E. D., R. A. Denne, J. S. Kaiser, and D. P. Eickhoff, 2014, Impact on sedimentation into the north-central deepwater Gulf of Mexico as a result of the Chicxulub event: *Gulf Coast Association of Geological Societies Journal*, 3, 41–50.

Scott, E., R. Denne, J. Kaiser, and D. Eichkoff, 2016, Immediate and post-event effects of the K/Pg boundary Chicxulub impact on the northern Gulf of Mexico: *New Understanding of the Petroleum Systems of the Continental Margins of the World*: Society of Economic Paleontologists and Mineralogists Gulf Coast Section, 32nd Annual Research Conference Proceedings, 96–109, doi:10.5724/gcs.12.32.

Scott, K. R., W. E. Hayes, and R. P. Fietz, 1961, Geology of the Eagle Mills Formation: *Gulf Coast Association of Geological Societies Transactions*, 11, 1–14.

Scott, R. W., 1984, Significant fossils of the Knowles Limestone, Lower Cretaceous, Texas, in W. P. S. Ventress, D. G. Bebout, B. F. Perkins, and C. H. Moore, eds., *The Jurassic of the Gulf Rim*: Society of Economic Paleontologists and Mineralogists Gulf Coast Section, 3rd Annual Research Conference Proceedings, 333–346.

Scott, R. W., 1990, Models and stratigraphy of mid-Cretaceous reef communities, Gulf of Mexico: in B. H. Lidz, ed., *Concepts in Sedimentology and Paleontology*, 2, Society of Economic Paleontologists and Mineralogists, 1–102.

Scott, R. W., D. G. Benson, R. W. Morin, B. L., Shaffer, and F. E. Oboh-Ikuenobe, 2002, Integrated Albian-Lower Cenomanian chronostratigraphy standard, Trinity River section Texas, in R. W. Scott, ed., *U. S. Gulf Coast Cretaceous Stratigraphy and Paleoecology*: Society of Economic Paleontologists and Mineralogists Gulf Coast Section, 23rd Bob F. Perkins Memorial Conference, 277–334.

Scott, R., W. Whitney, R. Hojnacki, X. Lin, and Y. Wang, 2016, Albian stratigraphy of the San Marcos Platform, Texas: why the person formation correlates with Upper Fredericksburg group not Washita Group, in *Mesozoic of the Gulf Rim and Beyond: New Progress in Science and Exploration of the Gulf of Mexico Basin: Society of Economic Paleontologists and Mineralogists Gulf Coast Section, 35th Annual Gulf Section SEPM Foundation Perkins Rosen Research Conference*, 536–545.

Sempere, J. C., J. Lin, H. S. Brown, H. Schouten, and G. M. Purdy, 1993, Segmentation and morphotectonic variations along a slow-spreading center: the Mid-Atlantic Ridge (24°00′N–30°40′N): *Marine Geophysical Researches*, 15, 153–200.

Seni, S. J., and M. P. A. Jackson, 1983, Evolution of salt structures, East Texas diapir province, part 1: sedimentary record of halokinesis: *American Association of Petroleum Geologists Bulletin*, 67, 1219–1244.

Sharman, G. R., J. A. Covault, D. F. Stockli, A F.-J. Wroblewski, and M. A. Bush, 2016, Early Cenozoic drainage reorganization of the United States Western Interior: Gulf of Mexico routing system: *Geology*, 44, 187–190. doi10.1130/G38765.1.

Shellhouse, K., 2017, *The Cretaceous–Paleogene Boundary Deposit in LaSalle Parish, Louisiana*: Master's thesis, The University of Louisiana at Lafayette, 175 p.

Shideler, G. L., 1986, *Regional Geologic Cross-Section Series of Neogene-Quaternary Deposits, Louisiana Continental Shelf*: Geological Society of America, Map and Chart Series MC-54, 8 p.

Shultz, A. W., 2010: Facies and environments of the Lobo trend, Webb and Zapata counties, Texas: examples from core, in J. Long, W. L. Stapp, R. W. Debus, and A. N. Smith, eds., *Contributions to the Geology of South Texas*: South Texas Geological Society, 229–240.

Sluijs A., L. van Roij, G. J. Harrington, *et al.*, 2014, Warming, euxinia and sea level rise during the Paleocene–Eocene Thermal Maximum on the Gulf Coastal Plain: implications for ocean oxygenation and nutrient cycling: *Climate of the Past*, 10, 1421–1439.

Smit, J., and J. Hertogen, 1980, An extraterrestrial event at the Cretaceous–Tertiary boundary: *Nature*, 285, 198.

Smith, T., 2013, Unleasing the mad dog: *GeoExPro*, 10, 22–27.

Smith, T., 2015, Puzzling salt structures: *GeoExPro*, 12, 20–24.

Smith, W. H. F., and D. T. Sandwell, 1997, Global seafloor topography from satellite altimetry and ship depth soundings, *Science*, 277, 1957–1962.

Snedden, J. W., and R. S. Jumper, 1990, Shelf and shoreface reservoirs, Tom Walsh-Owen Field, Texas, in J. H. Barwis, J. G. McPherson, and J. R. J. Studlick, eds., *Sandstone Petroleum Reservoirs: Casebooks in Earth Sciences*, Springer, 415–436.

Snedden, J. W., and D. G. Kersey, 1982, Depositional environments and gas reproduction trends Olmos Sandstone, Upper Cretaceous, Webb County, Texas: *Gulf Coast Association of Geological Societies Transactions*, 32, 497–518.

Snedden, J. W., and C. Liu, 2011, Recommendations for a uniform chronostratigraphic designation system for Phanerozoic depositional sequences: *American Association of Petroleum Geologists Bulletin* 95, 1095–1122, doi:10.1306/01031110138.

Snedden, J. W., and D. F. Stockli, 2019, Paleogeographic and depositional reconstruction of Oxfordian Aeolian sandstone reservoirs in Mexico offshore areas: comparison to the Norphlet Aeolian play of the Northern Gulf of Mexico: The Geological Society, London, Petroleum Geology of Mexico and the Northern Caribbean Conference (abstract).

Snedden, J. W., R. W. Tillman, and S. J. Culver, 2011, Genesis and evolution of a mid-shelf, storm-built sand ridge, New Jersey continental shelf, U.S.A.: *Journal of Sedimentary Research*, 81, 534–552, doi:10.2110/jsr.2011.26.

Snedden, J. W., W. E. Galloway, T. L. Whiteaker, and P. E. Ganey-Curry, 2012, Eastward shift of deep-water fan axes during the Miocene in the Gulf of Mexico: possible causes and models: *Gulf Coast Association of Geological Societies Journal*, 1, 131–144.

Snedden, J., D. Eddy, G. Christeson, *et al.*, 2013, A new temporal model for eastern Gulf of Mexico Mesozoic deposition: *Gulf Coast Association of Geological Societies Transactions*, 63, 609–612.

Snedden, J. W., I. O. Norton, G. L. Christeson, and J. C. Sanford, 2014, Interaction of deepwater deposition and a mid-ocean spreading center, eastern Gulf of Mexico basin, USA: *Gulf Coast Association of Geological Societies Transactions*, 64, 371–383.

Snedden, J., W. E. Galloway, P. Ganey-Curry, and M. Blum, 2015, The geologic history of submarine fans in the deepwater Gulf of Mexico: Mesozoic to modern: *Gulf Coast Association of Geological Societies Transactions*, 65, 521–527.

Snedden, J. W., A. C. Bovay, and J. Xu, 2016a, New models of Early Cretaceous source-to-sink pathways in the Eastern Gulf of Mexico, in C. M. Lowery, J. W. Snedden, and M. D. Blum, eds., *Mesozoic of the Gulf Rim and Beyond: New Progress in Science and Exploration of the Gulf of Mexico Basin*: Gulf Coast Section SEPM, Perkins Rosen Research Conference, 380–415.

Snedden, J. W., J. Virdell, T. L. Whiteaker, and P. Ganey-Curry, 2016b, A basin-scale perspective on Cenomanian–Turonian (Cretaceous) depositional systems, greater Gulf of Mexico (USA): *Interpretation*, 4, SC1–SC22, doi:10.1190/INT-2015-0082.1.

Snedden, J. W., W. E., Galloway, K. T. Milliken, *et al.*, 2018a, Validation of empirical source-to-sink scaling relationships in a continental-scale system: the Gulf of Mexico basin Cenozoic record: *Geosphere*, 14, 1–17, doi:10.1130/GES01452.1.

Snedden, J. W., L. D. Tinker and J. Virdell, 2018b, Southern Gulf of Mexico Wilcox source-to-sink: investigating and predicting Paleogene Wilcox reservoirs in Eastern Mexico deep-water areas: *American Association of Petroleum Geologists Bulletin*, 102, 2045–2074.

Snedden, J. W., I. Norton, M. Hudec, A. Eljalafi, and F. Peel, 2018c, Paleogeographic reconstruction of the Louann salt basin in the Gulf of Mexico: *American Association of Petroleum Geologists, Annual Convention and Exhibition* (abs).

Snyder, F., and R. Ysaccis, 2018, New offshore exploration opportunities within the Salina Del Istmo Basin, Mexico: *American Association of Petroleum Geologists, Search and Discovery* 11143, 41 p, . doi:10.1306/11143snyder.

Sømme, T. O., W. Helland-Hansen, and D. Granjeon, 2009a, Impact of eustatic amplitude variations on shelf morphology, sediment dispersal, and sequence stratigraphic interpretation: icehouse versus greenhouse systems: *Geology*, 37, 587–590, doi:10.1130/G31134Y.1.

Sømme, T. O., W. Helland-Hansen, O. J. Martinsen, and J. B. Thurmond, 2009b, Relationships between morphological and sedimentological parameters in source-to-sink systems: a basis for predicting semi-quantitative characteristics in subsurface systems: *Basin Research*, 21, 361–387, doi:10.1111/j.1365-2117.2009.00397.x.

Sosa Patron, A. A., J. G. Cardenas Lopez, C. C. Lara, *et al.*, 2009, Integrated geological interpretation and impact on the definition of Neogene Plays in the Isthmus Saline Basin, Mexico, in C. Bartolini and J.R. Roman Ramos, eds., *Petroleum Systems in the Southern Gulf of Mexico*: American Association of Petroleum Geologists, Memoir 90, 29–48.

Stabler, C., 2016, Triple-porosity diagram proposed to characterize complex carbonate reservoirs – examples from Mexico, in C. M. Lowery, J. W. Snedden, and N. C. Rosen, eds., *Mesozoic of the Gulf Rim and Beyond: New Progress in Science and Exploration of the Gulf of Mexico Basin*: Society of Economic Paleontologists and Mineralogists Gulf Coast Section, 35th Annual Research Conference.

Steffens, G. S., E. K. Biegert, H. S. Sumner, and D. Bird, 2003, Quantitative bathymetric analysis of selected siliciclastic margins: receiving basin configurations for deepwater fan systems: *Marine and Petroleum Geology*, 20, 547–561, doi:10.1016/j.marpetgeo.2003.03.007.

Steier, A., and P. Mann, 2019, Late Mesozoic gravity sliding and Oxfordian hydrocarbon reservoir potential of the northern Yucatan margin: *Marine and Petroleum Geology*, 103, 681–701.

Stephens, B. P., 2009, Basement controls on subsurface geologic patterns and coastal geomorphology across the northern Gulf of Mexico: implications for subsidence studies and coastal restoration: *Gulf Coast Association of Geological Societies Transactions*, 59, 729–751.

Stephens, B. P., 2010, Basement controls on subsurface geologic patterns and near-surface geology across the northern Gulf of Mexico: a deeper perspective on coastal Louisiana: *American Association of Petroleum Geologists, Search and Discovery* 30129.

Stephenson, S. N., G. G. Roberts, M. J. Hoggard, and A. C. Whittaker, 2014, A Cenozoic uplift history of Mexico and its surroundings from longitudinal river profiles: *Geochemistry, Geophysics, Geosystems*, 15, 4734–4758.

Stern, R. J., and W. R. Dickinson, 2010, The Gulf of Mexico is a Jurassic backarc basin: *Geosphere*, 6, 739–754, doi:10.1130/GES00585.1.

Stern, R. J., E. Y. Anthony, M. Ren, *et al.*, 2011, Southern Louisiana salt dome xenoliths: first glimpse of Jurassic (ca. 160 Ma) Gulf of Mexico crust: *Geology*, 39, 315–318.

Stoneburner, R. K., 2015, The discovery, reservoir attributes, and significance of the Hawkville Field and the Eagle Ford Trend: implications for future development: *Gulf Coast Association of Geological Societies Transactions*, 65, 377–387.

Strong, M. A., 2013, *Investigation and Characterization of Features on a Cretaceous–Paleogene Seismic Horizon in Northern Louisiana*: Master's thesis, University of Louisiana at Lafayette, 94 p.

Strong, M. A., and G. L. Kinsland, 2014, Chicxulub impact tsunami deposits at the K–Pg boundary in northern Louisiana?: *American Association of Petroleum Geologists, Search and Discovery* 330379.

Styzen, M. J., 1996, A chart in two sheets of the Late Cenozoic chronostratigraphy of the Gulf of Mexico. *The Gulf Coast Section of the Society of Economic Paleontologists and Mineralogists, GCSSEPM Foundation 16th Annual Research Conference.*

Suter, J. R., and H. L. Berryhill, 1985. Late Quaternary shelf-margin deltas, northwest Gulf of Mexico: *American Association of Petroleum Geologists Bulletin*, 69, 77–91.

Swanson, S. M., C. B. Enomoto, K. O. Dennen, B. J. Valentine, and C. D. Lohr, 2013, Geologic model for the assessment of undiscovered hydrocarbons in Lower to Upper Cretaceous carbonate rocks of the Fredericksburg and Washita groups, U.S. Gulf Coast Region: *Gulf Coast Association of Geological Societies Transactions*, 63, 423–437.

Sweet, M. L., and M. D. Blum, 2012, Paleocene–Eocene Wilcox submarine canyons and thick deepwater sands of the Gulf of Mexico: very large systems in a greenhouse world, not a Messinian-like crisis: *Gulf Coast Association of Geological Societies Transactions*, 61, 443–450.

Sweet, M. L., and M. D. Blum, 2016, Connections between fluvial to shallow marine environments and submarine canyons: implications for sediment transfer to deep water: *Journal of Sedimentary Research*, 86, 1147–1162, doi: 10.2110/jsr.2016.64.

Swift, T. E., and P. Mladenka, 1997, Technology tackles low-permeability sand in South Texas: *Oil & Gas Journal*, 95, 68–72.

Syvitski, J. P. M., and J. D. Milliman, 2007, Geology, geography, and humans battle for dominance over the delivery of fluvial sediment to the coastal ocean: *Journal of Geology*, 115, 1–19.

Tada, R., M. A. Iturralde-Vinent, T. Matsui, *et al.*, 2003, K/T boundary deposits in the Paleo-western Caribbean basin, in C. Bartolini, R. T. Buffler, and J. Blickwede, eds., *The Circum-Gulf of Mexico and the Caribbean: Hydrocarbon Habitats, Basin Formation, and Plate Tectonics*: American Association of Petroleum Geologists Memoir 79, 582–604.

Talling, P. J., 2013, Hybrid submarine flows comprising turbidity current and cohesive debris flow: deposits, theoretical and experimental analyses, and generalized models: *Geosphere*, 9, 460–488, doi:10.1130/GES00793.1.

Taylor, T. R., M. R. Giles, L. A. Hathon, *et al.*, 2010, Sandstone diagenesis and reservoir quality prediction: models, myths, and reality: *American Association of Petroleum Geologists Bulletin*, 94, 1093–1132. doi:10.1306/04211009123.

Tew, B. H., R. M. Mink, E. A. Mancini, S. D. Mann, and D. C. Kopaska-Merkel, 1991, *Regional Geologic Framework and Petroleum Geology of the Smackover Formation, Alabama Coastal Waters and Adjacent Federal Waters Area*, Geological Survey of Alabama and State Oil and Gas Board, Energy and Coastal Geology Division, Report, 84 p.

Thacher, C., J. Stefani, C. Wu, *et al.*, 2013, Subsalt imaging and 4D reservoir monitoring evaluation of Tahiti field, Gulf of Mexico: Society of Petroleum Engineers, Annual Meeting, 4875–4879.

Thomas, W. A., 2011, The Iapetan rifted margin of southern Laurentia: *Geosphere*, 7, 97–120, doi: 10.1130/GES00574.1.

Tian, Y., W. B. Ayers, and W. D. McCain, Jr., 2012, Regional analysis of stratigraphy, reservoir characteristics, and fluid phases in the Eagle Ford Shale, South Texas: *Gulf Coast Association of Geological Societies Transactions*, 62, 471–483.

Treviño García, F. J., 2012, *Caracterización inicial de un campo marino en aguas someras del Sur del Golfo de México*: Master's thesis, Instituto Politecnico National, 193 p.

Trudgill, B. D., M. G. Rowan, P. Weimer, *et al.*, 1995, The structural geometry and evolution of the salt-related Perdido Fold Belt, Alaminos Canyon, Northwestern Deep Gulf of Mexico: *Society of Economic Paleontologists and Mineralogists Gulf Coast Section 16th Annual Research Conference*, 275–284.

Trudgill, B. D., M. G. Rowan, J. C. Fiduk, *et al.*, 1999, The Perdido fold belt, northwestern deep Gulf of Mexico, part 1: structural geometry, evolution and regional implications: *American Association of Petroleum Geologists Bulletin*, 83, 88–113.

Tsikos, H., H.C. Jenkyns, B. Walsworth-Bell, *et al.*, 2004, Carbon-isotope stratigraphy recorded by the Cenomanian–Turonian oceanic anoxic event: correlation and implications based on three key localities: *Journal of the Geological Society of London*, 161, 711–719.

Tye, R. S., 1992, Fluvial–sandstone reservoirs of the Travis Peak Formation, East Texas Basin, in A. D. Miall, and N. Tyler, eds., *The Three-Dimensional Facies Architecture of Terrigenous Clastic Sediments and Its Implications for Hydrocarbon Discovery and Recovery* : Society of Economic Paleontologists and Mineralogists, Concepts in Sedimentology and Paleontology 3, 172–178.

Tyler, N., and W. A. Ambrose, 1984, *Facies Architecture and Production Characteristics of Strandplain Reservoirs in the Frio Formation*: The University of Texas at Austin, Bureau of Economic Geology, Report of Investigations 146, 42 p.

Tyler, N., and W. A. Ambrose, 1986, *Depositional Systems and Oil and Gas Plays in the Cretaceous Olmos Formation, South Texas*: The University of Texas at Austin, Bureau of Economic Geology, Report of Investigations 152, 49 p.

Umbarger, K. F., 2018, *Late Triassic North American Paleodrainage networks and sediment dispersal of the Chinle Formation: a quantitative approach using detrital zircons*: MS thesis, University of Kansas, 131 p.

Umbarger, K. F., and J. W. Snedden, 2016, Delineation of post-KPg carbonate slope deposits as a sedimentary record of the Paleogene linkage of De Soto canyon and the Suwannee strait, northern Gulf of Mexico: *Interpretation*, 4, SC51–SC60.

United States Department of Energy, Minerals Management Service, 1987, *Correlation of Cenozoic Sediments, Gulf of Mexico Outer Continental Shelf, Part 1: Galveston Area Offshore Texas Through Vermilion Area Offshore Louisiana*, OCS Report MMS 87-0026.

USGS, 2005, Assessment of undiscovered oil and gas resources of the North Cuba Basin, Cuba, 2004: United States Geological Survey, World Assessment of Oil and Gas Fact Sheet, 2 p, https://pubs.usgs.gov/fs/2005/3009/pdf/fs2005_3009.pdf.

USGS, 2014, Geology and assessment of unconventional resources of northeastern Mexico: US Geological Survey, Open-File Report 2015 1112, 1 p., doi:10.3133/ofr20151112.

Valdés, M. D., L. V. Rodriguez, and E. C. Garcia, 2009, *Geochemical Integration and Interpretation of Source Rocks, Oils, and Natural Gases in Southeastern Mexico*: American Association of Petroleum Geologists, Memoir 90, 337–368.

Vallejo, V. V. A., E. F. Solis, A. Olivares, *et al.*, 2012, Drilling a deep-water well in a subsalt structure in Mexico: Deep Offshore Technology International, paper 145, 1–17.

Van Avendonk, H. J., D. R. Eddy, G. Christeson, *et al.*, 2013, Structure and early evolution of the Northwestern Gulf of Mexico: new constraints from marine seismic refraction data: *American Association of Petroleum Geologists, Search and Discovery* 90163.

Van Avendonk, H., G. Christeson, I. Norton, and D. Eddy, 2015, Continental rifting and sediment infill in the northwestern Gulf of Mexico: *Geology*, 43, 631–634, doi:10.1130/G36798.1.

Vásquez, R. O., S. P. J. Cossey, D. S. van Nieuwenhuise, *et al.*, 2014, New insights into the stratigraphic framework and depositional history of the Paleocene and Eocene Chicontepec Formation, onshore Eastern Mexico: *American Association of Petroleum Geologists, Search and Discovery* 30334.

Vega, F. J., and T. F. Lawton, 2011, Upper Jurassic (Lower Kimmeridgian-Olvido) carbonate strata from the La Popa Basin diapirs, NE Mexico: *Boletín de la Sociedad Geológica Mexicana*, 63(2), 313–321.

Veltman, W., E. Velez, and V. Lujan, 2012, A fresh look for natural fracture characterization using advance borehole acoustics techniques: *American Association of Petroleum Geologists, Search and Discovery* 40915, 17 p.

Vernon, R. C., 1971, *Possible Future Petroleum Potential of Pre-Jurassic, Western Gulf Basin*: American Association of Petroleum Geologists, Memoir 15, 954–979.

Vila-Concejo, A., D. L. Harris, H. E. Power, A. M. Shannon, and J. M. Webster, 2013, Sediment transport and mixing depth on a coral reef sand apron: *Geomorphology*, 222, 143–150, doi:10.1016/j.geomorph.2013.09.034.

Waite, L. E., 2009, Edwards (Stuart City) shelf margin of South Texas: new data, new concepts: *American Association of Petroleum Geologists, Search and Discovery* 10177.

Walles, F. E., 1993, Tectonic and diagenetically induced seal failure within the south-western Great Bahamas Bank: *Marine and Petroleum Geology*, 10, 14–28, doi:10.1016/0264-8172(93)90096-B.

Walsh, J. P., P. L. Wiberg, R. Aalto, C. A. Nittrouer and S. A. Kuehl, 2016, Source-to-sink research: economy of the Earth's surface and its strata: *Earth-Science Reviews*, 153, 1–6. doi:10.1016/j.earscirev.2015.11.010.

Wang, F. P., U. Hammes, and L. Qinghui, 2013, Overview of Haynesville shale properties and production, in U. Hammes and J. Gale, eds., *Geology of the Haynesville Gas Shale in East Texas and West Louisiana*: American Association of Petroleum Geologists Bulletin, Memoir 105, 155–177, doi:10.1306/13441848M1053527.

Ward, W. C., G. Keller, W. Stinnesbeck, and T. Adatte, 1995, Yucatan subsurface stratigraphy: implications and constraints for the Chicxulub impact: *Geology*, 23 873–876, doi:10.1130/0091-7613(1995)023<0873:YNSSIA>2.3.CO;2.

Warwick, P. E., 2017, *Geologic Assessment of Undiscovered Conventional Oil and Gas Resources in the Lower Paleocene Midway and Wilcox Groups, and the Carrizo Sand of the Claiborne Group of the Northern Gulf Coast Region*: U.S. Geological Survey, Open-File Report 2017-111, 60 p.

Warzeski, E. R., 1987, Revised stratigraphy of the Mural Limestone: a lower Cretaceous carbonate shelf in Arizona and Sonora, in W. R. Dickinson and M. F. Klute, eds., *Mesozoic Rocks of Southern Arizona and Adjacent Areas*: Arizona Geological Society, Digest 18, 335–363.

Watkins, J. S., B. E. Bradshaw, S. Huh, R. Li, and J. Zhang, 1996a, Structure and distribution of growth faults in the northern Gulf of Mexico OCS: *Gulf Coast Association of Geological Societies Transactions*, 46, 63–77.

Watkins, J. S., W. R. Bryant, and R. T. Buffler, 1996b, *Structural Framework Map of the Northern Gulf of Mexico*, Gulf Coast Association of Geological Societies, Special Publication 80, 95–98.

Weber, R. D., and B. W. Parker, 2016, Pre-Albian biostratigraphical and paleoecological observations from the De Soto Canyon area; Gulf of Mexico, USA, in C. M. Lowery, J. W. Snedden, and M. D. Blum, eds., *Mesozoic of the Gulf Rim and Beyond: New Progress in Science and Exploration of the Gulf of Mexico Basin*: Gulf Coast Section SEPM, Perkins Rosen Research Conference paper, 154–172.

Webster, R. E., D. Luttner, and L. Liu, 2008, Fairway James Lime Field, East Texas: still developing after 48 years: *American Association of Petroleum Geologists, Search and Discovery* 110061.

Weeks, L. G., 1958, Habitat of oil and some factors that control it, in L. Weeks, ed., *Habitat of Oil: A Symposium*: American Association of Petroleum Geologists, 1–61.

Weeks, W. B., 1938, South Arkansas stratigraphy with emphasis on the older Coastal Plain beds: *American Association of Petroleum Geologists Bulletin*, 22, 958–964.

Weimer, P., 1990, Sequence stratigraphy, facies geometries, and depositional history of the Mississippi Fan, Gulf of Mexico: *American Association of Petroleum Geologists Bulletin*, 74, 425–453.

Weimer, P., and R. T. Buffler, 1992, Structural geology and evolution of the Mississippi Fan Fold Belt, Deep Gulf of Mexico: *American Association of Petroleum Geologists Bulletin*, 76, 225–251.

Weimer, P., P. Varnai, F. M. Budhijanto, *et al.*, 1998, Sequence stratigraphy of Pliocene and Pleistocene turbidite systems, northern Green Canyon and Ewing Bank (offshore Louisiana), northern Gulf of Mexico: *American Association of Petroleum Geologists Bulletin*, 82, 918–960.

Weimer, P., E. Zimmerman, R. Bouroullec, *et al.*, 2016, Temporal and spatial evolution of reservoirs, Northern Deepwater Gulf of Mexico: *Gulf Coast Association of Geological Societies Transactions*, 66, 539–555.

Weimer, P., R. Bouroullec, and O. Serrano, 2017, Petroleum geology of the Mississippi Canyon, Atwater Valley, western DeSoto Canyon, and western Lloyd Ridge protraction areas, northern deep-water Gulf of Mexico: traps, reservoirs, and tectono-stratigraphic evolution: *American Association of Petroleum Geologists Bulletin*, 101, 1073–1108.

Weise, B. R., 1980, *Wave-Dominated Delta Systems of the Upper Cretaceous San Miguel Formation, Maverick Basin, South Texas*: The University of Texas at Austin, Bureau of Economic Geology, Report of Investigations 107, 39 p.

Weislogel, A. L., B. Hunt, A. Lisi, T. Lovell, and D. M. Robinson, 2015, Detrital zircon provenance of the eastern Gulf of Mexico subsurface: constraints on Late Jurassic paleogeography and sediment dispersal of North America, in T. H. Anderson, A. N. Didenko, C. L. Johnson, A. I. Khanchuk, and J. H. MacDonald Jr., eds., *Late Jurassic Margin of Laurasia: A Record of Faulting Accommodating Plate Rotation*: Geological Society, Special Publication 513, 89–105, doi:10.1130/2015.2513(02).

Weislogel, A. L., K. Wiley, S. Bowman, and D. Robinson, 2016, Triassic–Jurassic provenance signatures in the nascent Eastern Gulf of Mexico region from detrital zircon geochronology, in C. M. Lowery, J. W. Snedden, and M. D. Blum, eds., *Mesozoic of the Gulf Rim and Beyond: New Progress in Science and Exploration of the Gulf of Mexico Basin*: Society of Economic Paleontologists and Mineralogists Gulf Coast Section, Perkins Rosen Research Conference, 252–270.

Weissert, H., A. Lini, K. B. Föllmi, and O. Kuhn, 1998, Correlation of Early Cretaceous carbon isotope stratigraphy and platform drowning events: a possible link?: *Palaeogeography, Palaeoclimatology, Palaeoecology*, 137, 189–203.

Wenger, L. M., L. R. Goodoff, O. P. Gross, S. C. Harrison, and K. C. Hood, 1994, Northern Gulf of Mexico: an integrated approach to source, maturation, and migration, in N. Scheidermann, P. Cruz, and R. Sanchez, eds., *Geologic Aspects of Petroleum Systems: 1st Joint AAPG-AMGP Hedberg Research Conference*, 5 p.

Whidden, K. J., J. Pitman, O. Pearson, *et al.*, 2018, The 2017 USGS Assessment of undiscovered oil and gas resources in the Eagle Ford Shale and associated Cenomanian Strata, Texas (abs.): *American Association of Petroleum Geologists, Annual Convention & Exhibition.*

White, B. R., and J. R. Sawyer, 1966, Black Lake Field: before and after: *Gulf Coast Association of Geological Societies Transactions*, 16, 219–225.

White, C., and J. Snedden, 2016, Seismic stratigraphic analysis of the Yoakum/Lavaca Canyon System, South Texas, USA (abs.): *American Geophysical Union, Fall Meeting 2016*, abstract #EP43B-0956.

White, C., and J. Snedden,in prep., Seismic stratigraphic analysis of the Yoakum/Lavaca Canyon System, South Texas, USA.

White, G., S. Blanke, and C. Clawson II, 1999, Evolutionary model of the Jurassic sequences of the East Texas Basin: implications for hydrocarbon exploration: *Gulf Coast Association of Geological Societies Transactions*, 49, 488–498.

Wiley, K. S., 2017, *Provenance of Syn-rift Clastics in the Eastern Gulf of Mexico: Insight from U-Pb Detrital Zircon Geochronology and Thin Sections*: MS thesis, West Virginia University, 194 p.

Williams-Rojas, C. T., and N. F. Hurley, 2001, Geologic controls on reservoir performance in Muspac and Catedral Gas Fields, Southeast Mexico, in C. Bartolini, R. T. Buffler, and A. Cantú-Chapa, eds., *The Western Gulf of Mexico Basin: Tectonics, Sedimentary Basins, and Petroleum Systems*: American Association of Petroleum Geologists, Memoir 75, 443–472.

Williams-Rojas, C., E. Reyey-Tovar, L. Miranda Peralta, *et al.*, 2012, Hydrocarbon potential of the deepwater portion of the "Salina del Istmo" province, southeastern Gulf of Mexico, Mexico: *Gulf Coast Association of Geological Societies Transactions*, 62, 641–644.

Wilson, H. H., 1993, The age of salt in the Gulf of Mexico Basin: *Journal of Petroleum Geology*, 16, 125–152.

Wilson, J. L., 1975, *Carbonate Facies in Geologic History*: Springer-Verlag, 471 p.

Winker, C. D., 1982, Cenozoic shelf margins, northwestern Gulf of Mexico: *Gulf Coast Association of Geological Societies Transactions*, 32, 427–448.

Winker, C. D., 1984, Clastic shelf margins of the post-Comanchean Gulf of Mexico: implications for deep-water sedimentation: *Society of Economic Paleontologists and Mineralogists Gulf Coast Section, 5th Annual Research Conference*, 109–120.

Winker, C. D., and J. R. Booth, 2000, Sedimentary dynamics of the salt-dominated continental slope, Gulf of Mexico: integration of observations from the seafloor, near-surface, and deep subsurface, in P. Weimer, R. M. Slatt, J. Coleman, *et al.*, eds., *Deep-Water Reservoirs of the World*: Society of Economic Paleontologists and Mineralogists Gulf Coast Section 20th Annual Bob F. Perkins Research Conference, 1059–1086.

Winker, C. D., and R. T. Buffler, 1988, Paleogeographic evolution of early deep-water Gulf of Mexico and margins, Jurassic to Middle Cretaceous (Comanchean): *American Association of Petroleum Geologists Bulletin*, 72, 318–346.

Winker, C. D., R. A. Morton, T. E. Ewing, and D. D. Garcia, 1983, *Depositional Setting, Structural Style, and Sandstone Distribution in Three Geopressured Geothermal Areas, Texas Gulf Coast*: The University of Texas at Austin, Bureau of Economic Geology, Report of Investigations 134, 60 p.

Winston, G. O., 1976, Florida's Ocala Uplift is not an uplift: *American Association of Petroleum Geologists Bulletin*, 60(6), 992–994.

Winter, R. R., 2018, Coarse-grained deep water, slope and basin-floor systems: Influence of tectonic processes on internal and external architecture: unpublished PhD dissertation, The University of Texas at Austin, 163 p.

Withjack, M. O., R. W. Schlische, and P. E. Olsen, 1998, Diachronous rifting, drifting, and inversion on the passive margin of central eastern North America: an analog for other passive margins: *American Association of Petroleum Geologists Bulletin*, 82, 817–835.

Withjack, M. O., R. W. Schlische, and P. E. Olsen, 2002, Rift-basin structure and its influence on sedimentary systems, in R. Renault and G. Ashley, eds., *Sedimentation in Continental Rifts*: Society of Economic Paleontologists and Mineralogists, Special Publication, 73, 57–81, doi:10.2110/pec.02.73.0057.

Witt, C., S. Brichau, and A. Carter, 2012, New constraints on the origin of the Sierra Madre de Chiapas (south Mexico) from sediment provenance and apatite thermochronometry: *Tectonics*, 31(6), TC6001, doi:10.1029/2012TC003141.

Wood, G. D., and D. G. Benson Jr., 2000, The North American occurrence of the algal coenobium plaesiodictyon: paleographic, paleoecologic, and biostratigraphic importance in the Triassic: *American Association of Stratigraphic Palynologists*, 24, 8–20.

Woods, R. D., and J. W. Addington, 1973, Pre-Jurassic geologic framework Northern Gulf Basin: *Gulf Coast Association of Geological Societies Transactions*, 23, 92–108.

Woolf, K., 2012, *Regional Character of the Lower Tuscaloosa Formation Depositional Systems and Trends in Reservoir Quality*: Ph.D. dissertation, The University of Texas at Austin, 241 p.

Worrall, D. M., and S. Snelson, 1989, Evolution of the Northern Gulf of Mexico, with emphasis on Cenozoic growth faulting and the role of salt, in A. W. Bally and A. R. Palmer, eds., *The Geology of North America*: Geological Society of America, 97–138.

Wu, X., and Galloway, W. E., 2002, Upper Miocene depositional history of the central Gulf of Mexico basin: *Gulf Coast Association of Geological Societies Transactions*, 52, 1019–1030.

Wu, X., and Galloway, W. E., 2003, Upper Miocene depositional history and paleogeographic evolution of central Gulf of Mexico basin, in H. H. Roberts, N. C. Rosen, R. H. Fillon, and J. B. Anderson, eds., *Shelf Margin Deltas and Linked Downslope Petroleum Systems: Global Significance and Future Exploration Potential*: Society of Economic Paleontologists and Mineralogists Gulf Coast Section, 23rd Annual Gulf Coast Section SEPM Bob. F. Perkins Research Conference Proceedings, 139–165.

Xaio, L., J. W. Zhao, H. S. Liu, *et al.*, 2017, Ages and geochemistry of the basement granites of the Chicxulub impact crater: implications for peak ring formation: *Lunar and Planetary Science*, 48, 1–2.

Xu, J., J. W. Snedden D. F. Stockli, C. S. Fulthorpe, and W. E. Galloway, 2016a, Early Miocene continental-scale sediment supply to the Gulf of Mexico Basin based on detrital zircon analysis: *Geological Society of America Bulletin*, 129, 3–22, doi:10.1130/B31465.1.

Xu, J., J. W. Snedden, W. E. Galloway, K. T. Milliken, and M. D. Blum, 2016b, Channel-belt scaling relationship and application to early Miocene source-to-sink systems in the Gulf of Mexico basin: *Geosphere*, 13, 179–200.

Xu, J., D. F. Stockli, and J. W. Snedden, 2017, Enhanced provenance interpretation using combined U–Pb and (U–Th)/He double dating of detrital zircon grains from lower Miocene strata, proximal Gulf of Mexico Basin, North America: *Earth and Planetary Science Letters*, 475, 44–57.

Xue, L., 1997, Depositional cycles and evolution of the Paleogene Wilcox strata, Gulf of Mexico basin, Texas: *American Association of Petroleum Geologists Bulletin*, 81, 937–953.

Xue, L., and W. E. Galloway 1993, Sequence stratigraphic and depositional framework of the Paleocene lower Wilcox strata, northwest Gulf of Mexico Basin: *Gulf Coast Association of Geological Societies Transactions*, 43, 453–464.

Xue, L., and W. E. Galloway, 1995, High-resolution depositional framework of the Paleocene Middle Wilcox strata, Texas coastal plain: *American Association of Petroleum Geologists Bulletin*, 79, 205–230.

Yancey, T. E., 1996, Stratigraphy and depositional environments of the Cretaceous–Tertiary boundary complex and basal Paleocene section, Brazos River, Texas: *Gulf Coast Association of Geological Societies Transactions*, 46, 433–442.

Yancey, T. E., W. C. Elsik, and R. H. Sancay, 2003, The palynological record of Late Eocene climate change, northwest Gulf of Mexico, in D.R. Prothero, L.C. Ivany, and E.A. Nesbitt, eds., *From Greenhouse to Icehouse: The Marine Eocene–Oligocene Transition*: Columbia University Press, 252–268.

Yancey, T. E., M. T. Heizler, B. V. Miller, and R. N. Guillemette, 2018, Eocene–Oligocene chronostratigraphy of ignimbrite flareup volcanic ash beds on the Gulf of Mexico coastal plains: *Geosphere*, 14, 1232–1252. doi.org/10.1130/GES01621.1.

Young, K., 1977, Rocks of the Austin area, in K. Young, ed., *Guidebook to the Geology of Travis County*, University of Texas Student Geology Society, 16–25.

Young, K., 1985, The Austin Division of central Texas, in K. Young and C. M. Woodruff, *Austin Chalk in Its Type Area, Stratigraphy and Structure*, Austin Geological Society, Guidebook, 7, 3–52.

Ysaccis, R., G. Hernandez, R. Villa, *et al.*, 2006, Structural styles of the Canela and its surrounded areas, Tabasco– Southern Mexico: *American Association of Petroleum Geologists Search and Discovery* 90052.

Yurewicz, D. A., T. B. Marler, K. A. Meyerholtz, and F. X. Siroky, 1993, Early Cretaceous carbonate platform, north rim of the Gulf of Mexico, Mississippi and Louisiana, in J. A. T. Simo, R. W. Scott, and J.-P. Masse, eds., *Cretaceous Carbonate Platforms*: American Association of Petroleum Geologists Memoir 56, 81–96.

Yurewicz, D. A., R. J. Chuchla, M. Richardson, *et al.*, 1997, Hydrocarbon generation and migration in the Tampico–Misantla Basin and Sierra Madre Oriental, east-central Mexico: evidence from an exhumed oil field in the Sierra de el Abra, in *Sedimentation and Diagenesis of Middle Cretaceous Platform Margins, East Central Mexico: American Association of Petroleum Geologists Bulletin, Field Trip Guidebook*, Dallas Geological Society and Society of Economic Paleontologists and Mineralogists, 24 p.

Zarra, L., 2007, Chronostratigraphic framework for the Wilcox Formation (Upper Paleocene–Lower Eocene) in the deep-water Gulf of Mexico: biostratigraphy, sequences, and depositional systems, in L. Kennan, J. Pindell, and N.C. Rosen, eds., *The Paleogene of the Gulf of Mexico and Caribbean Basins: Processes Events and Petroleum Systems*: Society of Economic Paleontologists and Mineralogists Gulf Coast Section, 26th Annual Research Conference, 81–146.

Zeng, H., and T. F. Hentz, 2002, High-frequency sequence stratigraphy from seismic sedimentology: applied to Miocene, Vermilion Block 50, Tiger Shoal area, offshore Louisiana: *American Association of Petroleum Geologists Bulletin*, 88, 153–174.

Zhang, J., R. Steel, and W. A. Ambrose, 2016, Greenhouse shoreline migration: Wilcox deltas: *American Association of Petroleum Geologists Bulletin*, 100, 1803–1831.

Zhang, J., J. Covault, M. Pyrcz, *et al.*, 2018, Quantifying sediment supply to continental margins: applications to the Paleogene Wilcox Group, Gulf of Mexico: *American Association of Petroleum Geologists Bulletin*, 100, 1685–1702, doi: 10.1306/01081817308.

Index